# Vibration Damping, Control, and Design

# Mechanical Engineering Series

*Frank Kreith and Roop Mahajan - Series Editors*

## Published Titles

# Vibration Damping, Control, and Design

Edited by

## Clarence W. de Silva

The University of British Columbia
Vancouver, Canada

CRC Press
Taylor & Francis Group
Boca Raton London New York

CRC Press is an imprint of the
Taylor & Francis Group, an **informa** business

This material was previously published in *Vibration and Shock Handbook* © 2005 by CRC Press, LLC.

CRC Press
Taylor & Francis Group
6000 Broken Sound Parkway NW, Suite 300
Boca Raton, FL 33487-2742

First issued in paperback 2019

© 2007 by Taylor & Francis Group, LLC
CRC Press is an imprint of Taylor & Francis Group, an Informa business

No claim to original U.S. Government works

ISBN-13: 978-1-4200-5321-0 (hbk)
ISBN-13: 978-0-367-38926-0 (pbk)

**Library of Congress Cataloging-in-Publication Data**

Vibration damping, control, and design / editor, Clarence W. de Silva.
    p. cm. -- (Mechanical engineering series)
  Includes bibliographical references and index.
  ISBN-13: 978-1-4200-5321-0 (alk. paper)
  ISBN-10: 1-4200-5321-3 (alk. paper)
  1. Vibration. 2. Damping (Mechanics) I. De Silva, Clarence W. II. Title.

TA355.V5222 2007
620.3'7--dc22                                      2006100169

Visit the Taylor & Francis Web site at
http://www.taylorandfrancis.com

and the CRC Press Web site at
http://www.crcpress.com

# Preface

In individual chapters authored by distinguished leaders and experienced professionals in their respective topics, this book provides for engineers, technicians, designers, researchers, educators, and students, a convenient, thorough, up-to-date, and authoritative reference source on techniques, tools, and data for analysis, design, monitoring and control of vibration, noise, and acoustics. Vibration suppression, damping, and control; design for and control of vibration; system design, application, and control implementation; and acoustics and noise suppression are treated in the book. Important information and results are summarized as windows, tables, graphs, and lists throughout the chapters, for easy reference and information tracking. References are given at the end of each chapter, for further information and study. Cross-referencing is used throughout to indicate other places in the book where further information on a particular topic is provided.

In the book, equal emphasis is given to theory and practical application. Analytical formulations, design approaches, and control techniques are presented and illustrated. Examples and case studies are given throughout the book to illustrate the use and application of the included information. The material is presented in a format that is convenient for easy reference and recollection.

Mechanical vibration is a manifestation of the oscillatory behavior in mechanical systems as a result of either the repetitive interchange of kinetic and potential energies among components in the system, or a forcing excitation that is oscillatory. Such oscillatory responses are not limited to purely mechanical systems, and are found in electrical and fluid systems as well. In purely thermal systems, however, free natural oscillations are not possible, and an oscillatory excitation is needed to obtain an oscillatory response. Sound, noise, and acoustics are manifestations of pressure waves, sources of which are often vibratory dynamic systems.

Low levels of vibration mean reduced noise and improved work environment. Vibration modification and control can be crucial in maintaining high performance and production efficiency, and prolonging the useful life in industrial machinery. Consequently, a considerable effort is devoted today to studying and controlling the vibration generated by machinery components, machine tools, transit vehicles, impact processes, civil engineering structures, fluid flow systems, and aircraft. Noise and acoustic problems can originate from undesirable vibrations and fluid–structure interactions, as found, for example in automobile engines. Noises from engine, environment, and high-speed and high-temperature exhaust gases in a vehicle will not only cause passenger discomfort and public annoyance, but also will result in damaging effects to the vehicle itself. Noise suppression methods and devices, and sound absorption material and structures are crucial under such situations. Before designing or controlling a system for good vibratory or acoustic performance, it is important to understand, analyze, and represent the dynamic characteristics of the system. This may be accomplished through purely analytical means, computer analysis of analytical models, testing and analysis of test data, or by a combination of these approaches.

In recent years, educators, researchers, and practitioners have devoted considerable effort towards studying and controlling vibration and noise in a range of applications in various branches of engineering, particularly, civil, mechanical, aeronautical and aerospace, and production and manufacturing. Specific applications are found in machine tools, transit vehicles, impact processes, civil engineering structures, construction machinery, industrial processes, product qualification and quality control, fluid flow systems, ships, and aircraft. This book is a contribution towards these efforts. In view of these analytical methods, practical considerations, design issues, and experimental techniques are presented throughout the book, and in view of the simplified and snap-shot style presentation of formulas, data, and advanced theory, the book serves as a useful reference tool and an extensive information source for engineers and technicians in industry and laboratories, researchers, instructors, and students in the areas of vibration, shock, noise, and acoustics.

**Clarence W. de Silva**
Editor-in-Chief
Vancouver, Canada

# Acknowledgments

I wish to express my gratitude to the authors of the chapters for their valuable and highly professional contributions. I am very grateful to Michael Slaughter, Acquisitions Editor-Engineering, CRC Press, for his enthusiasm and support throughout the project. Editorial and production staff at CRC Press have done an excellent job in getting this volume out in print. Finally, I wish to lovingly acknowledge the patience and understanding of my family.

# Editor-in-Chief

**Dr. Clarence W. de Silva**, P.Eng., Fellow ASME, Fellow IEEE, Fellow Canadian Academy of Engineering, is Professor of Mechanical Engineering at the University of British Columbia, Vancouver, Canada, and has occupied the NSERC-BC Packers Research Chair in Industrial Automation since 1988. He has earned Ph.D. degrees from the Massachusetts Institute of Technology and the University of Cambridge, England. De Silva has also occupied the Mobil Endowed Chair Professorship in the Department of Electrical and Computer Engineering at the National University of Singapore. He has served as a consultant to several companies including IBM and Westinghouse in the U.S., and has led the development of eight industrial machines and devices. He is recipient of the Henry M. Paynter Outstanding Investigator Award from the Dynamic Systems and Control Division of the American Society of Mechanical Engineers (ASME), Killam Research Prize, Lifetime Achievement Award from the World Automation Congress, Outstanding Engineering Educator Award of IEEE Canada, Yasurdo Takahashi Education Award of the Dynamic Systems and Control Division of ASME, IEEE Third Millennium Medal, Meritorious Achievement Award of the Association of Professional Engineers of BC, and the Outstanding Contribution Award of the Systems, Man, and Cybernetics Society of the Institute of Electrical and Electronics Engineers (IEEE).

He has authored 16 technical books including *Sensors and Actuators: Control System Instrumentation* (Taylor & Francis, CRC Press, 2007); *Mechatronics—An Integrated Approach* (Taylor & Francis, CRC Press, Boca Raton, FL, 2005); *Soft Computing and Intelligent Systems Design—Theory, Tools, and Applications* (with F. Karry, Addison Wesley, New York, NY, 2004); *Vibration: Fundamentals and Practice* (Taylor & Francis, CRC Press, 2nd edition, 2006); *Intelligent Control: Fuzzy Logic Applications* (Taylor & Francis, CRC Press, 1995); *Control Sensors and Actuators* (Prentice Hall, 1989); 14 edited volumes, over 170 journal papers, 200 conference papers, and 12 book chapters. He has served on the editorial boards of 14 international journals, in particular as the Editor-in-Chief of the *International Journal of Control and Intelligent Systems*, Editor-in-Chief of the *International Journal of Knowledge-Based Intelligent Engineering Systems*, Senior Technical Editor of *Measurements and Control*, and Regional Editor, North America, of *Engineering Applications of Artificial Intelligence – the International Journal of Intelligent Real-Time Automation*. He is a Lilly Fellow at Carnegie Mellon University, NASA-ASEE Fellow, Senior Fulbright Fellow at Cambridge University, ASI Fellow, and a Killam Fellow. Research and development activities of Professor de Silva are primarily centered in the areas of process automation, robotics, mechatronics, intelligent control, and sensors and actuators as principal investigator, with cash funding of about $6 million.

# Contributors

**S. Akishita**
Ritsumeikan University
Kusatsu, Japan

**Su Huan Chen**
Jilin University
Changchun, People's Republic of
    China

**Kourosh Danai**
University of Massachusetts
Amherst, Massachusetts

**Clarence W. de Silva**
The University of British Columbia
Vancouver, British Columbia, Canada

**Ebrahim Esmailzadeh**
University of Ontario
Oshawa, Ontario, Canada

**Seon M. Han**
Texas Tech University
Lubbock, Texas

**Nader Jalili**
Clemson University
Clemson, South Carolina

**Takayuki Koizumi**
Doshisha University
Kyoto-Hu, Japan

**Robert G. Landers**
University of Missouri at Rolla
Rolla, Missouri

**L.Y. Lu**
National Kaohsiung First University
    of Science and Technology
Kaohsiung, Taiwan

**Kiyoshi Nagakura**
Railway Technical Research Institute
Tokyo-To, Japan

**Teruo Obata**
Teikyo University
Totigi-Ken, Japan

**Kiyoshi Okura**
Mitsuboshi Belting Ltd.
Hyogo-Ken, Japan

**Randall D. Peters**
Mercer University
Macon, Georgia

**H. Sam Samarasekera**
Sulzer Pumps (Canada), Inc.
Burnaby, British Columbia, Canada

**Y.B. Yang**
National Taiwan University
Taipei, Taiwan

**J.D. Yau**
Tamkang University
Taipei, Taiwan

# Contents

# 1

# Vibration Damping

Clarence W. de Silva
*The University of British Columbia*

## Summary

*Damping in vibrating systems occurs through the dissipation of mechanical energy. This chapter presents modeling, analysis, and measurement of mechanical damping. The types of damping covered include material internal damping (including viscoelastic damping and hysteretic damping), structural damping, fluid damping, interface damping, viscous damping, Coulomb friction, and Stribeck damping. Representations of various types of damping using equivalent viscous damping models are analyzed. Damping in rotating devices is also studied.*

## 1.1   Introduction

Damping is the phenomenon by which mechanical energy is dissipated (usually by conversion into internal thermal energy) in dynamic systems. Knowledge of the level of damping in a dynamic system is important in the utilization, analysis, and testing of the system. For example, a device with natural frequencies within the seismic range (that is, less than 33 Hz) and which has relatively low damping, could produce damaging motions under resonance conditions when subjected to a seismic disturbance. This effect could be further magnified by low-frequency support structures and panels with low damping. This example shows that knowledge of damping in constituent devices, components, and support structures is important in the design and operation of complex mechanical systems. The nature and the level of component damping should be known in order to develop a dynamic model of the system and its peripherals. Knowledge of damping in a system is also important in imposing dynamic environmental limitations on the system (that is, the maximum dynamic excitation the system can withstand) under in-service conditions. Furthermore, knowledge of a system's damping can be useful in order to make design modifications in a system that has failed the acceptance test.

However, the significance of knowledge of damping levels in a test object for the development of test excitation (input) is often overemphasized. Specifically, if the response spectrum method is used to represent the required excitation in a vibration test, then there is no need for the damping value used in the development of the required response spectrum specification to be equal to the actual damping in the test object. The only requirement is that the damping used in the specified response spectrum be equal to that used in the test response spectrum. The degree of dynamic interaction between the test object and the shaker table, however, will depend on the actual level of damping in these systems. Furthermore, when testing near the resonant frequency of a test object, it is desirable to know about the damping in the test object.

In characterizing damping in a dynamic system it is important, first, to understand the major mechanisms associated with mechanical energy dissipation in the system. Then a suitable damping model should be chosen to represent the associated energy dissipation. Finally, damping values (model parameters) should be determined, for example, by testing the system or a representative physical model, by monitoring system response under transient conditions during normal operation or by employing already available data.

## 1.2    Types of Damping

There is some form of mechanical energy dissipation in any dynamic system. In the modeling of systems, damping can be neglected if the mechanical energy that is dissipated during the time duration of interest is small in comparison to the initial total mechanical energy of excitation in the system. Even for highly damped systems, it is useful to perform an analysis with the damping terms neglected, in order to study several crucial dynamic characteristics, e.g., modal characteristics (undamped natural frequencies and mode shapes).

Several types of damping are inherently present in a mechanical system. If the level of damping that is available in this manner is not adequate for proper functioning of the system then external damping devices may be added either during the original design or during subsequent design modifications of the system. Three primary mechanisms of damping are important in the study of mechanical systems. They are:

1. Internal damping (of material)
2. Structural damping (at joints and interfaces)
3. Fluid damping (through fluid–structure interactions)

Internal (material) damping results from mechanical energy dissipation within the material due to various microscopic and macroscopic processes. Structural damping is caused by mechanical energy dissipation resulting from relative motions between components in a mechanical structure that has common points of contact, joints or supports. Fluid damping arises from the mechanical energy dissipation resulting from drag forces and associated dynamic interactions when a mechanical system or its components move in a fluid.

Two general types of external dampers may be added to a mechanical system in order to improve its energy dissipation characteristics. They are:

1. Passive dampers
2. Active dampers

Passive dampers are devices that dissipate energy through some kind of motion, without needing an external power source or actuators. Active dampers have actuators that need external sources of power. They operate by actively controlling the motion of the system that needs damping. Dampers may be considered as vibration controllers. In the present chapter, the emphasis will be on damping that is inherently present in a mechanical system.

### 1.2.1    Material (Internal) Damping

Internal damping of materials originates from the energy dissipation associated with microstructure defects, such as grain boundaries and impurities; thermoelastic effects caused by local temperature gradients resulting from nonuniform stresses, as in vibrating beams; eddy current effects in ferromagnetic materials; dislocation motion in metals; and chain motion in polymers. Several models have been employed to represent energy dissipation caused by internal damping. This variety of models is primarily a result of the vast range of engineering materials; no single model can satisfactorily represent the internal damping characteristics of all materials. Nevertheless, two general types of internal damping can be identified: viscoelastic damping and hysteretic damping. The latter term is actually a misnomer, because all types of internal damping are associated with hysteresis loop effects. The stress ($\sigma$) and strain ($\varepsilon$) relations at a point in a vibrating continuum possess a hysteresis loop, such as the one shown in Figure 1.1. The area of the hysteresis loop gives the energy dissipation per unit volume of the material, per stress cycle. This is termed the per-unit-volume damping capacity, and is denoted by $d$. It is clear that $d$ is given by the cyclic integral

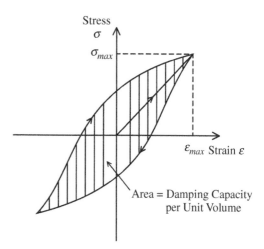

**FIGURE 1.1** A typical hysteresis loop for mechanical damping.

$$d = \oint \sigma \, d\varepsilon \tag{1.1}$$

In fact, for any damped device, there is a corresponding hysteresis loop in the displacement–force plane as well. In this case, the cyclic integral of force with respect to the displacement, which is the area of the hysteresis loop, is equal to the work done against the damping force. It follows that this integral (loop area) is the energy dissipated per cycle of motion. This is the *damping capacity* which, when divided by the material volume, gives the per-unit-volume damping capacity as before.

It should be clear that, unlike a pure elastic force (e.g., a spring force), a damping force cannot be a function of displacement ($q$) alone. The reason is straightforward. Consider a force $f(q)$ which depends on $q$ alone. Then, for a particular displacement point, $q$, of the component the force will be the same regardless of the direction of motion (i.e., the sign of $\dot{q}$). It follows that, in a loading and unloading cycle, the same path will be followed in both directions of motion. Hence, a hysteresis loop will not be formed. In other words, the net work done in a complete cycle of motion will be zero. Next consider a force $f(q, \dot{q})$ which depends on both $q$ and $\dot{q}$. Then, at a given displacement point, $q$, the force will depend on $\dot{q}$ as well. Hence, force in one direction of motion will be different from that in the opposite direction. As a result, a hysteresis loop will be formed, which corresponds to work done against the damping force (i.e., energy dissipation). We can conclude then that the damping force has to depend on a relative velocity, $\dot{q}$, in some manner. In particular, Coulomb friction, which does not depend on the magnitude of $\dot{q}$, does depend on the sign (direction) of $\dot{q}$.

### 1.2.1.1 Viscoelastic Damping

For a linear viscoelastic material, the stress–strain relationship is given by a linear differential equation with respect to time, having constant coefficients. A commonly employed relationship is

$$\sigma = E\varepsilon + E^* \frac{d\varepsilon}{dt} \tag{1.2}$$

which is known as the Kelvin–Voigt model. In Equation 1.2, $E$ is Young's modulus and $E^*$ is a viscoelastic parameter that is assumed to be time independent. The elastic term $E\varepsilon$ does not contribute to damping, and, as noted before, mathematically, its cyclic integral vanishes. Consequently, for the Kelvin–Voigt model, damping capacity per unit volume is

$$d_v = E^* \oint \frac{d\varepsilon}{dt} d\varepsilon \pi \tag{1.3}$$

For a material that is subjected to a harmonic (sinusoidal) excitation, at steady state, we have

$$\varepsilon = \varepsilon_{max} \cos \omega t \tag{1.4}$$

When Equation 1.4 is substituted in Equation 1.3, we obtain

$$d_v = \pi \omega E^* \varepsilon_{max}^2 \tag{1.5}$$

Now, $\varepsilon = \varepsilon_{max}$ when $t = 0$ in Equation 1.4, or when $d\varepsilon/dt = 0$. The corresponding stress, according to Equation 1.2, is $\sigma_{max} = E\varepsilon_{max}$. It follows that

$$d_v = \frac{\pi \omega E^* \sigma_{max}^2}{E^2} \tag{1.6}$$

These expressions for $d_v$ depend on the frequency of excitation, $\omega$.

Apart from the Kelvin–Voigt model, two other models of viscoelastic damping are also commonly used. They are, the Maxwell model given by

$$\sigma + c_s \frac{d\sigma}{dt} = E^* \frac{d\varepsilon}{dt} \tag{1.7}$$

and the standard linear solid model given by

$$\sigma + c_s \frac{d\sigma}{dt} = E\varepsilon + E^* \frac{d\varepsilon}{dt} \tag{1.8}$$

It is clear that the standard linear solid model represents a combination of the Kelvin–Voigt model and the Maxwell model, and is the most accurate of the three. But, for most practical purposes, the Kelvin–Voigt model is adequate.

### 1.2.1.2 Hysteretic Damping

It was noted above that the stress, and hence the internal damping force, of a viscoelastic damping material depends on the frequency of variation of the strain (and consequently the frequency of motion). For some types of material, it has been observed that the damping force does not significantly depend on the frequency of oscillation of strain (or frequency of harmonic motion). This type of internal damping is known as hysteretic damping.

Damping capacity per unit volume ($d_h$) for hysteretic damping is also independent of the frequency of motion and can be represented by

$$d_h = J\sigma_{max}^n \tag{1.9}$$

A simple model that satisfies Equation 1.9, for the case of $n = 2$, is given by

$$\sigma = E\varepsilon + \frac{\tilde{E}}{\omega} \frac{d\varepsilon}{dt} \tag{1.10}$$

which is equivalent to using a viscoelastic parameter, $E^*$, that depends on the frequency of motion in Equation 1.2 according to $E^* = \tilde{E}/\omega$.

Consider the case of harmonic motion at frequency $\omega$, with the material strain given by

$$\varepsilon = \varepsilon_0 \cos \omega t \tag{1.11}$$

Then, Equation 1.10 becomes

$$\sigma = E\varepsilon_0 \cos \omega t - \tilde{E}\varepsilon_0 \sin \omega t = E\varepsilon \cos \omega t + \tilde{E}\varepsilon_0 \cos\left(\omega t + \frac{\pi}{2}\right) \tag{1.12}$$

Note that the material stress has two components, as given by the right-hand side of Equation 1.12. The first component corresponds to the linear elastic behavior of a material and is in phase with the strain. The second component of stress, which corresponds to hysteretic damping, is 90° out of phase. (This stress component leads the strain by 90°.) A convenient mathematical representation is possible, by using the usual complex form of the response according to

$$\varepsilon = \varepsilon_0 e^{j\omega t} \tag{1.13}$$

Then, Equation 1.10 becomes

$$\sigma = (E + j\tilde{E})\varepsilon \tag{1.14}$$

It follows that this form of simplified hysteretic damping may be represented by using a complex modulus of elasticity, consisting of a real part which corresponds to the usual linear elastic (energy storage) modulus (or Young's modulus) and an imaginary part which corresponds to the hysteretic loss (energy dissipation) modulus.

By combining Equation 1.2 and Equation 1.10, a simple model for combined viscoelastic and hysteretic damping may be given by

$$\sigma = E\varepsilon + \left(E^* + \frac{\tilde{E}}{\omega}\right)\frac{d\varepsilon}{dt} \tag{1.15}$$

The equation of motion for a system whose damping is represented by Equation 1.15 can be deduced from the pure elastic equation of motion by simply substituting $E$ by the operator

$$E + \left(E^* + \frac{\tilde{E}}{\omega}\right)\frac{\partial}{\partial t}$$

in the time domain.

## Example 1.1

Determine the equation of flexural motion of a nonuniform slender beam whose material has both viscoelastic and hysteretic damping.

## Solution

The Bernoulli–Euler equation of bending motion on an undamped beam subjected to a dynamic load of $f(x, t)$ per unit length, is given by

$$\frac{\partial^2}{\partial x^2}EI\frac{\partial^2 q}{\partial x^2} + \rho A\frac{\partial^2 q}{\partial t^2} = f(x, t) \tag{1.16}$$

Here, $q$ is the transverse motion at a distance, $x$, along the beam. Then, for a beam with material damping (both viscoelastic and hysteretic) we can write

$$\frac{\partial^2}{\partial x^2}EI\frac{\partial^2 q}{\partial x^2} + \frac{\partial^2}{\partial x^2}\left(E^* + \frac{\tilde{E}}{\omega}\right)I\frac{\partial^3 q}{\partial t \partial x^2} + \rho A\frac{\partial^2 q}{\partial t^2} = f(x, t) \tag{1.17}$$

in which $\omega$ is the frequency of the external excitation $f(x, t)$ in the case of steady forced vibrations. In the case of free vibration, however, $\omega$ represents the frequency of free vibration decay. Consequently, when analyzing the modal decay of free vibrations, $\omega$ in Equation 1.17 should be replaced by the appropriate frequency ($\omega_i$) of modal vibration in each modal equation. Hence, the resulting damped vibratory system possesses the same normal mode shapes as the undamped system. The analysis of the damped case is very similar to that for the undamped system.

## 1.2.2 Structural Damping

Structural damping is a result of mechanical energy dissipation caused by friction due to the relative motion between components and by impacting or intermittent contact at the joints in a mechanical system or structure. Energy dissipation behavior depends on the details of the particular mechanical system. Consequently, it is extremely difficult to develop a generalized analytical model that would satisfactorily describe structural damping. Energy dissipation caused by rubbing is usually represented by a Coulomb friction model. Energy dissipation caused by impacting, however, should be determined from the coefficient of restitution of the two members that are in contact.

The most common method of estimating structural damping is by measurement. The measured values, however, represent the overall damping in the mechanical system. The structural damping component is obtained by subtracting the values corresponding to other types of damping, such as material damping, present in the system (estimated by environment-controlled experiments, previous data, and so forth) from the overall damping value.

Usually, internal damping is negligible compared to structural damping. A large proportion of mechanical energy dissipation in tall buildings, bridges, vehicle guideways, and many other civil engineering structures and in machinery, such as robots and vehicles, takes place through the structural damping mechanism. A major form of structural damping is the slip damping that results from energy dissipation by interface shear at a structural joint. The degree of slip damping that is directly caused by Coulomb (dry) friction depends on such factors as joint forces (for example, bolt tensions), surface properties and the nature of the materials of the mating surfaces. This is associated with wear, corrosion, and general deterioration of the structural joint. In this sense, slip damping is time-dependent. It is a common practice to place damping layers at joints to reduce undesirable deterioration of the joints. Sliding causes shear distortions in the damping layers, causing energy dissipation by material damping and also through Coulomb friction. In this way, a high level of equivalent structural damping can be maintained without causing excessive joint deterioration. These damping layers should have a high stiffness (as well as a high specific-damping capacity) in order to take the structural loads at the joint.

For structural damping at a joint, the damping force varies as slip occurs at the joint. This is primarily caused by local deformations at the joint, which occur with slipping. A typical hysteresis loop for this case is shown in Figure 1.2(a). The arrows on the hysteresis loop indicate the direction of relative velocity. For idealized Coulomb friction,

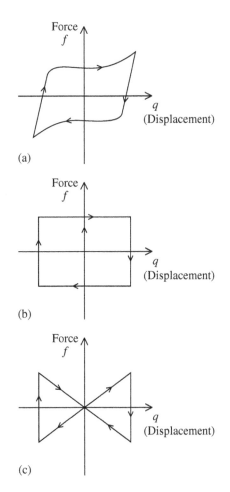

**FIGURE 1.2** Some representative hysteresis loops: (a) typical structural damping; (b) Coulomb friction model; and (c) simplified structural damping model.

the frictional force ($F$) remains constant in each direction of relative motion. An idealized hysteresis loop for structural Coulomb damping is shown in Figure 1.2(b). The corresponding constitutive relation is

$$f = c \, \text{sgn}(\dot{q}) \tag{1.18}$$

in which $f$ is the damping force, $q$ is the relative displacement at the joint and $c$ is a friction parameter. A simplified model for structural damping caused by local deformation may be given by

$$f = c|q|\text{sgn}(\dot{q}) \tag{1.19}$$

The corresponding hysteresis loop is shown in Figure 1.2(c). Note that the *signum function* is defined by

$$\text{sgn}(v) = \begin{cases} 1 & \text{for } v \geq 0 \\ -1 & \text{for } v < 0 \end{cases} \tag{1.20}$$

## 1.2.3 Fluid Damping

Consider a mechanical component moving in a fluid medium. The direction of relative motion is shown parallel to the $y$-axis in Figure 1.3. Local displacement of the element relative to the surrounding fluid is denoted by $q(x, y, t)$.

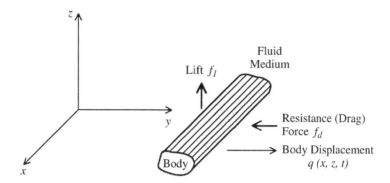

**FIGURE 1.3**  A body moving in a fluid medium.

The resulting drag force per unit area of projection on the $x$–$z$ plane is denoted by $f_d$. This resistance is the cause of mechanical energy dissipation in fluid damping. It is usually expressed as

$$f_d = \tfrac{1}{2} c_d \rho \dot{q}^2 \, \text{sgn}(\dot{q}) \qquad (1.21)$$

in which $\dot{q} = \partial q(x, z, t)/\partial t$ is the relative velocity. The drag coefficient, $c_d$, is a function of the Reynold's number and the geometry of the structural cross section. A net damping effect is generated by viscous drag produced by the boundary layer effects at the fluid–structure interface, and by pressure drag produced by the turbulent effects resulting from flow separation at the wake. The two effects are illustrated in Figure 1.4. Fluid density is $\rho$. For fluid damping, the damping capacity per unit volume associated with the configuration shown in Figure 1.3 is given by

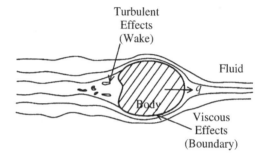

**FIGURE 1.4**  Mechanics of fluid damping.

$$d_f = \frac{\oint \int_0^{L_x} \int_0^{L_z} f_d \; dz \; dx \; dq(x, z, t)}{L_x L_z q_0} \qquad (1.22)$$

in which, $L_x$ and $L_z$ are cross-sectional dimensions of the element in the $x$ and $y$-directions, respectively, and $q_0$ is a normalizing amplitude parameter for relative displacement.

## Example 1.2

Consider a beam of length $L$ and uniform rectangular cross section that is undergoing transverse vibration in a stationary fluid. Determine an expression for the damping capacity per unit volume for this fluid–structure interaction.

## Solution

Suppose that the beam axis is along the $x$-direction and the transverse motion is in the $z$-direction. There is no variation in the $y$-direction, and hence, the length parameters in this direction cancel out.

$$d_f = \frac{\oint \int_0^{L} f_d \; dx \; dq(x, t)}{L q_0}$$

or

$$d_f = \frac{\int_0^{T} \int_0^{L} f_d \dot{q}(x, t) dx \; dt}{L q_0} \qquad (1.23)$$

in which $T$ is the period of the oscillations. Assuming constant $c_d$, we substitute Equation 1.21 into Equation 1.23:

$$d_f = \frac{1}{2} \frac{c_d \rho}{L q_0} \int_0^L \int_0^T |\dot{q}|^3 \, dt \, dx \tag{1.24}$$

For steady-excited harmonic vibration at frequency $\omega$ and shape function $Q(x)$ (or for free-modal vibration at natural frequency $\omega$ and mode shape $Q(x)$) we have

$$q(x, t) = q_{max} Q(x) \sin \omega t \tag{1.25}$$

In this case, with the change of variable $\theta = \omega t$, Equation 1.24 becomes

$$d_f = 2 c_d \rho \frac{q_{max}^3}{L q_0} \int_0^L |Q(x)|^3 \, dx \, \omega^2 \int_0^{\pi/2} \cos^3 \theta \, d\theta$$

or

$$d_f = \frac{4}{3} c_d \rho q_{max}^3 \omega^2 \frac{\displaystyle\int_0^L |Q(x)|^3 \, dx}{L q_0}$$

*Note.* The integration interval of $t = 0$ to $T$ becomes $\theta = 0$ to $2\pi$ or four times that from $\theta = 0$ to $\pi/2$.
If the normalizing parameter is defined as

$$q_0 = \frac{1}{L} q_{max} \int_0^L |Q(x)|^3 \, dx$$

then, we get

$$d_f = \frac{4}{3} c_d \rho q_{max}^2 \omega^2 \tag{1.26}$$

A useful classification of damping is given in Box 1.1.

# Box 1.1

# DAMPING CLASSIFICATION

| Type of Damping | Origin | Typical Constitutive Relation |
|---|---|---|
| Internal damping | Material properties | **Viscoelastic** $\sigma = E\varepsilon + E^* \dfrac{d\varepsilon}{dt}$ <br> **Hysteretic** $\sigma = E\varepsilon + \dfrac{\tilde{E}}{\omega} \dfrac{d\varepsilon}{dt}$ |
| Structural damping | Structural joints and interfaces | **Structural deformation** $f = c|q| \, \mathrm{sgn}(\dot{q})$ <br> **Coulomb** $f = c \, \mathrm{sgn}(\dot{q})$ <br> **General interface** $f = \begin{cases} f_s & \text{for } v = 0 \\ f_{sb}(v) \, \mathrm{sgn}(v) & \text{for } v \neq 0 \end{cases}$ |
| Fluid damping | Fluid–structure interactions | $f_d = \frac{1}{2} c_d \rho \dot{q}^2 \, \mathrm{sgn}(\dot{q})$ |

# 1.3 Representation of Damping in Vibration Analysis

It is not practical to incorporate detailed microscopic representations of damping in the dynamic analysis of systems. Instead, simplified models of damping that are representative of various types of energy dissipation are typically used. Consider a general $n$-degree-of-freedom mechanical system. Its motion can be represented by the vector $\mathbf{x}$ of $n$ generalized coordinates, $x_i$, representing the independent motions of the inertia elements. For small displacements, linear spring elements can be assumed. The corresponding equations of motion may be expressed in the vector matrix form

$$\mathbf{M}\ddot{\mathbf{x}} + \mathbf{d} + \mathbf{K}\mathbf{x} = \mathbf{f}(t) \tag{1.27}$$

in which $\mathbf{M}$ is the mass (inertia) matrix and $\mathbf{K}$ is the stiffness matrix. The forcing-function vector is $\mathbf{f}(t)$. The damping force vector $\mathbf{d}(\mathbf{x}, \dot{\mathbf{x}})$ is generally a nonlinear function of $\mathbf{x}$ and $\dot{\mathbf{x}}$. The type of damping used in the system model may be represented by the nature of $\mathbf{d}$ that is employed in the system equations. The various damping models that may be used, as discussed in the previous section, are listed in Table 1.1. Only the linear viscous damping term given in Table 1.1 is amenable to simplified mathematical analysis. In simplified dynamic models, other types of damping terms are usually replaced by an equivalent viscous damping term. Equivalent viscous damping is chosen so that its energy dissipation per cycle of oscillation is equal to that for the original damping. The resulting equations of motion are expressed by

$$\mathbf{M}\ddot{\mathbf{x}} + \mathbf{C}\dot{\mathbf{x}} + \mathbf{K}\mathbf{x} = \mathbf{f}(t) \tag{1.28}$$

In modal analysis of vibratory systems, the most commonly used model is proportional damping, where the damping matrix satisfies

$$\mathbf{C} = c_m\mathbf{M} + c_k\mathbf{K} \tag{1.29}$$

The first term on the right-hand side of Equation 1.29 is known as the inertial damping matrix. The corresponding damping force on each concentrated mass is proportional to its momentum. It represents the energy loss associated with a change in momentum (for example, during an impact). The second term is known as the stiffness damping matrix. The corresponding damping force is proportional to the rate of change of the local deformation forces at joints near the concentrated mass elements. Consequently, it represents a simplified form of linear structural damping. If damping is of the proportional type, it follows that the damped motion can be uncoupled into individual modes. This means that, if the damping model is of the proportional type, the damped system (as well as the undamped system) will possess real modes.

**TABLE 1.1**  Some Common Damping Models Used in Dynamic System Equations

| Damping Type | Simplified Model $d_i$ |
|---|---|
| Viscous | $\sum_j c_{ij}\dot{x}_j$ |
| Hysteretic | $\sum_j \dfrac{1}{\omega} c_{ij}\dot{x}_j$ |
| Structural | $\sum_j c_{ij}|x_j|\,\mathrm{sgn}(\dot{x}_j)$ |
| Structural Coulomb | $\sum_j c_{ij}\,\mathrm{sgn}(\dot{x}_j)$ |
| Fluid | $\sum_j c_{ij}|\dot{x}_j|\dot{x}_j$ |

### 1.3.1   Equivalent Viscous Damping

Consider a linear, single-DoF system with viscous damping, subjected to an external excitation. The equation of motion, for a unit mass, is given by

$$\ddot{x} + 2\zeta\omega_n\dot{x} + \omega_n^2 x = \omega_n^2 u(t) \tag{1.30}$$

If the excitation force is harmonic, with frequency $\omega$, we have

$$u(t) = u_0 \cos \omega t \tag{1.31}$$

Then, the response of the system at steady state is given by

$$x = x_0 \cos(\omega t + \phi) \tag{1.32}$$

in which the response amplitude is

$$x_0 = u_0 \frac{\omega_n^2}{\left[(\omega_n^2 - \omega^2) + 4\zeta^2\omega_n^2\omega^2\right]^{1/2}} \tag{1.33}$$

and the response phase lead is

$$\phi = -\tan^{-1}\frac{2\zeta\omega_n\omega}{(\omega_n^2 - \omega^2)} \tag{1.34}$$

The energy dissipation (i.e., damping capacity), $\Delta U$, per unit mass in one cycle is given by the net work done by the damping force, $f_d$; thus,

$$\Delta U = \oint f_d \, dx = \int_{-\phi/\omega}^{(2\pi - \phi)\omega} f_d \dot{x} \, dt \tag{1.35}$$

Since the viscous damping force, normalized with respect to mass (see Equation 1.30), is given by

$$f_d = 2\zeta\omega_n\dot{x} \tag{1.36}$$

the damping capacity, $\Delta U_v$, for viscous damping, can be obtained as

$$\Delta U_v = 2\zeta\omega_n \int_0^{2\pi/\omega} \dot{x}^2 \, dt \tag{1.37}$$

Finally, using Equation 1.32 in Equation 1.37 we get

$$\Delta U_v = 2\pi x_0^2 \omega_n \omega \zeta \tag{1.38}$$

For any general type of damping (see Table 1.1), the equation of motion becomes

$$\ddot{x} + \mathbf{d}(x, \dot{x}) + \omega_n^2 x = \omega_n^2 u(t) \tag{1.39}$$

The energy dissipation in one cycle (Equation 1.35) is given by

$$\Delta U = \int_{-\phi/\omega}^{(2\pi - \phi)/\omega} d(x, \dot{x})\dot{x} \, dt \tag{1.40}$$

Various damping force expressions, $d(x, \dot{x})$, normalized with respect to mass, are given in Table 1.2. For fluid damping, for example, the damping capacity is

$$\Delta U_f = \int_{-\phi/\omega}^{(2\pi - \phi)/\omega} c|\dot{x}|\dot{x}^2 \, dt \tag{1.41}$$

By substituting Equation 1.32 in Equation 1.41 for steady, harmonic motion we obtain

$$\Delta U_f = \tfrac{8}{3} c x_0^3 \omega^2 \tag{1.42}$$

**TABLE 1.2** Equivalent Damping Ratio Expressions for Some Common Types of Damping

| Damping Type | Damping Force, $d(x, \dot{x})$, per Unit Mass | Equivalent Damping Ratio, $\zeta_{eq}$ |
|---|---|---|
| Viscous | $2\zeta\omega_n\dot{x}$ | $\zeta$ |
| Hysteretic | $\dfrac{c}{\omega}\dot{x}$ | $\dfrac{c}{2\omega_n\omega}$ |
| Structural | $c\lvert x\rvert\,\mathrm{sgn}(\dot{x})$ | $\dfrac{c}{\pi\omega_n\omega}$ |
| Structural Coulomb | $c\,\mathrm{sgn}(\dot{x})$ | $\dfrac{2c}{\pi x_0\omega_n\omega}$ |
| Fluid | $c\lvert\dot{x}\rvert\dot{x}$ | $\dfrac{4}{3\pi}\left(\dfrac{\omega}{\omega_n}\right)x_0 c$ |

By comparing Equation 1.42 with Equation 1.38, the equivalent damping ratio for fluid damping is obtained as

$$\zeta_f = \frac{4}{3\pi}\left(\frac{\omega}{\omega_n}\right)x_0 c \tag{1.43}$$

in which $x_0$ is the amplitude of steady-state vibrations, as given by Equation 1.33. For the other types of damping listed in Table 1.1, expressions for the equivalent damping ratio can be obtained in a similar manner. The corresponding equivalent damping ratio expressions are given in Table 1.2. It should be noted that, for nonviscous damping types, $\zeta$ is generally a function of the frequency of oscillation, $\omega$, and the amplitude of excitation, $u_0$. It should be noted that the expressions given in Table 1.2 are derived assuming harmonic excitation. Engineering judgment should be exercised when employing these expressions for nonharmonic excitations.

For multi-DoF systems that incorporate proportional damping, the equations of motion can be transformed into a set of one-DoF equations (modal equations) of the type given in Equation 1.30. In this case, the damping ratio and natural frequency correspond to the respective modal values and, in particular, $\omega = \omega_n$.

## 1.3.2 Complex Stiffness

Consider a linear spring of stiffness $k$ connected in parallel with a linear viscous damper of damping constant $c$, as shown in Figure 1.5(a). Suppose that a force, $f$, is applied to the system, moving it through distance $x$ from the relaxed position of the spring. Then we have

$$f = kx + c\dot{x} \tag{1.44}$$

Suppose that the motion is harmonic, as given by

$$x = x_0 \cos \omega t \tag{1.45}$$

It is clear that the spring force, $kx$, is in phase with the displacement, but the damping force, $c\dot{x}$, has a 90° phase lead with respect to the displacement. This is because the velocity, $\dot{x} = -x_0\omega \sin \omega t = x_0\omega \cos(\omega t + \pi/2)$, has a 90° phase lead with respect to $x$. Specifically, we have

$$f = kx_0 \cos \omega t + cx_0\omega \cos\left(\omega t + \frac{\pi}{2}\right) \tag{1.46}$$

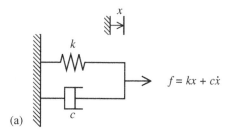

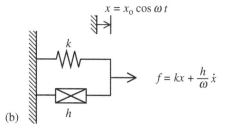

**FIGURE 1.5** Spring element in parallel with (a) a viscous damper and (b) a hysteretic damper.

This same fact may be represented by using complex numbers, where the in-phase component is considered as the real part and the 90° phase lead component is considered as the imaginary part with each component oscillating at the same frequency $\omega$. Then, we can write Equation 1.46 in the equivalent form

$$f = kx + j\omega cx \qquad (1.47)$$

This is exactly what we get by starting with the complex representation of the displacement

$$x = x_0\, e^{j\omega t} \qquad (1.48)$$

and substituting it in Equation 1.44. We note that Equation 1.47 may be written as

$$f = k^* x \qquad (1.49)$$

where $k^*$ is a "complex" stiffness, given by

$$k^* = k + j\omega c \qquad (1.50)$$

Clearly, the system itself and its two components (spring and damper) are real. Their individual forces are also real. The complex stiffness is simply a mathematical representation of the two force components (spring force and damping force), which are 90° out of phase, when subjected to harmonic motion. It follows that the linear damper may be "mathematically" represented by an "imaginary" stiffness. In the case of viscous damping this imaginary stiffness (and hence, the damping force magnitude) increases linearly with the frequency, $\omega$, of the harmonic motion. The concept of complex stiffness when dealing with discrete dampers is analogous to the use of complex elastic modulus in material damping, as discussed earlier in this chapter.

We have noted that, for hysteretic damping, the damping force (or damping stress) is independent of the frequency in harmonic motion. It follows that a hysteretic damper may be represented by an equivalent damping constant of

$$c = \frac{h}{\omega} \qquad (1.51)$$

which is valid for a harmonic motion (e.g., modal motion or forced motion) of frequency $\omega$. This situation is shown in Figure 1.5(b). It can be seen that the corresponding complex stiffness is

$$k^* = k + jh \qquad (1.52)$$

## Example 1.3

A flexible system consists of a mass, $m$, attached to the hysteretic damper and spring combination shown in Figure 1.5(b). What is the frequency response function of the system relating an excitation force, $f$, applied to the mass and the resulting displacement response, $x$? Obtain the resonant frequency of the system. Compare the results with the case for viscous damping.

## Solution

For a harmonic motion of frequency $\omega$, the equation of motion of the system is

$$m\ddot{x} + \frac{h}{\omega}\dot{x} + kx = f \qquad (1.53)$$

With a forcing excitation of $f = f_0\, e^{j\omega t}$ and the resulting steady-state response, $x = x_0\, e^{j\omega t}$, where $x_0$ has a phase difference (i.e., it is a complex function) with respect to $f_0$. Then, in the frequency domain, substituting the harmonic response $x = x_0\, e^{j\omega t}$ into Equation 1.53 we get

$$\left[ -\omega^2 m + \frac{h}{\omega} j\omega + k \right] x = f$$

resulting in the frequency transfer function

$$\frac{x}{f} = \frac{1}{[k - \omega^2 m + jh]} \tag{1.54}$$

Note that, as usual, this result is obtained simply by substituting $j\omega$ for $d/dt$. The magnitude of transfer function is at a maximum at resonance. This corresponds to a minimum value of

$$p(\omega) = (k - \omega^2 m)^2 + h^2$$

If we set $dp/d\omega = 0$, we get,

$$2(k - \omega^2 m)(-2\omega) = 0$$

Hence, the resonant frequency corresponds to the root of

$$k - \omega^2 m = 0$$

This gives the resonant frequency

$$\omega_r = \sqrt{\frac{k}{m}} \tag{1.55}$$

Note that, in the case of hysteretic damping, the resonant frequency is equal to the undamped natural frequency, $\omega_n$, and, unlike in the case of viscous damping, does not depend on the level of damping itself. For convenience consider the system response as the spring force

$$f_s = kx \tag{1.56}$$

Then, a normalized transfer function is obtained, as given by

$$\frac{f_s}{f} = G(j\omega) = \frac{1}{\left[1 - \omega^2 \dfrac{m}{k} + j\dfrac{h}{k}\right]} \tag{1.57}$$

or,

$$\frac{f_s}{f} = \frac{1}{[1 - r^2 + j\alpha]} \tag{1.58}$$

where

$$r = \frac{\omega}{\omega_n} \quad \text{and} \quad \alpha = \frac{h}{k} \tag{1.59}$$

which are the normalized frequency and the normalized hysteretic damping, respectively. The magnitude of the transfer function is

$$\left|\frac{f_s}{f}\right| = \frac{1}{\sqrt{(1 - r^2)^2 + \alpha^2}} \tag{1.60}$$

and the phase angle (phase lead) is

$$\angle f_s/f = -\tan^{-1} \frac{\alpha}{(1 - r^2)} \tag{1.61}$$

These results are sketched in Figure 1.6.

## 1.3.3 Loss Factor

We define the *damping capacity* of a device (damper) as the energy dissipated in a complete cycle of motion; specifically

$$\Delta U = \oint f_d \, dx \tag{1.62}$$

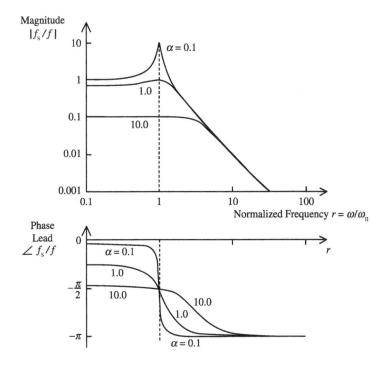

**FIGURE 1.6**   Frequency transfer function of a simple oscillator with hysteretic damping.

This is given by the area of the hysteresis loop in the displacement force plane. If the initial (total) energy of the system is denoted by $U_{max}$, then the *specific damping capacity, D,* is given by the ratio

$$D = \frac{\Delta U}{U_{max}} \tag{1.63}$$

The *loss factor, $\eta$,* is the specific damping capacity per radian of the damping cycle. Hence,

$$\eta = \frac{\Delta U}{2\pi U_{max}} \tag{1.64}$$

Note that $U_{max}$ is approximately equal to the maximum kinetic energy and also to the maximum potential energy of the device when the damping is low.

Equation 1.38 gives the damping capacity per unit mass of a device with viscous damping as

$$\Delta U = 2\pi x_0^2 \omega_n \omega \zeta \tag{1.65}$$

Here, $x_0$ is the amplitude and $\omega$ is the frequency of harmonic motion of the device, $\omega_n$ is the undamped natural frequency and $\zeta$ is the damping ratio. The maximum potential energy per unit mass of the system is

$$U_{max} = \frac{1}{2}\frac{k}{m}x_0^2 = \frac{1}{2}\omega_n^2 x_0^2 \tag{1.66}$$

Hence, from Equation 1.64, the loss factor for a viscous damped simple oscillator is given by

$$\eta = \frac{2\pi x_0^2 \omega_n \omega \zeta}{2\pi \times \frac{1}{2}\omega_n^2 x_0^2} = \frac{2\omega\zeta}{\omega_n} \tag{1.67}$$

For free decay of the system, we have $\omega = \omega_d \cong \omega_n$, where the latter approximation holds for low damping. For forced oscillation, the worst response conditions occur when $\omega = \omega_d \cong \omega_n$, which is what one must consider with regard to energy dissipation. In either case, the loss factor is approximately given by

$$\eta = 2\zeta \qquad (1.68)$$

For other types of damping, Equation 1.68 will still hold when the equivalent damping ratio, $\zeta_{eq}$, (see Table 1.2) is used in place of $\zeta$.

The loss factors of some common materials are given in Table 1.3. Definitions of useful damping parameters, as defined here, are summarized in Table 1.4. Expressions of loss factors for some useful damping models are given in Table 1.5.

TABLE 1.3  Loss Factors of Some Useful Materials

| Material | Loss Factor $\eta \cong 2\zeta$ |
|---|---|
| Aluminum | $2 \times 10^{-5}$ to $2 \times 10^{-3}$ |
| Concrete | 0.02 to 0.06 |
| Glass | 0.001 to 0.002 |
| Rubber | 0.1 to 1.0 |
| Steel | 0.002 to 0.01 |
| Wood | 0.005 to 0.01 |

TABLE 1.4  Definitions of Damping Parameters

| Parameter | Definition | Mathematical Formula |
|---|---|---|
| Damping capacity ($\Delta U$) | Energy dissipated per cycle of motion (area of displacement–force hysteresis loop) | $\oint f_d \, dx$ |
| Damping capacity per volume ($d$) | Energy dissipated per cycle per unit material volume (area of strain–stress hysteresis loop) | $\oint \sigma \, d\varepsilon$ |
| Specific damping capacity ($D$) | Ratio of energy dissipated per cycle ($\Delta U$) to the initial maximum energy ($U_{max}$) *Note*: for low damping, $U_{max}$ = maximum potential energy = maximum kinetic energy | $\dfrac{\Delta U}{U_{max}}$ |
| Loss factor ($\eta$) | Specific damping capacity per unit angle of cycle. *Note*: for low damping, $\eta = 2 \times$ damping ratio | $\dfrac{\Delta U}{2\pi U_{max}}$ |

TABLE 1.5  Loss Factors for Several Material Damping Models

| Material Damping Model | Stress–Strain Constitute Relation | Loss Factor ($\eta$) |
|---|---|---|
| Viscoelastic Kelvin–Voigt | $\sigma = E\varepsilon + E^* \dfrac{d\varepsilon}{dt}$ | $\dfrac{\omega E^*}{E}$ |
| Hysteretic Kelvin–Voigt | $\sigma = E\varepsilon + \dfrac{\tilde{E}}{\omega}\dfrac{d\varepsilon}{dt}$ | $\dfrac{\tilde{E}}{E}$ |
| Viscoelastic standard linear solid | $\sigma + c_s \dfrac{d\sigma}{dt} = E\varepsilon + E^*\dfrac{d\varepsilon}{dt}$ | $\dfrac{\omega E^*}{E}\dfrac{(1 - c_s E/E^*)}{(1 + \omega^2 c_s)}$ |
| Hysteretic standard linear solid | $\sigma + c_s \dfrac{d\sigma}{dt} = E\varepsilon + \dfrac{\tilde{E}}{\omega}\dfrac{d\varepsilon}{dt}$ | $\dfrac{\tilde{E}}{E}\dfrac{(1 - \omega c_s E/\tilde{E})}{(1 + \omega^2 c_s)}$ |

## 1.4  Measurement of Damping

Damping may be represented by various parameters (such as specific damping capacity, loss factor, $Q$-factor, and damping ratio) and models (such as viscous, hysteretic, structural, and fluid). Before attempting to measure damping in a system, we need to decide on a representation (model) that will adequately characterize the nature of mechanical energy dissipation in the system. Next, we should decide on the parameter or parameters of the model that need to be measured.

It is extremely difficult to develop a realistic yet tractable model for damping in a complex piece of equipment operating under various conditions of mechanical interaction. Even if a satisfactory damping modal is developed, experimental determination of its parameters could be tedious. A major difficulty arises because it usually is not possible to isolate various types of damping (for example, material, structural, and fluid) from an overall measurement. Furthermore, damping measurements must be conducted under actual operating conditions for them to be realistic.

If one type of damping (say, fluid damping) is eliminated during the actual measurement then it would not represent the true operating conditions. This would also eliminate possible interacting effects of the eliminated damping type with the other types. In particular, overall damping in a system is not generally equal to the sum of the individual damping values when they are acting independently. Another limitation of computing equivalent damping values using experimental data arises because it is assumed for analytical simplicity that the dynamic system behavior is linear. If the system is highly nonlinear, a significant error could be introduced into the damping estimate. Nevertheless, it is customary to assume linear viscous behavior when estimating damping parameters using experimental data.

There are two general ways by which damping measurements can be made: using a time–response record and using a frequency–response function of the system to estimate damping.

### 1.4.1  Logarithmic Decrement Method

This is perhaps the most popular time–response method that is used to measure damping. When a single-DoF oscillatory system with viscous damping (see Equation 1.30) is excited by an impulse input (or an initial condition excitation), its response takes the form of a time decay (see Figure 1.7), given by

$$y(t) = y_0 \exp(-\zeta \omega_n t) \sin \omega_d t \tag{1.69}$$

in which the damped natural frequency is given by

$$\omega_d = \sqrt{1 - \zeta^2} \, \omega_n \tag{1.70}$$

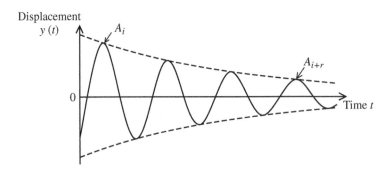

**FIGURE 1.7**   Impulse response of a simple oscillator.

If the response at $t = t_i$ is denoted by $y_i$, and the response at $t = t_i + 2\pi r/\omega_d$ is denoted by $y_{i+r}$, then, from Equation 1.69, we have

$$\frac{y_{i+r}}{y_i} = \exp\left(-\zeta \frac{\omega_n}{\omega_d} 2\pi r\right), \qquad i = 1, 2, \ldots, n$$

In particular, suppose that $y_i$ corresponds to a peak point in the time decay function, having magnitude $A_i$, and that $y_{i+r}$ corresponds to the peak point $r$ cycles later in the time history, and its magnitude is denoted by $A_{i+r}$ (see Figure 1.7). Even though the above equation holds for any pair of points that are $r$ periods apart in the time history, the peak points seem to be the appropriate choice for measurement in the present procedure, as these values would be more prominent than any arbitrary points in a response–time history. Then,

$$\frac{A_{i+r}}{A_i} = \exp\left(-\zeta \frac{\omega_n}{\omega_d} 2\pi r\right) = \exp\left[-\frac{\zeta}{\sqrt{1 - \zeta^2}} 2\pi r\right]$$

where Equation 1.70 has been used. Then, the logarithmic decrement $\delta$ is given by (per unit cycle)

$$\delta = \frac{1}{r} \ln\left(\frac{A_i}{A_{i+r}}\right) = \frac{2\pi\zeta}{\sqrt{1 - \zeta^2}} \tag{1.71}$$

or the damping ratio may be expressed as

$$\zeta = \frac{1}{\sqrt{1 + (2\pi/\delta)^2}} \tag{1.72}$$

For low damping (typically, $\zeta < 0.1$), $\omega_d \cong \omega_n$ and Equation 1.71 become

$$\frac{A_{i+r}}{A_i} \cong \exp(-\zeta 2\pi r) \tag{1.73}$$

or

$$\zeta = \frac{1}{2\pi r} \ln\left(\frac{A_i}{A_{i+r}}\right) = \frac{\delta}{2\pi} \qquad \text{for } \zeta < 0.1 \tag{1.74}$$

This is in fact the "per-radian" logarithmic decrement.

The damping ratio can be estimated from a free-decay record, using Equation 1.74. Specifically, the ratio of the extreme amplitudes in prominent $r$ cycles of decay is determined and substituted into Equation 1.74 to get the equivalent damping ratio.

Alternatively, if $n$ cycles of damped oscillation are needed for the amplitude to decay by a factor of two, for example, then, from Equation 1.74, we get

$$\zeta = \frac{1}{2\pi n} \ln(2) = \frac{0.11}{n} \qquad \text{for } \zeta < 0.1 \tag{1.75}$$

For slow decays (low damping), we have

$$\ln\left(\frac{A_i}{A_{i+1}}\right) \cong \frac{2(A_i - A_{i+1})}{(A_i + A_{i+1})} \tag{1.76}$$

Then, from Equation 1.74, we get

$$\zeta = \frac{A_i - A_{i+1}}{\pi(A_i + A_{i+1})} \qquad \text{for } \zeta < 0.1 \tag{1.77}$$

Any one of Equation 1.72, Equation 1.74, Equation 1.75, and Equation 1.77 could be employed in computing $\zeta$ from test data. It should be noted that the results assume single-DoF system behavior. For multi-DoF systems, the modal damping ratio for each mode can be determined using this method if the initial excitation is such that the decay takes place primarily in one mode of vibration.

In other words, substantial modal separation and the presence of "real" modes (not "complex" modes with nonproportional damping) are assumed.

## 1.4.2 Step–Response Method

This is also a time–response method. If a unit-step excitation is applied to the single-DoF oscillatory system given by Equation 1.30, its time–response is given by

$$y(t) = 1 - \frac{1}{\sqrt{1-\zeta^2}} \exp(-\zeta\omega_n t)\sin(\omega_d t + \phi)$$

(1.78)

in which $\phi = \cos\zeta$. A typical step–response curve is shown in Figure 1.8. The time at the first peak (peak time), $T_p$, is given by

$$T_p = \frac{\pi}{\omega_d} = \frac{\pi}{\sqrt{1-\zeta^2}\,\omega_n}$$

(1.79)

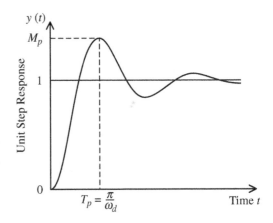

**FIGURE 1.8** A typical step–response of a simple oscillator.

The response at peak time (peak value), $M_p$, is given by

$$M_p = 1 + \exp(-\zeta\omega_n T_p) = 1 + \exp\left(\frac{-\pi\zeta}{\sqrt{1-\zeta^2}}\right)$$

(1.80)

The percentage overshoot, PO, is given by

$$PO = (M_p - 1) \times 100\% = 100\,\exp\left(\frac{-\pi\zeta}{\sqrt{1-\zeta^2}}\right)$$

(1.81)

It follows that, if any one parameter of $T_p$, $M_p$ or PO is known from a step–response record, the corresponding damping ratio, $\zeta$, can be computed by using the appropriate relationship from the following:

$$\zeta = \sqrt{1 - \left(\frac{\pi}{T_p\omega_n}\right)^2}$$

(1.82)

$$\zeta = \frac{1}{\sqrt{1 + \dfrac{1}{\left[\dfrac{\ln(M_p - 1)}{\pi}\right]^2}}}$$

(1.83)

$$\zeta = \frac{1}{\sqrt{1 + \dfrac{1}{\left[\dfrac{\ln(PO/100)}{\pi}\right]^2}}}$$

(1.84)

It should be noted that when determining $M_p$ the response curve should be normalized to unit steady-state value. Furthermore, the results are valid only for single-DoF systems and modal excitations in multi-DoF systems.

## 1.4.3 Hysteresis Loop Method

For a damped system, the force versus displacement cycle produces a hysteresis loop. Depending on the inertial and elastic characteristics and other conservative loading conditions (e.g., gravity) in the system, the shape of the hysteresis loop will change. But the work done by conservative forces (e.g., inertial, elastic, and gravitational) in a complete cycle of motion will be zero. Consequently, the net work done will be equal to the energy dissipated due to damping only. Accordingly, the area of the displacement–force hysteresis loop will give the damping capacity, $\Delta U$ (see Equation 1.62). The maximum energy in the system can also be determined from the displacement–force curve. Then, the loss factor, $\eta$, can be computed using Equation 1.64, and the damping ratio from Equation 1.68. This method of damping measurement may also be considered basically as a time domain method.

Note that Equation 1.65 is the work done against (i.e., energy dissipation in) a single loading–unloading cycle per unit mass. It should be recalled that $2\zeta\omega_n = c/m$, where $c =$ viscous damping constant and $m =$ mass. Accordingly, from Equation 1.65, the energy dissipation per unit mass and per hystereris loop is $\Delta U = \pi x_0^2 \omega c/m$. Hence, without normalizing with respect to mass, the energy dissipation per hysteresis loop of viscous damping is

$$\Delta U_v = \pi x_0^2 \omega c \qquad (1.85)$$

Equation 1.85 can be derived by performing the cyclic integration indicated in Equation 1.62 with the damping force $f_d = c\dot{x}$, harmonic motion $x = x_0\, e^{j\omega t}$ and the integration interval $t = 0$ to $2\pi/\omega$.

Similarly, in view of Equation 1.51, the energy dissipation per hysteresis loop of hysteretic damping is

$$\Delta U_h = \pi x_0^2 h \qquad (1.86)$$

Now, since the initial maximum energy may be represented by the initial maximum potential energy, we have

$$U_{max} = \tfrac{1}{2} k x_0^2 \qquad (1.87)$$

Note that the stiffness, $k$, may be measured as the average slope of the displacement–force hysteresis loop. Hence, in view of Equation 1.64, the loss factor for hysteretic damping is given by

$$\eta = \frac{h}{k} \qquad (1.88)$$

Then, from Equation 1.68, the equivalent damping ratio for hysteretic damping is

$$\zeta = \frac{h}{2k} \qquad (1.89)$$

## Example 1.4

A damping material was tested by applying a loading cycle of $-900$ to $900$ N and back to $-900$ N to a thin bar made of the material and measuring the corresponding deflection. The smoothed load vs. deflection curve obtained in this experiment is shown in Figure 1.9. Assuming that the damping is predominantly of the hysteretic type, estimate

1. The hysteretic damping constant
2. The equivalent damping ratio

## Solution

Approximating the top and the bottom segments of the hysteresis loop by triangles, we estimate the area of the loop as

$$\Delta U_h = 2 \times \tfrac{1}{2} \times 2.5 \times 900 \text{ N.mm}$$

Alternatively, we may obtain this result by counting the squares within the hysteresis loop. The deflection amplitude is

$$x_0 = 8.5 \text{ mm}$$

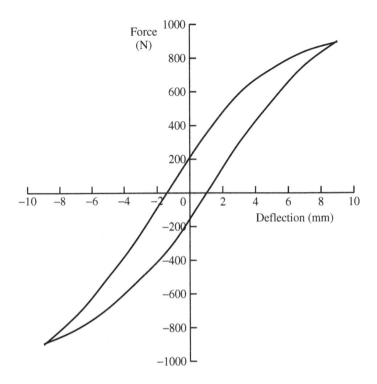

**FIGURE 1.9**   An experimental hysteresis loop of a damping material.

Hence, from Equation 1.86 we have

$$h = \frac{2 \times \dfrac{1}{2} \times 2.5 \times 900}{\pi \times 8.5^2} \text{ N/mm} = 9.9 \text{ N/mm}$$

The stiffness of the damping element is estimated as the average slope of the hysteresis loop; thus

$$k = \frac{600}{4.5} \text{ N/mm} = 133.3 \text{ N/mm}$$

Hence, from Equation 1.89, the equivalent damping ratio is

$$\zeta = \frac{9.9}{2 \times 133.3} \approx 0.04$$

### 1.4.4   Magnification Factor Method

This is a frequency–response method. Consider a single-DoF oscillatory system with viscous damping. The magnitude of its frequency–response function is

$$|H(\omega)| = \frac{\omega_n^2}{\left[(\omega_n^2 - \omega^2)^2 + 4\zeta^2 \omega_n^2 \omega^2\right]^{1/2}} \tag{1.90}$$

A plot of this expression with respect to $\omega$, the frequency of excitation, is given in Figure 1.10. The peak value of magnitude occurs when the denominator of the expression is at its minimum. This corresponds to

$$\frac{d}{d\omega}\left[(\omega_n^2 - \omega^2)^2 + 4\zeta^2 \omega_n^2 \omega^2\right] = 0 \tag{1.91}$$

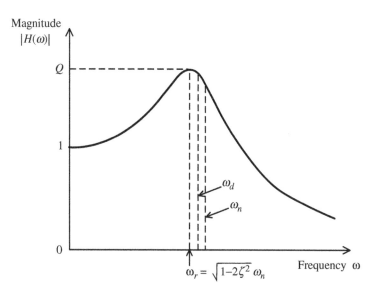

**FIGURE 1.10**  The magnification factor method of damping measurement applied to a single-DoF system.

The resulting solution for $\omega$ is termed the resonant frequency, $\omega_r$

$$\omega_r = \sqrt{1 - 2\zeta^2}\,\omega_n \tag{1.92}$$

It is noted that $\omega_r < \omega_d$ (see Equation 1.70), but for low damping ($\zeta < 0.1$), the values of $\omega_n$, $\omega_d$, and $\omega_r$ are nearly equal. The amplification factor, $Q$, which is the magnitude of the frequency–response function at resonant frequency, is obtained by substituting Equation 1.92 in Equation 1.90:

$$Q = \frac{1}{2\zeta\sqrt{1 - \zeta^2}} \tag{1.93}$$

For low damping ($\zeta < 0.1$), we have

$$Q = \frac{1}{2\zeta} \tag{1.94}$$

In fact, Equation 1.94 corresponds to the magnitude of the frequency–response function at $\omega = \omega_n$.

It follows that, if the magnitude curve of the frequency–response function (or a Bode plot) is available, then the system damping ratio, $\zeta$, can be estimated using Equation 1.94. When using this method, the frequency–response curve must be normalized so that its magnitude at zero frequency (termed *static gain*) is unity.

For a multi-DoF system modal damping values may be estimated from the magnitude of the Bode plot of its frequency–response function, provided that the modal frequencies are not too closely spaced and the system is lightly damped. Consider the logarithmic (to the base ten) magnitude plot shown in Figure 1.11. The magnitude is expressed in decibels (dB), which is calculated by multiplying the $\log_{10}$(magnitude) by a factor of 20. At the $i$th resonant frequency, $\omega_i$, the amplification factor, $q_i$ (in dB), is obtained by drawing an asymptote to the preceding segment of the curve and measuring the peak value from the asymptote. Then,

$$Q_i = (10)^{q_i/20} \tag{1.95}$$

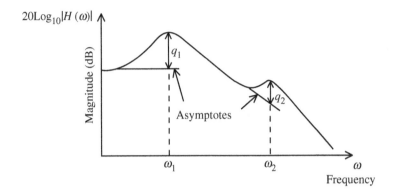

**FIGURE 1.11**   Magnification factor method applied to a multi-DoF system.

and the modal damping ratio

$$\zeta = \frac{1}{2Q_i}, \qquad i = 1, 2, \dots, n \tag{1.96}$$

If the significant resonances are closely spaced, curve-fitting to a suitable function may be necessary in order to determine the corresponding modal damping values. The Nyquist plot may also be used in computing damping using frequency domain data.

### 1.4.5   Bandwidth Method

The bandwidth method of damping measurement is also based on frequency–response. Consider the frequency–response function magnitude given by Equation 1.90 for a single-DoF, oscillatory system with viscous damping. The peak magnitude is given by Equation 1.94 for low damping. Bandwidth (half-power) is defined as the width of the frequency–response magnitude curve when the magnitude is $(1/\sqrt{2})$ times the peak value. This is denoted by $\Delta\omega$ (see Figure 1.12). An expression

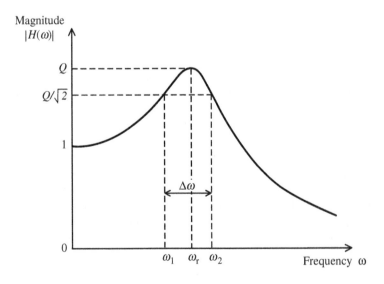

**FIGURE 1.12**   Bandwidth method of damping measurement in a single-DoF system.

for $\Delta\omega = \omega_2 - \omega_1$ is obtained below using Equation 1.90. By definition, $\omega_1$ and $\omega_2$ are the roots of the equation

$$\frac{\omega_n^2}{\left[(\omega_n^2 - \omega^2)^2 + 4\zeta^2\omega_n^2\omega^2\right]^{1/2}} = \frac{1}{\sqrt{2} \times 2\zeta} \tag{1.97}$$

for $\omega$. Equation 1.97 can be expressed in the form

$$\omega^4 - 2(1 - 2\zeta^2)\omega_n^2\omega^2 + (1 - 8\zeta^2)\omega_n^4 = 0 \tag{1.98}$$

This is a quadratic equation in $\omega_2$, having roots $\omega_1^2$ and $\omega_2^2$, which satisfy

$$(\omega^2 - \omega_1^2)(\omega^2 - \omega_2^2) = \omega^4 - (\omega_1^2 + \omega_2^2)\omega^2 + \omega_1^2\omega_2^2 = 0$$

Consequently,

$$\omega_1^2 + \omega_2^2 = 2(1 - 2\zeta^2)\omega_n^2 \tag{1.99}$$

and

$$\omega_1^2\omega_2^2 = (1 - 8\zeta^2)\omega_n^4 \tag{1.100}$$

It follows that

$$(\omega_2 - \omega_1)^2 = \omega_1^2 + \omega_2^2 - 2\omega_1\omega_2 = 2(1 - 2\zeta^2)\omega_n^2 - 2\sqrt{1 - 8\zeta^2}\,\omega_n^2$$

For small $\zeta$ (in comparison to 1), we have

$$\sqrt{1 - 8\zeta^2} \cong 1 - 4\zeta^2$$

Hence,

$$(\omega_2 - \omega_1)^2 \cong 4\zeta^2\omega_n^2$$

or, for low damping

$$\Delta\omega = 2\zeta\omega_n = 2\zeta\omega_r \tag{1.101}$$

From Equation 1.101 it follows that the damping ratio can be estimated from the bandwidth using the relation

$$\zeta = \frac{1}{2}\frac{\Delta\omega}{\omega_r} \tag{1.102}$$

For a multi-DoF system with widely spaced resonances, the foregoing method can be extended to estimate modal damping. Consider the frequency–response magnitude plot (in dB) shown in Figure 1.13.

Since a factor of $\sqrt{2}$ corresponds to 3 dB, the bandwidth corresponding to a resonance is given by the width of the magnitude plot at 3 dB below that resonant peak. For the *i*th mode, the damping ratio is given by

$$\zeta_i = \frac{1}{2}\frac{\Delta\omega_i}{\omega_i} \tag{1.103}$$

The bandwidth method of damping measurement indicates that the bandwidth at a resonance is a measure of the energy dissipation in the system in the neighborhood of that resonance. The simplified

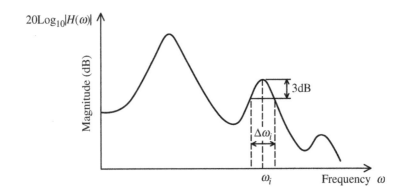

**FIGURE 1.13**  Bandwidth method of damping measurement in a multi-DoF system.

relationship given by Equation 1.103 is valid for low damping, however, and is based on linear system analysis. Several methods of damping measurement are summarized in Box 1.2.

## 1.4.6  General Remarks

There are limitations to the use of damping values that are experimentally determined. For example, consider time–response methods for determining the modal damping of a device for higher modes. The customary procedure is to first excite the system at the desired resonant frequency, using a harmonic exciter, and then to release the excitation mechanism. In the resulting transient vibration, however, there invariably will be modal interactions, except in the case of proportional damping. In this type of test, it is tacitly assumed that the device can be excited in the particular mode. In essence, proportional damping is assumed in modal damping measurements. This introduces a certain amount of error into the measured damping values.

Expressions used in computing damping parameters from test measurements are usually based on linear system theory. However, all practical devices exhibit some nonlinear behavior. If the degree of nonlinearity is high, the measured damping values will not be representative of the actual behavior of the system. Furthermore, testing to determine damping is usually performed at low amplitudes of vibration. The corresponding responses could be an order of magnitude lower than, for instance, the amplitudes exhibited under extreme operating conditions. Damping in practical devices increases with the amplitude of motion, except for relatively low amplitudes (see Figure 1.14 illustrating nonlinear behavior). Consequently, the damping values determined from experiments should be extrapolated when they are used to study the behavior of the system under various operating conditions. Alternatively, damping could be associated with a stress level in the device. Different components in a device are subjected to varying levels of stress, however, and it might be difficult to obtain a representative stress value for the entire device. One of the methods recommended for estimating damping in structures under seismic disturbances, for example, is by analyzing earthquake response records for structures that are similar to the one being considered. Some typical damping ratios that are applicable under operating basis earthquake (OBE) and safe-shutdown earthquake (SSE) conditions for a range of items are given in Table 1.6.

When damping values are estimated using frequency–response magnitude curves, accuracy becomes poor at very low damping ratios ($< 1\%$). The main reason for this is the difficulty in obtaining a sufficient number of points in the magnitude curve near a poorly damped resonance when the frequency–response function is determined experimentally. As a result, the magnitude curve is poorly defined in the neighborhood of a weakly damped resonance. For low damping ($< 2\%$), time–response methods are particularly useful. At high damping values, the rate of decay can be so fast that the measurements contain large errors. Modal interference in closely spaced modes can also affect measured damping results.

# Box 1.2

## DAMPING MEASUREMENT METHODS

| Method | Measurements | Formulas |
|---|---|---|
| Logarithmic decrement method | $A_i$ = first significant amplitude; $A_{i+r}$ = amplitude after $r$ cycles | Logarithmic decrement $$\delta = \frac{1}{r}\ln\frac{A_i}{A_{i+r}} \text{ (per cycle)}$$ $$\frac{\delta}{2\pi} = \frac{\zeta}{\sqrt{1-\zeta^2}} \text{ (per radian)}$$ or, $$\zeta = \frac{1}{\sqrt{1+(2\pi/\delta)^2}}$$ For low damping $$\zeta = \frac{\delta}{2\pi}$$ $$\zeta = \frac{A_i - A_{i+1}}{\pi(A_i + A_{i+1})}$$ |
| Step response method | $M_p$ = first peak value normalized r.t. steady-state value; PO = percentage overshoot (over steady-state value) | $$M_p = 1 + \exp\left[\frac{-\pi\zeta}{\sqrt{1-\zeta^2}}\right]$$ $$PO = 100\exp\left[\frac{-\pi\zeta}{\sqrt{1-\zeta^2}}\right]$$ |
| Hysteresis loop method | $\Delta U$ = area of displacement–force hysteresis loop; $x_0$ = maximum displacement of the hysteresis loop; $k$ = average slope of the hysteresis loop | Hysteretic damping constant $$h = \frac{\Delta U}{\pi x_0^2}$$ Loss factor $$\eta = \frac{h}{k}$$ Equivalent damping ratio $$\zeta = \frac{h}{2k}$$ |
| Magnification factor method | $Q$ = amplification at resonance, w.r.t. zero-frequency value | $$Q = \frac{1}{2\zeta\sqrt{1-\zeta^2}}$$ For low damping $$\zeta = \frac{1}{2Q}$$ |
| Bandwidth method | $\Delta\omega$ = bandwidth at $1/\sqrt{2}$ of resonant peak (i.e., half-power bandwidth); $\omega_r$ = resonant frequency | $$\zeta = \frac{\Delta\omega}{2\omega_r}$$ |

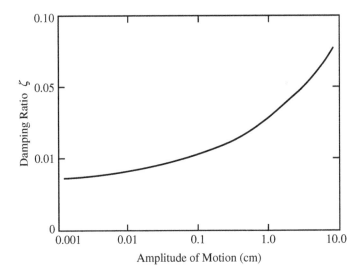

**FIGURE 1.14** Effect of vibration amplitude on damping in structures.

**TABLE 1.6** Typical Damping Values Suggested by ASME for Seismic Applications

| System | Damping Ratio ($\zeta$%) | |
|---|---|---|
| | OBE | SSE |
| Equipment and large diameter piping systems ($>$ 12 in. diameter) | 2 | 3 |
| Small diameter piping systems ($\leq$ 12 in. diameter) | 1 | 2 |
| Welded steel structures | 2 | 4 |
| Bolted steel structures | 4 | 7 |
| Prestressed concrete structures | 2 | 5 |
| Reinforced concrete structures | 4 | 7 |

## 1.5 Interface Damping

In many practical applications damping is generated at the interface of two sliding surfaces. This is the case, for example, in bearings, gears, screws, and guideways. Even though this type of damping is commonly treated under structural damping, due to its significance we will consider it again here in more detail, as a category of its own.

Interface damping was formally considered by DaVinci in the early 1500s and again by Coulomb in the 1700s. The simplified model used by them is the well-known Coulomb friction model as given by

$$f = \mu R \operatorname{sgn}(v) \tag{1.104}$$

where

$f$ = the frictional force that opposes the motion
$R$ = the normal reaction force between the sliding surfaces
$v$ = the relative velocity between the sliding surfaces
$\mu$ = the coefficient of friction

Note that the signum function "sgn" is used to emphasize that $f$ is in the opposite direction of $v$. This simple model is not expected to provide accurate results in all cases of interface damping. It is known that, apart from the loading conditions, interface damping depends on a variety of factors such as material properties, surface characteristics, nature of lubrication, geometry of the moving parts, and the magnitude of the relative velocity.

A somewhat more complete model for interface damping, incorporating the following characteristics, is shown in Figure 1.15:

1. Static and dynamic friction, with stiction and stick–slip behavior.
2. Conventional Coulomb friction (Region 1).
3. A drop in dynamic friction, with a negative slope, before increasing again. This is known as the "Stribeck effect" (Region 2).
4. Conventional viscous damping (Region 3).

These characteristics cover the behavior of interface damping that is commonly observed in practice. In particular, suppose that a force is exerted to generate a relative motion between two surfaces. For small values of the force, there will not be a relative motion, in view of friction. The minimum force, $f_s$, that is needed for the motion to start is the static frictional force. The force that is needed to maintain the motion will drop instantaneously to $f_d$, as the motion begins. It is as though initially the two surfaces were "stuck" and $f_s$ is the necessary breakaway force. Hence, this characteristic is known as stiction. The minimum force $f_d$ that is needed to maintain the relative motion between the two surfaces is called dynamic friction. In fact, under dynamic conditions, it is possible for "stick–slip" to occur where repeated sticking and breaking away cycles of intermittent motion take place. Clearly, such "chattering" motion corresponds to instability (for example, in machine tools). It is an undesirable effect and should be avoided.

After the relative motion begins, conventional Coulomb type damping behavior may dominate for small relative velocities, as represented in Region 1. For lubricated surfaces, at low relative velocities, there will be some solid-to-solid contact that generates a Coulomb-type damping force. As the relative speed increases, the degree of this solid-to-solid contact will decrease and the damping force will drop, as in Region 2 of Figure 1.15. This characteristic is known as the Stribeck effect. Since the slope of the friction curve is negative in Regions 1 and 2, this corresponds to the unstable region. As the relative velocity is further increased, in fully lubricated surfaces, viscous-type damping will dominate, as shown in Region 3 of Figure 1.15. This is the stable region. It follows that a combined model of interface damping may be expressed as

$$f = \begin{cases} f_s & \text{for } v = 0 \\ f_{sb}(v)\text{sgn}(v) + bv & \text{for } v \neq 0 \end{cases} \tag{1.105}$$

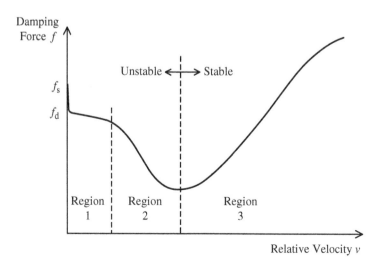

**FIGURE 1.15** Main characteristics of interface damping.

Note that $f_{sb}(v)$ is a nonlinear function of velocity that will represent both dynamic friction (for $v > 0$) and the Stribeck effect. Models that have been used to represent this effect include the following

$$f_{sb} = \frac{f_d}{1 + (v/v_c)^2}$$  (1.106)

$$f_{sb} = f_d\, e^{-(v/v_c)^2}$$  (1.107)

and

$$f_{sb} = (f_d + \alpha |v|^{1/2})\mathrm{sgn}(v)$$  (1.108)

Note that $f_d$ represents dynamic Coulomb friction and $v_c$ and $\alpha$ are modal parameters.

## Example 1.5

An object of mass $m$ rests on a horizontal surface and is attached to a spring of stiffness $k$, as shown in Figure 1.16. The mass is pulled so that the extension of the spring is $x_0$, and is moved from rest from that position. Determine the subsequent sliding motion of the object. The coefficient of friction between the object at the horizontal surface is $\mu$.

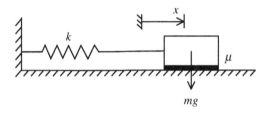

FIGURE 1.16  An object sliding against Coulomb friction.

## Solution

Note that, when the object moves to the left, the frictional force, $\mu mg$ acts to the right, and *vice versa*. Now, consider the first cycle of motion, stating from rest with $x = x_0$ moving to the left, coming to rest with the spring compressed and then moving to the right.

### First Half Cycle (Moving to Left)
The equation of motion is

$$m\ddot{x} = -kx + \mu mg$$  (i)

or

$$\ddot{x} + \omega_n^2 x = \mu g$$  (ii)

where $\omega_n = \sqrt{k/m}$ is the undamped material frequency. Equation ii has a homogeneous solution of

$$x_h = A_1 \sin(\omega_n t) + A_2 \cos(\omega_n t)$$  (iii)

and a particular solution of

$$x_p = \frac{\mu g}{\omega_n^2}$$  (iv)

Hence, the total solution is

$$x = A_1 \sin(\omega_n t) + A_2 \cos(\omega_n t) + \frac{\mu g}{\omega_n^2}$$  (v)

Using the initial conditions $x = x_0$ and $\dot{x} = 0$ at $t = 0$, we get $A_1 = 0$ and $A_2 = x_0 - (\mu g/\omega_n^2)$ Hence, Equation v becomes

$$x = \left(x_0 - \frac{\mu g}{\omega_n^2}\right) \cos(\omega_n t) + \frac{\mu g}{\omega_n^2}$$  (vi)

At the end of this half cycle we have $\dot{x} = 0$ or $\sin \omega_n t = 0$. Hence the corresponding time is $t = \pi/\omega_n$. Substituting this in Equation vi, the corresponding position of the object is (note: $\cos \pi = -1$)

$$x_{l1} = -\left(x_0 - \frac{2\mu g}{\omega_n^2}\right) \tag{vii}$$

**Second Half Cycle (Moving to Right)**
The equation of motion is

$$m\ddot{x} = -kx - \mu mg \tag{viii}$$

or

$$\ddot{x} + \omega_n^2 x = -\mu g \tag{ix}$$

The corresponding response is given by

$$x = B_1 \sin(\omega_n t) + B_2 \cos(\omega_n t) - \frac{\mu g}{\omega_n^2} \tag{x}$$

Using the initial conditions $x = -(x_0 - (2\mu g/\omega_n^2))$ and $\dot{x} = 0$ at $t = \pi/\omega_n$, we get $B_1 = 0$ and $B_2 = x_0 - (3\mu g/\omega_n^2)$. Hence, Equation x becomes

$$x = \left(x_0 - \frac{3\mu g}{\omega_n^2}\right) \cos(\omega_n t) - \frac{\mu g}{\omega_n^2} \tag{xi}$$

The object will come to rest ($\dot{x} = 0$) next at $t = 2\pi/\omega_n$, hence the position of the object at the end of the present half cycle would be

$$x_1 = x_0 - \frac{4\mu g}{\omega_n^2} \tag{xii}$$

The response for the next cycle is determined by substituting $x_1$ as given by Equation xii with Equation vi for the left motion and with Equation xi for the right motion. Then, we can express the general response as

$$\text{left motion in cycle } i: \ x = [x_0 - (4i - 3)\Delta] \cos \omega_n t + \Delta \tag{xiii}$$

$$\text{right motion in cycle } i: \ x = [x_0 - (4i - 1)\Delta] \cos \omega_n t - \Delta \tag{xiv}$$

where

$$\Delta = \frac{\mu g}{\omega_n^2} \tag{xv}$$

Note that the amplitude of the harmonic part of the response should be positive for that half cycle of motion to be possible. Hence, we must have

$x_0 > (4i - 3)\Delta$ for left motion in cycle $i$
$x_0 > (4i - 1)\Delta$ for right motion in cycle $i$

Also note from Equation xiii and Equation xiv that the equilibrium position for the left motion is $+\Delta$ and for the right motion is $-\Delta$. A typical response curve is sketched in Figure 1.17.

## 1.5.1 Friction in Rotational Interfaces

Friction in gear transmissions, rotary bearings, and other rotary joints has a somewhat similar behavior. Of course, the friction characteristics will depend on the nature of the devices and also the loading conditions, but, experiments have shown that the frictional behavior of these devices may be represented by the interface damping model given here. Typically, experimental results are presented as

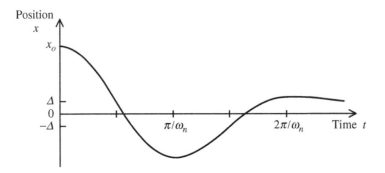

**FIGURE 1.17** A typical cyclic response under Coulomb friction.

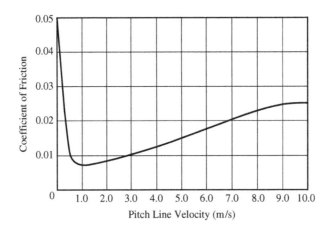

**FIGURE 1.18** Frictional characteristics of a pair of spur gears.

curves of coefficient of friction (frictional force/normal force) vs. relative velocity of the two sliding surfaces. In the case of rotary bearings, the rotational speed of the shaft is used as the relative velocity, while for gears; the pitch line velocity is used. Experimental results for a pair of spur gears are shown in Figure 1.18.

What is interesting to notice from the result is the fact that, for this type of rotational device, the damping behavior may be approximated by two straight line segments in the velocity–friction plane; the first segment having a sharp negative slope and the second segment having a

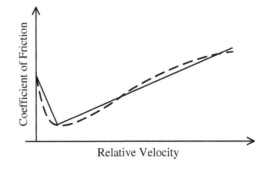

**FIGURE 1.19** A friction model for rotatory devices.

moderate positive slope that represents the equivalent viscous damping constant, as shown in Figure 1.19.

## 1.5.2  Instability

Unstable behavior or self-excited vibrations, such as stick–slip and chatter, that are exhibited by interacting devices such as metal removing tools (e.g., lathes, drills, and milling machines) may be easily explained using the interface damping model. In particular, it is noted that the model has a region of negative slope (or negative damping constant), which corresponds to low relative velocities, and a region of positive slope, which corresponds to high relative velocities. Consider the single-DoF model:

$$m\ddot{x} + b\dot{x} + kx = 0 \tag{1.109}$$

without an external excitation force. Initially the velocity is $\dot{x} = 0$. But, in this region, the damping constant, $b$, will be negative and hence the system will be unstable. Hence, a slight disturbance will result in a steadily increasing response. Subsequently, $\dot{x}$ will increase above the critical velocity where $b$ will be positive and the system will be stable. As a result, the response will steadily decrease. This growing and decaying cycle will be repeated at a frequency that primarily depends on the inertia and stiffness parameters ($m$ and $k$) of the system. Chatter is caused in this manner in interfaced devices.

## Bibliography

Blevins, R.D. 1977. *Flow-Induced Vibration*, Van Nostrand Reinhold, New York.

den Hartog, J.P. 1956. *Mechanical Vibrations*, McGraw-Hill, New York.

de Silva, C.W., Dynamic beam model with internal damping, rotatory inertia and shear deformation, *AIAA J.*, 14, 5, 676–680, 1976.

de Silva, C.W., Optimal estimation of the response of internally damped beams to random loads in the presence of measurement noise, *J. Sound Vib.*, 47, 4, 485–493, 1976.

de Silva, C.W., An algorithm for the optimal design of passive vibration controllers for flexible systems, *J. Sound Vib.*, 74, 4, 495–502, 1982.

de Silva, C.W. 2005. *MECHATRONICS — An Integrated Approach*, Taylor & Francis, CRC Press, Boca Raton, FL.

de Silva, C.W. 2006. *VIBRATION — Fundamentals and Practice*, 2nd ed., Taylor & Francis, CRC Press, Boca Raton, FL.

Ewins, D.J. 1984. *Modal Testing: Theory and Practice*. Research Studies Press Ltd, Letchworth, England.

Inman, D.J. 1996. *Engineering Vibration*, Prentice Hall, Englewood Cliffs, NJ.

Irwin, J.D. and Graf, E.R. 1979. *Industrial Noise and Vibration Control*, Prentice Hall, Englewood Cliffs, NJ.

Van de Vegte, J. and de Silva, C.W., Design of passive vibration controls for internally damped beams by modal control techniques, *J. Sound Vib.*, 45, 3, 417–425, 1976.

# 2

# Damping Theory

Randall D. Peters
*Mercer University*

**Summary**

*This introductory chapter synthesizes the many, though largely disjointed attributes of friction as they relate to damping (also see Chapter 1). Among other means, events selected from the history of physics are used to show that damping models have suffered from the inability of physicists to describe friction from first principles. To support fundamental arguments on which the chapter is based, evidence is provided for a claim that important nonlinear properties have been mostly missing from classical damping models. The chapter illustrates how the mechanisms of internal friction responsible for hysteretic damping in solids can lead to serious errors of interpretation. Such is the case even though hysteretic damping often masquerades as a linear phenomenon. One attempt to correct common model deficiencies is the author's work toward a "universal damping model," that is described in Section 2.17. Section 2.17 is developed in a "canonical" damping form. It shows the value of a direct, as opposed to an indirect, involvement of energy in model development. To keep a better perspective on how the treatment of damping is likely to evolve in the future, the last section of the chapter addresses some of the remarkable complexities of damping that are only beginning to be discovered. The manner in which technology has played a role in some of these discoveries is addressed in Chapter 3.*

# 2.1   Preface

The sheer volume of published material on the subject is a testament to the difficulty of selecting topics for inclusion in a chapter on damping. Viscoelasticity alone is the basis for several voluminous engineering handbooks. The present chapter is purposely different from similarly titled chapters of other reference books. There is little repetition of well-known and proven classical methods, for which the reader is referred to excellent other sources, such as de Silva (2006) and Chapter 1 of the present handbook. They provide solution techniques for many of the routine problems of engineering. The goal of the present chapter is to provide assistance with problems that are not routine, problems that are being encountered more frequently as technology advances. It is thought that this goal is best served by revisiting fundamental issues of the physics responsible for damping.

Once a multibody system has come to steady state, its damping treatment can be far less formidable than its description during approach to steady state. When dealing with limit cycles involving aeroelasticity and joints in helicopters, nonlinearity has a profound influence on the transient behavior. Attempts to model it have been largely unsuccessful, forcing the empirical selection of elastomers to reduce the vibration. (In the old days hydraulic dampers were used; Hodges, 2003). At a much lower level of sophistication, our understanding is quite limited on some common phenomena, such as the negative damping character of sound generated by a violin or a clarinet. Historically, when technology "hit the wall" because of too much theoretical handwaving, it became apparent that fundamental assumptions needed to be examined. In physics, a complete alteration of conventional wisdom was sometimes necessary, one of the best examples being the events that gave birth to quantum mechanics. Hopefully, from the multitude of seemingly disparate (but assumed by the author to be connected) observations which follow, the purpose for the architecture of this chapter can be partially realized. The enormous complexity of damping in general makes it unrealistic to hope for complete success.

Physics played a prominent role in developing the classical foundations of damping, starting in the 19th century. Subsequently, engineers uncovered many features of the subject that physicists never even thought about. In recent years, however, physics has been circumstantially forced to reconsider damping fundamentals. With the advent of personal computing, and an increased awareness of the importance of nonlinearity, new discoveries point to serious limitations of the classical foundation. The field of mechanics was severely limited until it began tackling problems of nonlinearity (not of damping type), and became concerned with previously ignored features giving unique system properties. Just as these unique properties could only be solved by techniques more sophisticated than the equations of linear type, there is mounting evidence that nonlinear damping may be the key to understanding some bewildering engineering cases.

It is important to try to identify the major mechanisms responsible for energy dissipation. This is easier said than done, since a host of different friction processes are usually at work. Moreover, the description of friction from first principles remains a daunting task. Thus we are forced to work with phenomenological models. There are also conflicts of nomenclature, with a given word meaning two different things from one profession to another. Thus, much of this chapter will attempt to define carefully terms while focusing on the physics, the treatment of which follows naturally along the lines of historical developments.

Engineers tend to be interested in higher frequencies and higher amplitudes of vibration than are scientists. A perfect damping model would be unconcerned with such differences of application; however, such a model is far from being realized. Because small-amplitude, long-period (low and slow) oscillations provide a valuable means for studying many processes of damping in general, much of this chapter focuses in that direction.

From the multitude of choices available to writers on the subject of damping, this author has selected a single (hopefully) unifying theme — nonlinear damping, especially as found in low and slow oscillations. Because it is a field still in its infancy, many of the ideas that follow are more speculative than one would prefer; however, they deserve discussion because of their perceived importance. To this author's knowledge, damping has not been previously treated in the manner of this chapter. Concerning the earliest relevant paper (Peters, 2001a, 2001b), the following was indicated by oft-cited Prof. A.V. Granato: "I don't know of anyone thinking about internal friction along the lines you have mentioned."

There are two important elements to the unifying theme of nonlinear dissipation: (i) the influence of nonlinear damping on multibody systems in their approach to steady state, and (ii) the close connection between damping and mechanical noise. When vibration decay is not exponential because of nonlinearity, there are significant ramifications and they are only beginning to be appreciated.

The novel features of this chapter are possible because of dramatic improvements in both sensing and data collection/analysis in the last decade. Demonstrating that a decay is not purely exponential requires both (i) a good linear sensor and (ii) the means to study readily long-time records when the damping is small (high Q). The first prerequisite has been met through the use of this author's patented fully differential capacitive sensor. The second has been realized with the availability of good, inexpensive analog-to-digital (A/D) converters having user-friendly, yet powerful Windows-based software. In addition to the "preview" software that comes with Dataq's A/D converter, a proven means for identifying nonexponential decay has been the analysis of records imported to Microsoft Excel. Details of these novel methods will be provided in the various sections that follow.

There are many examples in the engineering literature of nonlinear damping; even Coulomb damping is nonlinear because the friction force involves the algebraic sign of the velocity rather than the velocity itself, as in linear viscous damping. What has been realized for the first time in the course of writing this chapter is the following. As will be shown in the subsequent material, a decay process is not usually a pure exponential. Whatever the reason for a pure exponential, whether fundamentally linear (viscous) or nonlinear (hysteretic present model), the quality factor $Q$ for such a pure exponential decay is constant. When there is a second mechanism, such as amplitude-dependent damping (even if it is the only mechanism), the $Q$ now becomes time dependent. This is significant to mode coupling for the following reason. When a pair of modes couple because of elastic nonlinearity (a process that is impossible assuming linear dynamics), the strength of the coupling is proportional to the product of the individual amplitudes of the pair.

Consequently, variability in $Q$ can influence the evolution to steady state. It is a factor in determining which modes ultimately survive and/or dominate. Moreover, the distribution of the modes which remain depends on initial conditions, including the intensity of excitation.

Long ago, musicians learned to deal with nonlinearity, due in part to properties of the ear that are responsible for aural harmonics. A pair of purely harmonic signals can beat in the ear to produce a "sound" that does not exist when sensed with a linear detector. For example, consider a strong and undistorted 500 Hz signal sounded simultaneously with a pure 1003-Hz sound. The ear will hear a 3-Hz beat due to the superposition of the ear's aural second harmonic of the first with the fundamental of the second. However, there's more to this story. Conductors call for *fortissimo* and *pianissimo* sounds, not only because of the ear's nonlinearity, but also because of nonlinearities inherent to musical instruments. For example, it is easy with a good microphone and LabView (see de Silva, 2006) to demonstrate that the timbre of stringed instruments is intensity dependent. Not only is the mix of harmonics, as displayed in a fast Fourier transform (FFT) power spectrum, different according to volume, but their distribution also changes with time.

Noise is not typically treated in an engineering discussion of damping; however, mechanical noise is an important part of the technical material included in this chapter. Believing that there is a great deal of connectivity among vibration, damping, and noise, evidence will be provided in support of a premise — that the most important and least understood form of internal mechanical damping (material = hysteretic = "universal") is closely allied with the most important and least understood form of noise ($1/f$ = flicker = pink). If this premise is true, then the foundations of damping physics need reconstruction on several counts. Evidence in support of the premise will be provided through tidbits of experimental discoveries from a host of independent investigations. It is hoped that the unusual and lengthy introduction that follows will be beneficial in this regard. Historical elements serve to synthesize the many parts and are offered without apology. Following the introduction, some practical and novel equations of damping will eventually be provided. Even if readers find little identification with the philosophies that birthed them, it is hoped they will at least carefully examine the equations that are presented here in Section 2.17 for the first time.

## 2.2   Introduction

### 2.2.1   General Considerations of Damping

The etymology of the word "damping" is difficult to determine. It is obviously allied with the word damper, commonly defined as a "device that decreases the amplitude of electronic, mechanical, aerodynamic or acoustical oscillations," used for centuries, for example, to describe the sound attenuator pedal on the piano. Perhaps the German word *dampfen* (to choke) has had an influence in the evolution of the word. One can only wonder if water, as a moistening agent, played any role. Certainly, liquid water is important to some cases of energy dissipation in oscillators. Moreover, friction determined by the viscosity of a fluid (gas or liquid) is an important type of damping. A curious piece of history, in the celebrated work of Stokes, is why his expression "index of friction" did not take precedence over our modern word, viscosity. Peculiar terminology is also encountered to describe damping, such as the engineering device known as a dashpot, which is a mechanical damper. The vibrating part is attached to a piston that moves in a liquid-filled chamber.

We will see that the number of adjectives used to describe various types of damping is extensive. This multiplicity of terms to describe the loss of oscillatory energy to heat is no doubt an indicator of the complexity of damping phenomena in general. We will attempt (i) to identify similarities and differences among various types of damping, while (ii) explaining some of the physics responsible for the characteristics observed. Conceptual ideas and techniques of both theory and experiment will be provided, targeting the lowest level of sophistication for which semimeaningful results can be obtained. The reader should be aware that a "grand-unified" theory of damping does not exist, nor is it likely that one will ever be created.

Damping causes a portion of the energy of an oscillator, otherwise periodically exchanged between potential and kinetic forms, to be irreversibly converted to heat, sometimes by way of acoustical noise. Whether by suitable choice of materials during design of passive equipment, or by using feedback in active control of a sophisticated system, control of damping is important since mechanical vibrations can be detrimental or even catastrophic. An oft-quoted example of catastrophe is the Tacoma Narrows bridge, which collapsed in high winds on November 7, 1940. Like the vibration of a clarinet reed, this disaster is probably best described by the term negative damping, which can drive parametric oscillations.

The optimal amount of damping for a given system might fall anywhere in a wide range from great to extremely small, depending on system needs. The engineering world frequently wants oscillations to be as close to critically damped as possible. Physics experiments, such as those searching for the elusive gravitational wave (centered at the Laser Interferometer Gravitational Observatories, or Laser Interferometer Gravitational Wave Observatories [LIGO], in the United States; GEO600 in Germany [involving the British]; VIRGO in Italy [with the French], and TAMA in Japan), want damping in some of their components to be as small as possible. Frequency standards the world over require very small damping to insure high precision for timekeeping.

For the specific components of a system, a successful design frequently requires identification of the specific mechanisms primarily responsible for the dissipation of energy. Even after identifying the dominant sources, the theoretical difficulty of their treatment can also range from great to small, depending on the type of damping. For dashpot fluid damping, adequate models have existed for decades. For material damping, on the other hand, theories of internal friction are numerous and largely lacking in self-consistency.

The fundamental mechanisms responsible for damping are in most cases nonlinear; however, the oscillator's motion can itself be approximated in many cases by a linear second-order differential equation. If the potential energy is quadratic in the displacement, then the undamped linear equation of motion is that of the simple harmonic oscillator, because its solution is a combination of the sine and cosine (harmonic) functions. This undamped equation comprises the sum of two terms, one being a displacement and the other term an acceleration. The constant parameter multiplying each term of the pair depends on the nature of the system. For example, in the case of a mass–spring oscillator, the acceleration is multiplied by the mass, and the displacement by the spring constant. Thus, the equation corresponds to Newton's Second Law applied to a Hooke's Law (idealized) spring. In an electronic L–C oscillator, the "displacement" corresponds to the charge on the capacitor (divided by C) and the "acceleration" corresponds to the second time derivative of the capacitor's charge (multiplied by inductance L).

The usual means to describe damping, which is always present with oscillation, is to add a velocity term to the aforementioned displacement and acceleration. Although the damping could derive from several causes, there is usually a single dominant process. For example, the damping of current in a series-connected resistor, inductor, capacitor (RLC) circuit may depend mostly on Joule heating in the resistor R, in spite of the fact that there must also be energy loss in the form of radiation. Thus, the equation of motion includes a first-time derivative of the capacitor's charge (current) multiplied by R, in accord with Ohm's law.

Whether radiation is important for damping of the RLC circuit depends on the amount of coupling to the environment. If the circuit communicates with a final amplifier connected to an antenna, then radiation may become more important than Joule losses. The frequency of oscillation is a key parameter in this case, and also for damping problems in general. Unfortunately for some common systems, theoretical efforts to account properly for the effects of frequency have proven largely unsuccessful — except for models of phenomenological type developed by empiricism.

## 2.2.2   Specific Considerations

The mass–spring oscillator is the textbook example of harmonic motion, for which one of the most sophisticated mechanical oscillators ever built is the LaCoste version of vertical seismometer. Significant portions of the experimental data presented in this chapter were generated with an instrument designed around the LaCoste zero-length spring (LaCoste, 1934). The instrument used for this data collection was part of the World Wide Standardized Seismograph Network (WWSSN) during the 1960s. The spring of this seismometer is responsible for hysteretic damping of the instrument, rather than viscous damping as commonly assumed. Contrary to popular belief, air damping is not important for this seismograph at its nominal operating period, which is typically greater than 15 sec. Since every long-period pendulum apparently exhibits similar behavior, we thus find strong synergetic evidence in support of an old (mostly unheeded) claim that hysteretic damping (friction force independent of frequency) is universal (Kimball and Lovell, 1927). Their claim in 1927 to have discovered a universal form of internal friction (damping) is strengthened since the same behavior is seen in three distinctly different systems: (i) a mass–spring oscillator (as demonstrated by Gunar Streckeisen, details given later); (ii) a pendulum whose restoration depends on the Earth's gravitational field (demonstrated by several independent groups); and (iii) a rotating rod strained by a transverse deflection (1927 experiments of Kimball and Lovell).

The assumption of universality for hysteretic damping is a key point of this chapter. It will be shown that the damping of even a vibrating gas column (Ruchhardt's experiment to measure the ratio of heat capacities) is likely also hysteretic. The models that are described represent a departure from common theories of damping. Interestingly, the author's model has similarities to ordinary sliding friction, as given to us by Charles Augustin Coulomb. It effectively modifies the Coulomb coefficient of kinetic friction to yield an effective energy-dependent internal friction coefficient. The energy dependence is necessary to obtain exponential decay, as opposed to the linear decay of Coulomb damping. Just as with conventional Coulomb damping, its form is nonlinear, involving the algebraic sign of the velocity. We will see that the damping capacity predicted by the model permits an equivalent viscous form. Yet the underlying physics is related to creep of secondary type as opposed to the primary creep of viscoelasticity.

It is this author's opinion that much of the existing theory of damping is not the best means for modeling dissipation. The difficulties arise from approximating oscillator decay with linear mathematics. Although most individuals recognize the oft-stated caveat that viscous damping is an approximation to the actual physics of dissipation, they do not recognize some of the many serious limitations of the approximation. The situation is similar to the place in which we found ourselves at the beginning of the era labeled "deterministic chaos." The "butterfly effect" (Lorenz, 1972) has radically altered the thinking of many, but only in relationship to large-amplitude motions of a pendulum, where the instrument is no longer isochronous because of nonlinearity. As an archetype of chaos, the pendulum must be rigid and capable of "winding" (displacement greater than $\pi$) before chaos is possible. Nonlinearity is a prerequisite for the chaos, but it is not sufficient, since there are many examples of highly nonlinear but nonchaotic motions. For example, amplitude jumps of nonlinear oscillators, during a frequency sweep of an external drive, have been known for many years. They were observed before chaos was recognized, in systems like the Duffing and Van der Pol oscillators. Yet chaos, with its sensitive dependence on initial conditions (responsible for the butterfly effect), was not contemplated at the time. As with most significant advances, Lorenz's discovery was by accident, as he modeled convection in the atmosphere. The author's confrontation with complexity that derives from mesoscale structures in metals was likewise unexpected. "Strange phenomena" (as Richard Feynman would probably have labeled them) were encountered while using his patented fully differential capacitive sensor to study various mechanical systems, mainly oscillators.

As with chaos, the pendulum may ultimately serve as an archetype of complexity. When operated at low energy, especially through a combination of long period and small amplitude, the free decay of the physical pendulum departs radically from the predictions based on linear equations of motion. Such complexity can be easily demonstrated when the pendulum is fabricated from soft alloy metals. For example, Figure 2.1 illustrates the decay of a rod pendulum constructed with ordinary (heavy-gauge lead–tin) solder of the type used for joining electrical conductors (Peters, 2002a, 2002b, 2002c).

The "jerkiness" (discontinuities) in the record of Figure 2.1 is in no way related to amplitude jumps of the type previously mentioned; rather, these are jumps of the Portevin–Le Chatelier (PLC) type (Portevin and Le Chatelier, 1923). They are a fundamental, yet "dirty" phenomenon that physics has chosen for decades to try and ignore (even though materials science and engineering took early note of the PLC effect). The most obvious and profound thing that can be said about Figure 2.1 is the following: the presence of PLC jerkiness means that the concept of a potential energy function is not really valid, since the requirement for its definition is that a closed integral of the force with respect to displacement must vanish.

No matter the form of hysteresis, which is the cause for damping, it disallows the curl of the force to be zero, so that potential energy is never formally meaningful for a macroscopic oscillator (since there is always damping). In those cases where the damping is essentially continuous (not true for the example of Figure 2.1), the assumption of a potential energy function retains some computational meaning. For oscillators influenced by the PLC effect, this is no longer true. The resulting properties are important to a variety of technology issues, such as sensor performance, since noise is no longer the simple thermal form predicted by the fluctuation–dissipation theorem (used to characterize white, i.e., Johnson, noise).

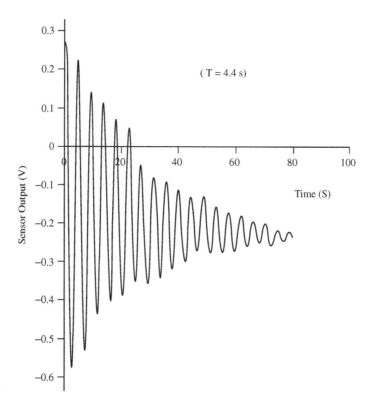

**FIGURE 2.1** Free-decay of a rod pendulum fabricated from solder.

Practical means for dealing with systems influenced by "stiction" have been known by engineers for decades. Because of metastabilities in the assumed potential function, the system is prone to latching (stuck in a localized potential well). One means for mastering the metastabilities (unstick the part designed to move) is to "dither" the system. The process has become more sophisticated in the last decade, which saw a major growth of interest in stochastic resonance. In the definition by Bulsara and Gammaitoni (1996):

> A stochastic resonance is a phenomenon in which a nonlinear system is subjected to a periodic modulated signal so weak as to be normally undetectable, but it becomes detectable due to resonance between the weak deterministic signal and stochastic noise.

The phenomenon is related to dithering (Gammaitoni, 1995). It is a case where the signal-to-noise ratio (SNR) can be increased by the counterintuitive act of raising the level of noise. Such a gain in SNR is not possible with a harmonic potential.

In recent studies of granular materials, "tapping" has become a popular means to study behavior that violates the fundamental theorem of calculus. Years ago, this author used tapping as a means to accelerate creep in wires under tension. Evidently, hammering the table on which the extensometer rested caused vibrational excitations of the wire that stimulated length changes of discontinuous PLC type. Because of the broad spectral character of an impulse, various eigenmodes of the wire (de Silva, 2006) could be thus readily excited. After "hammering-down" under load, a silver wire could by this same means be stimulated, after partial load removal, to exhibit length contractions. Since the total number of atoms is fixed, the process must involve exchange of atoms between the surface of the sample and its volume.

Extensometer studies of wires at elevated temperature have also displayed strange behavior. A polycrystalline silver wire of diameter 0.1 mm and approximate length 30 cm was found to exhibit large fluctuations when heated in air to within 100 K of its melting point, using a vertical furnace (Peters, 1993a, 1993b). The large fluctuation in length at these temperatures (reminiscent of critical phenomena and visible to the naked eye) may be associated with oxide states of the metal, since the experiments were not performed in vacuum. When cycled in temperature, fluctuations in the length of a gold wire were found to exhibit dramatic hysteresis. With influence from Prof. Tom Erber of Illinois Tech University, it was postulated in (Peters, 1993a) that there may be some mesoscale quantization of fundamental type responsible for the thermal hysteresis (hysteron).

Mechanical hysteresis resulting from mesoanelastic defect structures is evidently ubiquitous. Piezotranslators, which are used as actuators in atomic force microscopes and other nanotechnology applications, are afflicted with high levels of hysteresis when operating open loop. This behavior is consistent with the anomalously large damping that was observed with a pendulum (reported elsewhere in this document) in which there was a steel/PZT interface for the knife-edge.

Even the common strain gauge exhibits complex hysteresis behavior. The normally large hysteresis that is observed in preliminary cycling of a gauge is typically reduced by significant amounts after repeated cycling, a type of work hardening (if strained well below failure limits). It should be noted that the hysteresis of all the discussions in this chapter is not to be confused with backlash (as in a gear train).

All of these experiments are in keeping with the premise that *mesoanelastic complexity* determines the nature of hysteretic damping. It is seen that there are a plethora of examples where strain (and thus damping) of a sample is not simple, is not smooth, but more like the complex behavior of granular materials. In the case of polycrystalline metals, the same grains that are made visible by methods of acid etching decoration are evidently responsible for mesoscale (nonsmooth) internal friction damping. To assume that damping is quantal at the atomic scale, rather than the mesoscale, is without experimental justification. Nevertheless, this is a popular assumption with which estimates of the noise floor of an instrument such as a seismometer is estimated, to calculate SNR.

### 2.2.3  The Pendulum as an Instrument for the Study of Material Damping

Because of its early contributions to physics, which in those days was called natural philosophy, one might be tempted to believe that the pendulum is only important to (i) the history of science or (ii) teaching of fundamental principles. A single observation should be sufficient to resist this temptation — (as already noted) the pendulum has in the last 30 years become the primary archetype for the new science of chaos. Additionally, many of the data sets of this document, which show significant and previously unpublished results, were generated with a pendulum.

To a student of elementary physics, the choice of a pendulum may seem unsophisticated. Yet, to the author, who has spent 15 intense years trying to understand harmonic oscillators, the pendulum is the most versatile instrument with which to understand damping. It has been central to the development of science in general. It was studied by Galileo, Huygens, Newton, Hooke, and all the best-known scientists of the Renaissance period. It served to establish collision laws, conservation laws, the nature of Earth's gravitational field and, most of all, it was the basis for Newton's two-body central force theory. This theory was foundational to the development of classical mechanics, which is central to all of physics and engineering. Historian Richard Westfall has remarked: "Without the pendulum, there would be no Principia" (Westfall, 1990).

In 1850, Sir George Gabriel Stokes published a foundational paper (Stokes, 1850). His treatment of pendulum damping permitted the understanding, decades later, of a number of important phenomena in physics and engineering. For example, his studies were foundational to the Navier–Stokes equations of fluid mechanics. Moreover, viscous flow known as Stokes' Law was the basis for Millikan's famous oil drop experiment that determined the charge of the electron.

Stokes noted in his paper that, "… pendulum observations may justly be ranked among those most distinguished by modern exactness." He also noted

> The present paper contains one or two applications of the theory of internal friction to problems which are of some interest, but which do not relate to pendulums. … the resistance thus determined proves to be proportional, for a given fluid and a given velocity, not to the surface, but to the radius of the sphere. … Since the index of friction of air is known from pendulum experiments, we may easily calculate the terminal velocity of a globule [water] of given size. … The pendulum thus, in addition to its other uses, affords us some interesting information relating to the department of meteorology.

The last statement of this quotation speaks to some of the errors in the "common theory" of his day. In similar manner, some of the common-to-physics damping models of today are erroneously applied. Those who hold the viscous damping linear model in unwarranted regard, fail to recognize the limitations under which it is valid. There are frequent misapplications for reason of experimental deficiencies. We can all profit by taking seriously the following well-known words of Kelvin:

> When you can measure what you are speaking about, and express it in numbers, you know something about it. But when you cannot measure it, when you cannot express it in numbers, your knowledge is of a meager and unsatisfactory kind. It may be the beginning of knowledge but you have scarcely in your thoughts advanced to the state of science.   William Thomson, Lord Kelvin (1824 to 1907)

Simple (viscous) flow of the Stokes' Law type is possible only according to the restrictive conditions that Stokes spelled out in his paper. We now specify those conditions for viscous flow according to the nondimensional parameter given us late in the 19th century by Osborne Reynolds. Specifically, Stokes' Law is valid only for $Re = \rho v L / \eta < 60$ (approximately, for spheres); where $\rho$ is the density of the retarding fluid, $v$ is the speed of the object relative to the fluid, $L$ is a characteristic dimension of the object, and $\eta$ is the viscosity of the fluid. The requirement is not generally met for oscillators, and recent experiments have shown that contributions to the damping from air drag proportional to the square of the velocity cannot generally be ignored (Nelson and Olssen, 1986). This is just one example of how two or more damping types must sometimes be folded into an adequate model of dissipation. A novel method for combining all the common forms of damping in one mathematical expression is provided in this document. Additionally, it is shown how to calculate analytically the history of the amplitude of free-decay for such cases.

Considering the importance of Stokes' work, it is surprising that some of his requests for further experiments were apparently never seriously considered. On page 75 of his paper, one reads the following: "Moreover, experiments on the decrement of the arc of vibration are almost wholly wanting." Having noted this, Stokes appealed to experimentalists to generate such data. In the 19th century, collecting the data he requested would have been labor-intensive and therefore the experiments were probably never attempted. Sensors and data processing of the modern age now make them straightforward, but the pendulum has by now been viewed by too many as a relic rather than the important instrument described by Stokes. Much of the author's efforts have been directed at showing that the pendulum is still an important research instrument. For example, one physical pendulum of simple design was the basis for the generalized model of damping (modified Coulomb) that is here presented. Another has been used to illustrate surprisingly rich complexities of the motion that results from the ubiquitous defects of its structure (Peters 2002a, 2002b, 2002c). Thus studying the complex motions of "low and slow" physical pendula could yield significant new insight into the defect properties of materials — a field where relatively little first-principles progress has been made.

## 2.2.4   "Plenty of Room at the Bottom"

Richard Feynman gave a now-famous talk in 1959 titled, "There's plenty of room at the bottom" (presented at the American Physical Society's annual meeting at CalTech). Drawing on observations from biology, he spoke of a solid-state physics world involving "… strange phenomena that occur in complex situations." In the 44 years since Feynman's prophetic comments, there have been spectacular achievements in very large-scale integrated (VLSI) electronics, microelectromechanical systems (MEMS), and even nanotechnology. Progress in the mechanical (including sensor) realm has been much slower than in electronics; consequently, our present processing power far exceeds our acquisition (and actuator) capabilities.

One of the major obstacles to miniaturization involves dramatic change to physical properties that can occur as the size of a system shrinks below the mesoscale toward the atomic. For example, VLSI electronics is already beginning to be impacted by quantum properties of the atom, as component size continues to decrease in accord with Moore's law (Moore, 1965). Among other things, Feynman predicted that lubrication would no longer be "classical" at such a scale. On a related note, a paper by Nobel Laureate Edward M. Purcell (Purcell, 1977) draws a striking contrast between our macroscopic world and that of micro-organisms. At low Reynolds number, inertia becomes unimportant, and mechanics is dominated by viscous effects. The adoption of a new paradigm will be necessary for engineers to deal with these differences.

In the article "Plenty of room indeed" (Roukes, 2001), it is noted that there is an anticipated "dark side" of efforts to build truly useful micro- and nano-sized devices. Gaseous atoms and molecules constantly adsorb and desorb from device surfaces. This process is known to exchange momentum with the surface, even permitting scientific study of the gas–solid interface (Peters, 1990). The smaller the device, the less stable it will be because of adsorption/desorption. As Roukes has noted, this instability may pose a real disadvantage in various futuristic electromechanical signal-processing applications (Cleland and Roukes, 2002).

There is direct evidence, provided in the present chapter, that we need to be more concerned with noise: (i) the evacuated pendulum where it is speculated that outgassing influenced its free-decay, and (ii) the seismometer free-decay that showed both amplitude and phase noise and evidence for nonlinear damping. Concerning (i), when the vacuum chamber pressure is reduced, the preexisting steady state (normal rate balance between adsorption and desorption) becomes disturbed, so that there is a complex emission of gases from the surface of the pendulum. The emission is not likely to be spatially uniform, but more like the jets seen on Halley's comet when photographed by the Giotto spacecraft in 1986. In case (ii), the noise is seen to derive from mesoanelastic complexity of the structure of the pendulum itself rather than involving gases.

Miniaturized devices have the potential to serve as on-chip clocks, and the importance of phase noise to clocks is well documented. There is another, more subtle issue that points out the importance of phase noise. One of the best means for improving SNR is the technique of phase-sensitive detection, first employed by Robert Dicke at Princeton to improve solar experiments. The performance of miniaturized electromechanical sensors using "lock-in" amplifier methods may be influenced significantly by mechanical phase noise.

Phase noise of miniaturized devices is still mostly speculative. In addition to the mechanism just mentioned (adsorption/desorption), there is the matter that constitutes the theme of this chapter, defect organization. It is not possible to grow materials without dislocations and/or other disturbances to crystalline order, such as vacancies, interstitials, or substitutional impurities. Thus, "when mother nature fills the vacuum she abhors, she rarely does so with perfection."

Long before defects organize to the point of incipient failure (at much larger strains), they still influence vibration. They may even be a primary source of $1/f$ noise. Electronic noise of $1/f$ type is known to involve defects by means of trapping states, and these states derive from crystalline defects sometimes involving the surface. The interaction of the surface and the volume of a solid are important. For example, consider pure copper single crystals of the type used by the author in his doctoral work. A practical joke suggested by Vic Pare (that we never conducted because of the cost of these samples) would be to have a 98-lb weakling bend one by hand, then ask an NFL linebacker to straighten it back out! The striking irreversibility is the result of work hardening as dislocations develop at the surface and propagate into the bulk where they entangle.

In the case of polycrystalline materials, the memory features of hysteresis may be important according to the method of their fabrication. Wires are typically produced by pulling through successively smaller dies. This "swaging" may be conducive to the exchange of monolayer groups of atoms between the volume and the surface during fluctuation length changes. The fluctuation–dissipation theorem does not hold or, if it does, only in terms of larger entities than the atom. Thus, there are many yet-to-be-quantified elements of noise in the vibration of miniaturized devices. Feynman was right when he spoke of strange phenomena of the solid state.

Technology of the future is expected to be confronted increasingly with damping problems that must address issues of scaling — to deal with some factors discussed in this chapter, which, to the author's knowledge, have not been previously published. Until small (MEMS) oscillators become more common to the engineering world, we must study the mechanisms responsible for their damping by other than traditional means. One approach is similar to experimental techniques for the verification of the kinetic theory of gases. As noted by Present in his textbook (Present, 1958), there are two ways that Brownian motion can be studied: either (i) with small objects and an unsophisticated detector, or (ii) with larger objects and a very sensitive detector. It is the latter that provided some of our present knowledge of damping at the mesoscale. The fully differential capacitive transducer, whose patent label is "symmetric differential capacitive" (SDC), is a robust new technology that is sensitive, linear, and user-friendly (Peters, 1993a, 1993b). As with other sensitive detectors that have been used to predict the properties of small objects by studying larger ones, small-energy studies of various macroscopic pendula are demonstrating some of the "strange phenomena" of complex type predicted by Feynman.

From the author's perspective, we of the physics community have been guilty of two significant errors: (i) oversimplification of many problems by assuming a linear equation of motion based on viscous damping, and (ii) losing sight of fundamental issues by working with inappropriate, overly complicated damping models. The goal of this chapter is to assist progress toward a healthier balance between these extremes. It is hoped that readers will be thus better equipped to identify, and then dismantle, some of the impediments to the development of future technologies.

## 2.3  Background

### 2.3.1  Terminology

The large number of mechanisms capable of energy dissipation has resulted in a host of adjectives to describe damping phenomena in mechanical systems. They include (nonexhaustive list): viscous, eddy current, Coulomb, sliding, friction, structural, fluid, thermoelastic, internal friction, viscoelastic, material, solid, phonon–phonon, phonon–electron, and hysteretic. For present purposes, damping types will be grouped according to one of the following three categories: (i) fluid (including viscous), (ii) Coulomb, and (iii) hysteretic. Although hysteretic damping has come to be associated in engineering circles with a particular form of material damping in solids, it should be noted that all forms of damping involve hysteresis (for which the Greek meaning of the word is "to come late"). In a plot of periodic stress vs. strain, which is a straight line for displacements of a nondissipative, idealized substance, hysteresis causes the line to open into a loop. The size of the loop — more specifically the area inside this hysteresis loop — is a measure of the amount of nonrecoverable work done per cycle because of the damping. An actual force of friction is not readily recognizable in those cases that are labeled "internal friction." The word friction is used in a generic sense, meaning any process responsible for conversion of the oscillator's coherent motion into incoherent thermal activity.

With each of viscous, eddy current, and Coulomb damping, a force external to the oscillator is responsible for the dissipation of energy. The external force is associated respectively with (i) laminar fluid flow, (ii) induced currents, and (iii) surface friction. The surface friction case is not necessarily the trivial textbook presentation involving a coefficient of kinetic friction and a normal force. The cases just mentioned, along with thermoelastic damping; which is of internal rather than external origin, are much easier to treat theoretically than other cases. Viscous damping and eddy current damping (over the full range of the motion) are adequately described by a velocity term, which yields a linear equation of motion. Coulomb damping, however, is not proportional to velocity, but rather depends only on the algebraic sign of the velocity. The equation of motion is consequently nonlinear. Additionally, and unlike most other forms, Coulomb damping is not exponential. The turning points lie along a straight line when the motion is plotted vs. time. Similarly, if eddy current damping exists only over a small part of the motion, the decay is linear rather than exponential (Singh et al., 2002).

### 2.3.2  General Technical Features

Historically, viscous damping has been the model of choice because the resulting equation of motion is mathematically attractive and, for the RLC circuit, the form is appropriate. For mechanical oscillators, it is not generally appropriate, since viscous damping amounts to some part of the system moving in an external Newtonian fluid that removes energy because of a friction force that is proportional to velocity.

The defects responsible for material damping, such as dislocations, are also responsible for creep, so that high strength and high damping tend to be incompatible attributes. Magnesium alloys tend to be better than many other metals in this regard. Hardness of a material is neither a prerequisite for toughness nor for small damping, as recognized by those familiar with the mechanical properties of cast iron.

On a different scale, defects determine "how things break"; concerning which Marder and Fineberg have stated the following:

> the strength of solids calculated from an excessively idealized starting point comes out completely wrong; it is not determined by performance under ideal conditions, but instead by the survival of the most vulnerable spot under the most adverse of conditions.   (Marder and Fineberg, 1996)

Three famous scientists are primarily responsible for the highly popular viscous damping model of the simple harmonic oscillator; they are Lord Kelvin (Thomson and Tait, 1873), G.G. Stokes and

H.A. Lorentz. Stokes is best known for his equations of fluid dynamics that also include the name Navier. Stokes' Law, which describes the terminal velocity of a raindrop, was developed through his treatment of the damping of a pendulum. Not only does his law provide a basis for the simplest approximation for damping of a macroscopic oscillator, it was used by Robert Millikan to determine the charge of the electron. It should be noted that harmonic oscillation in a fluid (even at low Reynolds number) is much more complicated than steady-flow viscous friction. This topic is treated in Chapter 3, Section 3.9.

The first individual to use the term "simple harmonic oscillator" was probably Lord Kelvin. Such an oscillator is a key tool of experimental physics and also the foundation for much of theoretical physics. It is the basis for communication via electromagnetic waves and even esoteric theories of superfluids and superconductors.

Much of the underpinnings of theory involving harmonic oscillation derive from the work of Hendrik Anton Lorentz (1853–1928). Lorentz is well known for a variety of classical physics contributions, such as (i) the transformation of special relativity associated with Einstein and (ii) the force law for the acceleration of charged particles, both of which bear his name. Before the existence of electrons was proved, Lorentz proposed that light waves were due to oscillations of an electric charge in the atom. For his development of a mathematical theory of the electron, he received the Nobel Prize in 1902. The importance of his contributions is further realized by noting that it is common practice to describe the lineshape of atomic spectra by the term Lorentzian. The Lorentzian is equivalent to the resonance response of the driven viscous-damped simple harmonic oscillator.

It is easy to show how resistance in an electric network is responsible for damping; however, it is a challenge to understand anelastic processes of mechanical damping in terms of viscosity. From comments of his Ph.D. dissertation, it has been said that even Lorentz was never apparently satisfied with the velocity damping term in his equation — not knowing just how to relate it to the underlying physics. It is also clear from Stokes' paper that he recognized the need for caution in the use of his law of viscous friction. It appears that both Lorentz and Stokes were very careful compared with the carelessness with which the viscous model has been employed by many individuals in recent years.

The failure of solids influenced by "hysteretic" damping to be adequately described by the methods of viscoelasticity is not widely appreciated. It is unfortunate that too few people have expanded their view of damping to include other important types, such as derive from the anelasticity of solids. It is important in this work to recognize some subtle differences, for example, inelastic (not elastic) is not to be equated with anelastic (other than elastic).

## 2.3.3 Active vs. Passive Damping

With improvements in cost/performance of electronics, active damping is increasingly popular. Using force-feedback with integration/differentiation circuitry (opamps), a mechanical oscillator can sometimes be tailored for a specific purpose. A sophisticated example of this technology is the broadband seismometer that began to replace earlier version (passive) instruments roughly 35 years ago. The Sprengnether–LaCoste spring instrument that was used for some of the experiments reported in this document has been superceded by force-feedback units such as the Streckeisen STS-1 and STS-2.

In lieu of feedback, another way actively to influence the damping of a mechanical oscillator is to connect the sensor to an amplifier having a negative input resistance. The seismometers marketed by Lennartz Electronics in Germany use this in a patented technique to improve the performance of ordinary, off-the-shelf electrodynamic geophones.

Active damping depends on the nature of the transfer function of the composite system (electronics plus mechanical). The characteristics of the transfer function are determined by the location of its poles and zeros in the complex plane. Seismometers operate nominally near 0.707 of critical damping. This is done for two reasons: (i) the instrument is easier to adjust and (ii) the interpretation of earthquake records is simpler. Of course, to increase damping is to decrease sensitivity because of the fluctuation–dissipation theorem.

The force-feedback technique is not practical for some situations, regardless of cost. Additionally, it must be recognized that the method is not the answer to all problems, since electronics cannot compensate for a poor mechanical design. The description of commercial products is in some cases highly exaggerated, giving the impression that almost any sensor can perform flawlessly in this manner. Some accelerometers have employed dithering to offset the effects of "stiction" in bearings. The dithering was necessary because the potential energy function is not truly harmonic, being afflicted with the consequences of nonlinear damping. Even with sensing schemes that do not use a bearing, the effects of nonlinearity persist. In "Seismic Sensors and their Calibration" (Bormann and Bergmann, 2002), Erhard Wielandt, in talking about transient disturbances in the spring of a seismometer, says the following:

> Most new seismometers produce spontaneous transient disturbances, quasi-miniature earth-quakes caused by stress in the mechanical components.

In other words, internal friction from defects at the mesoscale cause behavior that is in some ways similar to ordinary sliding friction, where the static coefficient is greater than the kinetic coefficient. The postulate of Bantel and Newman is consistent with this idea (Bantel and Newman, 2000) when they refer to their observations as being consistent with a "stick–slip" model of internal friction.

It is seen then that one must use a detector that responds faithfully to the signal around which the servo-network functions. The linearity and sensitivity of that sensor are of paramount importance, since the basis for force-feedback design is linear system theory. For some less-challenging cases, the design approach is straightforward, since software packages like MATLAB® (see de Silva, 2006) have built in functions to describe behavior.

### 2.3.4 Magnetorheological Damping

A recent approach to damping control, that is quite different from the servo-networks mentioned above, is one that uses an magnetorheological (MR) fluid. It takes advantage of the large variation in viscosity of certain compound fluids according to the size of an applied magnetic field. J. David Carlson (Carlson, 2002) describes how an MR sponge damper is activated during the spin cycle of a washing machine to keep it from "walking out of the room." The peak in the Lorentzian (resonance response) of the machine is shown in his article to be substantially lowered by supplying current to the electromagnet of the damper.

### 2.3.5 Portevin–LeChatelier Effect

Physics, engineering, geoscience, and mathematics have all contributed greatly to a better understanding of damping phenomena; however, there has been little cross-discipline exchange of ideas and lessons learned. Some of the impediments to strong interdisciplinary programs derive from (i) the complexity of damping problems in general and (ii) the tendency for physics and mathematics research to be, on the one hand, less pragmatic and, on the other hand, highly specialized — focused on specific energy dissipation mechanisms. A good example of (i) involves the PLC effect, discovered in 1923. Why physics mostly ignored this early example of "dirty science" by two of their own number is not easily understood, although the birthing of quantum mechanics around this time may have been a factor. Had history turned in a different direction, perhaps we would already be able to explain from first principles the most important, but still barely understood, form of noise known as $1/f$, or flicker, or pink noise. Even though R.B. Johnson (well known for his discovery of white electronics noise in a resistor) was one of the first to see this form of noise, it still is not explained from first principles — although recent discoveries suggest an intimate connection with fractal geometry involving self-similarity. Such geometry is associated with the mesoscale of materials where the grain, rather than the atom, is the basic element of statistical mechanics.

For alloys, the PLC effect appears to be, in some ways, what the Barkhausen effect is to magnetic systems. In the case of ferrous materials, the noise which derives from the mesoscale has long been recognized; however, similar noise of mechanical type has not been seriously studied. This oversight is even more puzzling when one considers the admonition by G. Venkataraman, as recorded in the proceedings of a Fermi conference, for scientists to get involved in what he felt should become an important new field (Venkataraman, 1982).

## 2.3.6 Noise

Noise is purposely discussed in this chapter (also see the chapters in Section IX of this handbook) because it has been a, largely, missing component of efforts to understand the physics of damping. A feel for the importance of noise to damping research is to be gleaned from a comment by Kip Thorne in his foreword to the English translation of a book by V.B. Braginsky et al. (1985). Mainly because of instrumental needs of the Laser Interferometer Gravitational Wave Observatories (LIGO), Thorne writes,

> The central problem of such experiments is to construct an oscillator that is as perfectly simple harmonic as possible, and the largest obstacle to such construction is the oscillator's dissipation. If dissipation were perfectly smooth, it would not be much of an obstacle, but the fluctuation–dissipation theorem of statistical mechanics guarantees that any dissipation is accompanied by fluctuating forces. The stronger the dissipation, the larger the fluctuating forces, and the more seriously they mask the signals that the experimenter seeks to detect.

This comment by Thorne suggests a frequently important impediment to dialogue between engineering and physics — concern for different issues. LIGO is trying to minimize damping, whereas many engineering problems are concerned with just the opposite — making the damping as large as possible without compromising strength. More detailed discussions of noise are provided later.

## 2.3.7 Viscoelasticity

Within the world of polymers, damping is frequently described by the expression "viscoelasticity." This word, around which handbooks have been written (e.g., Lakes, 1998), is a combination of the two words, viscous and elastic. We like to think of ideal fluids as being viscous in the manner described by Newton. Likewise, ideal solids that obey Hooke's Law (stress proportional to strain) are described as elastic. Unfortunately, nature contains neither ideal solids nor ideal fluids. Real springs do not obey Hooke's Law, but rather are influenced by "anelasticity" (other than elastic) which gives rise to hysteresis in the stress–strain relationship. Real fluids usually have some (if not near total) degree of non-Newtonian character. Thus an envisioned "mixing" of fluid-like and solid-like character has dominated the thinking of those who, through the decades, attempted to develop theoretical models of damping.

It should be noted that the springs and dashpots used in models of viscoelasticity do not actually exist. They serve as a phenomenological means for (hopefully) understanding the elementary processes which their arrangement is designed to mimic. Consider, for example, high polymers, in which the interwoven structure of the long-chain molecules is one of extensive mechanical interference. (One popular visualization is that of an entanglement of a huge number of long, writhing snakes.) An increase of temperature is met with overall length reduction (negative temperature coefficient of expansion for the so-called entropy spring), which stands in stark contrast with metals. Such behavior is clearly important to damping since, as noted by Gross years ago, "…thermal movement interferes with the orientation and disorientation of the molecules and ultimately causes delay in the expansion and contraction of the specimen" (Gross, 1952).

## 2.3.8 Memory Effects

In this same article, Gross is one of the first to mention "memory" properties of creep. He describes a by-then old demonstration in which a "firmly suspended metal or plastic wire is twisted first in one direction for a long time and then in the other direction for a short time. Immediately after release, the deflection will be in the direction of the last twisting, but it decreases rapidly. Presently, a reversal occurs, and the wire begins to turn in the other direction, corresponding to the first twisting — the memory of the recent short-term handling has been obliterated by that of the more remote but longer lasting and therefore more impressive one!" Perhaps this old demonstration (sometimes today called the anelastic after-effect) is not so startling to those familiar with more modern shape-memory-alloys, which are expected by many to play increasingly important roles in the applied science of damping.

### 2.3.9   Early History of Viscoelasticity

Those who provided seminal influence in the development of the theories of viscoelasticity during the 19th century were some of the most famous names in physics, like Maxwell and Kelvin. Maxwell is best known for the electromagnetic equations associated with his name. He is far less known for two other significant contributions: (i) kinetic theory of gases and (ii) viscoelasticity — both of which are important to theories of oscillator damping. Maxwell's interest in the problem of viscoelasticity is first documented in a paper during his teen years, titled "The Equilibrium of Elastic Solids." Through his development with Boltzmann of the kinetic theory of gases, Maxwell showed a counterintuitive property of the viscosity of a gas. The viscosity does not decrease significantly as the pressure is reduced, until the mean free path between collisions of the molecules begins to approach dimensions of the chamber holding the gas. Important even to modern innovations such as MEMS oscillators, his surprising prediction was quickly verified by experiment. Maxwell's model of viscoelasticity combines a purely elastic spring with a purely viscous dashpot (fluid damper in which the friction force is proportional to the velocity).

Kelvin, probably the first to include a viscous damping term in the equation of motion of the simple harmonic oscillator, developed a similar model of viscoelasticity. Each of the two models is usually represented in literature (without original references) as containing a single spring and a single dashpot. They differ in that one connects the pair in series (Maxwell), while the other connects them in parallel (Kelvin–Voigt).

Both the Maxwell model and the Kelvin–Voigt model have been found by engineers to be less useful than the standard linear model (SLM) of anelasticity, largely advanced in the 20th century by Clarence Zener (Zener, 1948). In the three-component Zener model, a spring is connected in series with a parallel combination of spring and dashpot. Curiously, Zener is widely associated with electronics because of the common diode named after him, but fewer people know of his work in anelasticity. No doubt, his understanding of anelasticity helped him to better understand the complex processes at work in his diode.

### 2.3.10   Creep

The prevailing models of anelasticity appear frequently in the literature, but mostly in relationship to primary creep. Some of the papers exceptional to this rule are those by Berdichevsky (Berdichevsky et al., 1997). Recent work of a more heuristic type has shown that the equations of viscoelasticity are also able to accommodate secondary creep, in which the decay of strain rate with time has disappeared (Peters, 2001a, 2001b).

The importance of creep (and relaxation) physics to damping warrants some discussion. When a sample is subjected to a constant stress, the strain evolves through three phases of creep: (i) primary, (ii) secondary, and (iii) tertiary. An example of the first two of these phases is shown in Figure 2.2.

In the primary stage, the sample is deformed by anelastic processes involving defects of the crystalline structure. Influence of the disordering mechanisms is progressively reduced as the sample undergoes work hardening (such as pinning of dislocations). Work hardening would result in a purely exponential creep, in the absence of thermal effects which strive to undo the hardening (via diffusion processes). (At zero Kelvin, the creep would eventually cease, if described by a single time constant.) In the secondary stage, a balance between work hardening and thermal softening is attained, in which the strain vs. time has converted from exponential to linear. This balance cannot continue forever, if the stress is larger than

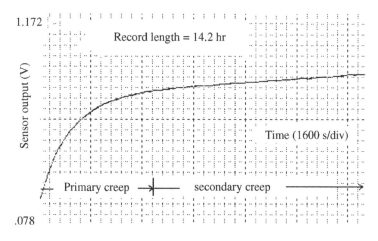

**FIGURE 2.2**  Example of creep in the spring of a vertical seismometer.

a threshold associated with failure (the elastic limit), and thus a final complex fracture of the sample finally occurs as the sample passes through the tertiary stage. Although one might want to divorce the issues of tertiary creep from considerations of damping, there is clearly a link between damping and failure, involving defects. We will return to this point later.

In Figure 2.2, creep resulted from the instrument having been severely disturbed (relocated accidentally by plumbers working in the building). As with any long-period mechanical oscillator, it is necessary for this instrument to stabilize after a major rebalancing. Primary creep is seen to have endured for about 5 h and what is labeled secondary creep in the figure does not continue indefinitely with the implied constant rate. Thus, the instrument will typically stabilize after one or several days, when the period of oscillation is of the order of 20 sec.

The total amount of creep in Figure 2.2 deserves mention. In the indicated 14.2 h, the mass of the seismometer moved a vertical distance of only 0.25 mm, which can be ascertained from the ordinate axis using the sensor calibration constant of 2000 V/m.

## 2.3.11  Stretched Exponentials

Systems typically demonstrate a more complex behavior than can be simply described by the SLM of viscoelasticity. In 1847, Kohlrausch (1847) discovered that the decay of the residual charge on a glass Leyden jar followed a *stretched exponential* law. The functional form that he discovered is often associated with a broad distribution of relaxation times, and has been found to describe a remarkably wide range of physical processes. To describe damping that is of the stretched exponential type, Kelvin chains or Maxwell elements in parallel have been used. Although an improved fit to the data can be realized by this means, the technique results in a high number of material parameters which have to be identified.

## 2.3.12  Fractional Calculus

A promising alternative to multiexponential decay models is to replace classical rheological dashpots by "fractional" elements. It is claimed that with only a few parameters, material behavior of many viscoelastic media can be described over large ranges of time and/or frequency (Hilfer, 2000). It may also be possible with fractional derivatives to treat the discontinuities that are sometimes present in decay (Asa, 1996). The disadvantages of fractional calculus are (i) the increasing computational/storage requirements and (ii) the esoteric mathematics, which is alien to the training of most.

## 2.3.13   Modified Coulomb Damping Model

Published here formally for the first time, with details described later, the heuristic "modified Coulomb" model is an alternative to all of the aforementioned damping models. It is thought to be closely related to secondary creep (Peters, 2001a, 2001b) and (like fractional calculus) accomplishes good fits with a small number of parameters. Developed from energy considerations, its equations are expressed in canonical form involving the quality factor $Q$.

## 2.3.14   Relaxation

Formally, relaxation is defined by the behavior of a sample subjected to a constant strain. Because of the mechanisms just discussed in relationship to creep, the stress relaxes exponentially toward zero (in the simplest approximation). In practice, the definition just given can be misleading since the word relaxation is used to describe a host of processes in which some quantity decays exponentially in time — for example, the relaxation of strain at constant stress in the Kelvin–Voigt model of viscoelasticity.

Some of the viscoelastic models using dashpots and springs have been quite successful in the limited regime of their applicability. For example, the Zener (Debye) model, which will be discussed again later, has been used for years to describe a particular form of damping in solids, which derives from relaxations associated with dislocations. Seminal experimental work of this type was conducted by Berry and Nowick in the 1950s (Berry and Norwich, 1958). A well-known theoretical model to describe dislocation damping was developed by Granato and Lucke (Granato and Lucke, 1956). The Granato model is that of a vibrating string (bowed Frank–Read source), where the end points of the "string" are points on the dislocation line that have been pinned. Recent theory shows that the Granato model is not always adequate; that "dislocation interactions may alter substantially the dislocation component of the spectrum observed during internal friction experiments." (Greaney et al., 2002) (excellent introductory material on this subject is to be found online at http://mid-ohio.mse.berkeley.edu/alex/rachel/rachel/rachel.html).

Bordoni (1954) performed experiments that led to his observation of relaxation-type internal friction processes where the acoustic attenuation is seen to peak at certain temperatures. The so-called Bordoni peaks occur at low temperatures or at ultrasonic frequencies. These losses, which are maximum when dislocation relaxations can take place in step with the driving frequency, were first observed in the FCC metals: lead, copper, aluminum, and silver.

Dislocation damping as just described is characterized by a temperature-dependent relaxation that exhibits Arrhenius behavior. By plotting the internal friction vs. reciprocal temperature, one may estimate the activation energy of the process responsible for the damping. The following quotation from the introduction of the Berry paper assists in defining some of the many expressions used historically to describe damping:

> Internal friction is often loosely described as the ability of a solid to damp out vibrations. More strictly, it is a measure of the vibrational energy dissipated by the operation of specific mechanisms within the solid. Internal friction arises even at the smallest stress levels if Hooke's Law does not properly describe the static stress–strain curve of the material. The nonelastic behavior which Zener has called anelasticity arises when the strain in the material is dependent on variables other than stress.

In a recent private communication, Prof. Granato has indicated the following:

> Dislocations do follow the Zener (or Debye) form fairly well for the damping, but not for the elastic modulus. This is because the response to a stress is given as a Fourier series. The higher order terms in the series have little effect on the damping, but a strong effect on the modulus at high frequencies. This makes the modulus fall off more slowly than with the reciprocal frequency.

## 2.4   Hysteresis — More Details

Hysteresis and creep are common to many systems, such as electromechanical actuators, especially when used at high drive levels. Their transfer function is influenced by "rate-independent memory effects." The state of the actuator depends not only on the present value of the input signal but also on the nature of their past amplitudes, especially the extremum values, but not on rates of the past (Visintin, 1996). This statement is in support of the author's secondary creep model of hysteretic damping, where the amplitude of the previous turning point determines the magnitude of the internal friction force for the half-cycle that follows. One of the most dramatic examples of a memory effect is the demonstration mentioned above, by Gross in the 1950s, concerning a twisted wire.

Damping complexities derive from the defect structures that are found in real materials and which give rise to hysteresis, which in the Greek language means to "come late." Although, almost everybody seems to appreciate magnetic hysteresis at some level, too few individuals (at least in physics) have been trained in the mechanisms of mechanical hysteresis responsible for damping. Dislocations, for example, are usually an add-on chapter to a solid-state physics text — even though they are known to be indispensable with regard to actual, as opposed to idealized, properties of materials.

In the case of ferrous materials, the magnetization of a specimen lags behind the field generated by an electric current, to which the specimen responds. In the case of real springs that do not obey Hooke's Law $F = -kx$, the displacement $x$ lags behind the spring's restoring force $F$. It is convenient to express the resulting hysteresis in terms of "intrinsic" variables instead of $x$ and $F$. Thus, the strain $\varepsilon$ (fractional change in the spring's length if it were a wire in tension) lags the stress $\sigma$ (force per unit area). Usually in engineering practice, the stress is reckoned with respect to the external force (negative of the spring $F$), so that the equivalent to Hooke's Law is $\sigma = E\varepsilon$, where $E$ is an elastic modulus descriptive of the material from which the spring is fabricated. In the case of a straight wire, $E$ would be Young's modulus but, for coil springs, $E$ is determined primarily by the shear modulus. Some of the ways in which hysteresis can be represented for a freely decaying oscillator are shown in Figure 2.3. The generalized coordinate $q$ would be spring elongation for the force case shown, or it would be strain when the ordinate quantity is stress. The graph of velocity vs. displacement is referred to as a phase-space plot. It is commonly used in describing chaotic systems and, if "strobed" at the frequency of the oscillator, becomes the Poincaré section. Notice that the circulation is of opposite sign when using external force as opposed to spring force, in addition to the curves occupying different quadrants. It is important to recognize this difference, particularly when discussing negative damping where the oscillation amplitude builds in time, as illustrated in the right hand part of the figure.

Although not very common in mechanical oscillators, it is possible to realize negative damping. One example is that of an optically driven pendulum, because of the LiF crystals that were placed in its support structure (containing a high density of color centers produced by radiation) (Coy and Molnar, 1997). An interesting feature of this pendulum was its unwillingness to entrain to the driving laser.

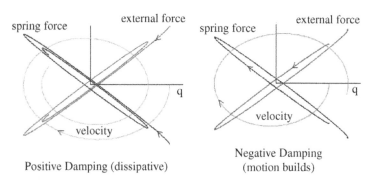

**FIGURE 2.3**   Three different ways to represent hysteresis damping for an oscillator in free-decay. Cases of both positive and negative damping are illustrated.

There are also examples of negative damping from aerodynamics, such as flutter. Since buildings and bridges can experience negative damping in catastrophic manner (Tacoma Narrows bridge as an example), it is not a subject to be ignored.

Another example of hysteresis that is very much like negative damping (though not usually labeled as such) is to be found in a heat engine (Peters, 2001a, 2001b). The motion is not simple harmonic; rather, the speed with which the hysteresis curve is traversed (in pressure vs. volume) increases as the size of the hysteresis loop increases. A larger loop (greater work done by the gas) results in higher revolutions per minute (r/min) of the engine as opposed to a larger amplitude of the motion at constant period. The gas pressure provides a force similar to the Hooke's Law force of the spring in a mass/spring oscillator.

It is usually assumed that hysteresis loops are "smooth," which is not necessarily true. For example, in the case of magnetic hysteresis, the "jerky" parts known as the Barkhausen effect (Barkhausen, 1919) are well known. The equivalent jerky behavior in metallic alloys is known as the Portevin–LeChatelier effect (Portevin and LeChatelier, 1923). Although we have historically avoided these cases that appear to be intractable in a mathematics sense (not obeying the fundamental theorem of calculus), their presence is undeniable testimony of the complex nature of hysteresis.

## 2.5 Damping Models

### 2.5.1 Viscous Damped Harmonic Oscillator

As first seen by students in a textbook, the equation of motion for a damped, driven harmonic oscillator is likely as follows:

$$m\ddot{x} + c\dot{x} + kx = F(t) \tag{2.1}$$

where $m$ is mass, $k$ is spring constant, $c$ is a "constant" of viscous damping, and $F(t)$ is the external force driving the oscillator. It is convenient to work with a coefficient of performance, or quality factor $Q$, and rewrite Equation 2.1 in canonical form as

$$\ddot{x} + \frac{\omega_0}{Q}\dot{x} + \omega_0^2 x = \omega_0^2\frac{F(t)}{k}, \text{ with } \omega_0^2 = \frac{k}{m} \tag{2.2}$$

For $F(t) = 0$ and an assumed solution, $x = A \exp(p\theta)$ with $\theta = \omega_0 t$, the differential equation becomes algebraic (quadratic) in $x(\theta)$, with the roots given by

$$p = -\frac{1}{2Q} \pm \sqrt{\frac{1}{4Q^2} - 1} \tag{2.3}$$

Depending on the value of $Q$, the motion is either overdamped (nonoscillatory), critically damped, or underdamped. Here, we restrict our attention to the last case corresponding to $Q > 1/2$, in which the square root term of Equation 2.3 is imaginary. Moreover, we are mostly concerned with systems in which $Q \gg 1$.

### 2.5.2 Definition of Q

The quality factor $Q$ is in general defined as $2\pi$ times the ratio of the energy of the oscillator to the energy lost to friction per cycle. For viscous damping (and hysteretic damping, later discussed), the $Q$ is independent of the amplitude of oscillation. For other types of damping, we will see that the $Q$ is not constant. In the case of the viscous damped oscillator, $Q = \omega_0/2\beta$ where $\beta$ appears in the solution as an amplitude decay "constant." The parameter $\beta$ is not really constant, as discussed in Chapter 3, Section 3.9.

$$x = A\, e^{-\beta t} e^{\pm j\omega_0 t\sqrt{1 - 1/4Q^2}} \tag{2.4}$$

Since $x$ is real, we use the real part of Equation 2.4 and employ Euler's identity to obtain

$$x(t) = A \, e^{-\beta t} \cos(\omega_1 t - \phi), \text{ with } \omega_1 = \sqrt{\omega_0^2 - \beta^2} \tag{2.5}$$

where $\phi$ is a constant determined by the initial conditions.

### 2.5.3  Damping "Redshift"

It is seen that the frequency of oscillation depends on the damping constant, $\beta$; however, the fractional change $\Delta\omega/\omega_0$ is almost always negligibly small. For example, the reduction in frequency is only 1.4% for $Q = 3$, which is close to critical damping of $Q = 0.5$. At these small values of $Q$, the lifetime of a freely decaying oscillator is so short that the frequency is ill-defined because of the Heisenberg uncertainty principle. At larger $Q$s, where the frequency is well-defined, the shift is negligible; i.e., at $Q = 100$, the fractional shift is only $1.3 \times 10^{-5}$. In the case of internal friction (hysteretic) damping, there is no redshift anyway because the oscillator is isochronous.

### 2.5.4  Driven System

When $F(t)$ is not zero, but rather corresponds to harmonic drive at angular frequency $\omega$ and amplitude $A$, the response involves the sum of Equation 2.5 (transient) and a particular solution (steady state).

$$x_p(t) = A_p \cos(\omega t - \delta), \text{ with } \delta = \tan^{-1}\left(\frac{2\omega\beta}{\omega_0^2 - \omega^2}\right). \tag{2.6}$$

The system resonates (amplitude a maximum) at $\omega \to \omega_R = \sqrt{\omega_0^2 - 2\beta^2}$, and the variation of the amplitude with $\omega$ at steady state at any drive frequency $\omega$ is given by

$$A_p = \frac{A\omega^2}{\sqrt{(\omega_0^2 - \omega^2)^2 + 4\omega^2\beta^2}} \tag{2.7}$$

The resonance response curve described by Equation 2.7 is called the Lorentzian. More frequently in physics, the term is used to describe pressure-broadened line widths (Milonni and Eberly, 1988). As noted previously, Lorentz was never apparently content with the damping term, $2\beta \, dx/dt$. In his Ph.D. dissertation concerned with the damping of electron oscillators through electromagnetic radiation, he was not able to satisfactorily describe the damping from first principles. Although we might be tempted to say that this failure derived from his classical (prequantum mechanics) description of the problem, such a viewpoint is an oversimplification.

### 2.5.5  Damping Capacity

#### 2.5.5.1  Viscous Damping

The loss per cycle, called the damping capacity, is computed for the viscous damping case as follows (per unit mass):

$$d_v = 2\beta \oint \dot{x} \, dx = 2\beta\omega A^2 \int_0^{2\pi} \sin^2\theta \, d\theta = 2\pi\beta\omega A^2 \tag{2.8}$$

where $A$ is the amplitude of the oscillation. Because the total energy per unit mass is $\omega^2 A^2/2$, we see that $Q = \omega/(2\beta)$.

### 2.5.5.2   Hysteretic Damping, Linear Approximation

The equation of motion in this case is given by $m\ddot{x} + h/\omega\dot{x} + kx = 0$ where $h$ is a constant. The energy loss in one cycle is given by

$$-\Delta E = md_h = \frac{h}{\omega}\int_0^T \dot{x}^2\,dt = \frac{h}{\omega}\omega A^2\int_0^{2\pi}\cos^2\theta\,d\theta = \pi hA^2 \rightarrow d_h = \frac{\pi}{m}hA^2 \qquad (2.9)$$

so that $Q = m\omega^2/h$.

### 2.5.5.3   Hysteretic Damping, Modified Coulomb Model

The nonlinear equation of motion introduced in this chapter to describe hysteretic damping is as follows:

$$\ddot{x} + c\sqrt{\frac{2E}{k}}\,\mathrm{sgn}(\dot{x}) + \omega^2 x = 0 = \ddot{x} + \frac{\pi\omega}{4Q_h}\sqrt{\omega^2 x^2 + \dot{x}^2}\,\mathrm{sgn}(\dot{x}) + \omega^2 x = \ddot{x} + cA_{\text{prev}}\mathrm{sgn}(\dot{x}) + \omega^2 x \qquad (2.10)$$

where, in the last expression, the subscript "prev" implies amplitude at the last (previous) turning point of the motion. This particular form for the damping term (Peters, 2002a, 2002b, 2002c), thought to result from secondary as opposed to primary creep (Peters, 2001a, 2001b), is not as computationally useful as the middle expression involving the $Q$. The damping capacity is given by

$$-\Delta E = md_h = 4cmA\int_0^{\pi/2} A\cos\theta\,d\theta \longrightarrow d_h = 4cA^2 \qquad (2.11)$$

yielding $Q = \pi\omega^2/(4c)$, so that the constant in the nonlinear model is related to the linear approximation constant through

$$c = \pi h/(4m)$$

## 2.5.6   Coulomb Damping

One of the simplest friction models is that in which a Hooke's Law spring is connected on one end to a mass that slides on a level table. The other end of the spring is connected to a stationary wall. The friction force of the mass against the table is of the type first described quantitatively by Charles Augustin Coulomb (1736–1806), although Leonardo da Vinci is probably the first to consider it scientifically. The equation of motion and its solution, for the free-decay of an oscillator damped by Coulomb friction, is given by

$$m\ddot{x} + f\,\mathrm{sgn}(\dot{x}) + kx = 0$$

Solution                                                                                                                    (2.12)

$$x(t) = [x_0 - (2n+1)\Delta_x]\cos\omega t + (-1)^n\Delta_x$$

The equation is nonlinear because of the sign of the velocity term, but it is easily integrated numerically; additionally, it is one of the few nonlinear equations for which an analytic solution is known and is given above (for more details, the reader is referred to Peters and Pritchett, 1997). The integer, $n$, specifies the number of half-cycle turning points from $t = 0$, and $\Delta_x$ is the decrement (linear, not logarithmic, having units of m) per half-cycle. There are occasions to use Equation 2.12; for example, problems in civil engineering where relative motion of members (slipping) occurs at a structural joint. The work against friction in one cycle can be obtained from energy considerations and is given by

$$f(4x_0 - 8\Delta_x) = \frac{1}{2}kx_0^2 - \frac{1}{2}k(x_0 - 4\Delta_x)^2 \qquad (2.13)$$

which, for small decrement, yields

$$\Delta_x = \frac{f}{k} = \frac{f}{m\omega^2} \qquad (2.14)$$

Damping characteristics for the models presently treated are summarized in Box 2.1.

# Box 2.1

## Damping Characteristics

| Type | Equation of Motion | Damping Capacity | $Q$ |
|------|-------------------|------------------|-----|
| Viscous | $\ddot{x} + 2\beta\dot{x} + \omega_0^2 x = 0$ | $2\pi\beta\omega A^2 m$ | $\dfrac{\omega}{2\beta}$ |
| Hysteretic (linear approximation) | $\ddot{x} + \dfrac{h}{m\omega}\dot{x} + \omega^2 x = 0$ | $\pi h A^2$ | $\dfrac{m\omega^2}{h}$ |
| Hysteretic (modified Coulomb) | $\ddot{x} + c_h A\,\mathrm{sgn}(\dot{x}) + \omega^2 x = 0$ | $4 c_h A^2 m$ | $\dfrac{\pi\omega^2}{4 c_h}$ |
| Coulomb | $\ddot{x} + \dfrac{f}{m}\,\mathrm{sgn}(\dot{x}) + \omega^2 x = 0$ | $4 f A$ | $\dfrac{\pi m \omega^2 A}{4 f}$ |
| Amplitude dependent | $\ddot{x} + c_f A^2\,\mathrm{sgn}(\dot{x}) + \omega^2 x = 0$ | $4 c_f A^3 m$ | $\dfrac{\pi\omega^2}{4 c_f A}$ |

## 2.5.7 Thermoelastic Damping

A microphone with Labview was used to analyze vibratory data of an aluminum rod. A rod of 1 m length can be excited to ear-piercing intensities by holding it at its center between thumb and finger of one hand, and stroking along the length with the other hand that is coated with violin-bow rosin. The decay of this "singing rod", which is a common part of physics demonstration equipment, was found to be in agreement with the following theoretical expression for thermoeleastic damping (Landau and Lifshitz, 1965):

$$\frac{1}{Q_{\mathrm{Th.d}}} = \frac{\kappa T \alpha^2 \rho \omega}{9 C^2} \tag{2.15}$$

where $\omega$ is the vibrational angular frequency, $T$ is the temperature, $\rho$ is the density of the bar, $C$ is the heat capacity per unit volume, $\alpha$ is its thermal expansion coefficient, and $\kappa$ is the thermal conductivity. The expression assumes adiabatic vibrations and there is no thermoelastic dissipation in pure shear oscillations (e.g., torsional oscillations of a bar) because the volume does not change and hence there is no local oscillation of the temperature. Notice, in particular, that the $Q$ is inversely proportional to frequency, unlike viscous damping that is proportional to the frequency, or hysteretic damping that is proportional to the square of the frequency. Thermoelastic damping is important for high-frequency compressional oscillations in materials with significant thermal coefficients, and especially for metals because of their large thermal conductivity.

The demonstration of comparable behavior in polymers (entropy spring, but opposite sign compared with metals) is quite easy. Stretch a rubber band between the hands and immediately touch it to the forehead. The increase in temperature is easily sensed. Conversely, releasing the tension in the band cools it enough to be sensed by placing the band to a part of the face that is sensitive to temperature change. Equation 2.15 does not apply to polymers.

## 2.6 Measurements of Damping

### 2.6.1 Sensor Considerations

The challenge to any measurement is to accomplish the task without significantly altering the system under study (see de Silva, 2006). For measurements on mechanical oscillators of the type described in this document, two types of sensor are generally superior to every other kind: (i) optical and (ii) capacitive. Optical sensors are probably the least perturbative but they do not readily yield themselves to large dynamic range with good linearity (and small quantization errors for digital type). Inductive sensors, such as the linear variable differential transformer (LVDT), are known from seismology to be inherently more noisy (up to 100 times) because of ferromagnetic granularity. Additionally, transformers are not amenable to miniaturization, and the components are inherently less stable. It is therefore a mystery why the widespread use of the fully differential inductive sensor (LVDT) continues when we have available the superior fully differential capacitive sensor, which is electrically equivalent (apart from its reactance type) and capable of miniaturization to the MEMS level. The challenge with really small capacitive sensors is the increase in output reactance of the device as they approach femtoFarad levels of individual capacitors.

All measurements reported in this document were taken with the fully differential unit whose patent name is "symmetric differential capacitive" (see Peters, 1993a, 1993b). It is especially useful for studying mechanical oscillators of macroscopic size and, morphed to various forms, it recently has found application in MEMS. It is capable of great sensitivity when configured in the form of an array, as shown in Figure 2.4.

Various lines in Figure 2.4 correspond to narrow insulator strips, such as the single vertical line in the set that connects to the amplifier. In the cross-connected static set, the plates labeled "1" are electrically distinct from the others labeled "2". The total-plate arrangement constitutes a symmetric AC bridge, and the central position of the moving set ($x = 0$ as shown in the figure) corresponds to bridge balance with $V_0 = 0$. Displacement away from balance gives a voltage output that is linear between $-w/4$ and $w/4$, as illustrated in the graph at the bottom of Figure 2.4.

The oscillator frequency is typically tens of kHz, and the amplifier is of instrumentation type (Horowitz and Hill, 1989). Unlike a bridge null detector, the linear response through $x = 0$ is realized

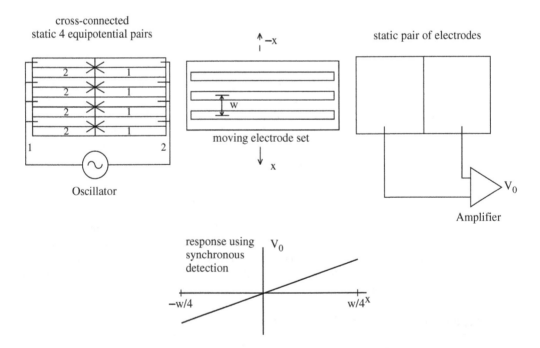

**FIGURE 2.4**  Illustration of a fully differential capacitive transducer array. For clarity, the three electrode-sets are shown separated from their operating positions (parallel with a small separation gap, with the moving electrodes in the middle).

when synchronous detection is employed. This can be accomplished with a lock-in amplifier, but the most recent Cavendish balance to employ the sensor uses diodes (Tel-Atomic Inc., online at http://www. telatomic.com/sdct1.html).

A tutorial ("detailed explanation") of the SDC sensor using diodes can also be found at this website.

In Figure 2.4, four individual SDC units have been shown connected in parallel. The total number, $N$, of individual units in an array depends on the characteristic width, $w$, for which the total range of detectable motion is $w/2$. If the requirement on range is small, then $N$ can in principle be made very large, which is desirable for the following reason. The sensitivity of this position sensor is inversely proportional to $w$ if output capacitance of the device is not a factor. As $w$ is reduced, however, the degrading influence of increased output reactance (capacitive) is more significant than the improved sensitivity that would result if the sensor could be connected to an amplifier with infinite input impedance. Since the instrumentation amplifier's input capacitance is not negligible, shrinking $w$ is beneficial only if the output reactance can by some means be kept low. This is accomplished with the array of individual units. In principle, the output capacitance could be held reasonably constant as $N$ approaches 100, by using photolithographic techniques and small spacing between the parallel electrodes. The concept has been deemed feasible because of existing technologies as well as the following: although not in the form of an array, Auburn University has fabricated a mesoscale accelerometer around the SDC sensor. The prototype was built on printed wire board (PWB) under U.S. Army contract (Dean, 2002).

No doubt the popular silicon-based MEMS accelerometers marketed by Analog Devices utilize the impedance advantages of an array, employing a large number of "fingers" in a force-feedback arrangement. Although employed mostly otherwise, the first case of a fully differential capacitive transducer using force-feedback was one based on simultaneous action of actuator and sensor functions in a single unit of nonlinear type (Peters et al., 1991).

## 2.6.2 Common-Mode Rejection

In attempts to measure damping, one can be confronted with difficulties of mode mixing. For example, the historical Cavendish experiment, using optical detection, has been traditionally difficult unless the instrument is placed in a very quiet location to avoid pendulous swinging of the boom. The high-frequency pendulous motion (of the order of 1 Hz) as a "noise" becomes superposed on the low-frequency torsional signal. The computerized Cavendish balance sold by Tel-Atomic overcomes this problem by means of a mechanical common-mode rejection feature. An SDC sensor placed near one boom end is connected in electrical phase opposition to a second SDC sensor placed near the other end of the boom. The boom itself serves as the moving electrode for both sensors. Neither sensor has a first-order response to boom motion parallel to its long axis. Pendulous motion perpendicular to the boom orientation is largely canceled.

## 2.6.3 Example of Viscous Damping

The aforementioned Sprengnether–LaCoste spring seismometer is well-suited to the demonstration of viscous damping, when damping is imposed in the following manner: the instrument was built with a Faraday Law (velocity) detector; i.e., a coil that moves with the mass of the instrument, in the field of a stationary magnet. As originally employed, the coil was connected to the amplifier of a recorder. In the present configuration, however, the velocity detector is not employed, since its sensitivity is severely limited at low frequencies. Instead, an SDC array of the type shown in Figure 2.4 is used to measure the position of the mass (a pair of lead weights, total mass 11 kg). If the instrument is operated with the coil open-circuit, there is no induced current. By connecting a resistor across the coil (through very fine copper wires that go to terminals on the case), mass motion induces a current. The induced current opposes the motion through Lenz's Law, resulting in damping. The damping depends on the size of the

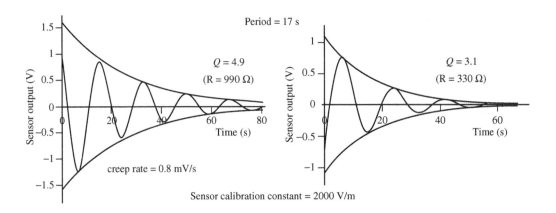

**FIGURE 2.5**   Examples of induced current damping of a vertical seismometer using two different resistors.

current and is thus an inverse function of the resistor's magnitude through Ohm's Law. The phenomenon is illustrated in Figure 2.5.

As compared with the "undamped" instrument, whose $Q$ is approximately 80 at a period of 17 sec, it is seen that the addition of a 990-ohm resistor lowered the $Q$ by more than an order of magnitude to 4.9. A 330-ohm resistor reduced it even further to 3.1. The amount of damping is also governed by the resistance of the coil winding, which is 480 ohm.

The envelopes that have been fitted to the decay curves were the basis for estimating the $Q$. The decay data were imported to Excel by first outputting the Dataq DI-154RS A/D generated record as a *.dat (CSV) file. The fits were produced by trial and error using the drag and autofill functions. Notice that the 990-ohm resistor (first) case is not as pure an exponential decay as the other case because of creep. The rate of creep is greater at large initial amplitudes of the motion.

## 2.6.4   Another Way to Measure Damping

Curve fitting (full nonlinear, in general) is the best way to estimate damping parameters, especially if the decay is not exponential. For more routine cases, simpler methods can be used. Among the host of ways that have been defined to specify the damping of an oscillator, one of the most common uses the logarithmic decrement. The solution to Equation 2.1 with zero right-hand side is given by

$$x(t) = x_0\, e^{-\beta t} \cos(\omega t + \phi). \tag{2.16}$$

The full-cycle turning points, $x_N = x_0\, e^{-\beta N T}$, with $N = 0, 1, 2, \dots$ can be used to compute the logarithmic decrement through

$$\beta T = \frac{1}{N} \ln \frac{x_0}{x_N} \tag{2.17}$$

Unfortunately, an estimate based on Equation 2.17 can be difficult due to the presence of either or both of two problems: (i) mean position offset in the decay record or (ii) asymmetry of the decay, where the turning points on one side of equilibrium decay at a different rate than those on the other side. Case (ii) occurs more often than one might expect; it is frequently a consequence of material complexity and not the result of nonlinearity in the electronics of the detector. It is important, however, to be sure that the detector is either linear or that corrections for the nonlinearity be utilized before estimating the damping.

A method to provide partial compensation uses half-cycle turning points $n = 2N$, and works with a minimum of three such points.

$$\beta T = -2 \ln[1 - (x_{n-1} - x_{n+1})/(x_{n-1} - x_n)] \tag{2.18}$$

Advantage is taken of random error reduction by using Equation 2.18 on a set of turning points (optimal number sometimes being about a dozen). The calculations are straightforward in a spreadsheet such as Excel by means of the autofill function.

## 2.7 Hysteretic Damping

### 2.7.1 Equivalent Viscous (Linear) Model

The few mechanical oscillators governed by Equation 2.1 tend to be those in which there is an external control, such as eddy current damping. For oscillators in which the damping derives from internal friction of its members, the following linear approximate form of the hysteretic damping model has been used:

$$m\ddot{x} + \frac{h}{\omega}\dot{x} + kx = F \tag{2.19}$$

It should be noted that hysteresis is the cause for all damping; however, engineers have come to use the term "hysteretic damping" for systems described by Equation 2.19. This equation differs in two important ways from Equation 2.1. For the viscous damped oscillator, $Q$ is proportional to the frequency, but for the hysteretic damped oscillator, $Q$ is proportional to the square of the frequency. Also, viscous damping changes the frequency of the oscillator, since $\omega_1 < \omega_0$ and, for resonance, the frequency is even lower. However, the hysteretic oscillator is isochronous, requiring only a single frequency $\omega = \sqrt{k/m} \rightarrow \omega_r$ to describe all features of the motion. For example, it is easy to show that the oscillator resonates at this frequency. Off resonance, the response is not the standard Lorentzian. To show this, assume steady state and use the phasor method given to us by Steinmetz, 1893 (complex exponential form for the variables); i.e., $F = F_0\, e^{j\omega t}$ and $x = x_0\, e^{j\omega t}$ to get the frequency transfer function

$$\frac{kx}{F} = \frac{1}{1 - \omega^2 \dfrac{m}{k} + j\dfrac{h}{k}} = \frac{1}{1 - r^2 + j\alpha} = Z, \text{ with } r = \frac{\omega}{\omega_r} \text{ and } \alpha = \frac{h}{k} = \frac{1}{Q} \tag{2.20}$$

for which the real and imaginary parts are given by

$$\text{Re } Z = \frac{1 - r^2}{(1 - r^2)^2 + \alpha^2}, \qquad \text{Im } Z = \frac{-\alpha}{(1 - r^2)^2 + \alpha^2} \tag{2.21}$$

which is expressible in polar form as

$$Z = |Z|\, e^{j\delta}, \text{ where } |Z| = \frac{1}{\sqrt{(1 - r^2)^2 + \alpha^2}} \text{ and } \delta = -\tan^{-1}\frac{\alpha}{1 - r^2} \tag{2.22}$$

It is interesting to compare the steady-state response of the driven, hysteretic damped oscillator with that of the driven, viscous damped oscillator; i.e., Equation 2.22 compared with normalized Equation 2.7. A Bode plot comparison (log–log, for the amplitude case) is provided in Figure 2.6. At small values of the damping parameter $\alpha$ (large $Q$), there is insignificant difference between the two cases. At large values, however, the difference is significant.

### 2.7.2 Examples from Experiment of Hysteretic Damping

The vertical seismometer that was used for several of the present studies is known to decay according to hysteretic damping. In Section 2.16.4 titled "Failure of Viscoelasticity", details are provided of the work by Gunar Streckeisen (1974) that showed this to be true. Decay curves of the instrument are

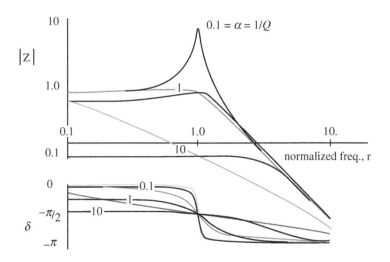

**FIGURE 2.6** Bode plot comparison of steady-state driven system with (i) hysteretic damping (dark curves) and (ii) viscous damping (light curves).

frequently a near-perfect exponential, once corrected for secular drift of the record. Sometimes, this drift is the result of creep in the spring of the instrument, but it may also be the result of other factors, such as (i) temperature change, or (ii) barometric pressure variation, or even (iii) tidal influence. The temperature sensitivity is due to the difference of thermal coefficients of the materials from which the instrument is constructed, and the pressure variation is a buoyancy effect. Tidal influence is the smallest of the three, which causes minute accelerations of the crust of the Earth with a period of about 12 h.

In the discussions which follow, two different decay records are provided. In both cases, the initial amplitude of oscillation is quite large, being a significant fraction of 1 mm, and the period for the two cases is different — the first case being 17 sec and the second one 21 sec. The first case time record, shown in Figure 2.7, contains 9800 points. Once a 12 $\mu$V/s (upward) drift was removed, the decay (left curve) is seen to be "nearly textbook" exponential.

The adjective "nearly" is appropriate because there is a 12% difference in the decay constants defining the upper and lower turning points (0.0022 top, 0.0025 bottom), which were determined by trial and error "eyeball" exponential fits using Excel. In this author's experience, such is the norm for virtually all mechanical oscillations; perfectly symmetric exponential decays have rarely been seen in the hundreds of cases studied.

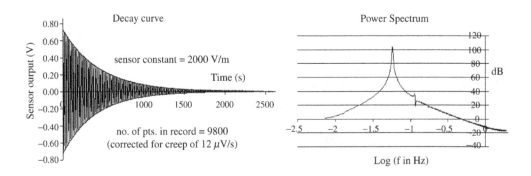

**FIGURE 2.7** Free-decay of a vertical seismometer due to hysteretic damping. The period of oscillation is 17 sec.

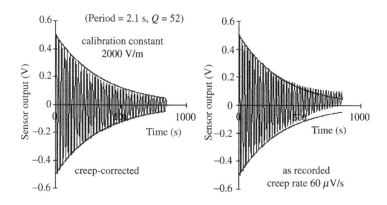

**FIGURE 2.8** Free-decay of a seismometer due to hysteretic damping.

Because there are roughly 150 oscillations in the time record of Figure 2.7, it is not possible to resolve individual turning points of the motion, but the oscillations are very nearly that of a pure damped sinusoid, as noted from the right-side graph of the figure. This spectrum was generated with a 4096-point FFT, comprising the first 1090 sec of the time record. The second harmonic is the only distortion observed, and it is about 65 dB below the fundamental. For the case presented in Figure 2.7, $Q = 80$.

Another example of free-decay hysteretic damping is provided in Figure 2.8. As usual, the record was afflicted by drift, possibly from creep in the spring, in this case a constant rate of 60 $\mu$V/s, as observed in the graph on the right. All of these graphs were produced with Excel, as noted earlier in the discussion of induced current damping. As with the decay curve of Figure 2.7, the creep-corrected graph on the left was generated by adding a secular term to the raw data. Once corrected, the decay is a near-perfect exponential of hysteretic type. We will see other examples (from pendulum studies) in which two damping mechanisms are simultaneously active in a decay.

The $Q$ values corresponding to Figure 2.7 and Figure 2.8 are consistent with hysteretic damping; i.e., 80 for the 17-sec oscillation and 52 for the 21-sec oscillation. As noted elsewhere in this document, $Q \alpha \omega^2$ for hysteretic damping as opposed to an exponent of 1 for viscous damping. Of course, one must collect data over a very much larger range of frequencies to verify this, as was done by Streckeisen (1974).

## 2.8 Failure of the Common Theory

Many mechanical oscillator studies in decades past, mainly by engineers, have shown that the so-called decay constant $\beta$ is proportional to $\omega^{-1}$ instead of being constant (e.g., Bert, 1973). The damping for these cases came to be called "material", "structural", or "hysteretic." A common way to obtain the correct frequency dependence was to divide the velocity by frequency and call the result an "equivalent viscous" form of damping. The adjective "equivalent" draws attention to the fact that internal friction in a solid cannot really result from fluid effects. Moreover, elsewhere in this document, there is plenty of support for the position that the linear equations of viscous damping type cannot produce truly meaningful (predictive) models when doing modal analysis on multibody systems.

An important early work by Kimball and Lovell (1927) is evidently the first experiment to show that internal friction ("force") of many solids is virtually independent of frequency. In other words, their elegant technique, in which a rotating rod is deflected by a transverse force, was the first to demonstrate the "universality" of hysteretic damping. Although both researchers were physicists at General Electric in the time of Steinmetz, few physicists of the 21st century know of this important work. As with the important contributions of Portevin and LeChatelier, their study of systems influenced by "dirty physics" was evidently ignored in favor of the "clean" new quantum mechanics of that era.

It is interesting that a bell made of lead does not tinkle at room temperature, but it can be made to do so at 77 K, by immersion in liquid nitrogen. This demonstration, which is often employed in physics "circuses," shows clearly that the internal friction of lead at audio frequencies can be reduced substantially by lowering the temperature. An important lesson to be learned from these observations is that damping, in general, is a complex function of temperature, frequency, conductivity, …(who knows where to terminate this list). Not only is a multitude of state variables necessary for a complete description of dissipation, but the previous history of stress–strain cycling may also be critical. Such is the nature of defect structures responsible for damping.

## 2.9   Air Influence

Even when operating an oscillator in high vacuum, there is a significant remanent damping that derives from internal friction. This fact is illustrated in Figure 2.9, which provides data for two different "simple" pendula. They are simple in the sense that the bob mass is concentrated near the bottom of the pendulum structure. In the figure, decay time (reciprocal of the decay constant, $\beta$) has been plotted against the natural log of the pressure in mtorr. Pressure reduction was done with a high-quality roughing pump, and the pressure was measured with (i) a mechanical gauge in the range 8 torr $< P <$ 760 torr and (ii) a thermocouple vacuum gauge for $0 < P < 100$ mtorr. In the range from 100 mtorr to 8 torr, the pressure could not be accurately measured with either of these gauges. Similarly, pressures below 1 mtorr could not be presently measured, but in similar other experiments with this pump, and using an ion gauge, it was easy to pump below 0.01 mtorr.

The period of each pendulum was very close to 1 sec, and the starting amplitude of the motion for every case was about 25 mrad. The heavier pendulum used a pair of pointed steel supports resting on single-crystalline silicon wafers to provide the axis of rotation. At the bottom of the pendulum was attached a solid lead ball whose mass was approximately 1 kg. The lighter pendulum was supported by a steel knife-edge resting on hard ceramic flats, and a large (10.3 cm dia.) lightweight (143 g) hollow metal sphere was attached at the bottom to provide as much air drag as possible. The motion was measured with an SDC sensor feeding the computer through a Dataq DI-154RS A/D converter.

Although air damping is evident in Figure 2.9, it is not as influential as one might expect, at least for the heavy pendulum. Moreover, at atmospheric pressure, it was easy to demonstrate the importance of nonlinear drag. As also noted in Nelson and Olssen (1986), this form of fluid friction caused a significant amplitude-dependent damping.

The remanent damping, once air influence is eliminated (pressure below 1 mtorr), is substantial relative to atmospheric damping, for both pendula. Removing the air increased the $Q$ from 7500 to

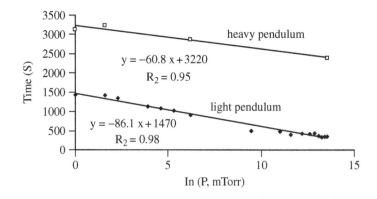

**FIGURE 2.9**   Pendulum damping as a function of pressure in a vacuum chamber.

10,100 for the heavy pendulum and, for the light pendulum, the increase was from 1000 to 4600. We thus see that even a pendulum designed to be heavily influenced by air drag also has significant damping that depends on the material from which the pendulum is fabricated or on the material upon which it rests.

The difference in internal friction damping between the heavy and light instruments was not expected to be so great. Although this might be due to the difference in axis-type (points for the heavy instrument and knife-edge for the light), no systematic effort was made to determine the primary source of the damping difference. In addition to different axis designs, the means for holding the instruments together was different. The light pendulum used a large-diameter solid brass wire between the axis and the lower mass, and the heavy pendulum used an aluminum tube.

Both of the pendula used to generate Figure 2.9 were relatively high-frequency instruments (period of 1 sec). The pivot was located, in each case, near the top end of the instrument. As such, they stand in stark contrast with the instruments that motivated this paper, where long-period pendula were used. A simple instrument to demonstrate some of the complexities of long-period instruments is a rod-pendulum of adjustable period (refer to Figure 2.1 above). The closer the axis to the center, the longer the period and the greater the influence of internal friction. It is easy to show that the sensitivity of a pendulum to external forces is proportional to the square of the period. Similarly, the ability to detect influence of internal configurational change is quadratic in the period.

# 2.10  Noise and Damping

## 2.10.1  General Considerations

Damping is inseparable from noise issues having nothing to do with undesirable sounds that might be produced by oscillation. In the simplest cases, the noise associated with damping can be described by the fluctuation–dissipation theorem. The viscous damped, thermally driven oscillator is a classic example of thermodynamic equilibrium, for which this theorem is applicable. The classic electronics analogous case is the Johnson noise of a resistor, described by the Nyquist (white) noise formula.

The largest obstacle to constructing a perfectly simple harmonic oscillator is the oscillator's dissipation. If damping were perfectly smooth, this would not be so great a challenge. However, the fluctuation–dissipation theorem of statistical mechanics guarantees that damping is accompanied by fluctuating forces. The larger the damping, the larger the fluctuating forces, i.e., the larger the noise. It is a standard problem in statistical mechanics to show that the magnitude of relative fluctuation is inversely proportional to the square root of the number of particles involved. In the case of internal friction noise, defects associated with mesoscale structures cause the effective number of particles responsible for the noise to be much smaller than the total number of atoms in a sample. Unfortunately, the fluctuation–dissipation theorem probably does not apply. It has been long known that it does not apply to the Barkhausen effect (Barkhausen, 1919). It has been recently demonstrated that it does not apply to structural glass (Grigera and Israeloff, 1999). The close relationship postulated by the author between the PLC effect and the Barkhausen effect implies that the fluctuation–dissipation theorem should also not generally apply to internal friction damping.

Internal friction noise is not white but rather more like $1/f$ (or flicker = pink) noise, a ubiquitous form that has not yet been explained from first principles. A frequently cited paper on self-organized criticality states the following:

> We shall see that the dynamics of a critical state has a specific temporal fingerprint, namely "flicker noise," in which the power spectrum $S(f)$ scales as $1/f$ at low frequencies. Flicker noise is characterized by correlations extended over a wide range of timescales, a clear indication of some sort of cooperative effect. Flicker noise has been observed, for example, in the light from quasars, the intensity of sunspots, the current through resistors, the sand flow in an hourglass, the flow of rivers such as the Nile, and even stock exchange price indices. Despite the ubiquity of flicker noise, its origin is not well understood. Indeed, one may say

that because of its ubiquity, no proposed mechanism to date can lay claim as the single general underlying root of $1/f$ noise. We shall argue that flicker noise is in fact not noise but reflects the intrinsic dynamics of self-organized critical systems. Another signature of criticality is spatial self-similarity. It has been pointed out that nature is full of self-similar "fractal" structures, though the physical reason for this is not understood.   (Bak, 1988)

It should be noted that controversy exists concerning this self-organized criticality paper, summarized in the following excerpt from Bak's book on $1/f$ noise called *How Nature Works: The Science of Self-Organized Criticality* (page 95):

In an earlier work (CFJ), performed while an undergraduate student in Aarhus, Denmark, (Kim Christensen) showed that our analysis of $1/f$ noise in the original sandpile article was not fully correct. Fortunately, we have since been able to recover from that fiasco in a joint project by showing that for a large class of models, $1/f$ noise does indeed emerge in the SOC state.

In the last few years, mathematicians have been drawn to "... an analogy, in which three areas of mathematics and physics, usually regarded as separate, are intimately connected. The analogy is tentative and tantalizing, but nevertheless fruitful. The three areas are eigenvalue asymptotics in wave physics, dynamical chaos, and prime number theory" (Berry and Keating, 1999). Some mathematicians speculate that a dynamical system (perhaps some form of a mesoanelastic pendulum, in the thinking of this author) could become a "machine" to generate prime numbers.

## 2.10.2   Example of Mechanical $1/f$ Noise

Shown in Figure 2.10 is an example of mechanical flicker noise made worse by creep that originates in the spring (LaCoste type) of a Sprengnether vertical seismometer. The data are from two separate time records, the first run preceding the second run by about a half-hour. Just before collecting the data of the first run, a clamping pin was removed from the seismometer. Used to constrain the mass from moving, this pin had been left in place overnight to determine the amount of electronics noise, including drift. The measured electronics noise (white $= 1/f^0$) was more than an order of magnitude smaller than the smallest (high frequency) noise components of mechanical (seismometer) type. The peak-to-peak amplitude of the oscillation in both cases was 0.5 mm (calibration constant for the sensor being 2000 V/m). The peak-to-peak amplitude for SNR $= 1$ for this system is of the order of 1 $\mu$m.

Although the spring force was not unloaded with the pin in place overnight, nevertheless, its removal caused a significant change to defect structures in the spring, as noted by the residuals between the data and their harmonic fits (magnified by a factor of ten).

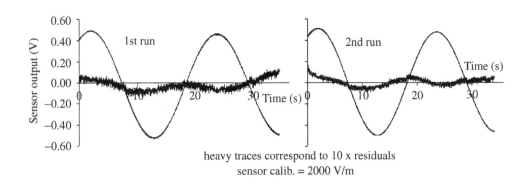

heavy traces correspond to 10 x residuals
sensor calib. = 2000 V/m

**FIGURE 2.10**   Evidence in support of $1/f$ mechanical noise in a seismometer.

The flicker character of the noise was demonstrated by computing power spectra of the residuals (not shown). The log–log plot, generated with the FFT (Cooley–Tukey FFT), showed $1/f$ frequency dependence for the second run. The larger noise of the first run was concentrated in the upper frequencies. The relaxation toward $1/f$ character suggests that flicker noise is a remanent of "work hardening." To demonstrate that the flicker noise was not due to the electronics, sensor output was recorded with the mass of the instrument locked. The electronics noise proved to be more than an order of magnitude smaller and "white" in character, probably mainly the result of A/D quantization (see Chapter 16).

Because the spring was not in equilibrium at the time the pin was removed (perhaps because of temperature change while the system was clamped), a great deal of initial molecular rearrangement occurred, involving atoms at grain boundaries. It is seen that the amount of fluctuations has noticeably decreased during the half-hour separating the two runs. Although the creep-noise would be undoubtedly much greater if the spring were relaxed altogether, it is not easy in such a case to quickly rebalance the seismometer to oscillate with a period in excess of 20 sec. These observations are in keeping with known properties of sensitive seismometers, as noted by Erhardt Wieland in "Instrumental self noise — transient disturbances" (ed. Borman and Bergmann, 2002):

> Most new seismometers produce spontaneous transient disturbances, quasi miniature earthquakes caused by stresses in the mechanical components. Although they do not necessarily originate in the spring, their waveform at the output seems to indicate a sudden and permanent (step-like) change in the spring force. Long-period seismic records are sometimes severely degraded by such disturbances. The transients often die out within some months or years; if not, and especially when their frequency increases, corrosion must be suspected. Manufacturers try to mitigate the problem with a low-stress design and by aging the components or the finished seismometer (by extended storage, vibrations, or alternate heating and cooling cycles). It is sometimes possible to relieve internal stresses by hitting the pier around the seismometer with a hammer, a procedure that is recommended in each new installation. (Wielandt, 2001)

Material damping noise appears to have features that are similar to Barkhausen noise — a magnetic phenomenon involving a system far-from-equilibrium. Such noise is associated with the granular nature of ferromagnetic domains and has consequence in the design of electronic instruments using iron alloys. For example, it is known (though not widely) that the popular LVDT is inherently less sensitive than a capacitive sensor of equivalent electrical type, because of its ferrous component (the rod-component that moves). As noted by Wielandt (2001), the capacitive sensor "…can be a hundred times better than that of the inductive type." Fully differential capacitive sensors, being electrically equivalent to the LVDT have still greater advantages borne of the higher symmetry. Additionally, by configuring the capacitive device as an array, it is possible for the sensitivity to also be greater.

Barkhausen noise and hysteretic damping noise may be much more similar than has been realized — involving granularity at the mesoscale, intermediate between micro- and macrophenomena. For such systems, first principle methods are very difficult to employ due to complexities that originate from a host of nonlinear interactions. For example, in the case of internal friction of solids, damping derives from stress–strain hysteresis determined by defect structures in the solid. Involving roughly $10^{12}$ atoms per "grain" in metal specimens, flicker noise evidently derives from self-similar structures of fractal geometry with a higher degree of spatial correlation than is true of white noise. The ubiquitousness of $1/f$ noise is consistent with the labeling of hysteretic damping as "universal," as first suggested by Kimball and Lovell (1927).

## 2.10.3 Phase Noise

The previous example was concerned with amplitude noise. It is also possible to see phase noise of mechanical type, as illustrated in Figure 2.11.

For the sensor calibration constant of 2000 V/m, it is seen that the initial amplitude of oscillation is 1.3 mrad, which is much too large to observe the discontinuities of mechanical Barkhausen type.

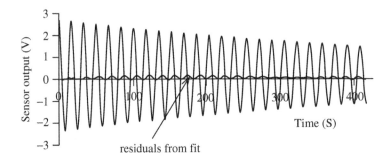

**FIGURE 2.11**    Illustration of phase noise in the free-decay of a vertical seismometer.

The phase noise is made obvious by comparing the decay to a "reference," i.e., by looking at the residuals from an Excel-generated fit to the data. By adjusting in a computer-generated damped sinusoid, (i) the initial values of amplitude and phase, (ii) the decay parameter, and (iii) the frequency, one can visually by trial and error come close to an optimum fit to the data. Having done so with Figure 2.11, the striking feature of the residuals (difference between data and fit) is the structure that looks something like "beats" but is not. The phase noise responsible for this behavior is thought to be consistent with $1/f$ mechanical noise. To visualize the noise, it is convenient to think in terms of a small randomizing (noise) vector whose tail is positioned at the head of the phasor used to generate the record. The component of the noise vector that is in the direction of the phasor generates amplitude noise as in Figure 2.10, whereas the perpendicular component is responsible for the phase noise of Figure 2.11.

Vibration phase noise imposes a serious limit on the performance of precision quartz crystal oscillators, since they are sensitive to acceleration. The phase noise in these oscillators can be observed by beating against a reference oscillator of known character; i.e., the reference serves the same purpose as the computer "fit" of Figure 2.11. To reduce the phase noise, crystals are isolated with low natural frequency vibration isolators (as described in the marketing literature of Wenzel Assoc., Austin, X).

# 2.11    Transform Methods

## 2.11.1    General Considerations

For linear systems, the Laplace and Fourier transforms (Laplace being more general) have been pre-eminent tools with which to study equations of motion (see Appendix 2A and Chapter 10). The author's transform experience (like most physicists) is mainly with Fourier transforms (FT). The discrete FT can be understood in terms of phasors (Peters, 1992). For linear differential equations, transforms are the means to convert differential forms to an equivalent algebraic form. Unfortunately, they cannot be directly employed on nonlinear equations due to the failure of superposition. Nevertheless, the linear approximations continue to be very valuable, so a chapter on damping deserves to mention some of their properties.

Ideas concerning the FFT were evidently originally treated by Gauss in the early 1800s, but the digital signal processing (DSP) "explosion" of the 1960s was largely due to the work of Cooley and Tukey (1965). For an interesting historical account about an "accident" in the publication of their paper, the reader is referred to Cipra (1993), who says the following about the FFT:

> The Fourier transform stands at the center of signal processing, which encompasses everything from satellite communications to medical imaging, from acoustics to spectroscopy. Fourier analysis, in the guise of x-ray crystallography, was essential to Watson and Crick's discovery of the double helix, and it continues to be important for the study of protein and viral structures. The Fourier transform is a fundamental tool, both theoretically and

computationally, in the solution of partial differential equations. As such, it is at the heart of mathematical physics, from Fourier's analytic theory of heat to the most modern treatments of quantum mechanics. Any kind of wave phenomenon — be it seismic, tidal, or electromagnetic — is a candidate for Fourier analysis. Many statistical processes, such as the removal of "noise" from data and computing correlations, are also based on working with Fourier transforms.

Concerning the last statement about noise, this author has used autocorrelation as a powerful means for identifying short-lived, low-frequency periodic signals in time records that do not readily show up in power spectra (FFTs). For example, they are the means for studying free-earth oscillations — eigenmodes excited by rapid relaxations of the Earth under tidal stressing (12 h periodic) (Peters, 2000). The FFT is used to generate the autocorrelation by means of the Wiener–Khintchine theorem (Press et al., 1986).

The great advantage of the FFT compared with the DFT has to do with degeneracy. The DFT proceeds to calculate the components of every "vector" in the reciprocal space (frequency reciprocal to time, units of "second", or wave number (spatial frequency) reciprocal to displacement, units of "meter") with disregard for the fact that many components have the same value, apart from a change of sign.

### 2.11.2 Bit Reversal

The key to the power of the FFT (central processor unit [CPU] time proportional to $n \log n$) compared with the discrete Fourier transform (DFT) (CPU time proportional to $n^2$) is the bit reversal scheme of the Cooley–Tukey algorithm. It is illustrated very simply by the following. Instead of a practically sized number of samples in the record to be transformed (minimum of $n = 1024$, typically), consider (for pedagogy) $n = 8$, distributed on the unit circle as shown in Figure 2.12.

Observe that the roots of unity in the complex plane, which have been numbered 0 to 7, divide the "pie" into eight equal pieces. (The algorithm requires that $n$ be expressible as a power of 2). The usual decimal counting scheme for the eight "vectors" is as indicated, traversing the phasor diagram (circle on left) sequentially. In the Cooley–Tukey algorithm, a choice is made to reverse the bits of the binary representation of the vector. Usually, the least significant bit is on the right and the most significant bit on the left, so that decimal counting is as shown on the right in the table, from 0 to 7. With bit reversal, "lsb" becomes the leftmost binary digit and the "msb" is the rightmost digit. Thus, for example, binary 110 (usually 6) becomes 3.

Using bit reversal, the phasor diagram is not traversed in the usual phasor (circulatory) sense, but rather in a "flip-flop" back and forth across the circle. By this means, there is no needless repetition in the calculation of "vector" components (real and imaginary values of a given term in the transform). For example, 5 is the simple negative of 1. It is much faster to reverse the sign on 1 to get 5 than to

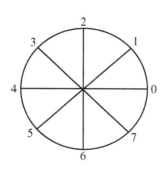

| decimal (bit-reversed) | binary number | decimal (usual) |
|---|---|---|
| 0 | 000 | 0 |
| 4 | 001 | 1 |
| 2 | 010 | 2 |
| 6 | 011 | 3 |
| 1 | 100 | 4 |
| 5 | 101 | 5 |
| 3 | 110 | 6 |
| 7 | 111 | 7 |

**FIGURE 2.12** Graphical illustration of why the Cooley–Tukey FFT algorithm is significantly faster than the original DFT.

needlessly calculate values for sine and cosine terms a second time. The saving in time is substantial as $n$ gets large, since there are then a great number of circulations of the phasor circle. For a 1K record, the FFT computes the transform 102.4 times faster than does the DFT. Additional details are provided in Peters (2003a, 2003b, 2003c).

### 2.11.3   Wavelet Transform

Recent work suggests that the wavelet transform (WT) may in the future replace the FT in some applications. It uses the Haar function, which is orthogonal on [0,1], as opposed to the orthogonality of the harmonic functions (sine and cosine) corresponding to [0,$2\pi$] (Strang, 1993). It is claimed that the WT is better able to address features of the Heisenberg uncertainty principle than the FFT.

### 2.11.4   Heisenberg's Famous Principle

The heart and soul of quantum mechanics is the Heisenberg uncertainty principle. As noted elsewhere in this chapter, it has things to say about damping models. According to well-known physicist Hans Bethe (1992), the principle has received "bad press":

> Many people believe that the uncertainty principle has made everything uncertain. It is quite the opposite. Without the uncertainty principle there could not exist any atoms, there could not be any certainty in the behavior of matter. So it is in fact a certainty principle.

Curiously, a failure figured in Heisenberg's discovery of the principle. During his thesis defense, in front of great theoretical physicist Arnold Sommerfeld (his director) and the famous experimentalist Wilhelm Wien, he proved unable to derive the magnifying power of a simple microscope. The scandal culminated with Professor Wien asking him to explain how a battery works, and he could not answer that question either. Knowing his extraordinary theoretical giftings, Sommerfeld gave him the highest possible grade to compensate for Wien's choice of an F. Thus, Heisenberg was awarded his doctorate.

Later, in an ironic turn of events, Heisenberg chose a microscope to illustrate features of the matrix quantum mechanics that he originated, and which corrected problems with the Bohr wave mechanics theory. His greatest source of embarrassment served to make Heisenberg famous!

## 2.12   Hysteretic Damping

### 2.12.1   Physical Basis

The model of simple harmonic oscillation with viscous damping assumes dissipation from an externally acting force. It is not suited to a conceptual understanding of hysteretic damping. To accommodate internal friction requires more than a single mass connected to the elastic component responsible for restoration. Two systems are pedagogically useful in this regard, one being a long-period physical pendulum (mechanical), and the other being the oscillator used by Ruchhardt to measure the ratio of heat capacities of a gas (mainly thermodynamic). Because of widespread confusion concerning the difference between viscous and hysteretic damping, both cases are presented here. The treatments are provided as evidence for the premise that hysteretic damping is the more important case for applied physics and engineering.

It is common knowledge that the damping of a mechanical oscillator results from the conversion of mechanical energy into thermal energy. One might expect, then, that a direct consideration of thermodynamics could yield conceptual understanding of the underlying physics. Although an ideal gas is rarely considered in this context, there is a classic experiment which speaks to its relevance. It is the ingenious technique used first in 1929 by Ruchhardt to measure $\gamma$, the ratio of heat capacity at constant pressure to that at constant volume (Zemansky, 1957).

## 2.12.2 Ruchhardt's Experiment

Consider a piston of mass $m$ moving in a cylinder of cross-sectional area $A$, alternately compressing and expanding a volume of ideal gas $V_0$ about the residual pressure $P_0$. Assume that there is no sliding friction between the piston and the cylinder. A small displacement $x$ of the mass results in volume change $\Delta V = V - V_0 = Ax$. There is a restoring force $F = A\Delta P$, where the pressure difference $\Delta P$ relates to $\Delta V$ through an assumed adiabatic process; i.e., the period of the motion is assumed too short for appreciable heat transfer into and out of the gas. Using $PV^\gamma = \text{constant}$, one obtains

$$\gamma P_0 V_0^{\gamma-1}\Delta V + V_0^\gamma \Delta P = 0 \tag{2.23}$$

from which one obtains

$$m\ddot{x} + \frac{\gamma P_0 A^2}{V_0}x = 0 \tag{2.24}$$

This is the equation of motion of a simple harmonic oscillator. There is no damping because of the assumed adiabatic process. By measuring the period $T = 2\pi/\omega = 2\pi\sqrt{V_0 m/\gamma P_0 A^2}$, one can estimate $\gamma$.

Historically, it appears that such measurements slightly underestimate $\gamma$, which can be understood as follows.

The ideal gas equation of state $PV = NkT$ yields, through differentiation

$$P_0 xA + V_0\frac{F}{A} = Nk\Delta T$$

$$m\ddot{x} + \frac{P_0 A^2}{V_0}x = \frac{NkA}{V_0}\Delta T(t) = F_d(t) \tag{2.25}$$

Notice the difference between Equation 2.24 and Equation 2.25. In Equation 2.25, damping is possible (a type of "negative drive" term) from temperature variations associated with heat transfer during traversal of the cycle. If it were possible for the oscillation to be isothermal ($\Delta T = 0$ at very low frequency, essentially quasistatic), then the frequency would be lower than that of the adiabatic case, since $\gamma > 1$ is missing from Equation 2.25. In the isothermal case, there would also be no damping, since the heat into the gas during compression would be balanced by that which leaves during expansion. The only way to get damping is for the paths of compression and expansion in a plot of pressure vs. volume to separate, i.e., for there to be hysteresis. Reality must correspond to something between the two extremes of adiabatic and isothermal, with experiment obviously favoring adiabatic. The process must depart somewhat from adiabatic, however, since there is damping, which Equation 2.25 shows to derive from temperature variations yielding hysteresis. It is interesting to look at the temperature variations relative to a "driving force," $F_d'(t)$. In the Ruchhardt experiment, there must be small variations $\Delta T'(t)$ that lag behind $x(t)$. (These are not the reversible temperature variations of the adiabat, onto which the $\Delta T'(t)$ are superposed.) By comparing with Equation 2.25, the right-hand zero of Equation 2.24 may be replaced with a damping force that can be written in terms of the velocity as

$$F'd(t) \propto \Delta T'(t) \rightarrow -\frac{c}{\omega}\dot{x} \tag{2.26}$$

where $c = \text{constant}$. Notice that the multiplier on the velocity is not simply a constant, but rather a constant divided by the angular frequency. The use of velocity is mathematically convenient, but the magnitude of the velocity (speed) is not expected to be a first order influence on the temperature changes of hysteresis type. The derivative of $x$ with respect to time not only shifts the phase by $90°$, which accommodates the lag with which heat is transferred, but it also introduces a frequency multiplier through the chain rule. Thus, to make damping proportional to the velocity would cause increased dissipative heat flow and thus increased damping as the frequency is increased. Since this does not happen, and lest we introduce a nonphysical term into the equation, it is necessary to divide by the frequency. Replacing the right-hand-side zero of Equation 2.24 with Equation 2.26, we obtain the modified equation of motion, with damping

$$m\ddot{x} + \frac{c}{\omega}\dot{x} + \frac{\gamma P_0 A^2}{V_0}x = 0 \tag{2.27}$$

Additional justification for the form of the damping term in Equation 2.27 can be realized by looking at cases where there is negative damping, i.e., $c < 0$. Such is true when the gas is caused to cycle as an engine. An illustrative case study was that of a low temperature Stirling engine (Peters, 2002a, 2002b, 2002c), in which reasonable agreement between theory and experiment was realized through the use of an equation based on the same arguments used to derive Equation 2.27.

It is seen that a straightforward modeling of Ruchhardt's experiment to include damping yields an equation of motion that is in the form of hysteretic damping. It appears that, for many systems in which the dissipation is dominated by internal friction, hysteretic damping is a near universal form.

### 2.12.3 Physical Pendulum

In the paper by Speake et al. (1999), one finds the following statement:

> the logarithmic decrement $(Q^{-1})$ varied as the inverse of the square of the frequency. We interpreted this as evidence that, in Cu–Be over this frequency range, the imaginary component of Young's modulus was independent of frequency, contrary to that which was predicted by the Maxwell model.

To fit their theory with experiment, they used a "modified" Maxwell model with a distribution of time constants that ranged from 30 sec to more than 4000 sec. Motivation for their continued modeling efforts derived partly from the observation by Kuroda (1995) that anelasticity was cause for some of the huge errors that have been present in estimates of the Newtonian gravitational constant, $G$, by the time-of-swing method.

Although it gives agreement with their particular experiment, the model of Speake et al. (1999) does not have the blessing of Occam's razor. Moreover, their claim that damping derived primarily from

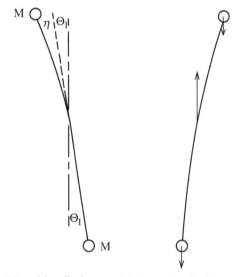

(a) positive displacement (b) negative displacement

**FIGURE 2.13** Idealized physical pendulum used to develop the modified Coulomb damping model.

their flex pivot of Cu–Be may not be true. Other studies suggest that the material defining the axis of a long-period pendulum is for many cases no more important (and sometimes much less important) to the damping than the material from which the pendulum proper is constructed. A model which also agrees with experiment of the type they conducted, but which is simpler, is now presented.

Illustrated in Figure 2.13 is an idealized long-period mechanical oscillator which could be labeled a "physical pendulum." The top and bottom masses are the same, $M$, assumed to be much greater than the mass of the connecting structure, which is represented by the curved line.

A primary mechanism for internal friction damping can be understood by looking at the external forces acting on the pendulum, which are pictured in the "negative displacement" (b) case. The upward "normal" force that acts through the pivot (usually a knife-edge) is opposed by the pair of bob-weights situated left and right, respectively, of the axis of rotation. As the pendulum swings alternately between positive and negative displacements, the structure undergoes periodic flexure. It should be pointed out that internal friction could still be operative throughout the structure even without net bending; i.e., there could be complementary pieces of the structure undergoing compression and tension. Even if the oscillator were in the weightlessness of space, a drive torque would result in dynamic reactionary forces that give rise to damping by this means.

Assume that the masses are separated a distance $2L$ and the axis of rotation is $\Delta L$ above the geometric center. Applying Newton's Second Law, with the lower mass causing a restoring torque and the upper mass a "destoring" torque, yields

$$\ddot{\theta}_1 + \frac{g}{2L}\left(1 + \frac{\Delta L}{L}\right)\theta_1 - \frac{g}{2L}\left(1 - \frac{\Delta L}{L}\right)\theta_2 = 0, \quad \theta_2 = \theta_1 + \eta \tag{2.28}$$

(*Note:* Equation 2.28 can be rewritten to accommodate larger displacements, where elastic nonlinearity gives rise to unusual behavior. The amplitude trend of the period is opposite to that of the gravitational nonlinearity, thus providing for improved isochronism. For details refer to Peters, 2003a, 2003b, 2003c).

The difference in displacement of the masses involves an elastic term proportional to $\theta_1$ and a dissipative term that depends on its time rate of change, i.e.

$$\eta = c\left(\theta_1 \cos \delta - \frac{\dot{\theta}_1}{\omega}\sin \delta\right), \quad \omega = \sqrt{g\frac{\Delta L}{L^2}} \tag{2.29}$$

where $c$ is a dimensionless constant. This result can be obtained by the complex exponential Steinmetz (phasor) method. The equation is consistent with the common assumption that stress and strain are related through a complex constant. The angle $\delta$ is the phase angle with which $\eta$ strain) lags behind $\theta_1$ (stress). To describe the motion of the lower mass, we can ignore the elastic part of $\eta$, since it does not contribute to the damping (or if the rod does not bend, assuming there still is damping as noted previously). We thus remove the subscript, and after some algebra obtain the result

$$\ddot{\theta} + \frac{\alpha}{\omega}\dot{\theta} + \omega^2 \theta = 0, \quad \alpha = \frac{gc}{2L}\sin \delta, \text{ for } \Delta L \ll L \tag{2.30}$$

which can also, in terms of $Q = 2\pi E/(-\Delta E)$, be expressed as

$$\ddot{\theta} + \frac{\omega}{Q}\dot{\theta} + \omega^2 \theta = 0, \quad Q = \frac{2L}{gc\delta}\omega^2, \quad \delta \ll 1 \tag{2.31}$$

If, as a material property, $\delta$ is independent of frequency, then $Q$ is quadratic in the frequency; i.e., the damping of the pendulum due to internal friction is inversely proportional to the square of the frequency — even though the internal friction (determined by $\delta$) is itself frequency-independent. It is important to note that the frequency dependence of internal friction is not to be equated with the frequency dependence of the $Q$ of the oscillator, even though internal friction is frequently stated as simply $1/Q$. This will be discussed in greater detail in Section 2.13.2.

### 2.12.3.1 Test of Q Dependencies

The dependence of $Q$ on frequency and length in Equation 2.31 was tested experimentally with a physical pendulum. Two Pb spheres, each of mass approximately 1 kg, were each drilled through a diameter to allow the insertion of the shaft of an aluminum alloy arrow (length approx. 70 cm) of the type commonly used with compound hunting bows. A second hole was drilled perpendicular to the first and tapped for a set screw. The shaft was sawed into two pieces, which were rigidly rejoined around a carbon–steel knife-edge using force fit and epoxy to machined protuberances above and below the knife-edges. The knife-edges extend perpendicularly outward on opposite sides of the arrow at its center.

### 2.12.3.2 Simple Method to Measure Damping

Although an SDC sensor could have been employed instead, the experiments to be described were performed with a measurement technique that warrants description because of its novel simplicity — yet it is reasonably accurate. To measure both period and damping, a small "flag" was super-glued to the top of the upper shaft. This flag was a small, thin, U-shaped piece of plastic in which the upper legs of the U were about 1 mm wide, with a spacing between centers of about 0.5 cm. An infra-red photogate of the type used in general education laboratories was mounted so the flag would trip the photogate during pendulum oscillation. Two different timing measurements were then performed, using a Pasco Smart Timer. In every run, the pendulum was displaced initially about 10° by hand and then released. There was no need for precision initialization.

In the pendulum mode of the timer, the period was directly measured. For this case, the photogate was positioned, relative to the U-shaped flag (for which one vertical arm is slightly longer than the other), so as to be interrupted only once by the pendulum per pass. In the time-interval mode, the flag was positioned so that both arms interrupted the photogate beam. The reciprocal of this time of interruption proved to be a reasonable measure of the instantaneous speed of the pendulum at the position of the photogate, which was that of maximum kinetic energy. The time intervals were recorded manually for traversals separated by one period, through five cycles of oscillation. These numbers were then typed into Excel and their reciprocals graphed. A trendline (using the option to print the slope) was applied to the near linearly declining graph. The decrement of this line (fractional decrease per cycle) proved to be a good approximation to the logarithmic decrement of the motion, which could have been estimated with exceptional precision by means of the other techniques mentioned in this chapter.

In the first set of experiments, the sphere on the lower shaft was maintained at a constant distance from the knife-edge, while the mass on the upper shaft was positioned at increasingly greater distances from the knife-edge to lengthen the period. Over the full range of periods considered, the distance between the two masses changed by a small amount around its nominal value of 67 cm. The results of this first study are shown in the left graph of Figure 2.14, where the log-decrement has been plotted vs. the square of the period. The $Q$ of the pendulum ($\pi/\Delta$) may be calculated for any value of the period using the indicated slope of 0.0004. For example, the $Q$ at a period of 10 sec was 76, this being near the shortest period considered. Near the other extreme of $T = 35$ sec, $Q = 6$. At the shortest possible periods, damping due to air drag would begin to become important.

The reasonable fit of the linear regression vs. period squared is consistent with the prediction by Equation 2.31 that $Q$ should be quadratic in the frequency.

The Equation 2.31 also indicates that the log-decrement should be proportional to the reciprocal of the distance, $L$, between the masses. To test this prediction, the period of the pendulum was measured as a function of mass separation, also using the smart timer. In generating the data for the right graph of Figure 2.14, the period was maintained constant at 20 sec. For every datum, the top sphere was always only slightly closer to the knife-edge than the lower sphere. At 0.049, the intercept of the trendline differs enough from zero, relative to the size of the error bars, to imply a systematic error. Possible sources of the error include: (i) the masses are of finite size, rather than being points as assumed by the model, and (ii) a nonnegligible mass from parts other than the spheres. Nevertheless, the data show a clear size dependence of the $Q$.

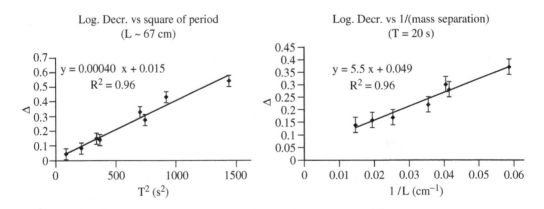

**FIGURE 2.14**   Results of experiments to test the dependencies of $Q$ on (i) frequency and (ii) length of pendulum.

The experiments just described do not permit one readily to isolate the source of the damping, which, for the cases in Figure 2.14, had the knife-edge resting on silicon wafers (integrated circuit stock material). It is not known to what extent the dissipation was dominated by strain in the knife-edge–silicon interface or by flexure of the aluminum arrow. Although the model that generated equation 2.31 assumed only the latter, there is nevertheless theoretical and experimental basis for model acceptance, regardless of the details of the damping.

### 2.12.3.3   Highly Dissipative, yet Hard Materials

The same pendulum was used to demonstrate some counterintuitive features of internal friction damping by replacing the Si wafers with various materials. When very soft material, such as lead, was the support for the knife-edges, there was a significant increase of the damping, as expected. It was also found, however, that cast iron increased the log-decrement (10-sec period) by more than 40%. The same was also true of ceramic PZT wafers of the type used to ignite gas grills by striking the wafer impulsively. Both the cast iron and the lead–zirconate–titanate samples are very hard, so the internal friction must derive from large defect densities in which atomic disorder is a sensitive function of stress. Some other hard interfaces, such as steel on glass, or steel on sapphire, did not show a difference from steel on Si, which suggests that the dominant source of damping for the pendulum in all these cases was flexure of the aluminum shaft.

The observation involving cast iron is consistent with its known excellent damping properties at higher frequencies — important, for example, to engine blocks. Some magnesium alloys have also been developed that have excellent damping characteristics without seriously sacrificing strength.

## 2.13   Internal Friction

### 2.13.1   Measurement and Specification of Internal Friction

Mechanical spectroscopy is a popular means for measuring internal friction of materials (Fantozzi, 1982). Typically, a torsion pendulum is used to stress harmonically a sample and the lag of the response (strain), relative to the stress, provides the loss tangent and thus the internal friction. In such experiments, it is widespread practice to report internal friction as $Q^{-1}$. There can be confusion because of this practice, depending on the nature of the measurement technique, i.e., whether one actually measures $Q$ as opposed to measuring something proportional to the stress–strain lag angle. If $Q$ is obtained from an oscillatory free decay, using the logarithmic decrement defined as follows, then there is no problem.

$$\Delta = \ln\frac{x_n}{x_{n+1}} = \beta T = \frac{\pi}{Q} \tag{2.32}$$

Here, $x_n$ and $x_{n+1}$ are adjacent turning point amplitudes separated by one period of the motion, $T$. In practice, it is very difficult to adjust a mechanical system to oscillate over a wide frequency range. The widest range known to the author, for a mass–spring system, involved the work of Gunar Streckeisen (1974), in which a vertical seismometer using the LaCoste spring was adjusted to have periods in the range between 7 and 140 sec. Because of the difficulties in attaining a wide range of eigenmodes, internal friction is typically determined with a specimen that does not oscillate. We now consider that case.

### 2.13.2   Nonoscillatory Sample

In the typical torsional pendulum used to measure internal friction, the sample is of very small mass. Such a pendulum was built, for example, around the original version of the fully differential capacitive sensor, to study magnetoelastic wires (Atalay and Squire, 1992). As with many delicate instruments, the Atalay and Squire instrument was of the type labeled "inverted." A silk fiber at the top of the specimen was used to provide minimal tension in the sample. They used one linear rotary differential capacitance transducer (LRDCT) (Peters, 1989) in the drive mode to provide a known stress to the delicate magnetoelastic sample and a second LRDCT to measure the strain magnitude and the angle by which it lags behind the stress because of an elasticity. As such, they were measuring the lag angle and not $Q$, as will now be shown.

Without an inertial term, the sample response $x$ to a periodic external force $F$ is governed by

$$F = Kx = (k + j\zeta)x = F_0 e^{j\omega t} \tag{2.33}$$

so that the transfer function is given by

$$\frac{x}{F} = k^{-1} - j\frac{\zeta}{k^2} \tag{2.34}$$

from which it is seen that the measurement does not yield $Q^{-1}$ but rather the lag angle $\zeta/k$, where $k$ is constant. Perhaps the measured angle, which is an indicator of the internal friction, has been called $Q^{-1}$ because $k = m\omega_0^2$ for an oscillator of frequency $\omega_0$, and $Q = m\omega_0^2/\zeta$ for the freely decaying oscillator. Bear in mind, however, that this expression for $k$ does not apply to the nonoscillatory measurement just described. There is a frequency square difference between such a measurement and what would be measured if an adjustable oscillator were being considered.

An example of the importance of this issue is found in the article by Lakes and Quakenbush (1996), in which one reads from the abstract the following statement:

> The damping, tan δ, followed a $\nu^{-n}$ dependence, with $n \approx 0.2$, over many decades of frequency $\nu$. This dependence corresponds to a stretched exponential relaxation function, and is attributed to a dislocation-point defect mechanism. It is not consistent with a self-organized criticality dislocation model which predicts tan $\delta \propto A^{-2}$. Dislocation damping in metals is relevant to development of high damping metals, the behavior of solders and of support wires in Cavendish balances.

The present arguments suggest that the experiment by Lakes and Quackenbush is (1996) not in strong disagreement with the SOC model; that the magnitude of the exponent difference between theory and experiment is really 0.2 and not 1.8 as they have indicated.

### 2.13.3   Isochronism of Internal Friction Damping

It is well known that, in the viscous damping free-decay case, the frequency of oscillation is lowered by damping according to

$$\omega_1 = \sqrt{\omega_0^2 - \beta^2} = \omega_0\sqrt{1 - (2Q_v)^{-2}} \tag{2.35}$$

and the resonance frequency of the driven oscillator is lowered even further (Marion and Thornton, 1998). It is not well known how difficult it is to measure this damping "red-shift," which brings in features of the Heisenberg uncertainty principle. Additionally, it is not well known that extensive damping experiments suggest that the frequency may not, for some systems, depend on the damping at all; i.e., the oscillator is isochronous. Isochronism cannot be realized with a linear homogeneous differential equation, but it can be realized with a nonlinear form that is obtained by modifying the damping term as follows:

$$\frac{\omega}{Q}\dot{x} \longrightarrow \frac{\pi}{4}\frac{\omega}{Q}\sqrt{\omega^2[x(t)]^2 + [\dot{x}(t)]^2}\,\mathrm{sgn}(\dot{x}) \tag{2.36}$$

where sgn($dx/dt$) is the algebraic sign of the velocity — it causes the equation of motion to be nonlinear even if the square root term were not present. For small damping, the square root term can be shown to be equal to the time-dependent amplitude of the motion multiplied by the angular frequency.

Other damping types are possible and are indicated in Peters (2002a, 2002b, 2002c) (…universal…) where evidence is also provided for harmonic distortion in the waveform because of the nonlinearity. It is shown in Peters and Pritchett (1997) that the oscillation is isochronous.

For large values of $Q$, the lag angle (radian measure) is given by $\delta = 1/Q$. Researchers usually measure $\delta$ and specify the magnitude of the internal friction as $Q^{-1}$. As noted previously, $Q$ is proportional to frequency for the viscous damped oscillator. Thus, for viscous damping, the internal friction is inversely proportional to the frequency.

For hysteretic damping we obtain the result

$$\tan \delta = \alpha = \frac{h}{k} \tag{2.37}$$

where the variables are defined in Equation 2.19. For small damping in which $\tan \delta = \delta = Q^{-1}$, we find that the internal friction for hysteretic damping is inversely proportional to the square of the frequency, since $h$ is constant and $k = m\omega^2$.

# 2.14 Mathematical Tricks — Linear Damping Approximations

## 2.14.1 Viscous Damping

In the Hooke's Law expression, $F = -kx$, it is common practice to approximate hysteresis of oscillatory motion by letting $k$ become a complex coefficient. This is also standard practice in a variety of fields, such as the description of lossy electromagnetic media. No doubt the practice has been further popularized by the standard approach of solving electrical engineering ac circuit problems by means of phasors, the technique developed by Steinmetz (1893).

We recognize in the expression $x(t) = x_0\, e^{j\omega t} = x_0 \cos \omega t + jx_0 \sin \omega t$ that harmonic variation is contained in the real part (or alternatively the imaginary part) of the complex exponential form. Using Newton's Second Law, and representing the spring constant by $k\, e^{j\delta}$ with $\delta \ll 1$ (small damping), we obtain the damped harmonic oscillator equation

$$m\ddot{x} + kx + (jk\delta)x = 0 \tag{2.38}$$

where the approximations $\cos \delta \to 1$ and $\sin \delta \to \delta$ have been employed.

However, since $\dot{x} = j\omega x$, and $\dfrac{k}{m} = \omega^2$, Equation 2.38 can be rewritten as

$$\ddot{x} + \omega\delta\dot{x} + \omega^2 x = 0 \tag{2.39}$$

We thus see that the damping constant $\omega\delta = \omega/Q = 2\beta$ permits us to express the logarithmic decrement $\Delta$ in terms of the angle $\delta$ with which $x$ lags $F$; i.e., $\Delta = \beta T = \pi\delta$. (Note that we are making no distinction here between the periods with and without damping, since the difference is small and hard to measure.) If $\beta$ were independent of frequency, then $\delta$ would be inversely proportional to the frequency, which is rarely realized in practice.

## 2.14.2 Hysteretic Damping

Equation 2.39 does not properly represent some of the most important engineering systems. For those labeled "hysteretic," we must use a different form for the complex spring constant. We assume that $F = -(k_{\text{complex}})x = (k + jh)x$ where $h$ is a real constant. Since $dx/dt = j\omega x$, this yields the equation of motion

$$\ddot{x} + \frac{h}{m\omega}\dot{x} + \omega^2 x = 0 \tag{2.40}$$

Since $h$ is assumed to be a true constant (independent of frequency), the lag angle between displacement and force is given by

$$\delta = \frac{h}{k} = \frac{1}{Q} = \frac{h}{m\omega^2} \tag{2.41}$$

which is seen to be inversely proportional to the square of the frequency. (Note that $\delta$ here is the same as $\alpha$ in Figure 2.6.) It should be noted that the complex form for the spring constant is not simply obtained using the common theory of viscoelasticity. Such theory requires a multitude of relaxation times (stretched exponentials) (Speake et al., 1999).

# 2.15    Internal Friction Physics

## 2.15.1    Basic Concepts

All damping derives from varying degrees of complexity because of the myriad interactions that are present, either internal of nonconservative type or external involving the environment. This is true even for systems that come closest to being governed by the textbook equations. For example, the author has attempted to produce ideal harmonic oscillators using viscous liquids for damping. Even they are complicated and do not strictly obey Stokes' Law of drag force proportional to the velocity. The nonlinear Navier–Stokes equation may be capable of describing them, but not in a simple form except to a first approximation that is not really very good relative to the precision that is possible with modern sensors.

Perhaps the closest to being an ideal viscous damped oscillator is that in which the damping force derives from eddy currents through Faraday's Law. A magnet is attached to the oscillator and, as it moves in proximity to a conductor, the time rate of change of magnetic flux gives rise to a retarding force that is proportional to velocity. Because there really is a force involved, and because of Lenz's Law, the damping term makes sense physically. This case might be completely ideal except for one factor — the magnet is part of a mechanical system that must possess structural integrity if it is to oscillate. Because of loads present in the structure (reactionary normal forces to the various weights), there will always be some creep. The creep is ultimately unavoidable, since there is apparently no stress threshold below which plastic deformation ceases to exist. It is important to realize that forces associated with inertial mass (Newton's Second Law) are just as important as the weights. Systems designed around an elastic member (such as a spring, in contrast to a simple pendulum) will experience damping in the weightlessness and the airlessness of space.

## 2.15.2    Dislocations and Defects

The extent to which mechanical defects, such as dislocations, have been ignored by large segments of the scientific community is surprising. The surprise is even greater when one considers the importance of defects in another field — that of electronics. Our present information age (world of computing) came into existence only after widespread recognition of the importance of the defects called impurities. The n-type and p-type semiconductor materials necessary to our modern age result from the substitution of silicon atoms with others of pentavalent and trivalent type in surprisingly small concentrations.

The strength of solids is very much less than as predicted by theories of an ideal (perfect) crystal. Dislocations are the primary culprits. Their influence on materials used in engineering has prompted the statement: "when mother nature fills the vacuum she abhors, she rarely does so with perfection." Unfortunately, few students exposed to fundamental science receive training in defect physics. Moreover, it is difficult to provide a self-consistent fundamental description of their properties, so very few scientists have more than a superficial knowledge of their importance.

"Viscoelasticity" is a misleading term. To combine the words viscous and elastic suggests that the state variables vary smoothly in time, i.e., as a fluid in the viscous part. Unfortunately, this is not true of hysteresis associated with either "domains" or with "grains." In the case of magnetic domains, it is quite easy to demonstrate nonsmooth (jerky) behavior that is called Barkhausen noise. Although the phenomenon was demonstrated by Barkhausen in 1919, only recent studies have begun to understand some of its complexities better (Urbach et al., 1995a,b).

A similar phenomenon, that must relate in some manner to the Barkhausen effect, is the PLC effect. Under applied stress, alloys frequently display discontinuous strain increase (jumps). The author has even demonstrated strain recovery of a similar type, catalyzed by "tapping." The polycrystalline metals that demonstrate these effects are obviously influenced by "granularity." They differ from the "granular materials" that have become a hot topic of recent research. Even pure polycrystalline metals exhibit these features. The German word to describe the deformation of tin under large stresses is *zinngeschrei* (=tin cries). Anyone who has ever bent large-diameter tungsten wire has experienced this phenomenon, since the nonsmooth strain can be both felt and heard.

There is still another type of material, thought to have great engineering potential in the future, that shows "granular" behavior — that of shape memory alloys (SMA). If an SMA specimen is cycled in temperature around the martensitic phase, it generates acoustic emissions (Amengual et al., 1987). For a figure taken from their work and other good pages about hysteresis, refer to the webpages of Prof. Sethna at http://www.lassp.cornell.edu/sethna/hysteresis/ReturnPointMemory.html. These emissions are probably related to the PLC effect and are characterized by surprising reproducibilities in spite of their complex behavior.

Thus, there is abundant experimental evidence against the overly simplistic view that hysteretic damping can be meaningfully described by simple, linear differential equations. The nonlinear terms necessary for a good mathematical treatment go beyond "chaos" to the world of "complexity." Chaos of deterministic type, though bewildering to many, is in many cases tractable (using equations that can be integrated numerically). Damping problems are much more complex than deterministic chaos. The challenges to our understanding derive in part from the long time that it has taken before there were any serious investigations of the mesoscale, the place where defect structures abide. If, as with Zener, we use the word anelasticity to describe systems that are "other than" elastic, then the term *mesoanelastic complexity* is an appropriate label for this poorly understood physics that is important and yet mostly unknown to many fields of both science and engineering.

## 2.16  Zener Model

### 2.16.1  Assumptions

The SLM of viscoelasticity provides a sound basis for some damping phenomena, yet it fails badly as an approximation for hysteretic damping. Its prominence in both the worlds of physics and engineering warrants the following detailed discussion so that the failure case may be properly documented.

Following the example of Zener, the following linear differential equation relates the stress, $\sigma$, the strain, $\varepsilon$, and their first time derivatives:

$$\sigma(t) + \tau_\varepsilon \dot{\sigma} = E_1(\varepsilon + \tau_\sigma \dot{\varepsilon}) \tag{2.42}$$

The $\tau$s are relaxation times (subscript $\varepsilon$ meaning at constant strain and subscript $\sigma$ at constant stress), and $E_1$ is the relaxed elastic modulus (ratio of stress to strain in a very slow process). Nominally, $\tau_\sigma > \tau_\varepsilon$, consistent with strain lagging stress. For periodic variations

$$\sigma(t) = \sigma_0\, e^{j\omega t}, \quad \varepsilon(t) = \varepsilon_0\, e^{j\omega t} \tag{2.43}$$

which, when substituted into Equation 2.42, yields

$$(1 + j\omega\tau_\varepsilon)\sigma_0 = E_1(1 + j\omega\tau_\sigma)\varepsilon_0 \tag{2.44}$$

The complex modulus of elasticity is defined by

$$E_C = \frac{1 + j\omega\tau_\sigma}{1 + j\omega\tau_\varepsilon} E_1 \tag{2.45}$$

and is seen to relate stress and strain according to

$$\sigma(t) = E_C \varepsilon(t) \tag{2.46}$$

From Equation 2.45, the real and imaginary parts of the modulus are found to be

$$\text{Real } (E_C) = \frac{1 + \omega^2 \tau_\varepsilon \tau_\sigma}{1 + \omega^2 \tau_\varepsilon^2} E_1 \tag{2.47}$$

$$\text{Imag } (E_C) = \frac{\omega(\tau_\sigma - \tau_\varepsilon)}{1 + \omega^2 \tau_\varepsilon^2} E_1 \tag{2.48}$$

The independent variable, or "frequency," for all cases is the convenient dimensionless parameter, $\omega\tau = \omega\sqrt{\tau_\varepsilon \tau_\sigma}$.

It is convenient to use polar form, so that

$$E_C = |E_C|\, e^{j\delta} \tag{2.49}$$

where $|E_C|$ is obtained by computing the square root of the sum of the squares of the real and imaginary parts. In this form, it is apparent that $\delta$ is a lag angle which determines the damping loss for the system. Moreover, from Equation 2.47 and Equation 2.48, it is seen to obey

$$\tan \delta = \frac{\omega(\tau_\sigma - \tau_\varepsilon)}{1 + \omega^2 \tau_\sigma \tau_\varepsilon} \tag{2.50}$$

### 2.16.2　Frequency Dependence of Modulus and Loss

The essential features of the Zener model are illustrated in Figure 2.15, where the "unrelaxed" high-frequency modulus obeys the relation $(E_1 E_2)/(E_1 + E_2) = E_1(\tau_\sigma/\tau_\varepsilon)$.

In viscous damping models, the damping is quantified by the product $\beta T$, which is equal to the logarithmic decrement. The logarithmic decrement is directly proportional to the period when the damping "constant" $\beta$ is truly constant. The graph in Figure 2.16 compares the logarithmic decrement computed by the standard model against a case where $\beta = $ constant. Also shown in the figure is a set of hysteresis curves for $\omega\tau = 10, 1,$ and $0.1$, respectively. Notice that the damping is large only for $\omega\tau$ near 1, in accord with the bottom plot of Figure 2.15. For that case, points (a) to (f) and back to (a) are shown, labels to illustrate work done by the stress in traversing the hysteresis loop. The algebraic sign of the work changes around the loop and the net work done in one cycle is just the area enclosed by the loop.

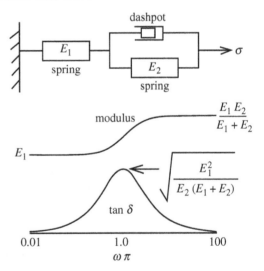

**FIGURE 2.15**　Zener Model of anelasticity. Bottom curves are "frequency" variation of modulus and loss respectively.

For damping based on the Zener (standard linear) model to agree with the simple viscous approximation, it is necessary that $\omega\tau \gg 1$; i.e., the period of the oscillator must be significantly shorter than the smaller of the relaxation times, as illustrated in the bottom graph of Figure 2.16.

### 2.16.3　Successes — Models of Viscoelasticity

Viscoelasticity, as an approximation for damping, is evidently quite adequate for some materials. The assumption of fluid character as a basis for hysteresis is expected to be closest to correct when

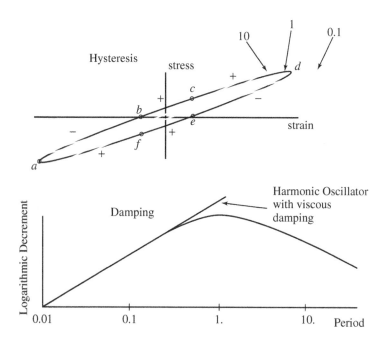

**FIGURE 2.16** Characteristics of the Zener model.

applied to those cases in which variations in strain are almost continuous. The materials of rheological type for which this appears to be most true are solids built from long chain polymers, i.e., various plastics. Such materials can yield surprising results, however. Shown in Figure 2.17 are results from a study that used a nylon monofilament sample (8-lb fishing line). The pair of torsional free-decay records corresponds to two different temperatures — 290 K (room temperature) and 390 K (above the glass transition temperature of the nylon). Although a significant increase in the period was observed as the temperature was increased above the glass transition temperature (changing from 18.2 to 27.8 sec), the logarithmic decrement was found to be almost unchanged. This was not in keeping with the expectation that softening of the material at the higher temperature would result in significantly greater damping. The effect is just the opposite of what was mentioned concerning cast iron, which, though very hard, does not have small damping. Here, a softening does not result in significantly increased damping.

Although there was some creep observed for both the decays of Figure 2.17, the creep was more pronounced in the higher temperature case. This is illustrated by the lower curve of the bottom graph, which is a computer fit in which the secular term necessary for best fit was removed to illustrate the creep. In both decay cases, the log decrement was calculated by importing the A/D data (Dataq DI-154RS) to Excel and then using trial and error adjustment of parameters to achieve the best fit.

Although the damping of glasses is normally treated using the theory of viscoelasticity, Granato (2002) has recently modeled these materials via defects. In his paper, Granato states the following: "As dislocations carry the deformation in crystals, interstitials are the basic microscopic elements carrying the deformation in glasses near and above the glass temperature."

## 2.16.4 Failure of Viscoelasticity

Unfortunately for the elegant theory of the Zener model that has been presented, there are many mechanical systems for which the $Q$ is not proportional to frequency, but rather proportional to the square of the frequency. The logarithmic decrement ($\Delta = \pi/Q$) has been measured for a host of

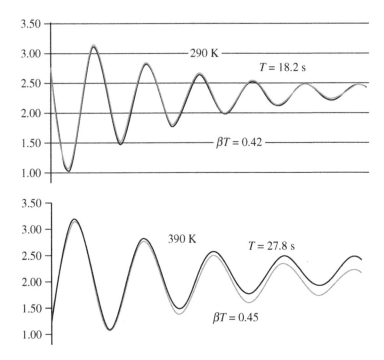

**FIGURE 2.17**   Torsional free-decay records of monofilament nylon at temperatures first below and then above the glass transition temperature. Although the modulus decreased dramatically at the higher temperature, the damping did not.

long-period mechanical oscillators, configured as some form of a pendulum. In all cases, these systems were described approximately by $\Delta = \beta T \alpha T^2$ rather than by $\beta T \alpha T$. Similar behavior has been noted in mechanical oscillators other than the pendulum — for example, in the geophysics research of Gunar Streckeisen and Erhard Wielandt, who are well known for the development of the widely employed STS-1 seismometer. During his pursuit of the Ph.D., Streckeisen (1974) measured the numerical damping (fraction of critical damping) of a vertical Sprengnether long-period seismometer 5100-V. After removing the magnet of the velocity transducer (to eliminate eddy currents and reduce viscous air damping, he found that the numerical damping was proportional to the square of the period between periods of 7 and 140 sec. He took about 30 measurements over this interval of periods, and showed that the damping increased from about 0.0008 to about 0.3 — a factor of roughly 400, not 20 as one would expect for viscous damping. To quote Wielandt (private communication), "the data are very clear."

# 2.17   Toward a Universal Model of Damping

## 2.17.1   Damping Capacity Quadratic in Frequency

The quadratic dependence on frequency of $Q$ (log decrement proportional to period squared) is equivalent to friction force being frequency-independent. In support for the claim of universality, it was noted in the Introduction (Section 2.2.2) that three very different systems showed this characteristic: (i) the vertical seismometer just discussed, (ii) various pendula, and (iii) the rotating rod direct measurement of internal friction first done by Kimball and Lovell (1927), who measured the transverse deflection of the end of a rod when it was rotated about a horizontal axis.

## 2.17.2 Pendula and Universal Damping

An example of one of the author's experiments that illustrate universal (hysteretic) damping is provided in Figure 2.14. Other works that illustrate hysteretic damping include those by Peter Saulson of Syracuse University, who has been frequently cited in the literature (see Saulson et al., 1994).

The pioneering work of Braginsky (important to LIGO) has already been mentioned in the context of small force measurements and noise. He and his Moscow group members argue that the internal friction in fused silica may be roughly independent of frequency from 0.1 Hz to 10 kHz (Braginsky et al., 1993).

An oft-cited paper speaking to the issues of hysteretic damping is an article by Quinn et al. (1992) concerned with material problems in the construction of long-period pendula. (The type of pendulum on which they based their studies was first described in the scientific literature 2 years earlier (Peters, 1990).) In a follow-on paper, Speake et al. (1999) state the following: "The analogues of anelasticity and its resultant $1/f$ noise are seen in a wide range of other processes (for example, dielectric and magnetic ones) described in terms of frequency-dependent susceptibilities."

The jerkiness (discontinuous change) that is the hallmark of the Barkhausen effect may have been first seen mechanically in the experiments that generated the metastable states paper. From a consideration of the chapter by James Brophy (Brophy, 1965), it was postulated in this 1990 paper that the jerky behavior of a mesodynamic pendulum is a type of mechanical Barkhausen effect.

## 2.17.3 Modified Coulomb Model — Background

The results that follow grew naturally out of the application of fully differential capacitive sensors to the study of mechanical oscillators. Efforts to model internal friction influence on long-period pendula uncovered something surprising to most — that the foundation for physics laid by Charles Augustin Coulomb may be much broader than had been realized. Most individuals in the physics community do not associate Coulomb's name with contributions other than to the laws of electrostatics. Engineers, however, have long used his name in the context of sliding friction, since, in fact, Coulomb gave us the empirical description

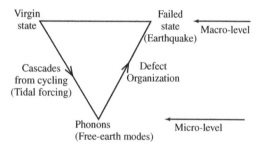

**FIGURE 2.18** Heuristic description of how materials fail — processes connected with damping.

which involves static and kinetic coefficients. Because of his interest in the civil engineering of soils (Heyman, 1997), Coulomb also provided something else — a basis for understanding granular flows and even some types of fracture. Concerning the latter, the Mohr criterion, applied to the Coulomb failure envelope, defines a "coefficient of internal friction," which is used to predict brittle failure (Gere and Timoshenko, 1996).

Coulomb friction is empirically simple, at least as a first approximation, since it depends only on the normal force between surfaces and the algebraic sign of the velocity when there is relative motion. Like so many problems of multibody type, a complete theory of sliding friction is far from being realized. Simplistic textbook efforts to explain energy loss, by picturing "hills and valleys" of the surface of two solids in contact, are useless. An example of such naivete can be realized by trying to understand the phenomenon of optical contacting. Two orthogonally oriented fused silica cylindrical fibers, allowed to touch, can experience atomic bonding forces that are surprisingly strong, being much greater than the weak attraction of the van der Waals type. Cleanliness of the surfaces is paramount for success in such a demonstration, which speaks to another issue — a connection between internal friction and surface physics.

The conversion of mechanical energy to thermal energy must involve nonlinear (avalanche or cascade) processes. A heuristic description of defect structural interactions that generate heat and eventual failure is the phonon triangle of Tom Erber (Illinois Tech University) shown in Figure 2.18. The author has extended Erber's triangle to include the larger-scale Earth in an attempt to explain earthquakes. Everybody recognizes that the bending of a wire does not take it from the virgin initial state to the failed state along the macroscopic upper leg.

There must first be a downward path to the microlevel, through cascading. These cascades can cause Barkhausen noise in the case of ferromagnetic samples, and acoustic emission in nonmagnetic metallic alloys (PLC effect). Failure requires the upward path of defect organization, the mechanisms of which are not yet understood. One of the first theories with possible implications to the organization leg is that of self-organized criticality. In the magnetic case, Erber has used a fluxgate magnetometer to improve failure predictions, since magnetic hysteresis is proving to be a sensitive indicator of mesoscale structure changes during cycling toward failure. Inferred from these studies is some yet-to-be discovered connectivity between noise, damping, and failure.

Surface friction is expected somehow to be connected with internal friction, the biggest difference being that the surface has many more defect states with which to redistribute energy. The larger density of states of the surface (reduced order) is probably an important factor in the difference between surface friction and the modified internal friction model which follows.

## 2.17.4 Modified Coulomb Damping Model — Equations of Motion

In the following damping model for internal friction, Coulomb's law of sliding friction is modified by assuming that the coefficient of friction is not constant, but rather involves the energy of oscillation $E$ in a power law; i.e.

$$m\ddot{x} + cm\left[\frac{2E}{k}\right]^{\lambda} \text{sgn}(\dot{x}) + kx = 0, \qquad E = \frac{1}{2}m\dot{x}^2 + \frac{1}{2}kx^2 \qquad (2.51)$$

where $c = $ constant that is different for each $\lambda$. For Coulomb (sliding) friction $\lambda = 0$. For amplitude-independent damping of hysteretic type, $\lambda = \frac{1}{2}$. For amplitude-dependent (such as large Reynolds number fluid) damping, $\lambda = 1$. In all cases, if $c \ll 1$ (small damping), the damping capacity is quadratic in the frequency, so that the internal friction $Q^{-1} \sim \omega^{-2}$. Equation 2.51 is easily implemented, in spite of its nonlinearity, which we will see later to be a cause for harmonics in the decay.

It is convenient to rewrite Equation 2.51 in canonical form so as to involve the $Q$ of the oscillator. For the case of hysteretic damping ($\lambda = \frac{1}{2}$), the equation becomes

$$\ddot{x} + \frac{\pi\omega}{4Q_h}\sqrt{\omega^2 x^2 + \dot{x}^2}\ \text{sgn}(\dot{x}) + \omega^2 x = 0 \qquad (2.52)$$

Similarly, for amplitude-dependent damping ($\lambda = 1$)

$$\ddot{x} + \frac{\pi}{4y_0 Q_{f0}}(\omega^2 x^2 + \dot{x}^2)\,\text{sgn}(\dot{x}) + \omega^2 x = 0 \qquad (2.53)$$

where $y_0$ is the initial value of the amplitude of $x$ (largest maximum of $x$), and $Q_f$ is found not to be constant, as in the case of hysteretic damping. Rather, in this case, the $Q$ increases as the amplitude decreases. On the other hand, the $Q$ of an oscillator influenced only by Coulomb ($\lambda = 0$, sliding) friction decreases with the amplitude, and the equation of motion in canonical form is given by

$$\ddot{x} + \frac{\pi\omega^2 y_0}{4Q_{c0}}\ \text{sgn}(\dot{x}) + \omega^2 x = 0 \qquad (2.54)$$

In Equation 2.53 and Equation 2.54, the subscript 0 is used to identify the initial value of the time varying $Q$. Equation 2.54 is equivalent to equation 2.12 with $Q_{c0}/y_0 = \pi/(4\Delta_x)$.

As will be illustrated with some examples, it is possible for an oscillator to be influenced simultaneously by all three types of friction. One may treat such a system with the following equation of motion

$$\ddot{x} + \left[ \frac{\pi\omega^2 y_0}{4Q_{c0}} + \frac{\pi\omega}{4Q_h}\sqrt{\omega^2 x^2 + \dot{x}^2} + \frac{\pi}{4y_0 Q_{f0}}(\omega^2 x^2 + \dot{x}^2) \right] \operatorname{sgn}(\dot{x}) + \omega^2 x = 0 \qquad (2.55)$$

At any instant during the decay, the total (time-dependent $Q$) is given by

$$\frac{1}{Q(t)} = \frac{1}{Q_c} + \frac{1}{Q_h} + \frac{1}{Q_f} \qquad (2.56)$$

in which it is seen that the smallest $Q$ in the set (largest damping term) is dominant in a manner reminiscent of capacitors connected in series.

It is instructive to look at the analytical solution for the time dependence of the amplitude (turning points, $y(t) = |x_{max}|$), when all the $Q$s $\gg 1$. Such a solution is obtained from energy considerations by noting first that the time rate of change of the energy is zero in the absence of friction, i.e.

$$\dot{E} = \frac{d}{dt}\left( \frac{1}{2}m\dot{x}^2 + \frac{1}{2}kx^2 \right) = \dot{x}(m\ddot{x} + kx) = 0, \qquad \text{no friction} \qquad (2.57)$$

With friction, $dE/dt$ is determined by the rate of doing work against the friction force; i.e., $dE/dt$ is proportional to $\omega yf$, where $f$ is the friction force. In the case of Coulomb friction, $f$ is constant (determined by $y_0$,) so $dE/dt$ is proportional to $E^{1/2}$. For hysteretic damping, $f$ is proportional to $y$, so $dE/dt$ is proportional to $E$. For fluid damping, $f$ is proportional to $y^2$, so $dE/dt$ is proportional to $E^{3/2}$. Thus, the general case is described by

$$\dot{E} = -\left( c_1 + c_2\sqrt{E} + c_3 E \right)\sqrt{E} \qquad (2.58)$$

Because the energy is proportional to $y^2$, we can write down the equation for the time varying amplitude as

$$\dot{y} = -c - by - ay^2 \qquad (2.59)$$

where $a$, $b$, and $c$ are constants. The solution to this first-order equation can be found in integral tables, and the result depends on the size of $c$ relative to the product $ab$. For present purposes, we will restrict ourselves to the case where Coulomb damping is not dominant, in which the solution involves an exponential. (For large $c$, one may develop the corresponding general case in terms of the tangent or its inverse.) The present result is as follows, using $r = (b^2 - 4ac)^{1/2}$, where $4ac < b^2$

$$\text{with } \alpha = 2ay_0 + b - r, \qquad \beta = 2ay_0 + b + r, \qquad p = \frac{\alpha}{\beta}e^{-rt}$$

$$y = \frac{b(p - 1) + r(p + 1)}{2a(1 - p)} \qquad (2.60)$$

In the case where $c = 0$, Equation 2.60 can be simplified to the following form, which is useful for curve fitting:

$$\frac{1}{y} = \left( \frac{a}{b} + \frac{1}{y_0} \right)e^{bt} - \frac{a}{b} \qquad (2.61)$$

For the case where $a = 0$, the better form for curve fitting is

$$y = \left( y_0 + \frac{c}{b} \right)e^{-bt} - \frac{c}{b} \qquad \text{(until } y = 0) \qquad (2.62)$$

Curve-fits based on the modified Coulomb damping model are summarized in Box 2.2.

# Box 2.2

# CURVE-FIT TO THE TURNING POINTS

If no damping

$$\dot{E} = \frac{d}{dt}\left(\frac{1}{2}m\dot{x}^2 + \frac{1}{2}kx^2\right) = \dot{x}(m\ddot{x} + kx) = 0, \qquad \text{no friction}$$

with damping ($E$ prop. to $y^2$, $\dot{E}$ prop. to $\omega y \cdot$ friction force)

$$\dot{E} = -\left(c_1 + c_2\sqrt{E} + c_3 E\right)\sqrt{E}$$

equivalent to ($c$ for Coulomb, $b$ for hysteretic, $a$ for fluid)

$$\dot{y} = -c - by - ay^2$$

general solution

$$\text{with } \alpha = 2ay_0 + b - r, \qquad \beta = 2ay_0 + b + r, \qquad p = \frac{\alpha}{\beta}e^{-rt}$$

$$y = \frac{b(p-1) + r(p+1)}{2a(1-p)}$$

special case, $c = 0$

$$\frac{1}{y} = \left(\frac{a}{b} + \frac{1}{y_0}\right)e^{bt} - \frac{a}{b}$$

special case, $a = 0$

$$y = \left(y_0 + \frac{c}{b}\right)e^{-bt} - \frac{c}{b} \qquad (\text{until } y = 0)$$

## 2.17.5  Model Output

Shown in Figure 2.19 is a case in which the decay is influenced by all three types of friction. Notice how the $Q$ rises initially, peaks at a value less than what would be true for hysteretic damping alone (constant $Q$ case), and then later declines. The initial rise is due to the amplitude-dependent damping term (size determined by coefficient $a$), and the later decline is due to the Coulomb damping term (determined by coefficient $b$).

The code in Table 2.1 that was used to generate Figure 2.19 has been reproduced here for two reasons: (i) to show the ease with which the modified Coulomb model may be numerically applied in general to a damping problem, and (ii) to illustrate an integration algorithm that has proven to be intuitive, simple, and powerful — the Cromer–Euler technique, which Alan Cromer first described as the "last point approximation (LPA)" (Cromer, 1981) in contrast to the unstable "first point approximation" given to us by Euler. Over the last 20 years, the author has employed the LPA in a host of applications that span from the generation of satellite ephemerides in the U.S. antisatellite program to both simple and several-body nonlinear problems of deterministic chaos type.

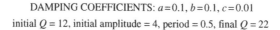

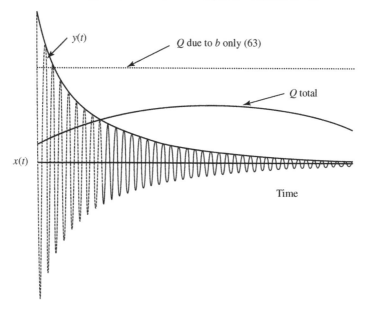

**FIGURE 2.19** Model generated results based on Equation 20.55 and Equation 2.60.

## 2.17.6 Experimental Examples

The code of Table 2.1 is useful in determining the nature of a given experimental case. Too frequently in the past, it has been naively assumed that the entire decay record was exponential. Particularly when longer records are collected, it is found that most damping is nonlinear. In the two experimental examples that follow to backup this claim, one is a near-perfect (nonlinear) particular case of amplitude-dependent (fluid) damping, and the other is a mixture of amplitude-dependent and hysteretic damping, but devoid of any Coulombic influence. Coulomb friction frequently tends to be either "all or nothing," depending on whether there is an unwanted mechanical contact that involves slippage. A notable exception is found in the case where a pendulum is influenced by eddy current damping in a narrow region of its total motion (Singh et al., 2002).

The pendulum that was used to generate the data displayed in Figure 2.14 was also used as follows. A large flat piece of plastic was attached to the bottom of the pendulum, so that its movement (normal vector to the surface in the direction of motion) would disturb a great amount of air in turbulent manner. As expected, there was a dominant initial (large level) amplitude-dependent damping, as shown in Figure 2.20.

The speed (maximum) was measured with a photogate as previously discussed. The expressions shown in the figure are consistent with Equation 2.59 and Equation 2.61. The data, which were collected by hand and typed into Excel, produced the "jagged" curve, and the computerized fit according to Equation 2.61 is the smooth curve of the pair. It is noteworthy that the quadratic drag of the air (determined by $a = 0.036$) is 40% greater than the viscous drag at the start of the decay. By cycle 37, the quadratic part has become much less significant than the constant $Q$ viscous part, having become roughly 60% smaller.

The fluid damping "soup-can" pendulum data of Figure 2.21 was generated with a can of Bush's black-eye peas. The container with enclosed unbroken contents, being a right circular cylinder of length 11 cm × diameter 7.4 cm, was suspended horizontally by a pair of knife edges under opposing end-lips

**TABLE 2.1**    QuickBasic Code to Calculate Amplitude History $y(t)$ and Integrate Equation of Motion to Obtain $x(t)$; Accommodates Three Common Forms of Friction

```
CLS
  REM: setup display
SCREEN 12: VIEW (0, 0) − (600, 470): WINDOW (−.2, − 5) − (1, 5)
  REM: assign constants and initialize variables
pi = 3.1416: dt = 0.002: t = 0
x0 = 4: x = x0: y0 = x0: xd = 0
Period = .5: omega = 2*pi/period: b = .1: a = .1: c = .01
  REM: print damping coefficients
PRINT "DAMPING COEFFICIENTS: a = "; a; ", b = "; b; ", c = "; c
r = SQR(b^2 − 4*a*c): alpha = 2*a*x0 + b − r
Beta = 2*a*x0 + b + r
  REM: Use a, b and c — set Q's to dampen (quadratic, linear, and constant resp.)
qf = omega/2/a/y0: qh = omega/2/b: qc = y0*omega/2/c
  REM: start integration loop
LOOP0:
t = t + dt
  REM: analytically compute amplitude (y = magnitude of x) at each time point
p = alpha*EXP(−r*t)/beta
y = (b*(p − 1) + r*(p + 1))/2/a/(1 − p)
  REM: integr. the eq. of motion to get x(t), using 3 fric. force/mass terms
  REM: The coeff.'s ff, fh & fc correspond to: quadratic in speed (fluid),
  REM: linear in speed (hysteretic), and independ. of speed (Coulomb) resp.
ff = (pi/4)*(1/y0)*(1/qf)*(omega^2*x^2 + xdot^2)
fh = (pi/4)*(omega/qh)*SQR(omega^2*x^2 + xdot^2)
fc = (pi/4)*omega^2*y0/qc
  REM: check algebraic sign − USE SIGN BUT NOT MAGNITUDE OF VELOCITY
IF xdot > 0 THEN GOTO SKIP
ff = − ff: fh = − fh: fc = − fc
SKIP: xdoubledot = − ff − fh − fc − omega^2*x
xdot = xdot + xdoubledot*dt: x = x + xdot*dt
  REM: calculate the energy and then the amplitude to evaluate Q
  REM: could instead use analytical result q = (pi/4)*omega^2*x/abs(ff + fh + fc)
Energy = .5*xdot^2 + .5*omega^2*x^2
Amplitude = SQR(2 * energy)/omega
  REM: calculate loss per period due to friction
loss = ABS(ff + fh + fc)*4*amplitude
q = 2*pi*energy/loss
IF t < 1.2*dt THEN PRINT "initial Q = "; 10*INT(q)/10;
IF t < 20 THEN GOTO SKIP2
PRINT ", initial Amplitude = "; x0; ", Period = "; period;
PRINT ", final Q = "; 10*INT(q)/10
  REM: DO GRAPH
SKIP2: PSET(.04*t, .5*q/omega): PSET(.04*t, .5*qh/omega), 4
PSET (.04*t, 4*x/y0): PSET(.04*t, 0): PSET(.04*t, 48*y/y0)
IF t > 20 OR y < 0 THEN GOTO pause
GOTO LOOP0
Pause: GOTO pause
RETURN: END: STOP
```

(Peters, 2002a, 2002b, 2002c). The motion of the can was measured with an SDC sensor connected to a Dataq A/D converter. Whereas experiments of similar type, with homogeneous fluid contents, have produced viscous decay records, the present case involved only friction of so-called "fluid" type; i.e., quadratic in the "velocity." To generate the figure, the A/D record was exported to the Microsoft software package, Excel. Fits to the data were then obtained by adjusting, through trial and error, the $a$, $b$, and $c$

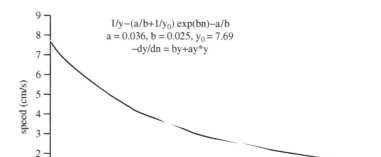

vmax vs cycle number n

$1/y - (a/b + 1/y_0) \exp(bn) - a/b$
$a = 0.036, b = 0.025, y_0 = 7.69$
$-dy/dn = by + ay*y$

**FIGURE 2.20** Decay of an air-damped pendulum as a function of cycle number *n*.

coefficients of a "fit" to the amplitude. For this case, the fit was easily accomplished because both *b* and *c* proved to be essentially zero.

The second case, involving an evacuated pendulum, was not a single pure type of damping, but can be seen in Figure 2.22 to have both hysteretic and amplitude-dependent contributions. Although fluid damping is amplitude-dependent in the same manner, with the damping term being proportional to the square of the amplitude, the word "fluid" is not used to describe this case since the system involved exclusively solid materials.

Not all decay records of this pendulum in vacuum yielded a mix of friction types as displayed in the figure. The effect was observed to be transient, and it is speculated that outgassing of components may have been a factor.

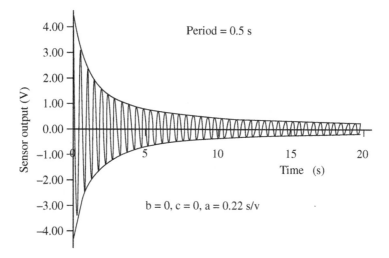

Period = 0.5 s

$b = 0, c = 0, a = 0.22$ s/v

**FIGURE 2.21** Example of fluid damping of a "soup-can" pendulum. The granular contents (black-eye peas and water) result in a friction force that is quadratic in the velocity.

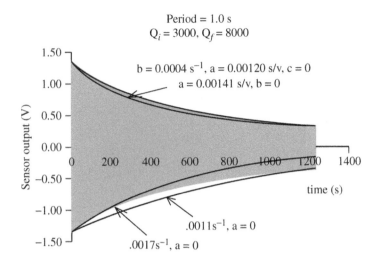

**FIGURE 2.22**   Example of a mix of two damping types, hysteretic and amplitude-dependent.

### 2.17.6.1   Numerical Integration

Instead of integrating the second-order equation of motion twice — first the acceleration, followed by the resulting velocity — more accurate results are obtained by integrating the equivalent pair of first-order equations.

For example, the equation of the simple harmonic oscillator with viscous damping is expressible as

$$\dot{p} = -q - kp \qquad \dot{q} = p \tag{2.63}$$

where the position variable has been represented by the generalized coordinate $q$ ($x$ elsewhere), and for the momentum $p = m\, dq/dt$, and here the mass, $m$, has been set to unity. Likewise, the spring constant has been set to unity. It is generally useful to distill a given problem to its most basic form when attempting to understand the physics. Constants that provide no useful information for trend analysis purposes are conveniently "normalized." Such is common practice, for example, in modeling chaotic systems.

The second-order set can always be reduced mathematically to a first-order pair; however, the pair results naturally from the use of the Hamiltonian as opposed to the Newtonian formulation of mechanics.

## 2.17.7   Damping and Harmonic Content

Equation 2.52 to Equation 2.54 are the nonlinear, modified Coulomb damping model forms that correspond, respectively, to (i) hysteretic, (ii) amplitude-dependent, and (iii) Coulomb damping. The damping term for each of the three cases can be expressed as follows:

$$\frac{f}{m} = \frac{\pi}{4}\frac{\omega}{Q}[v] \tag{2.64}$$

where $f$ is the friction force, and $[v]$ is the square wave whose fundamental in a Fourier series expansion is equal to the velocity of the oscillator times $4/\pi$; i.e., for a square wave $\pm h$, the amplitude of the fundamental is $\pm(4/\pi)h$. We see that all the damping types that have been considered in this chapter, when expressed in canonical form, correspond to a fundamental friction force $f = m\omega v/Q$. The simplicity of this result is probably why viscous damping has been viewed by so many physicists as "inviolate." One must be careful, however, because (as noted in the previous section) only for the case of hysteretic damping is $Q$ constant. For amplitude-dependent damping $Q_f = Q_{f0}(y_0/y)$ and for Coulomb damping

$Q_c = Q_{c0}(y/y_0)$. The time-dependent $Q$ of nonexponential cases will have significant influence on mode development in many-body systems because of elastic nonlinearity (necessary for mode coupling).

There is another important subtlety of Equation 2.64. When only the fundamental of [$v$] is retained, equivalent to viscous damping, $Q$ is proportional to frequency. When all odd harmonics are included (full square wave), $Q$ becomes proportional to frequency squared. This means that *harmonics in the friction force are responsible for the primary difference between hysteretic damping and viscous damping.* Something being presently considered is how, in an algorithmic sense, to modify Equation 2.52 to Equation 2.54 to provide for "dispersion," i.e., means for providing $Q$ dependence other than frequency squared. We posit the following: that hysteretic (exponential) damping is the idealized universal form of damping due to secondary creep. When there is an activation process of Zener (Debye) type, such as dislocation relaxation, then additional terms must be added to the hysteretic "background." It may be that this can be accommodated by a suitable removal of harmonics from the square wave of the hysteretic case, and it may happen that $Q$ is constant for systems that vary continuously. It is conjectured that the PLC effect, responsible for discontinuous changes, plays a role in those cases where $Q$ is not constant. Equations of motion based on the modified Coulomb damping model are summarized in Box 2.3.

---

# Box 2.3

# EQUATIONS OF MOTION BASED ON NONLINEAR DAMPING

---

Equation of motion in terms of energy

$$m\ddot{x} + cm\left[\frac{2E}{k}\right]^2 \text{sgn}(\dot{x}) + kx = 0, \qquad E = \frac{1}{2}m\dot{x}^2 + \frac{1}{2}kx^2$$

Hysteretic-only damping (exponential)

$$\ddot{x} + \frac{\pi\omega}{4Q_k}\sqrt{\omega^2 x^2 + \dot{x}^2}\,\text{sgn}(\dot{x}) + \omega^2 x = 0$$

Velocity-square (fluid) damping

$$\ddot{x} + \frac{\pi}{4y_0 Q_{f0}}(\omega^2 x^2 + \dot{x}^2)\,\text{sgn}(\dot{x}) + \omega^2 x = 0$$

Coulomb damping

$$\ddot{x} + \frac{\pi\omega^2 y_0}{4Q_{c0}}\,\text{sgn}(\dot{x}) + \omega^2 x = 0$$

All three damping types simultaneously active

$$\ddot{x} + \left[\frac{\pi\omega^2 y_0}{4Q_{c0}} + \frac{\pi\omega}{4Q_k}\sqrt{\omega^2 x^2 + \dot{x}^2} + \frac{\pi}{4y_0 Q_{f0}}(\omega^2 x^2 + \dot{x}^2)\right]\text{sgn}(\dot{x}) + \omega^2 x = 0$$

Quality factor

$$\frac{1}{Q(t)} = \frac{1}{Q_c} + \frac{1}{Q_k} + \frac{1}{Q_f}$$

# 2.18  Nonlinearity

## 2.18.1  General Considerations

Electrical nonlinearity is the type with which most engineers are familiar. It is the very basis for common nondigital forms of communication, such as that of frequency modulation type. A popular form of radio amateur communication is one in which the carrier and one of the two normal sidebands of a signal are suppressed before going to the antenna. At the receiver, the carrier is "regenerated" before going to the demodulator. The demodulator required for ultimate transduction by speaker is also a nonlinear device.

Nonlinearity of mechanical type is encountered throughout nature. The human ear, for example, is not linear, but rather characterized by both quadratic and cubic nonlinearities. If an intense, pure low frequency (inaudible) sound of frequency $f$ is present with a higher frequency audible one of frequency $F$, then one typically hears (in addition to $F$) tones at $F \pm f$ due to the quadratic nonlinearity and $F \pm 2f$ due to the cubic nonlinearity.

Very high frequency acoustics (ultrasound) is employed for studies of elasticity. The quasi-linear features of ultrasonic propagation have been the basis for measuring second-order elastic constants (determined by velocity of propagation) and internal friction (by attenuation of the beam, i.e., damping). A commonly employed ultrasonic technique that has been used to study both linear and nonlinear phenomena is the pulse-echo method. By using a thin specimen and extending the pulse width, the overlapped signal can add constructively or destructively and, in the former case, resonance is approached as the width gets very large (Peters, 1973). The pulse-echo method was the basis for this author's Ph.D. dissertation ("Temperature dependence of the nonlinearity parameters of copper single crystals," The University of Tennessee, 1968). The distinguished career of his professor, M.A. Breazeale, has focused on ultrasonic harmonic generation as a means to determine the shape of the interatomic potential of solids (Breazeale and Leroy, 1991). A longitudinal wave distorts because of the anharmonic potential (acoustic equivalence of optical frequency doubling with lasers in a KdP crystal). In like manner, phonon–phonon interactions are possible only because of nonzero elastic constants of order higher than second (second-order constants determining the harmonic potential). Because phonon–phonon interactions are part of damping, there must be consequences, at least for some cases, from nonlinear damping terms.

The unifying theme for this chapter is that damping is fundamentally nonlinear, in spite of the fact that linear approximations have prevailed in modeling and, for many purposes these linear models appear to be acceptable (Richardson and Potter, 1975). In their paper, Richardson and Potter state that "… an equivalent viscous damping component can always be derived, which will account for all of the energy loss from the system. Thus, in measuring the modal vibration parameters for the linear motion of a system, we don't care what the detailed damping mechanism really is."

Although their statement may be true for steady state, it is not expected to be true for the transient processes that lead to steady conditions of oscillation. As demonstrated elsewhere in this chapter, mixtures of different damping types are common among oscillators, and only with viscous or hysteretic damping is the Q independent of amplitude. Other cases may result, for example, from the decay being a combination of hysteretic damping and amplitude-dependent damping. An example used to illustrate this combination was an outgassing pendulum oscillating in vacuum. Similarly, a long, "simple" pendulum, oscillating in air, is found to require a pair of terms — viscous damping and "fluid" damping (Nelson and Olssen, 1986). In the Nelson and Olsson experiment, the drag was found, because of the size of the Reynolds number, to involve both first- and second-power velocity terms. Their case can, incidentally, be treated by the modified Coulomb, generalized damping model of this document.

The presence of either amplitude-dependent damping or Coulomb damping is expected to play a role in determining what modes of a multibody system are actually excited by external forcing. Concerning the latter, Coulomb friction is the basis for exciting chaotic vibrations in mechanical systems (Moon, 1987). Without the nonlinear friction, the excitation would be impossible. In similar manner (although chaotic motion may be present but not in an obvious way), friction from rosin on a violin bow is used to play the violin. Still another example of similar physics is the "singing rod" that was mentioned elsewhere as exhibiting thermoelastic damping.

Whatever combinations of normal modes are initially excited in a linear system are the only ones that can exist thereafter. Such is not the case, however, for many systems and, since nonlinearity is required for mode coupling, there must be nonlinearity in the equations of motion. There is no question about the existence and importance of elastic nonlinearity. Indeed, thermal expansion would be impossible in the absence of higher order elastic constants. The importance of nonlinear damping remains yet to be quantified, since models to include it have been few in number. For those who have found it advantageous to include the oldest and simplest type of nonlinearity in a damping model — Coulomb damping (sliding friction) — the improvements realized by their choice are unlikely to cause them to revisit the problem and try to solve it in terms of a viscous equivalent linear approximation.

There are many examples of damping of a single type other than viscous. In their efforts to improve the knowledge of the Newtonian gravitational constant $G = 6.67 \times 10^{-11}$ Nm$^2$/kg$^2$ (approx.), Bantel and Newman (2000) discovered a pure form of amplitude-dependent damping of internal friction type. They did their experiments at liquid helium temperature (4.2 K) and noted the following: "A striking feature noted in our data is the linearity of the amplitude dependence of $Q^{-1}$ for the three metal fiber materials," and also "Linearity implies that $Q$ may depend on frequency but not on amplitude, while in fact Fig.1 displays a significant amplitude dependence (and hence nonlinearity) of internal friction in all fibers tested." They also considered the temperature dependence of damping and note that there are two independent contributions in Cu−Be. One is linear and temperature-independent and the other amplitude-dependent and independent of temperature. Finally, it is worth noting their statement, "…our results are strongly suggestive of some kind of 'stick−slip' mechanism …," which lends strong support to the modified Coulomb internal friction damping model of the present document.

Repetition is felt to be warranted — such systems cannot always be reasonably described by an equivalent viscous form! For a case of amplitude-dependent $Q$, the equivalent form has no meaning unless the amplitude is fixed, i.e., it oscillates at steady state. Unfortunately, the evolution of the system to steady state is expected to depend on the damping form(s). Surely a model (not yet realized) that predicts what modes survive is worth much more than one which only characterizes the modes after they have reached steady state. The author and Prof. Dewey Hodges of Georgia Tech's Aerospace School are planning projects to try to develop such predictive capability. The present state of the art applied to structures suggests that a truly predictive model cannot ignore damping nonlinearity.

As demonstrated by Bantel and Newman (2000), the mixture of damping types that can co-exist in a system may change with temperature. Early experiments by Berry and Nowick (1958) also showed, as have many investigators subsequently, that damping generally depends on aging. It is naive to believe that aging would not also change the mix of damping types, when there is more than one type. Thus, an adequate damping model must be able to easily accommodate several damping types that are simultaneously active. A variety of engineering techniques have evolved to treat such problems. The most "successful" ones suffer from the fact that an excessive number of parameters or coupled equations must be adjusted by trial and error to yield decent agreement with experiment. This is reminiscent of the state of high-energy (nuclear) physics before the standard model. The hallmark of physics success has always been *simplification*. As noted by Albert Einstein: "All physics is either impossible or trivial. It is impossible until you understand it. Then it becomes trivial." It is hard to imagine, however, that certain damping physics could ever become trivial. Nevertheless, the simplifying nature of better conceptual understanding is a goal to strive for.

One of the remarkable things about the majority of damping models has been the absence of a direct consideration of energy in describing the dissipation process. After all, the most important quantity transformed by the damping is energy, so its inclusion is natural.

## 2.18.2 Harmonic Content

When the damping is nonlinear, the waveform of the oscillator in free-decay contains harmonics. The harmonic content is most obvious in the residuals (difference) after fitting a damped sinusoid to the record, as shown in Figure 2.23.

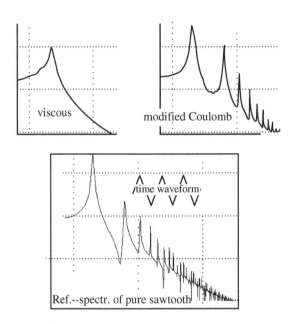

**FIGURE 2.23** Harmonic differences between the residuals of the modified Coulomb damping model and the classic viscous damping model. For reference purposes, a pure sawtooth is included in the figure.

Residuals are still present for the viscous case because the equation of motion was integrated numerically and compared against the classic exponentially decaying sinusoid (solution to the equation) that was used for fitting in all cases. There is always some degree of mismatch with the fit because of rounding errors in the computer. In Figure 2.23, the fundamental is smaller for the viscous case because the fit is inherently more perfect by about an order of magnitude in most of the "eye-ball" fits that were performed by Excel after importation of the data.

A test for harmonic content was performed on the seismometer (17-sec period) data displayed in Figure 2.11 illustrating phase noise. The power spectrum of the residuals for that case is shown in Figure 2.24.

The third harmonic is especially noticeable in this case. That the other harmonics are not so "cleanly" displayed may result from the significant phase noise of the record.

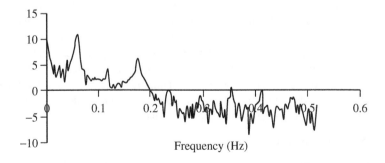

**FIGURE 2.24** Power spectrum of residuals, Sprengnether vertical seismometer free-decay, showing harmonic content.

By looking at the FFT of residuals, rather than the experimental record itself, one finds evidence for a combination of both mechanical and electronic noise. At lower frequencies, the noise (largely mechanical) is approximately $1/f$, while at higher frequencies the noise (largely electronic) begins to be more nearly "white" (frequency-independent) because of discretization errors of the resolution-limited 12-bit A to D converter.

In general, more spectral information can be gleaned from a consideration of the residuals than from the experimental data alone, particularly as one looks for harmonic distortion of mechanical type. Spectral "fingerprints" may prove ultimately useful in determining to what extent damping models of engineering type need to be implemented in full nonlinear form as opposed to an "equivalent viscous" form that is more convenient mathematically.

The importance of the harmonics observed in Figure 2.24 in determining system evolution is not completely known. It was noted earlier that they are expected to influence the evolution of a multibody system to steady state. Presently, it appears that they may serve to validate damping models. From one model type to another, there can be significant differences in the spectral character of the residuals, as shown in Figure 2.25. As compared with Figure 2.23, the fit with the modified Coulomb (hysteretic case) model has been tweaked to reduce the fundamental somewhat, but the odd harmonics remain significant. Observe that the spectrum of the residuals is almost the same for this model and the simplified structural model (see de Silva, 2000, p. 354). This is true even though the temporal variation of the friction force is dramatically different for the two, as seen from the lower time traces that were used to obtain the residuals (which are too small to be seen in the graphs).

From this author's perspective, the simplified structural model is unrealistic, since the friction force, given by $f = c|x|\,\mathrm{sgn}(\dot{x})$, vanishes for zero displacement (the absolute value of the displacement being used to get the hysteretic form of frequency dependence). This is seen in Figure 2.26, which compares hysteresis curves for several models. The modified Coulomb case shown is slightly different from Equation 2.52 that was used to generate Figure 2.25; Figure 2.26 was generated with the $A_{\mathrm{prev}}$ shown in Equation 2.10.

More studies of this type are obviously called for. The spectrum of residuals is a powerful means for the study of damping physics, and it needs to be more widely employed.

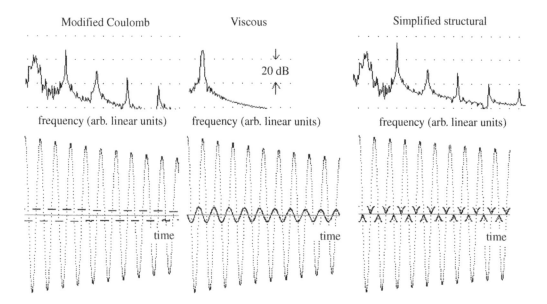

| Modified Coulomb | Viscous | Simplified structural |
|---|---|---|
| frequency (arb. linear units) | frequency (arb. linear units) | frequency (arb. linear units) |

20 dB

time · time · time

**FIGURE 2.25** Illustration of the spectral difference of the residuals for three different damping models. The corresponding temporal records used to generate the spectra are also shown underneath each case.

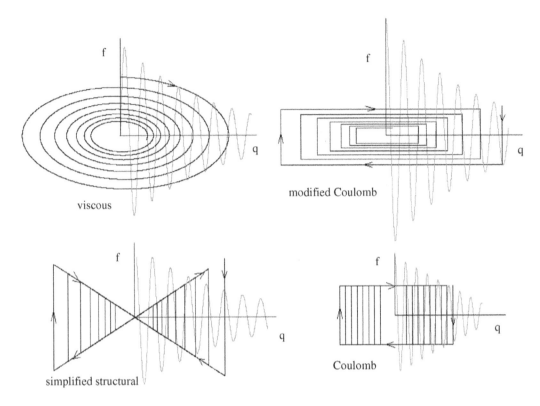

**FIGURE 2.26**  Comparison of hysteresis curves for some damping models.

## 2.18.3   Nonlinearity/Complexity and Future Technologies

Nonlinear damping models must improve if we are to overcome various technological barriers. One barrier is in the area of civil engineering. One of the pioneers of finite element modeling (FEM) is Prof. Emeritus Edward L. Wilson, of the University of California Berkeley. In Technical Note 19 (pertaining to "structural analysis programs") — a document published by his company Computers and Structures Inc — Dr. Wilson says the following:

> Linear viscous damping is a property of the computer model and is not a property of a real structure.

Expanding upon the statement, he notes:

> the use of linear modal damping, as a percentage of critical damping, has been used to approximate the nonlinear behavior of structures. The energy dissipation in real structures is far more complicated and tends to be proportional to displacements rather than proportional to the velocity. The use of approximate "equivalent viscous damping" has little theoretical or experimental justification...the standard "state of the art" assumption of modal damping needs to be re-examined and an alternative approach must be developed   [in reference to Rayleigh damping].

One of the hi-tech areas where modeling improvements are also sorely needed is that involving miniaturized mechanical systems. For example, MEMS devices have already encountered some of the "strange phenomena" of solid-state physics mentioned by Richard Feynman in his famous 1959 talk. To master or compensate for these phenomena, better understanding of the physics will be necessary.

## 2.18.4   Microdynamics, Mesomechanics, and Mesodynamics

At least three different broad fields of research have focused on problems associated with the structural defects that cause hysteresis. These are as follows.

### 2.18.4.1   Microdynamics

In the microdynamics world, the emphasis appears to have been primarily on "contact" friction. The 6th Microdynamics Workshop held at the Jet Propulsion Laboratory in 1999 produced the following statements (quoting Marie Levine's Program Overview): (1) "We have demonstrated that microdynamics exist. The next step is to qualify and quantify microdynamics through rigorous testing and analysis techniques." (2) Microdynamics is "defined as sub-micron nonlinear dynamics of materials, mechanisms (latches, joints, etc.) and other interface discontinuities."

In this workshop, it was noted that frequency-based computational methods *cannot* be used to model quasi-static, transient, and nonstationary disturbances. One of the flight operations they have recommended to minimize adverse effects of microdynamics is dithering.

### 2.18.4.2   Mesomechanics

Ostermeyer and Popov (1999) have the following to say about mesomechanics: "Real physical objects inherently possess discrete internal structures. Great efforts are needed to formulate continuum models of really granular bodies. The history of the last two centuries in a multitude of ways has been marked by highly successful attempts at formulating and analyzing the continuum models of the discrete world. In spite of great advances of continuum mechanics, a number of physical processes are amenable to simulation within the framework of continuum approaches only to a very limited extent. Among these are primarily all the processes whereby the medium continuity is impaired; i.e., those of nucleation and accumulation of damages and cracks and failure of materials and constructions."

Their paper speaks to one of the difficulties concerning granular materials that was mentioned earlier in this chapter — that the potential energy cannot be defined in the common manner. They introduce a temperature-dependent nonequilibrium interaction potential that is not constant in time due to the relaxation processes occurring in the system.

### 2.18.4.3   Mesodynamics

The author of this chapter is singlehandedly responsible for the use of the term "mesodynamics" in the context of mechanical oscillators. His research has been conducted independently of those doing mesomechanics; he came only recently to know of the latter. Whereas mesomechanics seems to have been largely concerned with failure, mesodynamics has been concerned with low-level hysteresis. It is probably closely related to the aforementioned microdynamics, except that the latter seems to have focused on surfaces (sliding friction), whereas mesodynamics is concerned with internal friction.

A group of individuals using "mesodynamics" to describe some of their computational physics is part of the Materials Science Division of Argonne National Laboratory. Their description of computational theory includes: (i) atomic-level simulation (using molecular dynamics); (ii) mesoscale simulation, i.e., "mesodynamics" (using FEM); and (iii) macroscale (continuum) simulation (FEM). Like the author of this chapter, they recognize that the mesoscale is not a continuum (meaning, for example, that the foundation of viscoelasticity is, for many cases, on shaky ground). They employ "dynamical simulation methods in which the microstructural elements (grain boundaries and grain junctions) are considered as the fundamental entities whose dynamical behavior determines microstructural evolution in space and time."

At the Theoretical Division of Los Alamos National Laboratory, Brad Lee Holian has been modeling mesodynamics via nonequilibrium molecular-dynamics (NEMD). In his paper, "Mesodynamics from Atomistics: A New Route to Hall-Petch," he notes that (i) the mesoscopic nonlinear elastic behavior must agree with the atomistic in compression; and (ii) the mesoscale cold curve in tension represents surface, rather than bulk cohesion, thereby decreasing inversely with grain size (Holian, 2003).

The complexity of mesodynamics, which this author has labeled "mesoanelastic complexity," is responsible for much of the aforementioned "strange phenomena." To those familiar with the Barkhausen effect and the PLC effect, they are less strange. It is thought that Richard Feynman, if he were still alive, would identify with mesodynamics because of material in his three-volume series (Feynman, 1970). For example, we have already noted his discussion of the Barkhausen effect, and he included in its entirety a reprint of the Bragg–Nye paper on bubbles which show two-dimensional defect structures such as dislocations, "grains," and "recrystallization" boundaries after stirring (Bragg et al., 1947).

Another famous individual, whose work related in an unexpected way to the material of this chapter, was Enrico Fermi. In one of the first dynamics calculations carried out on a computer, he and colleagues treated a chain of harmonic oscillators coupled together by a nonlinear term (Fermi, 1940). The continuum limit of their model is the remarkable nonlinear partial differential equation known as the Korteweg–deVries equation, whose solution is a soliton, used to advantage in optical fibers. Damping of solitons, whether of the KdV type or the Sine Gordon (kink/antikink) type, is not to be described by linear mathematics. Incidentally, the Sine Gordon soliton is used in modeling dislocations (Nabarro, 1987). The earliest theory to describe dislocation damping using kink/anti-kink pairs was that of Seeger (1956).

## 2.18.5   Example of the Importance of Mesoanelastic Complexity

As noted earlier in this chapter, once hysteretic damping was finally recognized to be important to the Cavendish experiment, better agreement with theory and experiment was possible. Curiously, Henry Cavendish may have been the first person to encounter a "strange" phenomenon (which he did not discuss) (Cavendish, 1798). In his first mass swing to perturb the balance, which used a "fiber" made of copper (silvered), there was an anomalously small period of oscillation that was only 55 sec. The period reported for subsequent trials was about 421 sec.

Whereas the Michell–Cavendish apparatus was a torsion balance, the instrument of Figure 2.27 is a physical pendulum. The perturbing masses, M, were hung from a bicycle wheel whose axle was suspended from the ceiling. The long-period pendulum was placed under a bell jar so that the instrument would not be driven by air currents. By rotating the wheel at constant angular

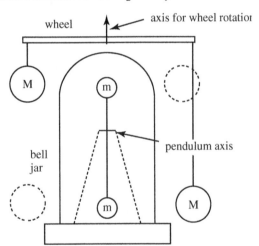

**FIGURE 2.27**   Physical pendulum used in the late 1980s to try and measure the Newtonian gravitational constant.

velocity, the driving force on the pendulum was harmonic. (In the figure, the position of each M one-half period later are shown by the dashed circles.) Knowing the amount of damping, as determined from large amplitude free-decay, it was easy to estimate the number of orbits of the bell jar, at the resonance frequency of the pendulum, required to excite motion to a level above noise in the sensor. Surprisingly, if it were initially at rest, no amount of drive by this means was able to get the pendulum oscillating! The reason involves metastabilities of the defect structures. The potential well is not harmonic (parabolic), but is rather modulated by "fine structure." When located in a deep metastability, the small gravitational force of the drive (in nanoNewtons) is not able to "unlatch" the system. If the pendulum had been dithered (a practice used in engineering) this problem could have been, at least partly, avoided. As it was, the pendulum rested on an isolation table of the type used in optics experiments.

More recently, a Hungarian research team has used a similar apparatus and postulated that the anomalies of their experiment derive from gravity being other than prescribed by Newton (Sarkadi and Badonyi, 2001). Although they claim that there is a "strong dependence of gravitational attraction on the mass ratio of interacting bodies," this author believes that additional experiments must be performed before such a claim has merit. It may be that the anomalous behavior of their pendulum is instead the result of mesoanelastic complexity, i.e., phenomena related to nonlinear damping.

The author's most recent research on damping complexity is based on the premise that the most important scale for the treatment of internal friction is the mesoscale, and not the atomic scale (Peters 2004). Experiments to support this position center around a study of the SMA NiTinol.

## 2.19   Concluding Remark

Much of the material of this part of the chapter on damping is clearly not appropriate to direct engineering application. It was deemed important to present some of the extensive background information responsible for birthing the practical equations of Section 2.17. In Chapter 3, the reader will find practical aids to the measurement of damping.

## Bibliography

Amengual, A., Manosa, L.L., Marco, F., Picornell, C., Segui, C.,, and Torra, V., Systematic study of the martensitic transformation in a Cu–Zn–Al alloy, reversibility versus irreversibility via acoustic emission, *Thermochim. Acta*, 116, 195–308, 1987.

Asa, F., Modal synthesis when modeling damping by use of fractional derivatives, *AIAA J.*, 34(5), 1051–1058, 1996.

Atalay, S. and Squire, P., Torsional pendulum system for measuring the shear modulus and internal friction of magnetoelastic amorphous wires, *Meas. Sci. Technol.*, 3, 735–739, 1992.

Bak, P., Tang, and Wiesenfeld, Self-organized criticality, *Phys. Rev. A*, 38, 364–374, 1988.

Bantel, M.K. and Newman, R.D., High precision measurement of torsion fiber internal friction at cryogenic temperatures, *J. Alloys Compd.*, 310, 233–242, 2000.

Barkhausen, H., *Phys. Z.*, 20, 401, 1919.

Berdichevsky, V., Hazzledine, P., and Shoykhet, B., Micromechanics of diffusional creep, *Int. J. Eng. Sci.*, 35, 10/11, 1003–1032, 1997.

Berry, M.V. and Keating, J.P., The Riemann zeros and eigenvalue asymptotics, *Siam Rev.*, 41, 2, 236–266, 1999.

Berry, B. and Nowick, A., Internal friction study of aluminum alloys containing 4 weight percent copper, *National Advisory Committee for Aeronautics, Technical Note 4225*, online at http://naca.larc.nasa.gov/reports/1958/naca-tn-4225, 1958.

Bert, C.W., Material damping: an introductory review of mathematical models. measures and experimental techniques, *J. Sound Vib.*, 29, 129–153, 1973.

Bethe, H., Lecture at Cornell University online info. at http://www.nd.edu/~bjanko/Copenhagen/Cop3.pdf, 1992.

Bordoni, P., Elastic and anelastic behavior of some metals at very low temperatures, *J. Acoust. Soc. Am.*, 26, 495, 1954.

Bormann P. and Bergmann E., eds. 2002. *New Manual of Observatory Practice*, Institute of Geophysics, University of Stuttgart, online at http://www.seismo.com/msop/nmsop/nmsop.html.

Bragg, Sir Lawrence, F.R.S., and Nye, J.F., A dynamical model of a crystal structure, *Proc. R. Soc. Lond., Ser. A, Math. Phys. Sci.*, 190, 1023, 474–481, 1947.

Braginsky, V.B., Mitrofanov, V.P., and Panov, V.I. 1985. *Systems with Small Dissipation*, The University of Chicago Press, Chicago.

Braginsky, V.B., Mitrofanov, V.P., and Okhrimenko, O.A., Isolation of test masses for gravitational wave antennae, *Phys. Lett. A*, 175, 82–84, 1993.

Breazeale, M.A. and Leroy, O. 1991. *Physical Acoustics: Fundamentals and Applications*, Kluwer Academic/ Plenum Publishers, New York.

Brophy, J. 1965. Fluctuations in magnetic and dielectric solids. In *Fluctuation Phenomena in Solids*, R.E. Burgess, Ed., Academic Press, New York.

Bulsara, A.R. and Gammaitoni, L., Tuning in to noise, *Phys. Today*, 49, 39–45, 1996.

Carlson, J.D., Controlling vibration with MR fluid damping, *Sensor Technol. Design*, 19, 2, 2002, online at http://www.sensorsmag.com/articles/0202/30/main.shtml.

Cavendish, H., Experiments to determine the density of the Earth, *Philos. Trans. R. Soc. Lond.*, 469–526, 1798.

Cipra, B., The FFT: making technology fly, *SIAM News*, 26, 3, 1993, online at http://www.siam.org/ siamnews/mtc/mtc593.htm.

Cleland, A. and Roukes, M., Noise processes in nanomechanical resonators, *J. Appl. Phys.*, 92, 5, 2758–2770, 2002.

Cooley, J. and Tukey, J., An algorithm for the machine calculation of complex Fourier Series, *Math. Comput.*, 19, 90, 297–301, 1965.

Coy, D. and Molnar, M., Optically driven pendulum, *Proc. NCUR XI*, 1621–1626, 1997.

Cromer, A., Stable solutions using the Euler approximation, *Am. J. Phys.*, 49, 5, 455–459, 1981.

Dean, R., Low-cost, high precision MEMS accelerometer fabricated in laminate online at http://www.eng. auburn.edu/ee/leap/MEMSFabricateTable.htm, 2002.

de Silva, C.W. 2006. *Vibration — Fundamentals and Practice*, 2nd ed., Taylor & Francis, CRC Press, Boca Raton, FL.

Fantozzi, G., Esnouf Benoit, W., and Richie, I., *Prog. Mater. Sci.*, 27, 311, 1982.

Fermi E., Pasta J., and Ulam S., Studies in nonlinear problems I, Los Alamos report (reproduced in R. Feynman, 1970: *Lectures on Physics*, Addison Wesley, Boston, 1940).

Fraden, J. 1996. *Handbook of Modern Sensors, Physics, Designs, & Applications*, 2nd ed., AIP Press (Springer), Secaucus, NJ.

Gammaitoni, L., Stochastic resonance and the dithering effect in threshold physical systems, *Phys. Rev. E*, 52, 469, 1995.

Gere, J., Timoshenko, S. 1996. *Mechanics of Materials*, Chapman & Hall, London.

Granato, A. 2002. High damping and the mechanical response of amorphous materials, Submitted to the Proceedings of the International Symposium on High Damping Materials in Tokyo, August 22, 2002 for publication in the Journal of Alloys and Compounds, private communication preprint.

Granato, A. and Lucke, K., Theory of mechanical damping due to dislocations, *J. Appl. Phys.*, 27, 583, 1956.

Greaney, P., Friedman, L.,, and Chrzan, D., Continuum simulation of dislocation dynamics: predictions for internal friction response, *Comput. Mater. Sci.*, 25, 387–403, 2002.

Grigera, T. and Israeloff, N., Observation of fluctuation–dissipation theorem violations in a structural glass, *Phys. Rev. Lett.*, 83, 24, 5038–5041, 1999.

Gross, B., The flow of solids, *Phys. Today*, 5, 8, 6–10, 1952.

Heyman, J. 1997. *Coulomb's Memoir on Statics: An Essay in the History of Civil Engineering*. Imperial College Press, London, ISBN: 1860940560.

Hilfer, P., ed. 2000. *Applications of Fractional Calculus in Physics*, World Scientific, London.

Hodges, D. 2003. Private communication (Georgia Tech School of Aerospace).

Holian, B. 2003. Mesodynamics from atomistics: a new route to Hall–Petch, private communication preprint.

Horowitz, H. and Hill, W. 1989. *Art of Electronics*, 2nd ed., Cambridge University Press.

Kimball, A. and Lovell, D., Internal friction in solids, *Phys. Rev.*, 30, 948–959, 1927.

Kohlrausch, R., *Ann. Phys.*, 12, 392, 1847, online information at http://www.ill.fr/AR-99/page/ 74magnetism.htm.

Kuroda, K., Does the time of swing method give a correct value of the Newtonian gravitational constant?, *Phys. Rev. Lett.*, 75, 2796–2798, 1995.

LaCoste, L., A new type long period vertical seismograph, *Physics*, 5, 178–180, 1934.

Lakes, R. 1998. *Viscoelastic Solids*, CRC Press, Boca Raton, FL.

Lakes, R. and Quackenbush, J., Viscoelastic behavior in indium tin alloys over a wide range of frequency and time, *Philos. Mag Lett.*, 74, 227–232, 1996.

Landau, L. and Lifshitz, E. 1965. *Theory of Elasticity*, Nauka, Moscow.

Lorenz, E. 1972. Predictability: Does the flap of a butterfly's wings in Brazil set off a tornado in Texas, presented to AAAS, Washington, DC.

Marder, M. and Fineberg, J., How things break, *Phys. Today*, 49, 24–29, 1996.

Marion, J. and Thornton, S. 1988. *Classical Mechanics of Particles and Systems*, 3rd ed., HBJ, Academic Press, New York, p. 114.

Milonni, P. and Eberly, J. 1988. *Lasers*. Wiley Interscience, Hoboken, NJ, p. 93.

Moon, F. 1987. *Chaotic Vibrations, An Introduction for Applied Scientists and Engineers*, Wiley Interscience, Hoboken, NJ.

Moore, G., Moore's law is described online at http://www.intel.com/research/silicon/mooreslaw.htm, 1965.

Nabarro, F. 1987. *Theory of Crystal Dislocations*, Dover, New York.

Nelson, R. and Olssen, M., The pendulum—rich physics from a simple system, *Am. J. Phys.*, 54, 112–121, 1986.

Ostermeyer, G. and Popov, V., Many-particle non-equilibrium interaction potentials in the mesoparticle method, *Phys. Mesomech.*, 2, 31–36, 1999.

Peters, R., Resonance generation of ultrasonic second harmonic in elastic solids, *J. Acoust. Soc. Am.*, 53, 6, 1673, 1973.

Peters, R., Linear rotary differential capacitance transducer, *Rev. Sci. Instrum.*, 60, 2789, 1989.

Peters, R., Metastable states of a low-frequency mesodynamic pendulum, *Appl. Phys. Lett.*, 57, 1825, 1990.

Peters, R., Fourier transform construction by vector graphics, *Am. J. Phys.*, 60, 439, 1992.

Peters, R., Fluctuations in the length of wires, *Phys. Lett. A*, 174, 3, 216, 1993a.

Peters, R., Full-bridge capacitive extensometer, *Rev. Sci. Instrum.*, 64, 8, 2250–2255, 1993b. This paper describes an SDC sensor with cylindrical geometry. Other geometries, including the more common planar one, are described online at http://physics.mercer.edu/petepag/sens.htm.

Peters, R. 2000. Autocorrelation Analysis of Data from a Novel Tiltmeter, abstract, *Amer. Geo. Union annual mtg.*, San Francisco.

Peters, R. 2001a. The Stirling engine refrigerator — rich pedagogy from applied physics, online at http://xxx.lanl.gov/html/physics/0112061.

Peters, R. 2001b. Creep and Mechanical Oscillator Damping, http://arXiv.org/html/physics/0109067/.

Peters, R. 2002a. The pendulum in the 21st century—relic or trendsetter, *Proc. The Int'l Pendulum Project*, University of New South Wales, Australia, Proceedings, October, 2002.

Peters, R. 2002b. The soup-can pendulum, *Proc. The Int'l Pendulum Project*, University of New South Wales, Australia, Proceedings, October. 2002.

Peters, R. 2002c. Toward a universal model of damping—modified Coulomb friction online at http://arxiv.org/html/physics/0208025.

Peters, R. 2003a. Graphical explanation of the speed of the Fast Fourier Transform, online at http://arxiv.org/html/math.HO/0302212.

Peters, R. 2003b. Nonlinear damping of the 'linear' Pendulum, online at http://arxiv.org/pdf/physics/03006081.

Peters, R. 2003c. Flex-Pendulum—basis for an improved timepiece, online at http://arxiv.org/pdf/physics/0306088.

Peters, 2004. Friction at the Mesoscale. In *Contemporary Physics*, P. Knight, Ed., Vol. 45, no. 6, 475–490, Imperial College, London 2004.

Peters, R. and Kwon, M., Desorption studies using Langmuir recoil force measurements, *J. Appl. Phys.*, 68, 1616, 1990.

Peters, R. and Pritchett, T., The not-so-simple harmonic oscillator, *Am. J. Phys.*, 65, 1067–1073, 1997.

Peters, R., Breazeale, M., and Pare, V., Temperature dependence of the nonlinearity parameters of Copper, *Phys. Rev.*, B1, 3245, 1970.

Peters, R., Cardenas-Garcia, J., and Parten, M., Capacitive servo-device for microrobotic applications, *J. Micromech. Microeng.*, 1, 103, 1991.

Portevin, A. and Le Chatelier, M., Tensile tests of alloys undergoing transformation, *C. R. Acad. Sci.*, 176, 507, 1923.

Present, R. 1958. *The Kinetic Theory of Gases.* McGraw-Hill, New York.

Press, W., Flannery, B., Teukolsky, S., and Vetterling, W. 1986. *Numerical Recipes—the Art of Scientific Computing.* Cambridge University Press.

Purcell, E., Life at low Reynolds number, *Am. J. Phys.*, 45, 3–11, 1977.

Richardson, M., and Potter, R. 1975. Viscous vs structural damping in modal analysis, *46th Shock and Vibration Symposium.*

Roukes, M., Plenty of room indeed, *Scientific American*, 285, 48–57, 2001.

Sarkadi, D. and Badonyi, L., A gravity experiment between commensurable masses, *J. Theor.*, 3–6, 2001.

Saulson, P., Stebbins, R., Dumont, F., and Mock, S., The inverted pendulum as a probe of anelasticity, *Rev. Sci. Instrum.*, 65, 182–191, 1994.

Seeger, A., On the theory of the low-temperature internal friction peak observed in metals, *Philos. Mag.*, 1, 1956.

Singh, A., Mohapatra, Y., and Kumar, S., Electromagnetic induction and damping, quantitative experiments using a PC interface, *Am. J. Phys.*, 70, 424–427, 2002.

Speake, C., Quinn, T., Davis, R., and Richman, S., Experiment and theory in anelasticity, *Meas. Sci. Technol.*, 10, 430–434, 1999. See also Quinn, Speake and Brown, 1992: Materials problems in the construction of long-period pendulums, *Philos. Mag.* A 65, 261–276, 1999.

Steinmetz, C.P., *Complex Number Technique, paper given at the International Electrical Congress*, Chicago, 1893.

Stokes, G., On the effect of the internal friction of fluids on the motion of pendulums, *Trans. Cambridge Philos. Soc.*, IX, 8, 1850, read December 9, 1850.

Strang, G., Wavelet transforms versus Fourier Transforms, *Bull. Am. Math. Soc.*, 28, 288–305, 1993.

Streckeisen, G. 1974. *Untersuchungen zur Messgenauigkeit langperiodischer Seismometer*, Diplomarbeit, Institut für Geophysik der ETH Zürich (communicated privately by E. Wielandt).

Tabor, M. 1989. The FUP Experiment. In *Chaos and Integrability in Nonlinear Dynamics: Introduction.* Wiley, New York.

Thomson, W., Tait, G. 1873. *Elements of Natural Philosophy, Part I.* The Clarendon Press, Oxford, (Thomson was later known as Lord Kelvin).

Urbach, J., Madison, R., and Markert, J., Reproducibility of magnetic avalanches in an Fe–Ni–Co alloy, *Phys. Rev. Lett.*, 75, 4694, 1995a.

Urbach, J., Madison, R., and Markert, J., Interface depinning, self-organized criticality, and the Barkhausen effect, *Phys. Rev. Lett.*, 75, 276, 1995b.

Venkataraman, G. 1982. Fluctuations and mechanical relaxation. In *Mechanical and Thermal Behavior of Metallic Materials*, Caglioti, G. and Milone, A., eds., pp. 278–414. North-Holland, Amsterdam.

Visintin, A. 1996. *Differential Models of Hysteresis.* Springer, Berlin.

Westfall, R. 1990. Making a world of precision: Newton and the construction of a quantitative physics. In *Some Truer Method. Reflections on the Heritage of Newton*, F. Durham and R.D. Purrington, eds., pp. 59–87. Columbia University Press, New York.

Wielandt, E. 2001. *Seismometry*, section Electronic Displacement Sensing, online at http://www.geophys. uni-stuttgart.de/seismometry/hbk_html/node1.html.

Zemansky, M.W. 1957. *Heat and Thermodynamics*, 4th ed., McGraw-Hill, New York, p. 127.

Zener, C. 1948. *Elasticity and Anelasticity of Metals*, Chicago Press, Chicago.

# 3

# Experimental Techniques in Damping

Randall D. Peters
*Mercer University*

**Summary**

*This chapter is a continuation of Chapter 1 and Chapter 2, and is concerned with practical experimental techniques for measuring damping. It begins with a discussion of the requirements placed on electronics. After demonstrating the importance of sensor linearity using a computer simulation, the issues of data acquisition and processing are addressed. The power of the Fast Fourier Transform is illustrated, not just for spectral analysis, but also (in "short time" form) for measuring the damping of each component when a system oscillates with multiple modes. Various sensor types are discussed in relation to their advantages and disadvantages for specific applications through the treatment of seven different systems studied in free decay. These seven cases differ with respect to factors such as (i) eigenfrequency, (ii) material type, and (iii) method of estimating the logarithmic decrement, and thus the Q of the decay. In the case of some solids, damping is shown to result largely from defects in the structures. A powerful test for nonlinear damping is demonstrated: simply looking at a graph of Q to see whether it changes with time. Two examples of driven oscillators are given. The first being very nearly linear, and the second being highly nonlinear, due to an anharmonic restoring force involving magnets plus several simultaneously acting damping mechanisms. The nonlinear system is used to illustrate difficulties in interpretation that can arise in driven systems due to phenomena such as frequency and/or amplitude jumps involving hysteresis. Then an illustration is given of how elastic-type nonlinearities may couple with damping-type nonlinearities in order to determine which modes of a complex system survive during the transient approach to steady state. Mechanical noise, another important feature of nonlinear damping, is also examined. The magnitude of the 1/f character in an evacuated pendulum is shown to decrease with time, as the oscillator is allowed to stabilize against creep. The final sections address the common misconception that viscous air friction is the most important form of mechanical oscillator damping. Cases are chosen to demonstrate that (i) internal friction is nearly always also important, if not the most important, and moreover, that (ii) fluid damping is rarely simple — involving the density as well as the viscosity of the fluid. It is shown that damping has a complicated frequency dependence, as opposed to the simple (overly idealized) form predicted by common theory.*

## 3.1 Electronic Considerations

### 3.1.1 Sensor Linearity

The importance of sensor linearity is often overlooked. It is naively assumed that one can simply employ a lookup table to provide calibration corrections. This assumption can result in serious misinterpretations of spectral data, especially in a multimode system. A classic example of artifacts (nonreal signals) that result from a nonlinear sensor is to be found in the ear. The phenomenon, known as aural harmonics, is well known to musicians and figures in the use of "fortissimo" and "pianissimo" in orchestral music. In this chapter we describe how the artifacts mentioned in Chapter 2 are generated. Figure 3.1 illustrates differences according to the nature of the nonlinearity.

The only "real" signals in Figure 3.1 are at frequencies $f_1$ and $f_2$. The number and type of other "unreal" (artifact) signals depends on the type of nonlinearity. The sensor response for the left graph (quadratic) is of the form $V = ax + bx^2$, whereas for the right graph $V = ax + bx^2 + cx^3$. The influence of terms other than $V = ax$ (ideal, linear output voltage) was generated by (i) simulating the pair of harmonic signals, (ii) inputting these signals to each simulated sensor, respectively, and (iii) performing a Fast Fourier Transform (FFT) on the output.

Although it is possible to understand mathematically the various artifacts using trigonometric identities, the phenomenon is much easier to demonstrate with a computer. For Figure 3.1, all

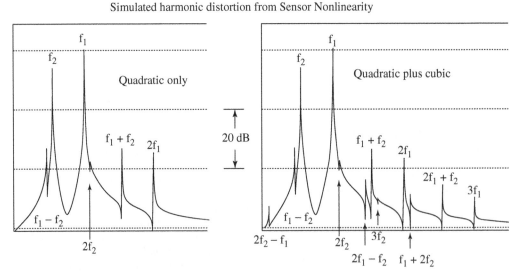

**FIGURE 3.1** Spectral illustration of sum and difference artifact frequencies according to nonlinear sensor type.

numerical operations were performed with code written by the author using QuickBasic. It was used to (i) simulate the harmonic signal that was written to a data file, after which it was (ii) read by the FFT algorithm based on the details supplied in *Numerical Recipes* (Press et al., 1986).

### 3.1.2 Frequency Issues

The choice of a sensor depends largely on the frequencies to be measured. For higher frequencies an excellent instrument for data collection is a digital (storage) oscilloscope, where a microphone can often be directly connected to the instrument. At lower frequencies, a serial-port analog-to-digital converter (ADC) is generally adequate and user-friendly. Examples of each will be provided. The majority of examples considered in this chapter involve low frequencies, where the eigenmode is typically described not in terms of frequency but rather the period (reciprocal of frequency).

### 3.1.3 Data Acquisition

In the absence of sophisticated data collection and analysis tools, the true character of damping is not readily discovered. Proper characterization is important, since a crude estimate of the damping, based on a single parameter (such as the viscous linear model), may be inappropriate if the oscillator is driven at places (either frequency or amplitude) other than where the parameter was measured. Some of the examples from the experiment that follows were selected to demonstrate the importance of nonlinearity. The probability that an oscillator, selected at random, might have a $Q$ that varies in time is proving to be more significant than anticipated. Were it not for dramatic improvements in numerical-type technology, this improved understanding of damping would not have been possible.

As with computer technology in general over the last decade, ADCs have become much more powerful. The Dataq model 700, for example, is superior (at lower frequencies) to many of the "plug-in" boards of the previous generation that were several times more expensive. The Dataq ADC operates through the USB port (Windows 98 and later), has 16-bit resolution and the software support is excellent. Especially useful for the present purposes are its ability to (i) easily perform data compression with which to view long records, (ii) quickly compute an FFT according to different, useful options, and (iii) easily output files to a spreadsheet.

## 3.2 Data Processing

### 3.2.1 Language Type

The author's experience with software began with early computers and even included the loading of the Fortran compiler of a PDP-11 using punch-tape. He has programmed computers (or hardware-specific processors) with (i) machine code, (ii) assembly language, (iii) Fortran, and (iv) Basic, and he has acquired a rudimentary knowledge of Pascal and C++. The drudgery of machine coding was a factor in his quest to better understand the Fourier transform (Peters, 1992, 2003a, 2003b, 2003c, 2003d). His philosophy with regard to numerical methods is similar to his view of hardware: choose the simplest package (lowest level of sophistication) consistent with the desired results for the problem at hand. The reader may be surprised to learn that QuickBasic (which some have modernized to Visual Basic for Windows) is his favorite language. Nearly all simulation results presented in this chapter were generated with the DOS version of QuickBasic.

### 3.2.2 Integration Technique

Too few have discovered the powerful integration scheme in which Cromer (1981) modified the unstable Euler algorithm. The difference between the two methods involves the sequencing (order) of updates to the state vectors in the discrete approximation of the integrals. The method was called the "last point approximation" (LPA) by Cromer, whereas the Euler technique would be called the "first point approximation" according to this nomenclature. The LPA was discovered by a high school student working for Cromer. She was attempting to simulate planetary motion with the Euler method and accidentally coded the LPA. The author first used the LPA to do intercept analyses for the U.S. antisatellite program — computing, among other things, orbital ephemerides. More recently he has used it in place of Runge Kutta techniques to do all kinds of mechanical system simulations, including nonlinear types with several DoF. A physics theorist at Texas Technical University, Professor Thomas Gibson, now regularly uses the LPA as part of the graduate-level course which he teaches in numerical methods.

### 3.2.3 Fourier Transform

With the Cooley–Tukey improvement to make it fast, the Fourier transform has become a tool of major software importance. Just as the integrated circuit dramatically changed hardware development, the FFT has had a profound influence on the evolution of scientific code.

Whereas many recognize the value of the FFT for viewing "raw" spectral data, few have discovered other powerful tools based in the FFT. For example, autocorrelation is unrivaled in its ability to uncover low-frequency signals of fairly short duration that are corrupted by noise. The number of cycles is not great enough (Heisenberg effect) for a well-defined line to be observed in the FFT by itself. Through the Wiener–Khintchin theorem, the autocorrelation overcomes this limitation. It is computed by multiplying the transform by its conjugate and then taking the inverse transform. The author has used this technique to study free oscillations of the earth (Peters, 2004).

#### 3.2.3.1 Short Time Fourier Transform

A powerful software tool is one in which the Fourier transform is not computed over the entire length of a record. Instead, the record is subdivided (usually with some degree of overlap between adjacent subsections), and the FFT is computed for each subsection. Because the data are generally of the temporal (rather than spatial) type, the technique is called the short time Fourier transform (STFT). For equivalent processing, where the independent variable has units of meters rather than seconds (as in optics applications), the technique could just as well be called the short space Fourier transform.

The STFT is especially useful when waveforms are not pure harmonic, as from a single-degree-of-freedom (single-DoF) oscillator. For systems with multiple modes, whether they derive from eigenmodes as recognized by most, or from mechanical noise as recognized by a few (generated as part of the internal friction of load bearing members); the STFT is a powerful means for isolating and thus determining the temporal history of individual spectral components.

The most common form of the STFT is the canned programs that are a part of software packages such as LabVIEW. With the Dataq software it is easy to accomplish the same thing manually, since one can readily step in time from place to place of a stored record, computing the FFT at any position. The intensity at a given position is obtained by clicking on the displayed spectral line of interest, which provides the value either in dB (Dataq version) or in volts. The amplitude history in dB of the line is thus obtained (equally spaced-in-time values) with a simple click of the mouse. In this way, the free decay of a single component of the system can be readily extracted from the total system response. For the present purposes, the individual intensities were copied by hand to paper and later typed into a spreadsheet for plotting. The process is not laborious, since the number of necessary points is typically less than two dozen.

An example of a manually generated STFT is provided in Figure 3.2. Unlike the methodology described above (operating on experimental data residing in a Dataq folder), the record of Figure 3.2 was generated by computer and written to an output file. The data correspond to three superposed, exponentially damped sinusoids.

From the upper graph (time record), it is not clear how the individual components are changing with time. The triplet of components becomes obvious in the frequency domain (lower left), and when the

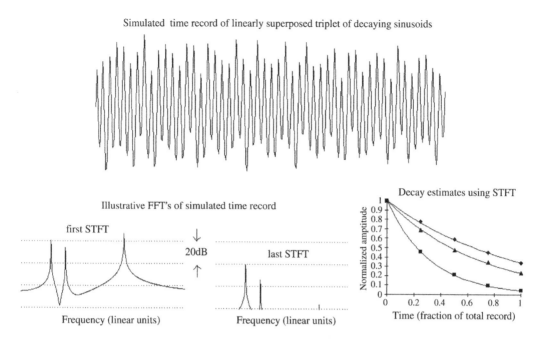

**FIGURE 3.2** Example of separating the time decay of superposed components using the STFT.

intensities of each line are plotted *versus* time (lower right) the exponential character of each becomes visible. To test the viability of this method, the individual damping parameters were evaluated from a given STFT graph and found to be in excellent agreement with the values supplied to the simulation.

### 3.2.3.2 Example Use of the STFT

The majority of the examples given in this chapter avoid low-energy oscillations as space does not allow mesoanelastic regimes to be treated at length. The following case was chosen to illustrate (i) the importance of the Portevin–LeChatelier (PLC) effect on damping (Portevin and Le Chatelier, 1923), and (ii) the power of the STFT in eliminating the influence of clutter. The STFT benefit was expected, since the difference between a noisy record and a multimode case like the previous example is in the number of modes. The PLC effect is significant for high-energy internal friction dissipation even though the jumps for which the effect is known are not obvious at these energies. Evidently, this is a consequence of the large number of events, for which the average effect is a fairly smooth decay.

The scales for the two graphs of Figure 3.3 are different by nearly three orders of magnitude, with the level of the sensor output for each case being indicated (mid-range values). Two features are evident from a direct visual inspection: (i) the change from a smooth to a jerky decay in going from high to low levels, and (ii) a 4% increase in the frequency of oscillation at the lower energy. The latter is recognizable from the vertical lines that have been added (every fifth peak). One of the more interesting (and surprising to most) features of the lower graph is that phase of the oscillation is not significantly altered as the result of mean position jumps.

The following information is provided for those skeptical of the comments concerning the lower trace of Figure 3.3. From the study of hundreds of low-level decays in different mechanical oscillators, the author has become confident that the jumps shown are not sensor (or other electronic) artifacts. Prejudice against this conclusion has been considerable over the last 14 years, in spite of the fact that a similar phenomenon was noted (and accepted) in magnetic materials many years ago, i.e., the Barkhausen effect (Barkhausen, 1919). Unfortunately, the related phenomenon in mechanical systems (PLC effect) is hardly known among physicists.

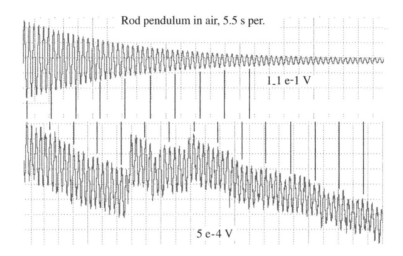

**FIGURE 3.3** Illustration of the character changes in decay in going from high to low energies of a pendulum.

The value of the STFT for the study of data such as that of Figure 3.3 is illustrated in Figure 3.4. Whereas Figure 3.3 showed only the first and the last portions of a long record, here the STFT was applied to the entire record using 27 different FFTs.

As seen in the upper left curve, there is a sharp decrease in the damping in STFT No. 5. Thus, the intensity was replotted in each of the intervals from 0 to 5 and 5 to 27. Both intervals show a near perfect linear trendline fit, indicating exponential decay. Thus the $Q$ was found to quickly change from 78 to 340 at an energy level in the region of $10^{-10}$ J. Although air damping was a factor in the early part of the record, it is not thought to be capable of causing the rapid change in $Q$ that was observed. A similar sharp

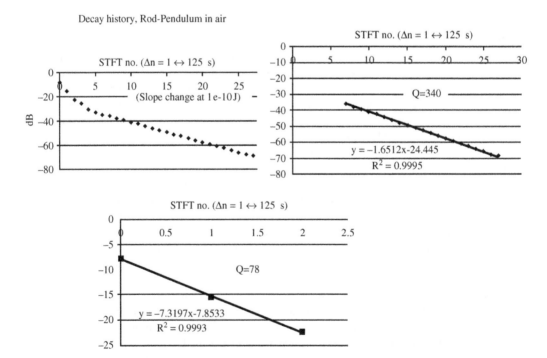

**FIGURE 3.4** Example of the use of STFT analysis applied to a dataset, the first and last portions of which are shown in Figure 3.3.

change in the damping of this same pendulum was seen at roughly $10^{-11}$ J with the pendulum swinging in a high vacuum. The slope change was equally rapid for the vacuum case, but the change in $Q$ was from 120 to 210. It is not known why the damping at low-level energy in a vacuum would have a lower $Q$ than in air. Perhaps the difference derives from a different placement of the knife-edges on the silicon flats. Some of the damping of this pendulum is the result of the knife-edges being fabricated from brass rather than a harder metal such as carbon steel.

## 3.3 Sensor Choices

Box 3.1 shows some representative sensors for damping measurements. The list is far from exhaustive; for a detailed description of each (plus discussion of other types), the reader is referred to Fraden (1996). Of the transducer types indicated, position sensors are generally the most versatile; but the present chapter also provides examples of the use of (i) velocity, (ii) microphone, and (iii) photogate measurements.

In addition to the need for linearity (discussed in Section 3.1 above), the ideal sensor will be noninvasive. In reality, it is not possible to perform a measurement that does not at some level perturb the system under study. The least perturbative types of direct measurement are optical and electrical — capacitive, followed by inductive.

Some of the advantages and disadvantages of the devices indicated in Box 3.1 are provided below.

---

# Box 3.1

## SOME SENSOR TYPES

Representative Sensors for Damping Measurements

| Position | Velocity | Pressure | Time Interval | Acceleration | Force/Strain |
|---|---|---|---|---|---|
| Capacitive | Faraday law (electromagnetic) | Microphone | Photogate | Accelerometer | Strain gauge |
| LVDT | | Pressure gauge | | | |
| Optical | | Capacitive | | | |
| Encoder | | Optoelectronic | | | |
| Shadow | | Piezoresistive | | | |
| Potentiometric | | | | | |

---

### 3.3.1 Direct Measurement

#### 3.3.1.1 Position Sensors

The inductive linear variable differential transformer (LVDT) is a sensor that is commonly used in engineering applications. Thus, it has been a natural choice for many position-sensing purposes; but it is both more invasive and noisier than capacitive sensors. Wielandt (2001) notes the following concerning the advantage of capacitive over inductive sensors: "Their sensitivity is ... typically a hundred times better than that of the inductive type."

Optical encoders are also readily available and have been used extensively. Because of their digital nature, based in a finite number of elements, their low-level resolution is poor compared to capacitive devices.

Optical sensing by shadow means is easy to employ — for example, using a solar cell of the type discussed later. The method is afflicted, however, by (i) an offset voltage, and (ii) the degrading influence of background light.

Potentiometers are very easy to use, but compared to other position sensors they are extremely invasive because of Coulomb friction in the slider and also the bearings that support it.

### 3.3.1.2  Velocity Sensor

The most important velocity sensor is that which functions on the basis of Faraday's law. Using a magnet and a coil, an electromagnetic force is generated in the wire of the coil when it experiences a changing magnetic flux. Prevalent in seismometers before the advent of broadband (feedback) instruments, its primary shortcoming is poor sensitivity at low frequencies.

### 3.3.1.3  Time Interval

Photogates have become the primary means for kinematic studies in introductory physics laboratories. Combined with compact, user-friendly timers, it is possible to measure both period and velocity. As illustrated later, they can be easily used to measure damping in slowly oscillating systems, but only in a limited amplitude range.

## 3.3.2  Indirect Measurement

In the cases of (i) pressure, (ii) acceleration, and (iii) force/stress sensing, the measurement is an indirect one. Consider, for example, the Ruchhardt experiment to measure the ratio of heat capacities of a gas (discussed in Chapter 2). The oscillation of the piston could be measured in several different ways. For instance, direct position sensing could be accomplished by attaching a small electrode to the piston and allowing it to move between stationary capacitor plates. Alternatively, a "flag" on the piston could be used to interrupt the light beam of a photogate. Depending on constraints, however, the easiest method might be an indirect measurement in which a pressure sensor monitors the gas through a catheter communicating with the cylinder of the apparatus.

Accelerometers can sometimes be connected directly to an oscillator, but only if the mass of the instrument is very small compared to the system being studied. As with the measurement of velocity, their sensitivity at low frequency is very poor (the second derivative of position yielding a response that is proportional to frequency-squared).

Strain gauges are easy to employ but also lack sensitivity (compared to position measurement), since they communicate with a very small portion of the oscillating sample (if noninvasive).

## 3.4  Damping Examples

### 3.4.1  Case 1: Vibrating Bar — Linear with Significant Noise

The simplest means to measure the $Q$ of an oscillator whose frequency is in the range of the human ear is to use a microphone connected to a digital oscilloscope. In this case, the microphone was an inexpensive dynamic type and the oscilloscope was a Tektronix TDS 3054. A better choice, had it been available, would be an electret microphone. The ring-down of a xylophone bar, following a strong (sharp) hammer strike, is shown in Figure 3.5.

The voltage *versus* time of the microphone output was saved to memory in the oscilloscope, from which the digital record was output to a floppy disk, using the CSV format. Data from the disk were read into columns A and B of an Excel spreadsheet using "Open file." An envelope fit was then performed on the turning points by placement of trial and error data into column C, using "autofill." A separate graph was generated for each value of the constant $b$ in the expression "$= 0.04 * \exp(-b * A1)$" typed into Cell C1. (The lower turning points were obtained by typing "$= -b1$" into Cell D1 and using autofill.

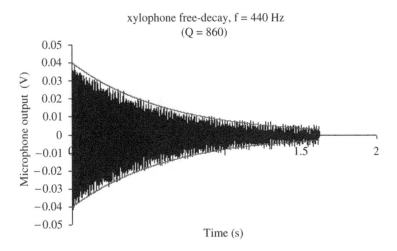

**FIGURE 3.5**   Free-decay record of a vibrating bar.

(Additional details concerning the use of Excel in this manner will be provided in the discussion of seismometer damping that follows.)

Although optimizing algorithms could be generated to perform such a fit (with a probable slight increase in accuracy), this visual technique is preferred here as it is more understandable, user-friendly, and its performance is proven. The total time required in Excel to generate Figure 3.5 using a 2K record (2048 points) is typically only a few minutes with a modern Pentium computer.[*]

Once a satisfactory fit was obtained ($b = 1.6$ in Figure 3.5), the $Q$ was estimated using

$$Q = \frac{\pi}{\Delta} = \frac{\pi f}{b} \tag{3.1}$$

There is a fair amount of electronic noise in Figure 3.5 because the microphone was connected directly to the oscilloscope. The smallest bandwidth of the oscilloscope, at 20 MHz, causes a large amount of Johnson (white) noise. Narrowing the bandpass by means of a preamplifier would improve the quality of the data dramatically. Such is typically true of signal to noise ratio (SNR) improvement by tailoring the electronics to the need.

## 3.4.2   Case 2: Vibrating Reed — Example of Nonlinear Damping

To illustrate another sensing technique, the system shown in Figure 3.6 was used.

A 90° twist was given to a hacksaw blade after heating with a torch and quenching. One end was clamped to the vertical post shown and a piece of cardboard was taped to the other end. An incandescent lamp is placed above the cardboard, which vibrates horizontally, and a solar panel below the cardboard is used as a sensor. The solar panel in this case is a commercial unit that comes with a cigarette lighter plug for charging automobile batteries. The output from the panel goes to the Tektronix digital scope also in the picture.

Unlike other sensing schemes described in this chapter, the solar panel output is not bipolar but instead has a constant voltage offset corresponding to the equilibrium position of the reed.

The frequency of oscillation is too low to operate the oscilloscope with a.c. coupling. Thus, it is important to make the d.c. offset as small as possible. This was accomplished by shielding nonactive parts

---

[*]A disclaimer is in order at this point. Present comments by the author should not be interpreted as an endorsement of Microsoft products in general. Although QuickBasic and Excel have both proven unusually beneficial to the work described in this chapter, they are the only software packages marketed by the company to have received a strong endorsement from the author.

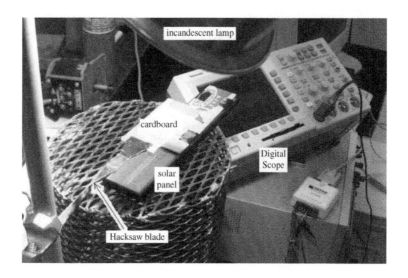

**FIGURE 3.6**   Setup for measurement of vibrating reed free-decay.

of the solar cell from the lamp. Although the d.c. offset could be removed with a voltage bucking battery, this was found to introduce unacceptable noise spikes. With the offset, the gain of the electronics was limited by the amount of vertical position shift allowed by the scope. The results of this study are illustrated in Figure 3.7.

The nonzero value of $b$ (0.018) in Figure 3.7 indicates the presence of amplitude-dependent damping (refer to Equation 2.61, Chapter 2). The nonlinear damping in this case probably derives from the air, rather than internal friction of the hacksaw blade. Its presence causes the $Q$ of the system to increase with time.

The $Q$ of the system is calculated from the expression

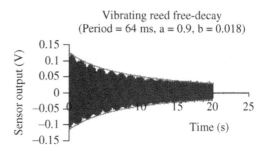

**FIGURE 3.7**   Vibrating reed decay with amplitude-dependent damping.

$$Q = \frac{\pi}{(b + ay)\tau} \tag{3.2}$$

where $\tau$ is the period of oscillation. At the start of the record ($y = 0.12$) the $Q$ is 390 and it approaches 2700 as the amplitude approaches zero.

### 3.4.3   Case 3: Seismometer

Since the ability of a seismometer to detect tremors is proportional to the square of the period of the instrument, they require a good low-frequency sensor. The most common sensor for the latest generation commercial instruments is a half-bridge (differential) capacitive type. Because of the greater sensitivity and linearity of the full-bridge symmetric differential capacitive (SDC) sensor mentioned in Chapter 2, it is well suited to these applications, being easy to employ. (The full-bridge character is described in a TEL-Atomic tutorial (Peters, 2002).) A significant advantage of the SDC symmetry (equivalent electrically to the inductive LVDT) is its relative insensitivity to construction imperfections, such as roughness of surface and nonparallelism of electrodes. Thus, construction can be done crudely without serious degradation of performance. For example, electrodes of the first prototype of the SDC sensor were fabricated from sheet

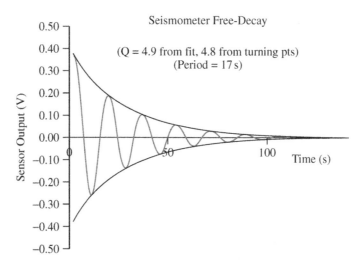

**FIGURE 3.8** Free-decay record of the Sprengnether vertical seismometer, period 17 sec.

copper that was cut with shears and subsequently flattened by hammer on a hard plane surface. This stands in stark contrast to the optically polished surfaces necessary for the use of some sensors.

The Sprengnether vertical seismometer discussed at several points in Chapter 2 was studied in a configuration for which $Q = 4.9$, as determined by an external 990 $\Omega$ resistor to provide induced-current damping. The resistor was connected across the coil that is part of the original equipment and which moves (with the mass of the instrument) in the field of a stationary magnet, i.e., a Faraday's law (velocity) detector. Excitation to initiate the free-decay study was accomplished by applying an alternating (square wave) current to the coil, reversing the direction of the current at each turning point of the motion of the mass. The fundamental (Fourier series) of a square-drive generated this way is shifted 90° from the mass motion, corresponding therefore to resonance. After cessation of the drive, data as shown in Figure 3.8 were collected with a Dataq DI-700 ADC (16-bit).

The graph in Figure 3.8 was generated with Excel after the Dataq record was saved to floppy disc as an *.dat (CSV) file. It was imported to Excel using "open file" with "comma delimiter." Once in Excel, these data were shifted one place to the right (from the default A column to the B column) to accommodate computer generation of a time-data column. The column of time values was generated according to the sample rate, the value of which is by default saved to the data file. To generate the time column, a 0 was placed in the first row, $n$ corresponding to the start of data. Dropping down one row in the A column, "$= A_n + 1/(\text{sample rate})$" was typed, to increment the time. Then the lower right hand corner "small solid square" of the box containing this time was grabbed and held with the left button of the mouse to autofill all the way to the last time point of the data. The computer-generated exponentials, which correspond to the turning points, were obtained by generating two additional columns. These were obtained by placing the cursor at a row corresponding to the time $A_n$ (in column C) and then typing "$= A_0 * \exp(-(\text{omega}/2/Q) * A_n)$." The value of $A_0$ is obvious from the data and a first estimate for $Q$ can be quickly obtained from about a dozen turning points (read with the Dataq software before the data are ever saved).

The technique is illustrated in Table 3.1. For example, in the case of Figure 3.8, $Q = 4.8$ from the 13 turning points. Thus the argument of the exponential was set to 0.0385. Using autofill, the columnar (upper) exponential was then quickly produced. Then a second (adjacent) column D was generated in similar manner, by taking the negative of the last point and then autofilling to the top row. (When one autofills downward, the rate with which Excel traverses the rows increases exponentially after the last row of data has been passed; thus, it is much easier to fill upwards rather than downwards.)

Once the pair of exponentials being fitted to the data have been graphed, along with the data, it is simple to adjust the curves by varying the argument (in this case, small changes around 0.0385) until a

**TABLE 3.1**    Estimation of *Q* from the Turning Points

Use of Excel to estimate logarithmic decrement from turning points of the motion

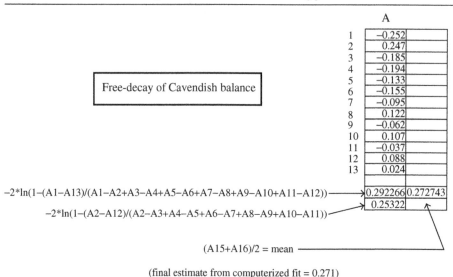

|   | A |
|---|---|
| 1 | −0.252 |
| 2 | 0.247 |
| 3 | −0.185 |
| 4 | −0.194 |
| 5 | −0.133 |
| 6 | −0.155 |
| 7 | −0.095 |
| 8 | 0.122 |
| 9 | −0.062 |
| 10 | 0.107 |
| 11 | −0.037 |
| 12 | 0.088 |
| 13 | 0.024 |

Free-decay of Cavendish balance

−2*ln(1−(A1−A13)/(A1−A2+A3−A4+A5−A6+A7−A8+A9−A10+A11−A12)) ⟶ 0.292266  0.272743

−2*ln(1−(A2−A12)/(A2−A3+A4−A5+A6−A7+A8−A9+A10−A11)) ⟶ 0.25322

(A15+A16)/2 = mean

(final estimate from computerized fit = 0.271)

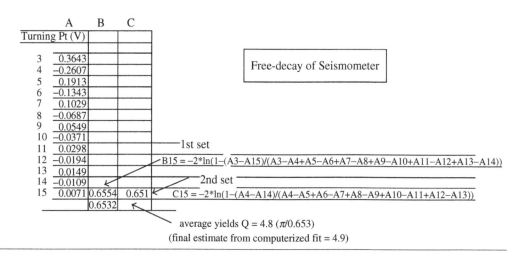

|   | A | B | C |
|---|---|---|---|
|   | Turning Pt (V) | | |
| 3 | 0.3643 | | |
| 4 | −0.2607 | | |
| 5 | 0.1913 | | |
| 6 | −0.1343 | | |
| 7 | 0.1029 | | |
| 8 | −0.0687 | | |
| 9 | 0.0549 | | |
| 10 | −0.0371 | | |
| 11 | 0.0298 | | |
| 12 | −0.0194 | | |
| 13 | 0.0149 | | |
| 14 | −0.0109 | | |
| 15 | 0.0071 | 0.6554 | 0.651 |
|   | | 0.6532 | |

Free-decay of Seismometer

1st set

B15 = −2*ln(1−(A3−A15)/(A3−A4+A5−A6+A7−A8+A9−A10+A11−A12+A13−A14))

2nd set

C15 = −2*ln(1−(A4−A14)/(A4−A5+A6−A7+A8−A9+A10−A11+A12−A13))

average yields Q = 4.8 (π/0.653)

(final estimate from computerized fit = 4.9)

good fit is obtained (using autofill each time). The fit is rapid and accurate when there are not too many parameters to vary, since the eye is well suited to this operation. Upon obtaining the best-fit by this means, the preliminary value of $Q = 4.8$ was altered to the final value of $Q = 4.9$.

Note that in Table 3.1 a new Excel worksheet was employed (column A no longer the time as in the discussion above). The voltages corresponding to the turning points (maximum and minimum) were each read by placing the cursor at an extremum and manually recording the value displayed in turn by the Dataq software. These values were then typed into Excel, as opposed to the "file open" method for importing large datasets, i.e., as used to generate Figure 3.7.

### 3.4.4   Case 4: Rod Pendulum with Photogate Sensor

One of the simplest ways to measure damping at larger levels is to use a photogate of the type common to general education physics laboratories. The infrared beam of the photogate is tripped by a "flag" attached to the oscillator. The rod pendulum pictured in Figure 3.9 was studied by this means. Figure 3.10 shows a closeup of the flag which is attached to the top of the pendulum and which passes through the photogate during oscillation.

As seen in the figures, various parts of the pendulum are clamped to a vertical steel rod. Both upper and lower masses are made of lead, each with a pair of holes drilled in it — one to pass the brass rod of the pendulum through and the other (after tapping) to hold a thumbscrew for securing the mass at different vertical positions on the rod.

This mechanical oscillator is a compound pendulum; the period can be made long as the knife-edge (also clamped to the rod) approaches the center of mass. At long periods, the instrument is not very responsive to external accelerations of the supporting frame; but it is sensitive to internal structural changes. Low-frequency instability is encountered as the upper parts of the instrument experience creep, particularly in the materials just above the knife-edge. The upper pendulum has similarities to an inverted pendulum, except that it is rigidly connected to the lower pendulum, and causes the oscillator to eventually exhibit double-well (Duffing) characteristics. This happens at larger amplitudes as the period is increased toward really long times. The tendency toward mesoanelastic complexity depends on the dimensions of the rod. As expected because of the well-known engineering properties of rods and tubes, a large diameter, thin-wall tube will behave differently from a solid rod made from the same amount of material (same total mass). This will be true if the tube does not experience localized (sharp) deformation prone to creasing.

Damping measurements with a photogate require that the time required for the flag to pass through the beam be fairly small — thus larger amplitudes of motion are required than with other sensors. Of course really large motion would result in a period increase, consistent with long-understood pendulum dynamics. For amplitudes within the acceptable range (which in practice is not overly restrictive), the velocity of the pendulum as it passes through the equilibrium position is inversely proportional to the time

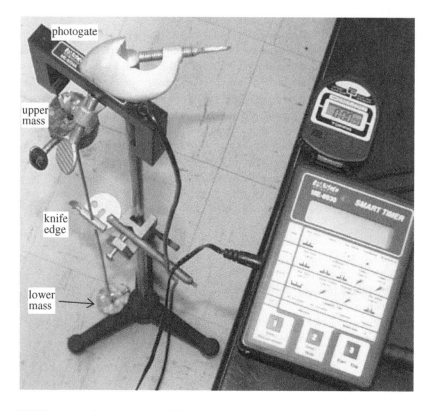

**FIGURE 3.9** Rod pendulum in which damping measurements are made with a photogate.

**FIGURE 3.10**  Top of the pendulum showing the upper mass and "flag" for tripping the photogate.

interval between interrupts of the photogate beam by the two vertical arms of the flag. If the period does not change with amplitude, then there is also an inverse relationship between the gate time and the amplitude. A plot of the inverse of these times *versus* cycle number is, for the constraints indicated, a reasonable approximation of the turning points of the free-decay.

In this case the single gate time interval measurements were made by a Pasco Smart Timer. It is a user-friendly instrument that also permits the period of the pendulum to be accurately measured by the flag (by using two different lengths of the flag arms). For experiments of this type, it may prove more convenient to measure the period with a stopwatch (infrequently as compared to the velocity). The sequentially increasing time intervals are read manually from the Smart Timer and recorded by hand, once per cycle. Of course, to do so requires that the period be long enough to permit these operations. The recorded values are conveniently analyzed by typing to a spreadsheet, which is then used to graph damping curves such as shown in Figure 3.11.

A pure exponential fit is not appropriate to the decay of Figure 3.11, in which the upper mass had been removed and a business card taped to the bottom of the pendulum to cause turbulent air damping (period near 1 sec). The fit shown, however, involving both linear and quadratic dampings, is seen to be quite reasonable. As in the case of the vibrating reed discussed earlier, this system is adequately described by the nonlinear damping equation 2.61 given in Chapter 2. For the data of Figure 3.11, $Q = 25$ initially and increases to 70 at the end of the record.

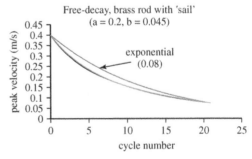

**FIGURE 3.11**  Free-decay of a pendulum as determined by photogate measurements.

Without the business card, and with the upper mass in place, the decay of this pendulum was found with the photogate measurement technique to be exponential, as expected for viscous damping with periods of about 5 sec. At periods in excess of about 10 sec, however, internal friction of the rod becomes more important than air damping. Although the decay is then still exponential at larger levels, the frequency dependence is not the same as required by linear air damping.

### 3.4.5 Case 5: Rod Pendulum Influenced by Material under the Knife-Edge

The data in Figure 3.11 were collected with the knife-edges resting on hard ceramic alumina flats. When supported by other materials, the damping of a rod pendulum can be influenced by anelastic flexure other than that of the rod. Hardness of the material does not guarantee low damping, as will be seen in the following examples. The data that follows were collected with a different pendulum, depicted in Figure 3.12.

The sensor in this case was an SDC unit, connected to the computer through the Dataq DI-700 A/D converter. The upper and lower masses are each approximately 1 kg and their separation distance on the aluminum hunting arrow from which the pendulum was fabricated was about 70 cm.

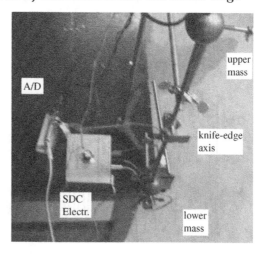

**FIGURE 3.12** Long period rod pendulum used to study the influence of different materials under the knife-edge.

#### 3.4.5.1 Lithium Fluoride Samples

The samples used to collect the data in Figure 3.13 were identical pairs, except that one pair had been irradiated with a huge dose of gamma rays. The resulting changes to the structure of the crystal are responsible not only for color centers as noted in the photograph in Figure 3.14, but also a dramatic change in the internal friction. It is clear from Figure 3.13 that internal friction in the LiF is the dominant source of damping of the rod pendulum that was used (Peters, 2003a, 2003b, 2003c, 2003d).

Lithium fluoride is used in thermoluminescent film badges (radiation monitors). When exposed to energetic radiation, atoms are "knocked" from their crystal lattice sites into metastable states corresponding to interstitial positions of the lattice. Upon ramping the temperature of the sample in an oven fitted with a photomultiplier tube, jumps from the metastable state are accompanied by the release of photons. The amount of light so generated is a measure of the dose that was received by the crystal. Because light flashes are observed with rather small changes in the temperature, it is reasonable to expect that mechanical strains might also cause a significant change to the defect state of such crystals. This postulate is confirmed by the data in Figure 3.13, which show a dramatic difference in the decay character of the pure (clear) crystals (bottom figure) and those which were extensively damaged by gammas (top figure).

In both of the decays in Figure 3.13 there is significant nonlinear damping, as evidenced in the early portions of each of the two records. The top case is nearly pure Coulombic, and the bottom case is partially amplitude dependent. This is revealed from estimates of the Q, shown in Figure 3.15.

The Q values in Figure 3.15 were computed from successive triplet-values of the turning points of the motion, read directly from the decay pattern displayed on the monitor by the Dataq software. The equation used is

$$Q = \frac{\pi}{-2\ln[1 - (\theta_n - \theta_{n+2})/(\theta_n - \theta_{n+1})]}, \quad n = 0, 1, 2, \ldots \tag{3.3}$$

Influence of Defects on damping from LiF crystals

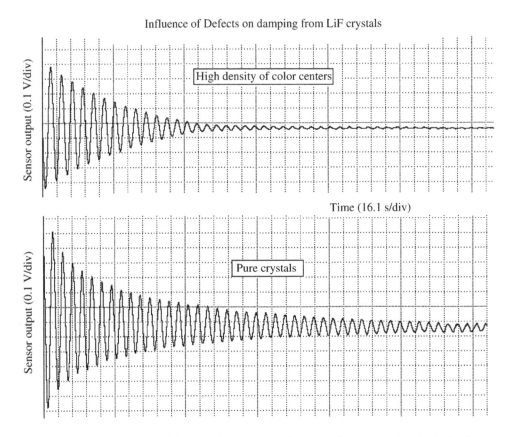

FIGURE 3.13    Illustration of damping difference according to specimen type under the knife-edge.

## 3.4.6  Hard Materials with Low *Q*

It is commonly (and mistakenly) thought that hard materials must necessarily also have low damping. The following two examples show that this is not necessarily so. Even though cast iron is very hard, it is also quite dissipative, which makes it an ideal material for engine blocks. Figure 3.16 shows a decay curve for the steel knife-edges of the pendulum resting on cast iron samples.

At the start of the record the damping with cast iron is nearly twice as great as that of steel-on-sapphire or steel-on-silicon, where the *Q* was found to be of the order of 80. This large damping measurement is consistent with the known excellent properties of cast iron for use in engine blocks, although the frequencies for such applications are much higher.

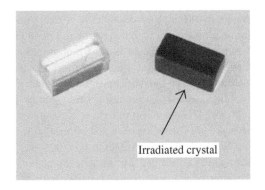

FIGURE 3.14    Photograph of LiF single crystals used to obtain the data in Figure 3.13.

Figure 3.17 is another very hard material which has large damping — the ceramic piezoelectric wafer formed from lead, zirconium, and titanium (PZT), which by means of a mechanical impulse is commonly used to generate an electric spark to ignite a gas grill. The secular decline of *Q* based on the short temporal record indicates Coulomb damping. It is consistent with the nearly straight-line turning points for the early part of the long-term record, also shown. The long-term record

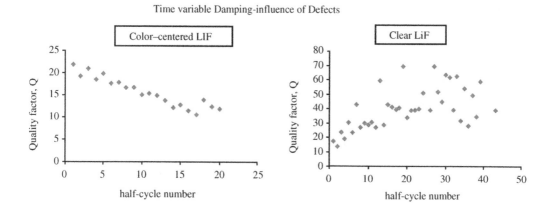

FIGURE 3.15 Temporal dependence of the Q, LiF crystal experiments.

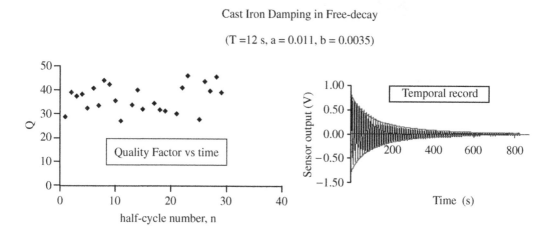

FIGURE 3.16 Data collected using cast iron samples.

is labeled as anomalous because it does not appear to be consistent with several simultaneously acting dissipation mechanisms. Instead, the strong Coulomb damping seen early on seems to disappear later, once the amplitude has dropped below a particular level. This suggests activation processes of a quantal type. It would be interesting to study the PZT wafers in a different pendulum configuration, and not operating "open-circuit" as in the present case, but rather with different resistors connected between the top and bottom of the wafers.

## 3.4.7 Anisotropic Internal Friction

With Polaroid material (H sheet) placed under the knife-edges it was found that the damping depends on the direction of the long-chain polymeric molecules. The direction of the molecules in a sample is readily determined by looking through the Polaroid at reflected light from a polished floor. When the reflection occurs close to the Brewster angle, only the horizontal component of the electric field is significant in the reflected light for unpolarized incident light. The direction of the molecules is thus determined by rotating the sample until the minimum of level of light is found. When this occurs the molecular chains are situated horizontally.

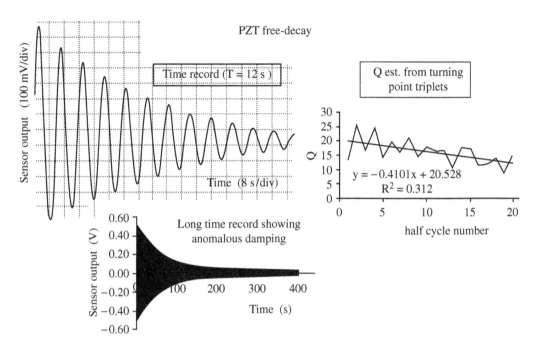

**FIGURE 3.17**   Data from an experiment involving PZT ceramic wafers.

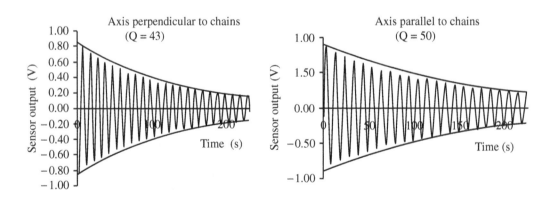

**FIGURE 3.18**   Free-decay curves showing anisotropy of the internal friction in polaroid material.

It was reasoned that the molecular properties of Polaroid might result in mechanical as well as optical anisotropies. This postulate proved to be true, as shown in Figure 3.18.

When oscillating on silicon at a period of 10 sec, previous studies have found that the instrument decays consistently with a $Q$ of 80 (uncertainty 3%). In the present study, half a dozen free-decay records were obtained for (i) edges parallel to the chains and (ii) edges perpendicular to the chains. The average $Q$ of oscillation was estimated at 50 for the parallel case and 43 for the perpendicular case. Reproducibility proved slightly better for the parallel case (4%) as compared to the perpendicular case (5%). Additional details are documented elsewhere (Peters, 2003a, 2003b, 2003c, 2003d).

### 3.4.7.1  Summary, Free-Decay $Q$ Estimation

All of the techniques so far described are methods based on free-decay, which is especially important for nonlinear systems. With linear systems it is also possible to use steady-state methods, as noted in Box 3.2 (last column; de Silva, 2006). Box 3.2 summarizes the techniques used in the present chapter to estimate the logarithmic decrement, $\beta T$, from which $Q = \pi/(\beta T)$.

---

# Box 3.2

# METHODS FOR QUANTIFYING DAMPING

Damping ($Q$ Estimation) Techniques ($Q = \pi/\beta T$, $T =$ Period)

| Logarithmic Decrement (Full-Cycle, $N$) | Turning Points (Half-Cycle, $n$) | Nonlinear Fit to Envelope | Time $\tau$ to $l/e$ ($0.3679 x_0$) | Short Time Fourier Transform | Bandwidth, Magnification Factor, Hysteresis Loop, Step-Response |
|---|---|---|---|---|---|
| $\beta T = \dfrac{1}{N} \ln \dfrac{x_0}{x_N}$ | $\beta T =$ $-2\ln\left[\dfrac{1-(x_{n-1}-x_{n+1})}{(x_{n-1}-x_{n+1})}\right]$ | $-\dot{y} = ay^2 + by$ $+ c$ | $\beta T = T/\tau$ | $-\beta T = \dfrac{T \ln 10}{20}$ [slope(dB/s)] | de Silva (2006), Chapter 7 |

---

The best method is to use a full nonlinear fit; the worst is to measure the time to $1/e$. The expression in Box 3.2 for the logarithmic decrement, using the STFT, is equivalent to

$$Q = 27.29 \frac{f}{\left|\dfrac{\mathrm{dB}}{\mathrm{s}}\right|} \tag{3.4}$$

where $f$ is the frequency in Hz and the STFT slope is specified in dB per s.

## 3.5  Driven Oscillators with Damping

This chapter has been mainly concerned with oscillators in free-decay. It is also possible to make quantitative predictions from measurements at steady state. Confidence in predictions, however, depends on the nature of the damping. Such data are of limited value for most nonlinear systems, unless supplemented with free-decay data.

### 3.5.1  MUL Apparatus

Some of the techniques applicable to driven systems are illustrated by the multipurpose undergraduate laboratory (MUL) apparatus shown in Figure 3.19, that has been used by students in the physics department at Mercer University.

For the purpose of measuring the Lorenz force (basis for defining the current unit, the ampere) a constant current is supplied through the posts to the pivoted-on-points brass wire on which a weight, $W$, is shown hanging on one of the horizontal arms of the wire. Current enters the wire through one post *via* the banana plug inserted into a drilled hole. It thereafter travels through the lower (invisible) shorter straight segment of the wire located between the poles of the drive magnet; and it finally exits through the banana plug on the opposite post. When carrying a current, the force on the wire from the part inside the magnet causes vertical deflection, the direction up or down being determined by the direction of the current. This results in a rotation about the pivot points (indented tops of the posts). The position is measured by the capacitive sensor, S (one of several variants of the SDC patent).

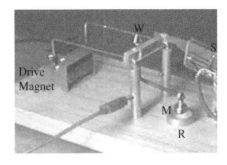

**FIGURE 3.19** Apparatus for studying resonance and the Lorenz force law.

The sensitivity of this current balance depends on the location of the center of mass of the oscillatory wire, which is determined in part by the position of the rare earth magnet, M, which hangs from a steel nut on the threaded part of the heavier brass rod having a 90° bend. The upper end of this threaded rod is held by a plexiglass member that also holds the ends of the oscillatory wire.

### 3.5.2 Driven Harmonic Oscillator

The MUL becomes a driven harmonic oscillator when the excitation current is a.c. rather than the d.c. used for the Lorenz force study. The damping is determined primarily by eddy currents in the aluminum ring, R, that lies on the wooden base underneath and in close proximity to magnet M.

The apparatus is useful for studying both free-decay and driven oscillation. Engineering students Brandon R. Bowden and James D. Sipe have programmed LabVIEW to generate both free-decay curves and resonances.

An example Lorentzian (resonance response) is given in Figure 3.20. (Additional information is found in a laboratory writeup (Peters, 1998).)

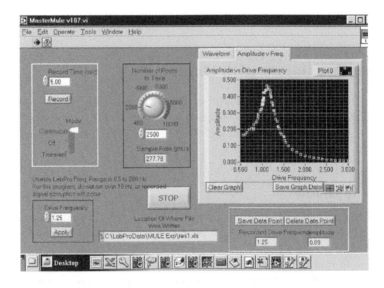

**FIGURE 3.20** Screens from the LabVIEW program used with the MUL to study both transient and resonance phenomena.

# 3.6 Oscillator with Multiple Nonlinearities

An oscillator can have significant nonlinearities of both the elastic and damping types. An example is the mechanical system pictured in Figure 3.21.

The instrument is a modified extensometer that was sold by TEL-Atomic and which was designed around the SDC sensor to measure Young's modulus and thermal expansion coefficients. The wire sample normally used with the instrument (along with a hollow power resistor that fits in the black clamp) has been removed, and two rare earth magnets have been employed. One magnet is superglued to the bottom of the pan where the weights are normally placed, as shown in Figure 3.22; and the other magnet is attached to the bottom of the inductor that is sitting on the top of the oscillator (cased instrument, Pasco) used for drive. The pair of magnets are positioned in close proximity so as to repel each other, thus supporting the mass of the moveable arm of the extensometer.

The study of nonlinear systems requires a linear sensor; i.e., any nonlinear contributions from the detector must be negligible. Figure 3.23 shows the calibration results for the instrument and its linear response for the range of amplitudes used in the study.

The potential energy of this oscillator was assumed to have the following form

$$U(x) = \frac{b}{x^n} + cx \tag{3.5}$$

and the parameters were estimated by measuring $x$ as small masses were placed on the pan. A linear regression fit to a log–log plot (using the sensor calibration constant of 550 V/m) yielded $b = 1.02 \times 10^{-5}$, $n = 1.526$, and $c = 0.304$ (system international units). Anharmonicity of the

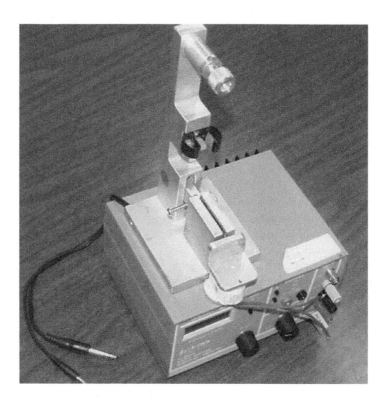

**FIGURE 3.21** Mechanical oscillator with multiple nonlinearities.

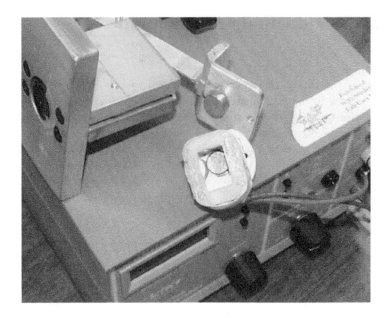

**FIGURE 3.22** Closeup picture of the oscillator in Figure 3.21 (nonoperational configuration), showing placement of the rare earth magnets.

potential is readily apparent in the plot shown in Figure 3.24, with the force of restoration being greater in compression ($x$ decreasing) than it is in extension. This feature is reminiscent of interatomic potentials, with anharmonicity being responsible for thermal expansion.

Because of the elastic nonlinearity, the mean position depends on the amplitude of the oscillation, as is evident in the free-decay curve in Figure 3.25.

The damping of this oscillator was also found to be nonlinear, as seen in Figure 3.26.

The oscillator exhibits hysteresis when driven at larger amplitudes, as shown in Figure 3.27, where it can be seen that the location of an amplitude jump depends on which way the oscillator is adjusted, either up or down in frequency. Such jumps (well known with oscillators with nonlinear elasticity) stand in stark contrast with the behavior of a linear oscillator, as can be seen by comparing Figure 3.27 with the screen picture in Figure 3.20.

A surprise from this study involves the frequency of oscillation. In general, the oscillator did not entrain to the drive. Moreover, the preferred frequencies were not necessarily the same as the free-decay frequency of 6.01 Hz. Some of the frequencies (measured with power spectra) are indicated in

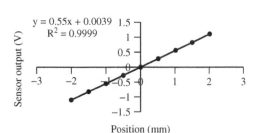

**FIGURE 3.23** Calibration data for the sensor used with the oscillator having multiple nonlinearities.

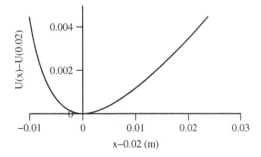

**FIGURE 3.24** Plot of the potential energy function of the oscillator.

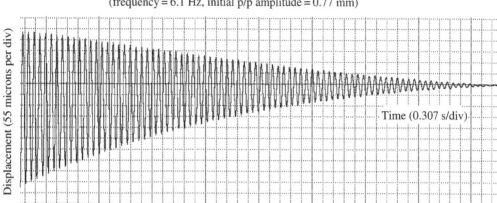

Free-decay of an Oscillator with multiple nonlinearities
(frequency = 6.1 Hz, initial p/p amplitude = 0.77 mm)

**FIGURE 3.25** Free-decay curve showing the mean position shift as a function of oscillator amplitude. (Decreasing sensor voltage corresponds to increasing $x$.) The frequency of oscillation is 6.01 Hz.

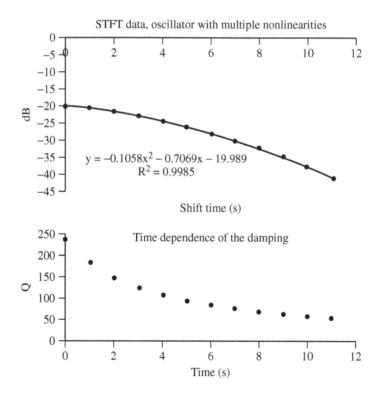

$$y = -0.1058x^2 - 0.7069x - 19.989$$
$$R^2 = 0.9985$$

**FIGURE 3.26** Free-decay character as determined using the short time Fourier transform.

Figure 3.27. The 3% frequency jumps observed in Figure 3.27 (in going from 5.46 Hz to 6.20 Hz) are not real but rather artifacts of the finite resolution of the 1024 point transforms that were employed.

Figure 3.27 demonstrates why nonlinear damping measurements should be done in free-decay. Figure 3.25 demonstrates how exponential fits make no sense for some oscillators. Fortunately, the STFT can be used to determine the amplitude dependence of the $Q$.

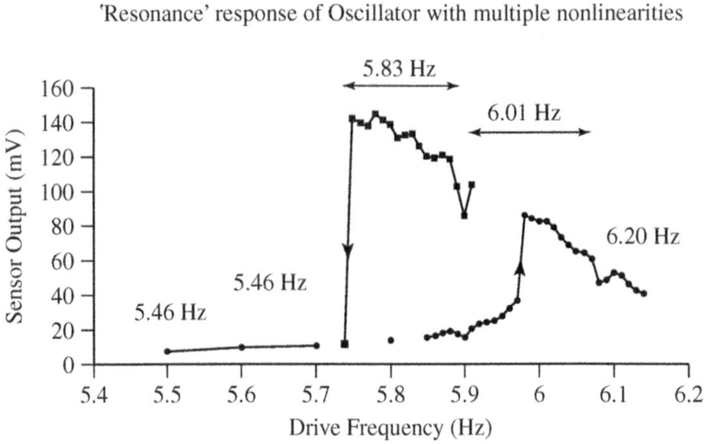

'Resonance' response of Oscillator with multiple nonlinearities

FIGURE 3.27   Resonance response (steady state) of the driven oscillator.


## 3.7   Multiple Modes of Vibration

### 3.7.1   The System

In engineering, multimode oscillations are common. Many, if not most, cases have mode mixing features even though they may in some cases be too small to be readily observed. The importance of nonlinearity to these problems is not widely appreciated, so a case to illustrate salient features is provided here. Free-decay records were obtained with an oscillator in the form of a vertically oriented (hanging) tungsten wire, of length 24 cm and diameter 0.31 mm. It was clamped at the top end, and at the bottom a rectangular plate was attached that was 11.3 cm long, 1.3 cm wide, and 0.8 mm thick. The plate was cut from double-sided copper circuit board. The board was positioned between the stationary plates of a capacitive sensor, as shown in Figure 3.28.

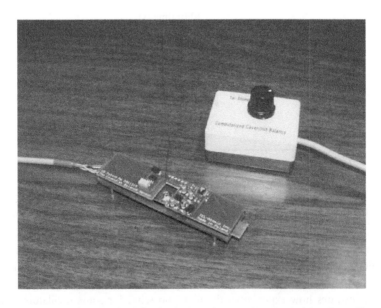

FIGURE 3.28   Photograph of the detector used to monitor the multimode oscillator.

For the picture, the apparatus was disassembled and the plate allowed to rest on the top of the bottom electrode set. Operationally, the plate was positioned midway between the upper and lower static electrode sets (separation distance of 4 mm); and there was no mechanical contact during oscillation. As can be seen, the top of the circuit board containing the upper electrode set contains more than a dozen electronic components; these are of the surface mount technology type. The detector is of the SDC type and this particular embodiment is manufactured in Poland for TEL-Atomic Inc., Jackson, MI, for use in the Computerized Cavendish Balance.

As can be seen in the picture, the wire was rather kinked instead of straight, which is expected to be a significant source of nonlinearity. For this reason, not to mention that it is very difficult to make larger diameter tungsten wires reasonably straight, no serious attempt was undertaken to remove the kinks.

## 3.7.2 Some Experimental Results

An example decay record generated with this apparatus is illustrated in Figure 3.29.

## 3.7.3 Short Time Fourier Transform

When multiple modes are present in a decay, as in Figure 3.29, it is not possible to readily estimate $Q$ for all of the various modes using time data. The decays can be estimated using the FFT, in a technique called the short time Fourier transform, which is built-in to various software packages related to acquisition systems, such as LabVIEW (see Appendix 15A). With the versatile software supplied with the Dataq A/D converter, it is straightforward to employ an equivalent manual technique. Using the number of points to define the FFT a value (always a power of 2 total) that is substantially smaller than the number of points in the record, a manual scan is performed in which one simply increments from start to finish, calculating a separate FFT at each position in time along the way. As an illustration of this powerful tool, Figure 3.30 shows spectra corresponding to the start and the finish of the data in Figure 3.29.

All the modes decay in time, and the rate of decay is especially large for those modes that correspond to sum and difference frequencies of the primary modes at 1.19 and 2.19 Hz. Table 3.2 gives the spectral intensities in dB for the two times considered. Where the rows are blank for the end of record case, the values were insignificantly small.

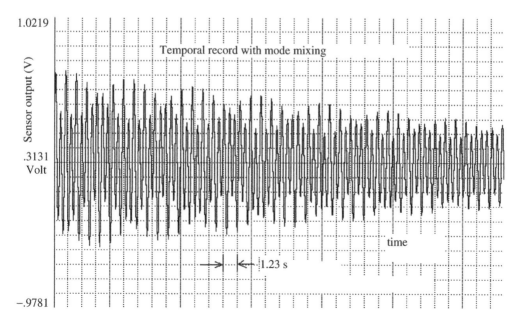

**FIGURE 3.29** Example free-decay of a multimode wire oscillator.

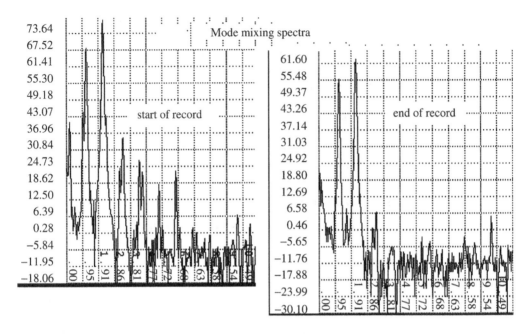

**FIGURE 3.30**  Beginning and end spectra corresponding to the temporal data from Figure 3.29. Ordinate values are spectral intensity in dB, abscissa values are frequency in Hz (linear scale).

The decibel values in the table are referenced to the bit-size (16 corresponding to 65536) of the ADC. In terms of the sensor output voltage, $V$, it is defined by Dataq as:

$$dB = 20 \log_{10}(32,768 \times V/FS) \tag{3.6}$$

where FS is the full-scale voltage as determined by the gain setting.

Elsewhere in this chapter, the decibel is calculated with a different reference. For example, for an FFT spectral line having real and imaginary components $R$ and $I$, respectively (voltage based), the intensity in dB is calculated using

$$dB = 20 \log_{10}\left[ \sqrt{R^2 + I^2} \Big/ \left(\frac{n}{2}\right) \right] \tag{3.7}$$

where $n$ is the number of points in the FFT. This is convenient for determining noise levels. For example, from later graphs showing electronics noise, the floor of the SDC sensor is found to be of the order of $-120$ dB, corresponding to a microvolt. The position resolution defined by this noise level is about 500 nm, i.e., the wavelength of visible light.

**TABLE 3.2**  Spectral Intensities for Some of the Lines Shown in Figure 3.30

| Frequency (Hz) | Start of Record (dB) | End of Record (dB) |
|---|---|---|
| 2.19 | 78.3 | 63.0 |
| 1.19 | 68.1 | 55.6 |
| 1.00 | 44.6 | |
| 0.19 | 40.8 | |
| 3.38 | 35.0 | 6.7 |
| 4.34 | 26.7 | |
| 4.53 | 22.4 | |
| 6.53 | 22.9 | |
| 5.53 | 17.8 | |

Of the two primary modes of this kinked-wire case study, the higher frequency (2.19 Hz) is the twisting mode and the lower frequency (1.19 Hz) is the swinging mode. The swinging mode is a little higher frequency than that which would result if the wire were completely flexible, yielding a near simple pendulum (1.02 Hz for 24 cm length). The swinging mode is two dimensional (pendulum equivalent called conical), but the sensor only responds (first order) to motion perpendicular to the long axis of the electrodes. It should also be noted that this motion is attenuated, relative to the twisting response, because of the mechanical common-mode rejection feature discussed in Chapter 2.

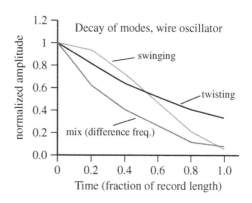

**FIGURE 3.31** Decay of modes of the wire oscillator, determined by the manual STFT.

The manual STFT was used on the data that generated Figure 3.29 to estimate the decay history of three different modes — both of the primary ones (twist and swing) and also the mode whose frequency is the difference between the frequencies of the primaries, i.e., 1 Hz. Figure 3.31 shows the results, where a Hanning window was used, and the total number of points in the record permitted five equally time-spaced FFTs, when working with a 1024 point transform.

Although the decay of the twisting mode is seen to be reasonably exponential, there was large beating between the modes (readily observed in Figure 3.29). Beating alone would not yield a mix signal whose frequency is 1.0 Hz. However, beating in a linear system can cause amplitude variations in the weaker swinging mode.

## 3.7.4 Nonlinear Effects — Mode Mixing

At least two signals in the spectra are the result of nonlinearity, i.e., the lines corresponding to the sum and difference of the frequencies of the primary pair — at 3.38 and 1.00 Hz, respectively. If the system of oscillator and detector were completely linear, then no such sum and difference cases would be possible. It is also to be noted that these mixtures are not the result of sensor nonlinearity, which as noted previously one must be careful to avoid.

The amplitude of a mix signal was expected to approximately obey the following relation:

$$A_m \propto A_1 A_2 \tag{3.8}$$

To test this premise, the STFT was used to estimate the amplitudes of each of the three components indicated in Equation 3.8. The amplitudes were all normalized, relative to the starting value for each case, and the results used to generate the graphs in Figure 3.32.

The amplitude of oscillation for a given mode, at the time of the transform, is found by using the peak value in dB of the intensity of the spectral line for that mode, according to

$$A \propto 10^{dB/20} \tag{3.9}$$

where the factor of 20 is used since the spectral intensities were calculated in terms of voltages. Although calibration constants (in V/m and V/rad) could be used to express the amplitude in meters or in radians, corresponding to the mode, nothing is gained by doing so for the present purposes.

The mixing index for these cases is defined by the expression

$$\text{index} = \frac{A_m}{\sqrt{A_1 A_2}} \tag{3.10}$$

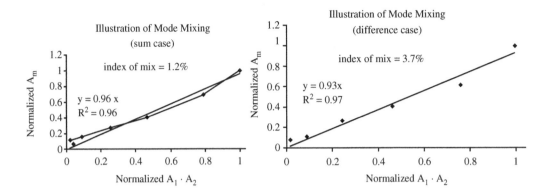

**FIGURE 3.32**   Evidence in support of nonlinear mixing according to Equation 3.8.


which is similar to expressions encountered in optics. It can be seen that the sum and difference frequencies are approximated reasonably well by theoretical expectation.


## 3.8   Internal Friction as Source of Mechanical Noise

Chapter 2 claims that internal friction is responsible not only for damping but also for significant mechanical noise of $1/f$ type. Figure 3.33 is provided in support of that claim.

The pendulum in these experiments (lead spheres near the ends of an aluminum tube with a pair of steel-points for the axis) was operated in a high vacuum to eliminate the influence of air. The electronics

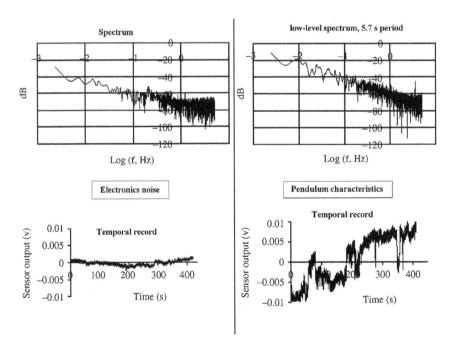

**FIGURE 3.33**   Power spectrum and associated temporal record showing mechanical $1/f$ noise (right pair). For reference, electronics noise is also provided (left pair).

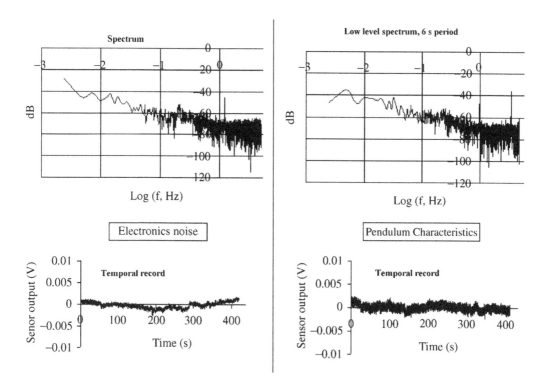

**FIGURE 3.34** Same as Figure 3.33, but after the pendulum had stabilized.

noise was obtained by removing the top mass and measuring the motion after oscillation reached a minimum (frequency approximately 1 Hz); some pendulous mode remains because of pump noise transmitted through the vacuum hose. Similarly, the pump vibrations excite a vibratory mode in the long-period pendulum (sharp spectral line at 3.5 Hz, right spectrum). In this vibratory mode the lead masses move in the same direction relative to the stationary axis (similar to the bending mode of the carbon dioxide molecule). It is interesting to note that there is no coherent oscillation to be seen above noise corresponding to the period of 5.7 sec. The mechanical noise is seen to include bistability, which is not uncommon for this type of system before hardening, following a significant force disturbance. The data in Figure 3.33 were collected after replacing the upper mass, which had been removed to measure the electronics noise, and after pumping to the operating pressure (below 5 $\mu$m Hg).

The mechanical noise is seen to be $1/f$ for $f < 1.5$ Hz, which is where electronics begins to contribute noticeably. Everywhere below 1 Hz, the electronics noise is an order of magnitude smaller than the mechanical noise.

After the pendulum had stabilized overnight and been allowed to oscillate through a number of free-decays (initialization by tilting the chamber), the data shown in Figure 3.34 was collected.

It can be seen that the mechanical noise has mostly settled out, leaving the remnant electronics noise.

## 3.9  Viscous Damping — Need for Caution

Recent experiments have shown important subtleties of viscous damping (Peters, 2003a, 2003b, 2003c, 2003d). It is true that the dissipation at a specified frequency can be adequately modeled by simply multiplying the velocity term in the differential equation by a coefficient. It is not proper, however, to call this coefficient a constant, since the damping coefficient is frequency dependent and also involves the density as well as the viscosity of the fluid in which oscillation takes place.

Some engineers have known about the history term, which is most simply treated in the case of a sphere executing simple harmonic motion. The friction force acting on the sphere in this case can be reasonably approximated by

$$f_{harmonic} = 6\pi\eta a\left(1 + C_H\frac{a}{\delta}\right)v, \quad \delta = \sqrt{\frac{2\eta}{\omega\rho}}, \quad (C_H \rightarrow 1 \text{ as } v \rightarrow 0) \tag{3.11}$$

where $\omega$ is the angular frequency of oscillation, $a$ is the radius of the sphere; and for the fluid, $\eta$ and $\rho$ are its viscosity and density, respectively. Only in the limit of zero frequency does the damping reduce to the form that one expects on the basis of Stokes' law of viscous friction (steady flow).

Using Equation 3.11 in the equation of motion for a pendulum yields for the $Q$

$$Q_v = \frac{I\omega}{6\pi\eta a\left(1 + \frac{a}{\delta}\right)L^2}, \quad \delta = \sqrt{\frac{2\eta}{\omega\rho}} \tag{3.12}$$

where $I$ is the moment of inertia, and $L$ is the distance from the axis to the center of the sphere. Typically, the ratio $a/\delta$ is significantly greater than unity so that the damping is governed by the surface area of the sphere rather than by its radius. These complexities of viscous damping are summarized in Box 3.3.

Reasonable experimental validation of the estimate for $Q$ was provided, as demonstrated in Figure 3.35.

The instrument in this case was a compound pendulum in which a mass was located above the axis of rotation, as well as the usual situation of mass below the axis. The water damping was provided through a small sphere at the bottom of the pendulum, immersed in water held by a rectangular container.

If the history term in Equation 3.12 is ignored there can be huge errors. For example, in the case of water damping, the damping can be underestimated by 1000 to 3000%, as shown in Figure 3.36.

At low frequencies, it is also important to correct for the influence of hysteretic damping of the pendulum. Figure 3.37 shows the large errors that occur when one fails to do so.

For some cases, buoyancy and added mass of the fluid are also quite significant to the frequency of oscillation, as shown in Figure 3.38.

# Box 3.3

## COMPLEXITIES OF VISCOUS DAMPING

Friction force is not a function only of viscosity $\eta$; it also depends on density $\rho$ and angular frequency $\omega$

$$f_{harmonic} = 6\pi\eta a\left(1 + C_H\frac{a}{\delta}\right)v, \quad \delta = \sqrt{\frac{2\eta}{\omega\rho}}, \quad (C_H \rightarrow 1 \text{ as } v \rightarrow 0)$$

resulting in a complicated frequency dependence for the $Q$ of viscous damping

$$Q_v = \frac{I\omega}{6\pi\eta a\left(1 + \frac{a}{\delta}\right)L^2}, \quad \delta = \sqrt{\frac{2\eta}{\omega\rho}}$$

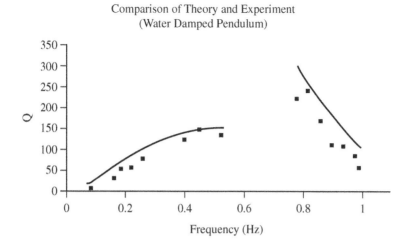

**FIGURE 3.35** Comparison of theory and experiment for a pendulum damped by water.

# 3.10 Air Influence

As seen from Figure 3.37, low-frequency motions are likely to be influenced more by internal friction than by any fluids that interact with the oscillator. The most important fluid is of course air, and a true delineation between external and internal effects requires that the oscillator be studied in a high vacuum. It is not enough to just remove most of the air, since the viscosity of gases is surprisingly constant until the mean free path between collisions becomes a significant fraction of chamber dimensions.

Theoretically, it is possible to roughly estimate air influence, although only in the simplest of geometries, such as a sphere. In such cases, Equation 3.11 could be used (with accounts for the history term, using appropriate values for the viscosity and density). It is also possible in some cases to estimate air influence experimentally, as in the example that follows.

## 3.10.1 Brass and Solder Rod Pendula

Because of its malleability, the internal friction of solder (lead–tin alloy) is large, compared to that of much harder brass. A pendulum of each material was studied, both having a length of about 50 cm and a

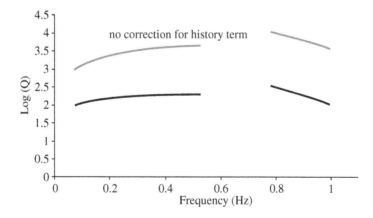

**FIGURE 3.36** Illustration of how huge errors can occur in damping estimates if one ignores the history term.

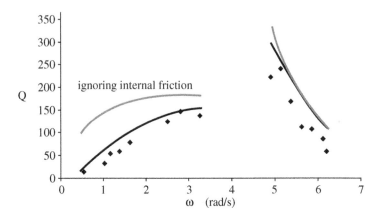

**FIGURE 3.37**  Illustration of significant low-frequency errors that result from a failure to recognize the hysteretic damping component of the pendulum.

diameter of about 3 mm. The technique used was the photogate method described in Section 3.4.4 (Case 4 above). Unlike the previous study, no lead masses were clamped on the rod — but it used the same adjustable knife-edge.

Figure 3.39 clearly shows that the internal friction for the solder pendulum is much greater than that of the brass pendulum.

A nonlinear fit was generated for each decay curve, from which the history of the quality factor was graphed as a function of velocity amplitude, as shown in Figure 3.40.

Consider the pair of brass curves in Figure 3.40. The large difference in $Q$ at 10 cm/sec (387 compared to 266) is in stark contrast with their near equality at 50 cm/sec. This is primarily a consequence of air drag that is quadratic in the velocity at the larger amplitude. It is more important to brass than to solder because of the small internal friction of the brass.

From the large difference in internal friction of the two materials, a first order correction for air influence on the solder pendulum is to simply subtract $1/Q$ of the brass from $1/Q$ (raw data) of the solder, to yield the reciprocal $Q$ (corrected) due to internal friction of the solder. This has been done in Figure 3.41.

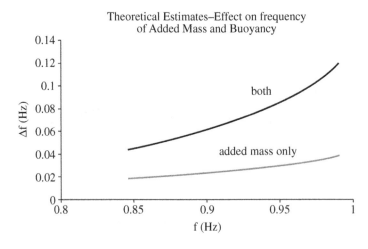

**FIGURE 3.38**  Example of how fluid properties influence the frequency as well as damping of an oscillator.

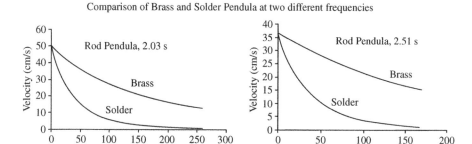

FIGURE 3.39 Free-decay curves for brass and solder pendula at two different frequencies, showing the larger internal friction of solder. The velocity is that of the peak value (amplitude) at the top of the pendulum, approx. 22 cm above the axis.

From Figure 3.41 it can be seen that the internal friction damping is not simply hysteretic (constant $Q$); rather it is a function of amplitude. It can also be seen, from the close proximity of the solid and dashed curves, that the air influence on the solder pendulum is much less than that of the internal friction. By contrast, air influence is of comparable magnitude to the internal friction in the case of the brass pendulum (or even larger, at large amplitude).

A minimum of two frequencies was considered for the study, since the frequency variation of the damping is different for external and internal frictions. (Note: although the period is a function of amplitude, the amount of nonisochronism is small compared to the damping changes and is ignored here.) The periods were matched for the two pendula at each of 2.03 and 2.51 sec. For hysteretic-only (internal friction) damping, the $Q$ at the shorter period should in theory be 1.53 times that of the longer period, for both brass and solder. If the damping were viscous only, the factor should be 1.24. In the case of solder at 10 cm/sec (corrected), the ratio is $1.66 = 131/71$, and for brass it is $1.46 = 387/266$. Although the ratio for solder is greater than the expected 1.53, the difference is within experimental

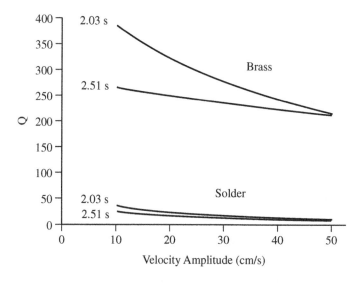

FIGURE 3.40 Illustration of amplitude-dependent damping in a rod pendulum made of (i) brass and (ii) solder. The two different matched periods of oscillation are indicated in seconds.

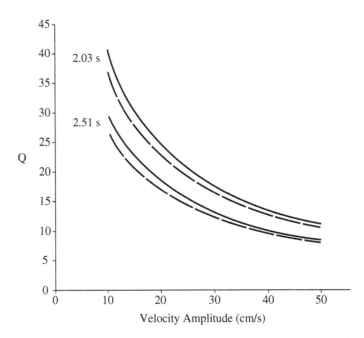

**FIGURE 3.41** Amplitude dependence of the estimated quality factor due only to internal friction in the solder pendulum. Dashed lines show the Q before correction for air damping.

uncertainty for individual Q values, which from other, more detailed experiments were in the neighborhood of 5 to 10%.

The ratio for brass (1.46) is between 1.24 and 1.53, as expected, because of the comparable influence of air and internal friction.

## References

Barkhausen, H., *Z. Phys.*, 20, 401, 1919.

Cromer, A., Stable solutions using the Euler approximation, *Am. J. Phys.*, 49(5), 455–459, 1981.

de Silva, C.W. 2006. *Vibration, Fundamentals and Practice*, 2nd ed., Taylor & Francis, CRC Press, Boca Raton, FL.

Fraden, J. 1996. *Handbook of Modern Sensors, Physics, Designs, & Applications*, 2nd ed., AIP Press (Springer), Secaucus, NJ.

Peters, R., Fourier transform construction by vector graphics, *Am. J. Phys.*, 60, 439, 1992.

Peters, R., Mercer physics laboratory experiment, described online at http://physics.mercer.edu/labs/sho.htm, 1998.

Peters, R., Tutorial description of the symmetric differential capacitive (SDC) sensor, online at http://www.telatomic.com/sdct2.html, 2002.

Peters, R., Graphical explanation of the speed of the Fast Fourier Transform, online at http://arxiv.org/html/math.HO/0302212, 2003a.

Peters, R., Nonlinear damping of the "linear" pendulum, online at http://arxiv.org/pdf/physics/03006081, 2003b.

Peters, R., Oscillator damping with more than one mechanism of internal friction dissipation, online at http://arxiv.org/html/physics/0302003/, 2003c.

Peters, R., Anisotropic internal friction damping, online at http://arxiv.org/html/physics/0302055, 2003d.

Peters, R., Folded pendulum measurements of the Earth's free oscillations, online at http://physics.mercer.edu/petepag/eigen.html, 2004.

Portevin, A. and Le Chatelier, M., Tensile tests of alloys undergoing transformation, *C. R. Acad. Sci.*, 176, 507, 1923.

Press, W., Flannery, B., Teukolsky, S., and Vetterling, W. 1986. *Numerical Recipes — The Art of Scientific Computing*, Cambridge University Press, Cambridge.

Wielandt, E., *Seismometry*, section: Electronic Displacement Sensing, online at http://www.geophys. uni-stuttgart.de/seismometry/hbk_html/node1.html, 2001.

# 4

# Structure and Equipment Isolation

**Y.B. Yang**
*National Taiwan University*

**L.Y. Lu**
*National Kaohsiung First University of Science and Technology*

**J.D. Yau**
*Tamkang University*

## Summary

*In this chapter, a brief review will be given of the concept of isolation for suppressing the vibrations in structures and equipment subjected either to harmonic or seismic ground excitations. The mechanism of various isolation devices, including the elastomeric bearing, sliding bearing, resilient-friction base isolator, and Electricite de France system, will first be described in Section 4.2, together with their mathematical models. In Section 4.3, a closed form solution will be derived for the dynamic response of a structure–equipment system isolated by bearings of the elastomeric type, subjected to harmonic motions. Such a solution enables us to interpret the various behaviors of the structure and equipment under excitation. The elastomeric bearings can help increase the fundamental period of the structure, thereby, reducing the accelerations transmitted to the superstructure. In Section 4.4 and Section 4.5, the seismic behavior of a structure–equipment system isolated by a sliding support, with and without resilient force, will be studied using a state-space incremental-integration approach. With the introduction of a frictional sliding interface, the motion of the structure–equipment system will be uncoupled from the ground excitation, and the influence of the latter will be mitigated. The residual base displacement caused by the sliding isolator can be reduced through inclusion of a resilient mechanism in the isolator. Nevertheless, the resilient mechanism can make the system more sensitive to the low-frequency components of excitation. In Section 4.6, issues related to design of base isolators will be discussed, along with the concepts underlying some design codes and guidelines. The notation used is listed at the end of the chapter.*

## 4.1  Introduction

Conventionally, structural designers are concerned about the safety of buildings, bridges, and other civil engineering structures that are subjected to earthquakes. The recent history of earthquakes reveals that strong earthquakes, such as the 1994 Northridge earthquake (U.S.A.), 1995 Kobe earthquake (Japan), and 1999 Chi-Chi earthquake (Taiwan), can cause some badly designed structures or buildings to fail or collapse, and also cause some well-designed structures to malfunction due to the damage or failure of the equipment housed in the structure or building. Both the failures of structures and equipment, also known as *structural* and *nonstructural* failures, respectively, can cause serious harm to the residents or personnel working in a building. For the case where the equipment is part of a key service system, such as in hospitals, power stations, telecommunication centers, high-precision factories, and the like, the lives and economic losses resulting from the malfunctioning of the equipment can be tremendous. Thus, the maintenance of the safety of structures and attached equipment during a strong earthquake is a subject of high interest in earthquake engineering. In this regard, *base isolation* has been proved to be an effective means for protecting the structures and attached equipment, which is made possible through reduction of the seismic forces transmitted from the ground to the superstructure (Yang et al., 2002).

For light secondary systems mounted on heavier primary systems, it was concluded that the response of the light secondary system, that is, the equipment, is affected by four major dynamic characteristics in earthquakes (Igusa and Der Kiureghian, 1985a, 1985b, 1985c; Yang and Huang, 1993). The first issue is *tuning*, which means that the natural frequency of the equipment is coincident with that of the structure. Such an effect may amplify the response of the equipment due to the fact that the light secondary system behaves as if it were a vibration absorber of the heavier primary system. The second issue is *interaction*, which is related to the feedback effect between the motions of the primary and secondary systems. Ignoring the feedback effect of interaction may result in an overestimation of the true response of the combined system. The third issue is *non-classical damping*, which may occur when the damping properties of the two systems are drastically different, such that the natural frequencies and mode shapes of the combined system can only be expressed in terms of complex numbers. Under such a circumstance, the conventional response spectrum analysis, based on modal superposition, becomes inapplicable. The last issue is *spatial coupling*, which relates to the effect of multiple support motions when the secondary system of interest is mounted at multiple locations. By considering the inelastic effect, Igusa (1990) proposed an equivalent linearization technique for investigating the response characteristics of an inelastic primary–secondary system with two degrees of freedom (DoF) under random vibrations. His results indicated that the existence of small nonlinearity is helpful for reducing the coupling system responses. With the concept of equivalent linearization, Huang et al. (1994) explored the response and reliability of a linear secondary system mounted on a yielding primary structure under white-noise excitations. It was concluded that the response of the secondary system could be reduced by increasing the equipment damping or by locating equipment at higher levels of the primary structure.

Owing to the fact that the mass and stiffness of a secondary system are much smaller than those of the primary structure, the interaction effect of the combined system, as well as the ill-conditioning in system matrices, may take place when one performs the dynamic analysis. To deal with this problem, some researchers chose to evaluate the response of the secondary systems from the floor motions. To avoid solving large eigenvalue problems and to account for the interaction between the building and equipment components, Villaverde (1986) applied the response spectrum technique to the analysis of a combined building–equipment system, by which the maximum response of light equipment mounted on the building under the earthquake is expressed in terms of the natural frequencies and mode shapes of the building and equipment. To take into account the equipment–structure interactions, Suarez and Singh (1989) proposed an analytical scheme for computing the dynamic characteristics of the combined system, using the modal properties to compute the floor spectra. Lai and Soong (1990) presented a statistical energy analysis technique for evaluating the response of coupling primary–secondary structural systems, based on the concept of power-balance equation, that is, the power input to the primary system is equal to the dissipated energy of the primary system plus the transferred energy to the secondary system. Using a mean-square condensation procedure, Chen and Soong (1994) considered the effect of interaction by calculating the multi-DoF response of a primary–secondary system under random excitations. Later on, Chen and Soong (1996) derived an exact solution for the mean-square response of a structure–equipment system under dynamic loads, indicating that there exists an optimal damping ratio for reducing the vibration of equipment attached to the primary structure. Gupta and coworkers investigated the response of a secondary system with multiple supports on a primary structure subjected to earthquakes, taking into account the interaction effect between the equipment and structure (Dey and Gupta, 1998, 1999; Chaudhuri and Gupta, 2002). Their results indicated that when the soil–structure interaction (SSI) is taken into account, the response of the equipment–structure system will be affected by the SSI, unless a very stiff soil condition is considered.

On the other hand, a number of research works have been conducted by implementing isolation systems at the base of the equipment–structure system, aiming to reduce the earthquake forces transmitted from the ground. Based on a theoretical and experimental investigation, Kelly and Tsai (1985) showed that seismic protection can be achieved effectively for lightweight equipment mounted on an isolated structure installed with elastic bearings at the base. A hybrid isolation system with base-isolated floors was proposed by Inaudi and Kelly (1993), for the protection of highly sensitive devices mounted on a structure subjected to support motions. Considering the effects of torsion and translation, Yang and Huang (1998) studied the seismic response of light equipment items mounted on torsional buildings supported by elastic bearings. Their results indicated that the response of an equipment–structure system can be effectively reduced through installation of base isolators, and that there exists an optimal location for mounting the equipment. Juhn et al. (1992) presented a series of experimental results for the secondary systems mounted on a sliding base-isolated structure. They concluded that the acceleration response of the secondary system may be amplified when the input motions are composed of low-frequency vibrations. In this case, the sliding bearings are not considered to be an effective isolation device, which implies that the base-isolated structure is not suitable for a construction site with soft soil.

Concerning the use of sliding bearings (supports) as base isolators, Lu and Yang (1997) investigated the response of an equipment item attached to a sliding primary structure under earthquake excitations. Their results showed that the response of the equipment can be effectively reduced through the installation of a sliding support at the structural base, in comparison with that of a structure with fixed base. To overcome the discontinuous nature of the sliding and nonsliding phases of a structural system with sliding base, a fictitious spring model was proposed by Yang and coworkers for simulating the mechanism of sliding and nonsliding (Yang et al., 1990, 2000; Yang and Chen, 1999). Such a model will be described in a later section of this chapter. Agrawal (2000) adopted the same fictitious spring model in studying the response of an equipment item mounted on a torsionally coupled structure with sliding support. His results indicated that sliding supports could effectively reduce the equipment response, compared to that of a fixed-base structure. However, in the tuning region, where the natural frequency of the equipment coincides with the fundamental frequency of the structure, the equipment response may be adversely amplified due to the increase in eccentricity of the torsionally coupled structure.

The problem of building isolation has recently received more attention than ever from researchers and engineers, due to the construction of high-precision factories worldwide. More and more stringent requirements have been employed in this regard for removing the ambient or man-made vibrations (Rivin, 1995; Steinberg, 2000). To allow sensitive electronic equipment to operate in a harsh environment, Veprik and Babitsky (2000) proposed an optimization procedure for the design of vibration isolators aimed at minimizing the response of the internal components of electronic equipment. As for the protection of high-tech equipment from micro- or ambient vibrations, Yang and Agrawal (2000) showed that passive hybrid floor isolation systems are more effective in mitigating the equipment response than passive or hybrid base isolation systems. Xu and coworkers studied the response of a batch of high-tech equipment mounted on a hybrid platform, which in turn is mounted on a building floor (Xu et al., 2003; Yang et al., 2003). Both their theoretical and experimental studies showed that the hybrid platform, which is composed of leaf springs, oil dampers, and an electron-magnetic actuator with velocity feedback control, is more effective in mitigating the velocity response of the high-tech equipment than the passive platform.

The objective of this chapter is to give an overview on the seismic behavior of various base isolators. The organization of this chapter can be summarized as follows. In Section 4.2, the mechanisms of various seismic isolators that are currently in use are introduced and explained. In Section 4.3, a structure–equipment system isolated by bearings of the elastomeric type is modeled by a three-DoF system composed of a spring and dashpot unit, for which a closed-form solution is obtained for the dynamic response of the isolated system subjected to harmonic earthquakes; remarks on the dynamic response of the system components are also made. In Section 4.4 and Section 4.5, the seismic behaviors of a structure–equipment system isolated by a sliding support, with and without resilient capability, will be investigated. Also presented are numerical methods based on the incremental-integration procedure for the analysis of structural systems with sliding-type isolators.

## 4.2   Mechanisms of Base-Isolated Systems

Figure 4.1 shows a simplified model for a structural system subjected to a support motion.

For this single-DoF system, the equation of motion can be written as

$$m\ddot{x} + c\dot{x} + kx = -m\ddot{x}_g \qquad (4.1)$$

where $m$ denotes the mass, $c$ the damping, $k$ the stiffness, $x$ the displacement of the system, and $\ddot{x}_g$ the ground acceleration. By assuming the system to be linearly elastic, the response $x(t)$ can be obtained using Duhamel's integral, as

$$x(t) = \frac{1}{\Omega_d} \int_0^t \ddot{x}_g(\tau) e^{-\zeta\Omega(t-\tau)} \sin \Omega_d(t-\tau) d\tau \qquad (4.2)$$

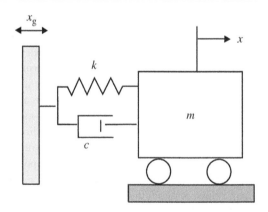

**FIGURE 4.1**   Model of a single-DoF system.

where the natural angular frequency, $\Omega$, damped natural frequency, $\Omega_d$, and damping ratio, $\zeta$, of the system are defined as follows:

$$\Omega = \sqrt{\frac{k}{m}} \qquad (4.3a)$$

$$\Omega_d = \Omega\sqrt{1 - \zeta^2} \qquad (4.3b)$$

$$\zeta = \frac{c}{2m\Omega} \qquad (4.3c)$$

Correspondingly, the natural period, $T$, and damped period, $T_d$, of the structure are

$$T = \frac{2\pi}{\Omega} = 2\pi\sqrt{\frac{m}{k}} \qquad (4.4a)$$

$$T_{\mathrm{d}} = \frac{2\pi}{\Omega_{\mathrm{d}}} = \frac{T}{\sqrt{1 - \zeta^2}} \qquad (4.4\mathrm{b})$$

For a given support acceleration, $\ddot{x}_{\mathrm{g}}$, the displacement, $x$, and acceleration, $\ddot{x}$, of the single-DoF system can be related to the natural period, $T$, and damping ratio, $\zeta$, of the system. Thus, for a specific earthquake, by first selecting a damping ratio, $\zeta$, and using Equation 4.2, one can compute the peak displacement $x$, for a structure with a period of vibration, $T$, with given values of $m$, $c$, and $k$. Repeating the above procedure for a wide range of periods, $T$, while keeping the damping ratio, $\zeta$, constant, one can obtain response curves similar to those shown in Figure 4.2. By varying the damping ratio, $\zeta$, one can construct the *displacement response spectra* and *pseudo-acceleration response spectra* for all single-DoF structures under a given earthquake, as schematically shown in Figures 4.2 and 4.3, respectively.

A general impression that is gained from Figure 4.2 and Figure 4.3 is that a structure with a shorter natural period has less displacement response when subjected to an earthquake, but it also has a larger acceleration response. Specifically, let us consider a structure of a constant damping ratio, $\zeta$, with its period increased from $T_1$ to $T_2$. As can be observed from the figures, the displacement of

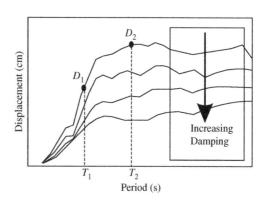

**FIGURE 4.2** Schematic of displacement response spectra.

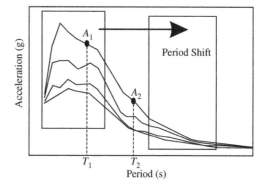

**FIGURE 4.3** Schematic of pseudo-acceleration response spectra.

the structure increases from $D_1$ to $D_2$, while the acceleration decreases from $A_1$ to $A_2$. Such a feature is known as the *period shift* effect. On the other hand, by increasing damping ratio of the structure, the displacement of the structure decreases significantly, as can be seen from Figure 4.2. The same is also true with the acceleration response, as can be seen from Figure 4.3. Moreover, a larger damping ratio also makes the structure less sensitive to the variation in ground vibration characteristics, as indicated by the smoother response curves for structures having higher damping ratios, in both figures. From the aforementioned two response spectra, one observes that the philosophy of base isolation is to lengthen the vibration period of the structure to be protected, using base isolators of some kind, by which the earthquake force transmitted to the structure can be greatly reduced. In the meantime, some additional damping must be introduced on the base isolators in order to control the relative displacements across the base isolators with tolerable limits.

To fulfill the function of lengthening the period of vibration of the structure to be protected, the base isolators that are inserted between the structure and its foundation must be flexible in the horizontal direction, but stiff enough in the vertical direction so as to carry the heavy loads transmitted from the superstructure. With such devices, the natural period of vibration of the structure will be significantly lengthened and shifted away from the dominant frequency range of the expected earthquakes. The following is a summary of the fundamental features of four types of isolators frequently used in engineering practice.

## 4.2.1 Elastomeric Isolation System

*Elastomeric bearing* is the type of base isolator most commonly known to researchers and engineers working on base isolation. It is usually composed of alternating layers of steel and hard rubber and, for this reason, it is also known as the *laminated rubber bearing*. This type of bearing is stiff enough to sustain the vertical loads, yet flexible under the lateral forces. The ability to deform horizontally enables the bearing to reduce significantly the structural base shear transmitted from the ground. While the major function of elastomeric bearings is to reduce the transmission of shear forces to the superstructure by lengthening the vibration period

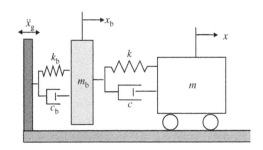

**FIGURE 4.4** Model for base-isolation systems with elastic bearing.

of the entire system, they must also provide sufficient rigidity under vertical loads. Let us consider a structure installed with elastomeric bearings, which is subjected to a support acceleration, $\ddot{x}_g$, as in Figure 4.4. By representing the isolated structure as a single-DoF system, based on the assumption that the superstructure is rigid in comparison with the stiffness of the elastic bearings, the equation of motion for the entire system can be written as

$$\begin{bmatrix} m & 0 \\ 0 & m_b \end{bmatrix} \begin{Bmatrix} \ddot{x} \\ \ddot{x}_b \end{Bmatrix} + \begin{bmatrix} c & -c \\ -c & c+c_b \end{bmatrix} \begin{Bmatrix} \dot{x} \\ \dot{x}_b \end{Bmatrix} + \begin{bmatrix} k & -k \\ -k & k+k_b \end{bmatrix} \begin{Bmatrix} x \\ x_b \end{Bmatrix} = -\begin{Bmatrix} m\ddot{x}_g \\ m_b\ddot{x}_g \end{Bmatrix} \tag{4.5}$$

where $m$, $c$, and $k$ denote the mass, damping, and stiffness of the superstructure, respectively, $m_b$, $c_b$, and $k_b$ denote the mass, damping, and stiffness of the base raft, respectively, and $x$ and $x_b$ denote the displacements of the superstructure and the base, respectively.

In reality, the reduction in the seismic forces transmitted to a superstructure through the installation of laminated rubber bearings is achieved at the expense of large relative displacements across the bearings. If substantial damping can be introduced into the bearings or the isolation system, then the problem of large displacements can be alleviated. It is for this reason that the laminated rubber bearing with a central lead plug inserted has been devised (Yang et al., 2002). To simulate the dynamic properties of the lead–rubber bearing (LRB) system, an equivalent linearized system has been proposed, for which the equation of motion is

$$\begin{bmatrix} m & 0 \\ 0 & m_b \end{bmatrix} \begin{Bmatrix} \ddot{x} \\ \ddot{x}_b \end{Bmatrix} + \begin{bmatrix} c & -c \\ -c & c+c_{eq} \end{bmatrix} \begin{Bmatrix} \dot{x} \\ \dot{x}_b \end{Bmatrix} + \begin{bmatrix} k & -k \\ -k & k+k_{eq} \end{bmatrix} \begin{Bmatrix} x \\ x_b \end{Bmatrix} = -\begin{Bmatrix} m\ddot{x}_g \\ m_b\ddot{x}_g \end{Bmatrix} \tag{4.6}$$

where $c_{eq}$ and $k_{eq}$ respectively represent the equivalent linearized damping and stiffness coefficients of the LRB system. The dynamic behavior of a structure–equipment system isolated by elastomeric bearings with linearized damping and stiffness coefficients, when subjected to harmonic and earthquake excitations, will be investigated analytically and numerically, respectively, in Section 4.3.

## 4.2.2 Sliding Isolation System

Another means for increasing the horizontal flexibility of a base-isolated structure is to insert a *sliding* or *friction surface* between the foundation and the base of the structure. The shear force transmitted to the superstructure through the sliding interface is limited by the static frictional force, which equals the product of the coefficient of friction and the weight of the superstructure. The coefficient of friction is usually kept as low as is practical. However, it must be high enough to provide a frictional force that can sustain strong winds and minor earthquakes without sliding. Since the sliding system has no dominant natural period, it is generally frequency-independent when the structure is subjected to earthquakes with a wideband frequency content. As mentioned previously, when a sliding structure is subjected to a ground motion, transitions may occur repeatedly between the sliding and nonsliding phases. To take into account such a phase transition, Yang et al. (1990) proposed the use of a fictitious spring between the structural base raft and the underlying ground to simulate the static–dynamic frictional force of the sliding device. With reference to Figure 4.5, the equation of motion for the structure with sliding base can be written as follows:

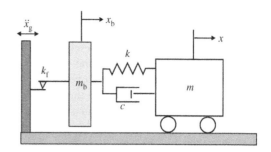

**FIGURE 4.5** Model for base-isolation systems with sliding support.

$$\begin{bmatrix} m & 0 \\ 0 & m_b \end{bmatrix} \begin{Bmatrix} \ddot{x} \\ \ddot{x}_b \end{Bmatrix} + \begin{bmatrix} c & -c \\ -c & c \end{bmatrix} \begin{Bmatrix} \dot{x} \\ \dot{x}_b \end{Bmatrix} + \begin{bmatrix} k & -k \\ -k & k \end{bmatrix} \begin{Bmatrix} x \\ x_b \end{Bmatrix} + \begin{Bmatrix} 0 \\ f_r \end{Bmatrix} = -\begin{Bmatrix} m\ddot{x}_g \\ m_b\ddot{x}_g \end{Bmatrix} \quad (4.7)$$

where $k_f$ is the stiffness of the fictitious spring and the frictional force, $f_r$, can be represented as

$$f_r = \begin{cases} k_f(x_b - x_{b0}) & \text{for non-sliding phase,} \\ \pm\mu(m + m_b)g & \text{for sliding phase} \end{cases} \quad (4.8)$$

with $x_{b0}$ indicating the initial elongation of the fictitious spring in the current nonsliding phase, $\mu$ the coefficient of friction, and $g$ the acceleration of gravity. The fictitious spring concept will be incorporated in the analysis of sliding structures in Section 4.4 of this chapter, when considering both harmonic and seismic excitations.

## 4.2.3 Sliding Isolation System with Resilient Mechanism

One particular problem with a sliding structure is the occurrence of residual displacements after earthquakes. To remedy such a drawback, the sliding surface is often made concave, so as to provide a recentering mechanism for the isolated structures. This is the idea behind the *friction pendulum system* (FPS), shown in Figure 4.6, which utilizes a spherical concave surface to produce a recentering force for the superstructure under excitations. To guarantee that a sliding structure can return to its original position, other mechanisms, such as high-tension springs and elastomeric bearings, can be used as an auxiliary system for providing the restoring forces. Previously, the sliding isolation systems have been successfully applied in the protection of important structures, such as nuclear power plants, emergency fire water tanks, large chemical storage tanks, and so on, from the damaging actions of severe earthquakes.

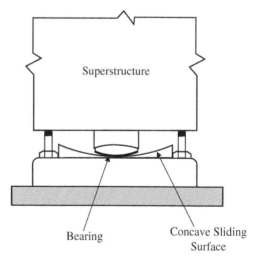

**FIGURE 4.6** Friction pendulum system.

To improve the performance of sliding isolators under strong earthquakes, Mostaghel (1984) and Mostaghel and Khodaverdian (1987) proposed the *resilient-friction base isolator* (RFBI) for controlling the transmission of shear force to the superstructures, while keeping the residual displacements within an allowable level. The RFBI device is basically made of a central rubber core and Teflon-coated steel plates, and offers a friction resistance for keeping the system in the nonsliding mode under wind excitations and small earthquakes, and a restoring force by the rubber ingredient for limiting the maximum sliding displacements. The equation of motion for a structure installed with RFBI, as shown in Figure 4.7, can be written as

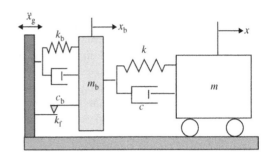

**FIGURE 4.7**  Model for base-isolation systems with RFBI device.

$$
\begin{bmatrix} m & 0 \\ 0 & m_b \end{bmatrix} \begin{Bmatrix} \ddot{x} \\ \ddot{x}_b \end{Bmatrix} + \begin{bmatrix} c & -c \\ -c & c + c_b \end{bmatrix} \begin{Bmatrix} \dot{x} \\ \dot{x}_b \end{Bmatrix} + \begin{bmatrix} k & -k \\ -k & k + k_b \end{bmatrix} \begin{Bmatrix} x \\ x_b \end{Bmatrix} + \begin{Bmatrix} 0 \\ f_r \end{Bmatrix} = - \begin{Bmatrix} m\ddot{x}_g \\ m_b\ddot{x}_g \end{Bmatrix} \quad (4.9)
$$

The interfacial frictional force, $f_r$, existing in the RFBI and appearing in Equation 4.9 serves as the outlet for energy dissipation. The behavior of a structure–equipment system supported by sliding isolators with resilient mechanism subjected to both harmonic and earthquake excitations will be investigated in Section 4.5.

### 4.2.4  Electricite de France System

To limit effectively the acceleration of base-isolated structures and internal secondary systems, such as those of nuclear power plants, when subjected to strong earthquakes, the *Electricite de France* (EDF) system was proposed by Gueraud et al. (1985). The design concept of an EDF system is to arrange the elastomeric bearing and sliding device at the base of a structure in series. For low-level ground motions, the EDF system will behave as an elastomeric bearing and return to the original position after support motions, while for strong earthquakes, the EDF system will behave as a sliding device. The EDF system may have a residual displacement after some major earthquakes. Because of the sliding mechanism of the EDF system, the maximum horizontal acceleration of the superstructure is kept within a certain range (Gueraud et al., 1985; Park et al., 2002), while the shear force transmitted to the superstructure through the frictional interface is smaller than the static frictional force. For the mathematical model shown for the EDF system in Figure 4.8, the equations of motion for the nonsliding and sliding phases can be written as

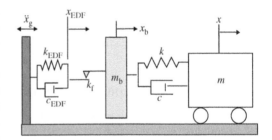

**FIGURE 4.8**  Model for base-isolation systems with EDF device.

(a)  nonsliding phase:

$$
\begin{Bmatrix} m\ddot{x} \\ m_b\ddot{x}_b \\ 0 \end{Bmatrix} + \begin{bmatrix} c & -c & 0 \\ -c & c & 0 \\ 0 & 0 & c_{EDF} \end{bmatrix} \begin{Bmatrix} \dot{x} \\ \dot{x}_b \\ \dot{x}_{EDF} \end{Bmatrix} + \begin{bmatrix} k & -k & 0 \\ -k & k + k_f & -k_f \\ 0 & -k_f & k_f + k_{EDF} \end{bmatrix} \begin{Bmatrix} x \\ x_b \\ x_{EDF} \end{Bmatrix}
$$

$$
= \begin{Bmatrix} -m\ddot{x}_g \\ -m_b\ddot{x}_g \\ 0 \end{Bmatrix} \quad (4.10a)
$$

(b) sliding phase:

$$\left\{\begin{array}{c} m\ddot{x} \\ m_b\ddot{x}_b \\ 0 \end{array}\right\} + \left[\begin{array}{ccc} c & -c & 0 \\ -c & c & 0 \\ 0 & 0 & c_{EDF} \end{array}\right]\left\{\begin{array}{c} \dot{x} \\ \dot{x}_b \\ \dot{x}_{EDF} \end{array}\right\} + \left[\begin{array}{ccc} k & -k & 0 \\ -k & k & 0 \\ 0 & 0 & k_{EDF} \end{array}\right]\left\{\begin{array}{c} x \\ x_b \\ x_{EDF} \end{array}\right\}$$

$$= \left\{\begin{array}{c} -m\ddot{x}_g \\ -m_b\ddot{x}_g \mp \mu(m + m_b)g \\ \pm \mu(m + m_b)g \end{array}\right\} \qquad (4.10b)$$

where $c_{EDF}$ and $k_{EDF}$, respectively, denote the damping and stiffness of the EDF system, and $x_{EDF}$ denotes the displacement of the system.

## 4.2.5 Concluding Remarks

To mitigate the transmission of earthquake forces to a structure, and the potentially earthquake-induced damage to the equipment attached to the structure, base isolation is an effective structural design philosophy. With the installation of base isolators, the natural period of vibration of the structure will be significantly lengthened and shifted away from the dominant frequency range of the expected earthquakes. In accordance, the earthquake force transmitted to the structure can be significantly reduced. In this section, the mechanisms of four types of base isolator frequently used in engineering practice are introduced. Since the base isolators, such as the elastomeric bearings or sliding isolations, have relatively flexible stiffness in the horizontal direction, the occurrence of residual displacements after earthquakes may cause certain problems on the structure to be protected. To remedy such a drawback and to further guarantee that a base-isolated structure can return to its original position, the RFBI is implemented for controlling the transmission of shear force to the superstructure, while keeping the residual displacement within an allowable level. On the other hand, to limit the acceleration level of internal secondary systems housed in a base-isolated structure under strong earthquakes, such as those of the nuclear power plants, the EDF system can be used as an alternative device for base isolation, even though some residual displacements may be induced after the earthquakes.

# 4.3 Structure–Equipment Systems with Elastomeric Bearings

Owing to the stringent requirements for normal functioning of high-tech facilities, such as printed circuit boards, semiconductor factories, and sensitive medical devices, the need to suppress excessive vibrations in sensitive structure–equipment systems has become an issue of great concern to structural designers. Besides, these high-tech facilities may suffer significant damages during a major earthquake. Using elastomeric isolation systems to reduce the earthquake forces transmitted from the ground is one of the most popular ways adopted by structural designers. In this section, the performance of elastomeric bearings in protecting structure–equipment systems against horizontal ground motions will be investigated.

## 4.3.1 Formulation of Base Isolation Systems with Elastic Bearing

By modeling the structure, internal equipment and the base of an isolated structure–equipment system as a lumped mass system, one can construct the mathematical model for the structure–equipment isolation system supported by an elastic bearing in Figure 4.9. The following is the

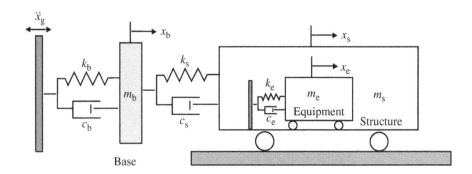

**FIGURE 4.9** Model of a structure–equipment isolation system with elastic bearing.

equation of motion for the base-isolated structure–equipment system when it is subjected to a ground acceleration, $\ddot{x}_g$:

$$
\begin{Bmatrix} m_e \ddot{x}_e \\ m_s \ddot{x}_s \\ m_b \ddot{x}_b \end{Bmatrix} + \begin{bmatrix} c_e & -c_e & 0 \\ -c_e & c_s + c_e & -c_s \\ 0 & -c_s & c_s + c_b \end{bmatrix} \begin{Bmatrix} \dot{x}_e \\ \dot{x}_s \\ \dot{x}_b \end{Bmatrix} + \begin{bmatrix} k_e & -k_e & 0 \\ -k_e & k_s + k_e & -k_s \\ 0 & -k_s & k_s + k_b \end{bmatrix} \begin{Bmatrix} x_e \\ x_s \\ x_b \end{Bmatrix}
$$
$$
= -\begin{Bmatrix} m_e \\ m_s \\ m_b \end{Bmatrix} \ddot{x}_g
\tag{4.11}
$$

where $m$ represents the mass, $c$ the damping coefficient, and $k$ the stiffness of the system. Also, the subscripts 'e', 's', and 'b' are associated with the DoF of the equipment, structure, and base, respectively. The notations employed in Figure 4.9 have been defined in Table 4.1. It should be mentioned that the elastic bearing stiffness, $k_b$, appearing in Equation 4.11, is a parameter relating to the boundary conditions of the system considered here. A small value of $k_b$ relative to the structural stiffness, $k_s$, means that the system is isolated by a set of soft bearings. In contrast, a large value of $k_b$ means that the structure is rigidly supported.

**TABLE 4.1**    Definition of Symbols

| Symbol | Definition |
|---|---|
| $c_e, c_s$ | Damping coefficients of equipment and superstructure |
| $k_e, k_s$ | Stiffness of equipment and superstructure |
| $k_b$ | Stiffness of elastic bearing or resilient stiffness of isolation system |
| $m_e, m_s, m_b$ | Masses of equipment, superstructure and base mat |
| $x_e(t), x_s(t), x_b(t)$ | Relative-to-the-ground displacements of equipment, superstructure and base mat |
| $\ddot{x}_g(t)$ | Ground acceleration |
| $\mu$ | Frictional coefficient of sliding isolation system |
| $\omega_b$ | Frequency of isolation system |
| $\omega_e, \omega_s$ | Frequencies of equipment and superstructure |
| $\omega_g$ | Frequency of ground excitation |
| $\zeta_e, \zeta_s$ | Damping ratios of equipment and structure |

### 4.3.2 Free Vibration Analysis

By neglecting the damping and forcing terms in Equation 4.11, the equation of motion for free vibration can be written as

$$
\begin{Bmatrix} m_e \ddot{x}_e \\ m_s \ddot{x}_s \\ m_b \ddot{x}_b \end{Bmatrix} + \begin{bmatrix} k_e & -k_e & 0 \\ -k_e & k_s + k_e & -k_s \\ 0 & -k_s & k_s + k_b \end{bmatrix} \begin{Bmatrix} x_e \\ x_s \\ x_b \end{Bmatrix} = \begin{Bmatrix} 0 \\ 0 \\ 0 \end{Bmatrix}
\tag{4.12}
$$

By solving the preceding equation, one can obtain the natural frequencies and vibration modes of the structure–equipment system with elastic bearings. As for the present problem, the horizontal stiffness of the elastic bearing is designed to be quite low compared with that of the superstructure. It follows that the superstructure in its entirety behaves essentially as a *rigid body* for the fundamental vibration mode shape of the combined system, which implies that the displacement responses for the equipment, structure, and base under free vibration can be approximately taken as the same, that is, $x_e = x_s = x_b = x$. By introducing such a condition into Equation 4.12, the equation of motion for the equivalent single-DoF base-isolated system can be written as

$$
(m_e + m_s + m_b)\ddot{x} + k_b x = 0
\tag{4.13}
$$

Equation 4.13 indicates that the fundamental frequency, $\omega_1$, of the base-isolated system can be approximated by $\omega_1 \approx \sqrt{k_b/(m_e + m_s + m_b)}$. Further, if the condition of fixed base is considered, that is, by letting the responses of the base be equal to zero, $x_b = \ddot{x}_b = 0$, the structure–equipment isolation system will be reduced to the case of a fixed-base system, such that the equation of motion becomes

$$
\begin{Bmatrix} m_e \ddot{x}_e \\ m_s \ddot{x}_s \end{Bmatrix} + \begin{bmatrix} k_e & -k_e \\ -k_e & k_s + k_e \end{bmatrix} \begin{Bmatrix} x_e \\ x_s \end{Bmatrix} = \begin{Bmatrix} 0 \\ 0 \end{Bmatrix}
\tag{4.14}
$$

as is well known.

### 4.3.3 Dynamics of Structure–Equipment Isolation Systems to Harmonic Excitations

The advantage of a closed-form solution is that it allows us to examine the key parameters involved in the problem considered. This is what will be sought herein. For the case of a harmonic ground excitation, $x_g$, with amplitude, $X_g$, that is, with $x_g = X_g\, e^{i\omega t}$, one may derive from Equation 4.11 the following:

$$
\begin{Bmatrix} m_e \ddot{x}_e \\ m_s \ddot{x}_s \\ m_b \ddot{x}_b \end{Bmatrix} + \begin{bmatrix} k_e & -k_e & 0 \\ -k_e & k_s + k_e & -k_s \\ 0 & -k_s & k_s + k_b \end{bmatrix} \begin{Bmatrix} x_e \\ x_s \\ x_b \end{Bmatrix} = \begin{Bmatrix} m_e \\ m_s \\ m_b \end{Bmatrix} X_g \omega^2\, e^{i\omega t}
\tag{4.15}
$$

Correspondingly, the steady-state responses of the system can be expressed as

$$
\begin{Bmatrix} x_e \\ x_s \\ x_b \end{Bmatrix} = \begin{Bmatrix} X_e \\ X_s \\ X_b \end{Bmatrix} e^{i\omega t}
\tag{4.16}
$$

where ($X_e$, $X_s$, and $X_b$) represent the amplitudes of the equipment, structure, and base, respectively. Substituting Equation 4.16 into Equation 4.15 yields

$$
\begin{bmatrix} k_e - m_e\omega^2 & -k_e & 0 \\ -k_e & k_s + k_e - m_s\omega^2 & -k_s \\ 0 & -k_s & k_s + k_b - m_b\omega^2 \end{bmatrix} \begin{Bmatrix} X_e \\ X_s \\ X_b \end{Bmatrix} = \begin{Bmatrix} m_e \\ m_s \\ m_b \end{Bmatrix} X_g \omega^2
\tag{4.17}
$$

from which the amplitudes ($X_e$, $X_s$, and $X_b$) for the system can be solved as follows:

$$X_e = \frac{X_s + X_g f_e^2}{1 - f_e^2} \tag{4.18a}$$

$$X_b = \frac{X_s + \varepsilon_b f_s^2 X_g}{1 + k_b/k_s - \varepsilon_b f_s^2} \tag{4.18b}$$

$$X_s = \frac{\left[m_s + \dfrac{m_e}{1 - f_e^2} + \dfrac{m_b}{1 + k_b/k_s - \varepsilon_b f_s^2}\right] X_g \omega^2}{\dfrac{k_b - m_b \omega^2}{1 + k_b/k_s - \varepsilon_b f_s^2} - \left(m_s + \dfrac{m_e}{1 - f_e^2}\right)\omega^2} \tag{4.18c}$$

where the amplitudes of the equipment and base, that is, $X_e$ and $X_b$, have been expressed in terms of the amplitude of the base, $X_s$. The parameters in Equation 4.18 are defined as

$$f_e = \omega/\omega_e \tag{4.19a}$$

$$f_s = \omega/\omega_s \tag{4.19b}$$

$$\varepsilon_b = m_b/m_s \tag{4.19c}$$

$$\omega_e = \sqrt{k_e/m_e} \tag{4.19d}$$

$$\omega_s = \sqrt{k_s/m_s} \tag{4.19e}$$

Finally, the state-steady absolute acceleration responses of the structure, equipment, and base can be expressed in terms of the ground acceleration $\ddot{x}_g$ as

$$a_s = \ddot{x}_s + \ddot{x}_g = -(X_s + X_g)\omega^2\, e^{i\omega t} = \frac{k_b \ddot{x}_g}{D(\omega)} \tag{4.20a}$$

$$a_e = \ddot{x}_e + \ddot{x}_g = -\frac{(X_s + X_g)\omega^2\, e^{i\omega t}}{1 - f_e^2} = \frac{k_b \ddot{x}_g}{(1 - f_e^2)D(\omega)} \tag{4.20b}$$

$$a_b = \ddot{x}_b + \ddot{x}_g = \frac{-[X_s + X_g + (k_b/k_s)X_g]\omega^2\, e^{i\omega t}}{1 + k_b/k_s - \varepsilon_b f_s^2} = \frac{(D^{-1}(\omega) + k_s^{-1})k_b \ddot{x}_g}{1 + k_b/k_s - \varepsilon_b f_s^2} \tag{4.20c}$$

$$D(\omega) = (k_b - m_b \omega^2) - (1 + k_b/k_s - \varepsilon_b f_s^2)\left(m_s + \frac{m_e}{1 - f_e^2}\right)\omega^2 \tag{4.20d}$$

As can be seen, the acceleration response of each component in the structure–equipment system depends mainly on the stiffness of the elastic bearing, $k_b$. In particular, the use of a smaller bearing stiffness, $k_b$, can result in significant reduction of the shear forces transmitted to the superstructure, as indicated by Equation 4.20a. This explains why an elastic bearing can be effectively used as an isolator for reducing the base shear of the structure–equipment system. In contrast, if the bearing stiffness, $k_b$, is made to be infinitely large, that is, by letting $k_b \to \infty$, the acceleration responses in Equation 4.20, reduce to

$$a_s = \frac{(1 - f_e^2)\ddot{x}_g}{(1 - f_e^2)(1 - f_s^2) - \varepsilon_e f_s^2} \tag{4.21a}$$

$$a_e = \frac{\ddot{x}_g}{(1 - f_e^2)(1 - f_s^2) - \varepsilon_e f_s^2} \tag{4.21b}$$

$$a_b = \ddot{x}_g \tag{4.21c}$$

with the use of L'Hospital's Rule, where $\varepsilon_e = m_e/m_s$. As can be seen from Equation 4.21c, the acceleration of the structural base is equal to the ground acceleration. Clearly, the present problem has been reduced to a two-DoF system with a rigid base, for which the solutions have been given in Equation 4.21a and Equation 4.21b.

Some important high-tech facilities, such as semiconductor factories and medical devices, are very sensitive to vibrations, especially to those caused by resonance. To consider the effect of resonance, we shall let the ground excitation frequency, $\omega$, coincide with the equipment frequency, $\omega_e$, that is, by letting $f_e = 1$ or $\omega_e = \omega$. For this case, the acceleration responses of the system in Equation 4.20 reduce to

$$a_s = 0 \qquad (4.22a)$$

$$a_e = \frac{-k_b \ddot{x}_g}{k_e[1 + (k_b - m_b \omega_e^2)/k_s]} \qquad (4.22b)$$

$$a_b = \frac{k_b \ddot{x}_g}{k_s + k_b - m_b \omega_e^2} \qquad (4.22c)$$

Because of the coincidence of the ground excitation frequency with the equipment frequency, the equipment behaves like a vibration absorber of the structure. For this reason, the response of the equipment is greatly amplified, as implied by Equation 4.22b, while the response of structure is completely suppressed, as indicated by Equation 4.22a. Moreover, if the frequency of the equipment is equal to the fundamental frequency of the structure–equipment isolation system, that is, $\omega_e(\approx \omega_1) = \sqrt{k_b/(m_e + m_s + m_b)}$, then the responses of the system in Equation 4.22 can further be expressed as follows:

$$a_s = 0 \qquad (4.23a)$$

$$a_e = \frac{-k_b \ddot{x}_g}{k_e[1 + (m_s + m_e)\omega_e^2/k_s]} \qquad (4.23b)$$

$$a_b = \frac{k_b \ddot{x}_g}{k_s + (m_s + m_e)\omega_e^2} \qquad (4.23c)$$

Since the equipment mass is generally much smaller than the structural mass, the preceding equation can be further reduced to

$$a_s = 0 \qquad (4.24a)$$

$$a_e = \frac{-k_b \ddot{x}_g}{k_e(1 + \omega_e^2/\omega_s^2)} \qquad (4.24b)$$

$$a_b = \frac{k_b \ddot{x}_g}{k_s(1 + \omega_e^2/\omega_s^2)} \qquad (4.24c)$$

As indicated by Equation 4.24b, the acceleration response of the equipment depends on the stiffness ratio, $k_b/k_e$, of the base to the equipment.

For the resonance condition of $\omega_e = \omega$, mentioned previously, let us consider the case when the structural frequency is equal to the equipment frequency, that is, $\omega_e = \omega_s$. For this case, the responses of the system in Equation 4.22 reduce to

$$a_s \approx 0 \qquad (4.25a)$$

$$a_e \approx \frac{-k_b \ddot{x}_g}{\varepsilon_e[k_s(1 - \varepsilon_b) + k_b]} \qquad (4.25b)$$

$$a_b = \frac{k_b \ddot{x}_g}{k_s(1 - \varepsilon_b) + k_b} \qquad (4.25c)$$

which indicates that the acceleration response of the equipment may be greatly amplified, as implied by the relatively small mass ratio $\varepsilon_e (= m_e/m_s)$ and large stiffness ratio, $k_b/k_s$, in Equation 4.25b. Such a phenomenon has been referred to as the *tuning of equipment*.

On the other hand, when the excitation frequency, $\omega$, coincides with the fundamental frequency, $\omega_1$, of the isolated system, that is, $\omega(\approx \omega_1) = \sqrt{k_b/(m_e + m_s + m_b)}$, resonant response may be induced on the structure–equipment isolation system. Considering that the first priority in design of high-tech equipment is to reduce the vibrations of the equipment, rather than the structure, by comparing the denominators in Equation 4.20a and Equation 4.20b, one may assume that the condition $|1 - f_e^2| \geq 1$ or $f_e = \omega/\omega_e \geq \sqrt{2}$ remains satisfied for a good design, which is equivalent to

$$\frac{\omega_e}{\omega_s} \leq \sqrt{\frac{k_b/k_s}{2[1 + (m_b + m_e)/m_s]}} \qquad (4.26)$$

Since the fundamental frequency, $\omega_s$, of a base-isolated structure is generally low in practice, the horizontal stiffness of the equipment attached to the structure must be designed to be soft enough such that Equation 4.26 can be satisfied. Certainly, this is one of the guidelines to be obeyed in the design of equipment for the sake of vibration reduction.

## 4.3.4 Illustrative Example

The forgoing formulations have been made by neglecting the damping of the structural system and by assuming the ground motion to be of the harmonic type. In practice, there is always some damping with the structural system, while the ground motion may be random in nature. To deal with such problems, the only recourse is to use numerical methods that are readily available. In this section, the Newmark $\beta$ method, proposed by Newmark (1959), with $\gamma = 1/2$ and $\beta = 1/4$, will be adopted for solving the second-order differential equation presented in Equation 4.11, which has the advantage of being numerically stable.

The example considered is the structure–equipment system isolated by elastomeric bearings, shown in Figure 4.9, with the data given in Table 4.2. As can be seen, the equipment has a frequency equal to five times the structural frequency, that is, $\omega_e = 5\omega_s$ ($= 8.34$ Hz). The 1940 El Centro earthquake (NS component) with a peak ground acceleration (PGA) of 341.55 gal is adopted as the ground excitation, as given in Figure 4.10. By an eigenvalue analysis, the natural frequencies solved for the base-isolated system are 2.46, 21.41, and 52.74 rad/sec. Because of the installation of elastic bearings on the structure–equipment system, the fundamental frequency of the system decreases significantly and is approximately equal to $\omega_1 \approx \sqrt{k_b(m_e + m_s + m_b)} = 2.51$ rad/sec, according to Section 4.3.2. From this example, one observes that the use of a single-DoF system to model a base-isolated system can give a generally good result for the first frequency of vibration.

As can be seen from Figure 4.11, for the structural acceleration of the system, the main-shock response of the fixed-base structure has been effectively eliminated due to installation of the elastic bearings. However, as indicated by Figure 4.12, because of the installation of soft bearings, the base displacement response of the isolated system is much larger and decays much slowly, even after the main shocks.

In Figure 4.13 the acceleration response of the equipment for the isolated and fixed-base cases are compared. As can be seen, the main-shock response of the fixed-base structure has been effectively suppressed through the installation of the elastic bearings. Furthermore, the equipment response appears to be almost identical to the structure response shown in Figure 4.11, due to the fact that the equipment

**TABLE 4.2**  System Parameters Used in Simulation (Section 4.3.4)

| Equipment | | Superstructure | | Isolation System | |
|---|---|---|---|---|---|
| Parameter | Value | Parameter | Value | Parameter | Value |
| Mass $m_e$ | 3 t ($= m_s/100$) | Mass $m_s$ | 300 t | Mass $m_b$ | 100 t ($= m_s/3$) |
| Horizontal stiffness $k_e$ | 8258 kN/m | Horizontal stiffness $k_s$ | 33,030 kN/m | Horizontal stiffness $k_b$ | 2546 kN/m |
| Damping | 15.74 kN m/s | Damping | 314.79 kN m/s | Damping | 50.46 kN m/s |
| Frequency $\omega_e = \sqrt{k_e/m_e}$ | 52.47 rad/sec | Frequency $\omega_s = \sqrt{k_s/m_s}$ | 10.46 rad/sec | Frequency $\omega_b = \sqrt{k_b/m_b}$ | 5.05 rad/sec |

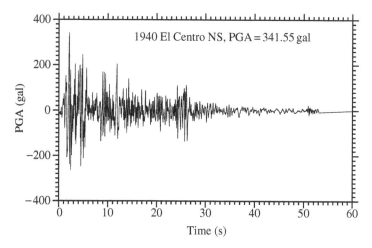

**FIGURE 4.10**  Waveform of 1940 El Centro earthquake (NS component).

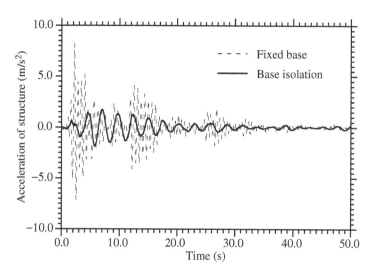

**FIGURE 4.11**  Comparison of structural accelerations.

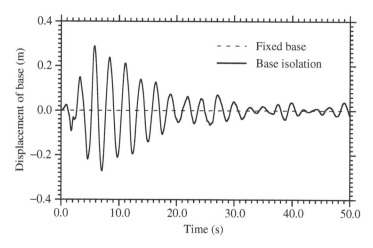

**FIGURE 4.12**  Comparison of base displacements.

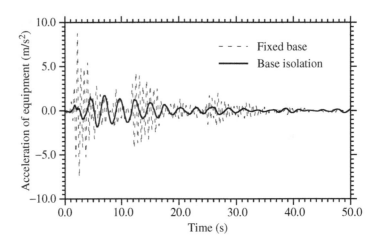

**FIGURE 4.13**   Comparison of equipment accelerations.

is rigidly attached to the structure, as implied by the relatively higher frequency of the equipment. As for the present example, the effectiveness of the elastic bearings in reducing the equipment response is ascertained.

For high-tech equipment, engineers may be concerned about the effect on equipment tuning induced by external excitations, such as earthquakes and traffic-induced vibrations. To investigate this effect on the structure–equipment isolation system considered, the maximum equipment acceleration was plotted as a function of the frequency ratio, $\omega_e/\omega_s$, in Figure 4.14. As can be seen, the response of the equipment is greatly amplified when it has a frequency close to the fundamental frequency of the structure–equipment isolation system, that is, when $\omega_e = 2.51$ rad/sec or $\omega_e/\omega_s = 0.24$. To avoid such a situation, it is suggested that isolators be mounted at both the structure base and equipment base. From Figure 4.14, one observes that the use of a small horizontal stiffness for the equipment will generally lead to greater equipment response due to tuning effect. However, as the stiffness of the equipment is further reduced, the equipment will reach another isolation state, in which the equipment response will be substantially suppressed, as indicated by the region with relatively small values of $\omega_e/\omega_s$. The margin for such a frequency ratio of $\omega_e/\omega_s$ can be obtained by substituting the parameters in Table 4.2 into

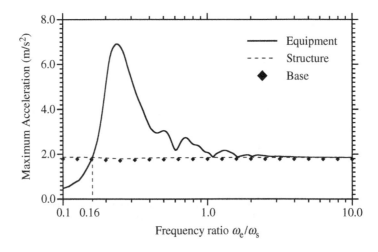

**FIGURE 4.14**   Maximum accelerations of structure–equipment isolation systems.

Equation 4.26, which yields a critical ratio of $\omega_e/\omega_s = 0.168$, very close to the value of 0.16 marked in Figure 4.14.

### 4.3.5 Concluding Remarks

This section investigates the dynamic response of a mathematical model of a structure–equipment system isolated by elastomeric bearings and subjected to ground excitations. Based on the closed-form solution of a structure–equipment isolation system subjected to harmonic support motions, one observes that the coincidence of the ground excitation frequency with the equipment frequency will make the equipment behave like a vibration absorber of the structure, of which the acceleration response will be greatly amplified. For the case that the first priority in design is to reduce the vibration of the equipment rather than that of the structure, Equation 4.26 provides a guideline for the design of equipment, which has been verified in the numerical example.

## 4.4 Sliding Isolation Systems

Sliding isolation can be an effective means for the seismic protection of structural systems. By implementing sliding isolators under the base mat of a structure, the transmission of ground excitation to the structure can be greatly reduced. Currently, applications of sliding isolation systems can be found elsewhere (Naeim and Kelly, 1999). A *sliding isolator* usually consists of a slider with frictional surfaces. For this reason, it is also referred to as a *friction isolator*. When subjected to an earthquake, the slider will slide along the frictional contact surfaces whenever the horizontal seismic force exceeds the maximum frictional force of the support, which, by Coulomb's theory, is equal to the normal contact force multiplied by the static (or dynamic) coefficient of friction of the sliding surfaces. Because of this, the seismic force transmitted to the superstructure is generally less than the maximum frictional force of the isolator. Obviously, the maximum frictional force is an important parameter for the design of a sliding isolation system, because it decides when the system starts to slide and how large the shear force is to be transmitted to the superstructure.

The motion of a sliding structure consists of two different states, namely, the sliding state and the stick (or nonsliding) state. At any instant of motion, the structure can only belong to one of the two states. Although in each state the sliding structure can be modeled as a linear system, the governing equations of motion for the two states are different. As a result, the overall behavior of the sliding structure is nonlinear. Such nonlinearity has resulted in the occurrence of subharmonic resonance in the frequency response of a sliding structure (Mostaghel et al., 1983; Westermo and Udwadia, 1983), making the dynamic response much more complicated. In some applications, a sliding isolation system has been designed with an automatic recentering mechanism (or resilient mechanism), so that the structure can slide back to its original position after the earthquake (Mokha et al., 1991). This type of sliding systems has been called the *resilient sliding isolation system*, which will be investigated in Section 4.5. The implementation of a recentering mechanism offers some advantages, but will inevitably introduce some disadvantages, as will be discussed in Section 4.5.

The purpose of this section is to investigate the seismic behavior of a sliding isolated structure and also the behavior of an equipment item mounted on the structure. No consideration will be made for the recentering mechanism. The nonlinear dynamic equation for a structure with an underneath friction element is first formulated. Next, two numerical approaches will be presented for solving the nonlinear equation, the *shear balance method* and *fictitious spring method*. Finally, using some assumed data, the harmonic response and seismic behavior of a sliding structure, together with the equipment mounted on it, will be presented. In this section, the frictional coefficient of the sliding system is assumed to be of the Coulomb type, that is, a time-invariant constant. For simplification, no distinction will be made between the static and dynamic frictional coefficients, or between the dynamic and maximum static frictional force.

### 4.4.1 Mathematical Modeling and Formulation

#### 4.4.1.1 Equation of Motion

A sliding isolated structure with an attached equipment item, as schematically shown in Figure 4.15, can be represented as a mass–spring–dashpot system of three DoF, as shown in Figure 4.16, for which the notations employed have been defined in Table 4.1. When the structural system is excited by an earthquake, the equation of motion can be written as

$$\mathbf{M}\ddot{\mathbf{x}}(t) + \mathbf{C}\dot{\mathbf{x}}(t) + \mathbf{K}\mathbf{x}(t) = -\mathbf{M}\mathbf{L}_1\ddot{x}_g(t) + \mathbf{L}_2 f(t) \tag{4.27}$$

where the vector $\mathbf{x}$ denotes the dynamic responses of the whole structural system

$$\mathbf{x}(t) = \begin{Bmatrix} x_e(t) \\ x_s(t) \\ x_b(t) \end{Bmatrix} \tag{4.28}$$

The mass, damping, and stiffness matrices in Equation 4.27 are defined as

$$\mathbf{M} = \begin{bmatrix} m_e & 0 & 0 \\ 0 & m_s & 0 \\ 0 & 0 & m_b \end{bmatrix},$$

$$\mathbf{C} = \begin{bmatrix} c_e & -c_e & 0 \\ -c_e & c_e + c_s & -c_s \\ 0 & -c_s & c_s \end{bmatrix}, \tag{4.29}$$

$$\mathbf{K} = \begin{bmatrix} k_e & -k_e & 0 \\ -k_e & k_e + k_s & -k_s \\ 0 & -k_s & k_s \end{bmatrix}$$

and the force distribution vectors as

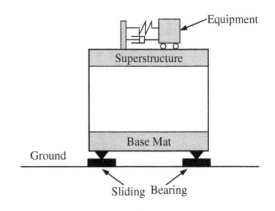

**FIGURE 4.15** Schematic for an isolated structure–equipment system with sliding bearing.

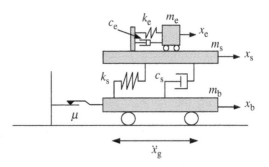

**FIGURE 4.16** Model for an isolated structure–equipment system with sliding support.

$$\mathbf{L}_1 = \begin{Bmatrix} 1 \\ 1 \\ 1 \end{Bmatrix}, \quad \mathbf{L}_2 = \begin{Bmatrix} 0 \\ 0 \\ 1 \end{Bmatrix} \tag{4.30}$$

Note that in Equation 4.27, the isolator frictional force, $f(t)$, which is not a constant, is moved to the right-hand side of the equation. This nonlinear force requires a special treatment in an analysis procedure, as will be explained later on.

For a systematic treatment, the above equation of motion can be further written in a state space form as shown below (Meirovitch, 1990):

$$\dot{\mathbf{z}}(t) = \mathbf{A}\mathbf{z}(t) + \mathbf{E}\ddot{x}_g(t) + \mathbf{B}f(t) \tag{4.31}$$

where the state vector $\mathbf{z}(t)$ and the system matrix $\mathbf{A}$ are defined as

$$
\mathbf{z}(t) = \begin{bmatrix} \dot{\mathbf{x}}(t) \\ \mathbf{x}(t) \end{bmatrix}, \quad \mathbf{A} = \begin{bmatrix} -\mathbf{M}^{-1}\mathbf{C} & -\mathbf{M}^{-1}\mathbf{K} \\ \mathbf{I} & 0 \end{bmatrix} \tag{4.32}
$$

and the excitation and friction distribution vectors as

$$
\mathbf{E} = \begin{bmatrix} -\mathbf{L}_1 \\ 0 \end{bmatrix}, \quad \mathbf{B} = \begin{bmatrix} \mathbf{M}^{-1}\mathbf{L}_2 \\ 0 \end{bmatrix} \tag{4.33}
$$

### 4.4.1.2 Conditions for Stick and Sliding States

As mentioned above, the motion of a sliding structure at any instant has two possible states, namely, the stick (or nonsliding) and sliding states. The following are the conditions that must be satisfied by the sliding structure:

(1) In stick state

$$
|f(t)| < f_{\max} = \mu W \tag{4.34a}
$$

$$
\dot{x}_b(t) = 0 \tag{4.34b}
$$

where $\mu$ is the coefficient of friction and $W$ is the total weight of the structure. According to the preceding equations, the frictional force in the stick state is an unknown with a magnitude less than the maximum frictional force, $f_{\max}$, which equals the product of $\mu$ and $W$, while the sliding velocity of the structure is simply zero. Whenever the frictional force satisfies Equation 4.34a, the sliding system remains in the stick state, otherwise it changes into the sliding state.

(2) In sliding state

$$
f(t) = -\operatorname{sgn}(\dot{x}_b(t))f_{\max} = -\operatorname{sgn}(\dot{x}_b(t))\mu W \tag{4.35a}
$$

$$
\dot{x}_b(t) \neq 0 \tag{4.35b}
$$

where the function $\operatorname{sgn}(x)$ denotes the sign of the variable $x$. According to Equation 4.35a and Equation 4.35b, the frictional force in the sliding state has a magnitude equal to the maximum frictional force, but directed in a sense opposite to that of the sliding velocity. On the other hand, the sliding velocity of the isolator remains as an unknown.

## 4.4.2 Methods for Numerical Analysis

Two numerical methods commonly used for the analysis of sliding isolated structural systems, the shear balance method and fictitious spring method, will be introduced in this section. By employing the discrete-time state-space formula, both methods can be cast in an incremental form that is suitable for the analysis of sliding systems with multiple DoF.

### 4.4.2.1 Shear Balance Method

Consider the state-space equation, Equation 4.31, and assume that both the ground acceleration and frictional force vary linearly within each time interval, as shown in Figure 4.17. Equation 4.31 may be written in the following incremental form (Meirovitch, 1990)

$$
\mathbf{z}[k+1] = \mathbf{A}_d\mathbf{z}[k] + \mathbf{E}_0\ddot{x}_g[k] + \mathbf{E}_1\ddot{x}_g[k+1] + \mathbf{B}_0 f[k] + \mathbf{B}_1 f[k+1] \tag{4.36}
$$

where the symbol $x[k]$ denotes that the variable $x$ is evaluated at the $k$th time step. The other coefficient matrices in equation 4.36 are defined as

$$A_d = e^{A\Delta t} = \sum_{i=0}^{\infty} \frac{\Delta t^i}{i!} A^i \qquad (4.37)$$

$$B_0 = \left[ (A)^{-1}A_d + \frac{1}{\Delta t}(A)^{-2}(I - A_d) \right] B \qquad (4.38a)$$

$$B_1 = \left[ -(A)^{-1} + \frac{1}{\Delta t}(A)^{-2}(A_d - I) \right] B \qquad (4.38b)$$

where $\Delta t$ denotes the size of the time step considered for analysis. The matrices $E_0$ and $E_1$ can be computed in a way similar to $B_0$ and $B_1$, except that the matrix $B$ in Equation 4.38a and Equation 4.38b should be replaced by the matrix $E$. In some applications, the system matrix $A$

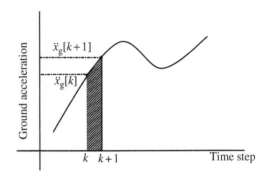

**FIGURE 4.17**   Force integration scheme with linear interpolation.

may not be invertible, that is, $A^{-1}$ does not always exist. If this is the case, one may compute $B_0$ and $B_1$ (and similarly $E_0$ and $E_1$) by the following formulas:

$$B_0 = A_d^* B, \quad B_1 = (\hat{A}_d - A_d^*)B \qquad (4.39)$$

where

$$A_d^* = \sum_{i=0}^{\infty} \frac{(\Delta t)^{i+1}}{i!(i+2)} A^i, \quad \hat{A}_d = \sum_{i=0}^{\infty} \frac{(\Delta t)^{i+1}}{(i+1)!} A^i \qquad (4.40)$$

Note that, on the right-hand side of Equation 4.36, the only unknown at the $k$th time step is the frictional force $f[k + 1]$; therefore, $f[k + 1]$ must be determined before the next time step response $z[k + 1]$ is computed. Wang et al. (1998) proposed the shear balance method for computing the frictional force $f[k + 1]$. By this method, the sliding structure is first assumed to be in the stick state at the $(k + 1)$th step, for which the condition given in Equation 4.34b must be satisfied

$$\dot{x}_b[k + 1] = Dz[k + 1] = 0 \qquad (4.41)$$

where $D$ is a relation matrix, equal to $D = [0\ 0\ 1\ 0\ 0\ 0]$ for the model shown in Figure 4.16. Substituting $z[k + 1]$ in Equation 4.36 into Equation 4.41, one may solve for the *estimated frictional force* at the $(k + 1)$th time step as

$$\bar{f}[k + 1] = -(DB_1)^{-1}D(A_d z[k] + B_0 f[k] + E_0 \ddot{x}_g[k] + E_1 \ddot{x}_g[k + 1]) \qquad (4.42)$$

where $\bar{f}[k + 1]$ with an overbar signifies that the frictional force is an estimate obtained by assuming the sliding structure to be at the stick state. Such a value may not be the actual one if the system is not in the stick state. The physical meaning for $\bar{f}[k + 1]$ is that it represents the *balanced shear force* required at the $(k + 1)$th time step for the structure to remain in the stick state. Therefore, the sign of $\bar{f}[k + 1]$ indicates the direction of the resistant force provided by the isolation system. In spite of the fact that $\bar{f}[k + 1]$ may not be the actual frictional force, it plays an important role for determining the actual state (stick or sliding) and the actual frictional force of the sliding isolated structure, as will be described below based on Equation 4.34a and Equation 4.35a.

(1) The system is in the "stick state" if $|\bar{f}[k + 1]| < f_{max}$ and the frictional force is

$$f[k + 1] = \bar{f}[k + 1] \qquad (4.43)$$

(2) The system is in the "sliding state" if $|\bar{f}[k + 1]| \geq f_{max}$ and the frictional force is

$$f[k + 1] = \text{sgn}(\bar{f}[k + 1])f_{max} \qquad (4.44)$$

As can be seen, the term $-\text{sgn}(\dot{x}_b[k+1])$ in Equation 4.35a is replaced by $\text{sgn}(\bar{f}[k+1])$ in Equation 4.44. Such a replacement is justified since the sign of $\bar{f}[k+1]$ indicates the direction of the resistant force at the $(k+1)$th time step. Once the correct frictional force, $f[k+1]$, is determined by using either Equation 4.43 or Equation 4.44, it can be substituted into Equation 4.36 to obtain the response $\mathbf{z}[k+1]$ for the next time step. The computational flow-chart for the shear balance method has been given in Figure 4.18.

### 4.4.2.2 Fictitious Spring Method

The fictitious spring method was first proposed by Yang et al. (1990) for the analysis of a sliding structure. Later, Lu and Yang (1997) reformulated the method into a state-space form for the analysis of equipment

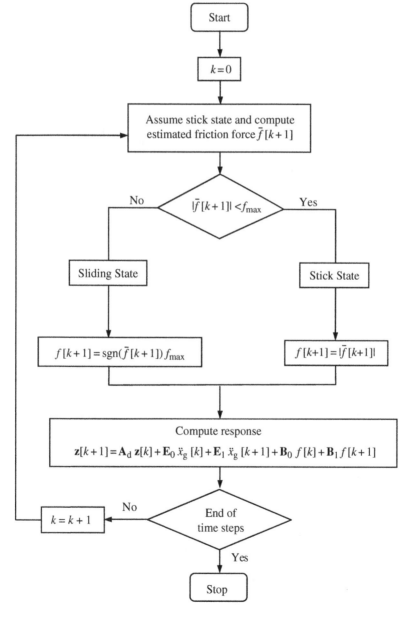

**FIGURE 4.18** Computational flow-chart for shear balance method.

mounted on a sliding structure. By this method, a fictitious spring, $k_f$, is introduced between the base mat and the ground, as in Figure 4.19, to represent the mechanism of sliding or friction. The stiffness, $k_f$, of the fictitious spring is taken as zero for the sliding state and as a very large value for the stick state. With the introduction of the fictitious spring, the stiffness matrix, $\mathbf{K}$, in Equation 4.29 should be modified as follows:

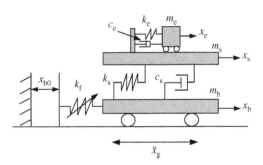

$$\mathbf{K} = \mathbf{K}(k_f) = \begin{bmatrix} k_e & -k_e & 0 \\ -k_e & k_e + k_s & -k_s \\ 0 & -k_s & k_s + k_f \end{bmatrix} \quad (4.45)$$

**FIGURE 4.19** Model for isolated structures with fictitious spring.

Accordingly, the state-space dynamic equation, Equation 4.31, should be modified as

$$\dot{z}(t) = \mathbf{A}(k_f)z(t) + \mathbf{E}\ddot{x}_g(t) + \mathbf{B}\tilde{f}(t) \quad (4.46)$$

where

$$\mathbf{A} = \mathbf{A}(k_f) = \begin{bmatrix} -\mathbf{M}^{-1}\mathbf{C} & -\mathbf{M}^{-1}\mathbf{K}(k_f) \\ \mathbf{I} & 0 \end{bmatrix} \quad (4.47)$$

Depending on the current state of the sliding system, the fictitious stiffness, $k_f$, and the modified friction term, $\tilde{f}(t)$, in Equation 4.46 may take one of the following two sets of values:

(1) In the stick state

$$k_f = \alpha k_s, \quad \tilde{f}(t) = k_f x_{b0} \quad (4.48)$$

(2) In the sliding state

$$k_f = 0, \quad \tilde{f}(t) = -\text{sgn}(\dot{x}_b(t))\mu W \quad (4.49)$$

In Equation 4.48, the symbol $\alpha$ represents a constant of very large value, and $x_{b0}$ the initial elongation of the fictitious spring in the current stick state (computation of $x_{b0}$ will be explained later). Note that the modified friction term, $\tilde{f}(t)$, may not be the actual frictional force. The actual frictional force can be determined as follows:

(1) In the stick state

$$f(t) = k_f(x_b(t) - x_{b0}) \quad (4.50)$$

(2) In the sliding state

$$f(t) = \tilde{f}(t) \quad (4.51)$$

According to Equation 4.50 and Equation 4.51, the actual frictional force of the isolation system in the stick state is equal to the internal force of the fictitious spring, while in the sliding state it is equal to the modified frictional force, $\tilde{f}(t)$. The frictional force computed from the preceding two equations should obey the conditions given in Equation 4.34a and Equation 4.35a as well.

With the conditions imposed for the stick and sliding states in Equation 4.48 and Equation 4.49, respectively, the equation of motion in Equation 4.46 actually represents two different sets of equations.

Specifically, Equation 4.46 and Equation 4.48 collectively describe the motion of the structure in the stick state, while Equation 4.46 and Equation 4.49 represent the motion of the structure in the sliding state. Owing to the fact that a sliding system may switch between the two states at certain instants, the behavior of the entire system should undoubtedly be regarded as a nonlinear one. Nevertheless, within each particular state, the behavior of the system as represented either by Equation 4.46 and Equation 4.48 or Equation 4.46 and Equation 4.49 is a linear one.

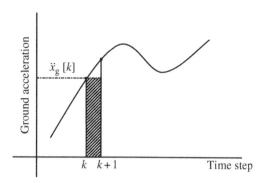

**FIGURE 4.20**   Constant force integration scheme.

In the following, a numerical solution scheme based on the concept of fictitious spring will be introduced. Let $\Delta t$ denote a time increment, which is usually taken as a very small value, and assume that the ground excitation and frictional force are constant within each time increment, $\Delta t$ (see Figure 4.20). Accordingly, the discrete-time solution of Equation 4.46 can be rewritten in an incremental form (Meirovitch, 1990) as

$$z[k+1] = A_d z[k] + E_d \ddot{x}_g[k] + B_d \tilde{f}[k] \tag{4.52}$$

where

$$A_d = A_d(k_f) = e^{A(k_f)\Delta t} = \sum_{i=0}^{\infty} \frac{\Delta t^i}{i!} A(k_f)^i \tag{4.53}$$

$$E_d = E_d(k_f) = A(k_f)^{-1}(A_d(k_f) - I)E \tag{4.54a}$$

$$B_d = B_d(k_f) = A(k_f)^{-1}(A_d(k_f) - I)B \tag{4.54b}$$

For the case where the system matrix $A$ is invertible, $B_d$ and $E_d$ may be computed instead using the following formulas:

$$E_d = E_d(k_f) = \left[ \sum_{i=0}^{\infty} \frac{\Delta t^i}{i!} A(k_f)^{i-1} \right] E \tag{4.55}$$

$$B_d = B_d(k_f) = \left[ \sum_{i=0}^{\infty} \frac{\Delta t^i}{i!} A(k_f)^{i-1} \right] B \tag{4.56}$$

Equation 4.52 is the solution of the sliding system given in incremental form, because the response, $z[k+1]$, can be computed from the solution of the previous step, $z[k]$. Note that, in Equation 4.53 and Equation 4.54, the coefficient matrices $A_d$, $E_d$, and $B_d$ have two possible sets of values, as the fictitious spring constant, $k_f$, may take different values for the sliding and stick states. Nevertheless, once the time step size, $\Delta t$, is chosen, the coefficient matrices $A_d$, $E_d$, and $B_d$, remain constant for each state. As such, they need only be calculated once at the beginning of the incremental procedure. The computational flow-chart for the fictitious spring method described above has been given in Figure 4.21.

The dynamic equation and its discrete-time solution for the sliding structure in the two states have been presented above. In the following, we shall describe how to determine the *transition time* for the sliding structure to switch from one state to the other. Once the transition time is determined, the original step size should be scaled down accordingly to reflect the transition point (Yang et al., 1990).

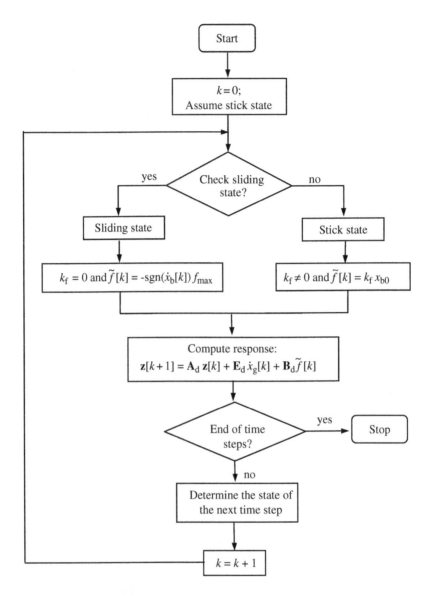

**FIGURE 4.21**   Computational flow-chart for fictitious spring method.

*Transition from stick to sliding state.* As stated in Equation 4.34a, the condition for a sliding isolated structure to remain in the stick state is that the static frictional force under the base mat be less than its maximum value, $f_{max}$. Once the frictional force exceeds this maximum value, the system starts to slide. With the fictitious spring method, the static frictional force is computed as the internal force of the fictitious spring, as in Equation 4.50. Based on the above considerations and Equation 4.50, the condition for the sliding system to transfer from the stick to the sliding state is

$$|f(t_0)| = |k_f(x_b(t_0) - x_{b0})| = f_{max} = \mu W \tag{4.57}$$

where $t_0$ denotes the transition time at which the structure starts to slide and $k_f$ is the spring constant given in Equation 4.48 for the stick state. Because a very large value has been used for the fictitious spring constant, $k_f$, in the stick state, the deviation in base displacement due to spring elongation is very small in this state, which can just be neglected. In practice, the transition time, $t_0$ may not occur precisely at the discrete points of time considered. It is likely that the spring force is less than $f_{max}$ at the current time step, say, the $k$th step, but exceeds $f_{max}$ at the following time step. If this is the case, numerical methods such as the bisection method should be employed to locate the transition time, $t_0$ within the time interval $(k\Delta t, (k + 1)\Delta t)$ considered based on Equation 4.57.

*Transition from sliding state to stick state.* The structure in the sliding state may return to the stick state whenever the following two conditions are satisfied. (1) The relative velocity of the base mat to the ground reaches zero, that is, $\dot{\xi}_b(t_0) = 0$ where $t_0$ is the transition time; (2) the estimated static frictional force, denoted by $\bar{f}(t_0)$, is less than the maximum static frictional force, that is, $\bar{f}(t_0) < f_{max}$. Here, the estimated static frictional force, $\bar{f}(t_0)$, is defined as the shear force required to balance the motion of the superstructure if the system is assumed to be in the stick state, similar to the one given in Equation 4.42 for $\bar{f}[k + 1]$. By letting the relative velocity and relative acceleration between the base mat and the ground be equal to zero, that is, $\dot{\xi}_b(t_0) = \ddot{\xi}_b(t_0) = 0$, the estimated frictional force can be calculated from the free-body diagram of the base mat as

$$\bar{f}(t_0) = m_b\ddot{x}_g(t_0) - k_s(x_s(t_0) - x_b(t_0)) - c_s\dot{x}_s(t_0) \tag{4.58}$$

For the sliding structure to transfer from the sliding to the stick state, both the aforementioned conditions must be satisfied simultaneously. Once the structure enters the stick state, the term $f(t_0)$ should be set equal to $\bar{f}(t_0)$ and used as the initial frictional force. For the sake of equilibrium, the initial base mat displacement, $x_{b0}$, should be computed as

$$x_{b0} = x_b(t_0) - (\bar{f}(t_0)/k_f) \tag{4.59}$$

where the value of $k_f$ is the one given for the stick state in Equation 4.48.

Concerning the two conditions mentioned above, it may happen that only the first condition, $\dot{\xi}_b = 0$ is satisfied, while the computed $\bar{f}(t_0)$ is still larger than $f_{max}$. If this is the case, the sliding system should not be regarded as a transition to the stick state. Rather, the situation should be regarded as an indication for reversing the direction of sliding in the next time step. Correspondingly, the frictional force, $f(t_0)$, should be set equal to the sliding frictional force, rather than the estimated one, $\bar{f}(t_0)$.

## 4.4.3 Simulation Results for Sliding Isolated Systems

### 4.4.3.1 Numerical Model and Ground Excitations

In this section, the dynamic behavior of the sliding isolated structure–equipment system shown in Figure 4.16 will be analyzed using the shear balance method. Although the sliding structure and equipment considered are both of single-DoF, there exists no difficulty for use of the method to solve problems with multi-DoF systems. In Table 4.3, the material properties adopted for the present model have been listed, which are intended to simulate a small, five-story, reinforced concrete frame. For the present purposes, two types of ground excitation are considered, namely, harmonic and earthquake excitations. The harmonic excitation is considered primarily for studying the frequency response of the sliding system, while the earthquake excitation is considered for the effect of earthquake intensity. For the

**TABLE 4.3** System Parameters Used in Simulation (Section 4.4.3)

| Equipment | | Superstructure | | Isolation System | |
|---|---|---|---|---|---|
| Parameter | Value | Parameter | Value | Parameter | Value |
| Mass $m_e$ | 3 t $(= m_s/100)$ | Mass $m_s$ | 300 t | Mass $m_b$ | 100 t $(= m_s/3)$ |
| Frequency $\omega_e$ | $5\omega_s$ or a variable | Frequency $\omega_s$ | 1.67 Hz | Frictional coefficient $\mu$ | 0.05, 0.1, 0.25 |
| Damping ratio $\zeta_e$ | 5% | Damping ratio $\zeta_s$ | 5% | — | — |

harmonic excitation, a sinusoidal ground acceleration of the following form is adopted:

$$\ddot{x}_g(t) = 0.5g \sin \omega_g t \tag{4.60}$$

where $\omega_g$ denotes the excitation frequency and $g$ is the gravitational acceleration. For the earthquake excitation, the 1940 El Centro earthquake (NS component) is considered, for which the waveform has been given in Figure 4.10. The PGA level of the earthquake will be adjusted for reasons of research. Concerning the effectiveness of response reduction, three quantities are chosen as the indices, namely, the base displacement (base drift), structural acceleration, and equipment acceleration. For all the figures shown in Section 4.4.3 below, the symbol $\mu$ in the legend is used to denote the frictional coefficient of the sliding isolation system and the word "fixed" denotes the response of the corresponding fixed-base structure.

### 4.4.3.2  Harmonic Response of Structure

*Time history.* For the isolated system subjected to a harmonic excitation of $\omega_g = 1$ Hz, the base displacements computed for different coefficients of friction were plotted in Figure 4.22 and Figure 4.23. As can be seen, the base displacements quickly reach the steady-state response within the first few cycles. Meanwhile, the use of a smaller frictional coefficient results in a larger permanent displacement before the steady state is reached. From the structural accelerations plotted in Figure 4.24, one observes that, for a sliding structure with a smaller coefficient of friction, the steady-state response is achieved in a faster way, accompanied by a larger reduction on structural acceleration. Of interest in Figure 4.24 is that the response of the fixed-base case shows a clear period of 1 sec, while in the sliding case, the response is contaminated by high-frequency signals caused by the sliding-stick transitions.

*Hysteretic behavior.* In order to understand the mechanical characteristics of a nonlinear device used for vibration control, it is common to present a diagram showing the force–deformation relation of the device, also referred to as the *hysteretic diagram* (Soong and Dargush, 1997). Figure 4.25 shows the hysteresis loops of the sliding isolation system (the sliding layer) for $\mu = 0.1$ and 0.25, when the system is subjected to a harmonic excitation of $\omega_g = 1$ Hz. In the figure, the horizontal and vertical axes, respectively, represent the base displacement and shear force, that is, the frictional force, under the mat. Just like many other frictional elements or devices, the shape of the hysteresis loop of a sliding bearing is rectangular. The height of the rectangle is equal to the maximum frictional force that depends on the coefficient of friction, while the width of the hysteresis loop is determined by the base-sliding displacement. As the coefficient of friction decreases, the height of the loop decreases, while the width increases. The total area of the hysteresis loop is equivalent to the portion of the energy dissipated by the sliding bearing.

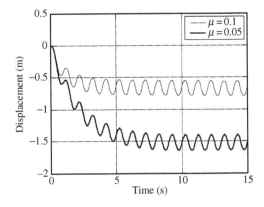

**FIGURE 4.22**  Comparison of base displacements ($\omega_g = 1$ Hz; $\omega_e = 5\omega_s$).

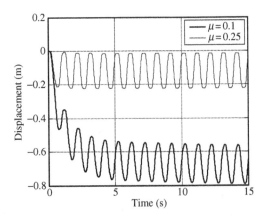

**FIGURE 4.23**  Comparison of base displacements ($\omega_g = 1$ Hz; $\omega_e = 5\omega_s$).

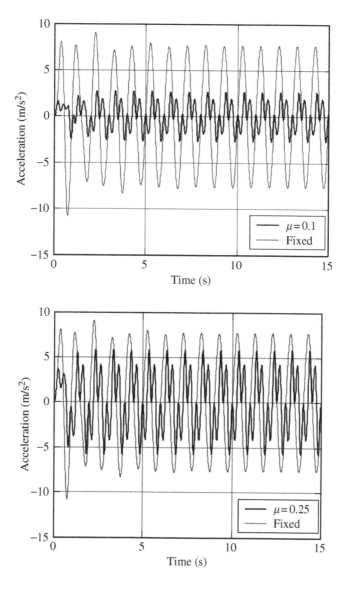

**FIGURE 4.24** Comparison of structural accelerations ($\omega_g = 1$ Hz; $\omega_e = 5\omega_s$).

*Frequency response.* Figure 4.26 shows the maximum structural acceleration with respect to the excitation frequency for four different frictional coefficients, $\mu = 0.05, 0.1, 0.25$ and $\infty$ (for the fixed-base case). Here, the maximum acceleration means the steady-state acceleration response. The following observations can be made from Figure 4.26: (1) Compared with the fixed-base case, the use of a smaller frictional coefficient can reduce the structural acceleration for the frequency range considered. (2) The sliding mechanism can effectively suppress the main resonant response, associated with the natural frequency of 1.67 Hz of the superstructure system. (3) As the coefficient of friction, $\mu$, decreases from $\infty$ to 0.05, the main resonant frequency associated with the structural natural frequency drifts from the fixed-base frequency of 1.67 Hz toward a higher value. (4) For the sliding cases of $\mu = 0.05, 0.1, 0.25$, there exist some minor peaks in the range of lower excitation frequencies, besides the main resonant peak. Such a phenomenon is called the *subharmonic resonance*. For a large frictional coefficient, say, with $\mu = 0.25$, the subharmonic resonant response may be even larger than that of the main resonance.

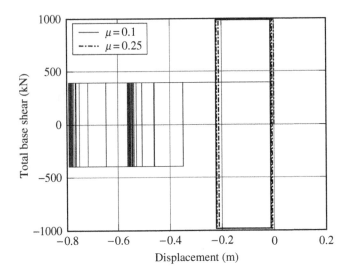

**FIGURE 4.25**   Hysteresis loops of a sliding bearing ($\omega_g = 1$ Hz; $\omega_e = 5\omega_s$).

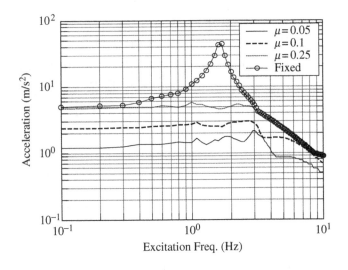

**FIGURE 4.26**   Maximum structural acceleration vs. ground excitation frequency ($\omega_e = 5\omega_s$).

Figure 4.27 shows the frequency responses of the maximum base displacement for various frictional coefficients. The following are observed: (1) The larger the frictional coefficient, the smaller the base drift is. (2) The base displacement has very large magnitudes in the lower excitation frequencies and decreases monotonically as the excitation frequency increases. (3) The extremely large base drift exhibited in the lower excitation frequency range is due to the initial permanent displacement observed in Figure 4.22.

#### 4.4.3.3   Harmonic Response of Equipment

*Time history.* Consider an equipment item of a natural frequency equal to five times of the structural frequency, that is, $\omega_e = 5\omega_s$. For the case of a harmonic excitation of $\omega_g = 1$ Hz, the accelerations solved for the equipment mounted on the structure with $\mu = 0.1$ and 0.25, along with the fixed-base case, have

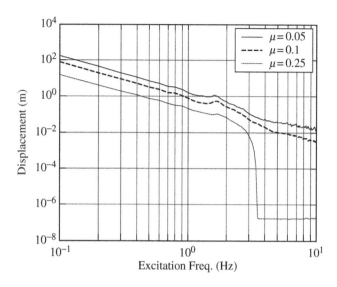

**FIGURE 4.27** Maximum base displacement vs. ground excitation frequency ($\omega_e = 5\omega_s$).

been plotted in Figure 4.28. As can be seen, the equipment quickly reaches the steady state within a few cycles of oscillation. The equipment response is effectively suppressed for the case with a smaller frictional coefficient. Additionally, the waveforms shown in Figure 4.28 for the equipment appear to be marginally higher than those of the primary structure shown in Figure 4.24, which can be attributed to the use of a relatively stiff equipment, that is, with $\omega_e = 5\omega_s$.

*Frequency responses.* For an equipment item of the frequency $\omega_e = 5\omega_s$ (= 8.34 Hz), the maximum acceleration response has been plotted as a function of the excitation frequency in Figure 4.29. By comparing Figure 4.29 with Figure 4.26, one observes that the frequency response curves of the equipment and primary structure are generally similar, except that a secondary resonant peak occurs around the equipment natural frequency of 8.34 Hz in Figure 4.29. The other observations from Figure 4.29 are as follows: (1) In comparison with the fixed-base case, the sliding isolation alleviates both the structural and equipment resonant peaks around the frequencies of 1.67 and 8.34 Hz, respectively. However, the level of alleviation is more apparent for the former than for the latter. (2) The equipment also exhibits the same subharmonic resonance behavior as that of the primary structure, in terms of the resonance peaks and frequencies. (3) As the frictional coefficient $\mu$ decreases from $\infty$ (for the fixed-base case) to 0.05, the main resonant frequency associated with the structure drifts toward a higher value. However, the resonant frequency associated with the equipment remains the same.

*Effect of equipment tuning.* The effect of equipment tuning refers to the case when the equipment frequency is tuned to the structural frequency, that is, $\omega_e = \omega_s$. Figure 4.30 shows the frequency response of the equipment when the equipment tuning occurs. Compared with Figure 4.29, this figure shows the following: (1) When equipment tuning occurs, the sliding isolation system can still mitigate the main resonant peak of the equipment, but the effectiveness of mitigation is drastically reduced. (2) Although the subharmonic resonance can still be observed, the relevant frequencies of the equipment are different from those of the primary structure. (3) The frequency of the maximum resonant response remains equal to the tuned equipment's natural frequency of 1.67 Hz, regardless of the change in the frictional coefficient, $\mu$, from $\infty$ to 0.05.

#### 4.4.3.4 Earthquake Response of Structure

*Time history.* For the isolated system subjected to the El Centro earthquake with a PGA of 0.5g, the structural acceleration and base displacement of the sliding system have been plotted in Figure 4.31 and Figure 4.32, respectively, together with the response for the fixed-base case in Figure 4.31. As can be

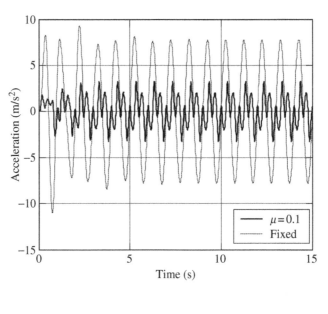

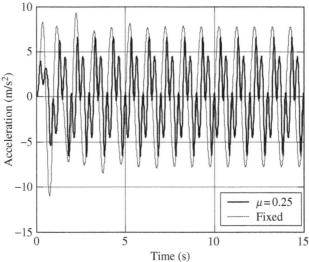

**FIGURE 4.28** Comparison of equipment accelerations ($\omega_g = 1$ Hz; $\omega_e = 5\omega_s$).

seen from Figure 4.31, the main-shock response occurring between 0 and 10 sec for the fixed-base structure has been effectively suppressed by the sliding isolators with $\mu = 0.1$ and 0.25. About 80 and 60% of the maximum structural acceleration have been suppressed by the isolators with $\mu = 0.1$ and 0.25, respectively. On the other hand, Figure 4.32 demonstrates that the better suppression effect for the case with $\mu = 0.1$ is achieved at the expense of a larger base displacement. It is interesting to note that the horizontal segments in the curves of Figure 4.32 actually represent the stick state of the sliding system, which is useful for unveiling the sliding-stick mechanism involved.

*Effect of earthquake intensity.* The maximum structural acceleration and base displacement vs. the PGA have been plotted in Figure 4.33 and Figure 4.34, respectively. As can be seen from Figure 4.33, the maximum structural acceleration for the fixed-base case is proportional to the earthquake intensity, while in all the sliding cases it remains essentially as a constant after the PGA reaches a certain level. In other words, the reduction in structural maximum response and the efficiency of isolation have increased

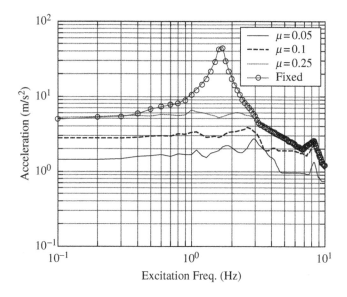

**FIGURE 4.29**  Maximum equipment acceleration vs. ground excitation frequency ($\omega_e = 5\omega_s$).

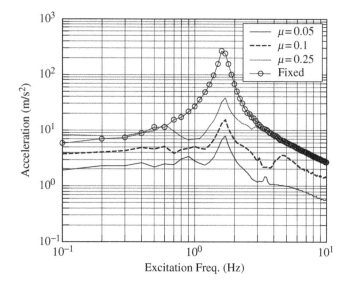

**FIGURE 4.30**  Maximum equipment acceleration vs. ground excitation frequency for equipment tuning ($\omega_e = \omega_s$).

with the increase in earthquake intensity. For example, for the case of $\mu = 0.1$, Figure 4.33 shows that the structural acceleration is reduced by around 60% at PGA $= 0.2g$, while it is reduced by more than 90% at PGA $= 1.0g$. However, as indicated by Figure 4.34, the above reduction in structural response has been achieved at the expense of increased base displacements. For the same PGA level, a sliding system with a smaller frictional coefficient has a better effect of vibration reduction, but this is accompanied by a larger base displacement.

*Residual base displacement.* The residual base displacement is defined as the permanent base displacement of the structure after it stops vibrating. This quantity is important in the study of sliding structures. Figure 4.35 shows the residual base displacement as a function of the earthquake PGA level. A first look at the figure reveals that no clear relation exists between the earthquake intensity and residual

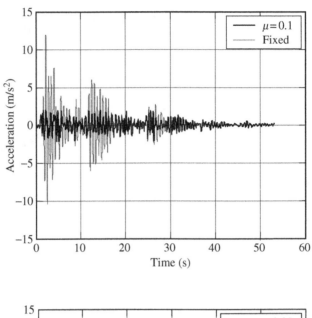

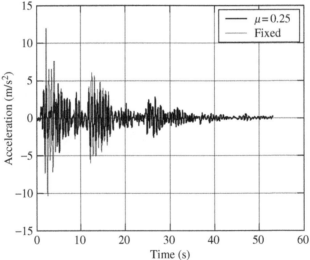

**FIGURE 4.31** Comparison of structural accelerations ($\omega_e = 5\omega_s$, PGA = 0.5$g$).

displacement, because a larger PGA may lead to a smaller residual base displacement. Nevertheless, after taking the average of the residual displacements over the PGA range of 0.1 to 1$g$, we obtain $x_{res} = 0.083$, 0.084, 0.084 m for the case of $\mu = 0.05$, 0.1, 0.25, respectively. These values indicate that a smaller frictional coefficient leads to a larger residual base displacement in general. However, when the frictional coefficient, $\mu$, approaches zero, the residual displacement approaches a constant equal to the permanent ground displacement.

### 4.4.3.5 Earthquake Response of Equipment

*Time history.* Consider an equipment item with a natural frequency equal to five times the structural frequency, that is, $\omega_e = 5\omega_s$ (= 8.34 Hz). For the isolated system subjected to the El Centro earthquake

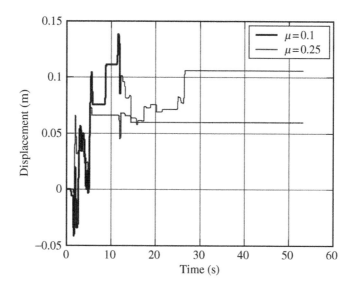

**FIGURE 4.32** Comparison of base displacements ($\omega_e = 5\omega_s$, PGA $= 0.5g$).

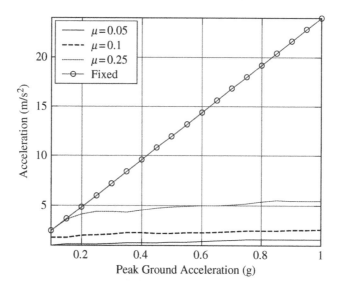

**FIGURE 4.33** Maximum structural acceleration vs. PGA ($\omega_e = 5\omega_s$).

with PGA $= 0.5g$, the time histories computed for the equipment acceleration for the cases with $\mu = 0.1$ and 0.25, along with the fixed-base case, have been plotted in Figure 4.36a and b. As can be seen, the main-shock response of the fixed-base structure occurring for the first 10 sec has been effectively suppressed through installation of the sliding isolator with $\mu = 0.1$ and 0.25. A higher level of reduction can be achieved if a smaller frictional coefficient is chosen.

*Effect of earthquake intensity.* Figure 4.37 shows the maximum equipment acceleration as a function of the PGA level. Because of the use of a relatively stiff equipment ($\omega_e = 5\omega_s$), the curves shown in Figure 4.37 are similar to those for the primary structure in Figure 4.33, but with slightly higher values. Therefore, the observations made previously for Figure 4.33 are applicable to Figure 4.37. The maximum response of equipment items with other frequencies will be discussed below.

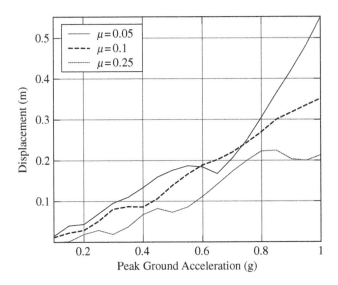

**FIGURE 4.34**   Maximum base displacement vs. PGA ($\omega_e = 5\omega_s$).

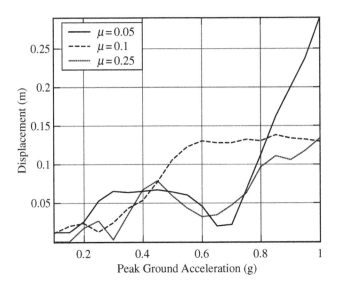

**FIGURE 4.35**   Residual base displacement vs. PGA ($\omega_e = 5\omega_s$).

*Effect of equipment tuning.* In order to study the equipment tuning effect, the maximum equipment acceleration has been plotted as a function of the equipment frequency in Figure 4.38. As can be seen, for all the values of $\mu$ considered, the equipment response is amplified when the equipment frequency moves close to the structural frequency of 1.67 Hz for the fixed-base case. Note that, since the resonant frequency of a sliding structure shifts to a higher value as the frictional coefficient decreases (see Figure 4.26), the frequency for which the most severe tuning effect occurs in Figure 4.38 also shifts from 1.67 Hz to a higher value as $\mu$ decreases. Nevertheless, it is concluded that, by choosing a smaller $\mu$, the amplification of the equipment response due to tuning effect can be effectively suppressed.

## 4.4.4   Concluding Remarks

The dynamic behavior of a sliding isolated structural system with an attached equipment item was investigated in this section. A sliding isolated structure is classified as a nonlinear dynamic system, as the

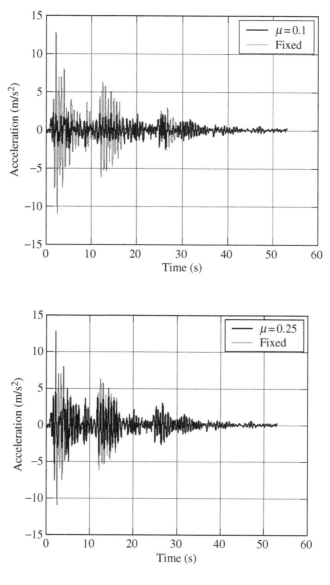

**FIGURE 4.36**  Comparison of equipment accelerations ($\omega_e = 5\omega_s$, PGA $= 0.5g$).

frictional forces induced on the sliding surface do not remain constant. To deal with such nonlinear systems, two analysis methods were formulated, the shear balance method and the fictitious spring method, both of which were presented in an incremental form that is suitable for direct implementation. Through the selection of a sliding isolated structure–equipment model, the responses of the structure and equipment subjected to both harmonic and earthquake excitations were analyzed. For the case of harmonic excitation, the results showed that the resonant responses of both the structure and attached equipment can be effectively suppressed, which remains good even when the equipment frequency is tuned to the structural frequency. For the case of seismic excitation, the results indicated that the level of reduction on the structural and equipment responses increases as the PGA level of the earthquake increases. Moreover, a sliding system with a smaller frictional coefficient has a higher isolation efficiency, at the expense of a larger base displacement.

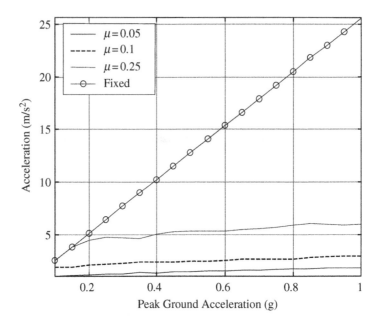

**FIGURE 4.37**   Maximum equipment acceleration vs. PGA ($\omega_e = 5\omega_s$).

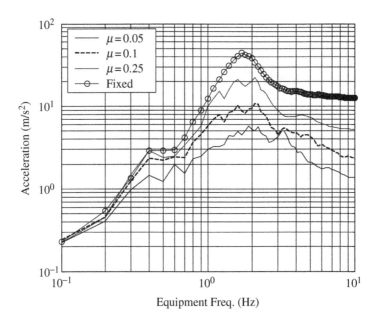

**FIGURE 4.38**   Maximum equipment acceleration vs. equipment frequency (PGA $= 0.5g$).

## 4.5   Sliding Isolation Systems with Resilient Mechanism

In Section 4.4, the relevant equations of analysis have been presented for a sliding isolated structural system, with no consideration made for the resilient (or recentering) mechanism. Because of this, rather large residual displacements may occur on the sliding isolation system, as have been numerically illustrated. If the concept of sliding isolators is to be applied to a real structure, it is important that the residual base displacements be controlled within certain limits, since they are not tolerable for some

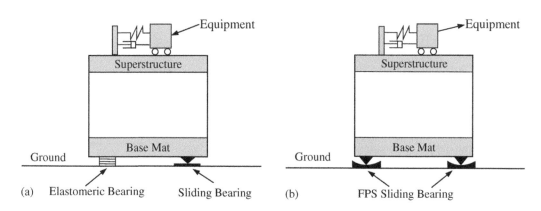

**FIGURE 4.39** Schematic for a structure–equipment system isolated by a sliding bearing with resilience capability: (a) combined isolation system; (b) friction pendulum system (FPS).

engineering applications. For example, residual base displacements may distort the networking of water and power lines, change the space between the isolated structure and adjacent buildings, and widen the gaps at building entrances. Therefore, in practice, a sliding isolation system is usually enhanced through inclusion of mechanisms that can provide some resilient force. However, for certain ground motions, the added resilient force may also present some negative effects not readily transparent to structural designers, as will be illustrated in the numerical studies later on.

As shown in Figure 4.39, there are at least two ways of implementing the resilient mechanism in a sliding isolation system. Figure 4.39a shows an isolation system that combines the elastomeric bearings with sliding bearings (Chalhoub and Kelly, 1990), in which the elastomeric bearings are used to provide the resilient force, and the sliding bearings to uncouple the structural system from the ground motion. On the other hand, the resilient mechanism can also be incorporated into each single sliding bearing, in a way similar to that in the RFBI described in Section 4.2.3 (Mostaghel and Khodaverdian, 1987) or the FPS shown in Figure 4.39b (Mokha et al., 1991). The FPS isolation system has been implemented in many existing buildings and bridges. A typical FPS bearing consists a spherical sliding surface and a slider, which usually has a smooth coating of very low friction. When an FPS device is implemented under a structure, the slider will slide on the spherical surface during an earthquake, and the gravitational load of the structure, together with the curved sliding surface, will provide the resilient force for the system to return to its original position. The resilient stiffness of an FPS bearing depends on the radius of curvature of the sliding surface and the structural weight carried by the bearing.

This section is aimed at investigating the behavior of a structure–equipment system isolated by a sliding system with resilient device. For convenience of discussion, a system with resilient device will be referred to as the *resilient sliding isolation* (RSI), and a sliding system without resilient device, as the one studied previously, as the *pure sliding isolation* (PSI).

## 4.5.1 Mathematical Modeling and Formulation

Both the RSI systems shown in Figure 4.39 can be represented by the mathematical model given in Figure 4.40, for which the symbols used have been defined in Table 4.1. The RSI model shown in Figure 4.40 differs from the PSI model shown Figure 4.16 in that a linear spring of stiffness, $k_b$, is added to simulate the resilient force of the isolator. Obviously, an RSI model can be considered as the composition of a friction element and a spring element in parallel. Owing to addition of resilient stiffness, the number of vibration frequencies of the system is increased by one. The newly introduced frequency, which depends on the resilient stiffness, is called the *isolation frequency*, which can be approximated by

$$\omega_b = \sqrt{k_b/(W/g)} \qquad (4.61)$$

where $W$ is the total weight of the isolated structure–equipment system. The isolation frequency commonly used in design is between 0.33 and 0.5 Hz, which implies a period of 2 to 3 sec (Naeim and Kelly, 1999).

For the type of combined sliding system shown in Figure 4.39(a), the actual value of resilient stiffness, $k_b$, is decided by the total horizontal stiffness of the elastomeric bearings implemented. On the other hand, for the FPS shown in Figure 4.39(b), the resilient stiffness, $k_b$, is approximated by the following equation for small isolator displacements:

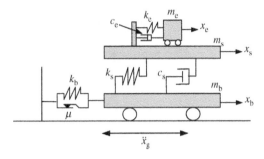

**FIGURE 4.40** Model for a structure–equipment system isolated by a sliding bearing with resilience capability.

$$k_b = \frac{W}{R} \qquad (4.62)$$

where $R$ denotes the radius of curvature of the sliding surface. By substituting Equation 4.62 into Equation 4.61, one can verify that the isolation period of an FPS is equal to the oscillation period of a pendulum; that is

$$T_b = \frac{2\pi}{\omega_b} = \frac{2\pi}{\sqrt{k_b/(W/g)}} = 2\pi\sqrt{R/g} \qquad (4.63)$$

When the sliding isolated system of Figure 4.40 is excited by a ground motion, its equation of motion may be written in exactly the same form as that of Equation 4.27; that is

$$\mathbf{M}\ddot{\mathbf{x}}(t) + \mathbf{C}\dot{\mathbf{x}}(t) + \mathbf{K}\mathbf{x}(t) = -\mathbf{M}\mathbf{L}_1\ddot{x}_g(t) + \mathbf{L}_2 f(t) \qquad (4.64)$$

All the variables used in the preceding equation are the same as those defined in Equation 4.28 to Equation 4.30, except that the stiffness matrix, $\mathbf{K}$, should be modified as

$$\mathbf{K} = \begin{bmatrix} k_e & -k_e & 0 \\ -k_e & k_e + k_s & -k_s \\ 0 & -k_s & k_s + k_b \end{bmatrix} \qquad (4.65)$$

Note that, in Equation 4.64, the frictional force, $f(t)$, which does not remain constant, is placed on the right-hand side, while the resilient stiffness, $k_b$, which remains constant, is absorbed by the stiffness matrix, $\mathbf{K}$, as in Equation 4.65. However, if the total shear force, $s(t)$ of the isolation system is of interest, it should be computed as the summation of the frictional force and resilient force (see Figure 4.40); that is

$$s(t) = k_b x_b(t) + f(t) \qquad (4.66)$$

The equation of motion as given in Equation 4.64 can be recast in the following form of the first-order state-space equation:

$$\dot{\mathbf{z}}(t) = \mathbf{A}\mathbf{z}(t) + \mathbf{E}\ddot{x}_g(t) + \mathbf{B}f(t) \qquad (4.67)$$

The definitions of the matrices $\mathbf{z}(t)$, $\mathbf{E}$, $\mathbf{A}$, and $\mathbf{B}$ are the same as those defined in Equation 4.32 and Equation 4.33, except that the system matrix, $\mathbf{A}$, should be modified to account for the addition of the resilient stiffness $k_b$ in the stiffness matrix, $\mathbf{K}$.

## 4.5.2 Methods for Numerical Analysis

If one compares the equation of motion for the RSI system in Equation 4.67 with that for the PSI system in Equation 4.31, one will conclude that the only source of nonlinearity in both equations comes from the same term, namely, the frictional force, $f(t)$. As a result, the two methods of solution mentioned in Section 4.4.2, the shear balance method (Wang et al., 1998) and fictitious spring method (Yang et al., 1990), remain valid for the analysis of the RSI systems, with no modification required. Moreover, owing to inclusion of the resilient stiffness, $k_b$, in the structural stiffness matrix, **K**, the system matrix, **A**, becomes nonsingular and invertible. This introduces some advantage in computation of relevant coefficient matrices, including the $\mathbf{B}_0$ and $\mathbf{B}_1$ matrices in Equation 4.38a and Equation 4.38b. In Section 4.5.3, the shear balance method will be employed to simulate the response of an RSI system.

## 4.5.3 Simulation Results for Sliding Isolation with Resilient Mechanism

### 4.5.3.1 Numerical Model and Ground Excitations

In this section, the dynamic behavior of a sliding system represented by the model shown in Figure 4.40 will be investigated. The data adopted in the analysis for the equipment, structure, and the isolator have been listed in Table 4.4. To facilitate comparison, some of the data are selected to be the same as those in Table 4.3. In particular, the isolation frequency chosen is $\omega_b = 0.4$ Hz, falling in the common range of 0.33 to 0.5 Hz. Again, two types of ground excitations are considered, namely, the harmonic and earthquake excitations. For the harmonic excitation, a waveform of ground acceleration identical to the one given in Equation 4.60 is used. And for the earthquake excitation, the 1940 El Centro earthquake with different levels of PGA will be used, of which the acceleration waveform has been given in Figure 4.10. The harmonic excitation is adopted mainly for studying the frequency response of the sliding isolated system, while the earthquake excitation is for studying the effect of earthquake intensity. The dynamic responses computed for the RSI system, including the structure and equipment, will be presented, with emphasis placed on comparison with the PSI system of the same parameters. Similar to what was done in Section 4.4.3, the symbol $\mu$ will be used to denote the frictional coefficient of the sliding isolation system in all figures, and the word "fixed" denotes the fixed-base structure.

### 4.5.3.2 Harmonic Response of Structure

*Time history.* Consider an RSI system subjected to a harmonic excitation of $\omega_g = 1$ Hz. The base displacement and structural acceleration of the RSI system have been plotted in Figure 4.41 and Figure 4.42, respectively. Clearly, both the base displacement and structural acceleration of the RSI system reach their steady-state harmonic responses in the first few cycles. Moreover, a smaller sliding frictional coefficient ($\mu = 0.1$) is more effective for suppressing the structural acceleration, as indicated by Figure 4.42. However, this is achieved only at the expense of a larger base displacement, as indicated by Figure 4.41. By comparing the result for the RSI system in Figure 4.41 with those for the PSI system in Figure 4.23, the effect of resilient mechanism in eliminating the permanent base displacement for the case with a small frictional coefficient of $\mu = 0.1$ can be clearly appreciated. In spite of the large difference in base displacement, the structural accelerations for the RSI and PSI systems shown in Figure 4.42 and Figure 4.24, respectively, appear to be quite similar, when interpreted in terms of the waveform and response amplitude. This implies that, for the harmonic excitation considered, the resilient mechanism in RSI has little influence on the isolation effectiveness.

*Hysteretic behavior.* In Figure 4.43, the hysteresis loops for RSI systems with $\mu = 0.1$ and 0.25 subjected to a harmonic excitation of $\omega_g = 1$ Hz have been plotted, in which the vertical axis represents

**TABLE 4.4** System Parameters Used in Simulation (Section 4.5.3)

| Equipment | | Superstructure | | Isolation System | |
|---|---|---|---|---|---|
| Parameter | Value | Parameter | Value | Parameter | Value |
| Mass $m_e$ | 3 t ($= m_s/100$) | Mass $m_s$ | 300 t | Mass $m_b$ | 100 t ($= m_s/3$) |
| Frequency $\omega_e$ | $5\omega_s$ or a variable | Frequency $\omega_s$ | 1.67 Hz | Frictional coefficient $\mu$ | 0.05, 0.1, 0.25 |
| Damping ratio $\zeta_e$ | 5% | Damping ratio $\zeta_s$ | 5% | Isolation frequency $\omega_b$ | 0.4 Hz |

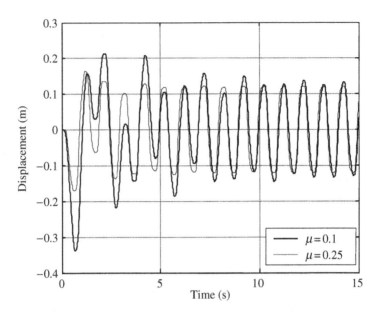

**FIGURE 4.41** Comparison of base displacements ($\omega_g = 1$ Hz, $\omega_e = 5\omega_s$).

the total shear, $s(t)$, of the bearing computed from Equation 4.66. It is interesting to note that, for an RSI system, the hysteresis loop is a parallelogram, of which the slope of the inclined upper and lower sides exactly represents the total resilient stiffness, $k_b$, and the height and width, respectively, are decided by the maximum frictional force and maximum base displacement. As the frictional coefficient decreases, the height of the parallelogram decreases, but the width increases. The total area of the hysteresis loop represents the portion of energy dissipated by the RSI system. Noteworthy is the fact that, when the resilient stiffness, $k_b$, reduces to zero, the hysteresis parallelogram reduces to a square as well, identical to the one shown in Figure 4.25 for the PSI system.

*Frequency responses.* The maximum accelerations of the steady-state response of the structure for four different frictional coefficients, that is, $\mu = 0.05$, 0.1, 0.25 and $\infty$ (fixed-base), have been plotted in Figure 4.44. From this figure, the following observations can be made: (1) Compared with the fixed-base case, the resonant peak occurring around the structural frequency, $\omega_s$, of 1.67 Hz was effectively suppressed by the RSI system, but a resonance of higher amplitude was induced in the lower frequency range (with frequencies lower than 0.6 Hz for the case studied). A further investigation reveals that the newly induced resonance is associated with the isolation frequency, $\omega_b$, of 0.4 Hz. Such an observation remains valid for all values of frictional coefficients, $\mu$. (2) The use of a lower frictional coefficient, $\mu$, will result in a smaller response in the high-frequency range for the RSI system, for example, with frequencies higher than 0.6 Hz, but a larger response for the low-frequency range. (3) Although both the RSI and PSI systems can effectively remove the resonant peak around the structural frequency of 1.67 Hz, the RSI system has the side effect of creating a low-frequency resonant peak at the isolation frequency, $\omega_b$. This implies that the RSI system is more sensitive to the excitation frequency.

The frequency responses of the maximum base displacement for the RSI system with various frictional coefficients have been plotted in Figure 4.45. When compared with the results for the PSI system in Figure 4.27, it is clear that the resilient mechanism of the RSI system considerably reduces the base displacement in the nonresonant excitation range, but it also amplifies the base displacement in the region when the excitation frequency is close to the isolation frequency, $\omega_b$. From Figure 4.44 and Figure 4.45, we observe that both the structural acceleration and base displacement of an RSI system may resonate at the isolation frequency, which is usually designed to be less than 0.5 Hz. This implies that an RSI system may be ineffective or unsafe for a ground motion with enriched low-frequency

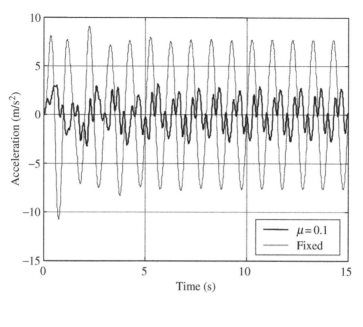

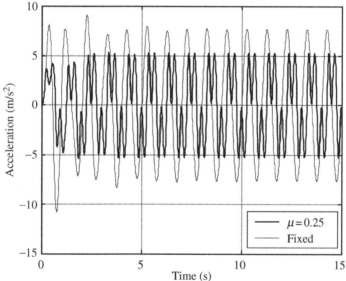

**FIGURE 4.42** Comparison of structural accelerations ($\omega_g = 1$ Hz, $\omega_e = 5\omega_s$).

(long-period) vibrations, such as the case with a near-fault earthquake containing a long-period, pulse-like waveform (Jangid and Kelly, 2001; Lu et al., 2003). Structural designers should be aware of such a side effect when designing an RSI system.

### 4.5.3.3  Harmonic Response of Equipment

*Time history.* Consider an equipment item with a frequency of $\omega_e = 5\omega_s$ (= 8.34 Hz), attached to the RSI system. The harmonic acceleration responses of the equipment for the case with $\mu = 0.1$ and 0.25 have been plotted in Figure 4.46a and b, respectively, together with those for the fixed-base case. As can be seen, the equipment acceleration has been effectively suppressed by the RSI with the smaller frictional coefficient ($\mu = 0.1$). Moreover, the acceleration waveforms shown in Figure 4.46 are similar to those of

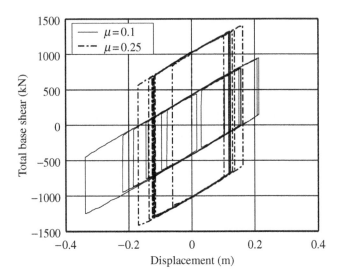

**FIGURE 4.43** Hysteresis loops of a sliding bearing with resilience capability ($\omega_g = 1$ Hz, $\omega_e = 5\omega_s$).

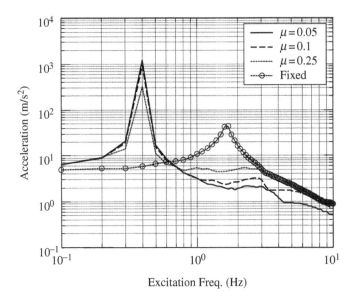

**FIGURE 4.44** Maximum structural acceleration vs. ground excitation frequency ($\omega_e = 5\omega_s$).

the PSI system shown in Figure 4.28. This implies that for the given excitation, the behavior of the equipment was not altered by introduction of the resilient mechanism in the RSI system.

*Frequency responses.* Figure 4.47 shows the acceleration frequency response curve of the attached equipment with a frequency of $\omega_e = 5\omega_s$ (8.34 Hz). A comparison of Figure 4.47 with Figure 4.44 indicates that the frequency responses of the equipment and primary structure are generally similar, except that a resonant peak associated with the equipment frequency around 8.34 Hz appears in Figure 4.47. Owing to such a similarity, the observations made previously for Figure 4.44 are applicable to Figure 4.47 for the attached equipment.

*Effect of equipment tuning.* Figure 4.48 shows the frequency response of the attached equipment for the case when the equipment frequency is tuned to the structural frequency, that is, with $\omega_e = \omega_s$. Similar to

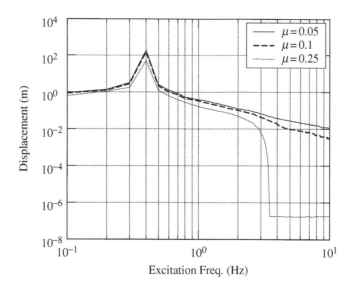

**FIGURE 4.45** Maximum base displacement vs. ground excitation frequency ($\omega_e = 5\omega_s$).

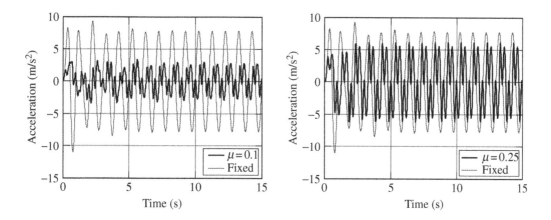

**FIGURE 4.46** Comparison of equipment accelerations ($\omega_g = 1$ Hz, $\omega_e = 5\omega_s$).

the structural frequency response shown in Figure 4.44, the equipment attached to the RSI system also resonates at the isolation frequency $\omega_b$ of 0.4 Hz. Such a resonance does not occur for the equipment attached to the PSI system (see Figure 4.30). Through comparison of the tuned case in Figure 4.48 with the detuned case in Figure 4.47, the following observations can be made: (1) Even when the equipment tuning occurs, an RSI system mitigates the equipment's resonant peak associated with the structural frequency at 1.67 Hz, although the effectiveness of isolation has been reduced. (2) The tuning effect has no influence on the resonant response associated with the isolation frequency of 0.4 Hz.

### 4.5.3.4 Earthquake Response of Structure

*Time history.* For an RSI system subjected to the El Centro earthquake with PGA $= 0.5g$, the structural acceleration and base displacement have been shown in Figure 4.49 and Equation 4.50, respectively. By comparing Figure 4.49 with Figure 4.31 for the corresponding PSI system, one observes that the structural accelerations of the RSI and PSI systems are generally similar, in terms of the response waveform and the response magnitude. Both systems reduce the maximum structural acceleration quite effectively, for example, by about 80% for $\mu = 0.1$. However, significant difference does exist between the

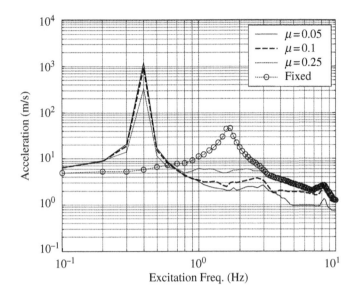

**FIGURE 4.47**  Maximum equipment acceleration vs. ground excitation frequency ($\omega_e = 5\omega_s$).

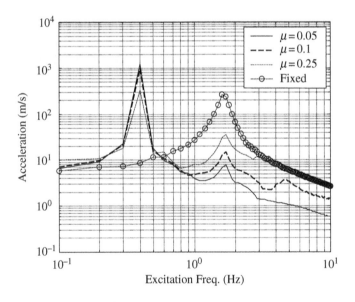

**FIGURE 4.48**  Maximum equipment acceleration vs. ground excitation frequency under tuning condition ($\omega_e = 5\omega_s$).

base displacements for the RSI system in Figure 4.50 and those for the PSI system in Figure 4.32. For example, for $\mu = 0.1$, the maximum base displacement experienced by the RSI system has been reduced by about 30%, while the residual base displacement has been reduced by about 70%, as can be seen by comparing Figure 4.50 with Figure 4.32. This implies that the resilient mechanism of the RSI system plays an important role in reducing the maximum and residual base displacements, especially the latter.

In spite of the observations made above, one should not forget that the frequency content of one earthquake may be different from an other. As was demonstrated in Figure 4.44 and Figure 4.45, an RSI system is generally sensitive to low-frequency excitations and may resonate at the isolation frequency.

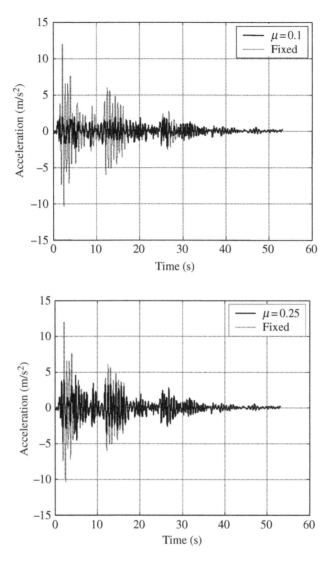

**FIGURE 4.49** Comparison of structural accelerations ($\omega_e = 5\omega_s$, PGA $= 0.5g$).

Therefore, if the RSI system is subjected to an earthquake containing more low-frequency components, unlike the El Centro earthquake, it is likely that the maximum structural responses induced exceed those of the PSI system.

*Effect of earthquake intensity.* The maximum structural acceleration and base displacement of the RSI system have been plotted with respect to the PGA in Figure 4.51 and Figure 4.52, respectively. These figures indicate that as the earthquake intensity increases from 0.1 to $1g$, the structural acceleration is reduced by an increasing amount by the RSI system, while the maximum base displacement also increases. By comparing Figure 4.51 and Figure 4.52 with Figure 4.33 and Figure 4.34 for the PSI system, one observes that both the RSI and PSI systems perform equally well for the El Centro earthquake, although the PSI system induces a slightly larger base displacement. On the other hand, unlike the response for the PSI system, the use of a smaller frictional coefficient for the RSI system does not always lead to a lower structural acceleration, as can be verified by comparing the responses for $\mu = 0.1$ and $0.05$ with a PGA greater than $0.8g$ in Figure 4.33 and Figure 4.51. This can be attributed to the large resilient force induced by the large base displacement under higher PGA levels.

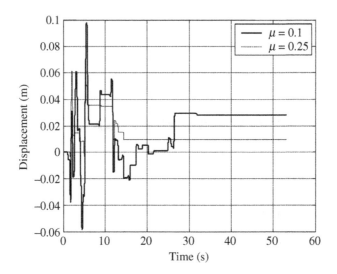

**FIGURE 4.50**  Comparison of base displacements ($\omega_e = 5\omega_s$, PGA = 0.5$g$).

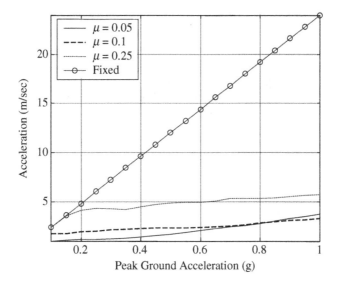

**FIGURE 4.51**  Maximum structural acceleration vs. PGA ($\omega_e = 5\omega_s$).

*Residual base displacement.* Figure 4.53 shows the residual base displacement of the RSI system vs. the PGA of the earthquake. For a given $\mu$, it is difficult to establish a relation between the earthquake intensity and residual displacement, because a larger PGA may result in a smaller residual base displacement in some cases. However, if one takes the average of residual displacements over the PGA range from 0.1 to 1$g$, the following can be computed: $x_{res} = 0.0065$, 0.011, and 0.014 m for $\mu = 0.05, 0.1$, and 0.25, respectively. These values indicate that a smaller frictional coefficient leads to a smaller residual base displacement, which can be attributed to the fact that for a SRI system with a smaller coefficient of friction, it is easier for the resilient mechanism to return the structure to its initial position after an earthquake. On the other hand, a comparison of Figure 4.53 with Figure 4.35 for the PSI system indicates that for the same value of $\mu$, the residual displacement was reduced substantially by the RSI system. This is certainly an advantage offered by the resilient mechanism of the RSI system.

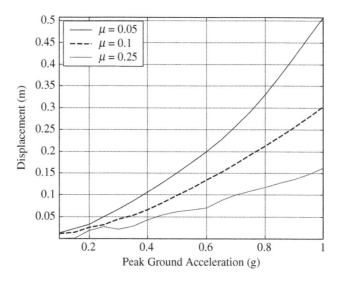

**FIGURE 4.52** Maximum base displacement vs. PGA ($\omega_e = 5\omega_s$).

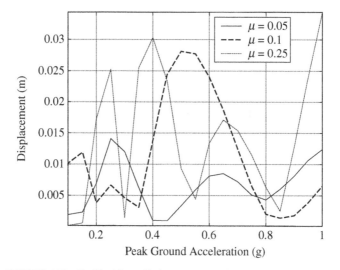

**FIGURE 4.53** Residual base displacement vs. PGA ($\omega_e = 5\omega_s$).

### 4.5.3.5 Earthquake Response of Equipment

*Time history.* Let us consider an equipment item of natural frequency equal to five times the structural frequency, that is, $\omega_e = 5\omega_s$ (= 8.34 Hz). The acceleration responses of the equipment mounted on the RSI system that were subjected to the El Centro earthquake with a PGA of 0.5g for $\mu = 0.1$ and 0.25 have been plotted in Figure 4.54a and b, respectively, along with those for the fixed-base cases. As can be seen, the main-shock response of the equipment appearing during the first 10 sec for the fixed-base system was effectively suppressed by the RSI system with $\mu = 0.1$ or 0.25. The level of reduction is more pronounced for the case with a smaller frictional coefficient, that is, with $\mu = 0.1$. By comparing Figure 4.54 with Figure 4.36 for the PSI system, one concludes that the effect of the resilient mechanism of the RSI system on the equipment response is insignificant for the earthquake and equipment frequency considered.

*Effect of earthquake intensity.* Figure 4.55 shows the maximum equipment acceleration vs. the PGA of the earthquake. This figure illustrates that for all values of the frictional coefficient, $\mu$, considered, an

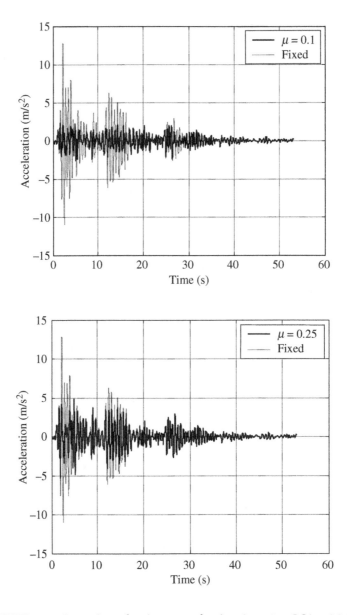

**FIGURE 4.54** Comparison of equipment accelerations ($\omega_e = 5\omega_s$, PGA $= 0.5g$).

increasing amount of reduction can be achieved by the RSI system as the earthquake intensity increases from 0.1 to 1$g$. Because relatively stiff equipment (i.e., with $\omega_e = 5\omega_s$) was assumed in the simulation, the curves shown in Figure 4.55 are similar to those of Figure 4.51 for the primary structure. Therefore, the observations made previously for Figure 4.51 apply here. The maximum response of equipment items with other natural frequencies will be discussed below. Moreover, a comparison of Figure 4.55 with Figure 4.37 (for the PSI system) reveals that the resilient mechanism can have some minor effect on the equipment response, but only when a smaller frictional coefficient (i.e., $\mu = 0.05$ or 0.1) is used and when the PGA level is high.

*Effect of equipment tuning.* In order to study the effect of equipment tuning, the maximum acceleration of the equipment has been plotted in Figure 4.56 for equipment frequencies ranging from 0.1 to 10 Hz.

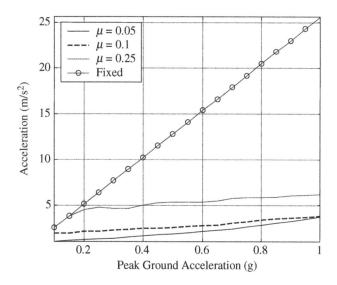

**FIGURE 4.55** Maximum equipment acceleration vs. PGA ($\omega_e = 5\omega_s$).

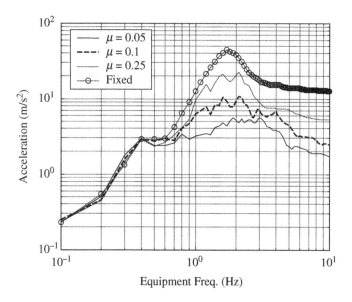

**FIGURE 4.56** Maximum equipment acceleration vs. equipment frequency (PGA = 0.5$g$).

As can be seen, for all the values of $\mu$ considered, the equipment response is amplified when the equipment frequency is close to the structural frequency, $\omega_s$, of 1.67 Hz, which means that the tuning effect tends to enlarge the equipment response. However, the use of a smaller $\mu$ can help in reducing the amplification of the equipment response resulting from the tuning effect. Finally, a comparison between Figure 4.56 and Figure 4.38 (for the PSI system) shows that the two diagrams are quite similar for an equipment item with a frequency higher than 1 Hz, but are different for that with a lower frequency. This implies that for the earthquake considered, the resilient mechanism of the RSI system has little effect on the response of the equipment with a higher stiffness.

### 4.5.4 Concluding Remarks

In this section, the behavior of a structure–equipment system isolated by an RSI system under both the harmonic and earthquake excitations has been investigated. Both the responses of the structure and equipment were studied, with special attention given to the effect of the resilient mechanism that characterizes an RSI system. The numerical results demonstrated that when subjected to a harmonic excitation, an RSI system is able to effectively suppress the resonant peaks associated with the structural frequency for both the structure and equipment, but it may also induce some resonant response near the isolation frequency due to the presence of resilient stiffness. Therefore, an RSI system is more sensitive to the frequency content of the ground excitation than a PSI system, especially to excitations of low-frequency components. As for the earthquake responses, the numerical results showed that the resilient mechanism of an RSI system can considerably reduce the residual base displacement. The resilient mechanism has a minor effect on the acceleration response of the structure and equipment, as long as no resonance is induced by the RSI system at the isolation frequency. By and large, both the RSI and PSI systems can be used as effective devices for reducing the acceleration responses of a structure and equipment.

## 4.6   Issues Related to Seismic Isolation Design

### 4.6.1   Design Methods

Having been developing for over 30 years, the technology of seismic isolation has matured. Many earthquake-prone countries, including the U.S., Japan, New Zealand, Taiwan, China, and European countries, have developed their own design codes, regulations, or guidelines (Fujita, 1998; Kelly, 1998; Martelli and Forni, 1998). Although most of the codes were developed based on the theory of structural dynamics, the design details outlined in the codes vary from one country to another. While a comprehensive explanation of the various design codes is not the purpose of this section, a brief overview of the concept underlying the design codes will be given. For more details, interested readers should refer to each code or to the books by Naeim and Kelly (1999) or Skinner et al. (1993). The design concept introduced herein is based on the series of Uniform Building Code (UBC, 1994, 1997).

Given the fact that base isolation devices are diverse, most design codes or regulations have been written in such a way as not to be specific with respect to the isolation systems. For instance, in the UBC (1997), no particular isolation system is identified as being acceptable; rather, it requires that every isolation system is stable for required displacement, has properties that do not degrade under repeated cyclic loadings, and provides increasing resistance with increasing displacement.

The design methods for base isolation can be classified as static analysis and dynamic analysis. The static analysis is applicable for stiff and regular buildings (in vertical and horizontal directions) that are constructed on soil of a relatively stiff condition. On the other hand, dynamic analysis is usually required for isolation systems with an irregular or long-period superstructure, or constructed on relatively soft soils. For a sophisticated design case, static analysis may be used in the preliminary design phase in order to draft or initiate the isolation design parameters, while dynamic analysis is employed in the final design phase for tuning or finalizing the design details of the isolation system. For simple design cases, static analysis alone is considered sufficient.

### 4.6.2   Static Analysis

For static analysis, a number of formulas have been specified in the design codes, so that engineers can easily calculate the following design parameters (shown in the design sequence): maximum isolator displacement, $D$; isolator total shear, $V_b$; total base shear, $V_s$, of superstructure; and seismic load, $F_i$, applied on each floor. These formulas were usually derived based on a simplified isolation model, assuming the isolation system can be linearized (even though most isolation systems are nonlinear) and the superstructure can be modeled as a rigid block. Such a simplified model is considered reasonable, since the displacements of an isolated structure are concentrated at the isolation level, which implies that the superstructure behaves as a rigid block. Based on such a model, only the first vibration mode with the superstructure treated as a rigid body has been considered in deriving the formulas. This explains why static analysis is suitable only for rigid and regular structures.

### 4.6.2.1 Computation of Maximum Isolator Displacement

An isolation design by static analysis usually starts with the calculation of the maximum isolator displacement, $D$, which depends on several factors:

$$D = D(Z, N, S, T_{ef}, \zeta_{ef}) \qquad (4.68)$$

where $Z$ denotes the earthquake zone factor, $N$ the near-fault factor, $S$ the soil condition factor, $T_{ef}$ the effective isolation period, and $\zeta_{ef}$ the effective isolation damping. For example, in the UBC (1994), the formula derived from the constant-velocity spectra over the period range of 1.0 to 3.0 sec has been given in the following form:

$$D = \frac{0.25ZNST_{ef}}{B} \qquad (4.69)$$

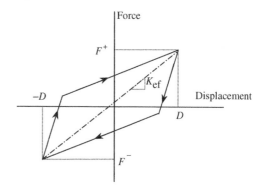

**FIGURE 4.57** Typical force–displacement diagram for an isolation system.

where $B$ is the damping factor, given as

$$B = B(\zeta_{ef}) \approx 0.25(1 - \ln \zeta_{ef}) \qquad (4.70)$$

In the above equations, the factors $Z$, $N$, and $S$ depend on conditions of the construction site of the isolated structure; however, the factors $T_{ef}$ and $\zeta_{ef}$ depend solely on the properties of the chosen isolation system. The factors $T_{ef}$ and $\zeta_{ef}$ are called the "effective" period and damping of the isolation system, because they are frequently obtained by linearizing a nonlinear isolation system. The way to linearize an isolation system will be explained below, along with the formulas for computing $T_{ef}$ and $\zeta_{ef}$. Suppose that for a nonlinear isolation system, the force-displacement relation (hysteresis loop) obtained from a component test is shown in Figure 4.57. The effective stiffness of this isolation system can be computed by

$$K_{ef} = \frac{F^+ - F^-}{2D} \qquad (4.71)$$

where $F^+$ and $F^-$, respectively, denote the largest positive and negative forces in the test. After the linearized stiffness is obtained from Equation 4.71, the corresponding effective quantities $T_{ef}$ and $\zeta_{ef}$ can be computed from the dynamic theory for a single DoF oscillation system; that is

$$T_{ef} = 2\pi\sqrt{\frac{W}{K_{ef}g}} \qquad (4.72)$$

$$\zeta_{ef} = \frac{1}{2\pi}\left(\frac{A}{K_{ef}D^2}\right) \qquad (4.73)$$

where $W$ is the structural weight, $g$ the gravitational acceleration, and $A$ the total area enclosed by the hysteresis loop in Figure 4.57.

#### 4.6.2.2 Computation of Maximum Isolator Shear

After the maximum isolator displacement, $D$, is obtained, the maximum isolator shear, $V_b$, can be estimated by the following formula:

$$V_b = K_{ef}D \tag{4.74}$$

Obviously, the above equation represents an equivalent static force exerted on the isolation system, when the system is displaced by an amount, $D$. In some design codes, $V_b$ has also been referred to as the design force beneath the isolation system.

#### 4.6.2.3 Computation of Total Base Shear

The total base shear, $V_s$, of the superstructure can be given as

$$V_s = \frac{K_{ef}D}{R_I} \tag{4.75}$$

where $R_I$ is a reduction factor (ductility factor) to account for structural ductility, which will be developed when the structure is subjected to an earthquake with intensity above the design level. In some codes, $V_s$ has also been referred to as the design force above the isolation system.

#### 4.6.2.4 Computation of Shear Force for Each Floor

Having computed the above total base shear, $V_s$, a formula is employed to distribute this total shear to each floor of the isolated structure. For instance, in the, UBC (1997), the shear force, $F_i$, exerted on each floor is computed by

$$F_i = V_s \frac{h_i w_i}{\sum_{j=1}^{n} h_j w_j} \tag{4.76}$$

where $n$ denotes the number of floors, $w_i$ the weight of the $i$th floor, and $h_i$ the height of the $i$th floor above the isolation level. Note that the sum of $F_i$ ($i = 1$ to $n$) must be equal to $V_s$.

The general procedure for static analysis was illustrated in Figure 4.58. Once the design parameters, $D$, $V_b$, $V_s$, and $F_i$, are all determined according to the code, they can be used in the detailed design of structural elements as well as of isolator elements. Nevertheless, in most applications, because the test data of the isolation system may not be available in the beginning of design, the values of $K_{ef}$, $T_{ef}$, and $\zeta_{ef}$, which are required in computing $D$, are not known to the designer. If this is the case, the design can begin with assumed values of $K_{ef}$, $T_{ef}$, and $\zeta_{ef}$, which may be obtained from experience or previous test data on similar isolators. After the preliminary design is completed, prototype isolators will be fabricated and tested. The actual values of $K_{ef}$, $T_{ef}$, and $\zeta_{ef}$, obtained from the tests will be used in the aforementioned code formulas to update the design parameters $D$, $V_b$, $V_s$. Moreover, one observes from Equation 4.71 that the linearized isolator stiffness, $K_{ef}$, is a function of the design parameter, $D$, itself, and so are $T_{ef}$, and $\zeta_{ef}$, obtained from Equation 4.72 and 4.73. In order to obtain $K_{ef}$, as well as $T_{ef}$, and $\zeta_{ef}$, an initial guess of $D$ is required at the beginning of design. As a result, the design procedure may have to be repeated iteratively until the difference between the final value of $D$ and the value $D_0$ computed in the last iteration is less than a preset tolerance. Such an iterative process is illustrated in Figure 4.58.

### 4.6.3 Dynamic Analysis

The dynamic analysis may be carried out in one of the two forms: response spectrum analysis and time-history analysis. Response spectrum analysis usually involves application of the concepts of response spectrum and modal superposition, and so on. Since these concepts primarily come from the dynamics of linear systems, the response spectrum analysis is only suitable for isolation systems with linear

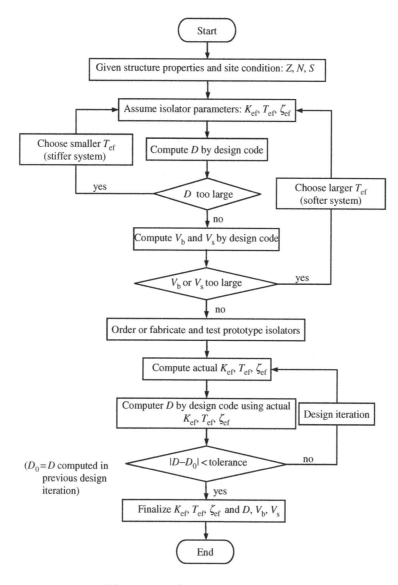

**FIGURE 4.58** Flow chart of static analysis.

properties. For the case when the isolation system or the superstructure appears to be highly nonlinear, a time-history analysis is generally required.

Because dynamic analysis depends generally on the usage of computer programs, relatively few formulas have been given in the dynamic analysis sections of design codes. Nevertheless, for a successful time-history analysis, the designer must prepare the following three basic elements: (1) a set of representative input ground motions, (2) accurate mathematic models for isolators and superstructures, and (3) a computer program that is capable of performing the nonlinear time-history analysis. These three elements are explained below.

### 4.6.3.1 Input Ground Motions

The response of an isolated system depends greatly on the chosen input ground motions, which are usually expressed in the form of ground accelerations. Each ground motion is called one record event.

The chosen events must be representative of the site conditions and soil characteristics. Design codes usually specify the minimum number of events required for analysis. Each ground motion event must be scaled so that all events are compatible with each other and also with the code specified target spectrum. In the UBC (1997), the scaling factor for each event is obtained in response spectra, and then applied to the time domain of the record data. In particular, site specific ground motions are required in the UBC for the following cases: (1) an isolated structure located on a soft soil, (2) an isolated structure located within certain distance (e.g., 10 km) of an active fault, (3) an isolated structure with very long period of vibration (e.g., greater than 3 sec).

### 4.6.3.2  Mathematic Models

Before any time-history analysis can be carried out, a mathematic model that can accurately reflect the mechanical behavior of the isolation system and the superstructure must be constructed. If the isolation system is nonlinear, the nonlinear parameters must be identified so that the constructed mathematic model can correctly describe the hysteretic behavior of the isolation system. In many cases, the isolation system is assumed to be nonlinear, but the superstructure linear. Establishing an accurate mathematic model is curial for obtaining reliable results in a time-history analysis.

### 4.6.3.3  Computer Programs

In practice, the task of time-history analysis is executed through the use of a computer program. The mathematic model properties mentioned above will be input to the program for analysis. The computer program selected should be capable of simulating the three-dimensional behavior of structures with selected nonlinear elements. To serve the purpose of isolation design and analysis, several structural analysis programs running on the platform of personal computers have been developed for easy access. Some of the widely used programs include (but are not limited to): ETABS (ETABS, 2004), SAP-2000 Nonlinear (SAP, 2000), and 3D-BASIS (Nagarajaiah et al., 1993). Most of these programs provide a set of imbedded mathematic models for the widely used isolator elements with linear or nonlinear parameters. The designers using these programs can easily build up the mathematic model for the isolated structure considered, specify the parameters of the isolator elements selected, and execute a nonlinear time-history analysis on a personal computer.

### 4.6.4  Concluding Remarks

In this section, the design concept of seismic isolation for structures was briefly reviewed. The design methods can be based either on static or dynamic analysis. The fundamental issues that should be considered in each design method were highlighted, along with some relevant formulas for computing the relevant parameters. It is believed that, with the concepts and procedures presented in this section, the readers should have a general knowledge of the procedure for base isolation design of structures and equipment.

# Acknowledgments

The authors are indebted to the graduate student, Cheng-Yan Wu, at the Department of Construction Engineering, National Kaohsiung First University of Science and Technology, for preparing some of the graphs presented in this chapter.

# Nomenclature

| Symbol | Quantity | Symbol | Quantity |
|---|---|---|---|
| $c, c_s$ | damping coefficients of superstructure | $V_b$ | total shear force of isolation system |
| $c_e$ | damping coefficients of equipment | $V_s$ | total base shear of superstructure |
| $c_{EDF}$ | damping of the EDF system | $w_i$ | weight of the $i$th floor |
| $D$ | maximum isolator displacement | $W$ | total weight of superstructure |
| $f_r$ | interfacial frictional force | $x, x_s(t)$ | relative-to-the-ground displacements |
| $F_i$ | seismic load applied on the $i$th floor | | of superstructure |
| $k, k_s$ | stiffness of superstructure | $x_b(t)$ | relative-to-the-ground displacements |
| $k_b$ | stiffness of isolation system | | of base mat |
| $k_e$ | stiffness of equipment | $x_e(t)$ | relative-to-the-ground displacements |
| $k_{EDF}$ | stiffness of the EDF system | | of equipment |
| $k_f$ | stiffness of the fictitious spring in sliding isolation | $\ddot{x}_g(t)$ | ground acceleration |
| | | $Z$ | zone factor |
| $K_{ef}$ | effective stiffness of isolation system | $\mu$ | frictional coefficient of sliding isolation system |
| $m, m_s$ | mass of superstructure | | |
| $m_b$ | mass of base mat | $\omega_b$ | frequency of isolation system |
| $m_e$ | mass of equipment | $\omega_e$ | frequencies of equipment |
| $n$ | number of building stories | $\omega_g$ | frequency of ground excitation |
| $N$ | near fault factor | $\zeta_e$ | damping ratios of equipment |
| $R_I$ | ductility factor | $\zeta_{ef}$ | effective damping ratios of isolation system |
| $S$ | soil factor | | |
| $T$ | natural period of superstructure | $\zeta, \zeta_s$ | damping ratios of superstructure |
| $T_d$ | damped period of superstructure | $\Omega_d$ | damped natural frequency |
| $T_{ef}$ | effective isolation period | $\Omega, \omega_s$ | frequencies of superstructure |

# References

Agrawal, A.K., Behaviour of equipment mounted over a torsionally coupled structure with sliding support, *Eng. Struct.*, 22, 72–84, 2000.

Chalhoub, M.S. and Kelly, J.M. 1990. Earthquake simulator test of a combined sliding bearing and rubber bearing isolation system, Report No. UCB/EERC-87/04, Earthquake Engineering Research Center, University of California, Berkeley, CA.

Chaudhuri, S.R. and Gupta, V.K., A response-based decoupling criterion for multiply-supported secondary systems, *Earthquake Eng. Struct. Dyn.*, 31, 1541–1562, 2002.

Chen, G. and Soong, T.T., Energy-based dynamic analysis of secondary systems, *J. Eng. Mech.*, ASCE, 120, 514–534, 1994.

Chen, G. and Soong, T.T., Exact solutions to a class of structure–equipment systems, *J. Eng. Mech.*, ASCE, 122, 1093–1100, 1996.

Dey, A. and Gupta, V.K., Response of multiply supported systems to earthquakes in frequency domain, *Earthquake Eng. Struct. Dyn.*, 27, 187–201, 1998.

Dey, A. and Gupta, V.K., Stochastic seismic response of multiply-supported secondary systems in flexible-base structures, *Earthquake Eng. Struct. Dyn.*, 28, 351–369, 1999.

ETABS, Integrated Analysis, Design and Drafting of Building Systems, Version 8, Software by Computers and Structures Inc., Berkeley CA, 2004.

Fujita, T., Seismic isolation of civil buildings in Japan, *Prog. Struct. Eng. Mater.*, 1, 295–300, 1998.

Gueraud, R., Noel-Leroux, J.P., Livolant, M., and Michalopoulos, A.P., Seismic isolation using sliding-elastomer bearing pads, *Nucl. Eng. Des.*, 84, 363–377, 1985.

Huang, C.D., Zhu, W.Q., and Soong, T.T., Nonlinear stochastic response and reliability of secondary systems, *J. Eng. Mech., ASCE*, 120, 177–196, 1994.

Igusa, T., Response characteristic of inelastic 2-DOF primary–secondary system, *J. Eng. Mech., ASCE*, 116, 1160–1174, 1990.

Igusa, T. and Der Kiureghian, A.D., Dynamic characteristic of two-degree-of-freedom equipment–structure systems, *J. Eng. Mech., ASCE*, 111, 1–19, 1985a.

Igusa, T. and Der Kiureghian, A.D., Dynamic response of multiply supported secondary systems, *J. Eng. Mech., ASCE*, 111, 20–41, 1985b.

Igusa, T. and Der Kiureghian, A.D., Generation of floor response spectra including oscillator–structure interaction, *Earthquake Eng. Struct. Dyn.*, 13, 661–676, 1985c.

Inaudi, J.A. and Kelly, J.M., Minimum variance control of base-isolation floors, *J. Struct. Eng., ASCE*, 119, 438–453, 1993.

Jangid, R.S. and Kelly, J.M., Base isolation for near-fault motion, *Earthquake Eng. Struct. Dyn.*, 30, 691–707, 2001.

Juhn, G., Manolis, G.D., Constantinou, M.C., and Reinhorn, A.M., Experimental study of secondary systems in base-isolated structure, *J. Struct. Eng., ASCE*, 118, 2204–2221, 1992.

Kelly, J.M., Seismic isolation of civil buildings in USA, *Prog. Struct. Eng. Mater.*, 1, 279–285, 1998.

Kelly, J.M. and Tsai, H.C., Seismic response of light internal equipment in base-isolated structures, *Earthquake Eng. Struct. Dyn.*, 13, 711–732, 1985.

Lai, M.L. and Soong, T.T., Statistical energy analysis of primary–secondary structural systems, *J. Eng. Mech., ASCE*, 116, 2400–2413, 1990.

Lu, L.Y. and Yang, Y.B., Dynamic response of equipment in structures with sliding support, *Earthquake Eng. Struct. Dyn.*, 26, 61–77, 1997.

Lu, L.Y., Shih, M.H., Tzeng, S.W., and Chang, C.S. 2003. Experiment of a sliding isolated structure subjected to near-fault ground motion, In *Proceedings of the Seventh Pacific Conference on Earthquake Engineering*, February 13–15, Christchurch.

Martelli, A. and Forni, M., Seismic isolation of civil buildings in Europe, *Prog. Struct. Eng. Mater.*, 1, 286–294, 1998.

Meirovitch, L. 1990. *Dynamics and Control of Structures*, Wiley, New York.

Mokha, A.S., Constantinous, M.C., Reinhorn, A.M., and Zayas, V.A., Experimental study of friction-pendulum isolation system, *J. Struct. Eng., ASCE*, 117, 1201–1217, 1991.

Mostaghel, N., 1984. *Resilient-friction Base Isolator*, Report No. UTEC 84-097, University of Utah, Salt Lake City, UT.

Mostaghel, N. and Khodaverdian, M., Dynamics of resilient-friction base isolator (R-FBI), *Earthquake Eng. Struct. Dyn.*, 15, 379–390, 1987.

Mostaghel, N., Hejazi, M., and Tanbakuchi, J., Response of sliding structures to harmonic support motion, *Earthquake Eng. Struct. Dyn.*, 11, 355–366, 1983.

Naeim, F. and Kelly, J.M. 1999. *Design of Seismic Isolated Structures: From Theory to Practice*, Wiley, New York.

Nagarajaiah, S., Li, C., Reinhorn, A.M., and Constantinou, M.C. 1993. 3D-BASIS-TABS: Computer program for nonlinear dynamic analysis of three dimensional base isolated structures, Technical report NCEER-93-0011, National Center for Earthquake Engineering Research, Buffalo, NY.

Newmark, N.M., A method of computation for structural dynamics, *J. Eng. Mech. Div., ASCE*, 85, 67–94, 1959.

Park, K.S., Jung, H.J., and Lee, I.W., A comparative study on aseismic performances of base isolation systems for multi-span continuous bridge, *Eng. Struct.*, 24, 1001–1013, 2002.

Rivin, E.I., Vibration isolation of precision equipment, *Precision Eng.*, 17, 41–56, 1995.

SAP 2000. Integrated Structural Analysis and Design Software, Software by Computers and Structures, Inc., Berkeley, CA.

Skinner, R.I., Robinson, W.H., and Mcverry, G.H. 1993. *An Introduction to Seismic Isolation*, Wiley, New York.

Soong, T.T. and Dargush, G.F. 1997. *Passive Energy Dissipation Systems in Structural Engineering*, Wiley, New York.

Steinberg, D.S. 2000. *Vibration Analysis for Electronic Equipment*, 3rd ed., Wiley, New York.

Suarez, L.E. and Singh, M.P., Floor spectra with equipment–structure–equipment interaction effects, *J. Eng. Mech., ASCE*, 115, 247–264, 1989.

UBC 1994. Uniform building code, *International Conference of Building Officials,* Whittier, CA.

UBC 1997. Uniform building code, *International Conference of Building Officials,* Whittier, CA.

Veprik, A.M. and Babitsky, V.I., Vibration protection of sensitive electronic equipment from harsh harmonic vibration, *J. Sound Vib.*, 238, 19–30, 2000.

Villaverde, R., Simplified seismic analysis of secondary systems, *J. Eng. Mech., ASCE*, 112, 588–604, 1986.

Wang, Y.P., Chung, L.L., and Liao, W.H., Seismic response analysis of bridges isolated with friction pendulum bearings, *Earthquake Eng. Struct. Dyn.*, 27, 1069–1093, 1998.

Westermo, B. and Udwadia, F., Period response of a sliding oscillator system to harmonic excitation, *Earthquake Eng. Struct. Dyn.*, 11, 135–146, 1983.

Xu, Y.L., Liu, H.J., and Yang, Z.C., Hybrid platform for vibration control of high-tech equipment in buildings subject to ground motion. Part 1. Experiment, *Earthquake Eng. Struct. Dyn.*, 32, 1185–1200, 2003.

Yang, J.N. and Agrawal, A.K., Protective systems for high-technology facilities against microvibration and earthquake, *Struct. Eng. Mech.*, 10, 561–575, 2000.

Yang, Y.B. and Chen, Y.C., Design of sliding-type base isolators by the concept of equivalent damping, *Struct. Eng. Mech.*, 8, 299–310, 1999.

Yang, Y.B. and Huang, W.H., Seismic response of light equipment in torsional buildings, *Earthquake Eng. Struct. Dyn.*, 22, 113–128, 1993.

Yang, Y.B. and Huang, W.H., Equipment–structure interaction considering the effect of torsion and base isolation, *Earthquake Eng. Struct. Dyn.*, 27, 155–171, 1998.

Yang, Y.B., Lee, T.Y., and Tsai, I.C., Response of multi-degree-of-freedom structures with sliding supports, *Earthquake Eng. Struct. Dyn.*, 19, 739–752, 1990.

Yang, Y.B., Hung, H.H., and He, M.J., Sliding and rocking response of rigid blocks due to horizontal excitations, *Struct. Eng. Mech.*, 9, 1–16, 2000.

Yang, Y.B., Chang, K.C., and Yau, J.D. 2002. Base isolation. In *Earthquake Engineering Handbook*, W.F. Chen and C. Scawthorn, Eds., CRC Press, Boca Raton, FL, chap. 17.

Yang, Z.C., Liu, H.J., and Xu, Y.L., Hybrid platform for vibration control of high-tech equipment in buildings subject to ground motion. Part 2. Analysis, *Earthquake Eng. Struct. Dyn.*, 32, 1201–1215, 2003.

# 5

# Vibration Control

Nader Jalili
*Clemson University*

Ebrahim Esmailzadeh
*University of Ontario*

## Summary

*The fundamental principles of vibration-control systems are formulated in this chapter (also see Chapter 7). There are many important areas directly or indirectly related to the main theme of the chapter. These include practical implementation of vibration-control systems, nonlinear control schemes, actual hardware implementation, actuator bandwidth requirements, reliability, and cost. Furthermore, in the process of designing a vibration-control system, in practice, several critical criteria must be considered. These include weight, size, shape, center-of-gravity, types of dynamic disturbances, allowable system response, ambient environment, and service life. Keeping these in mind, general design steps and procedures for vibration-control systems are provided.*

## 5.1 Introduction

The problem of reducing the level of vibration in constructions and structures arises in various branches of engineering, technology, and industry. In most of today's mechatronic systems, a number of possible devices such as reaction or momentum wheels, rotating devices, and electric motors are essential to the system's operation and performance. These devices, however, can also be sources of detrimental vibrations that may significantly influence the mission performance, effectiveness, and accuracy of operation. Therefore, there is a need for vibration control. Several techniques are utilized either to limit or alter the vibration response characteristics of such systems. During recent years, there has been considerable interest in the practical implementation of these vibration-control systems. This chapter presents the basic theoretical concepts for vibration-control systems design and implementation, followed by an overview of recent developments and control techniques in this subject. Some related practical developments in variable structure control (VSC), as well as piezoelectric vibration control of flexible structures, are also provided, followed by a summary of design steps and procedures for vibration-control systems. A further treatment of the subject is found in Chapter 7.

### 5.1.1   Vibration Isolation vs. Vibration Absorption

In vibration isolation, either the source of vibration is isolated from the system of concern (also called "force transmissibility"; see Figure 5.1a), or the device is protected from vibration of its point of attachment (also called "displacement transmissibility", see Figure 5.1b). Unlike the isolator, a vibration absorber consists of a secondary system (usually mass–spring–damper trio) added to the primary device to protect it from vibrating (see Figure 5.1c). By properly selecting absorber mass, stiffness, and damping, the vibration of the primary system can be minimized (Inman, 1994).

### 5.1.2   Vibration Absorption vs. Vibration Control

In vibration-control schemes, the driving forces or torques applied to the system are altered in order to regulate or track a desired trajectory while simultaneously suppressing the vibrational transients in the system. This control problem is rather challenging since it must achieve the motion tracking objectives while stabilizing the transient vibrations in the system. Several control methods have been developed for such applications: optimal control (Sinha, 1998); finite element approach (Bayo, 1987); model reference adaptive control (Ge et al., 1997); adaptive nonlinear boundary control (Yuh, 1987); and several other techniques including VSC methods (Chalhoub and Ulsoy, 1987; de Querioz et al., 1999; de Querioz et al., 2000).

As discussed before, in vibration-absorber systems, a secondary system is added in order to mimic the vibratory energy from the point of interest (attachment) and transfer it into other components or dissipate it into heat. Figure 5.2 demonstrates a comparative schematic of vibration control (both single-input control and multi-input configurations) on translating and rotating flexible beams, which could represent many industrial robot manipulators as well as vibration absorber applications for automotive suspension systems.

### 5.1.3   Classifications of Vibration-Control Systems

Passive, active, and semiactive (SA) are referred to, in the literature, as the three most commonly used classifications of vibration-control systems, either as isolators or absorbers (see Figure 5.3; Sun et al., 1995). A vibration-control system is said to be active, passive, or SA depending on the amount of

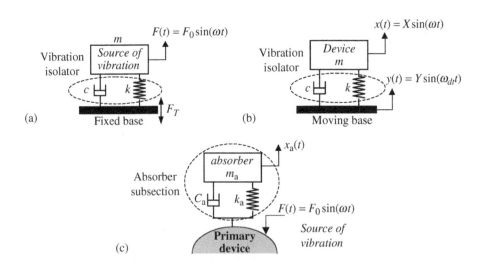

**FIGURE 5.1**   Schematic of (a) force transmissibility for foundation isolation; (b) displacement transmissibility for protecting device from vibration of the base and (c) application of vibration absorber for suppressing primary system vibration.

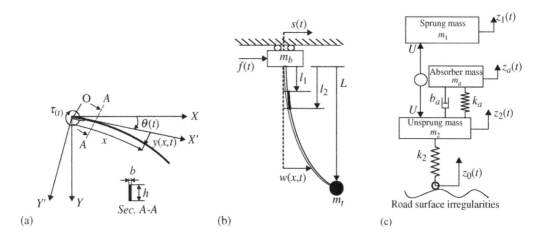

**FIGURE 5.2** A comparative schematic of vibration-control systems: (a) single-input simultaneous tracking and vibration control; (b) multi-input tracking and vibration control and (c) a two-DoF vehicle model with dynamic vibration absorber.

external power required for the vibration-control system to perform its function. A passive vibration control consists of a resilient member (stiffness) and an energy dissipater (damper) either to absorb vibratory energy or to load the transmission path of the disturbing vibration (Korenev and Reznikov, 1993; Figure 5.3a). This type of vibration-control system performs best within the frequency region of its highest sensitivity. For wideband excitation frequency, its performance can be improved considerably by optimizing the system parameters (Puksand, 1975; Warburton and Ayorinde, 1980; Esmailzadeh and Jalili, 1998a). However, this improvement is achieved at the cost of lowering narrowband suppression characteristics.

The passive vibration control has significant limitations in structural applications where broadband disturbances of highly uncertain nature are encountered. In order to compensate for these limitations, active vibration-control systems are utilized. With an additional active force introduced as a part of absorber subsection, $u(t)$ (Figure 5.3b), the system is controlled using different algorithms to make it more responsive to source of disturbances (Soong and Constantinou, 1994; Olgac and Holm-Hansen, 1995; Sun et al., 1995; Margolis, 1998). The SA vibration-control system, a combination of active and passive treatment, is intended to reduce the amount of external power necessary to achieve the desired performance characteristics (Lee-Glauser et al., 1997; Jalili, 2000; Jalili and Esmailzadeh, 2002), see Figure 5.3c.

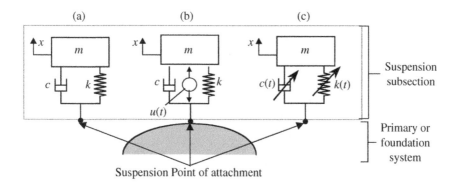

**FIGURE 5.3** A typical primary structure equipped with three versions of suspension systems: (a) passive; (b) active and (c) SA configurations.

### 5.1.4 Performance Characteristics of Vibration-Control Systems

In the design of a vibration-control system, it often occurs that the system is required to operate over a wideband load and frequency range that is impossible to meet with a single choice of required stiffness and damping. If the desired response characteristics cannot be obtained, an active vibration-control system may provide an attractive alternative vibration control for such broadband disturbances. However, active vibration-control systems suffer from control-induced instability in addition to the large control effort requirement. This is a serious concern, which prevents them from the common usage in most industrial applications. On the other hand, passive systems are often hampered by a phenomenon known as "detuning." Detuning implies that the passive system is no longer effective in suppressing the vibration it was designed for. This occurs due to one of the following reasons: (1) the vibration-control system may deteriorate and its structural parameters can be far from the original nominal design, (2) the structural parameters of the primary device itself may alter, or (3) the excitation frequency or the nature of disturbance may change over time.

A semiactive (also known as adaptive-passive) vibration-control system addresses these limitations by effectively integrating a tuning control scheme with tunable passive devices. For this, active force generators are replaced by modulated variable compartments such as variable rate damper and stiffness (see Figure 5.3c; Hrovat et al., 1988; Nemir et al., 1994; Franchek et al., 1995). These variable components are referred to as "tunable parameters" of the suspension system, which are retailored *via* a tuning control, thus resulting in semiactively inducing optimal operation. Much attention is being paid to these systems because of their low energy requirement and cost. Recent advances in smart materials, and adjustable dampers and absorbers have significantly contributed to applicability of these systems (Garcia et al., 1992; Wang et al., 1996; Shaw, 1998).

## 5.2 Vibration-Control Systems Concept

### 5.2.1 Introduction

With a history of almost a century (Frahm, 1911), the dynamic vibration absorber has proven to be a useful vibration-suppression device, widely used in hundreds of diverse applications. It is elastically attached to the vibrating body to alleviate detrimental oscillations from its point of attachment (see Figure 5.3). This section overviews the conceptual design and theoretical background of three types of vibration-control systems, namely the passive, active and SA configurations, along with some related practical implementations.

### 5.2.2 Passive Vibration Control

The underlying proposition in all vibration control or absorber systems is to adjust properly the absorber parameters such that the system becomes absorbent of the vibratory energy within the frequency interval of interest. In order to explain the underlying concept, a single-degree-of-freedom (single-DoF) primary system with a single-DoF absorber attachment is considered (Figure 5.4). The governing dynamics is expressed as

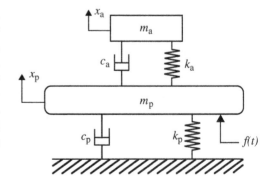

$$m_a \ddot{x}_a(t) + c_a \dot{x}_a(t) + k_a x_a(t)$$

$$= c_a \dot{x}_p(t) + k_a x_p(t) \qquad (5.1)$$

**FIGURE 5.4** Application of a passive absorber to single-DoF primary system.

$$m_p\ddot{x}_p(t) + (c_p + c_a)\dot{x}_p(t) + (k_p + k_a)x_p(t) - c_a\dot{x}_a(t) - k_a x_a(t) = f(t) \tag{5.2}$$

where $x_p(t)$ and $x_a(t)$ are the respective primary and absorber displacements, $f(t)$ is the external force, and the rest of the parameters including absorber stiffness, $k_a$, and damping, $c_a$, are defined as per Figure 5.4. The transfer function between the excitation force and primary system displacement in the Laplace domain is then written as

$$\text{TF}(s) = \frac{X_p(s)}{F(s)} = \left\{ \frac{m_a s^2 + c_a s + k_a}{H(s)} \right\} \tag{5.3}$$

where

$$H(s) = \{m_p s^2 + (c_p + c_a)s + k_p + k_a\}(m_a s^2 + c_a s + k_a) - (c_a s + k_a)^2 \tag{5.4}$$

and $X_a(s)$, $X_p(s)$, and $F(s)$ are the Laplace transformations of $x_a(t)$, $x_p(t)$, and $f(t)$, respectively.

### 5.2.2.1 Harmonic Excitation

When excitation is tonal, the absorber is generally tuned at the disturbance frequency. For this case, the steady-state displacement of the system due to harmonic excitation can be expressed as

$$\left| \frac{X_p(j\omega)}{F(j\omega)} \right| = \left| \frac{k_a - m_a \omega^2 + jc_a \omega}{H(j\omega)} \right| \tag{5.5}$$

where $\omega$ is the disturbance frequency and $j = \sqrt{-1}$. An appropriate parameter tuning scheme can then be selected to minimize the vibration of primary system subject to external disturbance, $f(t)$.

For complete vibration attenuation, the steady state, $|X_p(j\omega)|$, must equal zero. Consequently, from Equation 5.5, the ideal stiffness and damping of absorber are selected as

$$k_a = m_a \omega^2, \quad c_a = 0 \tag{5.6}$$

Notice that this tuned condition is only a function of absorber elements ($m_a$, $k_a$, and $c_a$). That is, the absorber tuning does not need information from the primary system and hence its design is stand alone. For tonal application, theoretically, zero damping in the absorber subsection results in improved performance. In practice, however, the damping is incorporated in order to maintain a reasonable trade-off between the absorber mass and its displacement. Hence, the design effort for this class of application is focused on having precise tuning of the absorber to the disturbance frequency and controlling the damping to an appropriate level. Referring to Snowdon (1968), it can be proven that the absorber, in the presence of damping, can be most favorably tuned and damped if adjustable stiffness and damping are selected as

$$k_{opt} = \frac{m_a m_p^2 \omega^2}{(m_a + m_p)^2}, \quad c_{opt} = m_a \sqrt{\frac{3k_{opt}}{2(m_a + m_p)}} \tag{5.7}$$

### 5.2.2.2 Broadband Excitation

In broadband vibration control, the absorber subsection is generally designed to add damping to and change the resonant characteristics of the primary structure in order to dissipate vibrational energy maximally over a range of frequencies. The objective of the absorber design is, therefore, to adjust the *absorber parameters* to minimize the peak magnitude of the frequency transfer function (FTF($\omega$) = $|\text{TF}(s)|_{s=j\omega}$) over the absorber parameters vector $\mathbf{p} = \{c_a \ k_a\}^T$. That is, we seek $\mathbf{p}$ to

$$\min_{\mathbf{p}} \left\{ \max_{\omega_{min} \leq \omega \leq \omega_{max}} \{|\text{FTF}(\omega)|\} \right\} \tag{5.8}$$

Alternatively, one may select the mean square displacement response (MSDR) of the primary system for vibration-suppression performance. That is, the absorber parameters vector, $\mathbf{p}$, is selected such that the MSDR

$$E\{(\bar{x}_p)^2\} = \int_0^\infty \{FTF(\omega)\}^2 S(\omega)d\omega \tag{5.9}$$

is minimized over a desired wideband frequency range. $S(\omega)$ is the power spectral density of the excitation force, $f(t)$, and FTF was defined earlier.

This optimization is subjected to some constraints in **p** space, where only positive elements are acceptable. Once the optimal absorber suspension properties, $c_a$ and $k_a$, are determined, they can be implemented using adjustment mechanisms on the spring and the damper elements. This is viewed as a SA adjustment procedure as it adds no energy to the dynamic structure. The conceptual devices for such adjustable suspension elements and SA treatment will be discussed later in Section 5.2.5.

### 5.2.2.3  Example Case Study

To better recognize the effectiveness of the dynamic vibration absorber over the passive and optimum passive absorber settings, a simple example case is presented. For the simple system shown in Figure 5.4, the following nominal structural parameters (marked by an overscore) are taken:

$$\bar{m}_p = 5.77 \text{ kg}, \quad \bar{k}_p = 251.132 \times 10^6 \text{ N/m}, \quad \bar{c}_p = 197.92 \text{ kg/sec}$$
$$\bar{m}_a = 0.227 \text{ kg}, \quad \bar{k}_a = 9.81 \times 10^6 \text{ N/m}, \quad \bar{c}_a = 355.6 \text{ kg/sec} \tag{5.10}$$

These are from an actual test setting, which is optimal by design (Olgac and Jalili, 1999). That is, the peak of the FTF is minimized (see thin lines in Figure 5.5). When the primary stiffness and damping increase 5% (for instance during the operation), the FTF of the primary system deteriorates considerably (the dashed line in Figure 5.5), and the absorber is no longer an optimum one for the present primary. When the absorber is optimized based on optimization problem 8, the retuned setting is reached as

$$k_a = 10.29 \times 10^6 \text{ N/m}, \quad c_a = 364.2 \text{ kg/sec} \tag{5.11}$$

which yields a much better frequency response (see dark line in Figure 5.5).

The vibration absorber effectiveness is better demonstrated at different frequencies by frequency sweep test. For this, the excitation amplitude is kept fixed at unity and its frequency changes every 0.15 sec from

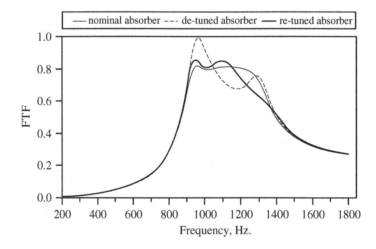

**FIGURE 5.5**  Frequency transfer functions (FTFs) for nominal absorber (thin-solid line), detuned absorber (thin-dotted line), and retuned absorber (thick-solid line) settings. (*Source:* From Jalili, N. and Olgac, N., *AIAA J. Guidance Control Dyn.*, 23, 961–970, 2000a. With permission.)

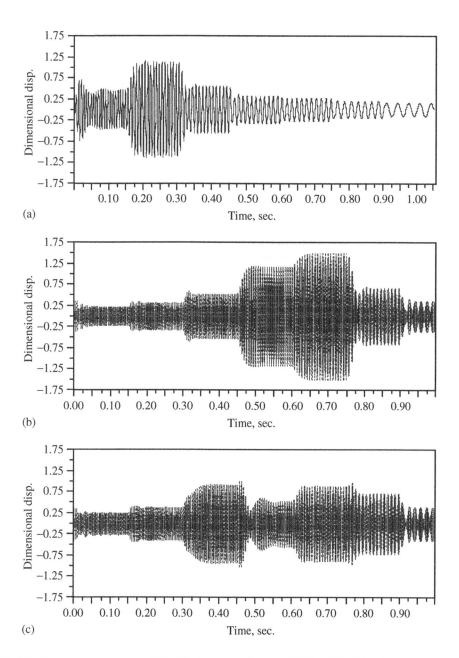

**FIGURE 5.6**  Frequency sweep each 0.15 with frequency change of 1860, 1880, 1900, 1920, 1930, 1950, and 1970 Hz: (a) nominally tuned absorber settings; (b) detuned absorber settings and (c) retuned absorber settings. (*Source:* From Jalili, N. and Olgac, N., *AIAA J. Guidance Control Dyn.*, 23, 961–970, 2000a. With permission.)

1860 to 1970 Hz. The primary responses with nominally tuned, with detuned, and with retuned absorber settings are given in Figure 5.6a–c, respectively.

## 5.2.3  Active Vibration Control

As discussed, passive absorption utilizes resistive or reactive devices either to absorb vibrational energy or load the transmission path of the disturbing vibration (Korenev and Reznikov, 1993; see Figure 5.7, top). Even with optimum absorber parameters (Warburton and Ayorinde, 1980;

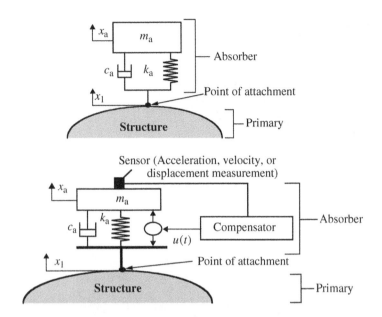

**FIGURE 5.7**   A general primary structure with passive (top) and active (bottom) absorber settings.

Esmailzadeh and Jalili, 1998a), the passive absorption has significant limitations in structural applications where broadband disturbances of highly uncertain nature are encountered.

In order to compensate for these limitations, active vibration-suppression schemes are utilized. With an additional active force, $u(t)$ (Figure 5.7, bottom), the absorber is controlled using different algorithms to make it more responsive to primary disturbances (Sun et al., 1995; Margolis, 1998; Jalili and Olgac, 1999). One novel implementation of the tuned vibration absorbers is the active resonator absorber (ARA) (Knowles et al., 2001b). The concept of the ARA is closely related to the concept of the delayed resonator (Olgac and Holm-Hansen, 1994; Olgac, 1995). Using a simple position (or velocity or acceleration) feedback control within the absorber subsection, the delayed resonator enforces that the dominant characteristic roots of the absorber subsection be on the imaginary axis, hence leading to resonance. Once the ARA becomes resonant, it creates perfect vibration absorption at this frequency. The conceptual design and implementation issues of such active vibration-control systems, along with their practical applications, are discussed in Section 5.3.

## 5.2.4   Semiactive Vibration Control

Semiactive (SA) vibration-control systems can achieve the majority of the performance characteristics of fully active systems, thus allowing for a wide class of applications. The idea of SA suspension is very simple: to replace active force generators with continually adjustable elements which can vary and/or shift the rate of the energy dissipation in response to instantaneous condition of motion (Jalili, 2002).

## 5.2.5   Adjustable Vibration-Control Elements

Adjustable vibration-control elements are typically comprised of variable rate damper and stiffness. Significant efforts have been devoted to the development and implementation of such devices for a variety of applications. Examples of such devices include electro-rheological (ER) (Petek, 1992; Wang et al., 1994; Choi, 1999), magneto-rheological (MR) (Spencer et al., 1998; Kim and Jeon, 2000) fluid dampers, and variable orifice dampers (Sun and Parker, 1993), controllable friction braces (Dowell and Cherry, 1994), and variable stiffness and inertia devices (Walsh and Lamnacusa, 1992;

Nemir et al., 1994; Franchek et al., 1995; Abe and Igusa, 1996). The conceptual devices for such adjustable properties are briefly reviewed in this section.

### 5.2.5.1 Variable Rate Dampers

A common and very effective way to reduce transient and steady-state vibration is to change the amount of damping in the SA vibration-control system. Considerable design work on SA damping was done in the 1960s to the 1980s (Crosby and Karnopp, 1973; Karnopp et al., 1974) for vibration control of civil structures such as buildings and bridges (Hrovat et al., 1983) and for reducing machine tool oscillations (Tanaka and Kikushima, 1992). Since then, SA dampers have been utilized in diverse applications ranging from trains (Stribersky et al., 1998) and other off-road vehicles (Horton and Crolla, 1986) to military tanks (Miller and Nobles, 1988). During recent years, there has been considerable interest in the SA concept in the industry for improvement and refinements of the concept (Karnopp, 1990; Emura et al., 1994). Recent advances in smart materials have led to the development of new SA dampers, which are widely used in different applications.

In view of these SA dampers, ER and MR fluids probably serve as the best potential hardware alternatives for the more conventional variable-orifice hydraulic dampers (Sturk et al., 1995). From a practical standpoint, the MR concept appears more promising for suspension, since it can operate, for instance, on vehicle battery voltage, whereas the ER damper is based on high-voltage electric fields. Owing to their importance in today's SA damper technology, we briefly review the operation and fundamental principles of SA dampers here.

#### 5.2.5.1.1 *Electro-Rheological Fluid Dampers*

ER fluids are materials that undergo significant instantaneous reversible changes in material characteristics when subjected to electric potentials (Figure 5.8). The most significant change is associated with complex shear moduli of the material, and hence ER fluids can be usefully exploited in SA absorbers where variable-rate dampers are utilized. Originally, the idea of applying an ER damper to vibration control was initiated in automobile suspensions, followed by other applications (Austin, 1993; Petek et al., 1995).

The flow motions of an ER fluid-based damper can be classified by shear mode, flow mode, and squeeze mode. However, the rheological property of ER fluid is evaluated in the shear mode (Choi, 1999). As a result, the ER fluid damper provides an adaptive viscous and frictional damping for use in SA system (Dimarogonas-Andrew and Kollias, 1993; Wang et al., 1994).

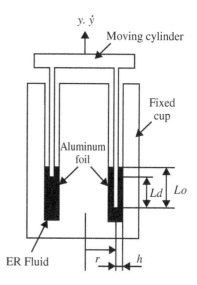

**FIGURE 5.8** A schematic configuration of an ER damper. (*Source:* From Choi, S.B., *ASME J. Dyn. Syst. Meas. Control*, 121, 134–138, 1999. With permission.)

#### 5.2.5.1.2 *Magneto-Rheological Fluid Dampers*

MR fluids are the magnetic analogies of ER fluids and typically consist of micron-sized, magnetically polarizable particles dispersed in a carrier medium such as mineral or silicon oil. When a magnetic field is applied, particle chains form and the fluid becomes a semisolid, exhibiting plastic behavior similar to that of ER fluids (Figure 5.9). Transition to rheological equilibrium can be achieved in a few milliseconds, providing devices with high bandwidth (Spencer et al., 1998; Kim and Jeon, 2000).

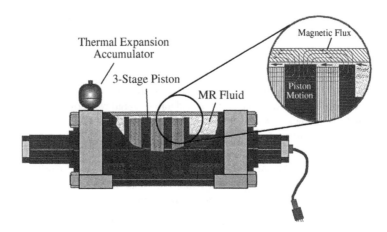

**FIGURE 5.9**  A schematic configuration of an MR damper. (*Source:* From Spencer, B.F. et al., *Proc. 2nd World Conf. on Structural Control*, 1998. With permission.)

### 5.2.5.2  Variable-Rate Spring Elements

In contrast to variable dampers, studies of SA springs or time-varying stiffness have also been geared for vibration-isolation applications (Hubard and Marolis, 1976), for structural controls and for vibration attenuation (Sun et al., 1995 and references therein). The variable stiffness is a promising practical complement to SA damping, since, based on the discussion in Section 5.2, both the absorber damping and stiffness should change to adapt optimally to different conditions. Clearly, the absorber stiffness has a significant influence on optimum operation (and even more compared to the damping element; Jalili and Olgac, 2000b).

Unlike the variable rate damper, changing the effective stiffness requires high energy (Walsh and Lamnacusa, 1992). Semiactive or low-power implementation of variable stiffness techniques suffers from limited frequency range, complex implementation, high cost, and so on. (Nemir et al., 1994; Franchek et al., 1995). Therefore, in practice, both absorber damping and stiffness are concurrently adjusted to reduce the required energy.

#### 5.2.5.2.1  Variable-Rate Stiffness (Direct Methods)

The primary objective is to directly change the spring stiffness to optimize a vibration-suppression characteristic such as the one given in Equation 5.8 or Equation 5.9. Different techniques can be utilized ranging from traditional variable leaf spring to smart spring utilizing magnetostrictive materials. A tunable stiffness vibration absorber was utilized for a four-DoF building (Figure 5.10), where a spring is threaded through a collar plate and attached to the absorber mass from one side and to the driving gear from the other side (Franchek et al., 1995). Thus, the effective number of coils, $N$, can be changed resulting in a variable spring stiffness, $k_a$:

$$k_a = \frac{d^4 G}{8D^3 N} \tag{5.12}$$

where $d$ is the spring wire diameter, $D$ is the spring diameter, and $G$ is modulus of shear rigidity.

#### 5.2.5.2.2  Variable-Rate Effective Stiffness (Indirect Methods)

In most SA applications, directly changing the stiffness might not be always possible or may require large amount of control effort. For such cases, alternatives methods are utilized to change the effective tuning ratio ($\tau = \sqrt{k_a/m_a}/\omega_{\text{primary}}$), thus resulting in a tunable resonant frequency.

In Liu et al. (2000), a SA flutter-suppression scheme was proposed using differential changes of external store stiffness. As shown in Figure 5.11, the motor drives the guide screw to rotate with slide block, $G$, moving along it, thus changing the restoring moment and resulting in a change of store

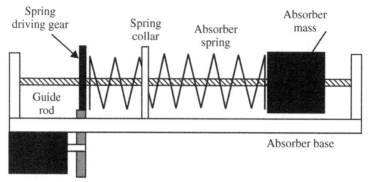

FIGURE 5.10 The application of a variable-stiffness vibration absorber to a four-DoF building. (*Source:* From Franchek, M.A. et al., *J. Sound Vib.*, 189, 565–585, 1995. With permission.)

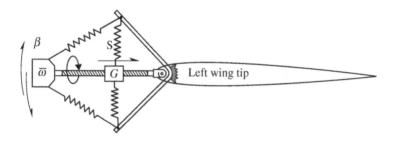

FIGURE 5.11 A SA flutter control using adjustable pitching stiffness. (*Source:* From Liu, H.J. et al., *J. Sound Vib.*, 229, 199–205, 2000. With permission.)

pitching stiffness. Using a double-ended cantilever beam carrying intermediate lumped masses, a SA vibration absorber was recently introduced (Jalili and Esmailzadeh, 2002), where positions of moving masses are adjustable (see Figure 5.12). Figure 5.13 shows an SA absorber with an adjustable effective inertia mechanism (Jalili et al., 2001; Jalili and Fallahi, 2002). The SA absorber consists of a rod carrying a moving block and a spring and damper, which are mounted on a casing. The position of the moving block, $r_v$, on the rod is adjustable which provides a tunable resonant frequency.

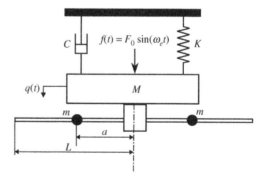

FIGURE 5.12 A typical primary system equipped with the double-ended cantilever absorber with adjustable tuning ration through moving masses, $m$. (*Source:* From Jalili, N. and Esmailzadeh, E., *J. Multi-Body Dyn.*, 216, 223–235, 2002. With permission.)

### 5.2.5.3 Other Variable-Rate Elements

Recent advances in smart materials have led to the development of new SA vibration-control systems using indirect influence on the suspension elements. Wang et al. used a SA piezoelectric network (1996) for structural-vibration control. The variable resistance and inductance in an external RL circuit are used as real-time adaptable control parameters.

Another class of adjustable suspensions is the so-called hybrid treatment (Fujita et al., 1991). The hybrid design has two modes, an active mode and a passive mode. With its aim of lowering the

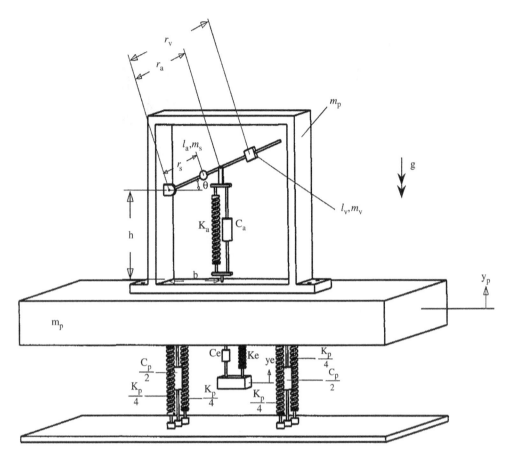

**FIGURE 5.13**   Schematic of the adjustable effective inertia vibration absorber. (*Source:* From Jalili, N. et al., *Int. J. Model. Simulat.*, 21, 148–154, 2001. With permission.)

control effort, relatively small vibrations are reduced in active mode, while passive mode is used for large oscillations. Analogous to hybrid treatment, the semiautomated approach combines SA and active suspensions to benefit from the advantages of individual schemes while eliminating their shortfalls (Jalili, 2000). By altering the adjustable structural properties (in the SA unit) and control parameters (in the active unit), a search is conducted to minimize an objective function subject to certain constraints, which may reflect performance characteristics.

## 5.3   Vibration-Control Systems Design and Implementation

### 5.3.1   Introduction

This section provides the basic fundamental concepts for vibration-control systems design and implementation. These systems are classified into two categories: vibration absorbers and vibration-control systems. Some related practical developments in ARAs and piezoelectric vibration control of flexible structures are also provided.

### 5.3.2   Vibration Absorbers

Undesirable vibrations of flexible structures have been effectively reduced using a variety of dynamic vibration absorbers. The active absorption concept offers a wideband of vibration-attenuation frequencies as well as real-time tunability as two major advantages. It is clear that the active control could be a destabilizing factor for the combined system, and therefore, the stability of the combined system (i.e., the primary and the absorber subsystems) must be assessed.

An actively tuned vibration-absorber scheme utilizing a resonator generation mechanism forms the underlying concept here. For this, a stable primary system (see Figure 5.7, top, for instance) is forced into a marginally stable one through the addition of a controlled force in the active unit (see Figure 5.7, bottom). The conceptual design for generating such resonance condition is demonstrated in Figure 5.14, where the system's dominant characteristic roots (poles) are moved and placed on the imaginary axis. The absorber

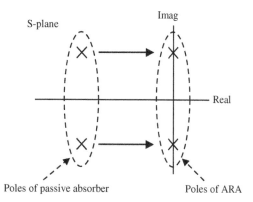

**FIGURE 5.14** Schematic of the active resonator absorber concept through placing the poles of the characteristic equation on the imaginary axis.

then becomes a resonator capable of mimicking the vibratory energy from the primary system at the point of attachment. Although there seem to be many ways to generate such resonance, only two widely accepted practical vibration-absorber resonators are discussed next.

#### 5.3.2.1 Delayed-Resonator Vibration Absorbers

A recent active vibration-absorption strategy, the delayed resonator (DR), is considered to be the first type of active vibration absorber when operated on a flexible beam (Olgac and Jalili, 1998; Olgac and Jalili, 1999). The DR vibration absorber offers some attractive features in eliminating tonal vibrations from the objects to which it is attached (Olgac and Holm-Hansen, 1994; Olgac, 1995; Renzulli et al., 1999), some of which are real-time tunability, the stand-alone nature of the actively controlled absorber, and the simplicity of the implementation. Additionally, this single-DoF absorber can also be tuned to handle multiple frequencies of vibration (Olgac et al., 1996). It is particularly important that the combined system, that is, the primary structure and the absorber together, is asymptotically stable when the DR is implemented on a flexible beam.

#### *5.3.2.1.1 An Overview of the Delayed-Resonator Concept*

An overview of the DR is presented here to help the reader. The equation of motion governing the absorber dynamics alone is

$$m_a \ddot{x}_a(t) + c_a \dot{x}_a(t) + k_a x_a(t) - u(t) = 0, \quad u(t) = g \ddot{x}_a(t - \tau) \tag{5.13}$$

where $u(t)$ represents the delayed acceleration feedback. The Laplace domain transformation of this equation yields the characteristics equation

$$m_a s^2 + c_a s + k_a - g s^2 \, e^{-\tau s} = 0 \tag{5.14}$$

Without feedback ($g = 0$), this structure is dissipative with two characteristic roots (poles) on the left half of the complex plane. For $g$ and $\tau > 0$, however, these two finite stable roots are supplemented by infinitely many additional finite roots. Note that these characteristic roots (poles) of Equation 5.14 are discretely located (say at $s = a + j\omega$), and the following relation holds:

$$g = \frac{|m_a s^2 + c_a s + k_a|}{|s^2|} e^{\tau a} \tag{5.15}$$

where $|\cdot|$ denotes the magnitude of the argument.

Using Equation 5.15, the following observation can be made:

- For $g = 0$: there are two finite stable poles and all the remaining poles are at $a = -\infty$.
- For $g = +\infty$: there are two poles at $s = 0$, and the rest are at $a = +\infty$.

Considering these and taking into account the continuity of the root loci for a given time delay, $\tau$, and as $g$ varies from 0 to $\infty$, it is obvious that the roots of Equation 5.14 move from the stable left half to the unstable right half of the complex plane. For a certain critical gain, $g_c$, one pair of poles reaches the imaginary axis. At this operating point, the DR becomes a perfect resonator and the imaginary characteristic roots are $s = \pm j\omega_c$, where $\omega_c$ is the resonant frequency and $j = \sqrt{-1}$. The subscript "c" implies the crossing of the root loci on the imaginary axis. The control parameters of concern, $g_c$ and $\tau_c$, can be found by substituting the desired $s = \pm j\omega_c$ into Equation 5.14 as

$$g_c = \frac{1}{\omega_c^2}\sqrt{(c_a\omega_c)^2 + (m_a\omega_c - k_a)^2}, \quad \tau_c = \frac{1}{\omega_c}\left\{\tan^{-1}\left[\frac{c_a\omega_c}{m_a\omega_c^2 - k_a}\right] + 2(\ell - 1)\pi\right\}, \quad \ell = 1, 2, \ldots$$

(5.16)

When these $g_c$ and $\tau_c$ are used, the DR structure mimics a resonator at frequency $\omega_c$. In turn, this resonator forms an *ideal absorber* of the tonal vibration at $\omega_c$. The objective of the control, therefore, is to maintain the DR absorber at this marginally stable point. On the DR stability, further discussions can be found in Olgac and Holm-Hansen (1994) and Olgac et al. (1997).

### 5.3.2.1.2    Vibration-Absorber Application on Flexible Beams

We consider a general beam as the primary system with absorber attached to it and subjected to a harmonic force excitation, as shown in Figure 5.15. The point excitation is located at $b$, and the absorber is placed at $a$. A uniform cross section is considered for the beam and Euler–Bernoulli assumptions are made. The beam parameters are all assumed to be constant and uniform. The elastic deformation from the undeformed natural axis of the beam is denoted by $y(x, t)$ and, in the derivations that follow, the dot ($\cdot$) and prime ($'$) symbols indicate a partial derivative with respect to the time variable, $t$, and position variable $x$, respectively.

Under these assumptions, the kinetic energy of the system can be written as

$$T = \frac{1}{2}\rho\int_0^L\left(\frac{\partial y}{\partial t}\right)^2 dx + \frac{1}{2}m_a\dot{q}_a^2 + \frac{1}{2}m_e\dot{q}_e^2$$

(5.17)

The potential energy of this system using linear strain is given by

$$U = \frac{1}{2}\text{EI}\int_0^L\left(\frac{\partial^2 y}{\partial x^2}\right)^2 dx + \frac{1}{2}k_a\{y(a, t) - q_a\}^2 + \frac{1}{2}k_e\{y(b, t) - q_e\}^2$$

(5.18)

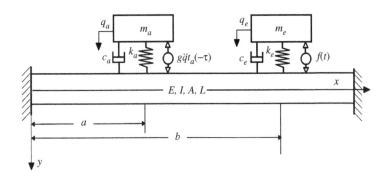

**FIGURE 5.15**   Beam–absorber–exciter system configuration. (*Source:* From Olgac, N. and Jalili, N., *J. Sound Vib.*, 218, 307–331, 1998. With permission.)

The equations of motion may now be derived by applying Hamilton's Principle. However, to facilitate the stability analysis, we resort to an assumed-mode expansion and Lagrange's equations. Specifically, $y$ is written as a finite sum "Galerkin approximation":

$$y(x, t) = \sum_{i=1}^{n} \Phi_i(x) q_{bi}(t) \tag{5.19}$$

The orthogonality conditions between these mode shapes can also be derived as (Meirovitch, 1986)

$$\int_0^L \rho \Phi_i(x) \Phi_j(x) dx = N_i \delta_{ij}, \quad \int_0^L EI \Phi_i''(x) \Phi_j''(x) dx = S_i \delta_{ij} \tag{5.20}$$

where $i, j = 1, 2, \ldots, n$, $\delta_{ij}$ is the Kronecker delta, and $N_i$ and $S_i$ are defined by setting $i = j$ in Equation 5.20.

The feedback of the absorber, the actuator excitation force, and the damping dissipating forces in both the absorber and the exciter are considered as non-conservative forces in Lagrange's formulation. Consequently, the equations of motion are derived.

Absorber dynamics is governed by

$$m_a \ddot{q}_a(t) + c_a \left\{ \dot{q}_a(t) - \sum_{i=1}^{n} \Phi_i(a) \dot{q}_{bi}(t) \right\} + k_a \left\{ q_a(t) - \sum_{i=1}^{n} \Phi_i(a) q_{bi}(t) \right\} - g \ddot{q}_a(t - \tau) = 0 \tag{5.21}$$

The exciter is given by

$$m_e \ddot{q}_e(t) + c_e \left\{ \dot{q}_e(t) - \sum_{i=1}^{n} \Phi_i(b) \dot{q}_{bi}(t) \right\} + k_e \left\{ q_e(t) - \sum_{i=1}^{n} \Phi_i(b) q_{bi}(t) \right\} = -f(t) \tag{5.22}$$

Finally, the beam is represented by

$$N_i \ddot{q}_{bi}(t) + S_i q_{bi}(t) + c_a \left\{ \sum_{i=1}^{n} \Phi_i(a) \dot{q}_{bi}(t) - \dot{q}_a(t) \right\} \Phi_i(a) + c_e \left\{ \sum_{i=1}^{n} \Phi_i(b) \dot{q}_{bi}(t) - \dot{q}_e(t) \right\} \Phi_i(b)$$

$$+ k_a \left\{ \sum_{i=1}^{n} \Phi_i(a) q_{bi}(t) - q_a(t) \right\} \Phi_i(a) + k_e \left\{ \sum_{i=1}^{n} \Phi_i(b) q_{bi}(t) - q_e(t) \right\} \Phi_i(b) + g \Phi_i(a) \ddot{q}_a(t - \tau)$$

$$= f(t) \Phi_i(b), \quad i = 1, 2, \ldots, n \tag{5.23}$$

Equation 5.21 to Equation 5.23 form a system of $n + 2$ second-order coupled differential equations.

By proper selection of the feedback gain, the absorber can be tuned to the desired resonant frequency, $\omega_c$. This condition, in turn, forces the beam to be motionless at $a$, when the beam is excited by a tonal force at frequency $\omega_c$. This conclusion is reached by taking the Laplace transform of Equation 5.21 and using feedback control law for the absorber. In short,

$$Y(a, s) = \sum_{i=1}^{n} \Phi_i(a) Q_{bi}(s) = 0 \tag{5.24}$$

where $Y(a, s) = \Im\{y(a, t)\}$, $Q_a(s) = \Im\{q_a(t)\}$ and $Q_{bi}(s) = \Im\{q_{bi}(t)\}$. Equation 5.24 can be rewritten in time domain as

$$y(a, t) = \sum_{i=1}^{n} \Phi_i(a) q_{bi}(t) = 0 \tag{5.25}$$

which indicates that the steady-state vibration of the point of attachment of the absorber is eliminated. Hence, the absorber mimics a resonator at the frequency of excitation and absorbs all the vibratory energy at the point of attachment.

### 5.3.2.1.3  Stability of the Combined System

In the preceding section, we have derived the equations of motion for the beam–exciter–absorber system in its most general form. As stated before, inclusion of the feedback control for active absorption is, indeed, an invitation to instability. This topic is treated next.

The Laplace domain representation of the combined system takes the form (Olgac and Jalili, 1998)

$$\mathbf{A}(s)\mathbf{Q}(s) = \mathbf{F}(s) \tag{5.26}$$

where

$$\mathbf{Q}(s) = \left\{ \begin{array}{c} Q_a(s) \\ Q_e(s) \\ Q_{b1}(s) \\ \vdots \\ Q_{bn}(s) \end{array} \right\}_{(n+2)\times 1}, \quad \mathbf{F}(s) = \left\{ \begin{array}{c} 0 \\ -F(s) \\ 0 \\ \vdots \\ 0 \end{array} \right\}_{(n+2)\times 1},$$

$$\mathbf{A}(s) = \begin{pmatrix} m_a s^2 + c_a s + k_a - g s^2\,e^{-\tau s} & 0 & -\Phi_1(a)(c_a s + k_a) & \cdots & -\Phi_n(a)(c_a s + k_a) \\ 0 & m_e s^2 + c_e s + k_e & -\Phi_1(b)(c_e s + k_e) & \cdots & -\Phi_n(b)(c_e s + k_e) \\ m_a \Phi_1(a)s^2 & m_e \Phi_1(b)s^2 & N_1 s^2 + cs + S_1(1 + j\delta) & \cdots & 0 \\ \vdots & \vdots & \vdots & \ddots & \vdots \\ m_a \Phi_n(a)s^2 & m_e \Phi_n(b)s^2 & 0 & \cdots & N_n s^2 + cs + S_n(1 + j\delta) \end{pmatrix}$$

$$\tag{5.27}$$

In order to assess the combined system stability, the roots of the characteristic equation, $\det(\mathbf{A}(s)) = 0$ are analyzed. The presence of feedback (transcendental delay term for this absorber) in the characteristic equations complicates this effort. The root locus plot observation can be applied to the entire system. It is typical that increasing feedback gain causes instability as the roots move from left to right in the complex plane. This picture also yields the frequency range for stable operation of the combined system (Olgac and Jalili, 1998).

### 5.3.2.1.4  Experimental Setting and Results

The experimental setup used to verify the findings is shown in Figure 5.16. The primary structure is a $3/8'$ in. $\times$ $1'$ in. $\times$ $12'$ in. steel beam (2) clamped at both ends to a granite bed (1). A piezoelectric actuator with a reaction mass (3 and 4) is used to generate the periodic disturbance on the beam. A similar actuator-mass setup constitutes the DR absorber (5 and 6). The two setups are located symmetrically at one quarter of the length along the beam from the center. The feedback signal used to implement the DR is obtained from the accelerometer (7) mounted on the reaction mass of the absorber structure. The other accelerometer (8) attached to the beam is present only to monitor the vibrations of the beam and to evaluate the performance of the DR absorber in suppressing them. The control is applied *via* a fast data acquisition card using a sampling of 10 kHz.

The numerical values for this beam–absorber–exciter setup are taken as

- *Beam*: $E = 210$ GPa, $\rho = 1.8895$ kg/m
- *Absorber*: $m_a = 0.183$ kg, $k_a = 10{,}130$ kN/m, $c_a = 62.25$ N sec/m, $a = L/4$
- *Exciter*: $m_e = 0.173$ kg, $k_e = 6426$ kN/m, $c_e = 3.2$ N sec/m, $b = 3L/4$

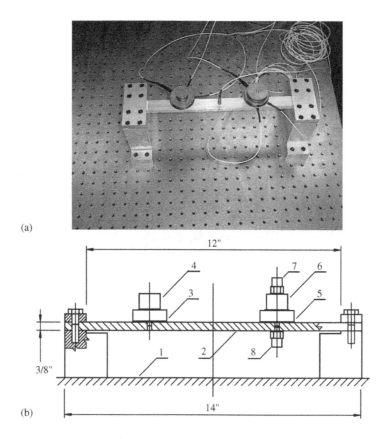

**FIGURE 5.16** (a) Experimental structure and (b) schematic depiction of the setup. (*Source:* From Olgac, N. and Jalili, N., *J. Sound Vib.*, 218, 307–331, 1998. With permission.)

### 5.3.2.1.5 Dynamic Simulation and Comparison with Experiments

For the experimental set up at hand, the natural frequencies are measured for the first two natural modes, $\omega_1$ and $\omega_2$. These frequencies are obtained much more precisely than those of higher-order natural modes. Table 5.1 offers a comparison between the experimental (*real*) and analytical (*ideal*) clamped–clamped beam natural frequencies.

The discrepancies arrive from two sources: first, the experimental frequencies are structurally damped natural frequencies, and second, they reflect the effect of partially clamped BCs. The theoretical frequencies, on the other hand, are evaluated for an undamped ideal clamped–clamped beam.

After observing the effect of the number of modes used on the beam deformation, a minimum of three natural modes are taken into account. We then compare the simulated time response vs. the experimental results of vibration suppression. Figure 5.17 shows a test with the excitation frequency $\omega_c = 1270$ Hz. The corresponding theoretical control parameters are $g_{c\ theory} = 0.0252$ kg and $\tau_{c\ theory} = 0.8269$ msec. The experimental control parameters for this frequency are found to be $g_{c\ exp.} = 0.0273$ kg and $\tau_{c\ exp.} = 0.82$ msec. The exciter disturbs the beam for 5 msec, then the DR tuning is triggered. The acceleration of the beam at the point of attachment decays exponentially. For all intents and purposes, the suppression takes effect in approximately 200 ms. These results match very closely with the experimental data, Figure 5.18. The only noticeable difference is in the frequency content of the exponential decay. This property is dictated by the dominant poles of the combined system. The imaginary part, however, is smaller in the analytical study. This difference is a nuance that does not affect the earlier observations.

**TABLE 5.1**    Comparison between Experimental and Theoretical Beam Natural Frequencies (Hz)

| Natural Modes | Peak Frequencies (Experimental) | Natural Frequencies (Clamped–Clamped) |
|---|---|---|
| First mode | 466.4 | 545.5 |
| Second mode | 1269.2 | 1506.3 |

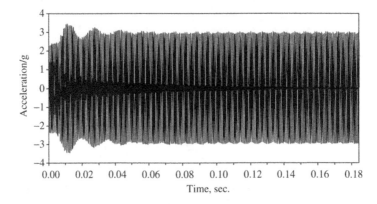

**FIGURE 5.17**    Beam and absorber response to 1270 Hz disturbance, analytical. (*Source:* From Olgac, N. and Jalili, N., *J. Sound Vib.*, 218, 307–331, 1998. With permission.)

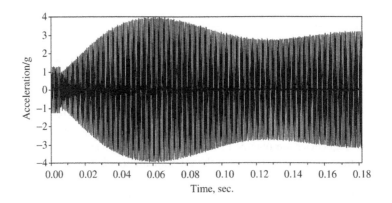

**FIGURE 5.18**    Beam and absorber response to 1270 Hz disturbance, experimental. (*Source:* From Olgac, N. and Jalili, N., *J. Sound Vib.*, 218, 307–331, 1998. With permission.)

### 5.3.2.2    Active Resonator Vibration Absorbers

One novel implementation of the tuned vibration absorbers is the ARA (Knowles et al., 2001b). The concept of the ARA is closely related to the concept of the delayed resonator (Olgac and Holm-Hansen, 1994; Olgac and Jalili, 1999). Using a simple position (or velocity or acceleration) feedback control within the absorber section, it enforces the dominant characteristic roots of the absorber subsection to be on the imaginary axis, and hence leading to resonance. Once the ARA becomes resonant, it creates perfect vibration absorption at this frequency.

A very important component of any active vibration absorber is the actuator unit. Recent advances in smart materials have led to the development of advanced actuators using piezoelectric ceramics, shape memory alloys, and magnetostrictive materials (Garcia et al., 1992; Shaw, 1998). Over the past two decades, piezoelectric ceramics have been utilized as potential replacements for conventional transducers.

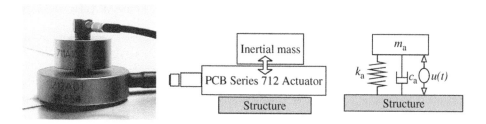

**FIGURE 5.19** A PCB series 712 PZT inertial actuator (left), schematic of operation (middle), and a simple single-DoF mathematical model (right). (Active Vibration Control Instrumentation, A Division of PCB Piezotronics, Inc., www.pcb.com.)

These materials are compounds of lead zirconate–titanate (PZT). The PZT properties can be optimized to suit specific applications by appropriate adjustment of the zirconate–titanate ratio. Specifically, a piezoelectric inertial actuator is an efficient and inexpensive solution for active structural vibration control. As shown in Figure 5.19, it applies a point force to the structure to which it is attached.

### 5.3.2.2.1 An Overview of PZT Inertial Actuators

PZT inertial actuators are most commonly made out of two parallel piezoelectric plates. If voltage is applied, one of the plates expands as the other one contracts, hence producing displacement that is proportional to the input voltage. The resonance of such an actuator can be adjusted by the size of the inertial mass (see Figure 5.19). Increasing the size of the inertial mass will lower the resonant frequency and decreasing the mass will increase it. The resonant frequency, $f_r$, can be expressed as

$$f_r = \frac{1}{2\pi}\sqrt{\frac{k_a}{m_a}} \tag{5.28}$$

where $k_a$ is the effective stiffness of the actuator, $m_a$, is defined as

$$m_a = m_{e_{PZT}} + m_{inertial} + m_{acc} \tag{5.29}$$

The PZT effective mass is $m_{e_{PZT}}$, $m_{inertial}$ is the inertial mass, and $m_{acc}$ is the accelerometer mass. Using a simple single-DoF system (see Figure 5.19), the parameters of the PZT inertial actuators can be experimentally determined (Knowles et al., 2001a). This "parameter identification" problem is an inverse problem. We refer interested readers to Banks and Ito (1988) and Banks and Kunisch (1989) for a general introduction to parameter estimation or inverse problems governed by differential equations.

### 5.3.2.2.2 Active Resonator Absorber Concept

The concept of ARA is closely related to that of the DR (Olgac, 1995; Olgac and Holm-Hansen, 1995; Olgac and Jalili, 1998). Instead of a compensator, the DR uses a simple delayed position (or velocity, or acceleration) feedback control within the absorber subsection for the mentioned "sensitization."

In contrast to that of the DR absorber, the characteristic equation of the proposed control scheme is rational in nature and is hence easier to implement when closed-loop stability of the system is concerned. Similar to the DR absorber, the proposed ARA requires only one signal from the absorber mass, absolute or relative to the point of attachment (see Figure 5.7 bottom). After the signal is processed through a compensator, an additional force is produced, for instance, by a PZT inertial actuator. If the compensator parameters are properly set, the absorber should behave as an ideal resonator at one or even more frequencies. As a result, the resonator will absorb vibratory energy from the primary mass at given frequencies. The frequency to be absorbed can be tuned in real time. Moreover, if the controller or the actuator fails, the ARA will still function as a passive absorber, and thus it is inherently fail-safe. A similar vibration absorption methodology is given by Filipović and Schröder (1999) for linear systems. The ARA, however, is not confined to the linear regime.

For the case of linear assumption for the PZT actuator, the dynamics of the ARA (Figure 5.7, bottom) can be expressed as

$$m_a\ddot{x}_a(t) + c_a\dot{x}_a(t) + k_ax_a(t) - u(t) = c_a\dot{x}_1(t) + k_ax_1(t) \tag{5.30}$$

where $x_1(t)$ and $x_a(t)$ are the respective primary (at the absorber point of attachment) and absorber mass displacements. The mass, $m_a$, is given by Equation 5.29 and the control, $u(t)$, is designed to produce designated resonance frequencies within the ARA.

The objective of the feedback control, $u(t)$, is to convert the dissipative structure (Figure 5.7, top) into a conservative or marginally stable one (Figure 5.7, bottom) with a designated resonant frequency, $\omega_c$. In other words, the control aims the placement of dominant poles at $\pm j\omega_c$ for the combined system, where $j = \sqrt{-1}$ (see Figure 5.14). As a result, the ARA becomes marginally stable at particular frequencies in the determined frequency range. Using simple position (or velocity or acceleration) feedback within the absorber section (i.e., $U(s) = \bar{U}(s)X_a(s)$), the corresponding dynamics of the ARA, given by Equation 5.30, in the Laplace domain become

$$(m_as^2 + c_as + k_a)X_a(s) - \bar{U}(s)X_a(s) = C(s)X_a(s) = (c_as + k_a)X_1(s) \tag{5.31}$$

The compensator transfer function, $\bar{U}(s)$, is then selected such that the primary system displacement at the absorber point of attachment is forced to be zero; that is

$$C(s) = (m_as^2 + c_as + k_a) - \bar{U}(s) = 0 \tag{5.32}$$

The parameters of the compensator are determined through introducing resonance conditions to the absorber characteristic equation, $C(s)$; that is, the equations $\text{Re}\{C(j\omega_i)\} = 0$ and $\text{Im}\{C(j\omega_i)\} = 0$ are simultaneously solved, where $i = 1, 2, \ldots, l$ and $l$ is the number of frequencies to be absorbed. Using additional compensator parameters, the stable frequency range or other properties can be adjusted in real time.

Consider the case where $U(s)$ is taken as a proportional compensator with a single time constant based on the acceleration of the ARA, given by

$$U(s) = \bar{U}(s)X_a(s), \quad \text{where } \bar{U}(s) = \frac{gs^2}{1 + Ts} \tag{5.33}$$

Then, in the time domain, the control force, $u(t)$, can be obtained from

$$u(t) = \frac{g}{T}\int_0^t e^{-(t-\tau)/T}\ddot{x}_a(\tau)d\tau \tag{5.34}$$

To achieve ideal resonator behavior, two dominant roots of Equation 5.32 are placed on the imaginary axis at the desired crossing frequency, $\omega_c$. Substituting $s = \pm j\omega_c$ into Equation 5.32 and solving for the control parameters, $g_c$ and $T_c$, one can obtain

$$g_c = m_a\left(\frac{c_a^2}{m_a^2\left(\omega^2 - \dfrac{k_a}{m_a}\right)} - \frac{k_a}{m_a\omega^2} + 1\right), \quad T_c = \frac{c_a\sqrt{k_a/m_a}}{\sqrt{m_ak_a}\left(\omega^2 - \dfrac{k_a}{m_a}\right)}, \quad \text{for } \omega = \omega_c \tag{5.35}$$

The control parameters, $g_c$ and $T_c$, are based on the physical properties of the ARA (i.e., $c_a$, $k_a$, and $m_a$) as well as the frequency of the disturbance, $\omega$, illustrating that the ARA does not require any information from the primary system to which it is attached. However, when the physical properties of the ARA are not known within a high degree of certainty, a method to autotune the control parameters must be considered. The stability assurance of such autotuning proposition will bring primary system parameters into the derivations, and hence the primary system cannot be totally decoupled. This issue will be discussed later in the chapter.

### 5.3.2.2.3  Application of ARA to Structural-Vibration Control

In order to demonstrate the effectiveness of the proposed ARA, a simple single-DoF primary system subjected to tonal force excitations is considered. As shown in Figure 5.20, two PZT inertial actuators are used for both the primary (model 712-A01) and the absorber (model 712-A02) subsections. Each system consists of passive elements (spring stiffness and damping properties of the PZT materials) and active compartment with the physical parameters listed in Table 5.2. The top actuator acts as the ARA with the controlled force, $u(t)$, while the bottom one represents the primary system subjected to the force excitation, $f(t)$.

The governing dynamics for the combined system can be expressed as

$$m_a\ddot{x}_a(t) + c_a\dot{x}_a(t) + k_a x_a(t) - u(t) = c_a\dot{x}_1(t) + k_a x_1(t) \tag{5.36}$$

$$m_1\ddot{x}_1(t) + (c_1 + c_a)\dot{x}_1(t) + (k_1 + k_a)x_1(t) - \{c_a\dot{x}_a(t) + k_a x_a(t) - u(t)\} = f(t) \tag{5.37}$$

where $x_1(t)$ and $x_a(t)$ are the respective primary and absorber displacements.

### 5.3.2.2.4  Stability Analysis and Parameter Sensitivity

The sufficient and necessary condition for asymptotic stability is that all roots of the characteristic equation have negative real parts. For the linear system, Equation 5.36 and Equation 5.37, when utilizing controller (Equation 5.34), the characteristic equation of the combined system (Figure 5.20, right) can be determined and the stability region for compensator parameters, $g$ and $T$, can be obtained using the Routh–Hurwitz method.

### 5.3.2.2.5  Autotuning Proposition

When using the proposed ARA configuration in real applications where the physical properties are not known or vary over time, the compensator parameters, $g$ and $T$, only provide partial vibration suppression. In order to remedy this, a need exists for an autotuning method to adjust the compensator

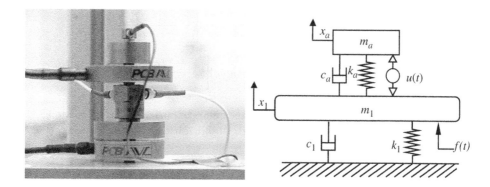

**FIGURE 5.20**  Implementation of the ARA concept using two PZT actuators (left) and its mathematical model (right).

**TABLE 5.2**  Experimentally Determined Parameters of PCB Series 712 PZT Inertial Actuators

| PZT System Parameters | PCB Model 712-A01 | PCB Model 712-A02 |
|---|---|---|
| Effective mass, $m_{ePZT}$ (gr) | 7.20 | 12.14 |
| Inertial mass, $m_{inertial}$ (gr) | 100.00 | 200.00 |
| Stiffness, $k_a$ (kN/m) | 3814.9 | 401.5 |
| Damping, $c_a$ (Ns/m) | 79.49 | 11.48 |

parameters, $g$ and $T$, by some quantities, $\Delta g$ and $\Delta T$, respectively (Jalili and Olgac, 1998b; Jalili and Olgac, 2000a). For the case of the linear compensator with a single time constant, given by Equation 5.33, the transfer function between primary displacement, $X_1(s)$, and absorber displacement, $X_a(s)$, can be obtained as

$$G(s) = \frac{X_1(s)}{X_a(s)} = \frac{m_a s^2 + c_a s + k_a - \dfrac{g s^2}{1 + Ts}}{c_a s + k_a} \tag{5.38}$$

The transfer function can be rewritten in the frequency domain for $s = j\omega$ as

$$G(j\omega) = \frac{X_1(j\omega)}{X_a(j\omega)} = \frac{-m_a \omega^2 + c_a \omega j + k_a + \dfrac{g\omega^2}{1 + T\omega j}}{c_a \omega j + k_a} \tag{5.39}$$

where $G(j\omega)$ can be obtained in real time by convolution of accelerometer readings (Renzulli et al., 1999) or other methods (Jalili and Olgac, 2000a). Following a similar procedure as is utilized in Renzulli et al. (1999), the numerator of the transfer function (Equation 5.39) must approach zero in order to suppress primary system vibration. This is accomplished by setting

$$G(j\omega) + \Delta G(j\omega) = 0 \tag{5.40}$$

where $G(j\omega)$ is the real-time transfer function and $\Delta G(j\omega)$ can be written as a variational form of Equation 5.39 as

$$\Delta G(\omega i) = \frac{\partial G}{\partial g}\Delta g + \frac{\partial G}{\partial T}\Delta T + \text{higher order terms} \tag{5.41}$$

Since the estimated physical parameters of the absorber (i.e., $c_a$, $k_a$, and $m_a$) are within the vicinity of the actual parameters, $\Delta g$ and $\Delta T$ should be small quantities and the higher-order terms of Equation 5.41 can be neglected. Using Equation 5.40 and Equation 5.41 and neglecting higher-order terms, we have

$$\Delta g = \text{Re}[G(j\omega)]\left[\frac{2Tc_a\omega^2 - k_a + k_aT^2\omega^2}{\omega^2}\right] + \text{Im}[G(j\omega)]\left[\frac{c_a - T^2\omega^2 c_a + 2k_a T}{\omega^2}\right],$$

$$\Delta T = \text{Re}[G(j\omega)]\left[\frac{c_a - T^2\omega^2 c_a + 2k_a T}{g\omega^2}\right] + \text{Im}[G(j\omega)]\left[\frac{k_a - 2Tc_a\omega^2 - k_aT^2\omega^2}{g\omega^3}\right] + \frac{T}{g}\Delta g \tag{5.42}$$

In the above expressions, $g$ and $T$ are the current compensator parameters given by Equation 5.35, $c_a$, $k_a$, and $m_a$ are the estimated absorber parameters, $\omega$ is the absorber base excitation frequency, and $G(j\omega)$ is the transfer function obtained in real time. That is, the retuned control parameters, $g$ and $T$, are determined as follows

$$g_{\text{new}} = g_{\text{current}} + \Delta g \quad \text{and} \quad T_{\text{new}} = T_{\text{current}} + \Delta T \tag{5.43}$$

where $\Delta g$ and $\Delta T$ are those given by Equation 5.42. After compensator parameters, $g$ and $T$, are adjusted by Equation 5.43, the process can be repeated until $|G(j\omega)|$ falls within the desired level of tolerance. $G(j\omega)$ can be determined in real time as shown in Liu et al. (1997) by

$$G(j\omega) = |G(j\omega)|e^{(j\phi(j\omega))} \tag{5.44}$$

where the magnitude and phase are determined assuming that the absorber and primary displacements are harmonic functions of time given by

$$x_a(t) = X_a \sin(\omega t + \phi_a), \quad x_1(t) = X_1 \sin(\omega t + \phi_1) \tag{5.45}$$

With the magnitudes and phase angles of Equation 5.44, the transfer function can be determined from Equation 5.44 and the following:

$$|G(j\omega)| = \frac{X_1}{X_a}, \quad \phi(j\omega) = \phi_1 - \phi_a$$

### 5.3.2.2.6 Numerical Simulations and Discussions

To illustrate the feasibility of the proposed absorption methodology, an example case study is presented. The ARA control law is the proportional compensator with a single time constant as given in Equation 5.34. The primary system is subjected to a harmonic excitation with unit amplitude and a frequency of 800 Hz. The ARA and primary system parameters are taken as those given in Table 5.2. The simulation was done using Matlab/Simulink® and the results for the primary system and the absorber displacements are given in Figure 5.21.

As seen, vibrations are completely suppressed in the primary subsection after approximately 0.05 sec, at which the absorber acts as a marginally stable resonator. For this case, all physical parameters are assumed to be known exactly. However, in practice these parameters are not known exactly and may vary with time, so the case with estimated system parameters must be considered.

To demonstrate the feasibility of the proposed autotuning method, the nominal system parameters $(m_a, m_1, k_a, k_1, c_a, c_1)$ were fictitiously perturbed by 10% (i.e., representing the actual values) in the simulation. However, the nominal values of $m_a$, $m_1$, $k_a$, $k_1$, $c_a$, and $c_1$ were used for calculation of the compensator parameters, $g$ and $T$. The results of the simulation using nominal parameters are given in Figure 5.22. From Figure 5.22, top, the effect of parameter variation is shown as steady-state oscillations of the primary structure. This undesirable response will undoubtedly be encountered when the experiment is implemented. Thus, an autotuning procedure is needed.

The result of the first autotuning iteration is given in Figure 5.22, middle, where the control parameters, $g$ and $T$, are adjusted based on Equation 5.43. One can see tremendous improvement in the primary system response with only one iteration (see Figure 5.22, middle). A second iteration is

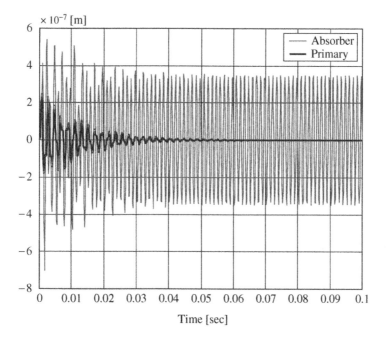

**FIGURE 5.21** Primary system and absorber displacements subjected to 800 Hz harmonic disturbance.

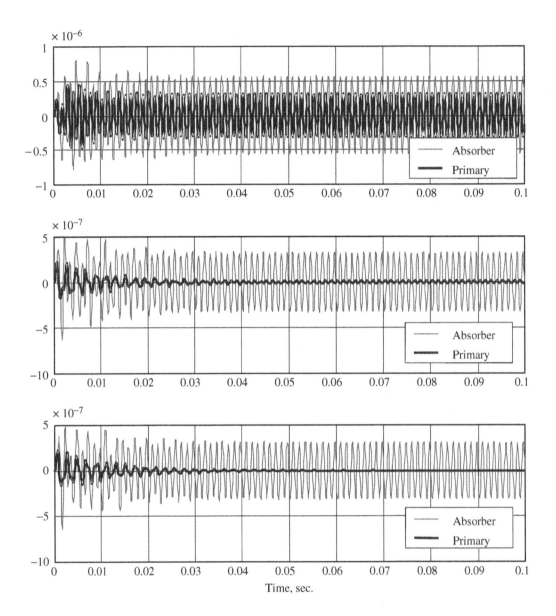

**FIGURE 5.22** System responses (displacement, m) for (a) nominal absorber parameters; (b) after first autotuning procedure and (c) after second autotuning procedure.

performed, as shown in Figure 5.22, bottom. The response closely resembles that from Figure 5.21, where all system parameters are assumed to be known exactly.

### 5.3.3 Vibration-Control Systems

As discussed, in vibration-control schemes, the control inputs to the systems are altered in order to regulate or track a desired trajectory while simultaneously suppressing the vibrational transients in the system. This control problem is rather challenging since it must achieve the motion-tracking objectives while stabilizing the transient vibrations in the system. This section provides two recent control methods developed for the regulation and tracking of flexible beams. The experimental implementations are also discussed. The first control method is a single-input vibration-control system discussed in

Section 5.3.3.1, while the second application utilizes a secondary input control in addition to the primary control input to improve the vibrational characteristics of the system (see Section 5.3.3.2).

### 5.3.3.1 Variable Structure Control of Rotating Flexible Beams

The vibration-control problem of a flexible manipulator consists in achieving the motion-tracking objectives while stabilizing the transient vibrations in the arm. Several control methods have been developed for flexible arms (Skaar and Tucker, 1986; Bayo, 1987; Yuh, 1987; Sinha, 1988; Lou, 1993; Ge et al., 1997; de Querioz et al., 1999). Most of these methods concentrate on a model-based controller design, and some of these may not be easy to implement due to the uncertainties in the design model, large variations of the loads, ignored high frequency dynamics, and high order of the designed controllers. In view of these methods, VSC is particularly attractive due to its simplicity of implementation and its robustness to parameter uncertainties (Yeung and Chen, 1989; Singh et al., 1994; Jalili et al., 1997).

#### 5.3.3.1.1 *Mathematical Modeling*

As shown in Figure 5.23, one end of the arm is free and the other end is rigidly attached to a vertical gear shaft, driven by a DC motor. A uniform cross section is considered for the arm, and we make the Euler–Bernoulli assumptions. The control torque, $\tau$, acting on the output shaft, is normal to the plane of motion. Viscous frictions and the ever-present unmodeled dynamics of the motor compartment are to be compensated *via* a perturbation estimation process, as explained later in the text. Since the dynamic system considered here has been utilized in literature quite often, we present only the resulting partial differential equation (PDE) of the system and refer interested readers to Junkins and Kim (1993) and Luo et al. (1999) for detailed derivations.

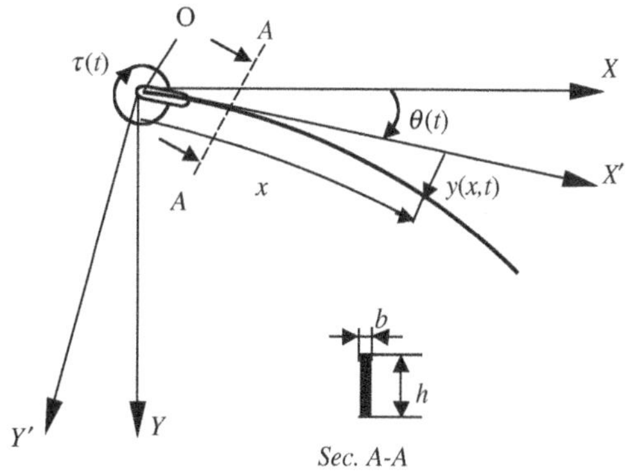

The system is governed by

**FIGURE 5.23** Flexible arm in the horizontal plane and kinematics of deformation. (*Source:* From Jalili, N., *ASME J. Dyn. Syst. Meas. Control*, 123, 712–719, 2001. With permission.)

$$I_h \ddot{\theta}(t) + \rho \int_0^L x \ddot{z}(x, t) \mathrm{d}x = \tau \qquad (5.46)$$

$$\rho \ddot{z}(x, t) + EI z''''(x, t) = 0 \qquad (5.47)$$

with the corresponding boundary conditions

$$z(0, t) = 0, \quad z'(0, t) = \theta(t), \quad z''(L, t) = 0, \quad z'''(L, t) = 0 \qquad (5.48)$$

where $\rho$ is the arm's linear mass density, $L$ is the arm length, $E$ is Young's modulus of elasticity, $I$ is the cross-sectional moment of inertia, $I_h$ is the equivalent mass moment of inertia at the root end of the arm, $I_t = I_h + \rho L^3/3$ is the total inertia, and the global variable $z$ is defined as

$$z(x, t) = x\theta(t) + y(x, t) \qquad (5.49)$$

Clearly, the arm vibration equation (Equation 5.47) is a homogeneous PDE but the boundary conditions (Equation 5.48) are nonhomogeneous. Therefore, the closed form solution is very tedious to obtain, if not impossible. Using the application of VSC, these equations and their associated boundary conditions can be converted to a homogeneous boundary value problem, as discussed next.

#### 5.3.3.1.2 *Variable Structure Controller*

The controller objective is to track the arm angular displacement from an initial angle, $\theta_d = \theta(0)$, to zero position, $\theta(t \rightarrow \infty) = 0$, while minimizing the flexible arm oscillations. To achieve the control insensitivity against modeling uncertainties, the nonlinear control routine of sliding mode control with an additional perturbation estimation (SMCPE) compartment is adopted here (Elmali and Olgac, 1992; Jalili and Olgac, 1998a). The SMCPE method, presented in Elmali and Olgac (1992), has many attractive features, but it suffers from the disadvantages associated with the truncated-model-base controllers. On the other hand, the infinite-dimensional distributed (IDD)-base controller design, proposed in Zhu et al. (1997), has practical limitations due to its measurement requirements in addition to the complex control law.

Initiating from the idea of the IDD-base controller, we present a new controller design approach in which an online perturbation estimation mechanism is introduced and integrated with the controller to relax the measurement requirements and simplify the control implementation. As utilized in Zhu et al. (1997), for the tip-vibration suppression, it is further required that the sliding surface enable the transformation of nonhomogeneous boundary conditions (Equation 5.48) to homogeneous ones. To satisfy vibration suppression and robustness requirements simultaneously, the sliding hyperplane is selected as a combination of tracking (regulation) error and arm flexible vibration as

$$s = \dot{w} + \sigma w \qquad (5.50)$$

where $\sigma > 0$ is a control parameter and

$$w = \theta(t) + \frac{\mu}{L} z(L, t) \qquad (5.51)$$

with the scalar, $\mu$, being selected later. When $\mu = 0$, controller (Equation 5.50) reduces to a sliding variable for rigid-link manipulators (Jalili and Olgac, 1998a; Yeung and Chen, 1988). The motivation for such sliding a variable is to provide a suitable boundary condition for solving the beam Equation 5.47, as will be discussed next and is detailed in Jalili (2001).

For the system described by Equation 5.46 to Equation 5.48, if the variable structure controller is given by

$$\tau = \psi_{est} + \frac{I_t}{1 + \mu} \left( -k \, \mathrm{sgn}(s) - Ps - \frac{\mu}{L} \ddot{y}(L, t) - \sigma(1 + \mu)\dot{\theta} - \frac{\sigma\mu}{L} \dot{y}(L, t) \right) \qquad (5.52)$$

where $\psi_{est}$ is an estimate of the beam flexibility effect

$$\psi = \rho \int_0^L x\ddot{y}(x, t)\mathrm{d}x \qquad (5.53)$$

$k$ and $P$ are positive scalars $k \geq 1 + \mu/I_t|\psi - \psi_{est}|$, $-1.2 < \mu < -0.45$, $\mu \neq -1$ and sgn( ) represents the standard signum function, then, the system's motion will first reach the sliding mode $s = 0$ in a finite time, and consequently converge to the equilibrium position $w(x, t) = 0$ exponentially with a time-constant $1/\sigma$ (Jalili, 2001).

### 5.3.3.1.3  *Controller Implementation*

In the preceding section, it was shown that by properly selecting control variable, $\mu$, the motion exponentially converges to $w = 0$ with a time-constant $1/\sigma$, while the arm stops in a finite time. Although the discontinuous nature of the controller introduces a robustifying mechanism, we have made the scheme more insensitive to parametric variations and unmodeled dynamics by reducing the required measurements and hence easier control implementation. The remaining measurements and ever-present modeling imperfection effects have all been estimated through an online estimation process. As stated before, in order to simplify the control implementation and reduce the measurement effort, the effect of all uncertainties, including flexibility effect ($\int_0^L x\ddot{y}(x, t)\mathrm{d}x$) and the ever-present unmodeled dynamics, is gathered into a single quantity named perturbation, $\psi$, as given by Equation 5.53. Noting Equation 5.46, the perturbation term can be expressed as

$$\psi = \tau - I_t\ddot{\theta}(t) \qquad (5.54)$$

which requires the yet-unknown control feedback $\tau$. In order to resolve this dilemma of causality, the current value of control torque, $\tau$, is replaced by the most recent control, $\tau(t - \delta)$, where $\delta$ is the small time-step used for the loop closure. This replacement is justifiable in practice since such an algorithm is implemented on a digital computer and the sampling speed is high enough to claim this. Also, in the absence of measurement noise, $\ddot{\theta}(t) \cong \ddot{\theta}_{cal}(t) = [\dot{\theta}(t) - \dot{\theta}(t - \delta)]/\delta$.

In practice and in the presence of measurement noise, appropriate filtering may be considered and combined with these approximate derivatives. This technique is referred to as "switched derivatives". This backward differences are shown to be effective when $\delta$ is selected to be small enough and the controller is run on a fast DSP (Cannon and Schmitz, 1984). Also, $\ddot{y}(L, t)$ can be obtained by attaching an accelerometer at arm-tip position. All the required signals are, therefore, measurable by currently available sensor facilities and the controller is thus realizable in practice. Although these signals may be quite inaccurate, it should be pointed out that the signals, either measurements or estimations, need not be known very accurately, since robust sliding control can be achieved if $k$ is chosen to be large enough to cover the error existing in the measurement/signal estimation (Yeung and Chen, 1989).

### 5.3.3.1.4   *Numerical Simulations*

In order to show the effectiveness of the proposed controller, a lightweight flexible arm is considered ($h \gg b$ in Figure 5.23). For numerical results, we consider $\theta_d = \theta(0) = \pi/2$ for the initial arm base angle, with zero initial conditions for the rest of the state variables. The system parameters are listed in Table 5.3. Utilizing assumed mode model (AMM), the arm vibration, Equation 5.47, is truncated to three modes and used in the simulations. It should be noted that the controller law, Equation 5.52, is based on the original infinite dimensional equation, and this truncation is utilized only for simulation purposes.

We take the controller parameter $\mu = -0.66$, $P = 7.0$, $k = 5$, $\varepsilon = 0.01$ and $\sigma = 0.8$. In practice, $\sigma$ is selected for maximum tracking accuracy taking into account unmodeled dynamics and actuator hardware limitations (Moura et al., 1997). Although such restrictions do not exist in simulations (i.e., with ideal actuators, high sampling frequencies and perfect measurements), this selection of $\sigma$ was decided based on actual experiment conditions.

The sampling rate for the simulations is $\delta = 0.0005$ sec, while data are recorded at the rate of only 0.002 sec for plotting purposes. The system responses to the proposed control scheme are shown in Figure 5.24. The arm-base angular position reaches the desired position, $\theta = 0$, in approximately 4 to 5 sec, which is in agreement with the approximate settling time of $t_s = 4/\sigma$ (Figure 5.24a). As soon as the system reaches the sliding mode layer, $|s| < \varepsilon$ (Figure 5.24d), the tip vibrations stop (Figure 5.24b), which demonstrates the feasibility of the proposed control technique. The control torque exhibits some residual vibration, as shown in Figure 5.24c. This residual oscillation is expected since the system

**TABLE 5.3**   System Parameters Used in Numerical Simulations and Experimental Setup for Rotating Arm

| Properties | Symbol | Value | Unit |
|---|---|---|---|
| Arm Young's modulus | $E$ | $207 \times 10^9$ | $N/m^2$ |
| Arm thickness | $b$ | 0.0008 | m |
| Arm height | $h$ | 0.02 | m |
| Arm length | $L$ | 0.45 | m |
| Arm linear mass density | $\rho$ | $0.06/L$ | kg/m |
| Total arm base inertia | $I_h$ | 0.002 | kg m$^2$ |
| Gearbox ratio | $N$ | 14:1 | — |
| Light source mass | — | 0.05 | kg |
| Position sensor sensitivity | — | 0.39 | V/cm |
| Motor back EMF constant | $K_b$ | 0.0077 | V/rad/sec |
| Motor torque constant | $K_t$ | 0.0077 | N m/A |
| Armature resistance | $R_a$ | 2.6 | $\Omega$ |
| Armature inductance | $L_a$ | 0.18 | mH |
| Encoder resolution | — | 0.087 | Deg/count |

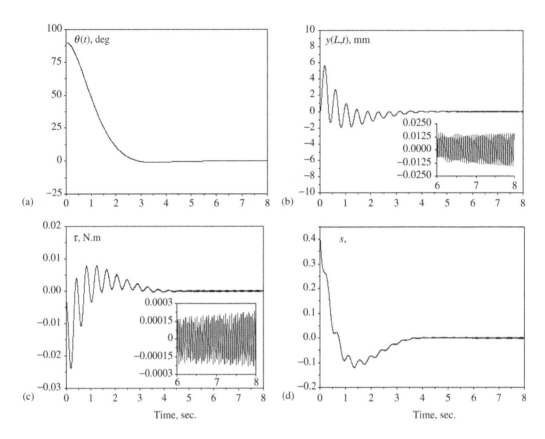

**FIGURE 5.24** Analytical system responses to controller with inclusion of arm flexibility, that is, $\mu = -0.66$: (a) arm angular position; (b) arm-tip deflection; (c) control torque and (d) sliding variable sec. (*Source:* From Jalili, N., *ASME J. Dyn. Syst. Meas. Control*, 123, 712–719, 2001. With permission.)

motiono is not forced to stay on $s = 0$ surface (instead it is forced to stay on $|s| < \varepsilon$) when saturation function is used. The sliding variable $s$ is also depicted in Figure 5.24d. To demonstrate better the feature of the controller, the system responses are displayed when $\mu = 0$ (Figure 5.25). As discussed, $\mu = 0$ corresponds to the sliding variable for the rigid link. The undesirable oscillations at the arm tip are evident (see Figure 5.25b and c).

### 5.3.3.1.5 Control Experiments

In order to demonstrate better the effectiveness of the controller, an experimental setup is constructed and used to verify the numerical results and concepts discussed in the preceding sections. The experimental setup is shown in Figure 5.26. The arm is a slender beam made of stainless steel, with the same dimensions as used in the simulations. The experimental setup parameters are listed in Table 5.3. One end of the arm is clamped to a solid clamping fixture, which is driven by a high-quality DC servomotor. The motor drives a built-in gearbox ($N = 14{:}1$) whose output drives an antibacklash gear. The antibacklash gear, which is equipped with a precision encoder, is utilized to measure the arm base angle as well as to eliminate the backlash. For tip deflection, a light source is attached to the tip of the arm, which is detected by a camera mounted on the rotating base.

The DC motor can be modeled as a standard armature circuit; that is, the applied voltage, $v$, to the DC motor is

$$v = R_a i_a + L_a \, di_a/dt + K_b \dot{\theta}_m \tag{5.55}$$

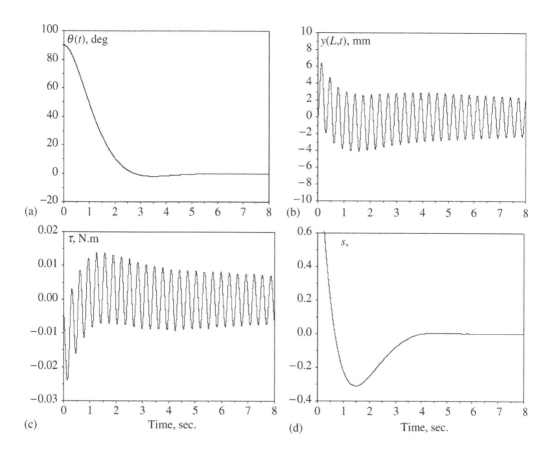

**FIGURE 5.25** Analytical system responses to controller without inclusion of arm flexibility, that is, $\mu = 0$: (a) arm angular position; (b) arm-tip deflection; (c) control torque and (d) sliding variable sec. (*Source:* From Jalili, N., *ASME J. Dyn. Syst. Meas. Control*, 123, 712–719, 2001. With permission.)

**FIGURE 5.26** The experimental device and setup configuration. (*Source:* From Jalili, N., *ASME J. Dyn. Syst. Meas. Control*, 123, 712–719, 2001. With permission.)

where $R_a$ is the armature resistance, $L_a$ is the armature inductance, $i_a$ is the armature current, $K_b$ is the back-EMF (electro-motive-force) constant, and $\theta_m$ is the motor shaft position. The motor torque, $\tau_m$ from the motor shaft with the torque constant, $K_t$, can be written as

$$\tau_m = K_t i_a \tag{5.56}$$

The motor dynamics thus become

$$I_e \ddot{\theta}_m + C_v \dot{\theta}_m + \tau_a = \tau_m = K_t i_a \tag{5.57}$$

where $C_v$ is the equivalent damping constant of the motor, and $I_e = I_m + I_L/N^2$ is the equivalent inertia load including motor inertia, $I_m$, and gearbox, clamping frame and camera inertia, $I_L$. The available torque from the motor shaft for the arm is $\tau_a$.

Utilizing the gearbox from the motor shaft to the output shaft and ignoring the motor electric time constant, $(L_a/R_a)$, one can relate the servomotor input voltage to the applied torque (acting on the arm) as

$$\tau = \frac{NK_t}{R_a} v - \left( C_v + \frac{K_t K_b}{R_a} \right) N^2 \dot{\theta} - I_h \ddot{\theta} \tag{5.58}$$

where $I_h = N^2 I_e$ is the equivalent inertia of the arm base used in the derivation of governing equations. By substituting this torque into the control law, the reference input voltage, $V$, is obtained for experiment.

The control torque is applied *via* a digital signal processor (DSP) with sampling rate of 10 kHz, while data are recorded at the rate 500 Hz (for plotting purposes only). The DSP runs the control routine in a single-input–single-output mode as a free standing CPU. Most of the computations and hardware commands are done on the DSP card. For this setup, a dedicated 500 MHz Pentium III serves as the host PC, and a state-of-the-art dSPACE® DS1103 PPC controller board equipped with a Motorola Power PC 604e at 333 MHz, 16 channels ADC, 12 channels DAC, as microprocessor.

The experimental system responses are shown in Figure 5.27 and Figure 5.28 for similar cases discussed in the numerical simulation section. Figure 5.27 represents the system responses when controller (Equation 5.52) utilizes the flexible arm (i.e., $\mu = -0.66$). As seen, the arm base reaches the desired position (Figure 5.27a), while tip deflection is simultaneously stopped (Figure 5.27b). The good correspondence between analytical results (Figure 5.24) and experimental findings (Figure 5.27) is noticeable from a vibration suppression characteristics point of view. It should be noted that the controller is based on the original governing equations, with arm-base angular position and tip deflection measurements only. The unmodeled dynamics, such as payload effect (owing to the light source at the tip, see Table 5.3) and viscous friction (at the root end of the arm), are being compensated through the proposed online perturbation estimation routine. This, in turn, demonstrates the capability of the proposed control scheme when considerable deviations between model and plant are encountered. The only noticeable difference is the fast decaying response as shown in Figure 5.27b and c. This clearly indicates the high friction at the motor, which was not considered in the simulations (Figure 24b and c). Similar responses are obtained when the controller is designed based on the rigid link only, that is, $\mu = 0$. The system responses are displayed in Figure 5.28. Similarly, the undesirable arm-tip oscillations are obvious. The overall agreement between simulations (Figure 24 and Figure 25) and the experiment (Figure 27 and Figure 28) is one of the critical contributions of this work.

### 5.3.3.2 Observer-Based Piezoelectric Vibration Control of Translating Flexible Beams

Many industrial robots, especially those widely used in automatic manufacturing assembly lines, are Cartesian types (Ge et al., 1998). A flexible Cartesian robot can be modeled as a flexible cantilever beam with a translational base support. Traditionally, a PD control strategy is used to regulate the movement of the robot arm. In lightweight robots, the base movement will cause undesirable vibrations at the arm tip because of the flexibility distributed along the arm. In order to eliminate

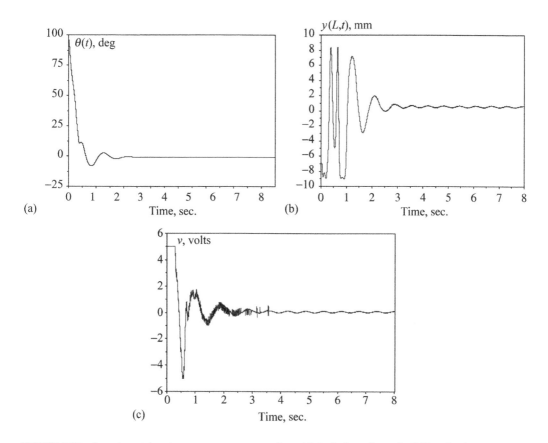

**FIGURE 5.27** Experimental system responses to controller with inclusion of arm flexibility, that is, $\mu = -0.66$: (a) arm angular position; (b) arm-tip deflection; (c) control voltage applied to DC servomotor. (*Source:* From Jalili, N., *ASME J. Dyn. Syst. Meas. Control*, 123, 712–719, 2001. With permission.)

such vibrations, the PD controller must be upgraded with additional compensating terms. In order to improve further the vibration suppression performance, which is a requirement for the high-precision manufacturing market, a second controller, such as a piezoelectric (PZT) patch actuator attached on the surface of the arm, can be utilized (Oueini et al., 1998; Ge et al., 1999; Jalili et al., 2002).

In this section, an observer-based control strategy is presented for regulating the arm motion (Liu et al., 2002). The base motion is controlled utilizing an electrodynamic shaker, while a piezoelectric (PZT) patch actuator is bonded on the surface of the flexible beam for suppressing residual arm vibrations. The control objective here is to regulate the arm base movement, while simultaneously suppressing the vibration transients in the arm. To achieve this, a simple PD control strategy is selected for the regulation of the movement of the base, and a Lyapunov-based controller is selected for the PZT voltage signal. The selection of the proposed energy-based Lyapunov function naturally results in velocity-related signals, which are not physically measurable (Dadfarnia et al., 2003). To remedy this, a reduced-order observer is designed to estimate the velocity related signals. For this, the control structure is designed based on the truncated two-mode beam model.

### 5.3.3.2.1 *Mathematical Modeling*

For the purpose of model development, we consider a uniform flexible cantilever beam with a PZT actuator bonded on its top surface. As shown in Figure 5.29, one end of the beam is clamped into

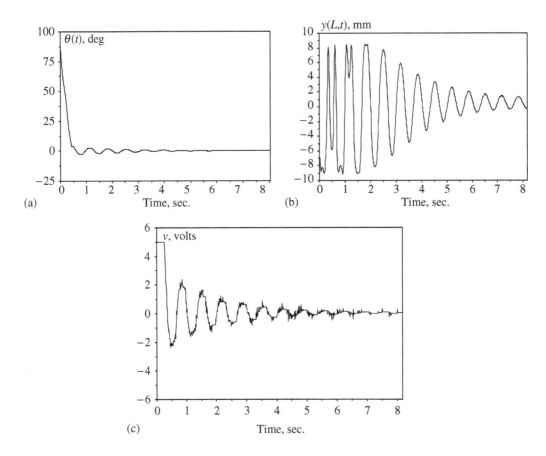

**FIGURE 5.28** Experimental system responses to controller without inclusion of arm flexibility, that is, $\mu = 0$: (a) arm angular position; (b) arm-tip deflection; (c) control voltage applied to DC servomotor. (*Source:* From Jalili, N., *ASME J. Dyn. Syst. Meas. Control*, 123, 712–719, 2001. With permission.)

a moving base with the mass of $m_b$, and a tip mass, $m_t$, is attached to the free end of the beam. The beam has total thickness $t_b$, and length $L$, while the piezoelectric film possesses thickness and length $t_b$ and ($l_2 - l_1$), respectively. We assume that the PZT and the beam have the same width, $b$. The PZT actuator is perfectly bonded on the beam at distance $l_1$ measured from the beam support. The force, $f(t)$, acting on the base and the input voltage, $v(t)$, applied to the PZT actuator are the only external effects.

To establish a coordinate system for the beam, the $x$-axis is taken in the longitudinal direction and the $z$-axis is specified in the transverse direction of the beam with midplane of the beam to be $z = 0$, as shown in Figure 5.30. This coordinate is fixed to the base.

The fundamental relations for the piezoelectric materials are given as (Ikeda, 1990)

$$\mathbf{F} = \mathbf{c}\mathbf{S} - \mathbf{h}\mathbf{D} \tag{5.59}$$

$$\mathbf{E} = -\mathbf{h}^{\mathrm{T}}\mathbf{S} + \boldsymbol{\beta}\mathbf{D} \tag{5.60}$$

where $\mathbf{F} \in \mathfrak{R}^6$ is the stress vector, $\mathbf{S} \in \mathfrak{R}^6$ is the strain vector, $\mathbf{c} \in \mathfrak{R}^{6 \times 6}$ is the symmetric matrix of elastic stiffness coefficients, $\mathbf{h} \in \mathfrak{R}^{6 \times 3}$ is the coupling coefficients matrix, $\mathbf{D} \in \mathfrak{R}^3$ is the electrical displacement vector, $\mathbf{E} \in \mathfrak{R}^3$ is the electrical field vector, and $\boldsymbol{\beta} \in \mathfrak{R}^{3 \times 3}$ is the symmetric matrix of impermittivity coefficients.

An energy method is used to derive the equations of motion. Neglecting the electrical kinetic energy, the total kinetic energy of the system is expressed as (Liu et al., 2002; Dadfarnia et al., 2004)

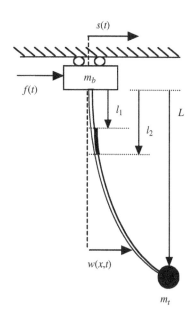

$$E_k = \frac{1}{2}m_b\dot{s}(t)^2 + \frac{1}{2}b\int_0^{l_1}\rho_b t_b(\dot{s}(t) + \dot{w}(x,t))^2 dx$$
$$+ \frac{1}{2}b\int_{l_1}^{l_2}(\rho_b t_b + \rho_p t_p)(\dot{s}(t) + \dot{w}(x,t))^2 dx$$
$$+ \frac{1}{2}b\int_{l_2}^{L}\rho_b t_b(\dot{s}(t) + \dot{w}(x,t))^2 dx$$
$$+ \frac{1}{2}m_t(\dot{s}(t) + \dot{w}(L,t))^2$$
$$= \frac{1}{2}m_b\dot{s}(t)^2 + \frac{1}{2}\int_0^{L}\rho(x)(\dot{s}(t) + \dot{w}(x,t))^2 dx$$
$$+ \frac{1}{2}m_t(\dot{s}(t) + \dot{w}(L,t))^2 \tag{5.61}$$

**FIGURE 5.29** Schematic of the SCARA/Cartesian robot (last link).

where

$$\rho(x) = [\rho_b t_b + G(x)\rho_p t_p]b \tag{5.62}$$
$$G(x) = H(x - l_1) - H(x - l_2)$$

and $H(x)$ is the Heaviside function, $\rho_b$ and $\rho_p$ are the respective beam and PZT volumetric densities. Neglecting the effect of gravity due to planar motion and the higher-order terms of quadratic in $w'$ (Esmailzadeh and Jalili, 1998b), the total potential energy of the system can be expressed as

$$E_p = \frac{1}{2}b\int_0^{l_1}\int_{-t_b/2}^{t_b/2}\mathbf{F}^T\mathbf{S}\,dy\,dx + \frac{1}{2}b\int_{l_1}^{l_2}\int_{-t_b/2}^{t_b/2}\mathbf{F}^T\mathbf{S}\,dy\,dx + \frac{1}{2}b\int_{l_1}^{l_2}\int_{t_b/2}^{(t_b/2)+t_p}[\mathbf{F}^T\mathbf{S} + \mathbf{E}^T\mathbf{D}]dy\,dx$$
$$+ \frac{1}{2}b\int_{l_2}^{L}\int_{-t_b/2}^{t_b/2}\mathbf{F}^T\mathbf{S}\,dy\,dx$$
$$= \frac{1}{2}\int_0^{L}c(x)\left[\frac{\partial^2 w(x,t)}{\partial x^2}\right]^2 dx + h_l D_y(t)\int_{l_1}^{l_2}\frac{\partial^2 w(x,t)}{\partial x^2}dx + \frac{1}{2}\beta_l(l_2 - l_1)D_y(t)^2 \tag{5.63}$$

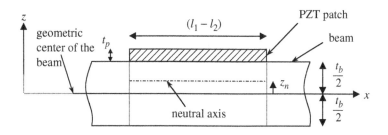

**FIGURE 5.30** Coordinate system.

where

$$c(x) = \frac{b}{3}\left\{\left(\frac{c_{11}^{b}t_{b}^{3}}{4}\right) + G(x)\left\{3c_{11}^{b}t_{b}z_{n}^{2} + c_{11}^{P}\left(t_{p}^{3} + 3t_{p}\left(\frac{t_{b}}{2} - z_{n}\right)^{2} + 3t_{p}^{2}\left(\frac{t_{b}}{2} - z_{n}\right)\right)\right\}\right\}$$

(5.64)

$$h_l = h_{12}t_p b(t_p + t_b - 2z_n)/2, \quad \beta_l = \beta_{22}bt_p$$

and

$$z_n = \frac{c_{11}^{P}t_p(t_p + t_b)}{c_{11}^{b}t_b + c_{11}^{P}t_p}$$

The beam and PZT stiffnesses are $c_{11}^{b}$ and $c_{11}^{P}$, respectively.

Using the AMM for the beam vibration analysis, the beam deflection can be written as

$$w(x,t) = \sum_{i=1}^{\infty} \phi_i(x)q_i(t), \quad P(x,t) = s(t) + w(x,t)$$

(5.65)

The equations of motion can now be obtained using the Lagrangian approach

$$\left[m_b + m_t + \int_0^L \rho(x)dx\right]\ddot{s}(t) + \sum_{j=1}^{\infty} m_j\ddot{q}_j(t) = f(t)$$

(5.66a)

$$m_i\ddot{s}(t) + m_{di}\ddot{q}_i(t) + \omega_i^2 m_{di}q_i(t) + h_l(\phi_i'(l_2) - \phi_i'(l_1))D_y(t) = 0$$

(5.66b)

$$h_l\sum_{j=1}^{\infty}\{(\phi_j'(l_2) - \phi_j'(l_1))q_j(t)\} + \beta_l(l_2 - l_1)D_y(t) = b(l_2 - l_1)v(t)$$

(5.66c)

where

$$m_{dj} = \int_0^L \rho(x)\phi_j^2(x)dx + m_t\phi_j^2(L), \quad m_j = \int_0^L \rho(x)\phi_j(x)dx + m_t\phi_j(L)$$

(5.67)

Calculating $D_y(t)$ from Equation 5.66b and substituting into Equation 5.66c results in

$$m_i\ddot{s}(t) + m_{di}\ddot{q}_i(t) + \omega_i^2 m_{di}q_i(t) - \frac{h_l^2(\phi_i'(l_2) - \phi_i'(l_1))}{\beta_l(l_2 - l_1)}\sum_{j=1}^{\infty}\{(\phi_j'(l_2) - \phi_j'(l_1))q_j(t)\}$$

$$= -\frac{h_l b(\phi_i'(l_2) - \phi_i'(l_1))}{\beta_l}v(t), \quad i = 1, 2, \ldots$$

(5.68)

which will be used to derive the controller, as discussed next.

### 5.3.3.2.2 Derivation of the Controller

Utilizing Equation 5.66a and Equation 5.68, the truncated two-mode beam with PZT model reduces to

$$\left[m_b + m_t + \int_0^L \rho(x)dx\right]\ddot{s}(t) + m_1\ddot{q}_1(t) + m_2\ddot{q}_2(t) = f(t)$$

(5.69a)

$$m_1\ddot{s}(t) + m_{d1}\ddot{q}_1(t) + \omega_1^2 m_{d1}q_1(t) - \frac{h_l^2(\phi_1'(l_2) - \phi_1'(l_1))}{\beta_l(l_2 - l_1)}$$

$$\times \{(\phi_1'(l_2) - \phi_1'(l_1))q_1(t) + (\phi_2'(l_2) - \phi_2'(l_1))q_2(t)\}$$

$$= -\frac{h_l b(\phi_1'(l_2) - \phi_1'(l_1))}{\beta_l}v(t)$$

(5.69b)

$$m_2\ddot{s}(t) + m_{d2}\ddot{q}_2(t) + \omega_2^2 m_{d2}q_2(t) - \frac{h_l^2(\phi_2'(l_2) - \phi_2'(l_1))}{\beta_l(l_2 - l_1)}$$
$$\times \{(\phi_1'(l_2) - \phi_1'(l_1))q_1(t) + (\phi_2'(l_2) - \phi_2'(l_1))q_2(t)\} \qquad (5.69c)$$
$$= -\frac{h_l b(\phi_2'(l_2) - \phi_2'(l_1))}{\beta_l} v(t)$$

The equations in Equation 5.69 can be written in the following more compact form

$$\mathbf{M\ddot{\Delta}} + \mathbf{K\Delta} = \mathbf{F}_e \qquad (5.70)$$

where

$$\mathbf{M} = \begin{bmatrix} \psi & m_1 & m_2 \\ m_1 & m_{d1} & 0 \\ m_2 & 0 & m_{d2} \end{bmatrix}, \quad \mathbf{K} = \begin{bmatrix} 0 & 0 & 0 \\ 0 & k_{11} & k_{12} \\ 0 & k_{12} & k_{22} \end{bmatrix}, \quad \mathbf{F}_e = \begin{Bmatrix} f(t) \\ \epsilon_1 v(t) \\ \epsilon_2 v(t) \end{Bmatrix}, \quad \mathbf{\Delta} = \begin{Bmatrix} s(t) \\ q_1(t) \\ q_2(t) \end{Bmatrix} \qquad (5.71)$$

and

$$\psi = m_b + m_t + \int_0^L \rho(x)dx, \quad \epsilon_1 = -\frac{h_l b}{\beta_l}(\phi_1'(l_2) - \phi_1'(l_1)), \quad \epsilon_2 = -\frac{h_l b}{\beta_l}(\phi_2'(l_2) - \phi_2'(l_1)),$$

$$k_{11} = \omega_1^2 m_{d1} - \frac{h_l^2}{\beta_l(l_2 - l_1)}(\phi_1'(l_2) - \phi_1'(l_1))^2,$$

$$\qquad (5.72)$$

$$k_{12} = -\frac{h_l^2}{\beta_l(l_2 - l_1)}(\phi_1'(l_2) - \phi_1'(l_1))(\phi_2'(l_2) - \phi_2'(l_1)),$$

$$k_{22} = \omega_2^2 m_{d2} - \frac{h_l^2}{\beta_l(l_2 - l_1)}(\phi_2'(l_2) - \phi_2'(l_1))^2$$

For the system described by Equation 5.70, if the control laws for the arm base force and PZT voltage generated moment are selected as

$$f(t) = -k_p\Delta s - k_d\dot{s}(t) \qquad (5.73)$$

$$v(t) = -k_v(\epsilon_1\dot{q}_1(t) + \epsilon_2\dot{q}_2(t)) \qquad (5.74)$$

where $k_p$ and $k_d$ are positive control gains, $\Delta s = s(t) - s_d$, $s_d$ is the desired set-point position, and $k_v > 0$ is the voltage control gain, then the closed-loop system will be stable, and in addition

$$\lim_{t \to \infty} \{q_1(t), q_2(t), \Delta s\} = 0$$

See Dadfarnia et al. (2004) for a detailed proof.

### 5.3.3.2.3  *Controller Implementation*
The control input, $v(t)$, requires the information from the velocity-related signals, $\dot{q}_1(t)$ and $\dot{q}_2(t)$, which are usually not measurable. Sun and Mills (1999) solved the problem by integrating the acceleration signals measured by the accelerometers. However, such controller structure may result in unstable closed-loop system in some cases. In this paper, a reduced-order observer is designed to estimate the velocity signals, $\dot{q}_1$ and $\dot{q}_2$. For this, we utilize three available signals: base displacement, $s(t)$, arm-tip deflection, $P(L, t)$, and beam root strain, $\epsilon(0, t)$; that is

$$y_1 = s(t) = x_1 \qquad (5.75a)$$

$$y_2 = P(L, t) = x_1 + \phi_1(L)x_2 + \phi_2(L)x_3 \qquad (5.75b)$$

$$y_3 = \epsilon(0, t) = \frac{t_b}{2}(\phi_1''(0)x_2 + \phi_2''(0)x_3) \qquad (5.75c)$$

It can be seen that the first three states can be obtained by

$$
\left\{\begin{array}{c} x_1 \\ x_2 \\ x_3 \end{array}\right\} = \mathbf{C}_1^{-1}\mathbf{y}
\tag{5.76}
$$

Since this system is observable, we can design a reduced-order observer to estimate the velocity-related state signals. Defining $\mathbf{X}_1 = [\, x_1 \quad x_2 \quad x_3 \,]^T$ and $\mathbf{X}_2 = [\, x_4 \quad x_5 \quad x_6 \,]^T$, the estimated value for $\mathbf{X}_2$ can be designed as

$$
\hat{\mathbf{X}}_2 = \mathbf{L}_r\mathbf{y} + \hat{\mathbf{z}}
\tag{5.77}
$$

$$
\dot{\hat{\mathbf{z}}} = \mathbf{F}\hat{\mathbf{z}} + \mathbf{G}\mathbf{y} + \mathbf{H}\mathbf{u}
\tag{5.78}
$$

where $\mathbf{L}_r \in R^{3 \times 3}$, $\mathbf{F} \in R^{3 \times 3}$, $\mathbf{G} \in R^{3 \times 3}$, and $\mathbf{H} \in R^{3 \times 2}$ will be determined by the observer pole placement. Defining the estimation error as

$$
\mathbf{e}_2 = \mathbf{X}_2 - \hat{\mathbf{X}}_2
\tag{5.79}
$$

the derivative of the estimation error becomes

$$
\dot{\mathbf{e}}_2 = \dot{\mathbf{X}}_2 - \dot{\hat{\mathbf{X}}}_2
\tag{5.80}
$$

Substituting the state-space equations of the system (Equation 5.77 and Equation 5.78) into Equation 5.80 and simplifying, we obtain

$$
\dot{\mathbf{e}}_2 = \mathbf{F}\mathbf{e}_2 + (\mathbf{A}_{21} - \mathbf{L}_r\mathbf{C}_1\mathbf{A}_{11} - \mathbf{G}\mathbf{C}_1 + \mathbf{F}\mathbf{L}_r\mathbf{C}_1)\mathbf{X}_1 + (\mathbf{A}_{22} - \mathbf{L}_r\mathbf{C}_1\mathbf{A}_{12} - \mathbf{F})\mathbf{X}_2 + (\mathbf{B}_2 - \mathbf{L}_r\mathbf{C}_1\mathbf{B}_1 - \mathbf{H})\mathbf{u}
\tag{5.81}
$$

In order to force the estimation error, $\mathbf{e}_2$, to go to zero, matrix $\mathbf{F}$ should be selected to be Hurwitz and the following relations must be satisfied (Liu et al., 2002):

$$
\mathbf{F} = \mathbf{A}_{22} - \mathbf{L}_r\mathbf{C}_1\mathbf{A}_{12}
\tag{5.82}
$$

$$
\mathbf{H} = \mathbf{B}_2 - \mathbf{L}_r\mathbf{C}_1\mathbf{B}_1
\tag{5.83}
$$

$$
\mathbf{G} = (\mathbf{A}_{21} - \mathbf{L}_r\mathbf{C}_1\mathbf{A}_{11} + \mathbf{F}\mathbf{L}_r\mathbf{C}_1)\mathbf{C}_1^{-1}
\tag{5.84}
$$

The matrix $\mathbf{F}$ can be chosen by the desired observer pole placement requirement. Once $\mathbf{F}$ is known, $\mathbf{L}_r$, $\mathbf{H}$, and $\mathbf{G}$ can be determined utilizing Equation 5.82, to Equation 5.84, respectively. The velocity variables, $\hat{\mathbf{X}}_2$, can now be estimated by Equation 5.77 and Equation 5.78.

### 5.3.3.2.4    Numerical Simulations

In order to show the effectiveness of the controller, the flexible beam structure in Figure 5.29 is considered with the PZT actuator attached on the beam surface. The system parameters are listed in Table 5.4.

First, we consider the beam without PZT control. We take the PD control gains to be $k_p = 120$ and $k_d = 20$. Figure 5.31 shows the results for the beam without PZT control (i.e., with only PD force control for the base movement). To investigate the effect of PZT controller on the beam vibration, we consider the voltage control gain to be $k_v = 2 \times 10^7$. The system responses to the proposed controller with a piezoelectric actuator based on the two-mode model are shown in Figure 5.32. The comparison between the tip displacement, from Figure 5.31 and Figure 5.32, shows that the beam vibration can be suppressed significantly utilizing the PZT actuator.

### 5.3.3.2.5    Control Experiments

In order to demonstrate better the effectiveness of the controller, an experimental setup is constructed and used to verify the numerical results. The experimental apparatus consists of a flexible beam with a PZT actuator and strain sensor attachments, as well as data acquisition,

**TABLE 5.4**  System Parameters Used in Numerical Simulations and
Experimental Setup for Translational Beam

| Properties | Symbol | Value | Unit |
|---|---|---|---|
| Beam Young's modulus | $c_{11}^b$ | $69 \times 10^9$ | N/m² |
| Beam thickness | $t_b$ | 0.8125 | mm |
| Beam and PZT width | $b$ | 20 | mm |
| Beam length | $L$ | 300 | mm |
| Beam volumetric density | $\rho_b$ | 3960.0 | kg/m³ |
| PZT Young's modulus | $c_{11}^P$ | $66.47 \times 10^9$ | N/m² |
| PZT coupling parameter | $h_{12}$ | $5 \times 10^8$ | V/m |
| PZT impermittivity | $\beta_{22}$ | $4.55 \times 10^7$ | m/F |
| PZT thickness | $t_P$ | 0.2032 | mm |
| PZT length | $l_2 - l_1$ | 33.655 | mm |
| PZT position on beam | $l_1$ | 44.64 | mm |
| PZT volumetric density | $\rho_P$ | 7750.0 | kg/m³ |
| Base mass | $m_b$ | 0.455 | kg |
| Tip mass | $m_t$ | 0 | kg |

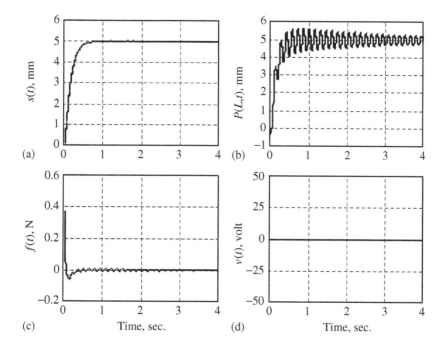

**FIGURE 5.31**  Numerical simulations for the case without PZT control: (a) base motion; (b) tip displacement;
(c) control force and (d) PZT voltage.

amplifier, signal conditioner and the control software. As shown in Figure 5.33, the plant consists
of a flexible aluminum beam with a strain sensor and a PZT patch actuator bound on each side of
the beam surface. One end of the beam is clamped to the base with a solid clamping fixture, which
is driven by a shaker. The shaker is connected to the arm base by a connecting rod. The
experimental setup parameters are listed in Table 5.4.

Figure 5.34 shows the high-level control block diagram of the experiment, where the shaker provides
the input control force to the base and the PZT applies a controlled moment on the beam. Two laser
sensors measure the position of the base and the beam-tip displacement. A strain-gauge sensor, which is
attached near the base of the beam, is utilized for the dynamic strain measurement. These three signals

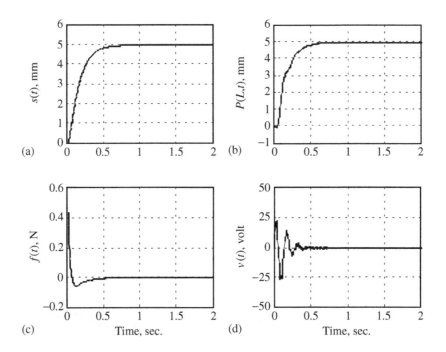

**FIGURE 5.32** Numerical simulations for the case with PZT control: (a) base motion; (b) tip displacement; (c) control force and (d) PZT voltage.

are fed back to the computer through the ISA MultiQ data acquisition card. The remaining required signals for the controller (Equation 5.66) are determined as explained in the preceding section. The data acquisition and control algorithms are implemented on an AMD Athlon 1100 MHz PC running under the RT-Linux operating system. The Matlab/Simulink environment and Real Time Linux Target are used to implement the controller.

The experimental results for both cases (i.e., without PZT and with PZT control) are depicted in Figure 5.35 and Figure 5.36, respectively. The results demonstrate that with PZT control, the arm vibration is eliminated in less than 1 sec, while the arm vibration lasts for more than 6 sec when PZT control is not used. The experimental results are in agreement with the simulation results except for some differences at the beginning of the motion. The slight overshoot and discrepancies at the beginning of the motion are due to the limitations of the experiment (e.g., the shaker saturation limitation) and unmodeled dynamics in the modeling (e.g., the friction modeling). However, it is still apparent that the PZT voltage control can substantially suppress the arm vibration despite such limitations and modeling imperfections.

## 5.4 Practical Considerations and Related Topics

### 5.4.1 Summary of Vibration-Control Design Steps and Procedures

In order to select a suitable vibration-control system, especially a vibration isolator, a number of factors must be considered.

#### 5.4.1.1 Static Deflection

The static deflection of the vibration-control system under the deadweight of the load determines to a considerable extent the type of the material to be used in the isolator. Organic materials, such as rubber

**FIGURE 5.33**   The experimental setup: (a) the whole system; (b) PZT actuator, ACX model No. QP21B; (c) dynamic strain sensor (attached on the other side of the beam), model No. PCB 740B02.

and cork, are capable of sustaining very large strains provided they are applied momentarily. However, if large strains remain for an appreciable period of time, they tend to drift or creep. On the other hand, metal springs undergo permanent deformation if the stress exceeds the yield stress of the material, but show minimal drift or creep when the stress is maintained below the yield stress.

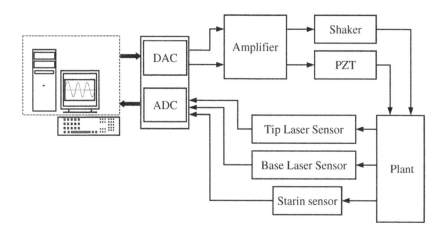

**FIGURE 5.34**   High-level control-block diagram.

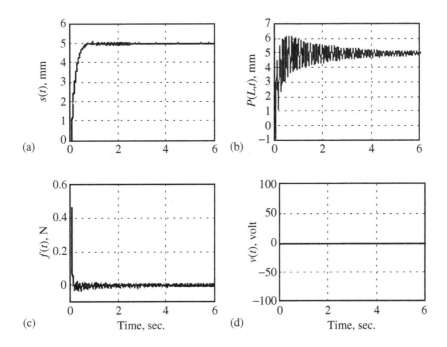

**FIGURE 5.35** Experimental results for the case without PZT control: (a) base motion; (b) tip displacement; (c) control force and (d) PZT voltage.

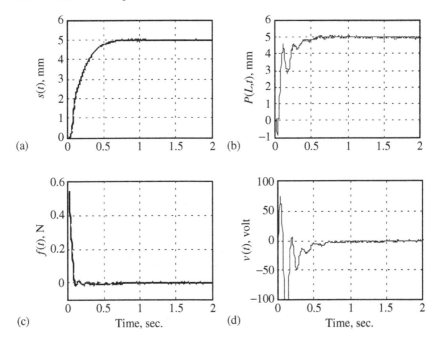

**FIGURE 5.36** Experimental results for the case with PZT control: (a) base motion; (b) tip displacement; (c) control force and (d) PZT voltage.

### 5.4.1.2 Stiffness in Lateral Directions

Resilient materials strained in compression are most useful when the load is relatively large and the static deflection is small. Such applications are difficult to design for a small load, unless the required static deflection is small. Otherwise, the small area and great thickness tend to cause a condition of instability. To a considerable extent, this limitation can be overcome by using sponge rubber, a material of lower modulus. In general, when the load is small, it is preferable to use rubber springs that carry the load in shear.

### 5.4.1.3   Environmental Conditions

It is highly common for vibration-control systems to be subjected to harsh environmental conditions. Especially in military applications, extreme ambient temperatures are encountered in addition to exposure to substances like ozone, rocket fuels, and so on. Organic materials are usually more susceptible to these environmental conditions than metal materials. However, owing to the superior mechanical properties of organic materials, such as lighter weight, smaller size, greater damping, and the ability to store large amounts of energy under shock, organic materials that are capable of withstanding the harsh conditions are being developed.

### 5.4.1.4   Damping Characteristics

In most of the vibration-control applications, the excitations cover a wide range of frequencies and may have random properties requiring the vibration-control systems to possess adequate damping. Elastomers possess very good damping properties when compared with metal springs, and they also eliminate the trouble of standing waves that occurs at high frequencies. If a metal spring is used in vibration-control applications requiring isolation of vibration at high frequencies, it is common to employ rubber pads in series with the metal spring, which also results in the damping of vibrations due to the addition of damping material.

### 5.4.1.5   Weight and Space Limitations

The amount of load-carrying resilient material is determined by the quantity of energy to be stored. In most of the cases, the vibration amplitude tends to be small relative to the static deflection, and the amount of material may be calculated by equating the energy stored in the material to the work done on the vibration control system.

### 5.4.1.6   Dynamic Stiffness

In the case of organic materials like rubber, the natural frequency calculated using the stiffness determined from a static-force deflection test of the spring is almost invariably lower than that experienced during vibration; that is, the dynamic modulus is greater than static modulus. The ratio between the dynamic and static modulus is generally between one and two. In many vibration-control applications, it is not feasible to mount the equipment directly upon the vibration-control system (isolator). Instead, a heavy, rigid block, usually made of concrete or heavy steel, supported by the isolator is employed.

## 5.4.2   Future Trends and Developments

During recent years, there has been considerable interest in the design and implementation of a variety of vibration-control systems. Recent developments in multivariable control design methodology and microprocessor implementation of modern control algorithms have opened a new era for the design of externally controlled passive systems for use in such systems: fuzzy reasoning (Yoshimura, 1998); adaptive algorithms (Venhovens, 1994); observer design (Hedrick et al., 1994); and many others.

Observing these developments combined with the substantial ongoing theoretical advances in the areas of adaptive and nonlinear controls (Astrom and Wittenmark, 1989; Alleyne and Hedrick, 1995), it is expected that the future will bring applications of these techniques in advanced vibration-control system design. For practical implementation, however, it is preferable to simplify these strategies, thus leading to simpler software implementations. Suboptimal policy neglecting some performance requirements can serve as an example of such simplifications.

# References

Abe, M. and Igusa, T., Semiactive dynamic vibration absorbers for controlling transient response, *J. Sound Vib.*, 198, 5, 547–569, 1996.

Alleyne, A. and Hedrick, J.K., Nonlinear adaptive control of active suspensions, *IEEE Trans. Control Syst. Technol.*, 3, 94–101, 1995.

Astrom, J.J. and Wittenmark, B. 1989. *Adaptive Control*, Addison-Wesley, Reading, MA.

Austin, S.A., The vibration damping effect of an electrorheological fluid, *ASME J. Vib. Acoust.*, 115, 136–140, 1993.

Banks, H.T. and Ito, K., A unified framework for approximation in inverse problems for distributed parameter systems, *Control Theor. Adv. Technol.*, 4, 73–90, 1988.

Banks, H.T. and Kunisch, K. 1989. *Estimation Techniques for Distributed Parameter Systems*, Birkhauser, Boston, MA.

Bayo, E., A finite-element approach to control the end-point motion of a single-link flexible robot, *J. Robotic Syst.*, 4, 63–75, 1987.

Cannon, R.H. Jr. and Schmitz, E., Initial experiments on the end-point control of a flexible one-link robot, *Int. J. Robotics Res.*, 3, 62–75, 1984.

Chalhoub, N.G. and Ulsoy, A.G., Control of flexible robot arm: experimental and theoretical results, *ASME J. Dyn. Syst. Meas. Control*, 109, 299–309, 1987.

Choi, S.B., Vibration control of flexible structures using ER dampers, *ASME J. Dyn. Syst. Meas. Control*, 121, 134–138, 1999.

Crosby, M. and Karnopp, D.C. 1973. The active damper—a new concept for shock and vibration control, *Shock Vibration Bulletin*, Part H, Washington, DC.

Dadfarnia, M., Jalili, N., Xian, B. and Dawson, D.M. 2003. Lyapunov-based piezoelectric control of flexible cartesian robot manipulators, In *Proceedings of the 22nd American Control Conference (ACC'03)*, Denver, CO.

Dadfarnia, M., Jalili, N., Liu, Z., and Dawson, D.M., An observer-based piezoelectric control of flexible cartesian robot manipulators: theory and experiment, *J. Control Eng. Practice*, 12, 1041–1053, 2004.

de Querioz, M.S., Dawson, D.M., Agrawal, M., and Zhang, F., Adaptive nonlinear boundary control of a flexible link robot arm, *IEEE Trans. Robotics Automat.*, 15, 779–787, 1999.

de Querioz, M.S., Dawson, D.M., Nagarkatti, S.P., and Zhang, F. 2000. *Lyapunov-Based Control of Mechanical Systems*, Birkhauser, Boston, MA.

Dimarogonas-Andrew, D. and Kollias, A., Smart electrorheological fluid dynamic vibration absorber, *Intell. Struct. Mater. Vib. ASME Des. Div.*, 58, 7–15, 1993.

Dowell, D.J., Cherry, S. 1994. Semiactive friction dampers for seismic response control of structures, Vol. 1, pp. 819–828. In *Proceedings of the Fifth US National Conference on Earthquake Engineering*, Chicago, IL.

Elmali, H. and Olgac, N., Sliding mode control with perturbation estimation (SMCPE): a new approach, *Int. J. Control*, 56, 923–941, 1992.

Emura, J., Kakizaki, S., Yamaoka, F., and Nakamura, M. 1994. Development of the SA suspension system based on the sky-hook damper theory, *SAE Paper No. 940863*.

Esmailzadeh, E. and Jalili, N., Optimal design of vibration absorbers for structurally damped Timoshenko beams, *ASME J. Vib. Acoust.*, 120, 833–841, 1998a.

Esmailzadeh, E. and Jalili, N., Parametric response of cantilever timoshenko beams with tip mass under harmonic support motion, *Int. J. Non-Linear Mech.*, 33, 765–781, 1998b.

Filipović, D. and Schröder, D., Vibration absorption with linear active resonators: continuous and discrete time design and analysis, *J. Vib. Control*, 5, 685–708, 1999.

Frahm, H., Devices for damping vibrations of bodies, US Patent #989958, 1911.

Franchek, M.A., Ryan, M.W., and Bernhard, R.J., Adaptive-passive vibration control, *J. Sound Vib.*, 189, 565–585, 1995.

Fujita, T., Katsu, M., Miyano, H., and Takanashi, S., Fundamental study of active-passive mass damper using *XY*-motion mechanism and hydraulic actuator for vibration control of tall building, *Trans. Jpn Soc. Mech. Engrs*, Part C, 57, 3532–3539, 1991.

Garcia, E., Dosch, J., and Inman, D.J., The application of smart structures to the vibration suppression problem, *J. Intell. Mater. Syst. Struct.*, 3, 659–667, 1992.

Ge, S.S., Lee, T.H., and Zhu, G., A nonlinear feedback controller for a single-link flexible manipulator based on a finite element method, *J. Robotic Syst.*, 14, 165–178, 1997.

Ge, S.S., Lee, T.H., and Zhu, G., Asymptotically stable end-point regulation of a flexible SCARA/cartesian robot, *IEEE/ASME Trans. Mechatron.*, 3, 138–144, 1998.

Ge, S.S., Lee, T.H., and Gong, J.Q., A robust distributed controller of a single-link SCARA/cartesian smart materials robot, *Mechatronics*, 9, 65–93, 1999.

Hedrick, J.K., Rajamani, R., and Yi, K., Observer design for electronic suspension applications, *Vehicle Syst. Dyn.*, 23, 413–440, 1994.

Horton, D.N. and Crolla, D.A., Theoretical analysis of a SA suspension fitted to an off-road vehicle, *Vehicle Syst. Dyn.*, 15, 351–372, 1986.

Hrovat, D., Barker, P., and Rabins, M., Semiactive versus passive or active tuned mass dampers for structural control, *J. Eng. Mech.*, 109, 691–705, 1983.

Hrovat, D., Margolis, D.L., and Hubbard, M., An approach toward the optimal SA suspension, *ASME J. Dyn. Syst. Meas. Control*, 110, 288–296, 1988.

Hubard, M. and Marolis, D. 1976. The SA spring: is it a viable suspension concept?, pp. 1–6. In *Proceedings of the Fourth Intersociety Conference on Transportation*, Los Angeles, CA.

Ikeda, T. 1990. *Fundamental of Piezoelectricity*, Oxford University Press, Oxford, New York.

Inman, D.J. 1994. *Engineering Vibration*, Prentice Hall, Englewood Cliffs, NJ.

Jalili, N., A new perspective for semi-automated structural vibration control, *J. Sound Vib.*, 238, 481–494, 2000.

Jalili, N., An infinite dimensional distributed base controller for regulation of flexible robot arms, *ASME J. Dyn. Syst. Meas. Control*, 123, 712–719, 2001.

Jalili, N., A comparative study and analysis of SA vibration-control systems, *ASME J. Vib. Acoust.*, 124, 593–605, 2002.

Jalili, N., Dadfarnia, M., Hong, F., and Ge, S.S. 2002. An adaptive non model-based piezoelectric control of flexible beams with translational base, pp. 3802–3807. In *Proceedings of the American Control Conference (ACC'02)*, Anchorage, AK.

Jalili, N., Elmali, H., Moura, J., and Olgac, N. 1997. Tracking control of a rotating flexible beam using frequency-shaped sliding mode control, pp. 2552–2556. In *Proceedings of the 16th American Control Conference (ACC'97)*, Albuquerque, NM.

Jalili, N. and Esmailzadeh, E., Adaptive-passive structural vibration attenuation using distributed absorbers, *J. Multi-Body Dyn.*, 216, 223–235, 2002.

Jalili, N. and Fallahi, B., Design and dynamics analysis of an adjustable inertia absorber for SA structural vibration attenuation, *ASCE J. Eng. Mech.*, 128, 1342–1348, 2002.

Jalili, N., Fallahi, B., and Kusculuoglu, Z.K., A new approach to SA vibration suppression using adjustable inertia absorbers, *Int. J. Model. Simulat.*, 21, 148–154, 2001.

Jalili, N. and Olgac, N., Time-optimal/sliding mode control implementation for robust tracking of uncertain flexible structures, *Mechatronics*, 8, 121–142, 1998a.

Jalili, N. and Olgac, N. 1998b. Optimum delayed feedback vibration absorber for MDOF mechanical structures, In *Proceedings of the 37th IEEE Conference on Decision Control (CDC'98)*, Tampa, FL.

Jalili, N. and Olgac, N., Multiple identical delayed-resonator vibration absorbers for multi-DoF mechanical structures, *J. Sound Vib.*, 223, 567–585, 1999.

Jalili, N. and Olgac, N., Identification and re-tuning of optimum delayed feedback vibration absorber, *AIAA J. Guidance Control Dyn.*, 23, 961–970, 2000a.

Jalili, N. and Olgac, N., A sensitivity study of optimum delayed feedback vibration absorber, *ASME J. Dyn. Syst. Meas. Control*, 121, 314–321, 2000b.

Junkins, J.L. and Kim, Y. 1993. *Introduction to Dynamics and Control of Flexible Structures*. AIAA Educational Series, Washington, DC.

Karnopp, D., Design principles for vibration-control systems using SA dampers, *ASME J. Dyn. Syst. Meas. Control*, 112, 448–455, 1990.

Karnopp, D.C., Crodby, M.J., and Harwood, R.A., Vibration control using SA force generators, *J. Eng. Ind.*, 96, 619–626, 1974.

Kim, K. and Jeon, D., Vibration suppression in an MR fluid damper suspension system, *J. Intell. Mater. Syst. Struct.*, 10, 779–786, 2000.

Knowles, D., Jalili, N., and Khan, T. 2001a. On the nonlinear modeling and identification of piezoelectric inertial actuators, In *Proceedings of the 2001 International Mechanical Engineering Congress and Exposition* (IMECE'01), New York.

Knowles, D., Jalili, N., and Ramadurai, S. 2001b. Piezoelectric structural vibration control using active resonator absorber, In *Proceedings of the 2001 International Mechanical Engineering Congress and Exposition* (IMECE'01), New York.

Korenev, B.G. and Reznikov, L.M. 1993. *Dynamic Vibration Absorbers: Theory and Technical Applications*, Wiley, Chichester, England.

Lee-Glauser, G.J., Ahmadi, G., and Horta, L.G., Integrated passive/active vibration absorber for multistory buildings, *ASCE J. Struct. Eng.*, 123, 499–504, 1997.

Liu, Z., Jalili, N., Dadfarnia, M., and Dawson, D.M. 2002. Reduced-order observer based piezoelectric control of flexible beams with translational base, In *Proceedings of the 2002 International Mechanical Engineering Congress and Exposition* (IMECE'02), New Orleans, LA.

Liu, J., Schönecker, A., and Frühauf, U. 1997. Application of discrete Fourier transform to electronic measurements, pp. 1257–1261. In *International Conference on Information, Communications and Signal Processing*, Singapore.

Liu, H.J., Yang, Z.C., and Zhao, L.C., Semiactive flutter control by structural asymmetry, *J. Sound Vib.*, 229, 199–205, 2000.

Luo, Z.H., Direct strain feedback control of flexible robot arms: new theoretical and experimental results, *IEEE Trans. Automat. Control*, 38, 1610–1622, 1993.

Luo, Z.H., Guo, B.Z., and Morgul, O. 1999. *Stability and Stabilization of Finite Dimensional Systems with Applications*, Springer, London.

Margolis, D., Retrofitting active control into passive vibration isolation systems, *ASME J. Vib. Acoust.*, 120, 104–110, 1998.

Meirovitch, L. 1986. *Elements of Vibration Analysis*, McGraw-Hill, New York.

Miller, L.R. and Nobles, C.M. 1988. The design and development of a SA suspension for military tank, *SAE* Paper No. 881133.

Moura, J.T., Roy, R.G., and Olgac, N., Frequency-shaped sliding modes: analysis and experiments, *IEEE Trans. Control Syst. Technol.*, 5, 394–401, 1997.

Nemir, D., Lin, Y., and Osegueda, R.A., Semiactive motion control using variable stiffness, *ASCE J. Struct. Eng.*, 120, 1291–1306, 1994.

Olgac, N., Delayed resonators as active dynamic absorbers, US Patent #5431261, 1995.

Olgac, N., Elmali, H., Hosek, M., and Renzulli, M., Active vibration control of distributed systems using delayed resonator with acceleration feedback, *ASME J. Dyn. Syst., Meas. Control*, 119, 380–389, 1997.

Olgac, N., Elmali, H., and Vijayan, S., Introduction to dual frequency fixed delayed resonator (DFFDR), *J. Sound Vib.*, 189, 355–367, 1996.

Olgac, N. and Holm-Hansen, B., Novel active vibration absorption technique: delayed resonator, *J. Sound Vib.*, 176, 93–104, 1994.

Olgac, N. and Holm-Hansen, B., Tunable active vibration absorber: the delayed resonator, *ASME J. Dyn. Syst., Meas. Control*, 117, 513–519, 1995.

Olgac, N. and Jalili, N., Modal analysis of flexible beams with delayed-resonator vibration absorber: theory and experiments, *J. Sound Vib.*, 218, 307–331, 1998.

Olgac, N. and Jalili, N., Optimal delayed feedback vibration absorber for flexible beams, *Smart Struct.*, 65, 237–246, 1999.

Oueini, S.S., Nayfeh, A.H., and Pratt, J.R., A nonlinear vibration absorber for flexible structures, *Nonlinear Dyn.*, 15, 259–282, 1998.

Petek, N.K., Shock absorbers uses electrorheological fluid, *Automot. Eng.*, 100, 27–30, 1992.

Petek, N.K., Romstadt, D.L., Lizell, M.B., and Weyenberg, T.R. 1995. Demonstration of an automotive SA suspension using electro-rheological fluid, *SAE* Paper No. 950586.

Puksand, H., Optimum conditions for dynamic vibration absorbers for variable speed systems with rotating and reciprocating unbalance, *Int. J. Mech. Eng. Educ.*, 3, 145–152, 1975.

Renzulli, M., Ghosh-Roy, R., and Olgac, N., Robust control of the delayed resonator vibration absorber, *IEEE Trans. Control Syst. Technol.*, 7, 683–691, 1999.

Shaw, J., Adaptive vibration control by using magnetostrictive actuators, *J. Intell. Mater. Syst. Struct.*, 9, 87–94, 1998.

Singh, T., Golnaraghi, M.F., and Dubly, R.N., Sliding-mode/shaped-input control of flexible/rigid link robots, *J. Sound Vib.*, 171, 185–200, 1994.

Sinha, A., Optimum vibration control of flexible structures for specified modal decay rates, *J. Sound Vib.*, 123, 185–188, 1988.

Skaar, S.B. and Tucker, D., Point Control of a one-link flexible manipulator, *J. Appl. Mech.*, 53, 23–27, 1986.

Snowdon, J.C. 1968. *Vibration and Shock in Damped Mechanical Systems*, Wiley, New York.

Soong, T.T. and Constantinou, M.C. 1994. *Passive and Active Structural Control in Civil Engineering*, Springer, Wien.

Spencer, B.F., Yang, G., Carlson, J.D., and Sain, M.K. 1998. Smart dampers for seismic protection of structures: a full-scale study, In *Proceedings of the Second World Conference on Structural Control*, Kyoto, Japan.

Stribersky, A., Muller, H., and Rath, B., The development of an integrated suspension control technology for passenger trains, *Proc. Inst. Mech. Engrs*, 212, 33–41, 1998.

Sturk, M., Wu, M., and Wong, J.Y., Development and evaluation of a high voltage supply unit for electrorheological fluid dampers, *Vehicle Syst. Dyn.*, 24, 101–121, 1995.

Sun, J.Q., Jolly, M.R., and Norris, M.A., Passive, adaptive, and active tuned vibration absorbers — a survey, *ASME Trans.*, 117, 234–242, 1995 (Special 50th Anniversary, Design issue).

Sun, D. and Mills, J.K. 1999. PZT actuator placement for structural vibration damping of high speed manufacturing equipment, pp. 1107–1111. In *Proceedings of the American Control Conference (ACC'99)*, San Diego, CA.

Sun, Y. and Parker, G.A., A position controlled disc valve in vehicle SA suspension systems, *Control Eng. Practice*, 1, 927–935, 1993.

Tanaka, N. and Kikushima, Y., Impact vibration control using a SA damper, *J. Sound Vib.*, 158, 277–292, 1992.

Venhovens, P.J., The development and implementation of adaptive SA suspension control, *Vehicle Syst. Dyn.*, 23, 211–235, 1994.

Walsh, P.L. and Lamnacusa, J.S., A variable stiffness vibration absorber for minimization of transient vibrations, *J. Sound Vib.*, 158, 195–211, 1992.

Wang, K.W., Kim, Y.S., and Shea, D.B., Structural vibration control via electrorheological-fluid-based actuators with adaptive viscous and frictional damping, *J. Sound Vib.*, 177, 227–237, 1994.

Wang, K.W., Lai, J.S., and Yu, W.K., An energy-based parametric control approach for structural vibration suppression via SA piezoelectric networks, *ASME J. Vib. Acoust.*, 118, 505–509, 1996.

Warburton, G.B. and Ayorinde, E.O., Optimum absorber parameters for simple systems, *Earthquake Eng. Struct. Dyn.*, 8, 197–217, 1980.

Yeung, K.S. and Chen, Y.P., A new controller design for manipulators using the theory of variable structure systems, *IEEE Trans. Automat. Control*, 33, 200–206, 1988.

Yeung, K.S. and Chen, Y.P., Regulation of a one-link flexible robot arm using sliding mode control technique, *Int. J. Control*, 49, 1965–1978, 1989.

Yoshimura, T., A SA suspension of passenger cars using fuzzy reasoning and the filed testing, *Int. J. Vehicle Des.*, 19, 150–166, 1998.

Yuh, J., Application of discrete-time model reference adaptive control to a flexible single-link robot, *J. Robotic Syst.*, 4, 621–630, 1987.

Zhu, G., Ge, S.S., Lee, T.H. 1997. Variable structure regulation of a flexible arm with translational base, pp. 1361–1366. In *Proceedings of the 36th IEEE Conference on Decision and Control*, San Diego, CA.

# 6

# Helicopter Rotor Tuning

Kourosh Danai

*University of Massachusetts*

**Summary**

*Before a helicopter leaves the plant, its rotors need to be tuned so that the helicopter vibration meets the required specifications during different flight regimes. For this, three different adjustments can be made to each rotor blade in response to the magnitude and phase of vibration. In this chapter (also see Chapter 7), the basic concepts for determining the blade adjustments are discussed, and three methods with fundamentally different approaches are described. A neural network-based method is described, which trains a feedforward network as the inverse model of the effect of the blade adjustments on helicopter vibrations, and uses the inverse model to determine the blade adjustments. Another is a probability-based method that maximizes the likelihood of success of the selected blade adjustments based on a stochastic model of the probability densities of the vibration components. The third method is an adaptive method that uses an interval model to represent the range of effect of blade adjustments on helicopter vibration, so as to cope with the nonlinear and stochastic nature of aircraft vibration. This method includes the a priori knowledge of the process by defining the initial coefficients of the interval model according to sensitivity coefficients between the blade adjustments and helicopter vibration, but then transforms these coefficients into intervals and updates them after each tuning iteration, to improve the model estimation accuracy. The details of rotor tuning are described through a case study, which demonstrates the application of the adaptive method.*

## 6.1 Introduction

Helicopter rotor tuning (track and balance) is the process of adjusting the rotor blades so as to reduce the aircraft vibration and the spread of rotors. Rotor tuning as applied to Sikorsky's Black Hawk (H-60) helicopters is performed as follows. For initial measurements, the aircraft is flown through six different regimes, during which measurements of rotor track and vibration (balance) are recorded. Rotor track is measured by optical sensors, which detect the vertical position of the blades. Vibration is measured at the frequency of once per blade revolution ( per rev) by two accelerometers, A and B, attached to the sides of the cockpit (see Figure 6.1, detail B). The vibration data are vectorially combined into two components: A + B, representing the vertical vibration of the aircraft, and A − B, representing its roll vibration. A sample of peak vibration levels for the six flight regimes, as well as the peak angular positions relative to a reference blade, are given in Table 6.1, along with a sample of track data.

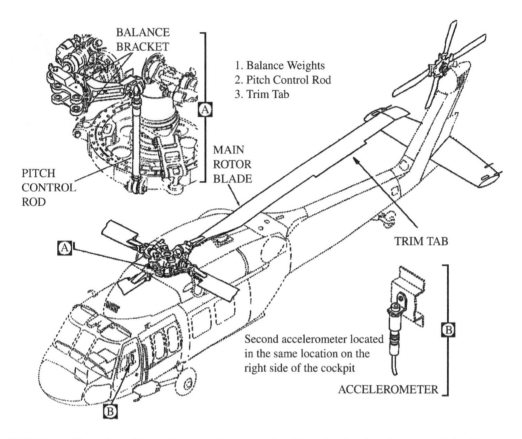

**FIGURE 6.1**  Illustration of the position of accelerometers A and B on the aircraft, and the rotor blade adjustments (push rod, trim tab, and hub weights).

**TABLE 6.1**  Typical Track and Balance Data Recorded during a Flight

| Flight Regime | Vibration | | | |
| --- | --- | --- | --- | --- |
| | A + B | | A − B | |
| | Magnitude (ips) | Phase (°) | Magnitude (ips) | Phase (°) |
| fpm | 0.19 | 332 | 0.38 | 272 |
| hov | 0.07 | 247 | 0.10 | 217 |
| 80 | 0.02 | 86 | 0.04 | 236 |
| 120 | 0.04 | 28 | 0.04 | 333 |
| 145 | 0.02 | 104 | 0.07 | 162 |
| vh | 0.10 | 312 | 0.12 | 211 |

| | Track (mm) | | | |
| --- | --- | --- | --- | --- |
| | Blade # | | | |
| | 1 | 2 | 3 | 4 |
| fpm | −2 | 3 | 1 | −2 |
| hov | −1 | 3 | 0 | −2 |
| 80 | 1 | 11 | 1 | −13 |
| 120 | 2 | 13 | −1 | −14 |
| 145 | 5 | 18 | −3 | −20 |
| vh | 2 | 13 | −1 | −14 |

The six flight regimes in Table 6.1 are: ground (fpm), hover (hov), 80 knots (80), 120 knots (120), 145 knots (145), and maximum horizontal speed (vh). The track data indicate the vertical position of each blade relative to a mean position.

In order to bring track and one per rev vibration within specification, three types of adjustments can be made to the rotor system: pitch control rod adjustments, trim tab adjustments, and balance weight adjustments (see Figure 6.1). Pitch control rods can be extended or contracted by a certain number of notches to alter the pitch of the rotor blades. Positive push rod adjustments indicate extension. Trim tabs, which are adjustable surfaces on the trailing edge of the rotor blades, affect the aerodynamic pitch moment of the air foils and consequently their vibration characteristics. Tab adjustments are measured in thousandths of an inch, with positive and negative changes representing upward and downward tabbing, respectively. Finally, balance weights can be either added to or removed from the rotor hub to tune vibrations through changes in the blade mass. Balance weights are measured in ounces, with positive adjustments representing the addition of weight. In the case of the Sikorsky H-60 helicopter, which has four main rotor blades, a total of 12 adjustments can be made to tune the rotors (i.e., three adjustments per blade). Among them, balance weights primarily affect the ground vibration, so they are not commonly used for in-flight tuning. Furthermore, since the symmetry of rotor blades in four-bladed aircraft produces identical effects for adjustments to opposite blades, the combined form of blade adjustments to opposite blade pairs can be used as inputs. Accordingly, the input vector can be defined as

$$\Delta \mathbf{x} = [\Delta x_1, \Delta x_2, \Delta x_3, \Delta x_4]^{\mathrm{T}} \tag{6.1}$$

where $\Delta x_1$ and $\Delta x_3$ denote the combined (condensed) trim tab adjustments ($\Delta T$) to blade combinations one/three and two/four, respectively, and $\Delta x_2$ and $\Delta x_4$ represent the combined pitch control rod adjustments ($\Delta P$) to blade combinations one/three and two/four, respectively. The relationships between the combined and individual adjustments are in the form:

$$\Delta x_1 = \Delta T_3 - \Delta T_1 \tag{6.2}$$

$$\Delta x_2 = \Delta P_3 - \Delta P_1 \tag{6.3}$$

$$\Delta x_3 = \Delta T_4 - \Delta T_2 \tag{6.4}$$

$$\Delta x_4 = \Delta P_4 - \Delta P_2 \tag{6.5}$$

Ideally, identical adjustments made to any two aircraft with different tail numbers should result in identical changes in vibration. In reality, however, significant inconsistencies in vibration changes may be present for identical adjustments to different tail numbers. This is perhaps due more to nonuniformity of flight conditions from weather or error in implementing the blade adjustments than factors such as dissimilarities between aircraft and rotor blades.

Virtually all of the current systems of rotor track and balance rely on the strategy shown in Figure 6.2, whereby the measurements of the flight just completed are used as the basis of search for the new blade adjustments. The search for blade adjustments is guided by the "process model" (see Figure 6.2), which represents the relationship between vibration changes and blade adjustments. A difficulty of rotor tuning is the excess of equations compared to degrees of freedom (four inputs to control 24 outputs), which translates into one-to-many mapping. Another difficulty is caused by the high level of noise present in the vibration measurements.

The traditional approach to rotor tuning uses linear relationships to define the process model

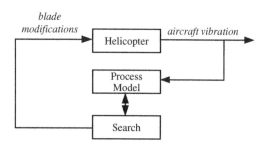

**FIGURE 6.2** Tuning strategy of the current methods.

and uses model inversion to streamline the search. The drawback of the traditional approach, therefore, is its neglect of the potential nonlinearity of track and balance, and the vibration noise, as well as its limited capacity to produce comprehensive solutions to facilitate model inversion due to its consideration of the most extreme vibration components. In an attempt to include the potential nonlinearity of the process, Taitel et al. (1995) trained a set of neural networks with actual track and balance data to map vibration measurements to blade adjustments as well as to evaluate the goodness of the solution. In effect, they developed an inverse model based on the solutions available in the historical track and balance data, and provided a forward model to evaluate the solution. The potential advantage of this method is that it can interpolate among the historical solutions to address potential nonlinearity and vibration noise. Its disadvantages are that it is only applicable to helicopters with extensive track and balance history, and that its solutions are constrained by those contained in the historical data.

Another deviation from the traditional approach is introduced by Ventres and Hayden (2000), who define the relationships between blade adjustments and vibration in frequency domain, and provide an extension of these relationships to higher order vibrations. They use an optimization method to search for the adjustments to reduce per rev vibration as well as higher-order vibrations. Accordingly, this approach has the capacity to provide a comprehensive solution, but it too neglects the potential nonlinearity between the blade adjustments and aircraft vibration as well as the noise in the measurements.

The most recent solutions to rotor tuning are those by Wang et al. (2005a, 2005b), which are designed to address both the stochastics of vibration and the potential nonlinearity of the tuning process. In the first solution, which is a probability-based method, the underlying model comprises two components: a deterministic component and a probability component. The method relies on the probability model to estimate the likelihood of the measured vibration satisfying the specifications and to search for blade adjustments that will maximize this likelihood. The likelihood measures in the probability model are computed according to the probability distribution of vibration derived from historical track and balance data. The second solution is an adaptive method that uses an interval model to cope with the potential nonlinearity of the process and to account for vibration noise. This method, which also incorporates learning to provide adaptation to the rotor tuning process, initializes the coefficients of the interval model according to the sensitivity coefficients between the blade adjustments and helicopter vibration. However, it modifies these coefficients after the first iteration to better represent the vibration measurements acquired. This method takes into account vibration data from all of the flight regimes during the search for the appropriate blade adjustments; therefore, it has the capacity to provide comprehensive solutions. The remainder of this chapter describes three of the methods discussed above to provide a representation of various solutions proposed for rotor tuning, followed by a case study to demonstrate the application of the adaptive method.

## 6.2 Neural Network-Based Tuning

As mentioned earlier, rotor tuning in four-bladed aircraft is performed by first specifying a condensed set of adjustments to reduce vibrations, and then expanding these adjustments into a detailed set to satisfy the track requirements. This same strategy is implemented in the system of neural networks shown in Figure 6.3 (Taitel et al., 1995). The first network in this system, called the selection net, determines the condensed blade adjustments (output) that will bring about a given change in vibration (input). To eliminate vibration, the negatives of the vibration measurements from the flight are utilized as inputs to this network. The validity of the condensed adjustments is then checked by predicting their effect on vibration via the condensed simulation net. Theoretically, these simulated vibration changes should be the negative of the vibration measurements from the aircraft so that their summation will be zero. However, owing to the inexactness of the neural network models and noise, the resultant vibration will most likely not equal zero. In cases where the resultant vibration is not within specifications (usually less than 0.20 inches per second [ips]), the condensed adjustments may be refined by feeding the resultant

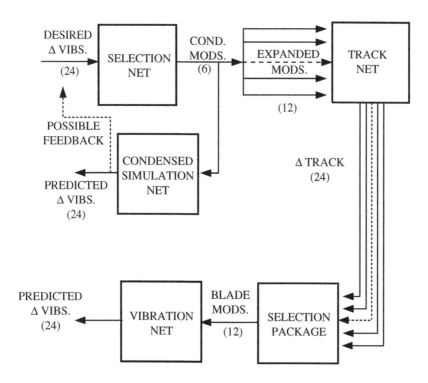

**FIGURE 6.3** Schematic of the rotor tuning system. The numbers inside parentheses represent the number of inputs or outputs of individual nets.

vibration back into the selection net. This feedback is depicted by the dashed feedback line in Figure 6.3. It should be noted that the condensed simulation net may also serve as a diagnostic tool by indicating behavior out of the norm. For example, an aircraft with vibrations significantly different from those predicted by this network may suffer from defective components.

Just as with the traditional approach, once the condensed solution has been specified, it needs to be expanded into a detailed form to satisfy the rotor track requirements. As previously mentioned, the condensed set of adjustments may be viewed as the constraint on detailed adjustments so as to ensure that the vibration solution is not compromised for track. Each one of these detailed sets of adjustments is a candidate for the final rotor tuning solution, and it is left to the track net and the selection package to determine which set of detailed adjustments provides the best tracking performance. For selection purposes, the track net simulates the changes in track due to a candidate set of detailed adjustments, and then adds these changes to the initial track measurements from the flight to estimate the resultant track. The set of detailed adjustments that yields the smallest estimated track (i.e., smallest maximum blade spread) is selected as the solution to the rotor tuning problem. The selected set of detailed adjustments is then checked via the vibration net, which, similar to the condensed simulation net, serves as an independent evaluator of the selected adjustments.

## 6.3 Probability-Based Tuning

The noted contribution of this method is its introduction of the likelihood of success as a criterion in the search for the blade adjustments (Wang et al., 2005). This method speculates the effectiveness of various adjustment sets in reducing the vibration and selects the set with the maximum probability of producing acceptable vibration. The concept of this method is explained in the context of a simple example. If the measured vibration from the current flight is denoted by $V_j(k-1)$ and the estimated vibration change according to the model is represented by $\Delta \hat{V}_j(k) = f(\Delta x)$ as a function of the blade adjustments, $\Delta x$, then the predicted vibration of the next flight, $\hat{V}_j(k)$, can be defined as

$$\hat{V}_j(k) = V_j(k-1) + \Delta \hat{V}_j(k) \qquad (6.6)$$

$$V_j(k) = \hat{V}_j(k) + \hat{e}_j(k) \qquad (6.7)$$

where $V_j(k)$ denotes the measured vibration for the next flight. In rotor tuning, the adjustments are selected according to the predicted vibration, $\hat{V}_j(k)$, whereas the objective is defined in terms of the measured vibration. The inclusion of the probability model here is to account for the inevitable uncertainty in the actual position of

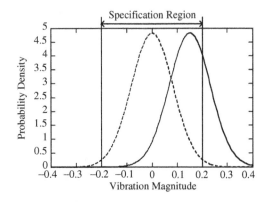

**FIGURE 6.4** Illustration of improved placement of the predicted vibration within the specification range.

the measured vibration. According to Equation 6.7, the mean value of the measured vibration is equal to the value of the predicted vibration plus the mean value of the prediction error. However, since the predicted vibration is a deterministic entity, the probability distribution of the measured vibration is the same as that of the prediction error. Accordingly, whereas the nominal value of the measured vibration can be controlled by the blade adjustment, its optimal position within the specification region should be determined according to its probability distribution. For a case where the prediction error, $\hat{e}_j(k)$, is zero-mean and normally distributed, as illustrated in Figure 6.4, placing the predicted vibration at the center of the specification range will be synonymous with maximizing the probability that the measured vibration will be within the range. The likelihood of success of blade adjustments can therefore be measured by the area under the probability density function of prediction error located within the specification region. The blade adjustment set that produces the highest likelihood will be the preferred adjustment.

The main difficulty with rotor tuning, however, is the limited number of DoFs, which precludes perfect positioning of the predicted vibration. This point is illustrated in Figure 6.5 for a case where two vibration components are to be positioned at the center of the specification region with only one adjustment. If one assumes that the effect of adjustment, $\Delta x$, on the change in the two vibration components, $\Delta \hat{V}_j(k)$, can be represented by a linear model, as

$$\Delta \hat{V}_j(k) = a_{ij} \Delta x$$

then the position of the predicted vibration components will be constrained to the line L in Figure 6.5. As illustrated in this figure, since it will be impossible to place the predicted vibration components at the center, a compromised position needs to be selected. In this method, the best compromised position for the predicted vibration is that which renders the largest probability of satisfying the specifications for the measured vibration. This position, for the two-component vibration example, is one that maximizes $P_r[(V_1, V_2) \in S] = \int_{(V_1,V_2) \in S} p(V_1, V_2) \mathrm{d}V_1 \, \mathrm{d}V_2$. The above formulation indicates that the placement of the predicted vibration requires knowledge of

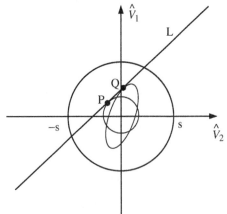

**FIGURE 6.5** Restricted placement of vibration components within the specification region for a two-dimensional case.

the joint probability density function, $p(V_1, V_2)$, of the vibration components. In the ideal case of independent vibration components with equal probability distributions, the loci of the points with equal probabilities $P_r[(V_1, \ldots, V_n) \in S]$ are surfaces of hyperspheres. Such ideal loci for the two-component vibration example of Figure 6.5 are circles centered at the origin (see Figure 6.5), which lead to point P as the best compromised position closest on line L to the center of the specification circle. Point P, however, does not represent the best position if the two vibration components are dependent or have unequal distributions. The loci of equal probabilities for this more general case are elliptical, as also shown in Figure 6.5, indicating point Q as the best position on line L for placing the predicted vibration. The inadequacy of the DoFs illustrated here is exacerbated in rotor tuning, where 24 correlated vibration components need to be positioned within the specification region using only four condensed blade adjustments. For the 24-component vector of measured vibration $\mathbf{V}(k) = [V_{c1}(k), V_{s1}(k), \ldots, V_{c12}(k), V_{s12}(k)]^T$, where $V_c$ and $V_s$ represent the cosine and sine components of each vibration measurement, respectively, the joint probability density function of measured vibration for the $k$th flight, $\mathbf{V}(k)$, can be characterized as an $N$-dimensional Gaussian function:

$$p(\mathbf{V}(k)) = \frac{1}{(2\pi)^{N/2} |\Phi|^{1/2}} \exp\left[ -\frac{1}{2} \hat{\mathbf{e}}(k)^T \Phi^{-1} \hat{\mathbf{e}}(k) \right] \tag{6.8}$$

$$\hat{\mathbf{e}}(k) = \mathbf{V}(k) - \mathbf{V}(k-1) - C\Delta\mathbf{x}(k) \tag{6.9}$$

where $\Phi$ represents the covariance matrix of the prediction error. Now, if $\Gamma = \{|\mathbf{V}_j| = \sqrt{V_{cj}^2 + V_{sj}^2} \le \alpha, j = 1, \ldots, 12\}$ denotes the specification region in 24-dimensional Euclidean space, the blade adjustments, $\Delta\mathbf{x}^*$, can be selected such that the probability that the measured vibration is within the acceptable range is maximized (see also Table 6.2). Formally,

$$\Delta\mathbf{x}^* = \arg_{\Delta\mathbf{x}} \max\left[ \Pr(\mathbf{V}(k) \in \Gamma) = \int_\Gamma p(\mathbf{V}(k)) d\mathbf{V}(k) \right] \tag{6.10}$$

**TABLE 6.2**  Summary of Probability-Based Tuning

For the input vector:

$$\Delta\mathbf{x} = [\Delta x_1, \Delta x_2, \Delta x_3, \Delta x_4]^T$$

where $\Delta x_1$ and $\Delta x_3$ denote the combined trim tab adjustments to blade combinations one to three and two to four, respectively, and $\Delta x_2$ and $\Delta x_4$ represent the combined pitch control rod adjustments to blade combinations one to three and two to four, respectively, the blade adjustments, $\Delta\mathbf{x}$, can be selected such that the probability that the measured vibration is within the acceptable range is maximized. Formally,

$$\Delta\mathbf{x}^* = \arg_{\Delta\mathbf{x}} \max[\Pr(\mathbf{V}(k) \in \Gamma) = \int_\Gamma p(\mathbf{V}(k)) d\mathbf{V}(k)]$$

where $\Pr(\mathbf{V}(k))$ denotes the probability of the measured vibration, $\Gamma$ denotes the specification region in 24-dimensional Euclidean space, and $p(\mathbf{V}(k))$ represents the joint probability density of the measured vibration for the $k$th flight characterized as an $N$-dimensional Gaussian function:

$$p(\mathbf{V}(k)) = \frac{1}{(2\pi)^{N/2} |\Phi|^{1/2}} \exp[ -\frac{1}{2} \hat{\mathbf{e}}(k)^T \Phi^{-1} \hat{\mathbf{e}}(k)]$$

with

$$\hat{\mathbf{e}}(k) = \mathbf{V}(k) - \mathbf{V}(k-1) - C\Delta\mathbf{x}(k)$$

representing the predicted error in vibration.

## 6.4   Adaptive Tuning

The schematic of this method is shown in Figure 6.6 (Wang et al., 2005). As in the other methods, it uses a process model as the basis of search for the appropriate blade adjustments, but instead of using a linear model, it uses an interval model to accommodate process nonlinearity and measurement noise. According to this model, the feasible region of the process is estimated first, to include the adjustments that will result in acceptable vibration estimates. This feasible region is then used to search for the blade adjustments that will minimize the modeled vibration. If the application of these adjustments does not result in satisfactory vibration, the interval model will

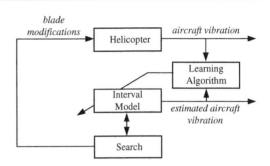

**FIGURE 6.6**   The strategy of the proposed tuning method.

be updated to better estimate the feasible region and improve the choice of blade adjustments for the next flight. Important parameters of adaptive tuning are summarized in Table 6.3.

### 6.4.1   The Interval Model

In order to account for the stochastics and nonlinearity of vibration, an interval model (Moore, 1979) is defined to represent the range of aircraft vibration caused by blade adjustments. The interval model used here has the form:

$$\Delta \vec{y}_j = \sum_{i=1}^{n} \vec{C}_{ji} \Delta x_i, \quad j = 1, \ldots, m \tag{6.11}$$

where each coefficient is defined as an interval:

$$\vec{C}_{ji} = [C_{Lji}, C_{Uji}]$$

In the above model, the variables with the two-sided arrow, $\leftrightarrow$, denote intervalled variables, $C_{Lji}$ and $C_{Uji}$ represent, respectively, the current values of the lower and upper bounds of the sensitivity coefficients between each input, $\Delta x_i$, and output, $\Delta \vec{y}_j$. The interval $\Delta \vec{y}_j$ denotes the estimated range of change of the $j$th output caused by the change to the current inputs, $\Delta x_1, \ldots, \Delta x_n$.

---

**TABLE 6.3**   Summary of Adaptive Tuning

In adaptive tuning, each vibration component is defined as

$$\Delta \vec{y}_j = \sum_{i=1}^{n} \vec{C}_{ji} \Delta x_i, \quad j = 1, \ldots, m$$

where each coefficient is defined as an interval:

$$\vec{C}_{ji} = [C_{Lji}, C_{Uji}]$$

with $C_{Lji}$ and $C_{Uji}$ representing, respectively, the current values of the lower and upper bounds of the sensitivity coefficients between each input, $\Delta x_i$, and output, $\Delta \vec{y}_j$. The blade adjustments are then sought by minimizing the objective function:

$$S = \frac{\sum_{e=1}^{N_e} \text{Distance}(x_c, x_e)}{\left( \prod_{s=1}^{N_s} \text{Distance}(x_c, x_s) \right)^{1/N_s}}$$

where $x_c$ represents a candidate set of blade adjustments within the feasible region, $x_e$ represents any set of blade adjustments within the selection region, $x_s$ denotes each of the previously selected blade adjustments, and $N_e$ and $N_s$ represent the number of the estimated feasible blade adjustments and the previously selected blade adjustments, respectively.

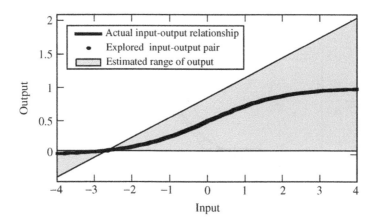

**FIGURE 6.7** Estimated range of output by the interval model using one reference input.

The fit provided by the interval model for a mildly nonlinear input/output relationship is illustrated in Figure 6.7, where the output range is estimated relative to one explored input.[1] According to Equation 6.11, the estimated range of the output becomes larger, and therefore less accurate, as the potential input is selected farther from the current input (producing a large $\Delta x_i$). This potential drawback of the interval model is considerably reduced when multiple inputs have been explored so that the interval model can take advantage of several inputs for estimating the output range. The estimated output, $\bar{\bar{y}}_j$, at a potential input, $x_i$, may be computed relative to any set of previously explored inputs, yielding different estimates of $\bar{\bar{y}}_j$ (due to different values of $\Delta x_i$). In order to cope with the multiplicity of estimates, $\bar{\bar{y}}_j$ is defined as the common range among all of the $\bar{\bar{y}}_j$ estimates (Yang, 2000). The estimation of $\bar{\bar{y}}_j$ using this commonality rule is illustrated in Figure 6.8, which indicates that using this estimation approach enables representation of the system nonlinearities in a piecewise fashion. It can be shown that the lack of commonality between the estimated ranges of output will cause a part of the input–output relationship to not be represented by the interval model. In such cases, however, the lack of compliance between the interval model and the *input–output* relationship can be corrected by adaptation of the coefficient intervals through learning.

## 6.4.2   Estimation of Feasible Region

The feasible region comprises all sets of blade adjustments that will reduce the aircraft vibration within specifications. The feasible region is estimated here by comparing the individually estimated $\bar{\bar{y}}_j$ values with their corresponding constraints, so as to decide whether the corresponding blade adjustments belong to the feasible region. In this method, even when the interval $\bar{\bar{y}}_j$ partly overlaps the vibration constraint, the corresponding blade adjustments are included in the estimated feasible region. The above procedure of estimating the feasible region based on individual outputs is then extended to multiple outputs by forming the conjunction of the estimated feasible regions from each output.

## 6.4.3   Selection of Blade Adjustments

The blade adjustments provide the coordinates of the feasible region, therefore, they need to provide a balanced coverage of the input space. As such, blade adjustment selection becomes synonymous with maximizing the distance of the selected blade adjustments from the previous blade adjustments, as well as

---

[1]An explored input represents an input for which the exact value of the output is available. In rotor tuning, an explored input would denote a blade adjustment that has been applied to the helicopter, and for which the corresponding vibration changes have been measured.

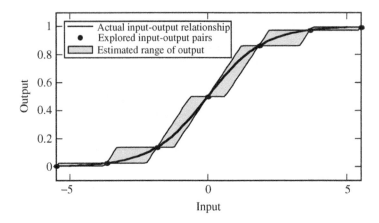

**FIGURE 6.8** Estimated range of output by the interval model using seven reference inputs.

bringing them closer to the center of the feasible region. This objective can be pursued by minimizing the following objective function:

$$S = \frac{\sum_{e=1}^{N_e} \text{Distance}(x_c, x_e)}{\left(\prod_{s=1}^{N_s} \text{Distance}(x_c, x_s)\right)^{1/N_s}} \tag{6.12}$$

where $x_c$ represents a candidate set of blade adjustments within the feasible region, $x_e$ represents any set of blade adjustments within the selection region, $x_s$ denotes each of the previously selected blade adjustments, and $N_e$ and $N_s$ represent, respectively, the numbers of the estimated feasible blade adjustments and the previously selected blade adjustments. Note that when the candidate set, $x_c$, is close to the previously selected blade adjustments, $\left(\prod_{s=1}^{N_s} \text{Distance}(x_c, x_s)\right)^{1/N_s}$ becomes small, and when the candidate set of blade adjustments, $x_c$, is far from the center of the feasible region, the value of $\sum_{e=1}^{N_e} \text{Distance}(x_c, x_e)$ becomes large. By minimizing $S$, the candidate blade adjustments are selected such that the above extremes are avoided.

### 6.4.4 Learning

Although an interval model defined according to the sensitivity coefficients may provide a suitable initial basis for tuning, it may not be the most representative of the rotor tuning process. As such, it may not be able to carry the search process to the end. A noted feature of the proposed method is its learning capability, which enables it to refine its knowledge base. To this end, the coefficients of the model are updated by considering new values for each of the upper and lower limits of individual coefficients. The objective is to make the range of the coefficients as small as possible while making sure that the interval model envelopes the acquired *input–output* data. The learning problem can be defined as

$$\text{Minimize } E = \sum_{m=1}^{K-1} \sum_{k>m}^{K} \{[y_L(m, k) - y(k)]^2 + [y_U(m, k) - y(k)]^2\} \tag{6.13}$$

subject to

$$y_U(m, k) \geq y(k) \tag{6.14}$$

$$y_L(m, k) \leq y(k) \tag{6.15}$$

$$C_{Ui} - \gamma \geq C_{Li} \tag{6.16}$$

where $K$ represents the total number of sample points collected so far, $y_L(m, k)$ and $y_U(m, k)$ represent, respectively, the lower and upper limits of the estimated output range at the $k$th sample point relative to the $m$th sample point, $y(k)$ denotes the actual output value at the $k$th sample point, and $C_{Ui}$ and $C_{Li}$ represent the upper and lower limits of the $i$th coefficient interval, respectively. The parameter $\gamma$ is a small positive number to control the range of the coefficients.

Most of the approaches that can be potentially used for adapting the coefficient intervals, such as gradient descent (Ishibuchi et al., 1993) or nonlinear programming, cannot be applied to rotor tuning due to their demand for rich training data and their impartiality to the initial value of the coefficients representing the *a priori* knowledge of the process. As an alternative, a learning algorithm is devised here to cope with the scarcity of track and balance data while staying true to the initial values of the coefficients. In this algorithm, the coefficients of the interval model, initially set pointwise at the sensitivity coefficients, are adapted after each flight in two steps: enlargement and shrinkage. First, the vibration measurements from all of the flights completed for the present tail number are matched against the estimated output ranges from the current interval model. If any of the measurements do not fit the upper or lower limits of the estimates, the coefficient intervals are enlarged in small steps, iteratively, and the output ranges are re-estimated at each iteration using the updated interval model. The enlargement of the coefficient intervals stops when the estimated output ranges include all of the measurements. At this point, even though the updated interval model provides a fit for the *input–output* data, it may be overcompensated. In order to rectify this situation, the coefficient intervals are shrunk individually by selecting new candidates for their upper and lower limits.

The shrinkage–enlargement learning algorithm has the form:

$$\Delta C_{Li} = -\eta \delta_L \Delta x_i(m, k) \tag{6.17}$$

$$\Delta C_{Ui} = -\eta \delta_U \Delta x_i(m, k) \tag{6.18}$$

where, during the enlargement phase, $\delta_L$ and $\delta_U$ are defined as

$$\delta_L = \begin{cases} \Delta y_L & \text{If } \Delta x_i(m, k) > 0 \text{ and } \Delta y_L > 0 \\ \Delta y_U & \text{If } \Delta x_i(m, k) < 0 \text{ and } \Delta y_U < 0 \\ 0 & \text{otherwise} \end{cases} \tag{6.19}$$

$$\delta_U = \begin{cases} \Delta y_U & \text{If } \Delta x_i(m, k) > 0 \text{ and } \Delta y_U < 0 \\ \Delta y_L & \text{If } \Delta x_i(m, k) < 0 \text{ and } \Delta y_L > 0 \\ 0 & \text{otherwise} \end{cases} \tag{6.20}$$

and during the shrinkage phase, they are defined as

$$\delta_L = \begin{cases} \Delta y_L & \text{If } \Delta x_i(m, k) > 0 \text{ and } \Delta y_L < 0 \\ \Delta y_U & \text{If } \Delta x_i(m, k) < 0 \text{ and } \Delta y_U > 0 \\ 0 & \text{otherwise} \end{cases} \tag{6.21}$$

$$\delta_U = \begin{cases} \Delta y_U & \text{If } \Delta x_i(m, k) > 0 \text{ and } \Delta y_U > 0 \\ \Delta y_L & \text{If } \Delta x_i(m, k) < 0 \text{ and } \Delta y_L < 0 \\ 0 & \text{otherwise} \end{cases} \tag{6.22}$$

with

$$\Delta x_i(m, k) = x_i(k) - x_i(m) \tag{6.23}$$

$$\Delta y_L = y_L(m, k) - y(k) \tag{6.24}$$

$$\Delta y_U = y_U(m, k) - y(k) \tag{6.25}$$

This procedure is repeated for each coefficient interval in an iterative fashion until the objective function $E$ (Equation 6.13) is minimized. The minimization of $E$ ensures limited adaptation of the coefficient intervals within the smallest possible range.

At the beginning of tuning, the limited number of *input–output* data available for learning will not provide a comprehensive representation of the process. Therefore, the coefficient intervals should not be shrunk drastically until enough *input–output* data have become available. For this, the length of each coefficient interval $[C_{Li}, C_{Ui}]$ is constrained by the minimal interval length for each tuning iteration as

$$\min L = \{C_{Ui}(0) - C_{Li}(0)\}(1 - \beta)^n \tag{6.26}$$

where $\beta \in [0, 1]$ controls the shrinkage rate of the coefficient interval, and $n$ denotes the number of tuning iterations. The coefficient interval cannot be shrunk when $\beta = 0$ and can be shrunk without limit when $\beta = 1$. Usually, $\beta$ is selected closer to 0.

## 6.5 Case Study

The utility of the Interval Model (IM) method is demonstrated in application to Black Hawks. Ideally, the performance of the proposed method should be evaluated side by side against that of the traditional method. However, such an evaluation would require tuning the aircraft with one method, undoing changes, and tuning the aircraft with another. Since such testing is prohibitively costly and infeasible, a compromised approach of evaluating the method in simulation is utilized. A process simulation model is therefore used to represent the block "helicopter" in Figure 6.6, with the block "forward model" represented by an interval model.

### 6.5.1 Simulation Model

Considering the potential nonlinearity of the effect of blade adjustments on the helicopter vibration and the high level of noise present in vibration measurements, multilayer neural networks offer the most suitable framework for modeling. A series of neural networks were trained with historical balance data to represent the relationships between vibration changes and blade adjustments, and the stochastic aspects of vibration were represented by the addition of random numbers to the outputs of the networks.

A total of 102 sets of vibration data were used to train and test the neural networks. The inputs to these networks were the combined blade adjustments of push rods and trim tabs to opposite blade pairs, and their outputs were the resulting vibration changes between two consecutive flights. Since the vibration data are vector quantities that are represented by both magnitude and phase components (see Table 6.1), the vibration data were transformed into Cartesian coordinates, so that each vector element would denote the change in the cosine or sine component of the $A + B$ or $A - B$ vibration of each of the six flight regimes (see Table 6.1). In this study, each neural network model consisted of four inputs and one output, so a total of 24 networks were trained to represent all of the vibration components. Alternatively, all of the vibration measurements may be represented by one neural network, but such a network is more difficult to train. Formally, the outputs of the neural networks, which represent the cosine and sine components of the vibration at different regimes, $v_{cj}(k)$ and $v_{sj}(k)$, respectively, are defined as

$$\hat{V}_{sj}(k) = V_{sj}(k - 1) + \Delta V_{sj}(k) + R_{sj}(k) \tag{6.27}$$

$$\hat{V}_{cj}(k) = V_{cj}(k - 1) + \Delta V_{cj}(k) + R_{cj}(k) \tag{6.28}$$

$$\Delta V_{sj}(k) = F_{sj}(\Delta x) \tag{6.29}$$

$$\Delta V_{cj}(k) = F_{cj}(\Delta x) \tag{6.30}$$

$$\hat{V}_j(k) = \sqrt{\hat{V}_{sj}(k)^2 + \hat{V}_{cj}(k)^2} \tag{6.31}$$

where the input vector $\Delta\mathbf{x} = \{\Delta x_1, \Delta x_2, \Delta x_3, \Delta x_4\}$ denotes the set of combined blade adjustments, each of the functionals, $F_{sj}$ and $F_{cj}$, represent the change in vibration between two consecutive flights as represented by a neural network, and $R_{sj}(k)$ and $R_{cj}(k)$ denote random numbers added to the outputs of the networks to account for measurement noise. Each of the networks consisted of two hidden layers, with four and eight processing elements in the first and second layers, respectively. To avoid overtraining, the 102 sets of data were divided into two equal subsets, one set to train the network and the other to test its performance. The random numbers, $R_{cj}$ and $R_{sj}$, were generated according to the Gaussian distribution $N(\mu, \sigma^2)$, with the mean $\mu$ and variance $\sigma^2$ defined as

$$\hat{\mu} = \frac{1}{M} \sum_{i=1}^{M} e_i \tag{6.32}$$

$$\hat{\sigma}^2 = \frac{1}{M-1} \sum_{i=1}^{M} (e_i - \hat{\mu})^2 \tag{6.33}$$

In the above formulation, $M$ represents the total number of data sets and $e_i$ denotes the difference between the measured and expected value of vibration, defined as

$$e_j(k) = V_j(k) - V_j(k-1) - \Delta V_j(k) \tag{6.34}$$

A sample of estimated vibration changes generated by the neural network model is compared side by side with the actual vibration changes in Figure 6.9. The results indicate close agreement between the predicted and actual vibration changes.

## 6.5.2   Interval Modeling

In application to the Black Hawks, a total of 24 interval models need to be constructed to approximate the changes in the cosine and sine components of the A + B and A − B vibrations at each of the six flight regimes. The interval models have the form

$$\vec{V}_{cj}(k) = V_{cj}(k-1) + \sum_{i=1}^{4} \vec{C}_{cji}(k-1)\Delta x_i(k) \tag{6.35}$$

$$\vec{V}_{sj}(k) = V_{sj}(k-1) + \sum_{i=1}^{4} \vec{C}_{sji}(k-1)\Delta x_i(k) \tag{6.36}$$

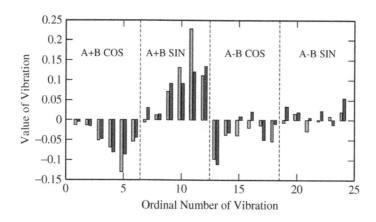

**FIGURE 6.9**   A sample set of simulated vibration changes shown side by side with the actual vibration changes.

$$\vec{V}_j(k) = \sqrt{\vec{V}_{sj}^2(k) + \vec{V}_{cj}^2(k)}, \quad i = 1, ..., 4 \text{ and } j = 1, ..., 12 \tag{6.37}$$

where the $\vec{V}_{cj}(k)$ and $\vec{V}_{sj}(k)$ represent, respectively, the estimated cosine and sine components of A + B or A − B vibration at each of the six flight regimes, $\vec{V}_j(k)$ denotes the magnitude of the vibration, and $\Delta x_i$ are the same as those in Equation 6.2 to Equation 6.5. For this study, the feasible region was defined to include all of the blade adjustments associated with vibration estimates that satisfied the specification: $\max\{\min(\vec{V}_1), ..., \min(\vec{V}_{12})\} \leq 0.2$. The above specification ensures that the lower limit of the estimated vibration range of the largest vibration component will be less than 0.2 ips (an industry standard). The selection of the lower limit here is to ensure that the feasible region is as large as possible, so as not to eliminate any potentially good candidate blade adjustments. The computation of the feasible region was based on the range $[−0.015, 0.015]$ for push rods and $[−0.035, 0.035]$ for trim tabs, within which 20,000 random sets of blade adjustments were evaluated for their feasibility. The blade adjustments associated with vibration ranges satisfying the specification were included in the feasible region.

As noted earlier, the proposed method uses the feasible region as the basis of search for the blade adjustments. For this study, the blade adjustments set that produced the smallest value for the objective function S (Equation 6.12) was selected to be applied to the helicopter. It should be noted that, given the stringent constraints on the vibration components, there were cases where the search algorithm could not find any feasible blade adjustments that would satisfy all of the constraints. In such cases, the set of blade adjustments that produced the smallest lower limit of the maximum estimated vibration was used as a compromised solution.

The interval model was updated after each tuning iteration. For shrinkage−enlargement learning, the parameter $\beta$ in Equation 6.26 was set to 1 and $\gamma$ to 0, so that the coefficient intervals could be shrunk without limits. Learning was performed separately for each tail number to customize the interval model to individual tail numbers; that is, the interval model was set to the sensitivity coefficients for each tail number and was adapted after the first tuning iteration. Accordingly, the interval model was actually a pointwise model for the first iteration and took the form of an interval model thereafter.

## 6.5.3 Performance Evaluation

The interval model (IM) method was tested on 39 tail numbers, for which actual track and balance data were available from the field. For each tail number, the IM method was applied iteratively until either the simulated vibrations were within their specifications, or an upper limit of five process iterations had been reached.

Since the stochastic aspects of vibration measurements impose randomness on the rotor tuning process, rotor tuning solutions cannot be evaluated by deterministic measures. This calls for the creation of performance measures that account for uncertainty. One such measure that assesses tuning efficiency is the *average tuning iteration number* (ATIN) which represents the average number of iterations taken for tuning each tail number. The number of flights used by the IM method for the 39 tail numbers is included in Table 6.4 along with those actually performed in the field. The results indicate that the IM method requires a smaller ATIN relative to that actually performed.

Another potentially significant aspect of the IM method is its adaptation capability, which enables it to transform a pointwise model into an IM, and to subsequently update it after the first iteration. Adaptation capability, however, may not be as significant in rotor tuning, which offers limited possibility for training. In order to evaluate the significance of learning in the performance of the IM method, the results in Table 6.4 were reproduced in Table 6.5 with the learning feature turned off. The ATINs indicate that with learning, the IM method requires fewer iterations for tuning each tail number, despite the small number of iterations taken to tune each tail number. This, in turn, indicates that the interval model enhances the performance of the IM method, since without learning, the model remains pointwise at the sensitivity coefficients. However, perhaps an equally interesting set of results in Table 6.5 are those indicating that even without learning, the IM method requires fewer iterations than actually

**TABLE 6.4**  The Number of Tuning Iterations Required by the Interval Model Method and Those Applied in the Field

| Tail # (39) | Number of Tuning Iterations | |
|---|---|---|
| | Actual | IM Method |
| 176 | 1 | 1 |
| 178 | 1 | 1 |
| 179 | 3 | 2 |
| 180 | 1 | 1 |
| 184 | 2 | 1 |
| ⋮ | ⋮ | ⋮ |
| 260 | 4 | 1 |
| ⋮ | ⋮ | ⋮ |
| 861 | 1 | 1 |
| Total | 71 | 48 |
| ATIN | 1.82 | 1.23 |

**TABLE 6.5**  The Number of Tuning Iterations Required by the IM Method (with and without Learning) along with Those Actually Applied in the Field

| Tail # (39) | Tuning Iteration Number | | |
|---|---|---|---|
| | Actual | IM Method | |
| | | With Learning | Without Learning |
| 185 | 3 | 2 | 3 |
| 186 | 3 | 2 | 3 |
| 208 | 2 | 2 | 3 |
| 245 | 3 | 2 | 3 |
| 260 | 4 | 2 | 3 |
| 802 | 3 | 2 | 3 |
| 822 | 3 | 2 | 3 |
| ⋮ | ⋮ | ⋮ | ⋮ |
| Total | 71 | 48 | 62 |
| ATIN | 1.82 | 1.23 | 1.59 |

performed in the field. Given that the adjustments associated with both sets of results were selected from the same model (i.e., sensitivity coefficients), the better performance of the IM method can only be attributed to its more effective search strategy that leads to more comprehensive solutions.

A preferred aspect of a system of rotor tuning is its ability to tune the aircraft within one iteration. This aspect of the method was evaluated by checking the number of tail numbers tuned within one iteration. For these results, in order to eliminate the difference between the simulation model and the helicopter, only the vibration estimates from simulation were used to evaluate the suitability of the adjustments. The results of this study are shown in Table 6.6, where the tail numbers tuned within one iteration are shown by a $\sqrt{}$ and those requiring more than one iteration are denoted by $\times$. The results indicate that the IM method satisfies this more stringent criterion better than the actual adjustments, further validating the claim that the IM method benefits from a more effective search engine.

Owing to the randomness of the vibration measurements, repeated applications of an adjustment set may lead to slightly different vibration measurements. This, in turn, may cause a variance in the number of iterations produced by adjustments when the resulting vibration is close to the specified threshold. It would be beneficial, therefore, to devise a measure for the probability of success of adjustments.

**TABLE 6.6**  Tally of the Tail Numbers Tuned within One Iteration According to Simulated Vibration

| Tail # (39) | Tuned within One Iteration | |
|---|---|---|
| | Actual | IM Method |
| 176 | ✓ | ✓ |
| 178 | ✗ | ✓ |
| 179 | ✓ | ✓ |
| ⋮ | ⋮ | ⋮ |
| 822 | ✗ | ✓ |
| 858 | ✗ | ✓ |
| 859 | ✓ | ✓ |
| 861 | ✓ | ✓ |
| Total | 19 | 30 |

The empirical measure, the acceptability index (AI), is defined here as

$$AI = \frac{1}{N} \sum_{l=1}^{N} s_l \qquad (6.38)$$

to denote the percentage of times an adjustment set will result in the vibration satisfying the specification. In the above equation, $N$ represents the total number of flights simulated to represent the repeated applications of the same adjustment set, and

$$s_l = \begin{cases} 1 & \text{if vibration of the } l\text{th simulation flight is acceptable} \\ 0 & \text{if vibration of the } l\text{th simulation flight is unacceptable} \end{cases}$$

The AIs computed for both the actual and selected adjustments at the first iteration are included in Table 6.7. The results indicate that the IM method provides adjustments with a higher probability of success as judged by the acceptability of vibration estimates from the simulation model. These results, which indicate that the selected adjustments from the IM method can more consistently tune the rotors within one iteration, imply the better positioning of the adjustments within the feasible region.

**TABLE 6.7**  The Values of Acceptability Index (Trial Mode) Computed for Both the Actual and Selected Adjustments at the First Flight

| Tail # (39) | Acceptability Index | |
|---|---|---|
| | Actual | IM Method |
| 176 | 0.92 | 0.87 |
| 178 | 0 | 0.54 |
| 179 | 0.61 | 0.52 |
| ⋮ | ⋮ | ⋮ |
| 260 | 0.40 | 0.89 |
| 261 | 0.09 | 0.95 |
| 263 | 0.18 | 0.12 |
| ⋮ | ⋮ | ⋮ |
| 822 | 0.00 | 0.64 |
| 857 | 0.62 | 0.55 |
| 858 | 0.64 | 0.93 |
| 859 | 0.94 | 0.67 |
| 861 | 0.96 | 0.74 |
| Average | 0.581 | 0.724 |

**TABLE 6.8** Comparison of the First Iteration Solutions of IM Method and Actual Solutions from Sikorsky's Production Line with the Cumulative Acceptable Adjustments

| Tail # | Modifications | | |
|---|---|---|---|
| | Actual Iteration 1 Modifications | CAM | IM Iteration 1 Modifications |
| 801 | 6, −4, −10, 11 | 2, −4, −4, 14 | 3, −5, −6, 12 |
| 802 | 5, 2, 0, 0 | 9, 0, −10, 10 | 8, −2, −5, 10 |
| 822 | 6, 0, −20, 0 | 10, −4, −23, 13 | 8, −4, −22, 6 |
| 858 | 7, 0, −14, 0 | 9, −2, −10, 3 | 9, −2, −15, 4 |

Another evaluation basis for the adjustments can be established by comparing them to the actual cumulative adjustments performed in the field. The cumulative adjustment set, $\sum x$, can be defined as

$$\sum x = \sum_{k=1}^{N} \Delta x_k \qquad (6.39)$$

where $N$ represents the total number of tuning iterations performed in the field for the tail number and $\Delta x_k$ denotes the adjustments applied at the $k$th iteration. A sample of actual first iteration adjustments, actual cumulative adjustments, and first iteration adjustments from the IM method is shown in Table 6.8. The results indicate that the adjustments from the IM method are closer to the actual cumulative adjustments than are the actual first iteration adjustments. Although the cumulative adjustments may not be the most desirable ones for the aircraft, they represent an acceptable set that has been proven in the field. The closeness of the IM method's solutions to the actual cumulative adjustments further validates its effectiveness.

# 6.6 Conclusion

A logical feature for future rotor tuning systems will be the capability to adjust the blades during the flight. For this, these systems will need to have the capability to learn from their mistakes. They will also need to be able to monitor the condition of the rotor system in-flight, so they will stop modifying the blade parameters when more drastic actions are necessary for saving the aircraft. As such, these systems will need to be used with strong operator interaction to prevent implementation of inappropriate adjustments, and must have the ability to explain the recommended adjustments to the operator.

# References

Ishibuchi, H., Tanaka, H., and Okada, H., An architecture of neural networks with interval weights and its applications to fuzzy regression analysis, *Fuzzy Sets Syst.*, 57, 27–59, 1993.

Moore, R.E. 1979. *Methods and Applications of Interval Analysis*, Society for Industrial and Applied Mathematics, Philadelphia.

Taitel, H., Danai, K., and Gauthier, D.G., Helicopter track and balance with artificial neural nets, *ASME J. Dyn. Syst. Meas. Contr.*, 117, 226–231, 1995.

Ventres, S. and Hayden, R.E. 2000. *Rotor tuning using vibration data only*, American Helicopter Society 56th Annual Forum, Virginia Beach, VA, May 2–4, 2000.

Wang, S., Danai, K., and Wilson, M., A probability-based approach to helicopter track and balance, *J. Am. Helicopter Soc.*, 50, 1, 56–64, 2005a.

Wang, S., Danai, K., and Wilson, M., An adaptive method of helicopter track and balance, *ASME J. Dyn. Syst. Meas. Contr.*, 2005b, March, in press.

Yang, D.Z. 2000. Knowledge-based interval modeling method for efficient global optimization and process tuning, Ph.D. Thesis. Department of Mechanical and Industrial Engineering, University of Massachusetts, Amherst.

# 7

# Vibration Design and Control

Clarence W. de Silva
*The University of British Columbia*

## Summary

*There are desirable and undesirable types and situations of mechanical vibration. Undesirable vibrations are those that cause human discomfort and hazards, structural degradation and failure, performance deterioration and malfunction of machinery and processes, and various other problems. This chapter discusses ways of either eliminating or reducing the undesirable effects of vibration. Specifically, some useful topics on design for vibration suppression and the control of vibration are addressed. General approaches to vibration mitigation may be identified from the dynamic systems point of view. Typically, a set of vibration specifications is given as simple threshold values (bounds) or frequency spectra, and the goal is to either design or control the system to meet these specifications. Frequency-domain techniques based on transfer functions such as transmissibility, and time-domain techniques using the state-space representation, optimal control and modal control are presented. Applications considered here include vibration isolation, balancing of rotating and reciprocating machinery, whirling suppression, and passive and active control of vibration.*

## 7.1 Introduction

Consider the schematic diagram of a vibratory system shown in Figure 7.1. Forcing excitations $\mathbf{f}(t)$ to the mechanical system $S$ cause the vibration responses $\mathbf{y}$. Our objective is to suppress $\mathbf{y}$ to a level that is acceptable. Clearly, there are three general ways of doing this:

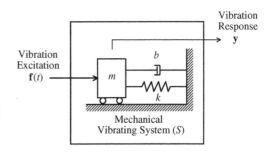

**FIGURE 7.1**  A vibrating mechanical system.

1. *Isolation.* Suppress the excitations of vibration. This method deals with $\mathbf{f}$.
2. *Design modification.* Modify or redesign the mechanical system so that for the same levels of excitation, the resulting vibrations are acceptable. This method deals with $S$.
3. *Control.* Absorb or dissipate the vibrations using external devices, through implicit or explicit sensing and control. This method deals with $\mathbf{y}$.

Within each of these three categories, several approaches can be used to achieve the objective of vibration mitigation. Essentially, each of these approaches involves designing (either complete through redesign or incremental design modification) of the system on the one hand, and controlling the vibration through external means (passive or active devices) on the other. Note that removal of faults (e.g., misalignments and malfunctions by repair or parts replacement) can also remove vibrations. This approach may fall into any of the three categories listed above.

The category of vibration isolation involves "isolating" a mechanical system ($S$) from vibration excitations ($\mathbf{f}$) so that the excitation signals are "filtered" out or dissipated prior to reaching the system. The use of properly designed suspension systems, mounts, and damping layers falls within this category. The category of design modification will involve making changes to the components and the structure of a mechanical system according to a set of specifications and design guidelines. Balancing of rotating machinery and structural modification through modal analysis and design techniques fall into this category. The category of control will involve either passive devices (which do not use external power), such as dynamic absorbers and dampers, or active control devices (which need external power for operation). In the passive case, the control device implicitly senses the vibration response and dissipates it (as in the case of a damper) or absorbs and stores its energy where it is slowly dissipated (as in the case of a dynamic absorber). In the active case, the vibrations $\mathbf{y}$ are explicitly sensed through *sensors* and *transducers*. The forces that should be acted on the system to counteract and suppress vibrations are determined by a controller, and the corresponding forces or torques are applied to the system through one or more *actuators*.

Note that there may be some overlap in the three general categories of vibration mitigation that were mentioned above. For example, the addition of a mount (category 1) may also be interpreted as a design modification (category 2) or as incorporating a passive damper (category 3). It should also be noted that the general approach, commonly known as *source alteration*, may fall into either category 1 or category 2. In this case, the purpose is to alter or remove the source of vibration. The source could either be external (e.g., road irregularities that result in vehicle vibrations) — a category 1 problem, or internal (imbalance or misalignment in rotating devices that results in periodic forces, moments, and vibrations) — a category 2 problem. It can be more difficult to alter external vibration sources (e.g., resurfacing the roadways) than to modify the internal sources (e.g., balancing of rotating machinery).

Furthermore, the external source of vibration may be quite random and may not be accessible for alteration at all (e.g., aerodynamic forces on an aircraft).

## 7.1.1 Shock and Vibration

Sometimes, response to shock loads is considered separately from response to vibration excitations for the purpose of design and control of mechanical systems. For example, shock isolation and vibration isolation are treated under different headings in some literature. This is actually unnecessary. Even though vibration analysis predominantly involves periodic excitations and responses, transient and random oscillations (vibrations) are also commonly found in practice. The frequency band of the latter two types of signals is much broader than that of a simple periodic signal. A shock signal is transient by definition, and has a very short duration (in comparison to the predominant time constants of the mechanical system to which the shock load is applied). Hence, it will possess a wide band of frequencies. Consequently, frequency-domain techniques are still applicable. Furthermore, time-domain techniques are particularly suited to dealing with transient signals in general and shock signals in particular. In that context, a shock excitation may be treated as an impulse whose effect is to instantaneously change the velocity of an inertia element. Then, in the time domain, a shock load may also be treated as an initial-velocity excitation of an otherwise free (unforced) system.

# 7.2 Specification of Vibration Limits

Design and control procedures of vibration have the primary objective of ensuring that under normal operating conditions, the system of interest does not encounter vibration levels that exceed the specified values. In this context, then, the ways of specifying vibration limits become important. This section will present some common ways of vibration specification.

## 7.2.1 Peak Level Specification

Vibration limits for a mechanical system may be specified in either the time domain or the frequency domain. In the time domain, the simplest specification is the peak level of vibration (typically, acceleration in units of $g$ — the acceleration due to gravity). Here, the techniques of isolation, design, or control should ensure that the peak vibration response of the system do not exceed the specified level. In this case, the entire time interval of operation of the system is monitored, and the peak values are checked against the specifications. Note that, in this case, it is the instantaneous peak value at a particular time instant that is of interest, and what is used in representing vibration is an instantaneous amplitude measure rather than an average amplitude or an energy measure.

## 7.2.2 Root-Mean-Square Value Specification

The root-mean-square (RMS) value of a vibration signal $y(t)$ is given by the square root of the average of the squared signal as

$$y_{\text{RMS}} = \left[ \frac{1}{T} \int_0^T y^2 \, dt \right]^{1/2} \tag{7.1}$$

Note that, by squaring the signal, its sign is eliminated and, essentially, the energy level of the signal is used. The period $T$, over which the squared signal is averaged, will depend on the problem and the nature of the signal. For a periodic signal, one period is adequate for averaging. For transient signals, several time constants (typically four times the larger time constant) of the vibrating system would be sufficient. For random signals, a value that is as large as feasible should be used.

In the method of RMS value specification, the RMS value of the acceleration response (typically, acceleration in gs) is computed using Equation 7.1 and is compared with the specified value. In this method, instantaneous bursts of vibration do not have a significant effect as they are filtered out because of the integration. It is the average energy or power of the response signal that is considered. The duration of exposure enters into the picture indirectly and in an undesirable manner. For instance, a highly transient vibration signal can initially have a damaging effect. However, the larger the $T$ that is used in Equation 7.1, the smaller the computed RMS value. Hence, in this case, the use of a large value for $T$ would lead to diluting or masking the damage potential. In practice, the longer the exposure to a vibration signal, the greater the harm caused. Hence, when using specifications such as peak and RMS values, they have to be adjusted according to the period of exposure. Specifically, a larger specification should be used for longer periods of exposure.

## 7.2.3  Frequency-Domain Specification

It is not realistic to specify the limitation to vibration exposure of a complex dynamic system by just a single threshold value. Usually, the effect of vibration on a system depends on at least the following three parameters:

1. Level of vibration (peak, RMS, power, etc.).
2. Frequency content (range) of excitation.
3. Duration of exposure to vibration.

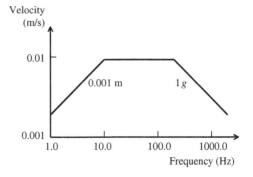

**FIGURE 7.2**  Operating vibration specification (nomograph) for a machine.

This is particularly true because the excitations that generate the vibration environment may not necessarily be a single-frequency (sinusoidal) signal and may be broadband and random. Furthermore, the response of the system to the vibration excitations will depend on its frequency transfer function, which determines its resonances and damping characteristics. Under these circumstances, it is desirable to provide specifications in a *nomograph* where the horizontal axis gives frequency (Hz) and the vertical axis could represent a motion variable such as displacement (m), velocity (m/s), or acceleration (m/s$^2$ or g). It is not important which of these motion variables represents the vertical axis of the nomograph. This is true because in the frequency domain

$$\text{Velocity} = j\omega \times \text{displacement}$$

$$\text{Acceleration} = j\omega \times \text{velocity}$$

and one form of motion may be easily converted into one of the remaining two motion representations. In each of the forms, assuming that the two axes of the nomograph are graduated in a logarithmic scale, the constant displacement, constant velocity, and constant acceleration lines are straight lines.

Consider a simple specification of machinery vibration limits as given by the following values:

$$\text{Displacement limit (peak)} = 0.001 \text{ m}$$

$$\text{Velocity limit} = 0.01 \text{ m/s}$$

$$\text{Acceleration limit} = 1.0g$$

This specification may be represented in a velocity vs. frequency nomograph (log–log) as in Figure 7.2.

Usually, such simple specifications in the frequency domain are not adequate. As noted above, the system behavior will vary depending on the excitation frequency range. For example, motion sickness in humans may be predominant in low frequencies in the range of 0.1 to 0.6 Hz and passenger discomfort in

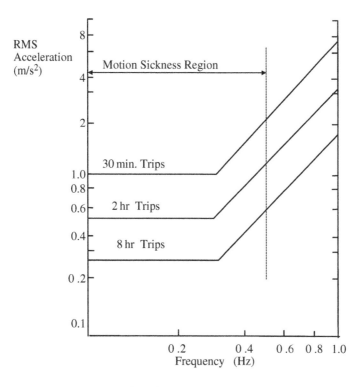

**FIGURE 7.3** A severe-discomfort vibration specification for ground transit vehicles.

ground transit vehicles may be most serious in the frequency range of 4 to 8 Hz for vertical motion and 1 to 2 Hz for lateral motion. In addition, for any dynamic system, particularly at low damping levels, the neighborhoods of resonant frequencies should be avoided and, hence, should be specified by low vibration limits in the resonant regions. Furthermore, the duration of vibration exposure should be explicitly accounted for in specifications. For example, Figure 7.3 presents a ride comfort specification for a ground transit vehicle, where lower vibration levels are specified for longer trips.

Finally, it should be noted that the specifications we are concerned with in the present context of design and control are upper bounds of vibration. The system should perform below (within) these specifications under normal operating conditions. Test specifications are lower bounds. The test should be conducted at or above these vibration levels so that the system would meet the test specifications. Some considerations of vibration engineering are summarized in Box 7.1.

## 7.3 Vibration Isolation

The purpose of vibration isolation is to "isolate" the system of interest from vibration excitations by introducing an *isolator* in between them. Examples of isolators are machine mounts and vehicle suspension systems. Two general types of isolation can be identified:

1. Force isolation (related to force transmissibility)
2. Motion isolation (related to motion transmissibility)

In force isolation, vibration forces that would be ordinarily transmitted directly from a source to a supporting structure (isolated system) are filtered out by an isolator through its flexibility (spring) and dissipation (damping) so that part of the force is routed through an inertial path. Clearly, the concepts of *force transmissibility* are applicable here. In motion isolation, vibration motions that are applied at a moving platform of a mechanical system (isolated system) are absorbed by an isolator through its flexibility and dissipation so that the motion that is transmitted to the system of interest is weakened.

# Box 7.1

# VIBRATION ENGINEERING

**Vibration mitigation approaches:**

- Isolation (buffers system from excitation)
- Design modification (modifies the system)
- Control (senses vibration and applies a counteracting force: passive/active)

**Vibration specification:**

- Peak and RMS values
- Frequency-domain specs on a nomograph
    *Vibration levels
    *Frequency content
    *Exposure duration

**Note:**

$|\text{Velocity}| = \omega \times |\text{Displacement}|$

$|\text{Acceleration}| = \omega \times |\text{Velocity}|$

**Limiting specifications:**
Operation (design) specifications: specify upper bounds
Testing specifications: specify lower bounds

The concepts of motion transmissibility are applicable in this case. The design problem in both cases is to select applicable parameters for the isolator so that the vibrations entering the system are below specified values within a frequency band of interest (the operating frequency range).

Let us revisit the main concepts of force transmissibility and motion transmissibility. Figure 7.4(a) gives a schematic model of force transmissibility through an isolator. Vibration force at the source is $f(t)$. In view of the isolator, the source system (with impedance $Z_m$) is made to move at the same speed as the isolator (with impedance $Z_s$). This is a parallel connection of impedances. Hence, the force $f(t)$ is split so that part of it is taken up by the inertial path (broken line) of $Z_m$. Only the remainder ($f_s$) is transmitted through $Z_s$ to the supporting structure, which is the isolated system. Force transmissibility is

$$T_f = \frac{f_s}{f} = \frac{Z_s}{Z_m + Z_s} \tag{7.2}$$

Figure 7.4(b) gives a schematic model of motion transmissibility through an isolator. Vibration motion $v(t)$ of the source is applied through an isolator (with impedance $Z_s$ and mobility $M_s$) to the isolated system (with impedance $Z_m$ and mobility $M_m$). The resulting force is assumed to transmit directly from the isolator to the isolated system and hence, these two units are connected in series. Consequently, we have the motion transmissibility:

$$T_m = \frac{v_m}{v} = \frac{M_m}{M_m + M_s} = \frac{Z_s}{Z_s + Z_m} \tag{7.3}$$

It can be seen that, according to these two models, we have

$$T_f = T_m \tag{7.4}$$

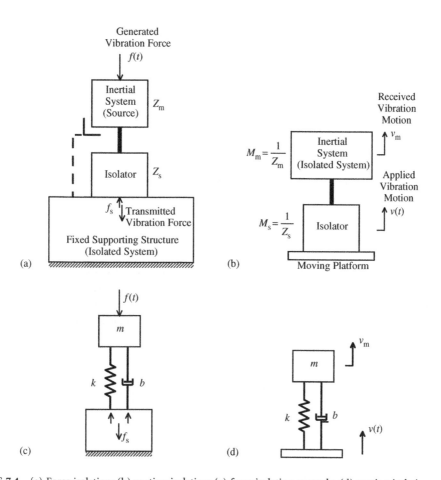

**FIGURE 7.4** (a) Force isolation; (b) motion isolation; (c) force isolation example; (d) motion isolation example.

As a result, the concepts of force transmissibility and motion transmissibility may be studied using just one common transmissibility function $T$.

Simple examples of force isolation and motion isolation are shown in Figure 7.4(c) and (d). For both cases, the transmissibility function is given by

$$T = \frac{k + bj\omega}{(k - m\omega^2 + bj\omega)} \tag{7.5}$$

where $\omega$ is the frequency of vibration excitation. Note that the model (Equation 7.5) is not restricted to sinusoidal vibrations. Any general vibration excitation may be represented by a Fourier spectrum, which is a function of frequency $\omega$. Then, the response vibration spectrum is obtained by multiplying the excitation spectrum by the transmissibility function $T$. The associated design problem is to select the isolator parameters $k$ and $b$ to meet the specifications of isolation.

Equation 7.5 may be expressed as

$$T = \frac{\omega_n^2 + 2\zeta\omega_n\omega j}{(\omega_n^2 - \omega^2 + 2\zeta\omega_n\omega j)} \tag{7.6}$$

where

$\omega_n = \sqrt{k/m}$ = undamped natural frequency of the system

$\zeta = \dfrac{b}{2\sqrt{km}}$ = damping ratio of the system

Equation 7.6 may be written in the nondimensional form:

$$T = \frac{1 + 2\zeta rj}{1 - r^2 + 2\zeta rj} \tag{7.7}$$

where the nondimensional excitation frequency is defined as

$$r = \omega/\omega_n$$

The transmissibility function has a phase angle as well as magnitude. In practical applications, the level of attenuation of the vibration excitation (rather than the phase difference between the vibration excitation and the response) is of primary importance. Accordingly, the transmissibility magnitude

$$|T| = \sqrt{\frac{1 + 4\zeta^2 r^2}{(1 - r^2)^2 + 4\zeta^2 r^2}} \tag{7.8}$$

is of interest. It can be shown that $|T| < 1$ for $r > \sqrt{2}$, which corresponds to the isolation region. Hence, the isolator should be designed such that the operative frequencies $\omega$ are greater than $\sqrt{2}\omega_n$. Furthermore, a threshold value for $|T|$ would be specified, and the parameters $k$ and $b$ of the isolator should be chosen so that $|T|$ is less than the specified threshold in the operating frequency range (which should be given). This procedure may be illustrated using an example.

## Example 7.1

A machine tool and its supporting structure are modeled as the simple mass–spring–damper system shown in Figure 7.5.

1. Draw a mechanical-impedance circuit for this system in terms of the impedances of the three elements: mass ($m$), spring ($k$), and viscous damper ($b$).
2. Determine the exact value of the frequency ratio $r$ in terms of the damping ratio $\zeta$, at which the force transmissibility magnitude will peak. Show that for small $\zeta$, this value is $r = 1$.
3. Plot $|T_f|$ vs. $r$ for the interval $r = [0, 5]$, with one curve for each of the five $\zeta$ values 0.0, 0.3, 0.7, 1.0, and 2.0, on the same plane. Discuss the behavior of these transmissibility curves.
4. From part (3), determine for each of the five $\zeta$ values and the excitation frequency range with respect to $\omega_n$, for which the transmissibility magnitude is:
   - Less than 1.05
   - Less than 0.5

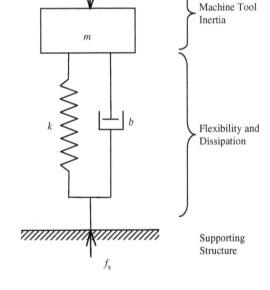

**FIGURE 7.5**  A simplified model of a machine tool and its supporting structure.

5. Suppose that the device in Figure 7.5 has a primary, undamped natural frequency of 6 Hz and a damping ratio of 0.2. It is necessary that the system has a force transmissibility magnitude of less than 0.5 for operating frequency values greater than 12 Hz. Does the existing system meet this requirement? If not, explain how you should modify the system to meet the requirement.

## Solution

1. Here, the elements $m$, $b$, and $f$ are in parallel with a common velocity $v$ across them, as shown in Figure 7.6. In the circuit, $Z_m = mj\omega$, $Z_b = b$, and $Z_k = k/j\omega$.

Force transmissibility

$$T_f = \frac{F_s}{F} = \frac{F_s/V}{F/V} = \frac{Z_s}{Z_s + Z_0}$$

$$= \frac{Z_b + Z_k}{Z_m + Z_b + Z_k} \qquad \text{(i)}$$

Substitute the element impedances. We obtain

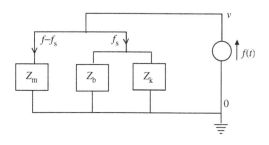

**FIGURE 7.6** The mechanical impedance circuit of the force isolation problem.

$$T_f = \frac{b + \dfrac{k}{j\omega}}{mj\omega + b + \dfrac{k}{j\omega}} = \frac{bj\omega + k}{-\omega^2 m + bj\omega + k} = \frac{j\omega b/m + k/m}{-\omega^2 + j\omega b/m + k/m} \qquad \text{(ii)}$$

The last expression is obtained by dividing the numerator and the denominator by $m$. Now, use the fact that

$$\frac{k}{m} = \omega_n^2 \quad \text{and} \quad \frac{b}{m} = 2\zeta\omega_n$$

and divide Equation ii throughout by $\omega_n^2$. We obtain

$$T_f = \frac{\omega_n^2 + 2\zeta\omega_n j\omega}{\omega_n^2 - \omega^2 + 2\zeta\omega_n j\omega} = \frac{1 + 2\zeta rj}{1 - r^2 + 2\zeta rj} \qquad \text{(iii)}$$

The transmissibility magnitude is

$$|T_f| = \sqrt{\frac{1 + 4\zeta^2 r^2}{(1 - r^2)^2 + 4\zeta^2 r^2}} \qquad (7.9)$$

where $r = \omega/\omega_n$ is the normalized frequency.

2. To determine the peak point of $|T_f|$, differentiate the expression within the square-root sign in Equation iv and equate to zero:

$$\frac{[(1 - r^2)^2 + 2\zeta^2 r^2]8\zeta^2 r - [1 + 4\zeta^2 r^2][2(1 - r^2)(-2r) + 8\zeta^2 r]}{[(1 - r^2)^2 + 4\zeta^2 r^2]^2} = 0$$

Hence,

$$4r\left\{[(1 - r^2)^2 + 2\zeta^2 r^2]2\zeta^2 + [1 + 4\zeta^2 r^2][(1 - r^2) - 2\zeta^2]\right\} = 0$$

which simplifies to

$$r(2\zeta^2 r^4 + r^2 - 1) = 0$$

The roots are

$$r = 0 \quad \text{and} \quad r^2 = \frac{-1 \pm \sqrt{1 + 8\zeta^2}}{4\zeta^2}$$

The root $r = 0$ corresponds to the initial stationary point at zero frequency. That does not represent a peak. Taking only the positive root for $r^2$ and then its positive square root, the peak point of the transmissibility magnitude is given by

$$r = \frac{\left[\sqrt{1 + 8\zeta^2} - 1\right]^{1/2}}{2\zeta} \qquad \text{(iv)}$$

For small $\zeta$, Taylor series expansion gives

$$\sqrt{1 + 8\zeta^2} \approx 1 + \frac{1}{2} \times 8\zeta^2 = 1 + 4\zeta^2$$

With this approximation, Equation iv equates to 1. Hence, for small damping, the transmissibility magnitude will have a peak at $r = 1$, and from Equation 7.9, its value is

$$|T_f| \approx \frac{\sqrt{1 + 4\zeta^2}}{2\zeta} \approx \frac{1 + \frac{1}{2} \times 4\zeta^2}{2\zeta}$$

or

$$|T_f| \approx \frac{1}{2\zeta} + \zeta \approx \frac{1}{2\zeta} \qquad (v)$$

3. The five curves of $|T_f|$ vs. $r$ for $\zeta = 0, 0.3, 0.7, 1.0,$ and 2.0 are shown in Figure 7.7. Note that these curves use the exact expression (see Equation 7.9).
   From the curves, we observe the following:

   1. There is always a nonzero frequency value at which the transmissibility magnitude will peak. This is the resonance.
   2. For small $\zeta$, the peak transmissibility magnitude is obtained at approximately $r = 1$. As $\zeta$ increases, this peak point shifts to the left (i.e., a lower value for peak frequency).
   3. The peak magnitude decreases with increasing $\zeta$.
   4. All the transmissibility curves pass through the magnitude value 1.0 at the same frequency $r = \sqrt{2}$.
   5. The isolation (i.e., $|T_f| < 1$) is given by $r > \sqrt{2}$. In this region, $|T_f|$ increases with $\zeta$.
   6. The transmissibility magnitude decreases for large $r$.

4. From the curves in Figure 7.7, we obtain:

   - For $|T_f| < 1.05$; $r > \sqrt{2}$ for all $\zeta$.
   - For $|T_f| < 0.5$; $r > 1.73, 1.964, 2.871, 3.77, 7.075$ for $\zeta = 0.0, 0.3, 0.7, 1.0,$ and 2.0, respectively.

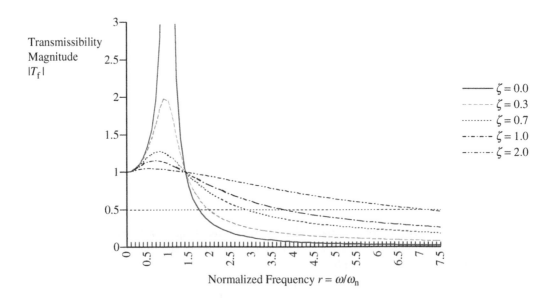

**FIGURE 7.7**  Transmissibility curves for a simple oscillator model.

5. We need

$$\sqrt{\frac{1 + 4\zeta^2 r^2}{(1 - r^2)^2 + 4\zeta^2 r^2}} < \frac{1}{2}$$

or

$$\frac{1 + 4\zeta^2 r^2}{(1 - r^2)^2 + 4\zeta^2 r^2} < \frac{1}{4}$$

or

$$4 + 16\zeta^2 r^2 < (1 - r^2)^2 + 4\zeta^2 r^2$$

or

$$r^4 - 2r^2 - 12\zeta^2 r^2 - 3 > 0$$

For $\zeta = 0.2$ and $r = 12/6 = 2$, the left-hand-side expression computes to

$$2^4 - 2 \times 2^2 - 12 \times (0.2)^2 \times 2^2 - 3 = 3.08 > 0$$

Hence, the requirement is met. In fact, since, for $r = 2$

$$\text{LHS} = 2^4 - 2 \times 2^2 - 12 \times 2^2 \zeta^2 - 3 = 5 - 48\zeta^2$$

it follows that the requirement would be met for

$$5 - 48\zeta^2 > 0$$

or

$$\zeta < \sqrt{\frac{5}{48}} = 0.32$$

If the requirement was not met (e.g., if $\zeta = 0.4$), the option would be to reduce damping.

## 7.3.1 Design Considerations

The *level of isolation* is defined as $1 - T$. It was noted that in the isolation region ($r > \sqrt{2}$) the transmissibility decreases (hence, the level of isolation increases) as the damping ratio $\zeta$ decreases. Thus, the best conditions of isolation are given by $\zeta = 0$. This is not feasible in practice, but we should maintain $\zeta$ as small as possible. For small $\zeta$ in the isolation region, Equation 7.8 may be approximated by

$$T = \frac{1}{(r^2 - 1)} \tag{7.10}$$

Note that $T$ is real, in this case, of $\zeta \cong 0$, and also is positive since $r > \sqrt{2}$. However, in general, $T$ may denote the magnitude of the transmissibility function. Substitute

$$r^2 = \omega^2/\omega_n^2 = \omega^2 m/k$$

We get

$$k = \frac{\omega^2 m T}{(1 + T)} \tag{7.11}$$

This equation may be used to determine the design stiffness of the isolator for a specified level of isolation $(1 - T)$ in the operating frequency range $\omega > \omega_0$ for a system of known mass (including the isolator mass). Often, the static deflection $\delta_s$ of spring is used in design procedures and is given by

$$\delta_s = \frac{mg}{k} \tag{7.12}$$

Substituting Equation 7.12 into Equation 7.11, we obtain

$$\delta_s = (1 + T)\frac{g}{\omega^2 T} \tag{7.13}$$

Since the isolation region is $\omega > \sqrt{2}\omega_n$ it is desirable to make $\omega_n$ as small as possible in order to obtain the widest frequency range of operation. This is achieved by making the isolator as soft as possible ($k$ as low as possible). However, there are limits to this in terms of structural strength, stability, and availability of springs. Then, $m$ may be increased by adding an inertia block as the base of the system, which is then mounted on the isolator spring (with a damping layer) or an air-filled pneumatic mount. The inertia block will also lower the centroid of the system, thereby providing added desirable effects of stability and reducing rocking motions and noise transmission. For improved load distribution, instead of just one spring of design stiffness $k$, a set of $n$ springs, each with stiffness $k/n$ and uniformly distributed under the inertia block, should be used.

Another requirement for good vibration isolation is low damping. Usually, metal springs have very low damping (typically $\zeta$ less than 0.01). On the other hand, higher damping is needed to reduce resonant vibrations that will be encountered during start-up and shutdown conditions when the excitation frequency will vary and pass through the resonances. In addition, vibration energy has to be effectively dissipated, even under steady operating conditions. Isolation pads made of damping material such as cork, natural rubber, and neoprene may be used for this purpose. They can provide damping ratios of the order of 0.01.

The basic design steps for a vibration isolator in force isolation are as follows:

1. The required level of isolation (1 to $T$) and the lowest frequency of operation ($\omega_0$) are specified. The mass of the vibration source ($m$) is known.
2. Use Equation 7.11 with $\omega = \omega_0$ to compute the required stiffness $k$ of the isolator.
3. If the component $k$ is not satisfactory, then increase $m$ by introducing an inertia block and recompute $k$.
4. Distribute $k$ over several springs.
5. Introduce a mounting pad of known stiffness and damping. Modify $k$ and $b$ accordingly and compute $T$ using Equation 7.8. If the specified $T$ is exceeded, then modify the isolator parameters as appropriate and repeat the design cycle.

Box 7.2 gives some relations that are useful in a design for vibration isolation.

## Example 7.2

Consider a motor and fan unit of a building ventilation system weighing 50 kg and operating in the speed range of 600 to 3600 rpm. Since offices are located directly underneath the motor room, a 90% vibration isolation is desired. A set of mounting springs, each having a stiffness of 100 N/cm, is available. Design an isolation system to mount the motor-fan unit on the room floor.

## Solution

For an isolation level of 90%, the required force transmissibility is $T = 0.1$. The lowest frequency of operation is $\omega = (600/60)2\pi$ rad/s. First, we try four mounting points. The overall spring stiffness is $k = 4 \times 100 \times 10^2$ N/m. Substitute in Equation 7.11.

$$4 \times 100 \times 100 = \frac{(10 \times 2\pi)^2 \times m \times 0.1}{1.1}$$

# Box 7.2

# Vibration Isolation

Transmissibility (force/force or motion/motion):

$$|T| = \sqrt{\frac{1 + 4\zeta^2 r^2}{(1 - r^2)^2 + 4\zeta^2 r^2}} \cong \frac{1}{(r^2 - 1)} \text{ for } r > 1 \text{ and small } \zeta$$

Properties:

1. $T_{peak} \cong \dfrac{\sqrt{1 + 4\zeta^2}}{2\zeta} \cong \dfrac{1}{2\zeta}$ for small $\zeta$

2. $T_{peak}$ occurs at $r_{peak} = \dfrac{[\sqrt{1 + 8\zeta^2} - 1]^{1/2}}{2\zeta} \cong 1$ for small $\zeta$

3. All $|T|$ curves coincide at $r = \sqrt{2}$ for all $\zeta$
4. Isolation region: $r > \sqrt{2}$
5. In isolation region:
   $|T|$ decreases with $r$ (i.e., better isolation at higher frequencies)
   $|T|$ increases with $\zeta$ (i.e., better isolation at lower damping)

Design formulas:

Level of isolation $= 1 - T$
Isolator stiffness:

$$k = \frac{\omega^2 mT}{(1 + T)}$$

where

$m$ = system mass
$\omega$ = operating frequency

Static deflection:

$$\delta_s = \frac{mg}{k} = (1 + T)\frac{g}{\omega^2 T}$$

which gives $m = 111.5$ kg. Since the mass of the unit is 50 kg, we should use an inertia block of mass 61.5 kg or more.

## 7.3.2 Vibration Isolation of Flexible Systems

The simple model shown in Figure 7.4(c) and (d) may not be adequate in the design of vibration isolators for sufficiently flexible systems. A more appropriate model for this situation is shown in Figure 7.8. Note that the vibration isolator has an inertia block of mass $m$ in addition to damped flexible mounts of stiffness $k$ and damping constant $b$. The vibrating system itself has a stiffness $K$ and damping constant $B$ in addition to its mass $M$.

In the absence of $K$, $B$, and the inertia block ($m$) as in Figure 7.4(c), the vibrating system becomes a simple inertia ($M$). Then, $y_a$ and $y$ are the same and the equation of motion is

$$M\ddot{y} + b\dot{y} + ky = f(t) \tag{7.14}$$

with the force transmitted to the support structure, $f_s$, given by

$$f_s = b\dot{y} + ky \qquad (7.15)$$

The force transmissibility in this case is

$$T_{\text{inertial}} = \frac{f_s}{f} = \frac{bs + k}{Ms^2 + bs + k} \qquad (7.16)$$

with $s = j\omega$

For the flexible system and isolator shown in Figure 7.8, the equation of motion:

$$M\ddot{y}_a + B(\dot{y}_a - \dot{y}) + K(y_a - y) = f(t) \quad (7.17)$$

$$m\ddot{y} + B(\dot{y} - \dot{y}_a) + K(y - y_a) + b\dot{y} + ky = 0 \qquad (7.18)$$

Hence, in the frequency domain, we have

$$(Ms^2 + Bs + K)y_a - (Bs + K)y = f \qquad (7.19)$$

$$[ms^2 + (B + b)s + K + k]y = (Bs + K)y_a$$
with $s = j\omega$ $\qquad (7.20)$

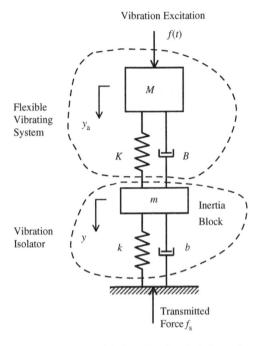

Vibration Excitation

$f(t)$

Flexible Vibrating System

$y_a$

$K$ $B$

$m$  Inertia Block

Vibration Isolator

$y$

$k$ $b$

Transmitted Force $f_s$

**FIGURE 7.8** A model for vibration isolation of a flexible system.

Substitute Equation 7.20 into Equation 7.19 for eliminating $y_a$. We obtain

$$\left\{(Ms^2 + Bs + K)\frac{[ms^2 + (B + b)s + K + k]}{(Bs + K)} - (Bs + K)\right\}y = f$$

which simplifies

$$\left\{\frac{Ms^2[ms^2 + (B + b)s + K + k] + (Bs + K)(ms^2 + bs + k)}{(Bs + K)}\right\}y = f \qquad (7.21)$$

The force transmitted to the supporting structure is still given by Equation 7.15. Hence, the transmissibility with the flexible system is

$$T_{\text{flexible}} = \frac{(Bs + K)(bs + k)}{\left\{Ms^2[ms^2 + (B + b)s + K + k] + (Bs + K)(ms^2 + bs + k)\right\}} \quad \text{with } s = j\omega \qquad (7.22)$$

From Equation 7.16 and Equation 7.22, the transmissibility magnitude ratio is

$$\frac{T_{\text{flexible}}}{T_{\text{inertial}}} = \left| \frac{(Bs + K)(Ms^2 + bs + k)}{Ms^2[ms^2 + (B + b)s + K + k] + (Bs + K)(ms^2 + bs = k)} \right| \quad \text{with } s = j\omega \qquad (7.23)$$

or

$$\frac{T_{\text{flexible}}}{T_{\text{inertial}}} = \left| \frac{(Ms^2 + bs + k)}{Ms^2(ms^2 + bs + k)/(Bs + K) + Ms^2 + ms^2 + bs + k} \right| \quad s = j\omega \qquad (7.24)$$

In the nondimensional form, we have

$$\frac{T_{\text{flexible}}}{T_{\text{inertial}}} = \left| \frac{1 - r^2 + 2j\zeta_b r}{-r^2(1 - r_m r^2 + 2j\zeta_b r)/(r_\omega^2 + 2j\zeta_a r_\omega r) + 1 - (1 + r_m)r^2 + 2j\zeta_b r} \right| \qquad (7.25)$$

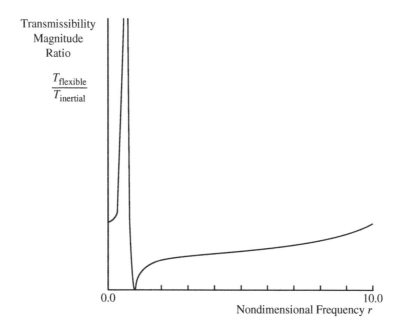

**FIGURE 7.9** The effect of system flexibility on the transmissibility magnitude in the undamped case (mass ratio = 1.0; natural frequency ratio = 10.0).

where

$$r = \frac{\omega}{\sqrt{k/M}}$$

$$r_{\mathrm{m}} = \frac{m}{M}$$

$$r_{\omega} = \frac{\sqrt{K/M}}{\sqrt{k/M}} = \sqrt{\frac{K}{k}}$$

$$\zeta_{\mathrm{a}} = \frac{B}{2\sqrt{KM}}$$

$$\zeta_{\mathrm{b}} = \frac{b}{2\sqrt{kM}}$$

Again, the design problem of vibration isolation is to select the parameters $r_{\mathrm{m}}$, $r_{\omega}$, $\zeta_{\mathrm{a}}$, and $\zeta_{\mathrm{b}}$ so that the required level of vibration isolation is realized for an operating frequency range of $r$.

A plot of Equation 7.25 for the undamped case with $r_{\mathrm{m}} = 1.0$ and $r_{\omega} = 10.0$ is given in Figure 7.9. Generally, the transmissibility ratio will be zero at $r = 1$ (the resonance of the inertial system) and there will be two values of $r$ (the resonances of the flexible system), for which the ratio will become infinity in the undamped case. The latter two neighborhoods should be avoided under steady operating conditions.

## 7.4 Balancing of Rotating Machinery

Many practical devices that move contain rotating components. Examples are wheels of vehicles, shafts, gear transmissions of machinery, belt drives, motors, turbines, compressors, fans, and rollers. An unbalance (imbalance) is created in a rotating part when its center of mass does not coincide with the axis of rotation. The reasons for this *eccentricity* include the following:

1. Inaccurate production procedures (machining, casting, forging, assembly, etc.)
2. Wear and tear
3. Loading conditions (mechanical)
4. Environmental conditions (thermal loads and deformation)
5. Use of inhomogeneous and anisotropic material (which does not have a uniform density distribution)
6. Component failure
7. Addition of new components to a rotating device

For a component of mass $m$, eccentricity $e$, and rotating at angular speed $\omega$, the centrifugal force that is generated is $me\omega^2$. Note the quadratic variation with $\omega$. This rotating force may be resolved into two orthogonal components, which will be sinusoidal with frequency $\omega$. It follows that harmonic forcing excitations are generated due to the unbalance, which can generate undesirable vibrations and associated problems.

Problems caused by unbalance include wear and tear, malfunction and failure of components, poor quality of products, and undesirable noise. The problem becomes increasingly important given the present trend of developing high-speed machinery. It is estimated that the speed of operation of machinery has doubled during the past 50 years. This means that the level of unbalance forces may have quadrupled during the same period, causing more serious vibration problems.

An unbalanced rotating component may be balanced by adding or removing material to or from the component. We need to know both the magnitude and location of the balancing masses to be added to, or removed. The present section will address the problem of component balancing for vibration suppression.

Note that the goal to remove the source of vibration (namely, the mass eccentricity) typically by adding one or more balancing mass elements. Two methods are available:

1. Static (single-plane) balancing
2. Dynamic (two-plane) balancing

The first method concerns balancing of planar objects (e.g., pancake motors, disks) whose longitudinal dimension about the axis of rotation is not significant. The second method concerns balancing of objects that have a significant longitudinal dimension. We will discuss both methods.

## 7.4.1  Static Balancing

Consider a disk rotating at angular velocity $\omega$ about a fixed axis. Suppose that the mass center of the disk has an eccentricity $e$ from the axis of rotation, as shown in Figure 7.10(a). Place a fixed coordinate frame

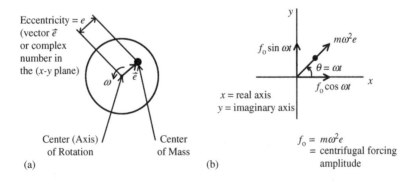

**FIGURE 7.10**   (a) Unbalance in a rotating disk due to mass eccentricity; (b) rotating vector (phasor) of centrifugal force due to unbalance.

$x$–$y$ at the center of rotation. The position $\vec{e}$ of the mass center in this coordinate frame may be represented as:

1. A position vector rotating at angular speed $\omega$,.
2. A complex number, with $x$-coordinate denoting the real part and $y$-coordinate denoting the imaginary part.

The centrifugal force due to the mass eccentricity is also a vector in the direction of $\vec{e}$, but with a magnitude $f_o = m\omega^2 e$, as shown in Figure 7.10(b). It is seen that harmonic excitations result in both $x$ and $y$ directions, given by $f_o \cos \omega t$ and $f_o \sin \omega t$, respectively, where $\theta = \omega t =$ orientation of the rotating vector with respect to the $x$-axis. To balance the disk, we should add a mass $m$ at $-\vec{e}$. However, we do not know the value of $m$ and the location of $\vec{e}$.

### 7.4.1.1 Balancing Approach

1. Measure the amplitude $V_u$ and the phase angle $\phi_1$ (e.g., by the signal from an accelerometer mounted on the bearing of the disk) of the unbalance centrifugal force with respect to some reference.
2. Mount a known mass (trial mass) $M_t$ at a known location on the disk. Suppose that its own centrifugal force is given by the rotating vector $\vec{V}_w$, and the resultant centrifugal force due to both the original unbalance and the final mass is $\vec{V}_r$.
3. Measure the amplitude $V_r$ and the phase angle $\phi_2$ of the resultant centrifugal force as in step 1, with respect to the same phase reference.

A vector diagram showing the centrifugal forces $\vec{V}_u$ and $\vec{V}_w$ due to the original unbalance and the trial-mass unbalance, respectively, is shown in Figure 7.11. The resultant unbalance is $\vec{V}_r = \vec{V}_u + \vec{V}_w$. Note that $-\vec{V}_u$ represents the centrifugal force due to the balancing mass. Therefore, if we determine the angle $\phi_b$ in Figure 7.11, it will give the orientation of the balancing mass. Suppose also that the balancing mass is $M_b$ and it is mounted at an eccentricity equal to that of the trial mass $M_t$. Then,

$$\frac{M_b}{M_t} = \frac{V_u}{V_w}$$

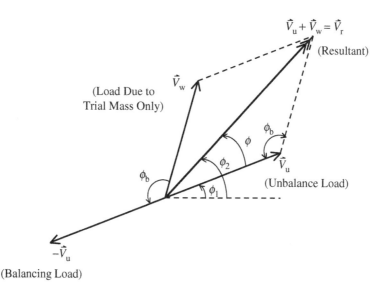

**FIGURE 7.11** A vector diagram of the single-plane (static) balancing problem.

We need to determine the ratio $V_u/V_w$ and the angle $\phi_b$. These values can be derived as follows:

$$\phi = \phi_2 - \phi_1 \tag{7.26}$$

The cosine rule gives

$$V_w^2 = V_u^2 + V_r^2 - 2V_uV_r \cos\phi \tag{7.27}$$

This will provide $V_w$ since $V_u$, $V_r$, and $\phi$ are known. Apply the cosine rule again:

$$V_r^2 = V_u^2 + V_w^2 - 2V_uV_w \cos\phi_b$$

Hence,

$$\phi_b = \cos^{-1}\left[\frac{V_u^2 + V_w^2 - V_r^2}{2V_uV_w}\right] \tag{7.28}$$

*Note*: One may think that since we measure $\phi_1$, we know exactly where $\vec{V}_u$ is. This is not the case because we do not know the reference line from which $\phi_1$ is measured. We only know that this reference is kept fixed (through strobe synchronization of the body rotation) during measurements. Hence, we need to know $\phi_b$, which gives the location of $-\vec{V}_u$ with respect to the known location of $\vec{V}_w$ on the disk.

## 7.4.2 Complex Number/Vector Approach

Again, suppose that the imbalance is equivalent to a mass of $M_b$ that is located at the same eccentricity (radius) $r$ as the trial mass $M_t$. Define complex numbers (mass location vectors in a body frame)

$$\vec{M}_b = M_b \angle \theta_b \tag{7.29}$$

$$\vec{M}_t = M_t \angle \theta_t \tag{7.30}$$

as shown in Figure 7.12.

Associated force vectors are

$$\vec{V}_u = \omega^2 r e^{j\omega t} M_b \angle \theta_b \tag{7.31}$$

$$\vec{V}_w = \omega^2 r e^{j\omega t} M_t \angle \theta_t \tag{7.32}$$

or

$$\vec{V}_u = \vec{A}\vec{M}_b \tag{7.33}$$

$$\vec{V}_w = \vec{A}\vec{M}_t \tag{7.34}$$

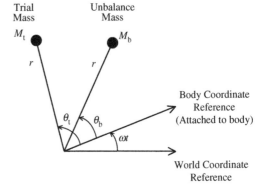

**FIGURE 7.12**   Rotating vectors of mass location.

where $\vec{A} = \omega^2 r e^{j\omega t}$ is the conversion factor (complex) from the mass to the resulting dynamic force (rotating). This factor is the same for both cases since $r$ is the same. We need to determine $\vec{M}_b$.

From Equation 7.33

$$\vec{M}_b = \frac{\vec{V}_u}{\vec{A}} \tag{7.35}$$

Substitute Equation 7.34:

$$\vec{M}_b = \frac{\vec{V}_u}{\vec{V}_w}\cdot\vec{M}_t \tag{7.36}$$

However, since

$$\vec{V}_r = \vec{V}_u + \vec{V}_w \tag{7.37}$$

we have

$$\vec{M}_{\mathrm{b}} = \frac{\vec{V}_{\mathrm{u}}}{(\vec{V}_{\mathrm{r}} - \vec{V}_{\mathrm{u}})} \cdot \vec{M}_{\mathrm{t}} \tag{7.38}$$

Since we know $\vec{M}_{\mathrm{t}}$ and we measure $\vec{V}_{\mathrm{u}}$ and $\vec{V}_{\mathrm{r}}$ to the same scaling factor, we can compute $\vec{M}_{\mathrm{b}}$. Locate the balancing mass at $-\vec{M}_{\mathrm{b}}$ (with respect to the body frame).

## Example 7.3

Consider the following experimental steps:

*Measured*: Accelerometer amplitude (oscilloscope reading) of 6.0 with a phase lead (with respect to strobe signal reference, which is synchronized with the rotating body frame) of 50°.
*Added*: Trial mass $M_{\mathrm{t}} = 20$ g at angle 180° with respect to a body reference radius.
*Measured*: Accelerometer amplitude of 8.0 with a phase lead of 60° (with respect synchronized strobe signal).

Determine the magnitude and location of the balancing mass.

## Solution

*Method 1*:

We have the data

$$\phi = 60 - 50° = 10°$$
$$V_{\mathrm{u}} = 6.0; \quad V_{\mathrm{r}} = 8.0$$

Hence, from Equation 7.27:

$$V_{\mathrm{w}} = \sqrt{6^2 + 8^2 - 2 \times 6 \times 8 \cos 10°} = 2.37$$

Balancing mass:

$$M_{\mathrm{b}} = \frac{6.0}{2.37} \times 20 = 50.63 \text{ g}$$

Equation 7.28 gives

$$\phi_{\mathrm{b}} = \cos^{-1}\left[\frac{6^2 + 2.37^2 - 8^2}{2 \times 6 \times 2.37}\right] = \cos^{-1}(-0.787) = 142° \text{ or } 218°$$

Pick the result $0° \le \phi_{\mathrm{b}} \le 180°$, as clear from the vector diagram shown in Figure 7.11. Hence,

$$\phi_{\mathrm{b}} = 142°$$

However,

$$\vec{M}_{\mathrm{t}} = 20 \angle 180° \text{ g}$$

It follows that

$$-\vec{M}_{\mathrm{b}} = 50.63 \angle (180° + 142°) \text{ g} = 50.63 \angle 322° \text{ g}$$

*Method 2*:

We have

$$\vec{M}_{\mathrm{t}} = 20 \angle 180° \text{ g}$$
$$\vec{V}_{\mathrm{u}} = 6.0 \angle 50°$$
$$\vec{V}_{\mathrm{r}} = 8.0 \angle 60°$$

Then, from Equation 7.38 we obtain

$$\bar{M}_b = \frac{6.0\angle 50°}{(8.0\angle 60° - 6.0\angle 50°)} 20\angle 180° \text{ g}$$

First, we compute

$$8.0\angle 60° - 6.0\angle 50° = (8.0\cos 60° + j8.0\sin 6.0°) - (6.0\cos 50° + j6.0\sin 50°)$$

$$= (8.0\cos 60° - 6.0\cos 50°) + j(8.0\sin 60° - 6.0\sin 50°) = 0.1433 + j2.332$$

$$= 2.336\angle 86.48°$$

Hence,

$$\bar{M}_b = \frac{6.0\angle 50°}{2.336\angle 86.48°} 20\angle 180° = \frac{6.0 \times 20}{2.336} \angle (50° + 180° - 86.48°) = 51\angle 143.5° \text{ g}$$

The balancing mass should be located at

$$-\bar{M}_b = 51\angle 323.5° \text{ g}$$

*Note*: This angle is measured from the same body reference as for the trial mass.

### 7.4.3 Dynamic (Two-Plane) Balancing

Instead of an unbalanced disk, consider an elongated rotating object supported at two bearings, as shown in Figure 7.13. In this case, in general, there may not be an equivalent single unbalance force at a single plane normal to the shaft axis. To show this, recall that a system of forces may be represented by a single force at a specified location and a couple (two parallel forces that are equal and opposite). If this single force (resultant force) is zero, we are left with only a couple. The couple cannot be balanced by a single force.

All the unbalance forces at all the planes along the shaft axis can be represented by an equivalent single unbalance force at a specified plane and a couple. If this equivalent force is zero, then to balance the couple we will need two equal and opposite forces at two different planes.

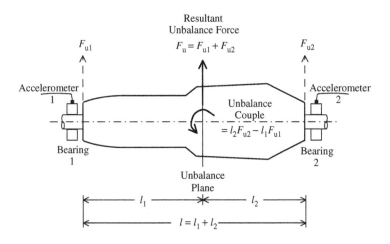

**FIGURE 7.13**  A dynamic (two-plane) balancing problem.

On the other hand, if the couple is zero, then a single force in the opposite direction at the same plane of the resultant unbalance force will result in complete balancing. However, this unbalance plane may not be reachable, even if it is known, for the purpose of adding the balancing mass.

In the present (two-plane) balancing problem, the balancing masses are added at the two bearing planes so that both the resultant unbalance force and couple are balanced, in general. It is clear from Figure 7.13 that even a sole unbalance mass $\bar{M}_b$ at a single unbalance plane may be represented by two unbalance masses $\bar{M}_{b1}$ and $\bar{M}_{b2}$ at the bearing planes 1 and 2. Likewise, in the presence of an unbalance couple, we can simply add two equal and opposite forces at the planes 1 and 2 so that its couple is equal to the unbalance couple. Hence, a general unbalance can be represented by the two unbalance masses $\bar{M}_{b1}$ and $\bar{M}_{b2}$ at planes 1 and 2, as shown in Figure 7.13. As for the single-plane balancing problem, the resultant unbalance forces at the two bearings (which would be measured by the accelerometers at 1 and 2) are

$$\bar{V}_{u1} = \bar{A}_{11}\bar{M}_{b1} + \bar{A}_{12}\bar{M}_{b2} \tag{7.39}$$

$$\bar{V}_{u2} = \bar{A}_{21}\bar{M}_{b1} + \bar{A}_{22}\bar{M}_{b2} \tag{7.40}$$

Suppose that a trial mass of $\bar{M}_{t1}$ (at a known location with respect to the body reference line) was added at plane 1. The resulting unbalance forces at the two bearings are

$$\bar{V}_{r11} = \bar{A}_{11}(\bar{M}_{b1} + \bar{M}_{t1}) + \bar{A}_{12}\bar{M}_{b2} \tag{7.41}$$

$$\bar{V}_{r21} = \bar{A}_{21}(\bar{M}_{b1} + \bar{M}_{t1}) + \bar{A}_{22}\bar{M}_{b2} \tag{7.42}$$

Next, suppose that a trial mass of $\bar{M}_{t2}$ (at a known location with respect to the body reference line) was added at plane 2, after removing $\bar{M}_{t1}$. The resulting unbalance forces at the two bearings are

$$\bar{V}_{r12} = \bar{A}_{11}\bar{M}_{b1} + \bar{A}_{12}(\bar{M}_{b2} + \bar{M}_{t2}) \tag{7.43}$$

$$\bar{V}_{r22} = \bar{A}_{21}\bar{M}_{b1} + \bar{A}_{22}(\bar{M}_{b2} + \bar{M}_{t2}) \tag{7.44}$$

The following subtractions of equations are now made.

Equation 7.41 minus Equation 7.39:

$$\bar{V}_{r11} - \bar{V}_{u1} = \bar{A}_{11}\bar{M}_{t1} \quad \text{or} \quad \bar{A}_{11} = \frac{\bar{V}_{r11} - \bar{V}_{u1}}{\bar{M}_{t1}} \tag{7.45}$$

Equation 7.42 minus Equation 7.40:

$$\bar{V}_{r21} - \bar{V}_{u2} = \bar{A}_{21}\bar{M}_{t1} \quad \text{or} \quad \bar{A}_{21} = \frac{\bar{V}_{r21} - \bar{V}_{u2}}{\bar{M}_{t1}} \tag{7.46}$$

Equation 7.43 minus Equation 7.39:

$$\bar{V}_{r12} - \bar{V}_{u1} = \bar{A}_{12}\bar{M}_{t2} \quad \text{or} \quad \bar{A}_{12} = \frac{\bar{V}_{r12} - \bar{V}_{u1}}{\bar{M}_{t2}} \tag{7.47}$$

Equation 7.44 minus Equation 7.40:

$$\bar{V}_{r22} - \bar{V}_{u2} = \bar{A}_{22}\bar{M}_{t2} \quad \text{or} \quad \bar{A}_{22} = \frac{\bar{V}_{r22} - \bar{V}_{u2}}{\bar{M}_{t2}} \tag{7.48}$$

Hence, generally

$$\bar{A}_{ij} = \frac{\bar{V}_{rij} - \bar{V}_{ui}}{\bar{M}_{tj}} \tag{7.49}$$

These parameters $A_{ij}$ are called *influence coefficients*.

Next, in Equation 7.39 and Equation 7.40 eliminate $\vec{M}_{b2}$ and $\vec{M}_{b1}$ separately to determine the other. Thus,

$$\vec{A}_{22}\vec{V}_{u1} - \vec{A}_{12}\vec{V}_{u2} = (\vec{A}_{22}\vec{A}_{11} - \vec{A}_{12}\vec{A}_{21})\vec{M}_{b1}$$

$$\vec{A}_{21}\vec{V}_{u1} - \vec{A}_{11}\vec{V}_{u2} = (\vec{A}_{21}\vec{A}_{12} - \vec{A}_{11}\vec{A}_{22})\vec{M}_{b2}$$

or

$$\vec{M}_{b1} = \frac{\vec{A}_{22}\vec{V}_{u1} - \vec{A}_{12}\vec{V}_{u2}}{(\vec{A}_{22}\vec{A}_{11} - \vec{A}_{12}\vec{A}_{21})} \tag{7.50}$$

$$\vec{M}_{b2} = \frac{\vec{A}_{21}\vec{V}_{u1} - \vec{A}_{11}\vec{V}_{u2}}{(\vec{A}_{21}\vec{A}_{12} - \vec{A}_{11}\vec{A}_{22})} \tag{7.51}$$

Substitute Equation 7.45 to Equation 7.48 into Equation 7.50 and Equation 7.51 to determine $\vec{M}_{b1}$ and $\vec{M}_{b2}$. Balancing masses that should be added are $-\vec{M}_{b1}$ and $-\vec{M}_{b2}$ in planes 1 and 2, respectively.

The single-plane and two-plane balancing approaches are summarized in Box 7.3.

## Example 7.4

Suppose that the following measurements are obtained.

*Without trial mass:*
Accelerometer at 1: amplitude = 10.0; phase lead = 55°.
Accelerometer at 2: amplitude = 7.0; phase lead = 120°

*With trial mass 20 g at location 270° of plane 1:*
Accelerometer at 1: amplitude = 7.0; phase lead = 120°
Accelerometer at 2: amplitude = 5.0; phase lead = 225°

*With trial mass 25 g at location 180° of plane 2:*
Accelerometer at 1: amplitude = 6.0; phase lead = 120°
Accelerometer at 2: amplitude = 12.0; phase lead = 170°

Determine the magnitude and orientation of the necessary balancing masses in planes 1 and 2 in order to completely balance (dynamic) the system.

## Solution

In the phasor notation, we can represent the given data as follows:

$$\vec{V}_{u1} = 10.0\angle55°; \quad \vec{V}_{u2} = 7.0\angle120°$$

$$\vec{V}_{r11} = 7.0\angle120°; \quad \vec{V}_{r21} = 5.0\angle225°$$

$$\vec{V}_{r12} = 6.0\angle120°; \quad \vec{V}_{r22} = 12.0\angle170°$$

$$\vec{M}_{t1} = 20\angle270° \text{ g}; \quad \vec{M}_{t2} = 25\angle180° \text{ g}$$

From Equation 7.45 to Equation 7.48, we have

$$\vec{A}_{11} = \frac{7.0\angle120° - 10.0\angle55°}{20\angle270°}; \quad \vec{A}_{21} = \frac{5.0\angle225° - 7.0\angle120°}{20\angle270°}$$

$$\vec{A}_{12} = \frac{6.0\angle120° - 10.0\angle55°}{25\angle180°}; \quad \vec{A}_{22} = \frac{12.0\angle170° - 7.0\angle120°}{25\angle180°}$$

These phasors are computed as below:

$$\vec{A}_{11} = \frac{(7.0\cos120° - 10\cos55°) + j(7\sin120° - 10\sin55°)}{20\angle270°} = \frac{-9.235 - j2.129}{20\angle270°} = \frac{9.477\angle193°}{20\angle270°}$$

$$= 0.474\angle-77°$$

---

# Box 7.3
# BALANCING OF ROTATING COMPONENTS

---

**Static or single-plane balancing** (balances a single equivalent dynamic force)

*Experimental approach:*

1. With respect to a body reference line of accelerometer signal at bearing, measure magnitude $(V)$ and phase $(\phi)$:
   - (a) Without trial mass: $(V_u, \phi_1)$ or $\vec{V}_u = V_u \angle \phi_1$
   - (b) With trial mass $M_t$: $(V_r, \phi_2)$ or $\vec{V}_r = V_r \angle \phi_2$.
2. Compute balancing mass $M_b$ and its location with respect to $M_t$.
3. Remove $M_t$ and add $M_b$ at determined location.

*Computation approach 1:*

$$V_w = [V_u^2 + V_r^2 - 2V_u V_w \cos(\phi_2 - \phi_1)]^{1/2}$$

$$\phi_b = \cos^{-1}\left[\frac{V_u^2 + V_w^2 - V_r^2}{2V_u V_w}\right] \text{ and } M_b = \frac{V_u}{V_w}M_t$$

Locate $M_b$ at $\phi_b$ from $M_t$.

*Computation approach 2:*

Unbalance mass phasor

$$\vec{M}_b = \frac{\vec{V}_u}{(\vec{V}_r - \vec{V}_u)}\vec{M}_t$$

where $\vec{M}_t = M_t \angle \theta_t$ (trial mass phasor).

Locate balancing mass at $-\vec{M}_b$.

**Dynamic or two-plane balancing** (balances an equivalent dynamic force and a couple)

*Experimental approach:*

1. Measure $\vec{V}_{ui}$ at bearings $i = 1, 2$, with a trial mass.
2. Measure $\vec{V}_{rij}$ at bearings $i = 1, 2$, with only one trial mass $\vec{M}_{tj}$ at $j = 1, 2$.
3. Compute unbalance mass phasor $\vec{M}_{bi}$ in planes $i = 1, 2$.
4. Remove trial mass and place balancing masses $-\vec{M}_{bi}$ in planes $i = 1, 2$.

*Computations:*

Influence coefficients: $\vec{A}_{ij} = (\vec{V}_{rij} - \vec{V}_{ui})/\vec{M}_{tj}$

Unbalance mass phasors:

$$\vec{M}_{b1} = \frac{\vec{A}_{22}\vec{V}_{u1} - \vec{A}_{12}\vec{V}_{u2}}{(\vec{A}_{22}\vec{A}_{11} - \vec{A}_{12}\vec{A}_{21})} \quad \text{and} \quad \vec{M}_{b2} = \frac{\vec{A}_{21}\vec{V}_{u1} - \vec{A}_{11}\vec{V}_{u2}}{(\vec{A}_{21}\vec{A}_{12} - \vec{A}_{11}\vec{A}_{22})}$$

---

$$\vec{A}_{21} = \frac{(5\cos 225° - 7\cos 120°) + j(5\sin 225° - 7\sin 120°)}{20\angle 270°} = \frac{-7.036 - j9.6}{20\angle 270°} = \frac{11.9\angle 234°}{20\angle 270°}$$

$$= 0.595\angle -36°$$

$$\vec{A}_{12} = \frac{(6\cos 120° - 10\cos 55°) + j(6\sin 120° - 10\sin 55°)}{25\angle 180°} = \frac{-8.736 - j3.0}{25\angle 180°} = \frac{9.237\angle 199°}{25\angle 180°}$$

$$= 0.369\angle 19°$$

$$\bar{A}_{22} = \frac{(12 \cos 170° - 7 \cos 120°) + j(12 \sin 170° - 7 \sin 120°)}{25\angle 180°} = \frac{-8.318 - j4.0}{25\angle 180°} = \frac{9.23\angle 205.7°}{25\angle 180°}$$

$$= 0.369\angle 25.7°$$

Next, the denominators of the balancing mass phasors (in Equation 7.50 and Equation 7.51) are computed as

$$\bar{A}_{22}\bar{A}_{11} - \bar{A}_{12}\bar{A}_{21} = (0.369\angle 25.7° \times 0.474\angle -77°) - (0.369\angle 19° \times 0.595\angle -36°)$$

$$= 0.1749\angle -51.3° - 0.2196\angle -17°$$

$$= (0.1749 \cos 51.3° - 0.2196 \cos 17°) - j(0.1749 \sin 51.3° - 0.2196 \sin 17°)$$

$$= -0.1 - j0.0723 = 0.1234\angle 216°$$

and, hence

$$-(\bar{A}_{22}\bar{A}_{11} - \bar{A}_{12}\bar{A}_{21}) = 0.1234\angle 36°$$

Finally, the balancing mass phasors are computed using Equation 7.50 and Equation 7.51 as

$$\bar{M}_{b1} = \frac{0.369\angle 25.7° \times 10\angle 55° - 0.369\angle 19° \times 7.0\angle 120°}{0.1234\angle 216°} = \frac{3.69\angle 80.7° - 2.583\angle 139°}{0.1234\angle 216°}$$

$$= \frac{(3.69 \cos 80.7° - 2.583 \cos 139°) + j(3.69 \sin 80.7° - 2.5838 \sin 139°)}{0.1234\angle 216°} = \frac{2.546 + j1.947}{0.1234\angle 216°}$$

$$= \frac{3.205\angle 37.4°}{0.1234\angle 216°} = 26\angle -178.6°$$

$$\bar{M}_{b2} = \frac{0.595\angle -36° \times 10\angle 55° - 0.474\angle -77° \times 7.0\angle 120°}{0.1234\angle 36°} = \frac{5.95\angle 19° - 3.318\angle 43°}{0.1234\angle 36°}$$

$$= \frac{(5.95 \cos 19° - 3.318 \cos 43°) - j(5.95 \sin 19° - 3.318 \sin 43°)}{0.1234\angle 36°} = \frac{3.2 + j0.326}{0.1234\angle 36°}$$

$$= \frac{1.043\angle 5.8°}{0.1234\angle 36°} = 8.45\angle -30.0°$$

Finally, we have

$$-\bar{M}_{b1} = 26\angle 1.4° \text{ g}; \qquad -\bar{M}_{b2} = 8.45\angle 150° \text{ g}$$

### 7.4.4　Experimental Procedure of Balancing

The experimental procedure for determining the balancing masses and locations for a rotating system should be clear from the analytical developments and examples given above. The basic steps are:

1. Determine the magnitude and the phase angle of accelerometer signals at the bearings with and without trial masses at the bearing planes.
2. Using this data, compute the necessary balancing masses (magnitude and location) at the bearing planes.
3. Place the balancing masses.
4. Check whether the system is balanced. If not, repeat the balancing cycle.

A laboratory experimental setup for two-plane balancing is shown schematically in Figure 7.14. A view of the system is shown in Figure 7.15. The two disks rigidly mounted on the shaft are driven by a DC motor. The drive speed of the motor is adjusted by the manual speed controller. The (two) shaft

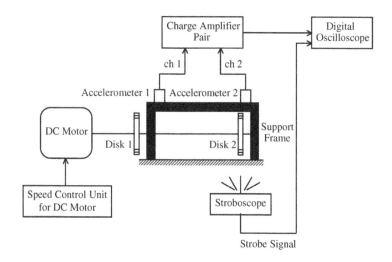

**FIGURE 7.14** Schematic arrangement of a rotor balancing experiment.

bearings are located very close to the disks, as shown in Figure 7.14. Two accelerometers are mounted on the top of the bearing housing so that the resulting vertical accelerations can be measured. The accelerometer signals are conditioned using the two-channel charge amplifier and read and displayed through two channels of the digital oscilloscope. The output of the stroboscope (tachometer) is used as the reference signal with respect to the phase angles of the accelerometer signals that are measured.

In Figure 7.15, the items of equipment are seen from left to right. The first item is the two-channel digital oscilloscope. The manual speed controller with control knob for the DC motor follows. Next is the pair of charge amplifiers for the accelerometers. The strobe-light unit (strobe-tacho) is placed on top of the common housing of the charge amplifier pair. The two-disk rotor system with the drive motor is shown as the last item to the right. Also, note the two accelerometers (seen as small vertical projections) mounted on the bearing frame of the shaft directly above the two bearings.

Because this reference always has to be fixed prior to reading the oscilloscope data, the strobe-tacho is synchronized with the disk rotation. This is achieved as follows (note that all the readings are taken with

**FIGURE 7.15** A view of the experimental setup for two-plane balancing at the University of British Columbia.

the same rotating speed, which is adjusted by the manual speed controller): First, make a physical mark (e.g., a black spot in a white background) on one of the disks. Aim the strobe flash at this disk. As the motor speed is adjusted to the required fixed value, the strobe flash is synchronized such that the mark on the disk "appears" stationary at the same location (e.g., at the uppermost location of the circle of rotation). This ensures not only that the strobe frequency is equal to the rotating speed of the disk, but also that the same phase angle reference is used for all readings of accelerometer signals.

The two disks have slots at locations whose radius is known, and whose angular positions in relation to a body reference line (a radius representing the 0° reference line) are clearly marked. Known masses (typically, bolts and nuts of known mass) can be securely mounted in these slots. Readings obtained through the oscilloscope are:

1. Amplitude of each accelerometer signal
2. Phase lead of the accelerometer signal with respect to the synchronized and reference-fixed strobe signal (note: a phase lag should be represented by a negative sign in the data)

The measurements taken and the computations made in the experimental procedure should be clear from Example 7.4.

## 7.5   Balancing of Reciprocating Machines

A reciprocating mechanism has a slider that moves rectilinearly back and forth along some guideway. A piston-cylinder device is a good example. Often, reciprocating machines contain rotatory components in addition to the reciprocating mechanisms. The purpose is to either covert a reciprocating motion to a rotary motions (as in the case of an automobile engine), or to convert a rotary motion to a reciprocating motions (as in the opto-slider mechanism of a photocopier). Irrespective of the reciprocating machine employed, it is important to remove the vibratory excitations that arise in order to realize the standard design goals of smooth operation, accuracy, low noise, reliability, mechanical integrity, and extended service life. Naturally, in view of their rotational asymmetry, reciprocating mechanisms with rotary components are more prone to unbalance than purely rotary components. Removing the "source of vibration" by proper balancing of the machine would be especially applicable in this situation.

### 7.5.1   Single-Cylinder Engine

A practical example of a reciprocating machine with integral rotary motion is the internal combustion (IC) engine of an automobile. A single-cylinder engine is sketched in Figure 7.16. Observe the nomenclature of the components. The reciprocating motion of the piston is transmitted through the connecting rod and crank into a rotary motion of the crankshaft. The crank, as sketched in Figure 7.16, has a counterbalance mass, the purpose of which is to balance the rotary force (centrifugal).

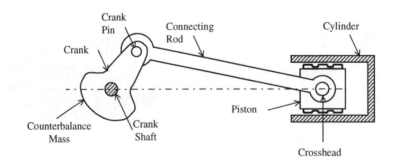

**FIGURE 7.16**   A single-cylinder reciprocating engine.

We will ignore this in our analysis because the goal is to determine the unbalance forces and ways to balance them.

Clearly, both the connecting rod and the crank have distributed mass and moment of inertia. To simplify the analysis, we approximate as follows:

1. Represent the crank mass by an equivalent lumped mass at the crank pin (equivalence may be based on either centrifugal force or kinetic energy).
2. Represent the mass of the connecting rod by two lumped masses, one at the crank pin and the other at the cross head (piston pin).

The piston itself has a significant mass, which is also lumped at the crosshead. Hence, the equivalent system has a crank and a connecting rod, both of which are considered massless, with a lumped mass $m_c$ at the crank pin and another lumped mass $m_p$ at the piston pin (crosshead).

Furthermore, under normal operation, the crankshaft rotates at a constant angular speed ($\omega$). Note that this steady speed is realized not by natural dynamics of the system, but rather by proper speed control (a topic which is beyond the scope of the present discussion).

It is a simple matter to balance the lumped mass $m_c$ at the crank pin. Simply place a countermass $m_c$ at the same radius in the radially opposite location (or a mass in inverse proportion to the radial distance form the crankshaft, but remaining in the radially opposite direction). This explains the presence of the countermass in the crank shown in Figure 7.16. Once complete balancing of the rotating inertia ($m_c$) is thus achieved, we still need to completely eliminate the effect of the vibration source on the crankshaft. To achieve this, we must compensate for the forces and moments on the crankshaft that result from:

1. The reciprocating motion of the lumped mass $m_p$
2. Time-varying combustion (gas) pressure in the cylinder

Both types of forces act on the piston in the direction of its reciprocating (rectilinear) motion. Hence, their influence on the crankshaft can be analyzed in the same way, except that the combustion pressure is much more difficult to determine.

The above discussion justifies the use of the simplified model shown in Figure 7.17 for analyzing the balancing of a reciprocating machine. The characteristics of this model are as follows:

1. A light crank OC of radius $r$ rotates at constant angular speed $\omega$ about O, which is the origin of the $x-y$ coordinate frame.
2. A light connecting rod CP of length $l$ is connected to the connecting rod at C and to the piston at P with frictionless pins. Since the rod is light and the joints are frictionless, the force $f_c$ supported by it will act along its length. (Assume that the force $f_c$ in the connecting rod is compressive, for the purpose of the sign convention). Connecting rod makes an angle $\phi$ with OP (the negative $x$ axis).
3. A lumped mass $m_p$ is present at the piston. A force $f$ acts at P in the negative $x$ direction. This may be interpreted as either the force due to the gas pressure in the cylinder or the inertia force $m_p a$ where $a$ is the acceleration $m_p$ in the positive $x$ direction. These two cases of forcing are considered separately.
4. A lateral force $f_l$ acts on the piston by the cylinder wall, in the positive $y$ direction.

Again, note that the lumped mass $m_c$ at C is not included in the model of Figure 7.17 because it is assumed to be completely balanced by a countermass in the crank. Furthermore, the lumped mass $m_p$ includes both the mass of the piston and also part of the inertia of the connecting rod.

There are no external forces at C. Furthermore, the only external forces at P are $f$ and $f_l$, where $f$ is interpreted as either the inertia force in $m_p$ or the gas force on the piston. Hence, there should be

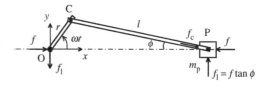

**FIGURE 7.17** The model used to analyze balancing of a reciprocating engine.

equal and opposite forces at the crankshaft O, as shown in Figure 7.17, to support the forces acting at P. Now, let us determine $f_l$.

Equilibrium at P gives

$$f = f_c \cos \phi$$

$$f_l = f_c \sin \phi$$

Hence,

$$f_l = f \tan \phi \tag{7.52}$$

This lateral force $f_l$ acting at both O and P, albeit in the opposite directions, forms a couple $\tau = xf_l$ or, in view of Equation 7.52:

$$\tau = xf \tan \phi \tag{7.53}$$

This couple acts as a torque on the crankshaft. It follows that, once the rotating inertia $m_c$ at the crank is completely balanced by a countermass, the load at the crankshaft is due only to the piston load $f$ and it consists of:

1. A force $f$ in the direction of the piston motion $(x)$
2. A torque $\tau = xf \tan \phi$ in the direction of rotation of the crankshaft $(z)$

As discussed below, the means of removing $f$ at the crankshaft will also remove $\tau$ to some extent. Hence, we will discuss only the approach of balancing $f$.

## 7.5.2 Balancing the Inertia Load of the Piston

First, consider the inertia force $f$ due to $m_p$. Here,

$$f = m_p a \tag{7.54}$$

where $a$ is the acceleration $\ddot{x}$, with the coordinate $x$ locating the position P of the piston (in other words OP $= x$). We notice from Figure 7.17 that

$$x = r \cos \omega t + l \cos \phi \tag{7.55}$$

However,

$$r \sin \omega t = l \sin \phi \tag{7.56}$$

Hence,

$$\cos \phi = \left[1 - \left(\frac{r}{l}\right)^2 \sin^2 \omega t\right]^{1/2} \tag{7.57}$$

which can be expanded up to the first term of Taylor series as

$$\cos \phi \cong 1 - \frac{1}{2}\left(\frac{r}{l}\right)^2 \sin^2 \omega t \tag{7.58}$$

This approximation is valid because $l$ is usually several times larger than $r$ and, hence, $(r/l)^2$ is much small than unity. Next, in view of

$$\sin^2 \omega t = \frac{1}{2}[1 - \cos 2\omega t] \tag{7.59}$$

we have

$$\cos \phi \cong 1 - \frac{1}{4}\left(\frac{r}{l}\right)^2 [1 - \cos 2\omega t] \tag{7.60}$$

Substitute Equation 7.60 into Equation 7.55. We get, approximately

$$x = r \cos \omega t + \frac{l}{4}\left(\frac{r}{l}\right)^2 \cos 2\omega t + l - \frac{l}{4}\left(\frac{r}{l}\right)^2 \tag{7.61}$$

Differentiate Equation 7.61 twice with respect to $t$ to get the acceleration

$$a = \ddot{x} = -r\omega^2 \cos \omega t - l\left(\frac{r}{l}\right)^2 \omega^2 \cos 2\omega t \tag{7.62}$$

Hence, from Equation 7.54, the inertia force at the piston (and its reaction at the crankshaft) is

$$f = -m_p r\omega^2 \cos \omega t - m_p l\left(\frac{r}{l}\right)^2 \omega^2 \cos 2\omega t \tag{7.63}$$

It follows that the inertia load of the reciprocating piston exerts a vibratory force on the crankshaft which has a *primary component* of frequency $\omega$ and a smaller *secondary component* of frequency $2\omega$, where $\omega$ is the angular speed of the crank. The primary component has the same form as that created by a rotating lumped mass at the crank pin. However, unlike the case of a rotating mass, this vibrating force acts only in the $x$ direction (there is no sin $\omega t$ component in the $y$ direction) and, hence, cannot be balanced by a rotating countermass. Similarly, the secondary component cannot be balanced by a countermass rotating at double the speed. To eliminate $f$, we use multiple cylinders whose connecting rods and cranks are connected to the crankshaft with their rotations properly phased (delayed), thus canceling out the effects of $f$.

## 7.5.3 Multicylinder Engines

A single-cylinder engine generates a primary component and a secondary component of vibration load at the crankshaft, and they act in the direction of piston motion ($x$). Because there is no complementary orthogonal component ($y$), it is inherently unbalanced and cannot be balanced using a rotating mass. It can be balanced, however, by using several piston-cylinder units with their cranks properly phased along the crankshaft. This method of balancing multicylinder reciprocating engines is addressed now.

Consider a single cylinder whose piston inertia generates a force $f$ at the crankshaft in the $x$ direction given by

$$f = f_p \cos \omega t + f_s \cos 2\omega t \tag{7.64}$$

Note that the primary and secondary forcing amplitudes $f_p$ and $f_s$, respectively, are given by Equation 7.63. Suppose that there is a series of cylinders in parallel, arranged along the crankshaft, and the crank of cylinder $i$ makes an angle $\alpha_i$ with the crank of cylinder 1 in the direction of rotation, as schematically shown in Figure 7.18(a). Hence, force $f_i$ on the crankshaft (in the $x$ direction, shown as vertical in Figure 7.18) due to cylinder $i$ is

$$f_i = f_p \cos(\omega t + \alpha_i) + f_s \cos(2\omega t + 2\alpha_i) \quad \text{for } i = 1, 2, \ldots, \text{ with } \alpha_1 = 0 \tag{7.65}$$

Not only do the cranks need to be properly phased, but the cylinders should also be properly spaced along the crankshaft to obtain the necessary balance. Consider two examples.

### 7.5.3.1 Two-Cylinder Engine

Consider the two-cylinder case, as shown schematically in Figure 7.18(b) where the two cranks are in radially opposite orientations (i.e., 180° out of phase). In this case, $\alpha_2 = \pi$. Hence,

$$f_1 = f_p \cos \omega t + f_s \cos 2\omega t \tag{7.66}$$

$$f_2 = f_p \cos(\omega t + \pi) + f_s \cos(2\omega t + 2\pi) = -f_p \cos \omega t + f_s \cos 2\omega t \tag{7.67}$$

It follows that the primary force components cancel out. However, they form a couple $z_0 f_p \cos \omega t$ where $z_0$ is the spacing of the cylinders. This causes a bending moment on the crankshaft, and it will not vanish

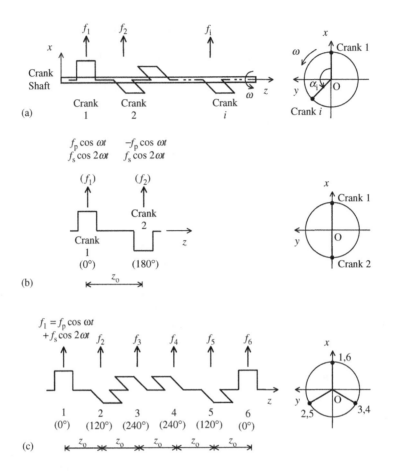

**FIGURE 7.18** (a) Crank arrangement of a multicylinder engine; (b) two-cylinder engine; (c) six-cylinder engine (balanced).

unless the two cylinders are located at the same point along the crankshaft. Furthermore, the secondary components are equal and additive to $2f_s \cos 2\omega t$. This resultant component acts at the midpoint of the crankshaft segment between the two cylinders. There is no couple due to the secondary components.

### 7.5.3.2 Six-Cylinder Engine

Consider the six-cylinder arrangement shown schematically in Figure 7.18(c). Here, the cranks are arranged such that $\alpha_2 = \alpha_5 = 2\pi/3$, $\alpha_3 = \alpha_4 = 4\pi/3$, and $\alpha_1 = \alpha_6 = 0$. Furthermore, the cylinders are equally spaced, with spacing $z_0$. In this case, we have

$$f_1 = f_6 = f_p \cos \omega t + f_s \cos 2\omega t \tag{i}$$

$$f_2 = f_5 = f_p \cos(\omega t + 2\pi/3) + f_s \cos(2\omega t + 4\pi/3) \tag{ii}$$

$$f_3 = f_4 = f_p \cos(\omega t + 4\pi/3) + f_s \cos(2\omega t + 8\pi/3) \tag{iii}$$

Now, we use the fact that

$$\cos \theta + \cos\left(\theta + \frac{2\pi}{3}\right) + \cos\left(\theta + \frac{4\pi}{3}\right) = 0 \tag{iv}$$

which may be proved either by straightforward trigonometric expansion or by using geometric interpretation (i.e., three sides of an equilateral triangle, the sum of whose components in any direction vanishes). The relation iv holds for any $\theta$, including $\theta = \omega t$ and $\theta = 2\omega t$. Furthermore,

$$\cos(2\omega t + 8\pi/3) = \cos\left(2\omega t + \frac{2\pi}{3}\right)$$

Thus, from Equation i to Equation iii, we can conclude that

$$f_1 + f_2 + f_3 + f_4 + f_5 + f_6 = 0 \tag{7.68}$$

This means that the lateral forces on the crankshaft that are exerted by the six cylinders will completely balance. Furthermore, by taking moments about the location of crank 1 of the crankshaft, we have

$$(z_0 + 4z_0)\left[f_p \cos\left(\omega t + \frac{2\pi}{3}\right) + f_s \cos\left(2\omega t + \frac{4\pi}{3}\right)\right]$$
$$+ (2z_0 + 3z_0)\left[f_p \cos\left(\omega t + \frac{4\pi}{3}\right) + f_s \cos\left(2\omega t + \frac{8\pi}{3}\right)\right] \tag{v}$$
$$+ 5z_0[f_p \cos \omega t + f_s \cos 2\omega t]$$

which also vanishes in view of relation iv. Hence, the set of six forces is in complete equilibrium and, as a result, there will be neither a reaction force nor a bending moment on the bearings of the crankshaft from these forces.

In addition, it can be shown that the torques $x_i f_i \tan \phi_i$ on the crankshaft due to this set of inertial forces $f_i$ will add to zero, where $x_i$ is the distance from the crankshaft to the piston of the $i$th cylinder and $\phi_i$ is the angle $\phi$ of the connecting rod of the $i$th cylinder. Hence, this six-cylinder configuration is in complete balance with respect to the inertial load.

## Example 7.5

An eight-cylinder in-line engine (with identical cylinders that are placed in parallel along a line) has its cranks arranged according to the phasing angles 0, 180, 90, 270, 270, 90, 180, and 0° on the crankshaft. The cranks (cylinders) are equally spaced, with spacing $z_0$. Show that this engine is balanced with respect to primary and secondary components of reaction forces and bending moments of inertial loading on the bearings of the crankshaft.

## Solution

The sum of the reaction forces on the crankshaft are

$$2\left[f_p \cos \omega t + f_s \cos 2\omega t + f_p \cos(\omega t + \pi) + f_s \cos(2\omega t + 2\pi) + f_p \cos\left(\omega t + \frac{\pi}{2}\right) + f_s(2\omega t + \pi)\right.$$
$$\left. + f_p \cos\left(\omega t + \frac{3\pi}{2}\right) + f_s \cos(2\omega t + 3\pi)\right] = 2\left[f_p \cos \omega t - f_p \cos \omega t - f_p \sin \omega t + f_p \sin \omega t\right.$$
$$\left. + f_s \cos 2\omega t + f_s \cos 2\omega t - f_s \cos 2\omega t - f_s \cos 2\omega t\right] = 0$$

Hence, both primary forces and secondary forces are balanced. The moment of the reaction forces about the crank 1 location of the crankshaft is

$$(z_0 + 6z_0)[f_p \cos(\omega t + \pi) + f_s \cos(2\omega t + 2\pi)] + (2z_0 + 5z_0)\left[f_p \cos\left(\omega t + \frac{\pi}{2}\right) + f_s \cos(2\omega t + \pi)\right]$$
$$+ (3z_0 + 4z_0)\left[f_p \cos\left(\omega t + \frac{3\pi}{2}\right) + f_s \cos(2\omega t + 3\pi)\right] + 7z_0[f_p \cos \omega t + f_s \cos 2\omega t]$$
$$= 7z_0[-f_p \cos \omega t + f_s \cos 2\omega t - f_p \sin \omega t - f_s \cos 2\omega t + f_p \sin \omega t - f_s \cos 2\omega t + f_p \cos \omega t + f_s \cos 2\omega t] = 0$$

Hence, both primary bending moments and secondary bending moments are balanced. Therefore, the engine is completely balanced.

The formulas applicable for balancing reciprocating machines are summarized in Box 7.4.

# Box 7.4

## Balancing of Reciprocating Machines

**Single cylinder engine:**
Inertia force at piston (and its reaction on crankshaft)

$$f = -m_p r \omega^2 \cos \omega t - m_p l \left( \frac{r}{l} \right)^2 \omega^2 \cos 2\omega t = f_p \cos \omega t + f_s \cos 2\omega t$$

where

$\omega$ = rotating speed of crank
$m_p$ = equivalent lumped mass at piston
$r$ = crank radius
$l$ = length of connecting rod
$f_p$ = amplitude of the primary unbalance force (frequency $\omega$)
$f_s$ = amplitude of the secondary unbalance force (frequency $2\omega$)

**Multicylinder engine:**
Net unbalance reaction force on crankshaft = $\sum_{i=1}^{n} f_i$
Net unbalance moment on crankshaft = $\sum_{i=1}^{n} z_i f_i$
where

$f_i = f_p \cos(\omega t + \alpha_i) + f_s \cos(2\omega t + 2\alpha_i)$
$\alpha_i$ = angular position of the crank of $i$th cylinder, with respect to a body (rotating) reference
    (i.e., crank phasing angle)
$z_i$ = position of the $i$th crank along the crankshaft, measured from a reference point on the
    shaft
$n$ = number of cylinders (assumed identical)

*Note*: For a completely balanced engine, both the net unbalance force and the net unbalance
moment should vanish.

Finally, it should be noted that, in the configuration considered above, the cylinders are placed in parallel along the crankshaft. These are termed in-line engines. Their resulting forces $f_i$ act in parallel along the shaft. In other configurations such as V6 and V8, the cylinders are placed symmetrically around the shaft. In this case, the cylinders (and their inertial forces, which act on the crankshaft) are not parallel. Here, a complete force balance may be achieved without having to phase the cranks. Furthermore, the bending moments of the forces can be reduced by placing the cylinders at nearly the same location along the crankshaft. Complete balancing of the combustion/pressure forces is also possible by such an arrangement.

## 7.5.4 Combustion/Pressure Load

In the balancing approach presented above, the force $f$ on the piston represents the inertia force due to the equivalent reciprocating mass. Its effect on the crankshaft is an equal reaction force $f$ in the lateral direction ($x$) and a torque $\tau = xf \tan \phi$ about the shaft axis ($z$). The balancing approach is to use a series of cylinders so that their reaction forces $f_i$ on the crankshaft from an equilibrium set so that no net reaction or bending moment is transmitted to the bearings of the shaft. The torques $\tau_i$ also can be balanced by the same approach, which is the case, for example, in the six-cylinder engine.

Another important force that acts along the direction of piston reciprocation is the drive force due to gas pressure in the cylinder (e.g., created by combustion of the fuel–air mixture of an internal combustion engine). As above, this force may be analyzed by denoting it as $f$. However, several important observations should be made first:

1. The combustion force $f$ is not sinusoidal of frequency $\omega$. It is reasonably periodic but the shape is complex and depends on the firing/fuel-injection cycle and the associated combustion process.
2. The reaction forces $f_i$ on the crankshaft, which are generated from cylinders $i$, should be balanced to avoid the transmission of reaction forces and bending moments to the shaft bearings (and hence, to the supporting frame — the vehicle). However, the torques $\tau_i$ in this case are in fact the drive torques. Obviously, they are the desired output of the engine and should not be balanced, unlike the inertia torques.

Therefore, although the analysis completed for balancing the inertia forces cannot be directly used here, we can employ similar approaches to the use of multiple cylinders for reducing the gas-force reactions. This is a rather difficult problem given the complexity of the combustion process itself. In practice, much of the leftover effects of the ignition cycle are suppressed by properly designed engine mounts. Experimental investigations have indicated that in a properly balanced engine unit, much of the vibration transmitted through the engine mounts is caused by the engine firing cycle (internal combustion) rather than the reciprocating inertia (sinusoidal components of frequency $\omega$ and $2\omega$). Hence, active mounts, where stiffness can be varied according to the frequency of excitation, are being considered to reduce engine vibrations in the entire range of operating speeds (e.g. 500 to 2500 rpm).

## 7.6 Whirling of Shafts

In the previous two sections, we studied the vibration excitations caused on rotating shafts and their bearings due to some form of mass eccentricity. Methods of balancing these systems to eliminate the undesirable effects were also presented. One limitation of the given analysis is the assumptions that the rotating shaft is rigid and, thus, does not deflect from its axis of rotation due to the unbalanced excitations. In practice, however, rotating shafts are made lighter than the components they carry (rotors, disks, gears, etc.) and will undergo some deflection due to the unbalanced loading. As a result, the shaft will bow out and this will further increase the mass eccentricity and associated unbalanced excitations and gyroscopic forces of the rotating elements (disks, rotors, etc.). The nature of damping of rotating machinery (which is rather complex and incorporates effects of rotation at bearings, structural deflections, and lateral speeds) will further affect the dynamic behavior of the shaft under these conditions. In this context, the topic of whirling of rotating shafts becomes relevant.

Consider a shaft that is driven at a constant angular speed $\omega$ (e.g., by using a motor or some other actuator). The central axis of the shaft (passing through its bearings) will bow out. The deflected axis itself will rotate, and this rotation is termed *whirling* or *whipping*. The whirling speed is not necessarily equal to the drive speed $\omega$ (at which the shaft rotates about its axis with respect to a fixed frame). However, when the whirling speed is equal to $\omega$, the condition is called *synchronous whirl*, and the associated deflection of the shaft can be quite excessive and damaging.

To develop an analytical basis for whirling, consider a light shaft supported on two bearings carrying a disk of mass $m$ in between the bearings, as shown in Figure 7.19(a). Note that C is the point on the disk at which it is mounted on the shaft. Originally, in the neutral configuration when the shaft is not driven ($\omega = 0$), the point C coincides with point O on the axis joining the two bearings. If the shaft were rigid, then the points C and O would continue to coincide during motion. The mass center (centroid or center of gravity for constant $g$) of the disk is denoted as G in Figure 7.19. During motion, C will move away from O due to the shaft deflection. The whirling speed (speed of rotation of the shaft axis) is the speed of rotation of the radial line OC with respect to a fixed reference. Denoting the angle of OC with respect to a fixed reference as $\theta$, the whirling speed is $\dot{\theta}$. This is explained in Figure 7.19(b) where an end view of the

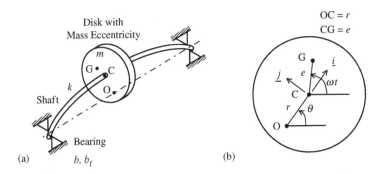

**FIGURE 7.19** (a) A whirling shaft carrying a disk with mass eccentricity; (b) end view of the disk and whirling shaft.

disk is given under deflected conditions. The constant drive speed $\omega$ of the shaft is the speed of the shaft spin with respect to a fixed reference, and is the speed of rotation of the radial line CG with respect to the fixed horizontal line shown in Figure 7.19(b). Hence, the angle of shaft spin is $\omega t$, as measured with respect to this line. The angle of whirl, $\theta$, is also measured from the direction of this fixed line, as shown.

## 7.6.1 Equations of Motion

Under practical conditions, the disk moves entirely in a single plane. Hence, its complete set of equations of motion consists of two equations for translatory (planar) motion of the centroid (with lumped mass $m$), and one equation for rotational motion about the fixed bearing axis. The latter equation depends on the motor torque that drives the shaft at constant speed $\omega$, and is not of interest in the present context. Thus, we will limit our development to the two translatory equations of motion. The equations may be written either in a Cartesian coordinate system $(x, y)$ or a polar coordinate system $(r, \theta)$. Here, we will use the polar coordinate system.

Consider a coordinate frame $(\mathbf{i}, \mathbf{j})$ that is fixed to the disk with its $\mathbf{i}$ axis lying along OC as shown in Figure 7.19(b). Note that the angular speed of this frame is $\dot{\theta}$ (about the $\mathbf{k}$ axis that is orthogonal to $\mathbf{i}$ and $\mathbf{j}$). Hence, as is well known, we have

$$\frac{d\mathbf{i}}{dt} = \dot{\theta}\mathbf{j} \quad \text{and} \quad \frac{d\mathbf{j}}{dt} = -\dot{\theta}\mathbf{i} \tag{7.69}$$

The position vector of the mass point G from O is

$$\vec{OG} = \mathbf{r}_G = \vec{OC} + \vec{CG} = r\mathbf{i} + e\cos(\omega t - \theta)\mathbf{i} + e\sin(\omega t - \theta)\mathbf{j} \tag{7.70}$$

The velocity vector $\mathbf{v}_G$ of the mass point G can be obtained simply by differentiating Equation 7.70 with the use of Equation 7.69. However, this can be simplified because $\omega$ is constant. Here, line CG has a velocity $e\omega$ that is perpendicular to it about C. This can be resolved along the axes $\mathbf{i}$ and $\mathbf{j}$. Hence, the velocity of G relative to C is

$$\mathbf{v}_{G/C} = -e\omega \sin(\omega t - \theta)\mathbf{i} + e\omega \cos(\omega t - \theta)\mathbf{j}$$

However, the velocity of point C is

$$\mathbf{v}_C = \frac{d}{dt}r\mathbf{i} = \dot{r}\mathbf{i} + r\frac{d\mathbf{i}}{dt} = \dot{r}\mathbf{i} + r\dot{\theta}\mathbf{j}$$

Hence, the velocity of G, which is given by $\mathbf{v}_G = \mathbf{v}_C + \mathbf{v}_{G/C}$, can be expressed as

$$\mathbf{v}_G = \dot{r}\mathbf{i} + r\dot{\theta}\mathbf{j} - e\omega \sin(\omega t - \theta)\mathbf{i} + e\omega \cos(\omega t - \theta)\mathbf{j} \tag{7.71}$$

Similarly, the acceleration of C is

$$\mathbf{a}_C = \frac{d}{dt}\mathbf{v}_C = \frac{d}{dt}[\dot{r}\mathbf{i} + r\dot{\theta}\mathbf{j}] = \ddot{r}\mathbf{i} + \dot{r}\dot{\theta}\mathbf{j} + \dot{r}\dot{\theta}\mathbf{j} + \dot{r}\dot{\theta}\mathbf{j} - r\dot{\theta}^2\mathbf{i} = (\ddot{r} - r\dot{\theta}^2)\mathbf{i} + (r\ddot{\theta} + 2\dot{r}\dot{\theta})\mathbf{j}$$

Also, since line CG rotates at constant angular speed $\omega$ about C, the point G has only a radial (centrifugal) acceleration $e\omega^2$ along GC. This can be resolved along $\mathbf{i}$ and $\mathbf{j}$ as before. Hence, the acceleration of G relative to C is

$$\mathbf{a}_{G/C} = -e\omega^2 \cos(\omega t - \theta)\mathbf{i} - e\omega^2 \sin(\omega t - \theta)\mathbf{j}$$

It follows that the acceleration of point G, given by $\mathbf{a}_G = \mathbf{a}_C + \mathbf{a}_{G/C}$, may be expressed as

$$\mathbf{a}_G = (\ddot{r} - r\dot{\theta}^2)\mathbf{i} + (r\ddot{\theta} + 2\dot{r}\dot{\theta})\mathbf{j} - e\omega^2 \cos(\omega t - \theta)\mathbf{i} - e\omega^2 \sin(\omega t - \theta)\mathbf{j} \qquad (7.72)$$

The forces acting on the disk are as follows:

Restraining elastic force due to lateral deflection of the shaft $= -kr\mathbf{i}$
Viscous damping force (proportional to the velocity of C) $= -b\dot{r}\mathbf{i} - br\dot{\theta}\mathbf{j}$

In addition, there is a frictional resistance at the bearing, which is proportional to the reaction and, hence, the shaft deflection is $r$ and also depends on the spin speed $\omega$. The following approximate model may be used:

$$\text{Bearing friction force} = -b_f r\omega\mathbf{j}$$

Here,

$k =$ lateral deflection stiffness of the shaft at the location of the disk
$b =$ viscous damping constant for lateral motion of the shaft
$b_f =$ bearing frictional coefficient

The overall force acting on the disk is

$$\mathbf{f} = -b\dot{r}\mathbf{i} - (br\dot{\theta} + b_f r\omega)\mathbf{j} \qquad (7.73)$$

The equation of rectilinear motion

$$\mathbf{f} = m\mathbf{a}_G \qquad (7.74)$$

on using Equation 7.72 and Equation 7.73, reduces to the following pair in the $\mathbf{i}$ and $\mathbf{j}$ directions:

$$-kr - b\dot{r} = m[\ddot{r} - r\dot{\theta}^2 - e\omega^2 \cos(\omega t - \theta)] \qquad (7.75)$$

$$-br\dot{\theta} - b_f r\omega = m[r\ddot{\theta} + 2\dot{r}\dot{\theta} - e\omega^2 \sin(\omega t - \theta)] \qquad (7.76)$$

These equations may be expressed as

$$\ddot{r} + 2\zeta_v \omega_n \dot{r} + (\omega_n^2 - \dot{\theta}^2)r = e\omega^2 \cos(\omega t - \theta) \qquad (7.77)$$

$$r\ddot{\theta} + 2(\zeta_v \omega_n r + \dot{r})\dot{\theta} + 2\zeta_f \omega_n \omega r = e\omega^2 \sin(\omega t - \theta) \qquad (7.78)$$

where the undamped natural frequency of lateral vibration is

$$\omega_n = \sqrt{\frac{k}{m}} \qquad (7.79)$$

and

$\zeta_v =$ viscous damping ratio of lateral motion
$\zeta_f =$ frictional damping ratio of the bearings

Equation 7.77 and Equation 7.78, which govern the whirling motion of the shaft-disk system, are a pair of coupled nonlinear equations, with excitations (depending on $\omega$) that are coupled with a motion variable ($\theta$). Hence, a general solution would be rather complex. A relatively simple solution is possible, however, under steady-state whirling.

## 7.6.2 Steady-State Whirling

Under steady-state conditions, the whirling speed $\dot{\theta}$ is constant at $\dot{\theta} = \omega_w$, hence, $\ddot{\theta} = 0$. Also, the lateral deflection of the shaft is constant, hence, $\dot{r} = \ddot{r} = 0$. Therefore, Equation 7.77 and Equation 7.78 become

$$(\omega_n^2 - \omega_w^2)r = e\omega^2 \cos(\omega t - \theta) \tag{7.80}$$

$$2\zeta_v \omega_n \omega_w r + 2\zeta_f \omega_n \omega r = e\omega^2 \sin(\omega t - \theta) \tag{7.81}$$

In Equation 7.80 and Equation 7.81, the left-hand side is independent of $t$. Hence, the right-hand side should also be independent of $t$. For this, we must have

$$\theta = \omega t - \phi \tag{7.82}$$

where $\phi$ is interpreted as the phase lag of whirl with respect to the shaft spin ($\omega$), and should be clear from Figure 7.19(b). It follows from Equation 7.82 that, for steady-state whirl, the whirling speed $\dot{\theta} = \omega_w$ is

$$\omega_w = \omega \tag{7.83}$$

This condition is called synchronous whirl because the whirl speed ($\omega_w$) is equal to the shaft spin speed ($\omega$). It follows that under steady-state conditions, we should have the state of synchronous whirl. The equations governing steady-state whirl are

$$(\omega_n^2 - \omega^2)r = e\omega^2 \cos \phi \tag{7.84}$$

$$2\zeta\omega_n \omega r = e\omega^2 \sin \phi \tag{7.85}$$

along with Equation 7.82 and, hence, Equation 7.83. Here, $\zeta = \zeta_v + \zeta_f$ is the overall damping ratio of the system. Note that the phase angle $\phi$ and the shaft deflection $r$ are determined from Equation 7.84 and Equation 7.85. In particular, squaring these two equations and adding to eliminate $\phi$, we obtain

$$r = \frac{e\omega^2}{\sqrt{(\omega_n^2 - \omega^2)^2 + (2\zeta\omega_n\omega)^2}} \tag{7.86}$$

which is of the form of magnitude of the frequency transfer function of a simple oscillator with an acceleration excitation. Divide Equation 7.85 by Equation 7.84 to get the phase angle:

$$\phi = \tan^{-1}\frac{2\zeta\omega_n\omega}{(\omega_n^2 - \omega^2)} \tag{7.87}$$

Using simple calculus (differentiate the square and equate to zero), we can show that the maximum deflection occurs at the critical spin speed $\omega_c$ given by

$$\omega_c = \frac{\omega_n}{\sqrt{1 - 2\zeta^2}} \tag{7.88}$$

This *critical speed* corresponds to a resonance. For light damping, we have approximately $\omega_c = \omega_n$. Hence, critical speed for low damping is equal to the undamped natural frequency of bending vibration of the shaft-rotor unit. The corresponding shaft deflection is (see Equation 7.86)

$$r_c = \frac{e}{2\zeta} \tag{7.89}$$

which is also a good approximation of $r$ at critical speed, with light damping. From Equation 7.84 and Equation 7.85, we can see that, at critical speed (with low damping), $\sin \phi = 1$ and $\cos \phi = 0$, which gives $\phi = \pi/2$. Also, note from Equation 7.86 that the steady-state shaft deflection is almost zero at low speeds and approaches $e$ at very high speeds. However, Equation 7.87 shows that, for small $\omega$, $\tan \phi$ is positive and small. We can see from Equation 7.85 that $\sin \phi$ is positive. This means $\phi$ itself is small for small $\omega$. For large $\omega$, we can see from Equation 7.86 that $r$ approaches $e$. Thus, we can see from

Equation 7.87 that tan $\phi$ is small and negative, whereas Equation 7.85 shows that sin $\phi$ is positive. Hence, $\phi$ approaches $\pi$ for large $\omega$.

It is seen from Equation 7.89 that, at critical speed, the shaft deflection increases with mass eccentricity and decreases with damping. This indicates that the approaches for reducing the damaging effects of whirling are:

1. Eliminate or reduce the mass eccentricity through proper construction practices and balancing.
2. Increase damping.
3. Increase shaft stiffness.
4. Avoid operation near critical speed.

There will be limitations to the use of these approaches, particularly making the shaft stiffer. Note also that our analysis did not include the mass distribution of the shaft. A Bernoulli–Euler type beam analysis has to be incorporated for a more accurate analysis of whirling for shafts whose mass cannot be accurately represented by a single parameter that is lumped at the location of the rotor. Formulas related to whirling of shafts are summarized in Box 7.5.

## Example 7.6

The fan of a ventilation system has a normal operating speed of 3600 rpm. The blade set of the fan weighs 20 kg and is mounted in the mid-span of a relatively light shaft that is supported on lubricated bearings at its two ends. The bending stiffness of the shaft at the location of the fan is $4.0 \times 10^6$ N/m. Equivalent damping ratio that acts on the possible whirling motion of the shaft is 0.05. Owing to fabrication error, the centroid of the fan has an eccentricity of 1.0 cm from the neutral axis of rotation of the shaft:

1. Determine the critical speed of the fan system and the corresponding shaft deflection at the location of the fan at steady state.
2. What is the steady-state shaft deflection at the fan during normal operation?

The fan was subsequently balanced using a mass of 5 kg. The centroid eccentricity was reduced to 2 mm by this means. What is the shaft deflection at the fan during normal operation now? Comment on the improvement that has been realized.

## Solution

1. The system is lightly damped. Hence, the critical speed is given by the undamped natural frequency; thus

$$\omega_c \cong \omega_n = \sqrt{\frac{k}{m}} = \sqrt{\frac{4 \times 10^6}{20}} \text{ rad/s} = 447.2 \text{ rad/s}$$

The corresponding shaft deflection is

$$r_c = \frac{e}{2\zeta} = \frac{1.0}{2 \times 0.05} \text{ cm} = 10.0 \text{ cm}$$

2. Operating speed $\omega = (3600/60) \times 2\pi$ rad/s $= 377$ rad/s. Using Equation 7.86, the corresponding shaft deflection, at steady state, is

$$r = \frac{1.0 \times (377)^2}{[(447.2^2 - 377^2)^2 + (2 \times 0.05 \times 447.2 \times 377)^2]^{1/2}} \text{ cm} = 2.36 \text{ cm}$$

After balancing, the new eccentricity $e = 0.2$ cm.
The new natural frequency (undamped) is

$$\omega_n = \sqrt{\frac{4 \times 10^6}{25}} \text{ rad/s} = 400 \text{ rad/s}$$

# Box 7.5

# WHIRLING OF SHAFTS

**Whirling:** A shaft spinning at speed $\omega$ about its axis, may bend due to flexure. The bent (bowed out) axis will rotate at speed $\omega_w$. This is called whirling.

**Equations of motion:**

$$\ddot{r} + 2\zeta_v\omega_n\dot{r} + (\omega_n^2 - \dot{\theta}^2)r = e\omega^2\cos(\omega t - \theta)$$

$$r\ddot{\theta} + 2(\zeta_v\omega_n r + \dot{r})\dot{\theta} + 2\zeta_f\omega_n\omega r = e\omega^2\sin(\omega t - \theta)$$

where $(r, \theta)$ are polar coordinates of shaft deflection at the mounting point of lumped mass.

$e$ = eccentricity of the lumped mass from the spin axis of shaft
$\dot{\theta} = \omega_w$ = whirling speed
$\omega$ = spin speed of shaft
$\omega_n = \sqrt{k/m}$ = natural frequency of bending vibration of shaft
$k$ = bending stiffness of shaft at lumped mass
$m$ = lumped mass
$\zeta_v$ = damping ratio of bending motion of shaft
$\zeta_f$ = damping ratio of shaft bearings

**Steady-state whirling (synchronous whirl):**
Here, whirling speed ($\dot{\theta}$ or $\omega_w$) is constant and equals the shaft spin speed $\omega$ (i.e., $\omega_w = \omega$ for steady-state whirling).

  Shaft deflection at lumped mass

$$r = \frac{e\omega^2}{\left[(\omega_n^2 - \omega^2)^2 + (2\zeta\omega_n\omega)^2\right]^{1/2}}$$

  Phase angle between shaft deflection ($r$) and mass eccentricity ($e$)

$$\phi = \tan^{-1}\frac{2\zeta\omega_n\omega}{(\omega_n^2 - \omega^2)}$$

where $\zeta = \zeta_v + \zeta_f$
**Note:** For small spin speeds $\omega$, we have small $r$ and $\phi$. For large $\omega$, we have $r \cong e$ and $\phi \cong \pi$
Critical speed:

$$\text{Spin speed } \omega = \frac{\omega_n}{\sqrt{1 - 2\zeta^2}} \cong \omega_n \text{ for small } \zeta$$

$$\phi = \pi/2$$

The corresponding shaft deflection during steady-state operation is

$$r = \frac{0.2 \times (377)^2}{[(400^2 - 377^2)^2 + (2 \times 0.05 \times 400 \times 377)^2]^{1/2}} \text{ cm} = 1.216 \text{ cm}$$

  Note that, even though the eccentricity has been reduced by a factor of five by balancing, the operating deflection of the shaft has been reduced only by a factor of less than two. The main reason for this is that the operating speed is close to the critical speed. Methods of improving the performance include changing the operating speed, using a smaller mass to balance the fan, using more damping, and making the shaft stiffer. However, some of these methods may not be feasible. Operating speed is determined by the task requirements. A location may not be available that is sufficiently distant to place a balancing mass that is appropriately small. Increased damping will increase heat generating, cause bearing problems, and will also reduce the operating speed. Replacement or stiffening of the shaft may require too much modification to the system and add cost. A preferable alternative would be to balance the fan by removing some mass. This will move the critical frequency (natural frequency) away from the operating speed rather than closer to it, while reducing the mass eccentricity at the same time. For example, suppose that a mass of 3 kg is removed from the fan, which results in an eccentricity of 2.0 mm. The new natural/critical frequency is

$$\sqrt{\frac{4 \times 10^6}{17}} \text{ rad/s} = 485.1 \text{ rad/s}$$

The corresponding shaft deflection during steady operation is

$$r = \frac{0.2 \times (377)^2}{[(485.1^2 - 377^2)^2 + (2 \times 0.05 \times 485.1 \times 377)^2]^{1/2}} \text{ cm} = 0.3 \text{ cm}$$

In this case, the deflection has been reduced by a factor of eight.

### 7.6.3 Self-Excited Vibrations

Equation 7.77 and Equation 7.78, which represent the general whirling motion of a shaft, are nonlinear and coupled. In these equations, the motion variables ($r$ and $\theta$) occur as (nonlinear) products of the excitation ($\omega$). Such systems are termed self-excited. Note that, in general (before reaching the steady state) the response variables $r$ and $\theta$ will exhibit vibratory characteristics in view of the presence of the excitation functions $\cos(\omega t - \theta)$ and $\sin(\omega t - \theta)$. Hence, a whirling shaft may exhibit self-excited vibrations. Because the excitation forces directly depend on the motion itself, it is possible that a continuous energy flow into the system could occur. This will result in a steady growth of the motion amplitudes and represents an *unstable* behavior.

A simple example of self-excited vibration is provided by a pendulum whose length is time variable. Although the system is stable when the length is fixed, it can become unstable under conditions of variable length. Practical examples of self-excited vibrations with possible exhibition of instability include the flutter of aircraft wings due to coupled aerodynamic forces, wind-induced vibrations of bridges and tall structures, galloping of ice-covered transmission lines due to air flow-induced vibrations, and chattering of machine tools due to friction-related excitation forces. Proper design and control methods, as discussed in this chapter, are important in suppressing self-excited vibrations.

## 7.7 Design through Modal Testing

Experimental modal analysis (EMA) involves extracting modal parameters (natural frequencies, modal damping ratios, mode shapes) of a mechanical system through testing (notably, through excitation-response data) and then developing a dynamic model of the system (mass, stiffness, and damping matrices) on that basis. The techniques of EMA are useful in modeling and model validation (i.e., verification of the accuracy of an existing model that was obtained, for example, through analytical modeling). In addition to these uses, EMA is also a versatile tool for design development. In the context of "design for vibration," EMA may be employed in the design and design modification of mechanical systems with the goal of achieving desired performance under vibrating conditions. This section will introduce this approach.

In applying EMA for design development of a mechanical system, three general approaches are employed:

1. Component modification
2. Modal response specification
3. Substructuring

The method of component modification allows us to modify (i.e., add, remove, or vary) physical parameters (inertia, stiffness, damping) in a mechanical system, and to determine the resulting effect on the modal response (natural frequencies, damping ratios, and mode shapes) of the system. The method of modal response specification provides the capability to establish the best changes, from the design viewpoint, in system parameters (inertia, stiffness, damping values, and associated directions) in order to realize a specified change in the modal response. In the techniques of substructuring, two or more subsystem models are combined using proper components of interfacing (interconnection), and the overall model of the integrated system is determined. Some of the subsystems used in this approach could be of analytical or computational origin (e.g., finite element models). It should be clear how these methods could be used in the design development of a mechanical system for proper vibration performance. The first method is essentially a trial and error technique of incremental design. Here, some appropriate parameters are changed and the resulting modal behavior is determined. If the resulting performance is not satisfactory, further changes are made in discrete steps until an acceptable performance (with regard to natural frequencies, response magnification factors, etc.) is achieved. The second method is clearly a direct design approach, where the design specifications are first developed in terms of modal characteristics, and then the design procedure will generate the size and type of the physical parameters to meet the specifications. In the third method, a suitable set of subsystems is first designed to meet performance characteristics of each subsystem. Then, these subsystems are linked through suitable mechanical interfacing components, and the performance of the overall system is determined to verify acceptance. In this manner, a complex system may be designed through the systematic design of its subsystems.

## 7.7.1 Component Modification

The method of component modification involves changing a mass, stiffness, or damping element in the system and determining the corresponding dynamic response, particularly the natural frequencies, modal damping ratios, and mode shapes. This is relatively straightforward because a single modal analysis or modal test (EMA) will give the required information. Because single step of component modification might not be acceptable as an appropriate design (e.g., a natural frequency might be too close to a significant frequency component of a vibration excitation), a number of modifications may be necessary. For such incremental procedures, modal analysis would be more convenient and cost effective than EMA because, in the latter case, physical modification and retesting would be needed, whereas the former involves the same computational steps as before, but with a new set of parameter values.

For example, consider an aluminum I beam that has a number of important modes of vibration, including bending and torsional modes. Figure 7.20(a) shows the fourth mode shape of vibration at natural frequency 678.4 Hz. The dotted line in Figure 7.20(b) shows the transfer function magnitude when the beam is excited at some location in the vertical direction and the response is measured in the vertical direction, at some other location, where neither of the locations are node points. The curve shows the first six natural frequencies.

Next, a lumped mass is added to the top flange at the shown location. The corresponding transfer function magnitude is shown by the solid curve in Figure 7.20(b). Note that all the natural frequencies have decreased due to the added mass, but the effect is larger for higher modes. Similarly, mode shapes also will change. If the new modes are not satisfactory (e.g., a particular natural frequency has not shifted enough) further modification and evaluation will be required.

Consider a mechanical vibrating system whose free response $\mathbf{y}$ is described by

$$\mathbf{M}\ddot{\mathbf{y}} + \mathbf{K}\mathbf{y} = 0 \tag{7.90}$$

Damping has been ignored for simplicity, but the following discussion can also be extended to a damped system (quite directly, for the case of proportional damping). If the mass matrix $\mathbf{M}$ and the stiffness matrix $\mathbf{K}$ are modified by $\delta\mathbf{M}$ and $\delta\mathbf{K}$, respectively, the corresponding response (as well as the natural frequencies and mode shapes) will be different from that of the original system. To illustrate, let the

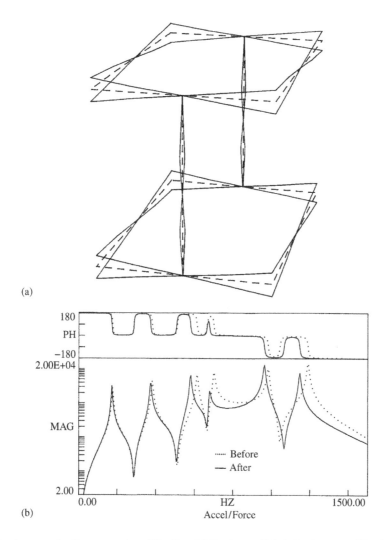

(a)

(b)

**FIGURE 7.20** An example of component modification: (a) the shape of Mode 4 prior to modification; (b) transfer function magnitude before and after modification.

modal matrix (the matrix whose columns are the independent mode shape vectors of the original system) be $\Psi$. Then, using the modal transformation

$$\mathbf{y} = \Psi\mathbf{q} \tag{7.91}$$

Equation 7.90 can be expressed in the canonical form, with modal generalized coordinates $\mathbf{q}$, as

$$\bar{\mathbf{M}}\ddot{\mathbf{q}} + \bar{\mathbf{K}}\mathbf{q} = 0 \tag{7.92}$$

where

$$\Psi^T\mathbf{M}\Psi = \bar{\mathbf{M}} = \text{diag}[M_1, M_2, \dots, M_n] \tag{7.93}$$

$$\Psi^T\mathbf{K}\Psi = \bar{\mathbf{K}} = \text{diag}[K_1, K_2, \dots, K_n] \tag{7.94}$$

If the same transformation (Equation 7.91) is used for the modified system

$$(\mathbf{M} + \delta\mathbf{M})\ddot{\mathbf{y}} + (\mathbf{K} + \delta\mathbf{K})\mathbf{y} = 0 \tag{7.95}$$

we obtain

$$(\bar{\mathbf{M}} + \boldsymbol{\Psi}^T \delta\mathbf{M}\boldsymbol{\Psi})\ddot{\mathbf{q}} + (\bar{\mathbf{K}} + \boldsymbol{\Psi}^T \delta\mathbf{K}\boldsymbol{\Psi})\mathbf{q} = 0 \qquad (7.96)$$

Since both $\boldsymbol{\Psi}^T \delta\mathbf{M}\boldsymbol{\Psi}$ and $\boldsymbol{\Psi}^T \delta\mathbf{K}\boldsymbol{\Psi}$ are not diagonal matrices in general, $\boldsymbol{\Psi}$ would not remain the modal matrix for the modified system. Furthermore, the original natural frequencies $\omega_i = \sqrt{K_i/M_i}$ will change due to the component modification. For the special case of proportional modifications ($\delta\mathbf{M}$ proportional to $\mathbf{M}$ and $\delta\mathbf{K}$ proportional to $\mathbf{K}$), the mode shapes will not change. However, in general, the natural frequencies will change.

The reverse problem is the modal response specification. Here, a required set of modal parameters ($\omega_{ir}$ and $\psi_{ir}$) is specified and the necessary changes $\delta\mathbf{M}$ and $\delta\mathbf{K}$ to meet the specifications must be determined. Note that the solution is not unique and is more difficult than the direct problem. In this case, a sensitivity analysis may initially be performed to determine the directions and magnitudes of the modal shift for a particular physical parameter shift. Then, the necessary magnitudes of physical shift to achieve the specified modal shift are estimated on that basis. The corresponding modifications are made and the modified system is analyzed/tested to check whether it is within specification. If not, further cycles of modification should be performed.

## Example 7.7

As an example of component modification, consider the familiar problem of a two-DoF system, as shown in Figure 7.21. The squared nondimensional natural frequencies $r_i^2 = (\omega_i/\omega_0)^2$ of the systems are given by

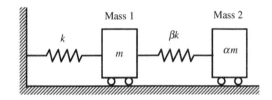

**FIGURE 7.21**    A two-degree-of-freedom example.

$$r_1^2, r_2^2 = \frac{1}{2\alpha}$$
$$\times \{\alpha + \beta + \alpha\beta\}\left\{1 \pm \sqrt{1 - \frac{4\alpha\beta}{(\alpha + \beta + \alpha\beta)^2}}\right\}$$

where $\omega_0 = \sqrt{k/m}$. We also showed that the mode shapes, as given by the ratio of the displacement of mass 2 to that of mass 1 at a natural frequency, are

$$\left(\frac{\psi_2}{\psi_1}\right)_i = \frac{1 + \beta - r_i^2}{\beta} \quad \text{for mode } i$$

Consider a system with $\alpha = 0.5$ and $\beta = 0.5$. By direct computation, we can show that $r_1 = 0.71$ and $r_2 = 1.41$. Estimate the modification of $\beta$ (the relative stiffness of the second spring) that would be necessary to shift the system natural frequencies to approximately $r_1 = 0.8$ and $r_2 = 2.0$. Check the corresponding shift in mode shapes.

## Solution

For $\alpha = 0.5$ and $\beta = 0.5$, direct substitution yields $r_1 = 0.71$ and $r_2 = 1.41$ with $(\psi_2/\psi_1)_1 = 2.0$ and $(\psi_2/\psi_1)_2 = -1.0$. Now, consider an incremental change in $\beta$ by 0.1. Then, $\beta = 0.6$. The corresponding natural frequencies are computed as

$$r_i^2 = \frac{1}{2 \times 0.5}\{0.5 + 0.6 + 0.5 \times 0.6\}\left\{1 \mp \left[1 - \frac{4 \times 0.5 \times 0.6}{(0.5 + 0.6 + 0.5 \times 0.6)^2}\right]^{1/2}\right\} = 0.528, 2.272$$

Hence,

$$r_1, r_2 = 0.727, 1.507$$

This step may be interpreted as a way of establishing the sensitivity of the system to the particular component modification. Clearly, the problem of modification is not linear. However, as a first approximation, assume a linear variation of $r_i^2$ with $\beta$, and make modifications according to

$$\frac{\delta\beta}{\delta\beta_0} = \frac{\delta r_i^2}{\delta r_{i0}^2} \tag{7.97}$$

where the subscript 0 refers to the initial trial variation ($\delta\beta_0 = 0.1$). Equation 7.97 is intuitively satisfying given the nature of the physical problem and the fact that, for a single-DoF problem, squared frequency varies with $k_0$. Then, we have

*For mode 1:*

$$\frac{\delta\beta}{0.1} = \frac{0.8^2 - 0.71^2}{0.727^2 - 0.71^2} = 5.634$$

or

$$\delta\beta = 0.56$$

*For mode 2:*

$$\frac{\delta\beta}{0.1} = \frac{2^2 - 1.41^2}{1.507^2 - 1.41^2} = 7.09$$

or

$$\delta\beta = 0.709$$

Therefore, we use $\delta\beta = 0.71$, which is the larger of the two. This corresponds to

$$\beta = 0.5 + 0.71 = 1.21$$

The natural frequencies are computed as usual:

$$r_1^2, r_2^2 = \frac{1}{2 \times 0.5}\{0.5 + 1.21 + 0.5 \times 1.21\}\left\{1 \mp \left[1 - \frac{4 \times 0.5 \times 1.21}{(0.5 + 1.21 + 0.5 \times 1.21)^2}\right]^{1/2}\right\} = 0.60, 4.03$$

or

$$r_1, r_2 = 0.78, 2.01$$

In view of the nonlinearity of the problem, this shift in frequencies is satisfactory. The corresponding mode shapes are

$$\left(\frac{\psi_2}{\psi_1}\right)_1 = \frac{(1 + 1.21) - 0.6}{1.21} = 1.33$$

$$\left(\frac{\psi_2}{\psi_1}\right)_2 = \frac{(1 + 1.21) - 4.03}{1.21} = -1.50$$

It follows that, as the stiffness of the second spring is increased, the motions of the two masses become closer in mode 1. Furthermore, in mode 2, the node point becomes closer to mass 1. Note the limitation of this particular component modification. As $\beta \to \infty$, the two masses become rigidly linked giving a frequency ratio of $r_1 = \sqrt{k/(m + \alpha m)}/\sqrt{k/m} = 1/\sqrt{1 + \alpha} = 1/\sqrt{1.5} = 0.816$, with $r_2 \to \infty$. Hence, it is unreasonable to expect a frequency ratio that is closer to this value of $r_1$ by a change in $\beta$ alone.

## 7.7.2 Substructuring

For large and complex mechanical systems with many components, the approach of substructuring can make the process of "design for vibration" more convenient and systematic. In this approach, the system is first divided into a convenient set of subsystems that are more amenable to testing and analysis. The subsystems are separately modeled and designed through the approaches of modal analysis and testing, along with any other convenient approaches (e.g., finite element technique). Note that the performance of the overall system depends on the interface conditions that link the subsystems, as well as the characteristics of the individual subsystems. Hence, it is not possible to translate the design specifications for the overall system into those for the subsystems without taking the interface conditions into account. The overall system is *assembled* from the designed subsystems by using *compatibility* requirements at the assembly locations together with dynamic equations of the interconnecting components such as spring–mass–damper units or rigid linkages. If the assembled system does not meet the design specifications, then modifications should be made to one or more of the subsystems and interfacing (assembly) linkages, and the procedure should be repeated. Thus, the main steps of using the approach of substructuring for vibration design of a complex system are as follows:

1. Divide the mechanical system into convenient subsystems (substructuring) and represent the interconnection points of subsystems by forces/moments.
2. Develop models for the subsystems through analysis, modal testing, and other standard procedures.
3. Design the subsystems so that their performance is well within the performance specifications provided for the overall system.
4. Establish the interconnecting (assembling) linkages for the subsystems, and obtain dynamic equations for them in terms of the linking forces/moments and motions (displacements/rotations).
5. Establish continuity (force balancing) and compatibility (motion consistency) conditions at the assembly locations.
6. Using matrix methods, eliminate the unknown variables and assemble the overall system.
7. Analyze (or test) the overall system to determine its vibration performance. If satisfactory, stop. If not, make modifications to the systems or assembly conditions and repeat step 4 to step 7.

As a simple example, consider two single-DoF systems that are interconnected by a spring linkage, as shown in Figure 7.22. The two subsystems may be represented by

$$\begin{bmatrix} m_1 & 0 \\ 0 & m_2 \end{bmatrix} \ddot{y} + \begin{bmatrix} k_1 & 0 \\ 0 & k_2 \end{bmatrix} y = 0$$

and the corresponding natural frequencies are

$$\omega_{s1} = \sqrt{k_1/m_1} \quad \text{and} \quad \omega_{s2} = \sqrt{k_2/m_2}$$

The overall interconnected system is given by

$$\begin{bmatrix} m_1 & 0 \\ 0 & m_2 \end{bmatrix} \ddot{y} + \begin{bmatrix} k_1 + k_c & -k_c \\ -k_c & k_2 + k_c \end{bmatrix} y = 0$$

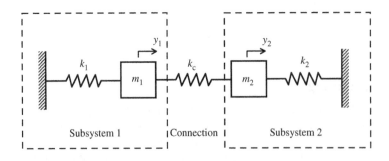

**FIGURE 7.22**  An example of substructuring.

Its natural frequencies are obtained by solving the equation

$$\det\begin{bmatrix} k_1 + k_c - \omega^2 m_1 & -k_c \\ -k_c & k_2 + k_c - \omega^2 m_2 \end{bmatrix} = 0$$

or

$$(k_1 + k_c - \omega^2 m_1)(k_2 + k_c - \omega^2 m_2) - k_c^2 = 0$$

which simplifies to

$$\omega^4 - \left[\frac{k_1 + k_c}{m_1} + \frac{k_2 + k_c}{m_2}\right]\omega^2 + \frac{k_1 k_2 + k_c(k_1 + k_2)}{m_1 m_2} = 0$$

The sum of the roots is

$$\omega_1^2 + \omega_2^2 = \frac{k_1 + k_c}{m_1} + \frac{k_2 + k_c}{m_2} > \omega_{s1}^2 + \omega_{s2}^2$$

The product of the roots is

$$\omega_1^2 \omega_2^2 = \frac{k_1 k_2 + k_c(k_1 + k_2)}{m_1 m_2} > \omega_{s1}^2 \omega_{s2}^2$$

This does not mean that both frequencies will increase due to the interconnection. Note that the limit on the lower frequency, as $k_c \rightarrow \infty$, is given by that of a single-DoF system with mass $m_1 + m_2$ and stiffness $k_1 + k_2$, which is, $\sqrt{(k_1 + k_2)/(m_1 + m_2)}$. This value can be larger or smaller than the natural frequency of a subsystem depending on the relative values of the parameters. Hence, even for this system, exact satisfaction of a set of design natural frequencies would be somewhat challenging because these frequencies depend on the interconnection as well as the subsystems.

Substructuring is a design development technique where complex designs can be accomplished through parallel and separate development of several subsystems and interconnections. Through this procedure, dynamic interactions among subsystems can be estimated and potential problems can be detected, which will allow redesigning of the subsystems or interfacing linkages prior to building the over prototype. Design approaches using EMA that may be used in vibration problems are summarized in Box 7.6.

# 7.8 Passive Control of Vibration

The techniques discussed in this chapter for reduction of the effects of mechanical vibration fall into the categories of vibration isolation and design for vibration. The third category, vibration control, is addressed now. Characteristic of vibration control is the use of a sensing device to detect the level of vibration in a system, and an actuation (forcing) device to apply a forcing function to the system to counteract the effects of vibration. In some such devices, the sensing and forcing functions are implicit and integrated together.

Vibration control may be subdivided into the following two broad categories:

1. Passive control
2. Active control

Passive control of vibration employs passive controllers. By definition, passive devices do not require external power for their operation. The two passive controllers of vibration discussed in the present section are vibration absorbers (or dynamic absorbers or Frahm absorbers, named after H. Frahm, who first employed the technique for controlling ship oscillations) and dampers. In both types of devices, sensing is implicit and control is achieved through a force generated by the device from its response to the vibration excitation. A dynamic absorber is a mass–spring-type mechanism with little or no damping,

# Box 7.6

## Test-Based Design Approaches for Vibration

1. **Component modification**:
Modify a component (mass, spring, damper) and determine modal parameters (natural frequencies, damping ratios, mode shapes).
   - Can determine sensitivity to component changes.
   - Can check whether a particular change is satisfactory.

2. **Modal response specification**:
Specify a desired modal response (natural frequencies, damping ratios, mode shapes) and determine the "best" component changes (mass, spring, damper) that will realize the modal specs.
   - Can be accomplished by first performing a sensitivity study (as in item 1).

3. **Substructuring**:
   (i)   Design subsystems to meet specs (analytically, experimentally, or by a mixed approach).
   (ii)  Establish interconnections between subsystems, and obtain continuity (force balance) and compatibility (motion consistency) at assembly locations.
   (iii) Assemble the overall system by eliminating unknown variables at interconnections.
   (iv)  Analyze or test the overall system. If satisfactory, stop. Otherwise, make changes to the subsystems or interconnections, and repeat the above steps.

which can "absorb" the vibration excitation through energy transfer into it, thereby reducing the vibrations of the primary system. The energy received by the absorber will be slowly dissipated due to its own damping. A damper is a purely dissipative device which, unlike a dynamic absorber, directly dissipates the energy received from the system rather than storing it. Hence, it is a more wasteful device, which also may exhibit problems related to wear and thermal effects. However, it has advantages over an absorber, having, for example, a wider frequency of operation.

## 7.8.1 Undamped Vibration Absorber

A dynamic vibration absorber (or a dynamic absorber, vibration absorber, or Frahm absorber) is a simple mass–spring oscillator with very low damping. An absorber that is tuned to a frequency of vibration of a mechanical system and is able to receive a significant portion of the vibration energy from the primary system at that frequency. In effect, the resulting vibration of the absorber applies an oscillatory force opposing the vibration excitation of the primary system and thereby virtually cancels the effect. In theory, the vibration of the system can be completely removed while the absorber itself undergoes vibratory motion. Since damping is quite low in practical vibration absorbers, we will first consider the case of an undamped absorber.

A vibration absorber may be used for vibration control in two common types of situations, as shown in Figure 7.23. Here, the primary system whose vibration needs to be controlled is modeled as an

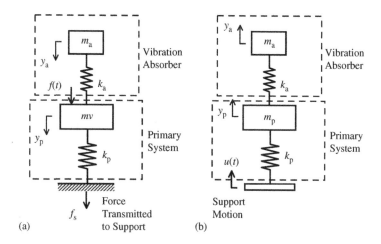

**FIGURE 7.23** Two types of applications of a vibration absorber: (a) reduction of the response to forcing excitation (or reducing the force transmitted to support structure); (b) reduction of the response to support motion.

undamped, single-DoF mass–spring system (denoted by the subscript p). An undamped vibration absorber is also a single-DoF mass–spring system (denoted by the subscript a). In the application shown in Figure 7.23(a), the objective of the absorber is to reduce the vibratory response $y_p$ of the primary system as a result of a vibration excitation $f(t)$. The force $f_s$ that is transmitted to the support structure due to the vibratory response of the system is given by

$$f_s = k_p y_p \tag{7.98}$$

Therefore, the objective of reducing $y_p$ may also be interpreted as one of reducing this transmitted vibratory force (a goal of vibration isolation). In the second type of application, represented in Figure 7.23(b), the primary system is excited by a vibratory support motion and the objective of the absorber is to reduce the resulting vibratory motions $y_p$ of the primary system. Note that, in both classes of application, the purpose is to reduce the vibratory responses. Hence, static loads (e.g., gravity) are not considered in the analysis.

Table 7.1 shows the development of the equations of motion for the two systems shown in Figure 7.23. Because we are interested mainly in the control of oscillatory responses to oscillatory excitations, the frequency-domain model is particularly useful. Note from Table 7.1 that the transfer function $f_s/f$ of System a is simply $k$ times the transfer function $y_p/f$, and is in fact identical to the transfer function $y_p/u$ of System b. The two problems are essentially identical and, thus, we need only address only one of them.

Before investigating the common transfer function for the two types of problems, let us look closely at the frequency-domain equations for the system shown in Figure 7.23(a). We have

$$(k_p + k_a - \omega^2 m_p)y_p - k_a y_a = f \tag{7.99}$$

$$(k_a - \omega^2 m_a)y_a = k_a y_p \tag{7.100}$$

along with Equation 7.98. Here, $m_p$ and $k_p$ are the mass and the stiffness of the primary system, $m_a$ and $k_a$ are the mass and the stiffness of the absorber, $f$ is the excitation amplitude, $\omega$ is the excitation frequency, $y_p$ is the primary mass response, and $y_a$ is the absorber response. Now note from Equation 7.100 that if $\omega = \sqrt{k_a/m_a}$ then $y_p = 0$. This means that if the absorber is tuned so that its natural frequency is equal to the excitation frequency (drive frequency), the primary system will not (ideally) undergo any vibratory motion, and is perfectly controlled. The reason for this should be clear from Equation 7.99 which, when $y_p = 0$ is substituted, gives $k_a y_a = -f$. In other words, a tuned absorber applies to the primary system a spring force that is exactly equal and opposite to the excitation force, thereby neutralizing the effect. The absorber mass moves, albeit 180° out of phase with the excitation.

**TABLE 7.1**    Equations for the Two Types of Absorber Applications

| | Absorber Application for the Reduction of Response to a: | |
|---|---|---|
| | Forcing Excitation | Support Motion |
| Time-domain equations | $m_p\ddot{y}_p = -k_p y_p - k_a(y_p - y_a) + f(t)$ <br> $m_a\ddot{y}_a = k_a(y_p - y_a)$ | $m_p\ddot{y}_p = k_p(u(t) - y_p) - k_a(y_p - y_a)$ <br> $m_a\ddot{y}_a = k_a(y_p - y_a)$ |
| Frequency-domain equations | $(-\omega^2 m_p + k_p + k_a)y_p = k_a y_a + f$ <br> $(-\omega^2 m_a + k_a)y_a = k_a y_p$ | $(-\omega^2 m_p + k_p + k_a)y_p = k_a y_a + k_p u$ <br> $(-\omega^2 m_a + k_a)y_a = k_a y_p$ |
| Matrix form | $\begin{bmatrix} k_p + k_a - \omega^2 m_p & -k_a \\ -k_a & k_a - \omega^2 m_a \end{bmatrix}\begin{bmatrix} y_p \\ y_a \end{bmatrix} = \begin{bmatrix} f \\ 0 \end{bmatrix}$ | $\begin{bmatrix} k_p + k_a - \omega^2 m_p & -k_a \\ -k_a & k_a - \omega^2 m_a \end{bmatrix}\begin{bmatrix} y_p \\ y_a \end{bmatrix} = k_p\begin{bmatrix} u \\ 0 \end{bmatrix}$ |
| Transfer-function matrix form | $\begin{bmatrix} y_p \\ y_a \end{bmatrix} = \dfrac{1}{\Delta}\begin{bmatrix} k_a - \omega^2 m_a & k_a \\ k_a & k_p k_a - \omega^2 m_p \end{bmatrix}\begin{bmatrix} f \\ 0 \end{bmatrix}$ | $\begin{bmatrix} y_p \\ y_a \end{bmatrix} = \dfrac{k}{\Delta}\begin{bmatrix} k_a - \omega^2 m_a & k_a \\ k_a & k_p k_a - \omega^2 m_p \end{bmatrix}\begin{bmatrix} u \\ 0 \end{bmatrix}$ |
| Vibration-control transfer function | $\dfrac{f_s}{f} = \dfrac{k_p y_p}{f} = \dfrac{k_p}{\Delta}(k_a - \omega^2 m_a)$ | $\dfrac{y_p}{u} = \dfrac{k_p}{\Delta}(k_a - \omega^2 m_a)$ |
| Characteristic polynomial | $\Delta = (k_p + k_a - \omega^2 m_p)(k_a - \omega^2 m_a) - k_a^2$ <br> $= m_p m_a \omega^4 - [k_a(m_p + m_a) + k_p m_a]\omega^2 + k_p k_a$ | |

The frequency of these motions will be $\omega$ (the same as that of the excitation) and the amplitude is proportional to that of the excitation ($f$) and inversely proportional to the stiffness of the absorber spring. It follows that a vibration absorber "absorbs" vibration energy from the primary system. Furthermore, note from Equation 7.98 that with a tuned absorber the vibration force transmitted to the support structure is (ideally) zero as well. All this information is observed without any mathematical manipulation of the equations of motion.

Note that we are dealing with vibratory excitations and responses. Therefore, static loading (such as gravity and spring preloads) is not considered (we investigate responses with respect to the static equilibrium configuration of the system). In summary, we are now able to state the characteristics of a vibration absorber (undamped) as follows:

1. It is effective only for a single excitation frequency (i.e., a sinusoidal excitation).
2. For the best effect, it should be "tuned" such that its natural frequency $\sqrt{k_a/m_a}$ is equal to the excitation frequency.
3. In the case of forcing vibration excitation, a tuned absorber can (ideally) make the vibratory response of the primary system and the vibratory force transmitted to the support structure zero.
4. In the case of a vibratory support motion, a tuned absorber can make the resulting response of the primary system zero.
5. It functions by acquiring vibration energy from the primary system and storing it (as kinetic energy of the mass or potential energy of the spring) rather than by directly dissipating the energy.
6. It functions by applying a vibration force to the primary system that is equal and opposite to the excitation force, thereby neutralizing the excitation.
7. The amplitude of motion of the vibration absorber is proportional to the excitation amplitude and is inversely proportional to the absorber stiffness. The frequency of the absorber motion is the same as the excitation frequency.

Now, consider the transfer function ($f_s/f$ or $y_p/u$) of an undamped vibration absorber, as given in Table 7.1. We have

$$G(\omega) = \frac{k_p(k_a - \omega^2 m_a)}{m_p m_a \omega^4 - [k_a(m_p + m_a) + k_p m_a]\omega^2 + k_p k_a} \tag{7.101}$$

It is convenient to use a nondimensional form in analyzing this frequency-transfer function. To that end, we define the following nondimensional parameters and frequency variable:

Fractional mass of the absorber $\mu = m_a/m_p$
Nondimensional natural frequency of the absorber $\alpha = \omega_a/\omega_p$
Nondimensional excitation (drive) frequency $r = \omega/\omega_p$

where

$\omega_a = \sqrt{k_a/m_a}$ = natural frequency of the absorber
$\omega_p = \sqrt{k_p/m_p}$ = natural frequency of the primary system

It is straightforward to divide the numerator and the denominator by $k_p k_a$ and then carry out simple algebraic manipulations to express the transfer function of Equation 7.101 in the nondimensional form as

$$G(r) = \frac{\alpha^2 - r^2}{r^4 - [\alpha^2(1 + \mu) + 1]r^2 + \alpha^2} \tag{7.102}$$

For this undamped system, there is no difference between the resonant frequencies (where the magnitude of the transfer function peaks) and the natural frequencies (roots of the characteristic equation that correspond to the "natural" or free time response oscillations). These are obtained by solving the equation

$$r^4 - [\alpha^2(1 + \mu) + 1]r^2 + \alpha^2 = 0 \tag{7.103}$$

which gives

$$r_1^2, r_2^2 = \frac{1}{2}[\alpha^2(1 + \mu) + 1] \mp \frac{1}{2}\sqrt{[\alpha^2(1 + \mu) + 1]^2 - 4\alpha^2} \tag{7.104}$$

These are squared frequencies, both of which are positive as clear from Equation 7.104. The actual, nondimensional natural frequencies are their square roots. The magnitude of the transfer function becomes infinite at either of these two natural/resonant frequencies. Furthermore, it is clear from Equation 7.102 that the transfer function magnitude becomes zero at $r = \alpha$, where the excitation frequency ($\omega$) is equal to the natural frequency of the absorber ($\omega_a$) as noted above. In the present undamped case, the transfer function $G(r)$ is real but it can be either positive or negative. The magnitude is thus the absolute value of $G(r)$, which is positive. The magnitude plot given in Figure 7.24 shows the resonant and control characteristics of a system with an undamped vibration absorber. Originally, the primary system had a resonance at $r = 1$ (i.e., $\omega = \omega_p$). When the absorber, which also has a resonance at $r = 1$, is added, the original resonance becomes an *antiresonance* with a zero response. Here, however, two new resonances are created, one at $r = 0.854$ and the other at $r = 1.171$, which are on either side of the tuned frequency ($r = 1$) of the absorber.

Owing to these two resonances, the effective region of the absorber is limited to a narrow frequency band centering its tuned frequency. Specifically, the absorber is not effective unless $|G| < 1$. The effective frequency band of a vibration absorber may be determined using this condition.

## Example 7.8

A high-precision, yet high-power positioning system uses a hydraulic actuator and a valve. The pressurized oil to this hydraulic servo system is provided by a gear-type rotary pump. The pump and the positioning system are mounted on the same workbench. The mass of the pump is 25 kg. The normal operating speed of the pump is 3600 rpm. During operation, it was observed that the pump exhibits a vertical resonance at this speed and it affects the accuracy of the position servo system. To control the vibrations of the pump at its operating speed, a vibration absorber of mass 1.25 kg tuned to the normal operating speed of the pump is attached, as shown schematically in Figure 7.25. Because the speed of the pump normally fluctuates during operation, we must determine the speed range within which the vibration absorber is effective. What are the new resonant frequencies of the system? (Neglect damping.)

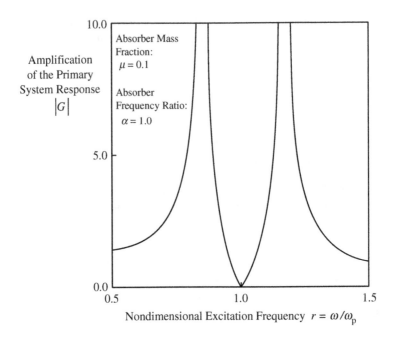

**FIGURE 7.24**   The effect of an undamped vibration absorber on the vibration response of a primary system.

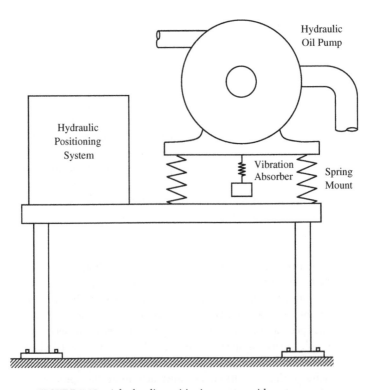

**FIGURE 7.25**   A hydraulic positioning system with a gear pump.

## Solution

For this problem, the fractional mass $\mu = 1.25/25.0 = 0.05$. Since the absorber is tuned to the resonant frequency of the pump, $\alpha = 1.0$. From Equation 7.103, the characteristic equation of the modified system becomes

$$r^4 - 2.05r^2 + 1 = 0$$

which has roots $r_1 = 0.854$ and $r_2 = 1.171$. It follows that the new resonances are at $0.854 \times 3600$ and $1.171 \times 3600$ rpm. These are 3074.4 and 4215.6 rpm, which should be avoided. From Equation 7.102, the system transfer function is

$$G(r) = \frac{1 - r^2}{(r^4 - 2.05r^2 + 1)}$$

The effective frequency band of the absorber corresponds to $|G(r)| < 1.0$. Since a sign reversal of $G(r)$ occurs at $r = 1$, we need to solve both

$$\frac{1 - r^2}{(r^4 - 2.05r^2 + 1)} = 1 \text{ and } -1$$

The first equation gives the roots $r = 0$ and $1.025$. The second equation gives the roots $r = 0.977$ and $1.45$.

Hence, the effective frequency band corresponds to $\Delta r = [0.977, 1.025]$. In terms of the operating speed of the pump, we have an effective band of 3517.2 to 3690 rpm. Thus, a speed fluctuation of about $\pm 80$ rpm is acceptable.

Finally, recall that the presence of the absorber generates two new resonances on either side of the resonance of the original system (to which the absorber is normally tuned). It is also clear from Equation 7.104 that these two resonances become farther and farther apart as the fractional mass $\mu$ of the vibration absorber is increased.

## 7.8.2 Damped Vibration Absorber

Damping is not the primary means by which vibration control is achieved in a vibration absorber. As noted above, the absorber acquires vibration energy from the primary system (and in turn, exerts a force on the system that is equal and opposite to the vibration excitation), thereby suppressing the vibratory motion. The energy received by the absorber has to be dissipated gradually and, hence, some damping should be present in the absorber. Furthermore, the two resonances that are created by adding the absorber have an infinite magnitude in the absence of damping. Hence, damping has the added benefit of lowering these resonant peaks.

The analysis of a vibratory system with a damped absorber is similar to but somewhat more complex than, that involving an undamped absorber. Furthermore, an extra design parameter — the damping ratio of the absorber — enters into the scene. Consider the model shown in Figure 7.26. Another version of application of a damped

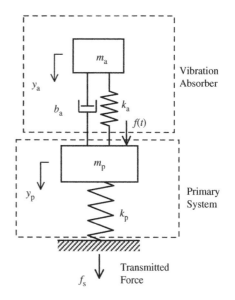

**FIGURE 7.26** Primary system with a damped vibration absorber.

absorber, which corresponds to Figure 7.23(b), may also be presented. However, because the two types of application have the same transfer function, it is sufficient to consider Figure 7.26 alone.

Again, the transfer function of vibration control may be taken as either $y_a/f$ or $f_s/f$, the latter being simply $k_p$ times the former. Although we will consider the dimensionless case of $f_s/f$, the results are equally valid for $y_p/f$, except that the responses must be converted from force to displacement by dividing by $k_p$.

There is no need to derive the transfer function anew for the damped system. Simply replace $k_a$ in Equation 7.101 by the complex stiffness $k_a + j\omega b_a$, which incorporates the viscous damping constant $b_a$ and the excitation frequency $\omega$. Hence, the transfer function of the damped system is

$$G(\omega) = \frac{k_p(k_a + j\omega b_a - \omega^2 m_a)}{m_p m_a \omega^4 - [(k_a + j\omega b_a)(m_p + m_a) + k_p m_a]\omega^2 + k_p(k_a + j\omega b_a)} \quad (7.105)$$

With the parameters defined as before, the nondimensional form of this transfer function is obtained by dividing throughout by $k_p k_a$ and then substituting the appropriate parameters. In particular, we use the fact that

$$\frac{b_a}{k_a} = \frac{2b_a}{2\sqrt{k_a m_a}}\sqrt{\frac{m_a}{k_a}} = \frac{2\zeta_a}{\omega_a} = \frac{2\zeta_a}{\omega_p}\frac{\omega_p}{\omega_a} = \frac{2\zeta_a}{\alpha\omega_p}$$

$$(7.106)$$

where the damping ratio $\zeta_a$ of the absorber is given by

$$\zeta_a = \frac{b_a}{2\sqrt{k_a m_a}} \quad (7.107)$$

as usual. Then, we follow the same procedure that used to derive Equation 7.102 from Equation 7.101 to get

$$G(r) = \frac{\alpha^2 - r^2 + 2j\zeta_a\alpha r}{r^4 - [(\alpha^2 + 2j\zeta_a\alpha r)(1 + \mu) + 1]r^2 + (\alpha^2 + 2j\zeta_a\alpha r)} \quad (7.108)$$

Note that this result is equivalent to simply replacing $\alpha^2$ by $\alpha^2 + 2j\zeta_a\alpha r$ in Equation 7.102.

It is important to note that the undamped natural frequencies are obtained by solving the characteristics equation with $\zeta_a = 0$. These are the same as before and given by the square roots of Equation 7.104. The damped natural frequencies are obtained by first setting $jr = \lambda$ (hence, $r^2 = -\lambda^2$ and $r^4 = \lambda^4$) and then solving the resulting characteristics equation (see the denominator of Equation 7.108).

$$\lambda^4 + 2\zeta_a\alpha(1 + \mu)\lambda^3 + (\alpha^2 + \alpha^2\mu + 1)\lambda^2 + 2\zeta_a\alpha\lambda + \alpha^2 = 0 \quad (7.109)$$

and then taking the imaginary parts of the roots of $\lambda$. These depend on $\zeta_a$ and are different from those obtained from Equation 7.104. The resonant frequencies correspond to the $r$ values where the magnitude of $G(r)$ will peak. Generally, these are not the same as the undamped or damped natural frequencies. However, for low damping (small $\zeta_a$ compared with 1), these three types of system characteristics frequencies are almost identical.

The magnitude of the transfer function (Equation 7.108) is plotted in Figure 7.27 for the case $\mu = 1.0$ and $\alpha = 1.0$, as in Figure 7.24, but for damping ratios $\zeta_a = 0.01$, 0.1, and 0.5. Note that the curve for $\zeta_a = 0.01$ is very close to that in Figure 7.24 for the undamped case. When $\zeta_a$ is large, as shown in the case of $\zeta_a = 0.5$, the two masses $m_p$ and $m_a$ tend to become locked together and appear to behave like a single mass. Then, the system tends to act like a single-DoF one, and the primary system is modified only in its mass (which increases). Consequently, only one resonant frequency is produced, which is smaller than that of the original primary system. Furthermore, as expected in this high-damping case, the effect of a vibration absorber is no longer present.

All three curves in Figure 7.27 pass through the two common points A and B, as shown. This is true for all curves corresponding to all values of $\zeta_a$, and particularly for the extreme cases of $\zeta_a = 0$ and $\zeta_a \to \infty$. Hence, these points can be determined as the points of intersection of the transfer function magnitude curves for the limiting cases $\zeta_a = 0$ and $\zeta_a \to \infty$.

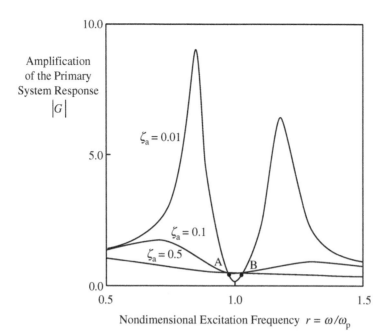

**FIGURE 7.27** Vibration amplification (transfer function magnitude) curves for damped vibration absorbers (absorber mass $\mu = 0.1$, absorber resonant frequency $\alpha = 1.0$).

Equation 7.102 gives $G(r)$ for $\zeta_a = 0$. Next, from Equation 7.108, we note that, as $\zeta_a \to \infty$, all the terms not containing $\zeta_a$ can be neglected. Hence,

$$G(r) = \frac{2j\zeta_a \alpha r}{-2j\zeta_a \alpha r(1 + \mu)r^2 + 2j\zeta_a \alpha r}$$

Cancel the common term and we get (for $r \neq 0$)

$$G(r) = \frac{1}{1 - (1 + \mu)r^2} \quad \text{for } \zeta_a \to \infty \tag{7.110}$$

Note that this is the normalized transfer function of a single-DoF system of natural frequency $1/\sqrt{1 + \mu}$. This result confirms the fact that as $\zeta_a \to \infty$, the two masses $m_p$ and $m_a$ become locked together and act as a single mass $(m_p + m_a)$ supported on a spring of stiffness $k_p$. Its natural frequency is $\sqrt{k_p/(m_p + m_a)}$ which, when normalized with respect to $\sqrt{k_p/m_p}$, becomes

$$\sqrt{\frac{k_p}{(m_p + m_a)} \frac{m_p}{k_p}} = \sqrt{\frac{m_p}{(m_p + m_a)}} = \frac{1}{\sqrt{1 + \mu}}$$

In determining the points of intersection between the functions (see Equation 7.102 and Equation 7.110), we should first note that at the first point of intersection (A), the function in Equation 7.102 is negative and positive in Equation 7.110, while the reverse is true for the second point of intersection (B). For either point, this means that the sign of one of the functions should be reversed before equating them. Thus,

$$\frac{\alpha^2 - r^2}{r^4 - [\alpha^2(1 + \mu) + 1]r^2 + \alpha^2} = -\frac{1}{1 - (1 + \mu)r^2}$$

which gives

$$(2 + \mu)r^4 - 2[\alpha^2(1 + \mu) + 1]r^2 + 2\alpha^2 = 0 \tag{7.111}$$

This is the equation whose roots (e.g., $r_1$ and $r_2$) give the points A and B. Then, we have the sum of the squared roots equal to the negative coefficient of $r^2$ in the quadratic (in $r^2$) Equation 7.111. Thus,

$$r_1^2 + r_2^2 = \frac{2[\alpha^2(1+\mu)+1]}{(2+\mu)} \tag{7.112}$$

In addition, the product of the squared roots is equal to the constant term in the quadratic ($r^2$) in Equation 7.111. Hence,

$$r_1^2 r_2^2 = \frac{2\alpha^2}{(2+\mu)} \tag{7.113}$$

### 7.8.2.1 Optimal Absorber Design

It has been pointed out (primarily by J.P. Den Hartog) that an optimal absorber design should not only have equal response magnitudes at the common points of intersection (i.e., equal ordinates of points A and B in Figure 7.27), but also that the resonances should occur at these points to achieve some balance and uniformity in the response amplification in the region surrounding the tuned frequency of the absorber. It is expected that these (intuitive) design conditions would give relations between the parameters $\alpha$, $\mu$, and $\zeta_a$, corresponding to an optimal absorber.

Consider the first requirement of equal transfer function magnitudes at A and B. As noted earlier, because these two points do not depend on $\zeta_a$, we use Equation 7.110 to satisfy the requirement. Thus, keeping in mind the sign reversal of the transfer function between A and B (i.e., as the transfer function passes through the resonance), we have

$$\frac{1}{1-(1+\mu)r_1^2} = -\frac{1}{1-(1+\mu)r_2^2}$$

which gives

$$r_1^2 + r_2^2 = \frac{2}{1+\mu} \tag{7.114}$$

Substituting this result (for equal ordinates) in the intersection-point condition (see Equation 7.112), we have

$$\frac{2}{1+\mu} = \frac{2[\alpha^2(1+\mu)+1]}{(2+\mu)}$$

On simplification, we get the simple result

$$\alpha = \frac{1}{1+\mu} \tag{7.115}$$

Next, we turn to the task of achieving peak magnitudes of the transfer function at the points of intersection (A and B). Generally, when one point peaks the other does not. As reported by Den Hartog, with straightforward but lengthy analysis, we obtain

$$\zeta_a^2 = \frac{\mu[3 - \sqrt{\mu/(\mu+2)}]}{8(1+\mu)^3} \tag{7.116}$$

for peak at the first intersection point, and

$$\zeta_a^2 = \frac{\mu[3 + \sqrt{\mu/(\mu+2)}]}{8(1+\mu)^3} \tag{7.117}$$

for peak at the second intersection point.

So, for design purposes, a balance is obtained by taking the average value of the results of Equation 7.116 and Equation 7.117 as

$$\zeta_a^2 = \frac{3\mu}{8(1+\mu)^3} \tag{7.118}$$

Thus, Equation 7.115 and Equation 7.118 correspond to an optimal vibration absorber. In addition, practical requirements and limitations need to be addressed in any design procedure. In particular, since $\mu$ is considerably less than unity (i.e., absorber mass is a small fraction of the primary mass), the absorber mass should undergo relatively large amplitudes at the operating frequency in order to receive the energy of the primary system. The absorber spring must be designed accordingly, while meeting the tuning frequency conditions that determine the ratio $m_a/k_a$.

## Example 7.9

The air compressor of a wind tunnel weighs 48 kg and normally operates at 2400 rpm. The first major resonance of the compressor unit occurs at 2640 rpm, with severe vibration amplitudes that are quite dangerous. Design a vibration absorber (damped) for installation on the mounting base of the compressor. What are the vibration amplifications of the compressor unit at the new resonances of the modified system? Compare these with the vibration amplitude of the original system in normal operation.

## Solution

As usual, we will tune the absorber to the normal operating speed (2400 rpm). Then, we have the nondimensional resonant frequency of the absorber:

$$\alpha = \frac{\omega_a}{\omega_p} = \frac{2400}{2640} = \frac{12}{13}$$

Now, for an optimal absorber, from Equation 7.115

$$\mu = \frac{1}{\alpha} - 1 = \frac{13}{12} - 1 = \frac{1}{12}$$

Hence, the absorber mass

$$m_a = 48 \times \frac{1}{12} \text{ kg} = 4.0 \text{ kg}$$

Then, from Equation 7.118, the damping ratio of the absorber is

$$\zeta_a = \left[ \frac{3/12}{8(1+1/12)^3} \right]^{1/2} = 0.157$$

Now,

$$\omega_a = \sqrt{\frac{k_a}{m_a}} = \sqrt{\frac{k_a}{4.0}} = \frac{2400}{60} \times 2\pi \text{ rad/s} = 88\pi \text{ rad/s}$$

Hence,

$$k_a = (88\pi)^2 \times 4.0 \text{ N/m} = 2.527 \times 10^5 \text{ N/m}$$

Also,

$$\zeta_a = \frac{1}{2} \frac{b_a}{\sqrt{m_a k_a}}$$

Then, we have

$$b_a = 2 \times 0.157\sqrt{4.0 \times 2.527 \times 10^5} \text{ N s/m} = 315.7 \text{ N s/m}$$

This gives us the damped absorber. Now, let us check its performance. We know that, in theory, the vibration amplitude at the operating speed should be almost zero now. However, two resonances are created around the operating point. Since damping is small, we use the undamped characteristic Equation 7.103 to compute these resonances:

$$r^4 - \left[ \frac{12^2}{13^2}\left(1 + \frac{1}{12}\right) + 1 \right]r^2 + \frac{12^2}{13^2} = 0$$

which gives

$$r^4 - \frac{25}{13}r^2 + \frac{12^2}{13^2} = 0$$

The roots of $r^2$ are 0.692 and 1.231. The (positive) roots of $r$ are 0.832 and 1.109.

These correspond to compressor speeds of (multiply $r$ by 2640 rpm) 2196 and 2929 rpm. Although they are approximately at $-10\%$ and $+20\%$ of the operating speed, the first resonance will be encountered during startup and shutdown conditions. To determine the corresponding vibration amplifications (force/force), use Equation 7.108 which, when the undamped characteristic equation is substituted into the denominator, becomes

$$G(r) = \frac{\alpha^2 - r^2 + 2j\zeta_a \alpha r}{[-2j\zeta_a \alpha r(1 + \mu)r^2 + 2j\zeta_a \alpha r]} = \frac{1 - j(\alpha^2 - r^2)/(2\zeta_a \alpha r)}{1 - (1 + \mu)r^2} \qquad (7.119)$$

Substitute the resonant frequencies $r_1 = 0.832$ and $r_2 = 1.109$. We get $|G(r_1)| = 4.223$ and $|G(r_2)| = 4.634$.

Without the absorber, we approximate the system by a simple undamped oscillator with transfer function

$$G_p(r) = \frac{1}{1 - r^2}$$

The corresponding vibration amplification at the operating speed is

$$|G_p(r_0)| = \frac{1}{|1 - 12^2/13^2|} = 6.76$$

It is observed that after adding the absorber, the resonant vibrations are smaller than even the operating vibrations of the original system. Hence, the design is satisfactory. Note that we used the force/force transfer functions. To get the displacement/force transfer functions we divide by $k_p$. However, we have

$$\sqrt{\frac{k_p}{m_p}} = \frac{2640}{60} \times 2\pi \text{ rad/s} = 88\pi \text{ rad/s}$$

Hence,

$$k_p = (88\pi)^2 \times 48 \text{ N/m} = 3.67 \times 10^6 \text{ N/m} = 3.67 \times 10^3 \text{ N/mm}$$

Thus, the amplitude of operating vibrations of the original system is

$$\frac{6.76}{3.67 \times 10^3}\text{mm/N} = 1.84 \times 10^{-3} \text{ mm/N}$$

The amplitudes of the resonant vibrations of the modified system are

$$\frac{4.223}{3.67 \times 10^3} \quad \text{and} \quad \frac{4.634}{3.67 \times 10^3} \text{ mm/N or } 1.15 \times 10^{-3} \text{ and } 1.26 \times 10^{-3} \text{ mm/N}$$

Vibration absorbers are simple and passive devices, which are commonly used in the control of narrowband vibrations (limited to a very small interval of frequencies). Applications are found in vibration suppression of transmission wires (e.g., a stockbridge damper, which simply consists of a piece of cable carrying two masses at its ends), consumer appliances, automobile engines, and industrial machinery. It should be noted that the concepts presented for a rectilinear vibration absorber may be directly extended to a rotary vibration absorber. Figure 7.28 provides a schematic representation of a rotary vibration absorber. This model corresponds to vibration force excitations (compare with Figure 7.23(a)). The case of rotational support-motion excitations (see Figure 7.23(b)), which has essentially the same transfer function, may also be addressed. Approaches of vibration control are summarized in Box 7.7.

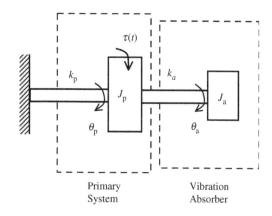

**FIGURE 7.28** The application of a rotary vibration absorber.

### 7.8.3 Vibration Dampers

As discussed above, vibration absorbers are simple and effective passive devices, which are used in vibration control. They have the added advantage of being primarily nondissipative. The main disadvantage of a vibration absorber is that it is only effective over a very narrow band of frequencies enclosing its resonant frequency (tuned frequency). When passive vibration control over a wide band of frequencies is required, a damper would be a preferable choice.

Vibration dampers are dissipative devices. They control vibration through direct dissipation of the vibration energy of the primary (vibrating) system. As a result, however, there will be substantial heat generation, and associated thermal problems and component wear. Consequently, methods of cooling (e.g., use of a fan, coolant circulation, and thermal conduction blocks) may be required in some special situations.

Consider a vibrating system modeled as an undamped single-DoF mass–spring system (simple oscillator). The magnitude of the excitation-response transfer function will have a resonance with a theoretically infinite magnitude in this case. Operation in the immediate neighborhood of such a resonance would be destructive. Adding a simple viscous damper, as shown in Figure 7.29(a), will correct the situation. The equation of motion (about the static equilibrium position) is

$$m\ddot{y} + b\dot{y} + ky = f(t) \tag{7.120}$$

with the dynamic force that is transmitted through the support base ($f_s$) given by

$$f_s = ky + b\dot{y} \tag{7.121}$$

Hence, the transfer function between the forcing excitation $f$ and the vibration response $y$ is

$$\frac{y}{f} = \frac{1}{k - \omega^2 m + j\omega b} \tag{7.122}$$

and that between the forcing excitation and the force transmitted to the support structure is

$$\frac{f_s}{f} = \frac{k + j\omega b}{k - \omega^2 m + j\omega b} \tag{7.123}$$

Using the nondimensional frequency variable $r = \omega/\omega_n$ where $\omega_n = \sqrt{k/m}$ is the undamped natural frequency of the system and the damping ratio $\zeta = b/(2\sqrt{km})$, we can express Equation 7.122 and

# Box 7.7

# VIBRATION CONTROL

**Passive control (no external power):**

1. Dampers
   - A dissipative approach (thermal problems, degradation)
   - Useful over a wide frequency band
2. Vibration absorbers (dynamic absorbers, Frahm absorbers)
   - Absorbs energy from vibrating system and applies counteracting force
   - Useful over a very narrow frequency band (near the tuned frequency)
   - Absorber executes large motions

*Undamped absorber design:*

$$\text{Transfer function of system with absorber} = \frac{\alpha^2 - r^2}{r^4 - [\alpha^2(1 + \mu) + 1]r^2 + \alpha^2}$$

where

$\mu$ = absorber mass/primary system mass
$\alpha$ = absorber natural frequency/primary system natural frequency
$r$ = excitation frequency/primary system natural frequency

The most effective operating frequency $r_{op} = \alpha$.
Avoid the two resonances.

*Optimal damped absorber design:*
   Mass ratio

$$\mu = \frac{1}{\alpha} - 1$$

   Damping ratio

$$\zeta_a = \frac{3\mu}{8(1 + \mu)^3}$$

**Active control (needs external power):**

1. Measure vibration response using sensors/transducers.
2. Apply control forces to vibrating system through actuators, according to a suitable control algorithm.

Equation 7.123 in the form

$$\frac{y}{f} = \frac{1}{k(1 - r^2 + 2j\zeta r)} \tag{7.124}$$

$$\frac{f_s}{f} = \frac{1 + 2j\zeta r}{(1 - r^2 + 2j\zeta r)} \tag{7.125}$$

When vibration control of the primary system is desired, we use the transfer function in Equation 7.124. However, when force transmissibility is the primary consideration, we use Equation 7.125.

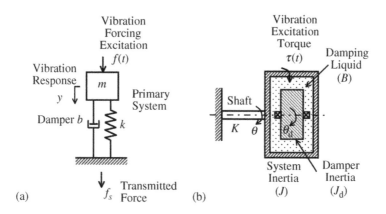

FIGURE 7.29   (a) A system with a linear viscous damper; (b) a rotary system with a Houdaille damper.

Furthermore, it is convenient to use the transfer function in Equation 7.124 in the nondimensional form:

$$\frac{ky}{f} = G(r) = \frac{1}{(1 - r^2 + 2j\zeta r)} \tag{7.126}$$

The magnitude of this transfer function is plotted in Figure 7.30 for several values of damping ratio. Note that the addition of significant levels of damping considerably lowers the resonant peak and flattens the overall response. This example illustrates the broadband nature of the effect of a damper. However, unlike a vibration absorber, it is not possible with a simple damper to bring the vibration levels to a theoretical zero. However, a damper is able to bring the response uniformly close to the static value (unity in Figure 7.30).

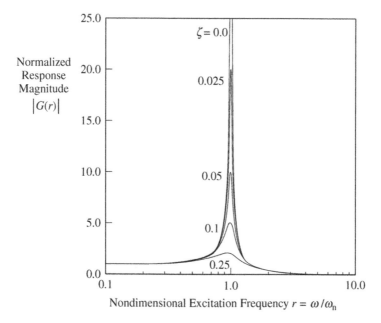

FIGURE 7.30   Frequency response of a system containing a linear damper.

Another common application of damper is connecting it through a free inertia element. For a rotational system, such an arrangement is know as the Houdaille damper, and is modeled as in Figure 7.29(b). The equations of motion are

$$J\ddot{\theta} + B(\dot{\theta} - \dot{\theta}_d) + K\theta = \tau(t) \tag{7.127}$$

$$J_d\ddot{\theta}_d + B(\dot{\theta}_d - \dot{\theta}) = 0 \tag{7.128}$$

In this case, the transfer function between the vibratory excitation torque $\tau$ and the response angle $\theta$ is given by

$$\frac{\theta}{\tau} = \frac{B + J_d j\omega}{KB - B(J + J_d)\omega^2 - J_d J j\omega^3 + K J_d j\omega} \tag{7.129}$$

Again, we use the normalized form of $K\theta/\tau$. Then, we obtain

$$\frac{K\theta}{\tau} = G(r) = \frac{2\zeta + jr\mu}{2\zeta[1 - (1 + \mu)r^2] + jr\mu(1 - r^2)} \tag{7.130}$$

where $r = \omega/\omega_n$, $\zeta = B/(2\sqrt{KJ})$, $\mu = J_d/J$, and $\omega_n = \sqrt{K/J}$.

Note the two extreme cases. When $\zeta = 0$, the system becomes the original undamped system, as expected. When $\zeta \to \infty$, the system becomes an undamped simple oscillator, but with a lower natural frequency of $r = 1/\sqrt{1 + \mu}$, instead of $r = 1$ that was present in the original system. This is to be expected because as $\zeta \to \infty$, the two inertia elements become locked together and act as a single combined inertia $J + J_d$. Clearly, in these two extreme systems, the effect of damping is not present. Optimal damping occurs somewhere in between, as is clear from the curves of response magnitude shown in Figure 7.31 for the case of $\mu = 0.2$.

Proper selection of the nature and values of damping is crucial for the effective use of a damper in vibration control. Damping in physical systems is known to be nonlinear and frequency dependent, as well as time-variant and dependent on the environment (e.g., temperature). Various models are available for different types of damping, but these are only approximate representations. In practice, such considerations as the type of damper used, the nature of the system, the specific application, and the

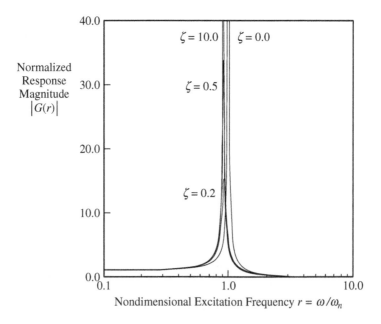

**FIGURE 7.31**  Response curves for a rotary system with Houdaille damper of inertia ratio $\mu = 0.2$.

speed of operation, determine which particular model (linear viscous, hysteric, Coulomb, Stribeck, quadratic aerodynamic, etc.) is suitable. In addition to the simple linear theory of viscous damper, specific properties of physical damping should be taken into consideration in practical designs.

## 7.9 Active Control of Vibration

Passive control of vibration is relatively simple and straightforward. Although it is robust, reliable, and economical, it has its limitations. Note that the control force that is generated in a passive device depends entirely on the natural dynamics. Once the device is designed (i.e., after the parameter values for mass, damping constant, stiffness, location, etc. are chosen), it is not possible to adjust the control forces that are naturally generated in real time. Furthermore, in a passive device there is no supply of power from an external source. Hence, even the magnitude of the control forces cannot be changed from their natural values. Since a passive device senses the response of the system as an integral process of the overall dynamics of the system, it is not always possible to directly target the control action at particular responses (e.g., particular modes). This can result in incomplete control, particularly in complex and high-order (e.g., distributed-parameter) systems. These shortcomings of passive control can be overcome using active control. Here, the system responses are directly sensed using sensor-transducer devices, and control actions of specific desired values are applied to desired locations/modes of the system.

### 7.9.1 Active Control System

Figure 7.32 presents a schematic diagram of an active control system. The mechanical dynamic system whose vibrations need to be controlled is the *plant* or *process*. The controller is the device that generates the signal (or command) according to some scheme (or control law) and controls vibrations of the plant. The plant and the controller are the two essential components of a *control system*. Usually, the plant must

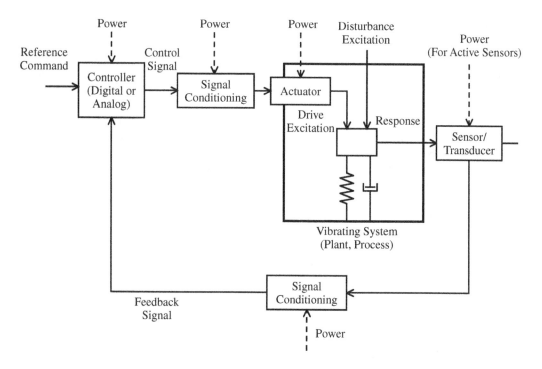

**FIGURE 7.32** A system for active control of vibration.

be monitored and its response must be measured using sensors providing feedback into the controller. Then, the controller compares the sensed signal with a desired response specified externally, and uses the error to generate a proper control signal. In this manner, we have a feedback control system. In the absence of a sensor and feedback, we have an open-loop control system. In *feed-forward control*, the excitation (i.e., input signal), not the response (i.e., output signal), is measured and used (i.e., fed forward into the controller) for generating the control signal. Both feedback and feedforward schemes may be used in the same control system.

The actuator that receives a control signal and drives the plant may be an integral part of the plant (e.g., the motor that drives the blade of a saw). Alternatively, it may have to be added specifically as an external component for the control actuation (e.g., a piezoelectric or electromagnetic actuator for controlling blade vibrations of a saw). In the former case, in particular, proper signal conditioning is needed to convert the control signal to a form that is compatible with the existing actuator. In the latter case, both the controller and the actuator must be developed in parallel for integration into the plant. In digital control, the controller is a digital processor. The control signal is in digital form and, typically, it has to be converted into the analog form prior to using in the actuator. Hence, digital-to-analog conversion (DAC) is a form of signal conditioning that is useful here. Furthermore, the analog signal that is generated may have to be filtered and amplified to an appropriate level for use in the actuator. It follows that filters and amplifiers are signal conditioning devices, which are useful in vibration control. In *software control*, the control signal is generated by a computer, which functions as the digital controller. In *hardware control*, the control signal is rapidly generated by digital hardware without using software programs. Alternatively, *analog control* may be used where the control signal is generated directly using analog circuitry. In this case, the controller is quite fast and it does not require DAC. Note that the actuator may need high levels of power. Furthermore, the controller and associated signal conditioning will require some power. The need for an external power source for control distinguishes active control from passive control.

In a feedback control system, sensors are used to measure the plant response, which enables the controller to determine whether the plant operates properly. A sensor unit that "senses" the response may automatically convert (transduce) this "measurement" into a suitable form. A piezoelectric accelerometer senses acceleration and converts it into an electric charge, an electromagnetic tachometer senses velocity and converts it into a voltage, and a shaft encoder senses a rotation and converts it into a sequence of voltage pulses. Hence, the terms *sensor* and *transducer* are used interchangeably to denote a sensor-transducer unit. The signal that is generated in this manner may need conditioning before feeding into the controller. For example, the charge signal from a piezoelectric accelerometer has to be converted to a voltage signal of appropriate level using a charge amplifier, and then it has to be digitized using an analog-to-digital converter (ADC) for use in a digital controller. Furthermore, filtering may be needed to remove measurement noise. Hence, signal conditioning is usually needed between the sensor and the controller as well as between the controller and the actuator. External power is required to operate active sensors (e.g., potentiometer) whereas passive sensors (e.g., electromagnetic tachometer) employ self-generation and do not need an external power source. External power may be needed for conditioning the sensor signals. Finally, as indicated in Figure 7.32, a vibrating system may have unknown disturbance excitations, which can make the control problem particularly difficult. Removing such excitations at the source level through proper design or vibration isolation is desirable, as discussed above. However, in the context of control, if these disturbances can be measured or some information about them is available, then they can be compensated for within the controller itself. This is, in fact, the approach of feedforward control.

## 7.9.2   Control Techniques

The purpose of a vibration controller is to excite (activate) a vibrating system in order to control its vibration response in a desired manner. In the present context of active feedback control, the controller uses measured response signals and compares them with their desired values in its task of determining an appropriate action. The relationship that generates the control action from a measured response (and a desired value for the response) is called a *control law*. Sometimes, a *compensator* (analog or digital, hardware or software) is employed to improve the system performance or to enhance the controller so that the task of control is easier. However, for our purpose, we may consider a compensator as an integral part of the controller and thus a distinction between the two is not made.

Various control laws, both linear and nonlinear, have been developed for practical applications. Many of them are suitable in vibration control. A comprehensive presentation of all such control laws is outside the scope of this book. We will give several linear control laws that are common and representative of what is available. These techniques are based on a linear representation (linear model) of the vibrating system (plant). Even when the overall operating range of a plant (e.g., robotic manipulator) is nonlinear, it is often possible to linearize the vibration response (e.g., link vibrations and joint vibrations of a robot) about a reference configuration (e.g., robot trajectory). These linear control techniques would be still suitable even though the overall dynamics of the system is nonlinear.

### 7.9.2.1 State-Space Models

In applying many types of control techniques, it is convenient to represent the vibrating system (plant) by a state-space model. This is simply a set of ordinary first-order differential equations, which could be coupled or nonlinear, and could have time-varying parameters (time-variant models). Here, we limit our discussion to linear and time-invariant state-space models. Such a model is expressed as

$$\dot{\mathbf{x}} = \mathbf{A}\mathbf{x} + \mathbf{B}\mathbf{u} \tag{7.131}$$

$$\mathbf{y} = \mathbf{C}\mathbf{x} + \mathbf{D}\mathbf{u} \tag{7.132}$$

where

$\mathbf{x} = [x_1, x_2, \ldots, x_n]^T$ = state vector (*n*th order column)
$\mathbf{u} = [u_1, u_2, \ldots, u_r]^T$ = input vector (*r*th order column)
$\mathbf{y} = [y_1, y_2, \ldots, y_m]^T$ = output vector (*m*th order column)
$\mathbf{A}$ = system matrix ($n \times n$ square)
$\mathbf{B}$ = input gain matrix ($n \times r$)
$\mathbf{C}$ = measurement gain matrix ($m \times n$)
$\mathbf{D}$ = feedforward gain matrix ($m \times r$)

Usually, for vibrating systems, it is possible to make $\mathbf{D} = 0$, and hence we will drop this matrix in the sequel. Furthermore, although a state variable $x_i$ need not have a direct physical meaning, an output variable $y_j$ should have some physical meaning and, in typical situations, should be measurable as well. The input variables are the "control variables" and are used for controlling the system (plant). The output variables are the "controlled variables," which correspond to the system response and are measured for feedback control.

It can be verified that the eigenvalues of the system matrix $\mathbf{A}$ occur in complex conjugates of the form $-\zeta_i\omega_i \pm j\sqrt{1 - \zeta_i^2}\,\omega_i$ in the damped oscillatory case, or as $\pm j\omega_i$ in the undamped case, where $\omega_i$ is the *i*th natural frequency of the system and $\zeta_i$ is the corresponding damping ratio (of the *i*th mode). The mathematical verification requires some linear algebra. An intuitive verification can be made since Equation 7.131 is an equivalent model for a system having the traditional mass–spring–damper model

$$\mathbf{M}\ddot{\mathbf{y}} + \mathbf{C}\dot{\mathbf{y}} + \mathbf{K}\mathbf{y} = \mathbf{f}(t) \tag{7.133}$$

where $\mathbf{M}$ = mass matrix, $\mathbf{C}$ = damping matrix, $\mathbf{K}$ = stiffness matrix, $\mathbf{f}(t)$ = forcing input vector, and $\mathbf{y}$ = displacement response vector. Where both models (Equation 7.131 and Equation 7.133), are equivalent they should have the same characteristic equation, which by its roots determines the natural frequencies and modal damping ratios. This is the case because we are simply looking at two different mathematical representations of the same system. Hence, the parameters of its dynamics, such as $\omega_i$ and $\zeta_i$, should remain unchanged. In fact, the state-space mode (Equation 7.131) is not unique, and different versions of state vectors and corresponding models are possible. Of course, all of them should have the same characteristics polynomial (and hence, the same $\omega_i$ and $\zeta_i$). One such state-space model may be derived from Equation 7.133 as follows:

Define the state vector as

$$x = \begin{bmatrix} y \\ \dot{y} \end{bmatrix} \quad \text{and} \quad \mathbf{u} = \mathbf{f}(t) \tag{7.134}$$

Since (for nonsingular **M**, as required), Equation 7.133 may be written as

$$\ddot{y} = -\mathbf{M}^{-1}\mathbf{K}\mathbf{y} - \mathbf{M}^{-1}\mathbf{C}\dot{\mathbf{y}} + \mathbf{M}^{-1}\mathbf{f}(t) \tag{7.135}$$

we have

$$\dot{\mathbf{x}} = \begin{bmatrix} \mathbf{0} & \mathbf{I} \\ -\mathbf{M}^{-1}\mathbf{K} & -\mathbf{M}^{-1}\mathbf{C} \end{bmatrix} \mathbf{x} + \begin{bmatrix} \mathbf{0} \\ \mathbf{M}^{-1} \end{bmatrix} \mathbf{u} \tag{7.136}$$

This is a state-space model that is equivalent to the conventional model (Equation 7.133), and can be shown to have the same characteristic equation. The development of a state-space model for a vibrating system can be illustrated using an example.

## Example 7.10

Consider a machine mounted on a support structure, modeled as in Figure 7.33. Using the excitation forces $f_1(t)$ and $f_2(t)$ as the inputs and the displacements $y_1$ and $y_2$ of the masses $m_1$ and $m_2$ as the outputs, develop a state-space model for this system.

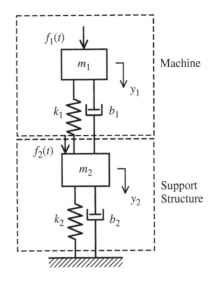

## Solution

Assume that the displacements are measured from the static equilibrium positions of the masses. Hence, the gravity forces do not enter into the formulation. Newton's Second law is applied to the two masses; thus

$$m_1\ddot{y}_1 = f_1 - k_1(y_1 - y_2) - b_1(\dot{y}_1 - \dot{y}_2)$$

$$m_2\ddot{y}_2 = f_2 - k_1(y_2 - y_1) - b_1(\dot{y}_2 - \dot{y}_1) - k_2 y_2 - b_2 \dot{y}_2$$

The following state variables are defined:

**FIGURE 7.33** A model of a machine mounted on support structure.

$$x_1 = y_1; \quad x_2 = \dot{y}_1; \quad x_3 = y_2; \quad x_4 = \dot{y}_2$$

Also, the input vector is $\mathbf{u} = [u_1 \ u_2]^T$ and the output vector is $\mathbf{y} = [y_1 \ y_2]^T$. Then, we have

$$\dot{x}_1 = x_2$$

$$m_1\dot{x}_2 = u_1 - k_1(x_1 - x_3) - b_1(x_2 - x_4)$$

$$\dot{x}_3 = x_4$$

$$m_2\dot{x}_4 = u_2 - k_1(x_3 - x_1) - b_1(x_4 - x_2) - k_2 x_3 - b_2 x_4$$

Accordingly, the state-space model is given by Equation 7.131 and Equation 7.132 with

$$
\mathbf{A} = \begin{bmatrix} 0 & 1 & 0 & 0 \\ -k_1/m_1 & -b_1/m_1 & k_1/m_1 & b_1/m_1 \\ 0 & 0 & 0 & 1 \\ k_1/m_2 & b_1/m_2 & (k_1+k_2)/m_2 & (b_1+b_2)/m_2 \end{bmatrix}, \quad \mathbf{B} = \begin{bmatrix} 0 & 0 \\ 1/m_1 & 0 \\ 0 & 0 \\ 0 & 1/m_2 \end{bmatrix},
$$

$$
\mathbf{C} = \begin{bmatrix} 1 & 0 & 0 & 0 \\ 0 & 0 & 1 & 0 \end{bmatrix}, \quad \text{and } \mathbf{D} = 0
$$

Also, note that the system can be expressed as

$$
\begin{bmatrix} m_1 & 0 \\ 0 & m_2 \end{bmatrix}\ddot{\mathbf{y}} + \begin{bmatrix} b_1 & -b_1 \\ -b_1 & (b_1+b_2) \end{bmatrix}\dot{\mathbf{y}} + \begin{bmatrix} k_1 & -k_1 \\ -k_1 & (k_1+k_2) \end{bmatrix}\mathbf{y} = \mathbf{f}(t)
$$

Its characteristic equation may be expressed as the determinant equation:

$$
\det\begin{bmatrix} m_1 s^2 + b_1 s + k & -b_1 s - k_1 \\ -b_1 s - k_1 & m_2 s^2 + (b_1+b_2)s + (k_1+k_2) \end{bmatrix} = 0
$$

It can be verified through direct expansion of the determinants that this equation is equivalent to the characteristic equation of the matrix $\mathbf{A}$, as given by $\det(\lambda\mathbf{I} - \mathbf{A}) = 0$, or

$$
\det\begin{bmatrix} \lambda & -1 & 0 & 0 \\ k_1/m_1 & \lambda + b_1/m_1 & -k_1/m_1 & -b_1/m_1 \\ 0 & 0 & \lambda & -1 \\ k_1/m_2 & -b_1/m_2 & -(k_1+k_2)/m_2 & \lambda - (b_1+b_2)/m_2 \end{bmatrix} = 0
$$

Note that, in the present context, $x$ and $y$ represent the vibration response of the plant and the control objective is to reduce these to zero. We will give some common control techniques that can achieve this goal.

### 7.9.2.2 Position and Velocity Feedback

In this technique, the position and velocity of each DoF is measured and fed into the system with sign reversal (negative feedback) and amplification by a constant gain. Because velocity is the derivative of position and since the gains are constant (i.e., proportional), this method falls into the general category of proportional-plus-derivative (PD or PPD) control. In this approach, it is tacitly assumed that the degrees of freedom are uncoupled. Then, control gains are chosen so that the DoF in the controlled system are nearly uncoupled, thereby justifying the original assumption. To explain this control method, suppose that a DoF of a vibrating system is represented by

$$
m\ddot{y} + b\dot{y} + ky = u(t) \tag{7.137}
$$

where $y$ is the displacement (position) of the DoF and $u$ is the excitation input that is applied. Now suppose that $u$ is generated according to the (active) control law

$$
u = -k_c y - b_c \dot{y} + u_r \tag{7.138}
$$

where $k_c$ is the position feedback gain and $b_c$ is the velocity feedback gain. The implication here is that the position $y$ and the velocity $\dot{y}$ are measured and fed into the controller which in turn generates $u$ according to Equation 7.138. Also, $u_r$ is some reference input that is provided externally to the controller.

Then, substituting Equation 7.138 into Equation 7.137, we obtain

$$m\ddot{y} + (b + b_c)\dot{y} + (k + k_c)y = u_r \tag{7.139}$$

The closed-loop system (the controlled system) now behaves according to Equation 7.139. The control gains $b_c$ and $k_c$ can be chosen arbitrarily (subject to the limitations of the physical controller, signal conditioning circuitry, the actuator, etc.) and may even be negative. In particular, by increasing $b_c$, the damping of the system can be increased. Similarly, by increasing $k_c$ the stiffness (and the natural frequency) of the system can be increased. Even though a passive spring and damper with stiffness $k_c$ and damping constant $b_c$ can accomplish the same task, once the devices are chosen it is not possible to conveniently change their parameters. Furthermore, it will not be possible to make $k_c$ or $b_c$ negative in this case of passive physical devices. The method of PPD control is simple and straightforward, but the assumptions of linear uncoupled DoF place a limitation on its general use.

### 7.9.2.3 Linear Quadratic Regulator Control

This is an optimal control technique. Consider a vibrating system that is represented by the linear state-space model:

$$\dot{x} = Ax + Bu \tag{7.131}$$

Assume that all the states **x** are measurable and all the system modes are controllable. Then, we use the constant-gain feedback control law:

$$u = Kx \tag{7.140}$$

The choice of parameter values for the feedback gain matrix **K** is infinite. Therefore, we can use this freedom to minimize the cost function:

$$J = \frac{1}{2} \int_t^\infty [x^T Q x + u^T R u] d\tau \tag{7.141}$$

This is the time integral of a quadratic function in both state and input variables, and the optimization goal may be interpreted as bringing **x** down to zero (regulating **x** to 0), but without spending a rather high control effort. Hence, the name linear quadratic regulation (LQR). In addition, **Q** and **R** are weighting matrices, with the former being at least positive semidefinite and the latter positive definite. Typically, **Q** and **R** chosen as diagonal matrices with positive diagonal elements whose magnitudes are determined by the degree of relative emphasis that should be given to various elements of **x** and **u**. It is well known that **K** that minimizes the cost function (Equation 7.141) is given by

$$K = -R^{-1}B^T K_r \tag{7.142}$$

where $K_r$ is the positive-definite solution of the matrix Riccati algebraic equation

$$K_r A + A^T K_r - K_r B R^{-1} B^T K_r + Q = 0 \tag{7.143}$$

It is also known that the resulting closed-loop control system is stable. Furthermore, the minimum (optimal) value of the cost function (Equation 7.141) is given by

$$J_m = \frac{1}{2} x^T K_r x \tag{7.144}$$

where **x** is the present value of the state vector. Major computational burden of the LQR method is in the solution (Equation 7.143). Other limitations of the technique arise due to the need to measure all the state variables (which may be relaxed to some extent). Although stability of the controlled system is guaranteed, the level of stability that is achieved (i.e., stability margin or the level of modal damping) cannot be directly specified. Further, robustness of the control system in the presence of model errors, unknown disturbances and so on, may be questionable. Besides, the cost function incorporates an integral over an infinite time duration, which does not typically reflect the practical requirement of rapid vibration control.

### 7.9.2.4 Modal Control

The LQR control technique has the serious limitation of not being able to directly achieve specified levels of modal damping, which may be an important goal in vibration control. The method of modal control that accomplishes this objective is *pole placement*, where poles (eignevalues) of the controlled system are placed at specified values. Specifically, consider the plant (Equation 7.131) and the feedback control law (Equation 7.140). Then, the closed-loop system is given by

$$\dot{x} = (\mathbf{A} + \mathbf{BK})\mathbf{x} \tag{7.145}$$

It is well known that if the plant $(\mathbf{A}, \mathbf{B})$ is controllable, then a control gain matrix $\mathbf{K}$ can be chosen that will arbitrarily place the eigenvalues of the closed-loop system matrix $\mathbf{A} + \mathbf{BK}$. Based on the given assumptions, the modal control technique assigns not only the modal damping but also the damped natural frequencies at specified values. The assumptions given above are quite stringent but they can be relaxed to some degree. However, a shortcoming of this method is the fact that it does not place a restriction on the control effort, for example, as the LQR technique does, in achieving a specified level of modal control.

## 7.10 Control of Beam Vibrations

Beam is a distributed-parameter system, which in theory has an infinite number of modes of vibration with associated mode shapes and natural frequencies. In this sense, it is an "infinite order" system with infinite DoF. Hence, the computation of modal quantities and associated control inputs can be quite complex. Fortunately, however, just a few modes may be retained in a dynamic model without sacrificing a great deal of accuracy, thereby facilitating simpler control. Some concepts of controlling vibrations in a beam are considered in this section. The present treatment is intended as an illustration of the relevant techniques and is not meant to be exhaustive. These techniques may be extended to other types of continuous system such as beams with different boundary conditions and plates. Because the control techniques that were outlined previously depend on a model, we will first illustrate the procedure of obtaining a state-space model for a beam.

### 7.10.1 State-Space Model of Beam Dynamics

Consider a Bernoulli–Euler-type beam with Kelvin–Voigt-type internal (material) damping. The beam equation may be expressed as

$$ELv(x, t) + E^* L \frac{\partial v(x, t)}{\partial t} + \rho A(x) \frac{\partial^2 v(x, t)}{\partial t^2} = f(x, t) \tag{7.146}$$

in which $L$ is the partial differential operator given by

$$L = \frac{\partial^2 I(x)}{\partial x^2} \frac{\partial^2}{\partial x^2} \tag{7.147}$$

and

  $f(x, t)$ = distributed force excitation per unit length of the beam
  $v(x, t)$ = displacement response at location $x$ along the beam at time $t$
  $I(x)$ = second moment of area of the beam cross section about the neutral axis
  $E$ = Young's modulus of the beam material
  $E^*$ = Kelvin–Voigt material damping parameter

Note that a general beam with nonuniform characteristics is assumed and, hence, the variations of $I(x)$ and $\rho A(x)$ with $x$ are retained in the formulation.

Using the approach of modal expansion, the response of the beam may be expressed by

$$v(x, t) = \sum_{i=1}^{\infty} Y_i(x) q_i(t) \tag{7.148}$$

where $Y_i(x)$ is the $i$th mode shape of the beam, which satisfies

$$LY_i(x) = \frac{\rho A(x)}{E} \omega_i^2 Y_i(x) \tag{7.149}$$

and $\omega_i$ is the $i$th undamped natural frequency. The orthogonality condition for this general example of a nonuniform beam is

$$\int_{x=0}^{l} \rho A Y_i Y_j dx = \begin{cases} 0 & \text{for } i \neq j \\ \alpha_j & \text{for } i = j \end{cases} \tag{7.150}$$

Suppose that the forcing excitation on the beam is a set of $r$ point forces $u_k(t)$ located at $x = l_k$, $k = 1, 2, \ldots, r$. Then, we have

$$f(x, t) = \sum_{k=1}^{r} u_k \delta(x - l_k) \tag{7.151}$$

where $\delta(x - l_i)$ is the *Dirac delta function*. Now, substitute Equation 7.148 and Equation 7.151 into Equation 7.146, use Equation 7.149, multiply throughout by $Y_j(x)$, and integrate over $x[0, l]$, using Equation 7.150. This gives

$$\ddot{q}_j + \gamma_j \dot{q}_j(t) + \omega_j^2 q_j = \frac{1}{\alpha_j} \sum_{k=1}^{r} u_k Y_j(l_k) \quad \text{for } j = 1, 2, \ldots \tag{7.152}$$

where

$$\gamma_j = \frac{E^*}{E} \omega_j^2 \tag{7.153}$$

Now, define the state variables $x_j$ according to

$$x_{2j-1} = \omega_j q_j, \quad x_{2j} = \dot{q}_j \quad \text{for } j = 1, 2, \ldots \tag{7.154}$$

Assuming that only the first $m$ modes are retained in the expansion, we then have the state equations

$$\dot{x}_{2j-1} = \omega_j x_{2j}, \quad \dot{x}_{2j} = -\omega_j x_{2j-1} - \gamma_j x_{2j} + \frac{1}{\alpha_j} \sum_{k=1}^{r} u_k Y_j(l_k) \quad \text{for } j = 1, 2, \ldots, m \tag{7.155}$$

This can be put in the matrix–vector form of a state-space model

$$\dot{\mathbf{x}} = \mathbf{A}\mathbf{x} + \mathbf{B}\mathbf{u} \tag{7.131}$$

where

$$\mathbf{A} = \begin{bmatrix} 0 & \omega_1 & & & & \\ -\omega_1 & -\gamma_1 & & & 0 & \\ & & \ddots & & & \\ & & & 0 & \omega_m \\ & 0 & & -\omega_m & -\gamma_m \end{bmatrix}_{n \times n} \tag{7.156}$$

and

$$
\mathbf{B} = \begin{bmatrix}
0 & & 0 \\
Y_1(l_1)/\alpha_1 & \cdots & Y_1(l_r)/\alpha_1 \\
\vdots & & \vdots \\
0 & & 0 \\
Y_m(l_1)/\alpha_m & \cdots & Y_m(l_r)/\alpha_m
\end{bmatrix}_{n \times r}
\tag{7.157}
$$

with $n = 2m$, where $m$ is the number of modes retained in the modal expansion. Note that, as the number of modes used in this model increases, both the accuracy and the computational effort that is needed for the control problem increase because of the proportional increase of the system order. At some point, the potential improvement in accuracy by further increasing the model size would be insignificant in comparison with added computational burden. Hence, a balance must be struck in this tradeoff.

## 7.10.2 Control Problem

The state-space model (Equation 7.131) for the beam dynamics, with matrices (Equation 7.156 and Equation 7.157), is known to be *controllable*. Hence, it is possible to determine a constant-gain feedback controller $\mathbf{u} = \mathbf{K}\mathbf{x}$ that minimizes a quadratic-integral cost function of the form in Equation 7.141. Also, a similar controller can be determined that places the eigenvalues of the system at specified locations thereby achieving not only specified levels of modal damping but also a specified set of natural frequencies. However, there is a practical obstacle to achieving such an active controller. Note that, in the model given in Equation 7.156 and Equation 7.157, the state variables are proportional to the modal variables $q_i$ and their time derivatives $\dot{q}_i$. They are not directly measurable. However, the displacements and velocities at a set of discrete locations along the beam can usually be measured. Let these locations ($s$) be denoted by $p_1, p_2, \ldots, p_s$. Thus, in view of the modal expansion (Equation 7.148), the measurements can be expressed as

$$
v(p_j, t) = \sum_{i=1}^{m} Y_i(p_j) q_i(t), \quad \dot{v}(p_j, t) = \sum_{i=1}^{m} Y_i(p_j) \dot{q}_i(t) \quad \text{for } j = 1, 2, \ldots, s
\tag{7.158}
$$

Now, define the output (measurement) vector $\mathbf{y}$ according to

$$
y = [v(p_1, t), \dot{v}(p_1, t), \ldots, v(p_s, t), \dot{v}(p_s, t)]^{\mathrm{T}}
\tag{7.159}
$$

In view of Equation 7.158 and the definitions of the state variable in Equation 7.154, we can write

$$
\mathbf{y} = \mathbf{C}\mathbf{x}
\tag{7.160}
$$

with

$$
C = \begin{bmatrix}
Y_1(p_1)/\omega_1 & 0 & \cdots & Y_m(p_1)/\omega_m & 0 \\
0 & Y_1(p_1) & \cdots & 0 & Y_m(p_1) \\
\vdots & \vdots & \cdots & \vdots & \vdots \\
Y_1(p_s)/\omega_1 & 0 & \cdots & Y_m(p_s)/\omega_m & 0 \\
0 & Y_1(p_s) & \cdots & 0 & Y_m(p_s)
\end{bmatrix}_{2s \times n}
\tag{7.161}
$$

Hence, an active controller is possible of the form:

$$
\mathbf{u} = \mathbf{H}\mathbf{y}
\tag{7.162}
$$

which is an output feedback controller. Therefore, in view of Equation 7.160, we have

$$\mathbf{u} = \mathbf{HCx} \tag{7.163}$$

This is not the same as complete state feedback $\mathbf{u} = \mathbf{Kx}$ where $\mathbf{K}$ can take any real value (and, hence, the LQR solution in Equation 7.142 and the complete pole placement solution cannot be applied directly). In Equation 7.163, only $\mathbf{H}$ can be arbitrarily chosen, and $\mathbf{C}$ is completely determined according to Equation 7.161. The resulting product $\mathbf{HC}$ will not usually correspond to either the LQR solution or the complete pole assignment solution. Still, the output feedback controller in Equation 7.162 can provide a satisfactory performance. However, a sufficient number of displacement and velocity sensors ($s$) have to be used in conjunction with a sufficient number of actuators ($r$) for active control. This will increase the system complexity and cost. Furthermore, due to added components and their active nature, the reliability of fault-free operation may degrade somewhat. A satisfactory alternative would be to use passive control devices such as dampers and dynamic absorbers, which is illustrated below. Note that in the matrices $\mathbf{B}$ and $\mathbf{C}$ given by Equation 7.157 and Equation 7.161, both the actuator locations $l_i$ and the sensor locations $p_j$ are variable. Hence, there exists an additional design freedom (or optimization parameters) in selecting the sensor and actuator locations in achieving satisfactory control.

## 7.10.3  Use of Linear Dampers

Now, consider the use of a discrete set of linear dampers for controlling beam vibration. Suppose that $r$ linear dampers with damping constants $b_j$ are placed at locations $l_j$, $j = 1, 2, \ldots, r$ along the beam, as schematically shown in Figure 7.34. The damping forces are given by

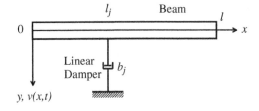

$$u_j = -b_j \dot{v}(l_j, t) \quad \text{for } j = 1, 2, \ldots, r \tag{7.164}$$

**FIGURE 7.34**  Use of linear dampers in beam vibration control.

Substituting the truncated modal expansion ($m$ modes)

$$\dot{v}(l_j, t) = \sum_{i=1}^{m} Y_i(l_j)\dot{q}_i(t) \tag{7.165}$$

we get, in view of Equation 7.154, the passive feedback control action

$$\mathbf{u} = -\mathbf{Kx} \tag{7.166}$$

with

$$\mathbf{K} = \begin{bmatrix} 0 & b_1 Y_1(l_1) & \cdots & 0 & b_1 Y_m(l_1) \\ \vdots & \vdots & \cdots & \vdots & \vdots \\ 0 & b_r Y_1(l_r) & \cdots & 0 & b_r Y_m(l_r) \end{bmatrix}_{r \times n} \tag{7.167}$$

By substituting Equation 7.166 into Equation 7.131, we have the closed-loop system equation

$$\dot{\mathbf{x}} = (\mathbf{A} - \mathbf{F})\mathbf{x} = \mathbf{A}_c \mathbf{x} \tag{7.168}$$

where $\mathbf{F} = \mathbf{BK}$ and is given by

$$\mathbf{F} = \begin{bmatrix} 0 & 0 & \cdots & 0 & 0 \\ 0 & \sum b_i Y_{11}(l_i)/\alpha_1 & \cdots & 0 & \sum b_i Y_{1m}(l_i)/\alpha_1 \\ \vdots & \vdots & \cdots & \vdots & \vdots \\ 0 & 0 & \cdots & 0 & 0 \\ 0 & \sum b_i Y_{m1}(l_i)/\alpha_m & \cdots & 0 & \sum b_i Y_{mm}(l_i)/\alpha_m \end{bmatrix}_{n \times n} \tag{7.169}$$

with

$$Y_{ij}(x) = Y_i(x)Y_j(x) \tag{7.170}$$

In this case, the controller design involves the selection of the damping constants $b_i$ and the damper locations $l_j$ to achieve the required performance. This may be achieved, for example, by seeking to make the eigenvalues of the closed-loop system matrix $\mathbf{A}_c$ reach a set of desired values. This achieves the desired modal damping and natural frequency characteristics. However, given that the structure of the $\mathbf{F}$ matrix is fixed, as seen in Equation 7.169, this is not equivalent to complete state feedback (or complete output feedback). Hence, generally, it is not possible to place the poles of the system at the exact desired locations.

### 7.10.3.1 Design Example

In realizing a desirable modal response of a beam using a set of linear dampers, one may seek to minimize a cost function of the form

$$J = \mathrm{Re}(\boldsymbol{\lambda} - \boldsymbol{\lambda}_d)^T \mathbf{Q} \, \mathrm{Re}(\boldsymbol{\lambda} - \boldsymbol{\lambda}_d) + \mathrm{Im}(\boldsymbol{\lambda} - \boldsymbol{\lambda}_d)^T \mathbf{R}(\boldsymbol{\lambda} - \boldsymbol{\lambda}_d) \tag{7.171}$$

where $\boldsymbol{\lambda}$ are the actual eigenvalues of the closed-loop system matrix ($\mathbf{A}_c$), and $\boldsymbol{\lambda}_d$ are the desired eigenvalues that will give the required modal performance (damping ratios and natural frequencies). "Re" denotes the real part and "Im" denotes the imaginary part. Weighting matrices $\mathbf{Q}$ and $\mathbf{R}$, which are real and diagonal with positive diagonal elements, should be chosen to relatively weight various eigenvalues. This allows the emphasis of some eigenvalues over others, with real parts and the imaginary parts weighting separately.

Various computational algorithms are available for minimizing the cost function (Equation 7.171). Although the precise details are beyond the scope of this book, we will present an example result. Consider a uniform simply supported 12 × 5 American Standard beam, with the following pertinent specifications: $E = 2 \times 10^8$ kPa ($29 \times 10^6$ psi), $\rho A = 47$ kg/m (2.6 lb/in.), length $l = 15.2$ m (600 in.), $I = 9 \times 10^{-5}$ m$^4$ (215.8 in.$^4$). The internal damping parameter for the $j$th mode of vibration is given by

$$E^*(\omega_j) = (g_1/\omega_j) + g_2 \tag{7.172}$$

in which $\omega_j$ is the $j$th undamped natural frequency given by

$$\omega_j = (j\pi/l)^2 \sqrt{EI/\rho A} \tag{7.173}$$

The numerical values used for the damping parameters are $g_1 = 88 \times 10^4$ kPa ($12.5 \times 10^4$ psi) and $g_2 = 3.4 \times 10^4$ kPa s ($5 \times 10^3$ psi s). For the present problem, $Y_i(x) = \sqrt{2}\sin(j\pi x/l)$ and $\alpha_j = \rho Al$ for all $j$.

First, $\omega_j$ and $\gamma_j$ are computed using Equation 7.173 and Equation 7.153, respectively, along with Equation 7.172. Next, the open-loop system matrix $\mathbf{A}$ is formed according to Equation 7.156 and its eigenvalues are computed. These are listed in Table 7.2, scaled to the first undamped natural frequency ($\omega_1$). Note that in view of the very low levels of internal material damping of the beam, the actual natural frequencies, as given by the imaginary parts of the eigenvalues, are almost identical to the undamped natural frequencies.

Next, we attempt to place the real parts of the (scaled) eigenvalues all at $-0.20$ while exercising no constraint on the imaginary parts (i.e., damped natural frequencies) by using: (a) single damper, and

**TABLE 7.2**   Eigenvalues of the Open-Loop (Uncontrolled) Beam

| Mode | Eigenvalue (rad/sec) (Multiply by 26.27) |
|------|------------------------------------------|
| 1 | $-0.000126 \pm j1.0$ |
| 2 | $-0.000776 \pm j4.0$ |
| 3 | $-0.002765 \pm j9.0$ |
| 4 | $-0.007453 \pm j16.0$ |
| 5 | $-0.016741 \pm j25.0$ |
| 6 | $-0.033.75 \pm j36.0$ |

(b) two dampers. In the cost function (Equation 7.171), the first three modes are more heavily weighted than the remaining three. Initial values of the damper parameters are $b_1 = b_2 = 0.1$ lbf s/in. (17.6 N s/m) and the initial locations $l_1/l = 0.0$ and $l_2/l = 0.5$. At the end of the numerical optimization, using a modified gradient algorithm, the following optimized values were obtained:

1. Single-damper control

$$b_1 = 36.4 \text{ lbf s/in. } (6.4 \times 10^3 \text{ N s/m})$$

$$l_1/l = 0.3$$

The corresponding normalized eigenvalues (of the closed-loop system) are given in Table 7.3.

2. Two-damper control

$$b_1 = 22.8 \text{ lbf s/in. } (4.0 \times 10^3 \text{ N s/m})$$

$$b_2 = 12.1 \text{ lbf s/in. } (2.1 \times 10^3 \text{ N s/m})$$

$$l_1/l = 0.25, \quad l_2/l = 0.43$$

The corresponding normalized eigenvalues are given in Table 7.4.

It would be overly optimistic to expect perfect assignment all real parts at $-0.2$. However, note that good levels of damping have been achieved for all modes except for Mode 3 in the single-damper control and Mode 4 in the two-damper control. In any event, because the contribution of the higher modes towards the overall response, is relatively smaller, it is found that the total response (e.g., at point $x = l/12$) is well damped in both cases of control.

**TABLE 7.3**   Eigenvalues of the Beam with an Optimal Single Damper

| Mode | Eigenvalue (rad/s) (Multiply by 26.27) |
|------|------------------------------------------|
| 1 | $-0.225 \pm j0.985$ |
| 2 | $-0.307 \pm j3.955$ |
| 3 | $-0.037 \pm j8.996$ |
| 4 | $-0.119 \pm j15.995$ |
| 5 | $-0.355 \pm j24.980$ |
| 6 | $-0.158 \pm j35.990$ |

**TABLE 7.4**   Eigenvalues of the Beam with Optimized Two Dampers

| Mode | Eigenvalue (rad/s) (Multiply by 26.27) |
|------|------------------------------------------|
| 1 | $-0.216 \pm j0.982$ |
| 2 | $-0.233 \pm j3.974$ |
| 3 | $-0.174 \pm j8.997$ |
| 4 | $-0.079 \pm j15.998$ |
| 5 | $-0.145 \pm j24.999$ |
| 6 | $-0.354 \pm j35.989$ |

# Bibliography

Beards, C.F. 1996. *Engineering Vibration Analysis with Application to Control Systems*, Halsted Press, New York.

Cao, Y., Modi, V.J., de Silva, C.W., and Misra, A.K., On the control of a novel manipulator with slewing and deployable links, *Acta Astronaut.*, 49, 645–658, 2001.

Caron, M., Modi, V.J., Pradhan, S., de Silva, C.W., and Misra, A.K., Planar dynamics of flexible manipulators with slewing deployable links, *J. Guid. Control Dyn.*, 21, 572–580, 1998.

Chen, Y., Wang, X.G., Sun, C., Devine, F., and de Silva, C.W., Active vibration control with state feedback in woodcutting, *J. Vibr. Control*, 9, 645–664, 2003.

den Hartog, J.P., 1956. *Mechanical Vibrations*, Mc-Graw-Hill, New York.

de Silva, C.W., Optimal estimation of the response of internally damped beams to random loads in the presence of measurement noise, *J. Sound Vibr.*, 47, 485–493, 1976.

de Silva, C.W., An algorithm for the optimal design of passive vibration controllers for flexible systems, *J. Sound Vibr.*, 74, 495–502, 1982.

de Silva, C.W. and Wormley, D.N. 1983. *Automated Transit Guideways: Analysis and Design*, D.C. Heath & Co., Lexington, KY.

de Silva, C.W. 1989. *Control Sensors and Actuators*, Prentice Hall, Englewood Cliffs, NJ.

de Silva, C.W. 1995. *Intelligent Control: Fuzzy Logic Applications*, CRC Press, Boca Raton, FL.

de Silva, C.W. 2005. *Mechatronics—an Integrated Approach*, Taylor & Francis, CRC Press, Boca Raton, FL.

de Silva, C.W. 2006. *Vibration—Fundamentals and Practice*, 2nd ed., Taylor & Francis, CRC Press, Boca Raton, FL.

Goulet, J.F., de Silva, C.W., Modi, V.J., and Misra, A.K., Hierarchical control of a space-based deployable manipulator using fuzzy logic, *AIAA J. Guid. Control Dyn.*, 24, 395–405, 2001.

Irwin, J.D. and Graf, E.R. 1979. *Industrial Noise and Vibration Control*, Prentice Hall, Englewood Cliffs, NJ.

Karray, F. and de Silva, C.W. 2004. *Soft Computing and Intelligent System Design: Theory, Tools, and Applications*, Pearson, England.

MATLAB *Control Systems Toolbox*, The MathWorks, Inc., Natick, MA, 2004.

Van de Vegte, J. and de Silva, C.W., Design of passive vibration controls for internally damped beams by modal control techniques, *J. Sound Vibr.*, 45, 417–425, 1976.

# Appendix 7A

# MATLAB® Control Systems Toolbox

## 7A.1 Introduction

Modeling, analysis, design, data acquisition, and control are important activities within the field of vibration. Computer software tools and environments are available for effectively carrying out, both at the learning level and at the professional application level. Several such environments and tools are commercially available.

MATLAB[1] is an interactive computer environment with a high-level language and tools for scientific and technical computation, modeling and simulation, design, and control of dynamic systems. SIMULINK[1] is a graphical environment for modeling, simulation, and analysis of dynamic systems, and is available as an extension to MATLAB. The Control Systems Toolbox of MATLAB is suitable in the analysis, design, and control of mechanical vibrating systems.

---

[1]MATLAB and SIMULINK are registered trademarks and products of The MathWorks, Inc. LabVIEW is a product of National Instruments, Inc.

## 7A.2  MATLAB

MATLAB interactive computer environment is very useful in computational activities in Mechatronics. Computations involving scalars, vectors, and matrices can be carried out and the results can be graphically displayed and printed. MATLAB toolboxes are available for performing specific tasks in a particular area of study such as control systems, fuzzy logic, neural network, data acquisition, image processing, signal processing, system identification, optimization, model predictive control, robust control, and statistics. User guides, Web-based help, and on-line help from the parent company, Math-Works, Inc., and various other sources. What is given here is a brief introduction to get started in MATLAB for tasks that are particularly related to Control Systems and Mechatronics.

### 7A.2.1  Computations

Mathematical computations can be done by using the MATLAB command window. Simply type in the computations against the MATLAB prompt ">>" as illustrated next.

### 7A.2.2  Arithmetic

An example of a simple computation using MATLAB is given below:

>> x = 2; y = − 3;
>> z = x^2 − x * y + 4
z =
    14

In the first line, we have assigned values 2 and 3 to two variables $x$ and $y$. In the next line, the value of an algebraic function of these two variables is indicated. Then, MATLAB provides the answer as 14. Note that if you place a ";" at the end of the line, the answer will not be printed/displayed.

Table 7A.1 gives the symbols for common arithmetic operations used in MATLAB.

Following example shows the solution of the quadratic equation $ax^2 + bx + c = 0$:

>> a = 2; b = 3; c = 4;
>> x = (−b + sqrt(b^2 − 4*a*c))/(2*a)
x =
 − 0.7500 + 1.1990i

The answer is complex, where i denotes $\sqrt{-1}$. Note that the function sqrt( ) is used, which provides the positive root only. Some useful mathematical functions are given in Table 7A.2.

### 7A.2.3  Arrays

An array may be specified by giving the start value, increment, and the end value limit. An example is given below.

>> x = (0.9: − 0.1:0.42)
x =

0.9000   0.8000   0.7000   0.6000   0.5000

**TABLE 7A.1**   MATLAB Arithmetic Operations

| Symbol | Operation |
| --- | --- |
| + | Addition |
| − | Subtraction |
| * | Multiplication |
| / | Division |
| ^ | Power |

**TABLE 7A.2**   Useful Mathematical Functions in MATLAB

| Function | Description |
|----------|-------------|
| abs( ) | Absolute value/magnitude |
| acos( ) | Arc-cosine (inverse cosine) |
| acosh( ) | Arc-hyperbolic-cosine |
| asin( ) | Arc-sine |
| atan( ) | Arc-tan |
| cos( ) | Cosine |
| cosh( ) | Hyperbolic cosine |
| exp( ) | Exponential function |
| imag( ) | Imaginary part of a complex number |
| log( ) | Natural logarithm |
| log10( ) | Log to base 10 (common log) |
| real( ) | Real part of a complex number |
| sign( ) | Signum function |
| sin( ) | Sine |
| sqrt( ) | Positive square root |
| tan( ) | Tan function |

Note that MATLAB is case sensitive.

The entire array may be manipulated. For example, all the elements are multiplied by $\pi$ as below:

```
>> x = x*pi
x =
```
2.8274   2.5133   2.1991   1.8850   1.5708

The second and the fifth elements are obtained by

```
>> x([2 5])
ans =
```
2.5133   1.5708

Next, we form a new array $y$ using $x$, and then plot the two arrays, as shown in Figure 7A.1:

```
>> y = sin(x);
>> plot(x,y)
```

A polynomial may be represented as an array of its coefficients. For example, the quadratic equation $ax^2 + bx + c = 0$ as given before, with $a = 2$, $b = 3$, and $c = 4$, may be solved using the function "roots" as below:

```
>> p = [2   3   4];
>> roots(p)
ans =
```
   $-0.7500 + 1.1990i$
   $-0.7500 - 1.1990i$

The answer is the same as we obtained before.

## 7A.2.4   Relational and Logical Operations

Useful relational operations in MATLAB are given in Table 7A.3. Basic logical operations are given in Table 7A.4.

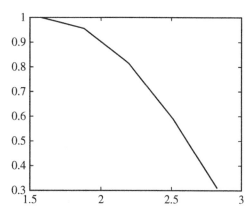

**FIGURE 7A.1**   A plot using MATLAB.

**TABLE 7A.3**   Some Relational Operations

| Operator | Description |
|---|---|
| < | Less than |
| <= | Less than or equal to |
| > | Greater than |
| >= | Greater than or equal to |
| == | Equal to |
| ~= | Not equal to |

**TABLE 7A.4**   Basic Logical Operations

| Operator | Description |
|---|---|
| & | AND |
| \| | OR |
| ~ | NOT |

Consider the following example:

```
>> x = (0:0.25:1)*pi
x =
    0   0.7854   1.5708   2.3562   3.1416
>> cos(x) > 0
ans =
    1   1   1   0   0
>> (cos(x) > 0)&(sin(x) > 0)
ans =
    0   1   1   0   0
```

In this example, first an array is computed. Then the cosine of each element is computed. Next it is checked whether the elements are positive. (A truth value of 1 is sent out if true and a truth value of 0 if false.) Finally, the "AND" operation is used to check whether both corresponding elements of two arrays are positive.

### 7A.2.5   Linear Algebra

MATLAB can perform various computations with vectors and matrices (see Appendix 3A and Appendix 6A). Some basic illustrations are given here.

A vector or a matrix may be specified by assigning values to its elements. Consider the following example:

```
>> b = [1.5   -2];
>> A = [2   1;   -1   1];
>> b = b'
b =
     1.5000
    -2.0000
>> x = inv(A) * b
x =
     1.1667
    -0.8333
```

In this example, first a second-order row vector and $2 \times 2$ matrix are defined. The row vector is transposed to get a column vector. Finally the matrix–vector equation $\mathbf{Ax} = \mathbf{b}$ is solved according to $\mathbf{x} = \mathbf{A}^{-1}\mathbf{b}$. The determinant and the eigenvalues of $\mathbf{A}$ are determined by

**TABLE 7A.5**   Some Matrix Operations in MATLAB

| Operation | Description |
|---|---|
| + | Addition |
| − | Subtraction |
| * | Multiplication |
| / | Division |
| ^ | Power |
| ' | Transpose |

**TABLE 7A.6**   Useful Matrix Functions in MATLAB

| Function | Description |
|---|---|
| det( ) | Determinant |
| inv( ) | Inverse |
| eig( ) | Eigenvalues |
| [,] = eig( ) | Eigenvectors and eigenvalues |

```
>> det(A)
ans =
    3
>> eig(A)
ans =
    1.5000 + 0.8660i
    1.5000 − 0.8660i
```

Both eigenvectors and eigenvalues of **A** computed as

```
>>[V,P] = eig(A)
V =
          0.7071              0.7071
       −0.3536 + 0.6124i   −0.3536 − 0.6124i
P =
    1.5000 + 0.8660i           0

          0            1.5000 − 0.8660i
```

Here, the symbol **V** is used to denote the matrix of eigenvectors. The symbol **P** is used to denote the diagonal matrix whose diagonal elements are the eigenvalues.

Useful matrix operations in MATLAB are given in Table 7A.5 and several matrix functions are given in Table 7A.6.

### 7A.2.6   M-Files

The MATLAB commands have to be keyed in on the command window, one by one. When several commands are needed to carry out a task, the required effort can be tedious. Instead, the necessary commands can be placed in a text file, edited as appropriate (using text editor), which MATLAB can use to execute the complete task. Such a file is called an M-file. The file name must have the extension "m" in the form *filename.m*. A toolbox is a collection of such files, for use in a particular application area (e.g., control systems, fuzzy logic). Then, by keying in the M-file name at the MATLAB command prompt, the file will be executed. The necessary data values for executing the file have to be assigned beforehand.

## 7A.3   Control Systems Toolbox

There are several toolboxes with MATLAB, which can be used to analyze, compute, simulate, and design control problems. Both time-domain representations and frequency-domain representations can be used. Also, both classical and modern control problems can be handled. The application is illustrated here through several control problems.

### 7A.3.1   MATLAB Modern Control Examples

Several examples in modern control engineering are given now to illustrate the use of MATLAB in control.

#### 7A.3.1.1   *Pole Placement of a Third-Order Plant*

A mechanical plant is given by the input–output differential equation $\dddot{x} + \ddot{x} = u$, where $u$ is the input and $x$ is the output. Determine a feedback law that will yield approximately a simple oscillator with a damped natural frequency of 1 unit and a damping ratio of $1/\sqrt{2}$.

To solve this problem, first we define the state variables as $x_1 = x$, $x_2 = \dot{x}_1$, and $x_3 = \dot{x}_2$. The corresponding state-space model is

$$\dot{\mathbf{x}} = \begin{bmatrix} \dot{x}_1 \\ \dot{x}_2 \\ \dot{x}_3 \end{bmatrix} = \underbrace{\begin{bmatrix} 0 & 1 & 0 \\ 0 & 0 & 1 \\ 0 & 0 & -1 \end{bmatrix}}_{A} \begin{bmatrix} x_1 \\ x_2 \\ x_3 \end{bmatrix} + \underbrace{\begin{bmatrix} 0 \\ 0 \\ 1 \end{bmatrix}}_{B} u$$

$$y = \underbrace{\begin{bmatrix} 1 & 0 & 0 \end{bmatrix}}_{C} \mathbf{x}$$

The open-loop poles and zeros are obtained using the following MATLAB commands:

```
>> A = [0 1 0; 0 0 1; 0 0 − 1];
>> B = [0; 0; 1];
>> C = [1 0 0];
>> D = [0];
>> sys_open = ss(A,B,C,D);
>> [nat_freq_open,damping_open,poles_open] = damp(sys_open)
>> pzmap(sys_open)
```

The open-loop poles are: $[0\ 0\ -1]^{\mathrm{T}}$.
The step response of the open-loop system is obtained using the command:

```
>> step(sys_open)
```

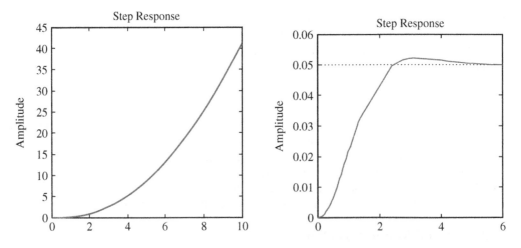

**FIGURE 7A.2**   (a) Step response of the open-loop system; (b) step response of the third-order system with pole-placement control.

The result is shown in Figure 7A.2(a). Clearly, the system is unstable.

With the desired damped natural frequency $\omega_d = 1$ and damping ratio $\zeta = 1/\sqrt{2}$, we get the undamped natural frequency $\omega_n = \sqrt{2}$ and, hence, $\zeta\omega_n = 1$. It follows that we need to place two poles at $-1 \pm j$. Also the third pole has to be far from these two on the left half plane (LHP); say, at $-10$. The corresponding control gain $K$ can be computed using the "place" command in MATLAB:

```
>> p = [-1 + j - 1 - j - 10];
>> K = place(A,B,p)
place:ndigits = 15
K =
    20.0000   22.0000   11.0000
```

The corresponding step response of the closed-loop system is shown in Figure 7A.2(b).

### 7A.3.1.2   Linear Quadratic Regulator for a Third-Order Plant

For the third-order plant in the previous example, we design a linear quadratic regulator (LQR), which has a state feedback controller, using MATLAB Control Systems Toolbox. The MATLAB command $K = lqr(A,B,Q,R)$ computes the optimal gain matrix $\mathbf{K}$ such that the state-feedback law $\mathbf{u} = -\mathbf{Kx}$ minimizes the quadratic cost function

$$J = \int_0^\infty (\mathbf{x}^T\mathbf{Qx} + \mathbf{u}^T\mathbf{Ru})dt$$

The weighting matrices $\mathbf{Q}$ and $\mathbf{R}$ are chosen to apply the desired weights to the various states and inputs. The MATLAB commands for designing the controller are

```
>> A = [0 1 0; 0 0 1; 0 0 - 1];
>> B = [0; 0; 1];
>> C = [1 0 0];
>> D = [0];
>> Q = [2 0 0; 0 2 0; 0 0 2];
>> R = 2;
>> Klqr = lqr(A,B,Q,R)
>> lqr_closed = ss(A-B*Klqr,B,C,D);
>> step(lqr_closed)
```

The step response of the system with the designed LQR controller is shown in Figure 7A.3.

### 7A.3.1.3   Modal Analysis Example

Consider the two-DoF mechanical system shown in Figure 7A.4. We now solve the modal analysis problem using MATLAB, for the numerical values

$$\alpha = 0.5, \quad \beta = 0.5, \quad m = 1 \text{ kg}; \quad k = 1 \text{ N/m}$$

For the given mass matrix $\mathbf{M}$ and the stiffness matrix $\mathbf{K}$, the solution steps for the alternative approach of modal analysis are

1. Determine $\mathbf{M}^{1/2}$ and $\mathbf{M}^{-1/2}$.
2. Solve for the eigenvalues $\lambda$ and the eigenvectors $\boldsymbol{\phi}$ of $\mathbf{M}^{-1/2}\mathbf{KM}^{-1/2}$. These eigenvalues are the squares of the natural frequencies of the original system.

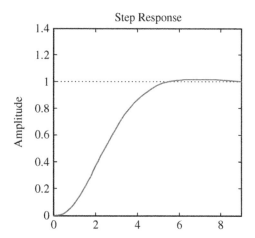

FIGURE 7A.3   Step response of the third-order system with LQR control.

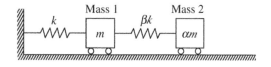

FIGURE 7A.4   A two-DoF system.

3. Determine the modal vectors $\psi$ of the original system using the transformation $\psi = M^{-1/2}\phi$.

The program code is given below:

```
%Modal Analysis Example
clear;
m = 1.0;
k = 1.0;
M = [m 0;  0 m/2 ];
K = [3/2 * k  − k/2;
      − k/2 k/2];
M_sqrt = M^0.5;
M_s_inv = inv(M_sqrt);
lemda = eig(M_s_inv * K * M_s_inv);
[U,D] = eig(M_s_inv * K * M_s_inv);
V = M_s_inv * U;
disp('Natural frequencies')
fprintf('omega1 = %10.3f %14.3f \n',sqrt(lemda(2,1)));
fprintf('\n omega2 = %10.3f%14.3f \n',sqrt(lemda(1,1)));
fprintf('\n')
fprintf('\nMode shapes \n')
fprintf('First mode   Second Mode \n')
for i = 1:2
    fprintf('%10.3f %14.3f \n',V(i,2)/V(1,2),V(i,1)/V(1,1));
end
```

The necessary results are obtained as shown below:

```
>> Natural frequencies
    omega1 = 1.414
    omega2 = 0.707
```

Mode Shapes

| First Mode | Second Mode |
| --- | --- |
| 1.000 | 1.000 |
| − 1.000 | 2.000 |

# 8

# Structural Dynamic Modification and Sensitivity Analysis

Su Huan Chen
*Jilin University*

**Summary**

*The matrix perturbation theory for structural dynamic modification and sensitivity analysis is presented in this chapter. The theory covers a broad spectrum of subjects, specifically, matrix perturbation of real modes of complex structures and matrix perturbation of complex modes. The contents include nine sections. Section 8.2 provides the preliminaries to matrix perturbation and sensitivity analysis. Section 8.3 presents the matrix perturbation method including first-order and second-order perturbation. Section 8.4 presents methods for design sensitivity analysis. In Section 8.5, high-accuracy modal superposition for sensitivity analysis of modes is given. Section 8.6 presents the sensitivity analysis of eigenvectors for free–free structures. In Section 8.7 and Section 8.8, matrix perturbations for repeated modes and closely spaced modes are discussed. In Section 8.9, the matrix perturbation approach for complex modes is presented.*

## 8.1 Introduction

In modern engineering problems, the dynamic design of structures is becoming increasingly important. In order to achieve an optimal design, we repeatedly have to modify the structural parameters and solve the generalized eigenvalue problem. The iterative vibration analysis can be very tedious for large and complex structures. Therefore, it is necessary to seek a fast computation method for sensitivity analysis and reanalysis. The matrix perturbation method is an extremely useful tool for this purpose.

The matrix perturbation method is concerned with how the natural frequencies and modal vectors change if small modifications are imposed on the parameters of structures. Engineering problems often involve many small modifications in the structural parameters, such as material property variations, manufacturing errors, iterative design of structural parameters, design sensitivity analysis, random eigenvalue analysis, robustness analysis of control systems, and so on.

In this chapter, it is assumed that the reader has an undergraduate knowledge in vibration theory and a working knowledge in the finite element method.

The contents of the chapter include the basic preliminaries: vibration equations of the finite element model, eigenvalue problem, modal vectors, orthogonality conditions, modal expansion theorem, and the power series expansion of eigensolutions. The chapter also covers such topics as: the perturbation method for distinct eigenvalues and corresponding eigenvectors; sensitivities of eigenvalues and eigenvectors; the high-accuracy modal superposition method for eigenvector derivatives; eigenvector derivatives for free–free structures; perturbation method for systems with repeated eigenvalues and close eigenvalues; and perturbation method of the complex modes of systems with real unsymmetric matrices.

## 8.2 Structural Dynamic Modification of Finite Element Model

The finite element method is an important tool to obtain numerical and computational solutions to problems in structural vibration analysis. By applying the finite element method to a structure, a discrete analysis model to idealize the continuum can be obtained. The finite equation of vibrations of a structure in the global coordinate system is

$$\mathbf{M\ddot{q} + C\dot{q} + Kq = Q} \tag{8.1}$$

where $\mathbf{M}$, $\mathbf{K}$, and $\mathbf{C}$ are the mass, stiffness, and damping matrices, respectively, $\mathbf{\ddot{q}}$, $\mathbf{\dot{q}}$, and $\mathbf{q}$ are the acceleration, velocity, and displacement vectors, respectively, and $\mathbf{Q}$ is the external load vector.

Neglecting the damping force and external load vector, Equation 8.1 becomes

$$\mathbf{M\ddot{q} + Kq = 0} \tag{8.2}$$

This is the natural vibration equation for the structure. Its solution (the natural vibration) is harmonic, and is given by

$$\mathbf{q = u}\cos(\omega t - \varphi) \tag{8.3}$$

where $\mathbf{u}$ is modal vector, and $\omega$ the natural frequency of the system. Substituting Equation 8.3 into Equation 8.2, the eigenproblem of structural vibration can be obtained as

$$\mathbf{Ku = \lambda Mu} \tag{8.4}$$

where $\lambda$ ($\lambda = \omega^2$) denote the eigenvalues of the system.

In structural vibration analysis, the natural frequencies and the corresponding modal vectors can be obtained by solving the eigenproblem (Equation 8.4). The solutions for $n$ eigenvalues and corresponding eigenvectors satisfy

$$\mathbf{KU} = \mathbf{MU\Lambda} \tag{8.5}$$

where $\mathbf{U}$, which is called the *modal matrix*, is an $(n \times n)$ matrix with its columns equal to the $n$ eigenvectors, and $\mathbf{\Lambda}$ is an $(n \times n)$ diagonal matrix consisting of the corresponding eigenvalues as the diagonal elements; specifically

$$\mathbf{U} = [\mathbf{u}_1, \mathbf{u}_2, ..., \mathbf{u}_n] \tag{8.6}$$

$$\mathbf{\Lambda} = \text{diag}(\lambda_i), \quad i = 1, 2, ..., n \tag{8.7}$$

An important relation for eigenvectors is that of $\mathbf{M}$ orthogonality and $\mathbf{K}$ orthogonality; that is, we have

$$\mathbf{u}_i^{\mathrm{T}} \mathbf{M} \mathbf{u}_j = \delta_{ij} \tag{8.8}$$

$$\mathbf{u}_i^{\mathrm{T}} \mathbf{K} \mathbf{u}_j = \lambda_i \delta_{ij} \tag{8.9}$$

where $\delta_{ij}$ is the Kronecker delta. For $n$ eigenpairs, Equation 8.8 and Equation 8.9 can be written as

$$\mathbf{U}^{\mathrm{T}} \mathbf{M} \mathbf{U} = \mathbf{I} \tag{8.10}$$

$$\mathbf{U}^{\mathrm{T}} \mathbf{K} \mathbf{U} = \mathbf{\Lambda} \tag{8.11}$$

Since the modal vectors are independent, an arbitrary displacement vector, $\mathbf{u}$, can be expressed as a linear combination of $\mathbf{u}_i$, $i = 1, 2, ..., n$; that is

$$\mathbf{u} = \sum_{r=1}^{n} c_r \mathbf{u}_r = \mathbf{UC} \tag{8.12}$$

where $c_r$ is a constant. Each constant $c_r$ can be determined by

$$c_r = \mathbf{u}_r^{\mathrm{T}} \mathbf{Mu}, \quad r = 1, 2, ..., n \tag{8.13}$$

This is known as the *expansion theorem*.

Suppose the physical parameter of a given structure is given a small modification. This will cause a small change in the matrices $\mathbf{K}_0$ and $\mathbf{M}_0$; that is

$$\mathbf{M} = \mathbf{M}_0 + \varepsilon \mathbf{M}_1, \quad \mathbf{K} = \mathbf{K}_0 + \varepsilon \mathbf{K}_1 \tag{8.14}$$

where $\varepsilon$ is a small parameter, $\mathbf{K}_0$ and $\mathbf{M}_0$ are the original mass and stiffness matrices, respectively, and $\varepsilon \mathbf{M}_1$ and $\varepsilon \mathbf{K}_1$ are the corresponding modifications. It is obvious that if $\mathbf{M}_0$ and $\mathbf{K}_0$ are symmetric, the matrices $\mathbf{M}_1$ and $\mathbf{K}_1$ are also symmetric.

If $\varepsilon \mathbf{M}_1$ and $\varepsilon \mathbf{K}_1$ are small, the changes of eigenvalues and eigenvectors of the structure are also small. According to the matrix perturbation theory, the eigensolutions of Equation 8.4 can be expressed in the form of a power series in $\varepsilon$; thus

$$\mathbf{u}_i = \mathbf{u}_{0i} + \varepsilon \mathbf{u}_{1i} + \varepsilon^2 \mathbf{u}_{2i} + \cdots \tag{8.15}$$

$$\lambda_i = \lambda_{0i} + \varepsilon \lambda_{1i} + \varepsilon^2 \lambda_{2i} + \cdots \tag{8.16}$$

where $\mathbf{u}_{0i}$ and $\lambda_{0i}$ are the eigensolutions of the original structure, $\lambda_{1i}$ and $\lambda_{2i}$ are the first- and the second-order perturbations of the eigenvalues, and $\mathbf{u}_{1i}$ and $\mathbf{u}_{2i}$ are the first- and the second-order perturbation of the eigenvectors.

Since the eigensolutions of the original structure, $\mathbf{u}_{0i}$ and $\lambda_{0i}$, are known, only the first- and the second-order perturbations of the eigensolutions are required without solving Equation 8.4.

## 8.3    Perturbation Method of Vibration Modes

The perturbation methods of vibration modes are well developed (Fox and Kapoor, 1968; Rogers, 1977; Chen and Wada, 1979; Hu, 1987; Chen, 1993). In this section, it is assumed that all eigenvalues of the original structure are distinct.

### 8.3.1    First-Order Perturbation of Distinct Modes

According to the expansion theorem, the first-order perturbation, $\mathbf{u}_{1i}$, can be expanded by the modal vectors, $\mathbf{u}_{0s}$, of the original structure as

$$\mathbf{u}_{1i} = \sum_{s=1}^{n} c_{1s}\mathbf{u}_{0s} \tag{8.17}$$

where

$$c_{1s} = \frac{1}{\lambda_{0i} - \lambda_{0s}}(\mathbf{u}_{0s}^{\mathrm{T}}K_1\mathbf{u}_{0i} - \lambda_{0i}\mathbf{u}_{0s}^{\mathrm{T}}\mathbf{M}_1\mathbf{u}_{0i}), \quad s \neq i \tag{8.18}$$

$$c_{1i} = -\tfrac{1}{2}\mathbf{u}_{0i}^{\mathrm{T}}\mathbf{M}_1\mathbf{u}_{0i} \tag{8.19}$$

The first-order perturbation of the eigenvalues is

$$\lambda_{1i} = \mathbf{u}_{0i}^{\mathrm{T}}K_1\mathbf{u}_{0i} - \lambda_{0i}\mathbf{u}_{0i}^{\mathrm{T}}\mathbf{M}_1\mathbf{u}_{0i} \tag{8.20}$$

### 8.3.2    Second-Order Perturbation of Distinct Modes

If the parameter modification is fairly large, in order to obtain high computing accuracy, the second-order perturbation must be used. According to the expansion theorem, the second-order perturbation, $\mathbf{u}_{2i}$, can be expanded by the modal vectors, $\mathbf{u}_{0s}$, of the original structure as

$$\mathbf{u}_{2i} = \sum_{s=1}^{n} c_{2j}\mathbf{u}_{0s} \tag{8.21}$$

where

$$c_{2s} = \frac{1}{\lambda_{0i} - \lambda_{0s}}(\mathbf{u}_{0s}^{\mathrm{T}}K_1\mathbf{u}_{1i} - \lambda_{0i}\mathbf{u}_{0s}^{\mathrm{T}}\mathbf{M}_1\mathbf{u}_{1i} - \lambda_{1i}\mathbf{u}_{0s}^{\mathrm{T}}\mathbf{M}_0\mathbf{u}_{1i} - \lambda_{1i}\mathbf{u}_{0s}^{\mathrm{T}}\mathbf{M}_1\mathbf{u}_{0i}), \quad s \neq i \tag{8.22}$$

$$c_{2s} = -\frac{1}{2}(\mathbf{u}_{1i}^{\mathrm{T}}\mathbf{M}_0\mathbf{u}_{1i} + \mathbf{u}_{0i}^{\mathrm{T}}\mathbf{M}_1\mathbf{u}_{1i} + \mathbf{u}_{1i}^{\mathrm{T}}\mathbf{M}_1\mathbf{u}_{0i}) \tag{8.23}$$

The second perturbation of the eigenvalues is

$$\lambda_{2i} = \mathbf{u}_{0i}^{\mathrm{T}}K_1\mathbf{u}_{1i} - \lambda_{0i}\mathbf{u}_{0i}^{\mathrm{T}}\mathbf{M}_1\mathbf{u}_{1i} - \lambda_{1i}\mathbf{u}_{0i}^{\mathrm{T}}\mathbf{M}_0\mathbf{u}_{1i} - \lambda_{1i}\mathbf{u}_{0i}^{\mathrm{T}}\mathbf{M}_1\mathbf{u}_{0i} \tag{8.24}$$

### 8.3.3    Numerical Examples

As illustrations of the matrix perturbation method, several numerical examples are given now.

### Example 8.1

Consider the five-degree-of-freedom (five-DoF) system shown in Figure 8.1. The physical parameters are given as

$$m_1 = m_2 = m_3 = m_4 = 1.0, \quad m_5 = 0.5, \quad k_1 = k_2 = k_3 = k_4 = k_5 = 1.0$$

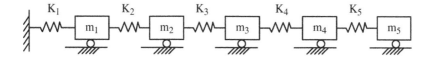

**FIGURE 8.1** Mass–spring system for Example 8.1.

In order to study the computing accuracy of first- and second-order perturbations, let us assume that the fifth mass undergoes a decrement of 5 to 30%, and the stiffness of the first spring undergoes a decrement of 5 to 30%.

The computed results for the natural frequencies are presented in Table 8.1, in which the initial solutions mean the eigensolutions of the original structure.

**TABLE 8.1** Comparison of Natural Frequencies

| Mode Number | | Changes of Structural Parameter (%) | | | | | |
|---|---|---|---|---|---|---|---|
| | | 5 | 10 | 15 | 20 | 25 | 30 |
| 1 | A | 0.3022 | 0.2922 | 0.2821 | 0.2724 | 0.2629 | 0.2536 |
| | B | 0.3128 | 0.3128 | 0.3128 | 0.3128 | 0.312 | 0.3128 |
| | C | 3.52 | 7.14 | 10.9 | 14.38 | 18.97 | 23.3 |
| | D | 0.3017 | 0.2903 | 0.2783 | 0.2658 | 0.2527 | 0.2388 |
| | C | 0.14 | 0.58 | 1.34 | 2.43 | 3.9 | 5.81 |
| | E | 0.3022 | 0.2922 | 0.2827 | 0.2740 | 0.2660 | 0.2588 |
| | C | 0.0033 | 0.068 | 0.24 | 0.58 | 0.93 | 2.08 |
| 2 | A | 0.8788 | 0.8512 | 0.8249 | 0.7998 | 0.7756 | 0.7523 |
| | B | 0.9079 | 0.98079 | 0.9079 | 0.98079 | 0.9079 | 0.98079 |
| | C | 3.31 | 6.66 | 10.1 | 13.52 | 17.06 | 20.68 |
| | D | 0.8775 | 0.8460 | 0.8133 | 0.7792 | 0.7435 | 0.7060 |
| | C | 0.15 | 0.61 | 1.41 | 2.58 | 4.14 | 6.15 |
| | E | 0.3789 | 0.8518 | 0.8267 | 0.8089 | 0.7835 | 0.7659 |
| | C | 0.0076 | 0.062 | 0.21 | 0.52 | 0.91 | 1.80 |
| 3 | A | 1.3732 | 1.3348 | 1.2989 | 1.2650 | 1.2332 | 1.2031 |
| | B | 1.1421 | 1.1421 | 1.1421 | 1.1421 | 1.1421 | 1.1421 |
| | C | 2.99 | 5.94 | 8.88 | 11.79 | 14.7 | 17.5 |
| | D | 1.371 | 1.3266 | 1.2806 | 1.2328 | 1.1832 | 1.1313 |
| | C | 0.15 | 0.62 | 1.41 | 2.54 | 4.05 | 5.96 |
| | E | 1.3733 | 1.3356 | 1.3015 | 1.2712 | 1.2449 | 1.2231 |
| | C | 0.0074 | 0.060 | 0.20 | 0.49 | 0.74 | 1.66 |
| 4 | A | 1.7355 | 1.6923 | 1.6520 | 1.6143 | 1.5790 | 1.5457 |
| | B | 1.7820 | 1.7820 | 1.7820 | 1.7820 | 1.7820 | |
| | C | 2.68 | 5.30 | 7.78 | 10.38 | 12.85 | 15.28 |
| | D | 1.7331 | 1.6828 | 1.6319 | 1.5773 | 1.5209 | 1.5209 |
| | C | 0.14 | 0.56 | 1.28 | 2.29 | 3.62 | 5.27 |
| | E | 1.7356 | 1.6933 | 1.6552 | 1.6216 | 1.5830 | 1.5695 |
| | C | 0.0070 | 0.057 | 0.19 | 0.45 | 0.68 | 1.53 |
| 5 | A | 1.9273 | 1.8825 | 1.8408 | 1.8017 | 1.7649 | 1.7304 |
| | B | 1.9753 | 1.9753 | 1.9753 | 1.9753 | 1.9753 | 1.9753 |
| | C | 2.49 | 4.93 | 7.31 | 9.63 | 11.92 | 14.16 |
| | D | 1.9248 | 1.8929 | 1.896 | 1.7696 | 1.7079 | 1.6492 |
| | C | 0.1300 | 0.51 | 1.15 | 2.06 | 3.23 | 4.69 |
| | E | 1.9274 | 1.8835 | 1.8439 | 1.8090 | 1.7790 | 1.7541 |
| | C | 0.0064 | 0.051 | 0.17 | 0.41 | 0.83 | 1.37 |

**TABLE 8.2** Comparison of Natural Frequencies

| Mode No. | A (Hz) | B (Hz) | F (%) | D (Hz) | G (%) | E (Hz) | H (%) |
|---|---|---|---|---|---|---|---|
| 1 | 27.78 | 25.45 | 8.39 | 29.27 | 5.36 | 28.14 | 1.29 |
| 2 | 109.1 | 107.7 | 1.28 | 110.39 | 1.18 | 109.35 | 0.28 |
| 3 | 157.4 | 153.2 | 2.67 | 159.53 | 1.35 | 158.79 | 0.88 |
| 4 | 230.5 | 233.2 | 1.17 | 231.24 | 0.32 | 231.02 | 0.022 |
| 5 | 320.7 | 325.6 | 1.53 | 320.58 | 0.04 | 320.89 | 0.04 |
| 6 | 391.1 | 393.7 | 0.66 | 392.06 | 0.24 | 319.35 | 0.06 |

As can be seen from the results, if the change of the structural parameter is 15%, the average change of the natural frequencies is 9.0%. Using the first-order perturbation, the average error of the frequencies is reduced to 1.32%.

If the change of the structural parameter is 30%, the average error of the natural frequencies is 18%. Using the first-order perturbation, the average error of natural frequencies is reduced to 1.6%.

The notation used in Table 8.1 and Table 8.2 is as follows:

- A: the exact solutions of the modified structure
- B: the initial solutions of the original structure
- C: percent error
- D: the first-order perturbation solutions
- E: the second-order perturbation solutions
- F: the percent errors of the initial solutions
- G: the percent errors of the first-order perturbation
- H: the percent errors of the second perturbation

## Example 8.2

Consider a truss structure (as shown in Figure 8.2) with 20 rods. The cross section area of the second rod is changed from 1.0 to 2.0 cm². The results calculated are listed in Table 8.2.

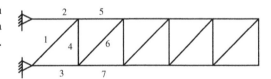

**FIGURE 8.2** Truss structure for Example 8.2.

## Example 8.3

Consider a torsional vibration system with five disks, as shown in Figure 8.3. The physical parameters of the system are as follows:

- $I_1 = 10.78$ kg cm sec²
- $I_2 = 82.82$ kg cm sec²
- $I_3 = 14.27$ kg cm sec²
- $I_4 = 29.56$ kg cm sec²
- $I_5 = 21.66$ kg cm sec²
- $K_1 = 10.48 \times 10^4$ kg cm/rad
- $K_2 = 34.30 \times 10^4$ kg cm/rad
- $K_3 = 24.40 \times 10^4$ kg cm/rad
- $K_4 = 40.60 \times 10^4$ kg cm/rad

The corresponding constrained system is shown in Figure 8.3b, in which the hung stiffness is $K_s = 4060$ kg cm/rad.

The exact eigensolutions of the constrained system are taken as the initial results. Using matrix

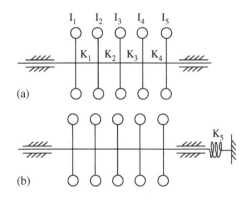

**FIGURE 8.3** Torsional vibration system for Example 8.3.

**TABLE 8.3** Comparison for Natural Frequencies

| | No. | | | | |
|---|---|---|---|---|---|
| | 1 | 2 | 3 | 4 | 5 |
| $\omega_0$ | 0.000000 | 62.554934 | 105.668169 | 177.680405 | 224.361444 |
| $\omega_x$ | 1.595950 | 62.604815 | 105.669360 | 177.704224 | 224.386235 |
| $\delta$ | | 0.079700 | 0.001070 | 0.013406 | 0.011000 |
| $\varepsilon\omega_1$ | −1.525707 | −0.049895 | −0.001132 | −0.023834 | −0.002508 |
| $\omega'_0$ | 0.080243 | 62.554920 | 105.448168 | 177.680390 | 224.383727 |
| $\delta'$ | | $2.24 \times 10^{-5}$ | $9.46 \times 10^{-7}$ | $8.22 \times 10^{-8}$ | $9.93 \times 10^{-5}$ |

**TABLE 8.4** Comparison for Eigenvectors

| No. | $u_0^i$ | $u_x^i$ | $\varepsilon u_1^i$ | $u'^i_0$ |
|---|---|---|---|---|
| 1 | 0.079283 | 0.097363 | −0.000080 | 0.079823 |
| | 0.079283 | 0.079342 | −0.000059 | 0.079283 |
| | 0.079283 | 0.079287 | −0.000004 | 0.079283 |
| | 0.079283 | 0.079198 | 0.000085 | 0.079283 |
| | 0.079283 | 0.079189 | 0.000153 | 0.079342 |
| 2 | −0.095647 | −0.095611 | 0.000036 | −0.095647 |
| | −0.057148 | −0.057064 | −0.000084 | −0.057148 |
| | 0.008612 | 0.008719 | −0.000107 | 0.008612 |
| | 0.099081 | 0.099192 | −0.000111 | 0.099081 |
| | 0.125223 | 0.125259 | 0.000036 | 0.125223 |
| 3 | 0.277894 | 0.277884 | 0.000010 | 0.277894 |
| | −0.041278 | −0.041284 | 0.000005 | −0.41279 |
| | −0.027509 | −0.027497 | −0.000013 | −0.027510 |
| | 0.009810 | 0.009840 | −0.000030 | 0.009810 |
| | 0.024263 | 0.024279 | −0.000016 | 0.024263 |
| 4 | −0.009038 | −0.009033 | −0.000005 | −0.009038 |
| | 0.020312 | −0.020309 | 0.000002 | 0.020312 |
| | −0.125560 | −0.125580 | 0.000024 | −0.125556 |
| | −0.098791 | 0.098735 | −0.000053 | −0.098788 |
| | 0.144375 | 0.144413 | −0.000038 | 0.144375 |
| 5 | 0.004823 | 0.004823 | 0.000001 | 0.004824 |
| | −0.020156 | −0.020154 | −0.0000002 | −0.020156 |
| | 0.217244 | 0.217227 | 0.000018 | 0.217245 |
| | −0.088717 | −0.088725 | 0.000008 | −0.088717 |
| | 0.052618 | 0.052652 | −0.000034 | 0.052618 |

perturbation, the eigensolutions of the free–free system can be obtained. The results are listed in Table 8.3 and Table 8.4.

The notation used in Table 8.3 is as follows:

- $\omega_0$: exact solution of the natural frequency of the free–free system (l/sec)
- $\omega'_0$: perturbation solutions of the natural frequency of the free–free system (l/sec)
- $\omega_x$: natural frequency of the constrained system (l/sec)

$\varepsilon\omega_1 = \omega'_0 - \omega_x$ the perturbation of the natural frequency

$$\delta = \frac{|\omega_0 - \omega_x|}{\omega_0} \ (\%)$$

$$\delta' = \frac{|\omega_0 - \omega'_0|}{\omega_0} \ (\%)$$

The notation used in Table 8.4 is as follows:

- $\mathbf{u}_0^i$: exact solution of eigenvectors of the free–free system
- $\mathbf{u}_0'^i$: perturbation solution of eigenvectors of the free–free system
- $\mathbf{u}_x^i$: eigenvector of the constrained system
- $\varepsilon \mathbf{u}_1^i$: first-order perturbation of eigenvectors

As can be seen from Table 8.3, the natural frequencies of the free–free system are increased by the hung elastic elements. For example, the frequency of the rigid mode is increased to 1.595950 (l/sec), and the frequency of the first elastic mode is increased by 0.8124%. By modifying the eigensolutions with the perturbation method, the frequency of the rigid mode is reduced to 0.079700 (l/sec), which is nearly equal to zero, and all the frequencies of the elastic modes become almost exact solutions. The results in Table 8.4 show that the mode shapes of the free–free system, $\mathbf{u}_0'^i$, are close to the exact solution, $u_0^i$.

# 8.4 Design Sensitivity Analysis of Structural Vibration Modes

In the optimization of structural analysis, the design sensitivity analysis of eigenvalues and eigenvectors plays an essential role. The designer can use this information directly in an interactive computer-aided design procedure as a valuable guide. Significant work has been done in this area (Haug et al., 1985; Adelmen and Haftka, 1986; Chen and Pan, 1986; Wang, 1991).

## 8.4.1 Direct Differential Method for Sensitivity Analysis

Design sensitivity analysis of eigenvalues and eigenvectors will reveal how the changes in some design parameters in the system affect the dynamic characteristics of the structure.

Let $\lambda_{i,j}$ and $\mathbf{u}_{i,j}$ denote the sensitivity of the eigenvalue, $\lambda_i$, and the eigenvector, $\mathbf{u}_i$, respectively, with respect to the design variables $b_j$ ($j = 1, 2, \ldots, L$), and let $\mathbf{K}_j$ and $\mathbf{M}_j$ denote the derivative of the stiffness and mass matrices, respectively, with respect to $b_j$. The design sensitivity of the eigenvalue is

$$\lambda_{i,j} = \mathbf{u}_i^T (\mathbf{K}_j - \lambda_i \mathbf{M}_j) \mathbf{u}_i \tag{8.25}$$

The sensitivity of the eigenvector, $\mathbf{u}_{i,j}$, can be expressed as the following series:

$$\mathbf{u}_{i,j} = \sum_{s=1}^{n} c_{ijs} \mathbf{u}_s \tag{8.26}$$

where

$$c_{ijs} = \frac{1}{\lambda_i - \lambda_s} \mathbf{u}_s^T (\mathbf{K}_j - \lambda_i \mathbf{M}_j) \mathbf{u}_i, \quad i \neq s, \quad i, s = 1, 2, \ldots, n, \quad i \neq s \tag{8.27}$$

$$c_{iji} = -\frac{1}{2} \mathbf{u}_i^T \mathbf{M}_j \mathbf{u}_i \tag{8.28}$$

## 8.4.2 Perturbation Sensitivity Analysis

Let $\Delta \mathbf{K}$ and $\Delta \mathbf{M}$ denote the increments of the stiffness and the mass matrices resulting from an incremental change of the design variable, $\Delta b_j$, and let $\Delta \lambda_i$ and $\Delta \mathbf{u}_i$ denote the corresponding perturbations of the eigenvalue and eigenvector, respectively. The direct differential method of design sensitivity analysis of vibration modes can now be put into perturbation form, approximately as

$$\lambda_{i,j} = \frac{\Delta \lambda_i}{\Delta b_j} \tag{8.29}$$

$$\mathbf{u}_{i,j} = \frac{\Delta \mathbf{u}_i}{\Delta b_j} \tag{8.30}$$

where $\Delta\lambda_i$ and $\Delta\mathbf{u}_i$ can be evaluated by the perturbation formulas presented in this chapter. In practical analysis, the design variables could be the cross-sectional area of the truss members, bending moment of inertia, equivalent torsional moment of inertia of a beam, the thickness of a plate, or other variable. In some complex structures, a mass, $m_r$, may be placed at a node point and moving in the direction of the $r$th DoF, or an elastic support with spring stiffness, $K_r$, may be placed at a certain node point. It is also possible that an elastic connector of stiffness, $K_j$, might exist between two components. They can also be considered as design variables. In finite element analysis, $\Delta\mathbf{K}$ and $\Delta\mathbf{M}$ are known to be the sum of the element increments, $\Delta\mathbf{K}^e$ and $\Delta\mathbf{M}^e$; thus

$$\Delta\mathbf{K} = \sum_e \Delta\mathbf{K}^e \tag{8.31}$$

$$\Delta\mathbf{M} = \sum_e \Delta\mathbf{M}^e \tag{8.32}$$

Hence, the sensitivity formulas of vibration modes as given above can be transformed into the finite element perturbation form (Chen and Pan, 1986)

$$\lambda_{i,j} = \frac{1}{\Delta b_j} \sum_e \bar{\mathbf{u}}_i^{\mathrm{T}} (\Delta\mathbf{K}^e - \lambda_i \Delta\mathbf{M}^e) \bar{\mathbf{u}}_i \tag{8.33}$$

and

$$\mathbf{u}_{i,j} = \frac{1}{\Delta b_j} \sum_e \left( \sum_{\substack{s=1 \\ s \neq i}}^{n} \frac{1}{\lambda_i - \lambda_s} \bar{\mathbf{u}}_s^{\mathrm{T}} (\Delta\mathbf{K}^e - \lambda_i \Delta\mathbf{M}^e) \bar{\mathbf{u}}_i \mathbf{u}_s - \frac{1}{2} \bar{\mathbf{u}}_i^{\mathrm{T}} \Delta\mathbf{M}^e \bar{\mathbf{u}}_i \mathbf{u}_i \right) \tag{8.34}$$

In these formulas, the overbar signifies that the eigenvector concerned contains only the components needed for the $e$th finite element. It is important to observe that, in Equation 8.33 and Equation 8.34, calculations are done on the element basis, and as a result, the calculations are greatly simplified.

Using the shorthand notations

$$\lambda_{i,j}^e = \frac{1}{\Delta b_j} \bar{\mathbf{u}}_i^{\mathrm{T}} \Delta\mathbf{K}^e - \lambda_i \Delta\mathbf{M}^e \bar{\mathbf{u}}_i \tag{8.35}$$

$$\mathbf{u}_{i,j}^e = \frac{1}{\Delta b_j} \left( \sum_{\substack{s=1 \\ s \neq i}}^{n} \frac{1}{\lambda_i - \lambda_s} \bar{\mathbf{u}}_s^{\mathrm{T}} \Delta\mathbf{K}^e - \lambda_i \Delta\mathbf{M}^e \bar{\mathbf{u}}_i \mathbf{u}_s - \frac{1}{2} \bar{\mathbf{u}}_i^{\mathrm{T}} \Delta\mathbf{M}^e \bar{\mathbf{u}}_i \mathbf{u}_i \right) \tag{8.36}$$

Equation 8.33 and Equation 8.34 can be written as

$$\lambda_{i,j} = \sum_e \lambda_{i,j}^e \tag{8.37}$$

and

$$\mathbf{u}_{i,j} = \sum_e \mathbf{u}_{i,j}^e \tag{8.38}$$

where $\lambda_{i,j}^e$ and $\mathbf{u}_{i,j}^e$ are the design sensitivity of the $e$th element for the eigenvalue $\lambda_i$ and the eigenvector $\mathbf{u}_i$, respectively. Let us consider the following important cases.

For a concentrated mass, $m_r$, placed at a node point and moved in the direction of the $r$th DoF, Equation 8.33 and Equation 8.34 become

$$\lambda_{i,r} = \frac{\Delta\lambda_i}{\Delta m_r} = -\lambda_i u_{ir}^2 \tag{8.39}$$

and

$$\mathbf{u}_{i,r} = \frac{\Delta \mathbf{u}_i}{\Delta m_r} = \sum_{\substack{s=1 \\ s \neq i}}^{n} \frac{-\lambda_i}{\lambda_i - \kappa_s} u_{sr} u_{ir} \mathbf{u}_s - \frac{1}{2} u_{ir}^2 \mathbf{u}_i \tag{8.40}$$

where $u_{ir}$ is the $r$th element of the $i$th eigenvector $\mathbf{u}_i$.

For an elastic connector with stiffness $K_j$ between the $r$th and the $l$th DoF of two components, Equation 8.33 and Equation 8.34 become

$$\lambda_{i,j} = \frac{\Delta \lambda_i}{\Delta k_j} = (u_{ir} - u_{il})^2 \tag{8.41}$$

and

$$\mathbf{u}_{i,j} = \frac{\Delta \mathbf{u}_i}{\Delta k_j} = \sum_{\substack{s=1 \\ s \neq i}}^{n} \frac{1}{\lambda_i - \lambda_s} (u_{sr} u_{ir} - u_{sl} u_{ir} - u_{sr} u_{il} + u_{sl} u_{il}) \mathbf{u}_s \tag{8.42}$$

For an elastic support with spring stiffness $K_r$ placed in the direction of the $r$th DoF, Equation 8.33 and Equation 8.34 become

$$\lambda_{i,r} = \frac{\Delta \lambda_i}{\Delta k_r} = u_{ir}^2 \tag{8.43}$$

and

$$\mathbf{u}_{i,r} = \frac{\Delta \mathbf{u}_i}{\Delta k_r} = \sum_{\substack{s=1 \\ s \neq i}}^{n} \frac{1}{\lambda_i - \lambda_s} u_{sr} u_{ir} \mathbf{u}_s \tag{8.44}$$

### 8.4.3 Numerical Example

The design sensitivity analysis of an automotive chassis is presented here as an illustration of the method.

### Example 8.4

The finite element model of an automobile chassis consists of 39 beam elements involving 30 nodal points and 180 DoF (Figure 8.4).

The design variables for the sensitivity analysis of eigenvalues in this example are the equivalent torsional moment of inertia, $J$, and the bending moment of inertia, $I_y$, of the beam element of

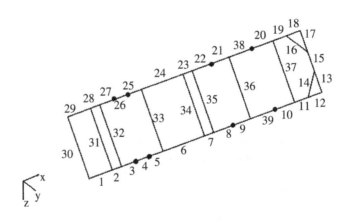

**FIGURE 8.4**   Finite element model of the automotive chassis for Example 8.4.

**TABLE 8.5** Sensitivities of the First Four Chassis Frequencies

| | | | | | | | | | |
|---|---|---|---|---|---|---|---|---|---|
| 1 | NE | 15 | 11 | 19 | 24 | 6 | 37 | 34 | 36 |
| | $\lambda^e_{1J}$ | 9.97 | 5.23 | 5.23 | 3.64 | 3.64 | 3.34 | 3.32 | 3.31 |
| | NE | 24 | 6 | 23 | 7 | 22 | 8 | 33 | 21 |
| | $\lambda^e_{1Iy}$ | 0.015 | $0.15 \times 10^{-2}$ | $0.33 \times 10^{-3}$ | $0.33 \times 10^{-3}$ | $0.28 \times 10^{-3}$ | $0.19 \times 10^{-3}$ | $0.11 \times 10^{-3}$ | $0.11 \times 10^{-3}$ |
| 2 | NE | 16 | 14 | 11 | 19 | 18 | 12 | 13 | 17 |
| | $\lambda^e_{2J}$ | $6.67 \times 10^{-2}$ | $0.67 \times 10^{-2}$ | $0.45 \times 10^{-2}$ | $0.45 \times 10^{-2}$ | $0.11 \times 10^{-2}$ | $0.11 \times 10^{-2}$ | $0.53 \times 10^{-3}$ | $0.53 \times 10^{-3}$ |
| | NE | 38 | 10 | 21 | 9 | 22 | 8 | 24 | 6 |
| | $\lambda^e_{2Iy}$ | 2.92 | 2.92 | 1.78 | 1.78 | 1.49 | 1.49 | 1.44 | 1.44 |
| 3 | NE | 15 | 19 | 11 | 37 | 31 | 30 | 32 | 1 |
| | $\lambda^e_{3J}$ | 95.9 | 47.2 | 47.2 | 26.5 | 25.4 | 25.2 | 23.8 | 12.5 |
| | NE | 10 | 38 | 9 | 21 | 8 | 22 | 6 | 24 |
| | $\lambda^e_{3Iy}$ | 2.96 | 2.96 | 1.97 | 1.97 | 1.74 | 1.74 | 1.70 | 1.70 |
| 4 | NE | 14 | 16 | 19 | 11 | 12 | 18 | 17 | 13 |
| | $\lambda^e_{4J}$ | 0.19 | 0.19 | 0.12 | 0.12 | 0.03 | 0.03 | 0.015 | 0.015 |
| | NE | 10 | 38 | 39 | 20 | 5 | 25 | 6 | 24 |
| | $\lambda^e_{4Iy}$ | 22.4 | 22.4 | 15.7 | 15.7 | 11.8 | 11.8 | 10.9 | 10.9 |

the structure. Results for the sensitivity of eigenvalues with respect to $J$ and $I_y$ are given in Table 8.5, in which NE denotes the number of the element. Only the highest eight values are given, and they are listed in descending order.

From Table 8.5, it is seen that for this particular chassis, the sensitivities of the first natural frequency, $\lambda^e_{1Iy}$, are much smaller than $\lambda^e_{1J}$. This indicates that there is very little effect of the change of bending moment of inertia, $I_y$, of the beams on the vibration of the chassis at its first natural frequency. Thus, we can conclude that the first mode is a torsional mode. Similarly, the results indicate that the third mode is also a torsional mode. On the other hand, the second and the fourth modes are recognized to be bending modes. This information is very useful to the designer when deciding on a change in the design. For example, if he wants to increase the first torsional frequency, the efficient way is for him to increase the equivalent torsional moments of inertia of beam elements 15, 11, 19, 24, 6, and so on.

It should be noted that only the first low-frequency modes are available and can be used as basis vectors of eigenvector derivatives in Equation 8.26. However, modal truncation induces errors, and the errors become significant if more high-frequency modes are truncated. An improvement to truncated modal superposition representation of eigenvector derivatives is presented in the next section.

### 8.4.4 Concluding Remarks

As can be seen from the numerical examples given above, the matrix perturbation method is an extremely useful tool for fast reanalysis of a modified structure. It is widely used in a range of structural modifications, such as the modification of various types of elements, local modification of structures, sensitivity analysis of vibration modes, and so on. Therefore, matrix perturbation plays an important role in dynamic analysis and optimization of structures.

# 8.5 High-Accuracy Modal Superposition for Sensitivity Analysis of Modes

The modal superposition method is often used to compute the derivatives of modal vectors. Because of the cost of generating computer solutions for a dynamic analysis, it is impractical to obtain all modes.

Therefore, only the first $L$ low-frequency modes are computed and are used as basis vectors of eigenvector derivatives. However, as noted above, modal truncation induces errors, which can be significant if more high-frequency modes are truncated. An explicit method to improve the truncated modal superposition representation of eigenvector derivatives is presented (Wang, 1991), in which a residual static mode is used to approximate the contribution due to unavailable high-frequency modes (method one).

In this section a more accurate modal superposition method (method two; Chen, 1993a; Liu and Chen, 1994a) than method one is given. In this method, the contribution of the truncated modes to the eigenvector derivatives is expressed exactly, as a convergent series that can be evaluated by a simple iterative procedure.

### 8.5.1   Method One

The modal sensitivity can be expressed as

$$u_{i,j} = \sum_{s=1}^{N} c_s u_s = \sum_{j=1}^{L} c_j \mathbf{u}_j + \mathbf{S}_R \tag{8.45}$$

where

$$\mathbf{S}_R = \sum_{j=L+1}^{N} c_j \mathbf{u}_j \tag{8.46}$$

Since $\lambda_i \ll \lambda_{L+1}$, Equation 8.46 can be approximated as

$$\mathbf{S}_R \approx \mathbf{S}_{RA} = \bar{\mathbf{H}}_0 - \bar{\mathbf{W}}_0 \tag{8.47}$$

where

$$\bar{\mathbf{H}}_0 = \mathbf{K}^{-1}(-\mathbf{K}_j + \lambda_{i,j}\mathbf{M} + \lambda_i \mathbf{M}_j) \tag{8.48}$$

$$\bar{\mathbf{W}}_0 = \sum_{j=1}^{L} \frac{1}{\lambda_j} \mathbf{u}_j^{\mathrm{T}}(-\mathbf{K}_j + \lambda_{i,j}\mathbf{M} + \lambda_i \mathbf{M}_j)\mathbf{u}_j \tag{8.49}$$

### 8.5.2   Method Two

The contribution of $u_{i,j}$, $\mathbf{S}_R$ due to truncated high-frequencies modes is as follows:

$$\mathbf{S}_R = \sum_{j=0}^{\infty} \lambda_i^j(\mathbf{H}_j - \mathbf{W}_j) \tag{8.50}$$

where

$$\mathbf{W}_j = \mathbf{U}_L \mathbf{\Lambda}_L^{-j-1} \mathbf{U}_L^{\mathrm{T}}(-\mathbf{K}_j + \lambda_{i,j}\mathbf{M} + \lambda_i \mathbf{M}_j) \tag{8.51}$$

$$\mathbf{U}_L = [\mathbf{u}_1, \mathbf{u}_2, ..., \mathbf{u}_L] \tag{8.52}$$

$\mathbf{H}_j$ can be obtained with the following iterative procedure:

$$\left. \begin{aligned} \mathbf{H}_0 &= \mathbf{K}^{-1}(-\mathbf{K}_j + \mathbf{K}_{i,j}\mathbf{M} + \lambda_i \mathbf{M}_j) \\ \mathbf{F}'_{j-1} &= \mathbf{M}\mathbf{H}_{j-1}, \quad j \geq 1 \\ \mathbf{H}_j &= \mathbf{K}^{-1}\mathbf{F}'_{j-1} \end{aligned} \right\} \tag{8.53}$$

Define $\mathbf{S}_R(k)$ as

$$\mathbf{S}_R(k) = \sum_{j=0}^{k} \lambda_i^j (\mathbf{H}_j - \mathbf{W}_j) \tag{8.54}$$

Using this definition, the given iterative process can be terminated if the following inequality

$$\|\mathbf{S}_R(k) - \mathbf{S}_R(k-1)\|_2 \le \varepsilon \tag{8.55}$$

is satisfied, where $\varepsilon$ is a specified accuracy requirement.

It should be noted that, if only the first term in the series (Equation 8.50) is retained with all the other terms neglected, then method two is reduced to method one. In addition, the series (Equation 8.50) can be used to estimate the errors induced by the modal truncation.

## 8.6 Sensitivity of Eigenvectors for Free–Free Structures

As can be seen from Equation 8.48 and Equation 8.53, both method one and method two fail to deal with the free–free structures with rigid-body modes because they involve the inversion of the stiffness matrix. However, we can transform the eigenproblem with a singular stiffness matrix into its equivalent eigenproblem with a nonsingular stiffness matrix, in the sense that these two eigenproblems have the same derivatives of eigenvalues and eigenvectors (Liu and Chen, 1994b).

Consider the eigenvalue problem

$$\bar{\mathbf{K}} \bar{\mathbf{u}}_i = \bar{\lambda}_i \mathbf{M} \bar{\mathbf{u}}_i \tag{8.56}$$

where

$$\bar{\mathbf{K}} \equiv \mathbf{K} - \mu \mathbf{M} \tag{8.57}$$

Here, $\mu$ is a nonzero scalar parameter and $\bar{\mathbf{K}}$ is nonsingular if $\mu \ne \lambda_i$ $(i = 1, 2, ..., n)$

It can be shown that

$$\bar{\lambda}_i = \lambda_i - \mu \tag{8.58}$$

$$\bar{\mathbf{u}}_i = \mathbf{u}_i, \quad i = 1, 2, ..., N \tag{8.59}$$

and

$$\frac{d\bar{\lambda}_i}{db} = \frac{d\lambda_i}{db} \tag{8.60}$$

$$\frac{d\bar{\mathbf{u}}_i}{db} = \frac{d\mathbf{u}_i}{db} \tag{8.61}$$

The derivatives $d\mathbf{u}_i/db$ can be obtained from the derivatives $d\bar{\mathbf{u}}_i/db$ of the eigenproblem of Equation 8.56, in which $\bar{\mathbf{K}}$ is nonsingular. In this context, both method one and method two, discussed in Section 8.5.1 and Section 8.5.2, can be applied to deal with the free–free structures with rigid-body modes.

To achieve a faster average convergent speed for all the first $m$ eigenvector derivatives, $\mu$ can be determined as

$$\begin{cases} \mu = \dfrac{\left( \sum\limits_{j=1}^{m} \lambda_j \right)}{m}, \quad j = 1, 2, ..., m \\ \mu \ne \lambda_j, \end{cases} \tag{8.62}$$

## 8.7   Matrix Perturbation Theory for Repeated Modes

### 8.7.1   Basic Equations

In this section, let us consider the case of repeated eigenvalues, namely, $\lambda_{0i} = \lambda_{0i+1} = \cdots = \lambda_{0i+m-1}$. The system is known as a *degenerate system*. In engineering, many complex and large structures, such as airplanes, rockets, tall towers, bridges, and ocean platforms, often have multiple or cluster eigenvalues. The matrix perturbation for the repeated modes is presented in Haug, et al. (1980), Chen and Pan (1986), Hu (1987), Mills-Curran (1988), Ojalvo (1988), Dailey (1989), Lim et al. (1989) and Shaw and Jayasuriya (1992).

Assume that $\lambda_0 = \lambda_{01} = \lambda_{02} = \cdots = \lambda_{0m}$; that is, $\lambda_0$ is a repeated eigenvalue with multiplicity equal to $m$, and $\mathbf{u}_{01}, \mathbf{u}_{02}, \cdots, \mathbf{u}_{0m}$ are the eigenvectors associated with $\lambda_0$. Then, a linear combination of $\mathbf{u}_{0j}$ $(j = 1, 2, \ldots, m)$, denoted as $\mathbf{U}_0$, will also be the eigenvector associated with $\lambda_0$:

$$\mathbf{U}_0 = \mathbf{U}_{0m}\alpha \tag{8.63}$$

where

$$\mathbf{U}_{0m} = [\mathbf{u}_{01}, \mathbf{u}_{02}, \ldots, \mathbf{u}_{0m}] \tag{8.64}$$

$$\alpha^{\mathrm{T}}\alpha = \mathbf{I} \tag{8.65}$$

and

$$\alpha = [\alpha_1, \alpha_2, \ldots, \alpha_m]^{\mathrm{T}} \tag{8.66}$$

Note that $\alpha$ is a constant matrix to be determined.

According to the matrix perturbation method, the eigenvalues and eigenvectors of the structure with repeated eigenvalues for the perturbed structure can be expressed as

$$\Lambda_m = \Lambda_0 + \varepsilon\Lambda_1 \tag{8.67}$$

$$\mathbf{U}_m = \mathbf{U}_{0m}\alpha + \varepsilon(\mathbf{U}_0\mathbf{C}_m + \mathbf{U}_A\mathbf{C}_A) = \mathbf{U}_{0m}\alpha + \varepsilon(\mathbf{U}_{0m}\alpha\mathbf{C}_m + \mathbf{U}_A\mathbf{C}_A) \tag{8.68}$$

where $\mathbf{U}_A$ is the $n \times (n - m)$ modal matrix containing all the eigenvectors except $\mathbf{U}_{0m}$, $\Lambda_m$ is the $m \times m$ eigenvalue diagonal matrix of the perturbed structure, $\Lambda_1$ is the $m \times m$ diagonal matrix with its diagonal elements equal to the first-order perturbations of eigenvalues, $\mathbf{C}_m$ is an $m \times m$ matrix to be determined, and $\mathbf{C}_A$ is an $(n - m) \times (n - m)$ matrix to be determined.

### 8.7.2   The First-Order Perturbation of Eigensolutions

$\Lambda_1$ and $\alpha$ can be computed from the following $(m \times m)$ eigenproblem:

$$\mathbf{W}\alpha = \alpha\Lambda_1, \quad \alpha^{\mathrm{T}}\alpha = \mathbf{I} \tag{8.69}$$

where

$$\mathbf{W} = \mathbf{U}_{0m}^{\mathrm{T}}(\mathbf{K}_1 - \lambda_0\mathbf{M}_1)\mathbf{U}_{0m} \tag{8.70}$$

Solving the $m \times m$ eigenproblem of Equation 8.69 can produce $\Lambda_1$ and $\alpha$.

If matrix $\mathbf{W}$ has no repeated eigenvalues, $\alpha$ can be uniquely determined; if matrix $\mathbf{W}$ has repeated eigenvalues, $\alpha$ can be determined using the higher order perturbation equations. Here, we assume that matrix $\mathbf{W}$ has no repeated eigenvalues; that is, $\lambda_{1i} \neq \lambda_{1j}$, $(i \neq j)$, where $\lambda_{1k}$ $(0 < k \leq m)$ are the elements of the diagonal matrix $\Lambda_1$.

The matrix $\mathbf{C}_A$ is

$$\mathbf{C}_A = (\Lambda_A - \lambda_0\mathbf{I})^{-1}\mathbf{U}_A^{\mathrm{T}}(\lambda_0\mathbf{M}_1 - \mathbf{K}_1)\mathbf{U}_{0m}\alpha \tag{8.71}$$

The elements of $C_m$ are

$$C_{ij}^m = \frac{R_{ij}}{\lambda_{jm}^{(1)} - \lambda_{im}^{(1)}}, \quad i \neq j \quad i,j = 1, 2, ..., m \tag{8.72}$$

where $R_{ij}$ are the elements of **R** given by

$$\mathbf{R} = -\boldsymbol{\alpha}^T\mathbf{U}_{0m}^T\mathbf{M}_1\mathbf{U}_{0m}\boldsymbol{\alpha}\boldsymbol{\Lambda}_1 + \boldsymbol{\alpha}^T\mathbf{U}_{0m}^T\mathbf{K}_1\mathbf{U}_A\mathbf{C}_A - \lambda_0\boldsymbol{\alpha}^T\mathbf{U}_{0m}^T\mathbf{M}_1\mathbf{U}_A\mathbf{C}_A - \boldsymbol{\alpha}^T\mathbf{U}_{0m}^T\mathbf{M}_0\mathbf{U}_A\mathbf{C}_A\boldsymbol{\Lambda}_1 \tag{8.73}$$

and

$$C_{ii}^m = \frac{1}{2}Q_{ii} \tag{8.74}$$

where $Q_{ii}$ is the diagonal elements of **Q**, given by

$$\mathbf{Q} = -\boldsymbol{\alpha}^T\mathbf{U}_{0m}^T\mathbf{M}_1\mathbf{U}_{0m}\boldsymbol{\alpha} \tag{8.75}$$

## 8.7.3 High-Accuracy Modal Superposition for the First-Order Perturbation of Repeated Modes

In Section 8.5, the high-accuracy modal superposition for the first-order perturbation of eigenvectors of distinct eigenvalues is given. In this section, we extend these methods to the situation with repeated modes.

### 8.7.3.1 Method One for Computing $\mathbf{U}_1$

Assuming $\mathbf{U}_{AL}$ and $\boldsymbol{\Lambda}_{AL}$ are the first $L$ modes and eigenvalues excluding the repeated modes, the first-order perturbation of eigenvectors is

$$\mathbf{U}_1 = \mathbf{U}_{0m}\boldsymbol{\alpha}\mathbf{C}_m + \mathbf{U}_{AL}\mathbf{C}_{AL} + \mathbf{S}_R \tag{8.76}$$

$$\mathbf{S}_R = \mathbf{U}_S - [\mathbf{U}_{0m}\vdots\mathbf{U}_{AL}]\mathrm{diag}(\lambda_0^{-1}, \boldsymbol{\Lambda}_{AL}^{-1})[\mathbf{U}_{0m}\vdots\mathbf{U}_{AL}]^T\mathbf{T} \tag{8.77}$$

where $\mathbf{U}_S$ is the static displacement obtained by

$$\mathbf{K}\mathbf{U}_S = \mathbf{T} \tag{8.78}$$

and

$$\mathbf{T} = \mathbf{M}_0\mathbf{U}_{0m}\boldsymbol{\alpha}\boldsymbol{\Lambda}_1 + \lambda_0\mathbf{M}_1\mathbf{U}_{0m}\boldsymbol{\alpha} - \mathbf{K}_1\mathbf{U}_{0m}\boldsymbol{\alpha} \tag{8.79}$$

In Equation 8.79, $\boldsymbol{\Lambda}_1$ and $\boldsymbol{\alpha}$ can be obtained from Equation 8.69.

The matrix $\mathbf{C}_{AL}$ is given by

$$\mathbf{C}_{AL} = (\boldsymbol{\Lambda}_{AL} - \lambda_0\mathbf{I})^{-1}\mathbf{U}_{AL}^T(\lambda_0\mathbf{M}_1 - \mathbf{K}_1)\mathbf{U}_{0m}\boldsymbol{\alpha} \tag{8.80}$$

and the elements of matrix $\mathbf{C}_m$ are

$$C_{ij}^m = \frac{R_{ij}}{\lambda_{jm}^{(1)} - \lambda_{im}^{(1)}}, \quad i \neq j, \quad i,j = 1, 2, ..., m \tag{8.81}$$

where **R** is given by

$$\mathbf{R} = \boldsymbol{\alpha}^T\mathbf{U}_{0m}^T\mathbf{M}_1\mathbf{U}_{0m}\boldsymbol{\Lambda}_1 - \boldsymbol{\alpha}^T\mathbf{U}_{0m}^T(\lambda_0\mathbf{M}_1 - \mathbf{K}_1)(\mathbf{U}_{AL}\mathbf{C}_{AL} + \mathbf{S}_R) - \boldsymbol{\alpha}^T\mathbf{U}_{0m}^T\mathbf{M}_0\mathbf{S}_R\boldsymbol{\Lambda}_1 \tag{8.82}$$

The diagonal elements of $\mathbf{C}_m$ are

$$C_{ii}^m = \frac{1}{2}Q_{ii} \tag{8.83}$$

where

$$\mathbf{Q} = -\boldsymbol{\alpha}^T \mathbf{U}_{0m}^T \mathbf{M}_1 \mathbf{U}_{0m} \boldsymbol{\alpha} - \boldsymbol{\alpha}^T \mathbf{U}_{0m}^T \mathbf{M}_0 \mathbf{S}_R - \mathbf{S}_R^T \mathbf{M}_0 \mathbf{U}_{0m} \boldsymbol{\alpha} \tag{8.84}$$

### 8.7.3.2   Method Two for Computing $\mathbf{U}_1$

The first-order perturbation of eigenvectors can be expressed as

$$\mathbf{U}_1 = \mathbf{U}_{0m} \boldsymbol{\alpha} \mathbf{C}_m + \mathbf{U}_{AL} \mathbf{C}_{AL} + \mathbf{S}_R \tag{8.85}$$

where $\mathbf{C}_{AL}$ can also be calculated using Equation 8.80; that is

$$\mathbf{C}_{AL} = (\boldsymbol{\Lambda}_{AL} - \lambda_0 \mathbf{I})^{-1} \mathbf{U}_{AL}^T (\lambda_0 \mathbf{M}_1 - \mathbf{K}_1) \mathbf{U}_{0m} \boldsymbol{\alpha}$$

and $\mathbf{S}_R$ is given by

$$\mathbf{S}_R = \sum_{j=0}^{\infty} \lambda_0^j (\mathbf{H}_j - \mathbf{W}_j) \tag{8.86}$$

where

$$\mathbf{W}_j = [\mathbf{U}_{0m} \vdots \mathbf{U}_{AL}] \boldsymbol{\Lambda}_0^{-j-1} [\mathbf{U}_{0m} \vdots \mathbf{U}_{AL}]^T \mathbf{T}, \quad j \geq 0 \tag{8.87}$$

$$\mathbf{T} = \mathbf{M}_0 \mathbf{U}_{0m} \boldsymbol{\alpha} \boldsymbol{\Lambda}_1 + \lambda_0 \mathbf{M}_1 \mathbf{U}_{0m} \boldsymbol{\alpha} - \mathbf{K}_1 \mathbf{U}_{0m} \boldsymbol{\alpha} \tag{8.88}$$

The iterative method for computing $\mathbf{H}_j$ is as follows:

$$\begin{aligned} \mathbf{H}_0 &= \mathbf{K}^{-1} \mathbf{T}, \\ \mathbf{F}_{j-1}' &= \mathbf{M} \mathbf{H}_{j-1}, \quad j \geq 1 \\ \mathbf{H}_j &= \mathbf{K}^{-1} \mathbf{F}_{j-1}', \end{aligned} \tag{8.89}$$

This iterative process can be terminated according to the accuracy requirement. If we define $\mathbf{S}_R(k)$ as

$$\mathbf{S}_R(k) = \sum_{j=0}^{k} \lambda_0^j (\mathbf{H}_j - \mathbf{W}_j) \tag{8.90}$$

the termination condition can be stated as

$$\|S_R(k) - S_R(k-1)\|_2 \leq \varepsilon, \quad j = 1, 2, \dots, m \tag{8.91}$$

where $\varepsilon$ is a specified accuracy requirement.

The computation method for $\mathbf{C}_m$ in Equation 8.85 is similar to that of Equation 8.81 to Equation 8.84. The only difference is that $\mathbf{S}_R$ in Equation 8.82 and Equation 8.84 can be replaced with $\mathbf{S}_R(k)$ in Equation 8.90.

## 8.8   Matrix Perturbation Method for Closely Spaced Eigenvalues

The vibration modes with close frequencies, that is, with clusters of frequencies, often occur in certain structural systems including large space structures, multispan beams, and in some nearly periodic structures and symmetric structures. Therefore, it is important here to present the perturbation method for vibration modes with close eigenvalues (Liu, 2000).

The perturbation analysis of close eigenvalues can be transformed into a problem with a repeated eigenvalue, which is equal to the average value of the close eigenvalues (Chen, 1993).

## 8.8.1 Method One of Perturbation Analysis for Close Eigenvalues

Consider vibration eigenproblem

$$\mathbf{K}_0[\mathbf{U}_0 \vdots \mathbf{U}_A] = \mathbf{M}_0[\mathbf{U}_0 \vdots \mathbf{U}_A]\text{diag}(\mathbf{\Lambda}_0, \mathbf{\Lambda}_A) \tag{8.92}$$

$$[\mathbf{U}_0 \vdots \mathbf{U}_A]^{\mathrm{T}}\mathbf{M}_0[\mathbf{U}_0 \vdots \mathbf{U}_A] = \mathbf{I} \tag{8.93}$$

where $\mathbf{K}_0$ and $\mathbf{M}_0$ are $n \times n$ real symmetric matrices, and $\mathbf{\Lambda}_0$ and $\mathbf{U}_0$ are the $m \times m$ diagonal matrix of close eigenvalues and the corresponding $n \times m$ modal matrix.

Using the spectral decomposition of $\mathbf{K}_0$, the problem can be expressed as

$$\mathbf{K}_0 = \bar{\mathbf{K}}_0 + \varepsilon\,\delta\mathbf{K}_0 \tag{8.94}$$

where

$$\bar{\mathbf{K}}_0 = \mathbf{M}_0(\lambda_0\mathbf{U}_0\mathbf{U}_0^{\mathrm{T}})\mathbf{M}_0 + \mathbf{M}_0(\mathbf{U}_A\mathbf{\Lambda}_A\mathbf{U}_A^{\mathrm{T}})\mathbf{M}_0 \tag{8.95}$$

$$\varepsilon\,\delta\mathbf{K}_0 = \mathbf{M}_0(\mathbf{U}_0(\varepsilon\,\delta\mathbf{\Lambda}_0)\mathbf{U}_0^{\mathrm{T}})\mathbf{M}_0 \tag{8.96}$$

$$\varepsilon\,\delta\mathbf{\Lambda}_0 = \mathbf{\Lambda}_0 - \lambda_0\mathbf{I} = \mathbf{\Lambda}_0 - \left(\frac{\displaystyle\sum_{i=1}^{m}\lambda_{0i}}{m}\right)\mathbf{I} \tag{8.97}$$

It can be seen that $\bar{\mathbf{K}}_0$ given by Equation 8.95 satisfies

$$\bar{\mathbf{K}}_0[\mathbf{U}_0 \vdots \mathbf{U}_A] = \mathbf{M}_0[\mathbf{U}_0 \vdots \mathbf{U}_A]\text{diag}(\lambda_0\mathbf{I}, \mathbf{\Lambda}_A) \tag{8.98}$$

$$[\mathbf{U}_0 \vdots \mathbf{U}_A]^{\mathrm{T}}\mathbf{M}_0[\mathbf{U}_0 \vdots \mathbf{U}_A] = \mathbf{I} \tag{8.99}$$

This indicates that $\lambda_0$ and $\mathbf{U}_0$ are the repeated eigenvalues and the corresponding eigenvector subspace with multiplicity $m$ of the eigenproblem (Equation 8.92), and $\mathbf{\Lambda}_A$ and $\mathbf{U}_A$ are also the eigensolution of eigenproblem (Equation 8.92).

If $\mathbf{\Lambda}_0 \rightarrow \lambda_0\mathbf{I}$, $\varepsilon\,\delta\mathbf{\Lambda}_0 \rightarrow \mathbf{0}$, and $\bar{\mathbf{K}}_0 \rightarrow \mathbf{K}_0$, and if the small parameter modifications $\varepsilon\mathbf{K}_1$ and $\varepsilon\mathbf{M}_1$ are introduced to the matrices $\mathbf{K}_0$ and $\mathbf{M}_0$, the eigenproblem with close eigenvalues becomes

$$(\mathbf{K}_0 + \varepsilon\mathbf{K}_1)\mathbf{U} = (\mathbf{M}_0 + \varepsilon\mathbf{M}_1)\mathbf{U}\mathbf{\Lambda} \tag{8.100}$$

$$\mathbf{U}(\mathbf{M}_0 + \varepsilon\mathbf{M}_1)\mathbf{U}^{\mathrm{T}} = \mathbf{I} \tag{8.101}$$

Substituting Equation 8.94 into Equation 8.100, we obtain

$$(\bar{\mathbf{K}}_0 + \varepsilon\bar{\mathbf{K}}_1)\mathbf{U} = (\mathbf{M}_0 + \varepsilon\mathbf{M}_1)\mathbf{U}\mathbf{\Lambda} \tag{8.102}$$

$$\mathbf{U}(\mathbf{M}_0 + \varepsilon\mathbf{M}_1)\mathbf{U}^{\mathrm{T}} = \mathbf{I} \tag{8.103}$$

where

$$\varepsilon\bar{\mathbf{K}}_1 = \varepsilon\,\delta\mathbf{K}_0 + \varepsilon\mathbf{K}_1 \tag{8.104}$$

$$\mathbf{\Lambda} = \lambda_0\mathbf{I} + \varepsilon\mathbf{\Lambda}_1 + \varepsilon^2\mathbf{\Lambda}_2 + \cdots \tag{8.105}$$

$$\mathbf{U} = \mathbf{U}_0\alpha + \varepsilon\mathbf{U}_1 + \varepsilon^2\mathbf{U}_2 + \cdots \tag{8.106}$$

Therefore, the eigenproblem of Equation 8.102 and Equation 8.103 can be considered to be a perturbed eigenproblem with the perturbation matrices equal to $(\delta \mathbf{K}_0 + \mathbf{K}_1)$ and $\mathbf{M}_1$, respectively. The eigensolutions, $\mathbf{\Lambda}$ and $\mathbf{U}$, can be obtained from Equation 8.102 and Equation 8.103 by using the perturbation method for repeated eigenvalues as discussed in Section 8.7. Accordingly, the perturbation problem of modes with close eigenvalues is transformed into one of the repeated eigenvalues.

The complete algorithm for $\mathbf{\Lambda}$ and $\mathbf{U}$ is given below.

(1)  Compute

$$\lambda_0 = \frac{\sum\limits_{i=1}^{m} \lambda_{0i}}{m}$$

(2)  Compute

$$\mathbf{W} = \mathbf{U}_0^T (\delta \mathbf{K}_0 + \mathbf{K}_1 - \lambda_0 \mathbf{M}_1) \mathbf{U}_0$$

(3)  Solve the eigenvalue problem

$$\mathbf{W}\boldsymbol{\alpha} = \boldsymbol{\alpha} \mathbf{\Lambda}_1$$

$$\boldsymbol{\alpha}^T \boldsymbol{\alpha} = \mathbf{I}$$

(4)  Compute the perturbed eigenvalues of the close eigenvalues

$$\mathbf{\Lambda} = \lambda_0 \mathbf{I} + \mathbf{\Lambda}_1$$

(5)  Compute the new eigenvectors $\mathbf{U}_0 \boldsymbol{\alpha}$ corresponding to $\lambda_0$.
(6)  Compute the matrix $\mathbf{C}_A$

$$\mathbf{C}_A = (\mathbf{\Lambda}_A - \lambda_0 \mathbf{I})^{-1} \mathbf{U}_A^T (\lambda_0 \mathbf{M}_1 - \mathbf{K}_1 - \delta \mathbf{K}_0) \mathbf{U}_0 \boldsymbol{\alpha}$$

(7)  Compute

$$\mathbf{R} = [R_{ij}]$$

$$\mathbf{R} = -\boldsymbol{\alpha}^T \mathbf{U}_0^T \mathbf{M}_1 \mathbf{U}_0 \boldsymbol{\alpha} \mathbf{\Lambda}_1 - \lambda_0 \boldsymbol{\alpha}^T \mathbf{U}_0^T \mathbf{M}_1 \mathbf{U}_A \mathbf{C}_A - \boldsymbol{\alpha}^T \mathbf{U}_0^T (\delta \mathbf{K}_0 + \mathbf{K}_1) \mathbf{U}_A \mathbf{C}_A - \boldsymbol{\alpha} \mathbf{U}_0^T \mathbf{M}_0 \mathbf{U}_A \mathbf{C}_A \mathbf{\Lambda}_1$$

(8)  Compute

$$\mathbf{C}_m = [C_{ij}^m]$$

$$C_{ij}^m = \frac{R_{ij}}{\lambda_{1j} - \lambda_{1i}}, \quad i \neq j, \quad i, j = 1, 2, \dots, m$$

$$C_{ii}^m = \frac{1}{2} Q_{ii}$$

$$\mathbf{Q} = -\boldsymbol{\alpha}^T \mathbf{U}_0^T \mathbf{M}_1 \mathbf{U}_0 \boldsymbol{\alpha}$$

(9)  Compute the perturbed eigenvectors $\mathbf{U}$

$$\mathbf{U} = \mathbf{U}_0\boldsymbol{\alpha} + \mathbf{U}_0\boldsymbol{\alpha}\mathbf{C}_m + \mathbf{U}_A\mathbf{C}_A$$

## 8.8.2  Method Two of Perturbation Analysis for Close Eigenvalues

Because of the importance of the problem in both theory and practice, we now present method two of perturbation analysis for close eigenvalues, which is equivalent to method one given above.

Using the spectral decomposition of $\mathbf{M}_0$, the problem can be expressed as

$$\mathbf{M}_0 = \bar{\mathbf{M}}_0 + \varepsilon\,\delta\mathbf{M}_0 \tag{8.107}$$

Then, the following equations hold:

$$\mathbf{K}_0[\mathbf{U}_0 \vdots \mathbf{U}_A] = \bar{\mathbf{M}}_0[\mathbf{U}_0 \vdots \mathbf{U}_A]\mathrm{diag}(\lambda_0\mathbf{I}, \boldsymbol{\Lambda}_A) \tag{8.108}$$

$$[\mathbf{U}_0 \vdots \mathbf{U}_A]^{\mathrm{T}}\mathbf{M}_0[\mathbf{U}_0 \vdots \mathbf{U}_A] = \mathbf{I} \tag{8.109}$$

where

$$\bar{\mathbf{M}} = \lambda_0^{-2}\mathbf{K}_0\mathbf{U}_0\mathbf{U}_0^{\mathrm{T}}\mathbf{K}_0 + \mathbf{K}_0[\mathbf{U}_A(\boldsymbol{\Lambda}_A^{-1})^2\mathbf{U}_A^{\mathrm{T}}]\mathbf{K}_0 \tag{8.110}$$

$$\varepsilon\,\delta\mathbf{M}_0 = \mathbf{K}_0[\mathbf{U}_0\varepsilon\,\delta(\boldsymbol{\Lambda}_0^{-1})^2\mathbf{U}_0^{\mathrm{T}}]\mathbf{K}_0 \tag{8.111}$$

$$\varepsilon\,\delta\boldsymbol{\Lambda}_0^{-2} = \boldsymbol{\Lambda}_0^{-2} - \lambda_0^{-2}\mathbf{I} \tag{8.112}$$

and

$$\lambda_0 = \frac{\displaystyle\sum_{i=1}^{m}\lambda_{0i}}{m} \tag{8.113}$$

It can be seen that $\bar{\mathbf{M}}_0$ and $\varepsilon\,\delta\mathbf{M}_0$ given by Equation 8.110 and Equation 8.111 satisfy Equation 8.108 and Equation 8.109; that is, $\lambda_0$ and $\mathbf{U}_0$ are the repeated eigenvalues and corresponding modal matrix of Equation 8.108 and Equation 8.109. $\boldsymbol{\Lambda}_A$ and $\mathbf{U}_A$ are the eigenvalue diagonal matrix and the corresponding modal matrix excluding $\boldsymbol{\Lambda}_0$ and $\mathbf{U}_0$, respectively.

If $\mathbf{K}_0$ and $\mathbf{M}_0$ are modified to $\mathbf{K}_0 + \varepsilon\mathbf{K}_1$ and $\mathbf{M}_0 + \varepsilon\mathbf{M}_1$, the eigenvalue problem becomes

$$(\mathbf{K}_0 + \varepsilon\mathbf{K}_1)\mathbf{U} = (\mathbf{M}_0 + \varepsilon\mathbf{M}_1)\mathbf{U}\boldsymbol{\Lambda} \tag{8.114}$$

$$\mathbf{U}(\mathbf{M}_0 + \varepsilon\mathbf{M}_1)\mathbf{U}^{\mathrm{T}} = \mathbf{I} \tag{8.115}$$

Substituting Equation 8.107 into Equation 8.114 yields

$$(\mathbf{K}_0 + \varepsilon\mathbf{K}_1)\mathbf{U} = (\bar{\mathbf{M}}_0 + \varepsilon\bar{\mathbf{M}}_1)\mathbf{U}\boldsymbol{\Lambda} \tag{8.116}$$

$$\mathbf{U}(\bar{\mathbf{M}}_0 + \varepsilon\bar{\mathbf{M}}_1)\mathbf{U}^{\mathrm{T}} = \mathbf{I} \tag{8.117}$$

where

$$\varepsilon\bar{\mathbf{M}}_1 = \varepsilon\,\delta\mathbf{M}_0 + \varepsilon\mathbf{M}_1 \tag{8.118}$$

Thus, Equation 8.116 and Equation 8.117 can be considered to be a perturbed eigenproblem with repeated eigenvalues, and the perturbation method for repeated eigenvalues can be used to obtain the perturbed eigensolutions of Equation 8.116 and Equation 8.117:

$$\boldsymbol{\Lambda} = \lambda_0\mathbf{I} + \varepsilon\boldsymbol{\Lambda}_1 + \cdots \tag{8.119}$$

$$\mathbf{U} = \mathbf{U}_0\alpha + \varepsilon\mathbf{U}_1 + \cdots \tag{8.120}$$

## Example 8.5

For the six-DoF mass–spring system shown in Figure 8.5, the stiffness and mass matrices $K_0$ and $M_0$ are given by

$$K_0 = \begin{bmatrix} 1500 & -1000 & & & & \\ -1000 & 1200 & -200 & & & \\ & -200 & 15,200 & -5000 & -5000 & -5000 \\ & & -5000 & 5000 & & \\ & & -5000 & & 5000 & \\ & & -5000 & & & 5000 \end{bmatrix} \text{(N/m)}$$

$$M_0 = \text{diag}(200, 300, 50, 20, 20, 20.004) \text{ (kg)}$$

The perturbation eigensolutions are computed for the following three cases:

Case 1

$$\varepsilon K_1 = \begin{bmatrix} 0 & 0 & 0 & 0 & 0 & 0 \\ 0 & 0 & 0 & 0 & 0 & 0 \\ 0 & 0 & 5 & 0 & -5 & 0 \\ 0 & 0 & 0 & 0 & 0 & 0 \\ 0 & 0 & -5 & 0 & 5 & 0 \\ 0 & 0 & 0 & 0 & 0 & 0 \end{bmatrix} \text{(N/m)}$$

$$\varepsilon M_1 = \text{diag}(0, \dots, 0) \text{ (kg)}$$

Case 2

$$\varepsilon K_1 = \begin{bmatrix} 5.00 & 0 & 0 & 0 & 0 & 0 \\ 0 & 0 & 0 & 0 & 0 & 0 \\ 0 & 0 & 0 & 0 & 0 & 0 \\ 0 & 0 & 0 & 0 & 0 & 0 \\ 0 & 0 & 0 & 0 & 0 & 0 \\ 0 & 0 & 0 & 0 & 0 & 0 \end{bmatrix} \text{(N/m)}$$

$$\varepsilon M_1 = \text{diag}(0, 0, 0, 0, 0, 0.5) \text{ (kg)}$$

Case 3

$$\varepsilon K_1 = 0, \ \varepsilon M_1 = \text{diag}(0, 0, 0, 0, 2.0, 0) \text{ (kg)}$$

The unperturbed eigensolutions have a single pair of close eigenvalues given by

$$\Lambda_0 = \text{diag}(249.966642, 250.000000)$$

$$U_0^T = \begin{bmatrix} 0.00000 & 0.00000 & -0.00012 & -0.091283 & -0.091283 & 0.182560 \\ 0.00000 & 0.00000 & 0.00000 & -0.158114 & 0.158114 & 0.00000 \end{bmatrix}$$

The other unperturbed eigensolutions are as follows:

$$\Lambda_A = \text{diag}(0.594885, 2.478725, 10.234656, 552.175102)$$

$\mathbf{U}_A$

$$= \begin{bmatrix} -0.058595 & -0.027542 & 0.028427 & -0.000001 \\ 0.032047 & -0.027659 & 0.039259 & 0.000127 \\ -0.006732 & 0.074593 & 0.058387 & -0.104793 \\ -0.007020 & 0.075340 & 0.058527 & 0.086699 \\ -0.007020 & 0.075340 & 0.058527 & 0.086699 \\ -0.007020 & 0.075340 & 0.058527 & 0.086667 \end{bmatrix}$$

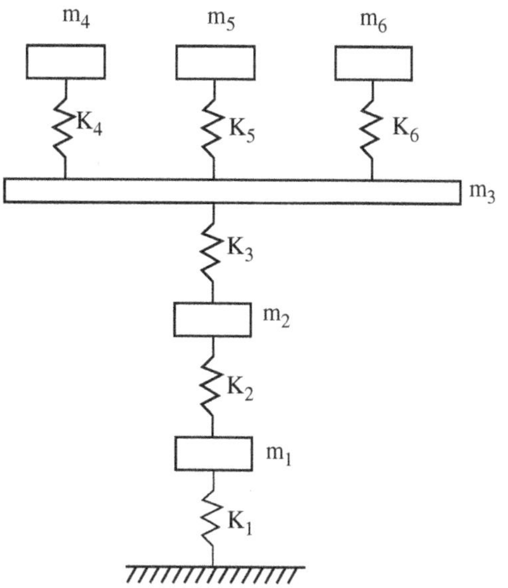

FIGURE 8.5 Six-DoF mass–spring system for Example 8.5.

The perturbed eigensolutions associated with the single pair of close eigenvalues for the three cases are summarized in Table 8.6. These results show that the perturbation analysis of distinct eigenvalues is not only inaccurate but also misleading when applied to close eigenvalues, and that the perturbed eigensolutions given by the present method are in good agreement with the exact solutions.

For example, in Case 3, the eigenvalue errors induced by the present method are reduced to 1.047950 and 0.000000, while the errors induced by the perturbation of distinct eigenvalues are 4.174100 and 3.025595. The eigenvectors obtained by the perturbation method of distinct eigenvalues are not only

**TABLE 8.6** Comparison of Eigensolutions with Close Eigenvalues

| | Exact | | Perturbation Method of Distinct Eigenvalues | | Perturbation Method of Close Eigenvalues | |
|---|---|---|---|---|---|---|
| | | | *Case 1* | | | |
| Eigenvalues | 249.973872 | 250.159351 | 250.008324 | 250.125000 | 249.973872 | 250.159451 |
| Eigenvectors | 0.000000 | 0.000000 | 0.000000 | 0.000000 | 0.000000 | 0.000000 |
| | 0.000000 | 0.000000 | 0.000000 | 0.000000 | 0.000000 | 0.000000 |
| | 0.000018 | −0.000066 | −0.000043 | 0.000079 | 0.000011 | 0.000005 |
| | 0.150492 | 0.103411 | −0.433574 | 0.039471 | 0.150501 | −0.103354 |
| | 0.014254 | −0.181965 | 0.251058 | 0.355699 | 0.014251 | 0.182015 |
| | −0.164753 | 0.078701 | 0.182585 | −0.395285 | −0.164745 | −0.078658 |
| | | | *Case 2* | | | |
| Eigenvalues | 243.878599 | 247.932109 | 244.759777 | 246.875000 | 243.725997 | 247.908461 |
| Eigenvectors | 0.000000 | 0.000000 | 0.000000 | 0.000000 | 0.000000 | 0.000000 |
| | 0.000000 | −0.000004 | 0.000002 | 0.000005 | 0.000000 | 0.000000 |
| | −0.000015 | 0.001506 | −0.000772 | −0.001976 | 0.000011 | 0.000006 |
| | −0.000622 | 0.182090 | 8.467433 | −5.098718 | 0.001103 | 0.183060 |
| | −0.155859 | −0.091163 | −8.648272 | −4.784467 | −0.157555 | −0.092069 |
| | 0.156472 | −0.090082 | 0.180521 | 9.882118 | 0.158052 | −0.090988 |
| | | | *Case 3* | | | |
| Eigenvalues | 234.474405 | 249.975015 | 245.800915 | 237.500000 | 233.326455 | 249.975015 |
| Eigenvectors | 0.000000 | 0.000000 | 0.000000 | 0.000000 | 0.000000 | 0.000000 |
| | 0.000016 | 0.000000 | −0.000008 | 0.000021 | 0.000000 | −0.000000 |
| | −0.005558 | −0.000015 | 0.003030 | −0.007905 | 0.000006 | 0.000010 |
| | −0.089507 | −0.158189 | 34.141638 | −19.920531 | −0.091153 | 0.158189 |
| | 0.175420 | 0.000158 | −34.321586 | −19.612210 | 0.182572 | −0.000158 |
| | −0.089778 | 0.158022 | 0.181586 | 39.528471 | −0.091415 | −0.158025 |

inaccurate but also misleading, while good agreement with the exact eigenvectors has been obtained by the present method.

### 8.8.3  Concluding Remarks

Perturbation analysis of vibration modes with close frequencies is presented in this section. It can be regarded as a general treatment of perturbation analysis, because the perturbation analysis of both distinct eigenvalues and repeated eigenvalues is contained in the present method. The results obtained by this method allow one to analyze the influence of parameter changes in a system on the dynamic characteristics of the system, which is very important for effective structural design.

## 8.9  Matrix Perturbation Theory for Complex Modes

In Section 8.2 to Section 8.8, the matrix perturbation for real modes of systems with real symmetric mass and stiffness matrices, $\mathbf{M}$ and $\mathbf{K}$, was given. However, in many engineering problems such as systems with nonproportional damping (see Chapter 1), dynamic systems under nonconservative forces, analysis of aero-elastic flutter, and structural vibration control systems, the system matrices are not symmetric and may not be diagonalizable. In this case, the matrix perturbation for real modes cannot be used, and we must use the matrix perturbation for complex modes (Murthy and Haftka, 1988; Chen, 1993; Liu, 1999; Adhikari and Friswell, 2001). In the following, we assume that the system is not defective; that is, the system has a complete eigenvector set to span the eigenspace. The discussion in this chapter is limited to the nondefective systems.

### 8.9.1  Basic Equations

The vibration equation of a linear system with $n$-DoFs is given by

$$\mathbf{M\ddot{q} + C\dot{q} + Kq = Q}(t) \tag{8.121}$$

where the matrices $\mathbf{M}$, $\mathbf{C}$, and $\mathbf{K}$, are assumed to be real and unsymmetric. The free vibration equation of the system is

$$\mathbf{M\ddot{q} + C\dot{q} + Kq = 0} \tag{8.122}$$

The corresponding right eigenvalues problem is

$$(\mathbf{M}s^2 + \mathbf{C}s + \mathbf{K})\mathbf{x} = 0 \tag{8.123}$$

and its adjoint eigenvalue problem is

$$(\mathbf{M}s^2 + \mathbf{C}s + \mathbf{K})^{\mathrm{T}}\mathbf{y} = 0$$

$$\mathbf{y}^{\mathrm{T}}(\mathbf{M}s^2 + \mathbf{C}s + \mathbf{K}) = 0 \tag{8.124}$$

It is common in literature to call $\mathbf{y}$ the *left eigenvector*, while $\mathbf{x}$ in the original system, a column vector, is called the *right eigenvector*.

Let us introduce a state vector

$$\mathbf{u} = \left\{ \begin{matrix} s\mathbf{x} \\ \mathbf{x} \end{matrix} \right\} = \mathbf{Tx} \tag{8.125}$$

where $\mathbf{T}$ is the state transformation matrix

$$\mathbf{T} = \left\{ \begin{matrix} s\mathbf{I} \\ \mathbf{I} \end{matrix} \right\} \tag{8.126}$$

Similarly, we introduce the state vector

$$\mathbf{v} = \left\{ \begin{array}{c} s\mathbf{y} \\ \mathbf{y} \end{array} \right\} = \mathbf{T}\mathbf{y} \tag{8.127}$$

Hence, Equation 8.123 and Equation 8.124 become

$$(\mathbf{A}s + \mathbf{B})\mathbf{u} = \mathbf{0} \tag{8.128}$$

$$(\mathbf{A}s + \mathbf{B})^{\mathsf{T}}\mathbf{v} = \mathbf{0} \tag{8.129}$$

or

$$\mathbf{v}^{\mathsf{T}}(\mathbf{A}s + \mathbf{B}) = \mathbf{0}$$

where

$$\mathbf{A} = \begin{bmatrix} -\mathbf{C} & -\mathbf{K} \\ \mathbf{I} & \mathbf{0} \end{bmatrix}$$

$$\mathbf{B} = \begin{bmatrix} \mathbf{M} & \mathbf{0} \\ \mathbf{0} & \mathbf{I} \end{bmatrix}$$

It is well known that the eigenvalues of the adjoint eigenproblem (Equation 8.129) are identical to that of the original eigenproblem (Equation 8.128). The *characteristic equation* is

$$\det(\mathbf{A} + s\mathbf{B}) = 0$$

This characteristic determinant is a polynomial of $2n$ order in $s$, and $2n$ eigenvalues $s_i$ ($i = 1, 2, ..., 2n$) can be found in the complex domain. The left and right modal vectors, $\mathbf{v}_i$ and $\mathbf{u}_i$, corresponding to $s_i$ satisfy

$$\mathbf{A}\mathbf{u}_i = s_i\mathbf{B}\mathbf{u}_i \tag{8.130}$$

and

$$\mathbf{A}^{\mathsf{T}}\mathbf{v}_i = s_i\mathbf{B}^{\mathsf{T}}\mathbf{v}_i \tag{8.131}$$

The *orthogonality conditions* are

$$\mathbf{v}_j^{\mathsf{T}}\mathbf{B}\mathbf{u}_i = \mathbf{0} \tag{8.132}$$

$$\mathbf{v}_j^{\mathsf{T}}\mathbf{A}\mathbf{u}_i = \mathbf{0} \tag{8.133}$$

The *normalization conditions* are

$$\mathbf{v}_i^{\mathsf{T}}\mathbf{B}\mathbf{u}_i = 1$$
$$\mathbf{u}_i^{\mathsf{T}}\mathbf{B}\mathbf{u}_i = 1 \tag{8.134}$$

Therefore, the orthogonality conditions can be written as

$$\mathbf{v}_j^{\mathsf{T}}\mathbf{B}\mathbf{u}_i = \delta_{ij}$$
$$\mathbf{v}_j^{\mathsf{T}}\mathbf{A}\mathbf{u}_i = s_i\delta_{ij} \tag{8.135}$$

## 8.9.2 Matrix Perturbation Method for Distinct Modes

If small changes are made on the structural parameters, the mass, damping, and stiffness matrices of the system also have small changes given by

$$\mathbf{M} = \mathbf{M}_0 + \varepsilon\mathbf{M}_1$$
$$\mathbf{C} = \mathbf{C}_0 + \varepsilon\mathbf{C}_1 \tag{8.136}$$
$$\mathbf{K} = \mathbf{K}_0 + \varepsilon\mathbf{K}_1$$

and we have

$$\mathbf{A} = \mathbf{A}_0 + \varepsilon\mathbf{A}_1 \tag{8.137}$$

$$\mathbf{B} = \mathbf{B}_0 + \varepsilon\mathbf{B}_1 \tag{8.138}$$

where $\varepsilon$ is a small parameter.

In the following, we first consider the case of distinct eigenvalues, $s_{0i}$, of the original system. According to the matrix perturbation theory, the eigenvalues and eigenvectors can be expressed as a power series in $\varepsilon$, that is

$$\mathbf{S} = \mathbf{S}_0 + \varepsilon\mathbf{S}_1 + \varepsilon^2\mathbf{S}_2 + \cdots \tag{8.139}$$

$$\mathbf{U} = \mathbf{U}_0 + \varepsilon\mathbf{U}_1 + \varepsilon^2\mathbf{U}_2 + \cdots \tag{8.140}$$

$$\mathbf{V} = \mathbf{V}_0 + \varepsilon\mathbf{V}_1 + \varepsilon^2\mathbf{V}_2 + \cdots \tag{8.141}$$

where $\mathbf{S}_0$, $\mathbf{U}_0$, and $\mathbf{V}_0$ are the eigensolutions of the original system; $\mathbf{S}_1$, $\mathbf{U}_1$, and $\mathbf{V}_1$ are the first-order perturbations of eigensolutions; and $\mathbf{S}_2$, $\mathbf{U}_2$, and $\mathbf{V}_2$ the second-order perturbations.

$\mathbf{U}_1$ can be expressed as a linear combination of the right eigenvectors of the original system as

$$\mathbf{U}_1 = \mathbf{U}_0\mathbf{C}^1 \tag{8.142}$$

where $\mathbf{C}^1$ is to be the determined matrix given by

$$C_{ij}^1 = \frac{1}{S_{0j} - S_{0i}}P_{ij}^1, \quad j \neq i, \quad i,j = 1, 2, \ldots \tag{8.143}$$

Also

$$\mathbf{S}_1 = \mathrm{diag}(P_{11}^1, P_{22}^1, \ldots) \tag{8.144}$$

where $P_{ij}^1$ are the elements of $\mathbf{P}^1$ given by

$$\mathbf{P}^1 = \mathbf{V}_0^{\mathrm{T}}(-\mathbf{A}_1\mathbf{U}_0 + \mathbf{B}_1\mathbf{U}_0\mathbf{S}_0) \tag{8.145}$$

The $\mathbf{V}_1$ can be expressed as the expansion of $\mathbf{V}_0$

$$\mathbf{V}_1 = \mathbf{V}_0\mathbf{D}^1 \tag{8.146}$$

where $\mathbf{D}^1$ is to be the determined coefficient matrix given by

$$D_{ij}^1 = \frac{1}{S_{0j} - S_{0i}}R_{ij}^1, \quad j \neq i, \quad i,j = 1, 2, \ldots \tag{8.147}$$

and $R_{ij}^1$ are the nondiagonal elements of $\mathbf{R}^1$

$$\mathbf{R}^1 = \mathbf{U}_0^{\mathrm{T}}(\mathbf{B}_1^{\mathrm{T}}\mathbf{V}_0\mathbf{S}_0 - \mathbf{A}_1^{\mathrm{T}}\mathbf{V}_0) \tag{8.148}$$

If the modification of the parameter is fairly large, the second-order perturbation must be used to obtain high accuracy. According to the expansion theorem, the second-order perturbation of eigenvectors, $\mathbf{U}_2$, can be expressed as

$$\mathbf{U}_2 = \mathbf{U}_0\mathbf{C}^2 \tag{8.149}$$

In a similar manner, $\mathbf{S}_2$ and the elements $C_{ij}^2$ can be obtained as

$$\mathbf{S}_2 = \mathrm{diag}(P_{11}^2, P_{22}^2, \ldots) \tag{8.150}$$

$$C_{ij}^2 = \frac{1}{S_{0j} - S_{0i}}P_{ij}^2, \quad j \neq i, i, \quad j = 1, 2, \ldots \tag{8.151}$$

where $P_{ij}^2$ are the nondiagonal elements of $\mathbf{P}^2$

$$\mathbf{P}^2 = \mathbf{V}_0^{\mathrm{T}}\mathbf{B}_0\mathbf{U}_1\mathbf{S}_1 + \mathbf{V}_0^{\mathrm{T}}\mathbf{B}_1\mathbf{U}_0\mathbf{S}_1 + \mathbf{V}_0^{\mathrm{T}}\mathbf{B}_1\mathbf{U}_1\mathbf{S}_0 - \mathbf{V}_0^{\mathrm{T}}\mathbf{A}_1\mathbf{U}_1 \tag{8.152}$$

$V_2$ can be expressed as

$$V_2 = V_0 D^2 \tag{8.153}$$

where $D^2$ is to be the determined coefficient matrix given by

$$D_{ij}^2 = \frac{1}{S_{0j} - S_{0i}} R_{ij}^2, \quad j \neq i, \quad i, j = 1, 2, \dots \tag{8.154}$$

and $R_{ij}^2$ are the nondiagonal elements of $R^2$

$$R^2 = U_0^T (B_0^T V_1 S_1 + B_1^T V_0 S_1 + B_1^T V_1 S_0 - A_1^T V_1) \tag{8.155}$$

If $j = i$, the coefficients $C_{ii}^1$, $D_{ii}^1$, $C_{ii}^2$, and $D_{ii}^2$ can be computed as

$$C_{ii}^1 = \frac{-1}{u_{0i}^T (B_0 + B_0^T) u_{0i}} \left( u_{0i}^T B_1 u_{0i} + \sum_{\substack{j=1 \\ j \neq i}}^{n} C_{ij} u_{0j}^T (B_0 + B_0^T) u_{0i} \right) \tag{8.156}$$

$$D_{ii}^1 = Q_{ii}^1 - C_{ii}^1 \tag{8.157}$$

where $Q_{ii}^1$ is the diagonal element of $Q_1$

$$Q_1 = -V_0^T B_1 U_0 \tag{8.158}$$

$$C_{ii}^2 = \frac{-u_{1i}^T B_0 u_{1i} + u_{0i}^T B_1 u_{1i} + u_{1i}^T B_1 u_{0i} - \sum_{\substack{j=1 \\ j \neq i}}^{n} C_{ij} u_{0j}^T (B_0 + B_0^T) u_{0i}}{u_{0i}^T (B_0 + B_0^T) u_{0i}} \tag{8.159}$$

$$D_{ii}^2 = Q_{ii}^2 - C_{ii}^2 \tag{8.160}$$

and $Q_{ii}^2$ is the diagonal element of $Q^2$

$$Q^2 = -V_0^T B_0 U_1 - V_1^T B_0 U_1 - V_1^T B_1 U_0 \tag{8.161}$$

## 8.9.3 High-Accuracy Modal Superposition for Eigenvector Derivatives

For a large-scale structure, only a small number of the first lower $L$ modes are extracted, and the higher modes are truncated in order to reduce the computational cost. The modal superposition method may not only give inaccurate result, but also may be misleading if the truncation is considerable. In this section, we give a high-accuracy modal superposition method for derivatives of the complex mode of nonsymmetric matrices.

### 8.9.3.1 Improved Modal Superposition

An improved modal superposition (IMS) to reduce the computation errors by modal truncation was proposed (Lim et al., 1989). The derivatives of modes can be expressed as

$$\frac{\partial u_i}{\partial b} = \bar{\alpha}_{ii} u_i + \bar{z}_i \tag{8.162}$$

$$\bar{z}_i = \sum_{\substack{j=1 \\ j \neq i}}^{L} \frac{v_j^T F_i}{S_i - S_j} u_j + A^{-1} F_i + \sum_{j=1}^{L} \frac{v_j^T F_i}{S_j} u_j \tag{8.163}$$

$$F_i = \left( \frac{\partial A}{\partial b} - \frac{\partial S_i}{\partial b} B - S_i \frac{\partial B}{\partial b} \right) u_i \tag{8.164}$$

where $\mathbf{A}^{-1}\mathbf{F}_i$ is the contribution of the truncated higher modes to the derivatives of modes, as given by

$$\bar{\alpha}_{ii} = -\frac{1}{2}\left(\mathbf{u}_i^{\mathrm{T}}\frac{\partial\mathbf{B}}{\partial\mathbf{b}}\mathbf{u}_i + \mathbf{u}_i^{\mathrm{T}}\mathbf{B}\bar{z}_i + \bar{z}_i^{\mathrm{T}}\mathbf{B}\mathbf{u}_i\right) \tag{8.165}$$

### 8.9.3.2 High-Accuracy Modal Superposition

Assume that the eigenvalues are ordered according to their modular magnitude, and satisfy the following condition:

$$|S_i| < |S_j|, \quad j > L \tag{8.166}$$

The derivatives of modes can be expressed as

$$\frac{\partial\mathbf{u}_i}{\partial\mathbf{b}} = \bar{\alpha}_{ii}\mathbf{u}_i + \bar{z}_i = \bar{\alpha}_{ii}\mathbf{u}_i + z_{iL} + z_{iH} \tag{8.167}$$

where

$$\bar{z}_i = z_{iL} + z_{iH} \tag{8.168}$$

$$z_{iL} = \sum_{\substack{j=1\\j\neq i}}^{L}\frac{\mathbf{v}_j^{\mathrm{T}}\mathbf{F}_i}{S_i - S_j}\mathbf{u}_j \tag{8.169}$$

$$z_{iH} = -\sum_{j=1}^{K}\left(\mathbf{A}^{-1}(\mathbf{B}\mathbf{A}^{-1})^{j-1} - \mathbf{U}_L((\mathbf{S}_L)^{-1})^j\mathbf{V}_L^{\mathrm{T}}\right)\mathbf{F}_i \tag{8.170}$$

Also, $\bar{\alpha}_{ii}$ can be obtained from Equation 8.165, and $K$ denotes the number of terms used in series (Equation 8.170).

It can be shown that for $K = 1$, Equation 8.167 is equivalent to Equation 8.162.

### 8.9.3.3 Numerical Example

### Example 8.6

Consider a 20-DoF system, as shown in Figure 8.6, with the parameters given by

$$m_1 = m_2 = \cdots = m_{19} = 2m, \quad m_{20} = m = 1.0 \text{ kg}$$

$$k_1 = k_2 = \cdots = k_{21} = 1.0 \times 10^3 \text{ N/m}$$

$$c_1 = c_2 = \cdots = c_7 = 3c, \quad c_8 = c_9 = \cdots = c_{14} = 2c$$

$$c_{15} = c_{16} = \cdots = c_{21} = c = 0.1 \text{ N sec/m}$$

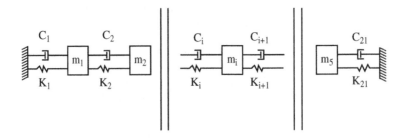

**FIGURE 8.6**　The 20-DoF system for Example 8.6.

**TABLE 8.7** Errors of Eigenvector Derivatives (%)

| | Modes used | TMS | IMS | HAMS | | |
|---|---|---|---|---|---|---|
| | | | | 2 | 3 | 4 |
| $\partial u_1/\partial b_1$ | 4 | 60.18 | 13.14 | 1.41 | 0.01 | 0.00 |
| | 8 | 26.66 | 1.01 | 0.03 | 0.00 | 0.00 |
| | 12 | 16.76 | 0.66 | 0.01 | 0.00 | 0.00 |
| $\partial u_1/\partial b_1$ | 4 | 101.00 | 50.94 | 27.72 | 7.64 | 1.51 |
| | 8 | 48.65 | 10.32 | 1.29 | 0.38 | 0.03 |
| | 12 | 23.98 | 3.37 | 0.35 | 0.06 | 0.01 |
| $\partial u_1/\partial b_1$ | 4 | 69.74 | 14.82 | 3.07 | 0.02 | 0.00 |
| | 8 | 26.60 | 1.39 | 0.23 | 0.00 | 0.00 |
| | 12 | 25.92 | 0.85 | 0.12 | 0.00 | 0.00 |
| $\partial u_1/\partial b_1$ | 4 | 90.63 | 43.79 | 21.33 | 9.01 | 2.00 |
| | 8 | 44.86 | 6.65 | 1.56 | 0.35 | 0.04 |
| | 12 | 23.76 | 3.23 | 0.37 | 0.11 | 0.02 |
| $\partial u_1/\partial b_1$ | 4 | 72.52 | 15.07 | 8.46 | 1.11 | 0.03 |
| | 8 | 37.95 | 1.22 | 0.57 | 0.03 | 0.00 |
| | 12 | 28.54 | 0.52 | 0.03 | 0.00 | 0.00 |
| $\partial u_1/\partial b_1$ | 4 | 63.33 | 9.19 | 1.92 | 0.42 | 0.00 |
| | 8 | 27.56 | 0.75 | 0.49 | 0.21 | 0.00 |
| | 12 | 25.00 | 0.47 | 0.15 | 0.01 | 0.00 |

For the purpose of comparison, the errors of the truncation modal superposition (TMS), the IMS, and the high-accuracy superposition (HAMS) in computing the derivatives of eigenvectors are listed in Table 8.7.

For the sake of simplicity, the errors of eigenvector derivatives are represented by

$$\left| \left( \frac{\partial u_i}{\partial b_j} \right)_\varepsilon - \left( \frac{\partial u_i}{\partial b_j} \right)_\alpha \right|$$

where $(\partial u_i/\partial b_j)_\varepsilon$ denotes the exact solution and $(\partial u_i/\partial b_j)_\alpha$ denotes those obtained by the three methods presented above. In the computation of the eigenvector derivatives, the parameters $m_1$ and $m_{10}$ are functions of the design variable $b_1$, the parameters $c_8$ and $c_{15}$ are functions of design variable $b_2$, and $K$ denotes the number of terms used in series (Equation 8.170).

The results in Table 8.7 confirm that the solution accuracy of the high-accuracy modal superposition is much higher than that of the TMS and the IMS. For example, if only the first four modes are used, the errors of $\partial u_2/\partial b_1$ are 101.00 and 50.94% for the truncated modal superposition and the IMS, and the errors are reduced to about 27.27, 7.64, and 1.51% for the high-accuracy modal superposition. If the first 12 modes are used, the error of $\partial u_1/\partial b_1$ is 16.76% for the truncated modal superposition, and the errors are reduced to 0.01, 0.00 and 0.00% for the case of $K = 2, 3, 4$ in the series (Equation 8.170), where the first four modes are used, respectively.

## 8.9.4 Matrix Perturbation for Repeated Eigenvalues of Nondefective Systems

### 8.9.4.1 Basic Equation

Consider a system having repeated eigenvalues, $S_0 = S_1 = \cdots = S_{0m}$, with multiplicity $m$, and the corresponding right and left modal matrices

$$\mathbf{U}_{0m} = [\, \mathbf{u}_{01} \quad \mathbf{u}_{02} \quad \cdots \quad \mathbf{u}_{0m} \,] \tag{8.171}$$

$$\mathbf{V}_{0m} = [\, \mathbf{v}_{01} \quad \mathbf{v}_{02} \quad \cdots \quad \mathbf{v}_{0m}\,] \tag{8.172}$$

the remaining eigenvalues being distinct.

The repeated eigenvalues satisfy the following equations:

$$\mathbf{A}_0\mathbf{U}_{0m} = \mathbf{B}_0\mathbf{U}_{0m}\mathbf{S}_0 \tag{8.173}$$

$$\mathbf{A}_0^{\mathrm{T}}\mathbf{V}_{0m} = \mathbf{B}_0^{\mathrm{T}}\mathbf{V}_{0m}\mathbf{S}_0 \tag{8.174}$$

$$\mathbf{V}_{0m}^{\mathrm{T}}\mathbf{B}_0\mathbf{U}_{0m} = \mathbf{I} \tag{8.175}$$

$$\mathbf{u}_{0i}^{\mathrm{T}}\mathbf{B}_0\mathbf{u}_{0i} = 1 \tag{8.176}$$

If small changes are made to the parameters, we have

$$\mathbf{A} = \mathbf{A}_0 + \varepsilon\mathbf{A}_1 \tag{8.177}$$

$$\mathbf{B} = \mathbf{B}_0 + \varepsilon\mathbf{B}_1 \tag{8.178}$$

The eigenvalues and eigenvectors of the perturbed system can be expressed as power series expansions in $\varepsilon$:

$$\mathbf{S}_m = \mathbf{S}_0 + \varepsilon\mathbf{S}_1 + \varepsilon^2\mathbf{S}_2 + \cdots \tag{8.179}$$

$$\mathbf{U}_m = \mathbf{U}_0 + \varepsilon\mathbf{U}_1 + \varepsilon^2\mathbf{U}_2 + \cdots \tag{8.180}$$

$$\mathbf{V}_m = \mathbf{V}_0 + \varepsilon\mathbf{V}_1 + \varepsilon^2\mathbf{V}_2 + \cdots \tag{8.181}$$

where

$$\mathbf{U}_0 = \mathbf{U}_{0m}\alpha \tag{8.182}$$

$$\mathbf{V}_0 = \mathbf{V}_{0m}\beta \tag{8.183}$$

and $\alpha_{m \times m}$ and $\beta_{m \times m}$ are to be determined coefficient matrices.

### 8.9.4.2    The First-Order Perturbation of Eigenvalues

The first-order perturbation diagonal matrix, $\mathbf{S}_1$, of the repeated eigenvalues and the coefficient matrix, $\alpha$, can be obtained from the equations:

$$\mathbf{W}\alpha = \alpha\mathbf{S}_1 \tag{8.184}$$

$$\mathbf{W}^{\mathrm{T}}\beta = \beta\mathbf{S}_1 \tag{8.185}$$

$$\mathbf{W} = \mathbf{V}_{0m}^{\mathrm{T}}(\mathbf{A}_1 - \mathbf{S}_0\mathbf{B}_1)\mathbf{U}_{0m} \tag{8.186}$$

and the normalization conditions

$$\alpha^{\mathrm{T}}\alpha = \mathbf{I} \tag{8.187}$$

$$\beta^{\mathrm{T}}\alpha = \mathbf{I} \tag{8.188}$$

If matrix $\mathbf{W}$ has no repeated eigenvalues, $\alpha$ and $\beta$ can be uniquely determined. If $\mathbf{W}$ has repeated eigenvalues, we must consider the higher order perturbation equations for determining $\alpha$ and $\beta$. Here, we assume that $S_{1i}$ are distinct eigenvalues, that is, $S_{1i} \neq S_{1j}$ ($i \neq j$), where $S_{1i}$ are the diagonal elements of $\mathbf{S}_1$.

### 8.9.4.3    The First-Order Perturbation of Eigenvectors

According to the modal expansion theorem, the first-order perturbation of the right and left eigenvectors, $\mathbf{U}_1$ and $\mathbf{V}_1$, can be expressed as

$$\mathbf{U}_1 = \mathbf{U}_{0m}\alpha\mathbf{C}_m^1 + \mathbf{U}_A\mathbf{C}_A^1 \tag{8.189}$$

$$\mathbf{V}_1 = \mathbf{V}_{0m}\beta\mathbf{D}_m^1 + \mathbf{V}_A\mathbf{D}_A^1 \tag{8.190}$$

where $\mathbf{C}_m^1$ and $\mathbf{C}_A^1$ are coefficient matrices which are to be determined, and $\mathbf{U}_A$ and $\mathbf{V}_A$ are the right and left modal matrices corresponding to the distinct eigenvalues:

$$\mathbf{C}_A^1 = (\mathbf{S}_A - \mathbf{S}_0)^{-1}\mathbf{V}_A^T(\mathbf{S}_0\mathbf{B}_1 - \mathbf{A}_1)\mathbf{U}_{0m}\boldsymbol{\alpha} \tag{8.191}$$

$$\mathbf{D}_A^1 = (\mathbf{S}_A - \mathbf{S}_0)^{-1}\mathbf{U}_A^T(\mathbf{S}_0\mathbf{B}_1^T - \mathbf{A}_1^T)\mathbf{V}_{0m}\boldsymbol{\beta} \tag{8.192}$$

The elements of matrix $\mathbf{C}_m^1$ can be computed by

$$C_{mij}^1 = \frac{R_{ij}^1}{\lambda_{1j} - \lambda_{1i}}, \quad i \neq j, \ i,j = 1,2,...,m \tag{8.193}$$

where $R_{ij}^1$ are the nondiagonal elements of $\mathbf{R}^1$:

$$\mathbf{R}^1 = \boldsymbol{\beta}^T\mathbf{V}_{0m}^T\mathbf{B}_0\mathbf{U}_A\mathbf{C}_A^1\mathbf{S}_1 + \boldsymbol{\beta}^T\mathbf{V}_{0m}^T\mathbf{B}_1\mathbf{U}_{0m}\boldsymbol{\alpha}\mathbf{S}_1 + \boldsymbol{\beta}^T\mathbf{V}_{0m}^T\mathbf{B}_1\mathbf{U}_A\mathbf{C}_A^1\mathbf{S}_0 - \boldsymbol{\beta}^T\mathbf{V}_{0m}^T\mathbf{A}_1\mathbf{U}_A\mathbf{C}_A^1 \tag{8.194}$$

$$C_{mii}^1 = \frac{-1}{\mathbf{u}_{0i}^T(\mathbf{B}_0 + \mathbf{B}_0^T)\mathbf{u}_{0i}}\left[\mathbf{u}_{0i}^T\mathbf{B}_1\mathbf{u}_{0i} + \sum_{\substack{j=1\\j\neq i}}^{m} c_{mij}^1\mathbf{u}_{0j}^T(\mathbf{B}_0 + \mathbf{B}_0^T)\mathbf{u}_{0i} + \sum_{j=m+1}^{n} c_{Aij}^1\mathbf{u}_{0j}^T(\mathbf{B}_0 + \mathbf{B}_0^T)\mathbf{u}_{0i}\right] \tag{8.195}$$

$$D_{mij}^1 = \frac{R_{ij}^2}{S_{1j} - S_{1i}}, \quad i \neq j, \ i,j = 1,2,...,m \tag{8.196}$$

where $R_{ij}^2$ are the nondiagonal elements of $\mathbf{R}^2$:

$$\mathbf{R}^2 = \boldsymbol{\alpha}^T\mathbf{U}_{0m}^T\mathbf{B}_0\mathbf{V}_A\mathbf{D}_A^1\mathbf{S}_1 + \boldsymbol{\alpha}^T\mathbf{U}_{0m}^T\mathbf{B}_1^T\mathbf{V}_{0m}\boldsymbol{\beta}\mathbf{S}_1 + \boldsymbol{\alpha}^T\mathbf{U}_{0m}^T\mathbf{B}_1^T\mathbf{V}_A\mathbf{D}_A^1\mathbf{S}_0 - \boldsymbol{\alpha}^T\mathbf{U}_{0m}^T\mathbf{A}_1^T\mathbf{V}_A\mathbf{D}_A^1 \tag{8.197}$$

$$D_{mii}^1 = Q_{ii}^2 - C_{mii}^1 \tag{8.198}$$

where $Q_{ii}^2$ are the diagonal elements of $\mathbf{Q}^2$:

$$\mathbf{Q}^2 = -\boldsymbol{\beta}^T\mathbf{V}_{0m}^T\mathbf{B}_1\mathbf{U}_{0m}\boldsymbol{\alpha} \tag{8.199}$$

## 8.9.5 Matrix Perturbation for Close Eigenvalues of Unsymmetric Matrices

Assume that $\mathbf{S}_0$ is a diagonal matrix with $m$ close eigenvalues; $\mathbf{U}_{0n\times m}$ and $\mathbf{V}_{0n\times m}$ are the corresponding right and left eigenvectors matrices; $\mathbf{S}_A$ is the remaining distinct eigenvalue diagonal matrix; $\mathbf{U}_{An\times(n-m)}$ and $\mathbf{V}_{An\times(n-m)}$ are the corresponding right and left eigenvector matrices. They satisfy the following equations:

$$\mathbf{A}_0[\mathbf{U}_0 \vdots \mathbf{U}_A] = \mathbf{B}_0[\mathbf{U}_0 \vdots \mathbf{U}_A]\text{diag}(\mathbf{S}_0, \mathbf{S}_A) \tag{8.200}$$

$$\mathbf{A}_0^T[\mathbf{V}_0 \vdots \mathbf{V}_A] = \mathbf{B}_0^T[\mathbf{V}_0 \vdots \mathbf{V}_A]\text{diag}(\mathbf{S}_0, \mathbf{S}_A) \tag{8.201}$$

$$[\mathbf{V}_0 \vdots \mathbf{V}_A]^T\mathbf{B}_0[\mathbf{U}_0 \vdots \mathbf{U}_A] = \mathbf{I} \tag{8.202}$$

$$\mathbf{u}_{0i}^T\mathbf{B}_0\mathbf{u}_{0i} = 1 \tag{8.203}$$

Construct the matrix $\bar{\mathbf{A}}_0$ as

$$\mathbf{A}_0 = \bar{\mathbf{A}}_0 + \varepsilon\tilde{\mathbf{A}}_0 \tag{8.204}$$

where

$$\bar{\mathbf{A}}_0 = \mathbf{B}_0[\mathbf{U}_0 \vdots \mathbf{U}_A]\text{diag}(S_0\mathbf{I}, \mathbf{S}_A)[\mathbf{V}_0 \vdots \mathbf{V}_A]^T\mathbf{B}_0 \tag{8.205}$$

$$\varepsilon\tilde{\mathbf{A}}_0 = \mathbf{B}_0\mathbf{U}_0(\varepsilon[\delta\mathbf{S}_0])\mathbf{V}_0^T\mathbf{B}_0 \tag{8.206}$$

$$\varepsilon[\delta\mathbf{S}_0] = \mathbf{S}_0 - S_0\mathbf{I} \tag{8.207}$$

$$S_0 = \frac{1}{m}\left(\sum_{k=1}^{n} S_{0i}\right) \tag{8.208}$$

Here, $S_{0i}$ are the close eigenvalues, and $S_0$ is the average of $S_{0i}$ ($i = 1, 2, \ldots, m$).

It can be shown that the following equations hold:

$$\bar{\mathbf{A}}_0 \mathbf{U}_0 = \mathbf{B}_0 \mathbf{U}_0 S_0 \mathbf{I} \tag{8.209}$$

$$\bar{\mathbf{A}}_0^T \mathbf{V}_0 = \mathbf{B}_0^T \mathbf{V}_0 S_0 \mathbf{I} \tag{8.210}$$

These equations indicate that $S_0$ is the repeated eigenvalue with multiplicity, $m$, for the eigenproblem defined by Equation 8.209 and Equation 8.210, and $\mathbf{U}_0$ and $\mathbf{V}_0$ are the corresponding right and left modal matrices, respectively.

If small modifications $\varepsilon \mathbf{A}_1$ and $\varepsilon \mathbf{B}_1$ are imposed on the matrices $\mathbf{A}_0$ and $\mathbf{B}_0$, then the eigenproblems of the perturbed system become

$$(\bar{\mathbf{A}}_0 + \varepsilon \bar{\mathbf{A}}_1)\mathbf{U} = (\mathbf{B}_0 + \varepsilon \mathbf{B}_1)\mathbf{U}S \tag{8.211}$$

$$(\bar{\mathbf{A}}_0 + \varepsilon \bar{\mathbf{A}}_1)^T \mathbf{V} = (\mathbf{B}_0 + \varepsilon \mathbf{B}_1)^T \mathbf{V}S \tag{8.212}$$

where

$$\varepsilon \bar{\mathbf{A}}_1 = \varepsilon \bar{\mathbf{A}}_0 + \varepsilon \mathbf{A}_1 \tag{8.213}$$

The eigensolutions of Equation 8.211 and Equation 8.212 are given by

$$\mathbf{U} = \mathbf{U}_0 \boldsymbol{\alpha} + \varepsilon \mathbf{U}_1 \tag{8.214}$$

$$\mathbf{V} = \mathbf{V}_0 \boldsymbol{\alpha} + \varepsilon \mathbf{V}_1 \tag{8.215}$$

$$\mathbf{S} = \mathbf{S}_0 + \varepsilon \mathbf{S}_1 \tag{8.216}$$

It should be noted that Equation 8.211 and Equation 8.212 are the eigenproblem for repeated eigenvalues. That is, the perturbation analysis for close eigenvalues has been transferred into that of repeated eigenvalues. Hence, the methods given by Section 8.9.4 can be used to compute $\mathbf{S}_1$, $\boldsymbol{\alpha}$, $\boldsymbol{\beta}$, $\mathbf{U}_1$, and $\mathbf{V}_1$ in Equation 8.214 to Equation 8.216.

# References

Adelmen, H.M. and Haftka, R.T., Sensitivity analysis of discrete structural systems, *AIAA J.*, 24, 823, 1986.

Adhikari, S. and Friswell, M.I., Eigenderivative analysis of asymmetric non-conservative systems, *Int. J. Numer. Methods Eng.*, 39, 1813, 2001.

Chen, S.H. 1993. *Matrix Perturbation Theory in Structural Dynamics*, International Academic Publishers, Beijing.

Chen, S.H. and Liu, Z.S., High accuracy modal superposition for eigenvector derivatives, *Chin. J. Mech.*, 25, 432, 1993a.

Chen, S.H. and Liu, Z.S., Perturbation analysis of vibration modes with close frequencies, *Commun. Numer. Methods Eng.*, 9, 427, 1993b.

Chen, S.H. and Pan, H.H., Design sensitivity analysis of vibration modes by finite element perturbation, *Proc. of the Fourth IMAC*, 38, 1986.

Chen, J.C. and Wada, B.K., Matrix perturbation for structural dynamics, *AIAA J.*, 15, 1095, 1979.

Dailey, R.L., Eigenvector derivatives with repeated eigenvalues, *AIAA J.*, 27, 486, 1989.

Fox, R.L. and Kapoor, M.P., Rates of change of eigenvalues and eigenvectors, *AIAA J.*, 12, 2426, 1968.

Haug, E.J., Komkov, V., and Choi, K.K. 1985. *Design Sensitivity Analysis of Structural Systems*, Academic Press, Orlando, FL.

Haug, E.J. and Rousselet, B., Design sensitivity analysis in structural dynamics, eigenvalue variations, *J. Struct. Mech.*, 8, 161, 1980.

Hu, H. 1987. *Natural Vibration Theory for Multi-DoF Structures*, Science Press, Beijing (in Chinese).

Lim, K.B., Juang, J.N., and Ghaemmaghani, P., Eigenvector derivatives of repeated eigenvalues using singular value decomposition, *J. Guidance Control Dyn.*, 12, 282, 1989.

Liu, J.K., Perturbation technique for non-self-adjoint systems with repeated eigenvalues, *AIAA J.*, 37, 222, 1999.

Liu, X.L., Derivation of formulas for perturbation analysis with modes of close eigenvalues, *Struct. Eng. Mech.*, 10, 427, 2000.

Liu, Z.S. and Chen, S.H., Contribution of the truncated modes to eigenvector derivatives, *AIAA J.*, 32, 1551, 1994a.

Liu, Z.S. and Chen, S.H., An accurate method for computing eigenvector derivatives for free–free structures, *Int. J. Comput. Struct.*, 52, 1135, 1994b.

Mills-Curran, W.C., Calculation of eigenvector derivatives for structures with repeated eigenvalues, *AIAA J.*, 26, 867, 1988.

Murthy, D.V. and Haftka, R.T., Derivatives of eigenvalues and eigenvectors of a general complex matrix, *Int. J. Numer. Methods Eng.*, 26, 293, 1988.

Ojalvo, E.U., Efficient computation of modal sensitivities for systems with repeated frequencies, *AIAA J.*, 26, 361, 1988.

Rogers, L.C., Derivatives of eigenvalues and eigenvectors, *AIAA J.*, 8, 943, 1977.

Shaw, J. and Jayasuriya, S., Modal sensitivities for repeated eigenvalues and eigenvalues derivatives, *AIAA J.*, 30, 850, 1992.

Wang, B.P., Improved approximate methods for computing eigenvector derivatives in structural dynamics, *AIAA J.*, 29, 1018, 1991.

# 9

# Vibration in Rotating Machinery

H. Sam Samarasekera
*Sulzer Pumps (Canada), Inc.*

**Summary**

*This chapter concerns vibration in rotating machinery. Although it is impractical to totally eliminate such vibrations, it is essential that they be controlled to within acceptable limits for safe and reliable operation of such machines. The two major categories of vibration phenomena that occur in rotating machinery are forced vibration and self-excited instability. Monitoring, diagnosis and control of these vibrations requires a sound understanding of rotor dynamics in machinery. Predicting the vibration behavior of a rotating machine by analytical means has become customary in many industries. With the advent of computer technology, several computer-based programs have been developed to accurately predict the behavior of rotating machinery. Significant strides in modeling techniques have also been made over the past century to accurately represent components such as shaft sections, disks, impellers, bearings, seals, rotor dampers, and rotor−stator interactions. This has enhanced the accuracy and reliability of both analytical and computational procedures. The chapter presents useful techniques of analysis, measurement, diagnosis, and control of vibration in rotating machinery.*

## 9.1 Introduction

Vibrations are an inherent part of all rotating machinery. Residual mass imbalance and dynamic interaction forces between the stationary and rotating components, which are practically impossible to eliminate, cause these vibrations. The challenge is to identify the source of vibration and control it to within reasonable limits. Because of economic advantages, the trend in industry has been to move towards high speed, high power, lighter and more compact machinery. This has resulted in machines operating above their first critical speeds, which was unheard of in the past. The new operating parameters have required concurrent development of vibration technology without which it is not possible to safely and reliably operate such machinery. Industry has also come to realize that *vibration* is an essential phenomenon, which could be used to assess the performance, durability, and reliability of rotating machinery.

Engineers at different levels approach the subject of vibration in rotating machinery differently. The machinery designer has to recognize the potential sources of vibration and control them to within acceptable levels. In the past few decades, owing to the advancement in computers and modeling techniques, better understanding of the dynamics of rotating machinery, including the identification of potential sources of vibration, has been realized. This has enabled designers to accurately predict the rotordynamic behavior of machinery, allowing it to reach higher operating speeds and larger energy capacities safely and reliably.

Approaching vibration from a different perspective, the maintenance engineer uses vibration standards and guidelines to monitor the health of equipment for their timely repair and refurbishment. Reliable vibration monitoring and diagnostics techniques have moved industry into *predictive* rather than *preventive* maintenance practices, which considerably reduce plant downtimes that rely on key rotating machinery. Premature replacement of machinery components has also been minimized. The resulting financial and economic benefits provide an added incentive for the study and understanding of vibration in rotating machinery.

The vibration specialist or troubleshooter has to use his knowledge of rotordynamics and his diagnostic capabilities to solve vibration problems in rotating machinery. In most cases, it is also important to have an understanding of the interfacial dynamics of the rotating machinery with the surrounding system in order to solve a vibration problem.

From a safety and reliability standpoint, the public must be concerned with vibration in rotating machinery. Their concerns are addressed through vibration standards and guidelines. These procedures have been developed for rotating machinery by numerous organizations, both at the national and international levels. Some of these standards are industry specific and some are equipment type specific, while a number of them try to cover a wide range of rotating machinery. The objective of most of these standards is to establish and control quality, safety, durability and reliable performance of rotating machinery for the benefit of those who use or operate it.

## 9.1.1 History of Vibration in Rotating Machinery

Although various types of rotating machinery have been in use for many centuries, understanding of their rotordynamic behavior did not begin until 1869 (Rankine, 1869). Since that time, there has been steady growth in the development and understanding of the vibration behavior of rotating machinery. A tabulation of major historical events that have contributed to this growth is presented in Table 9.1.

---

- All rotating machinery vibrates to some degree. For public safety and machine reliability, the vibrations have to be controlled to within acceptable limits.
- Modern trends towards more sophisticated, higher speed compact rotating machinery have contributed to the rapid development in vibration technology through a better understanding of their rotordynamics.
- Vibration technology is integrated into the areas of design, maintenance, and troubleshooting of rotating machinery.
- From a safety and reliability standpoint, the public is protected by the implementation of vibration standards and guidelines.
- The first publicly reported rotordynamic study was made in 1869.

**TABLE 9.1**  A Chronological Listing of Major Contributions that Have Led to the Development and
Understanding of Vibration in Rotating Machinery

| Year | Contributor | Description |
|---|---|---|
| 1869 | Rankine, W.J.M. | He examined the equilibrium of a frictionless, uniform shaft disturbed from its initial position. The resulting recorded article is recognized to be the first on the subject of rotor dynamics. He proposed that motion is stable below the first critical speed, is neutral or indifferent at the critical speed, and unstable above the critical speed |
| | | He also developed numerical formulae for critical speeds for the cases of a shaft resting freely on a bearing at each end and for an overhanging shaft fixed in direction at one end |
| 1883 | Greenhill, A.G. | He studied the effect of end thrust and torque on the stability of a long shaft and concluded that they were both unimportant. He also obtained formulae for the cases of an unloaded shaft resting on bearings at each end and fixed in direction at each end |
| Circa 1890 | Reynolds, O. | He extended the theory developed by Rankine and Greenhill for the case of a shaft loaded with pulleys |
| 1893 | Dunkerley, S. | He developed formulae for critical speeds for loaded shafts in terms of the diameter of the shaft, weights of pulleys, the manner in which the shaft is supported, and so on, and verified them by experiment |
| | | He postulated that any degree of unbalance will excite the shaft at the critical speed to very high amplitudes and that it is possible to operate above the first critical speed. The dependence of critical speed on the moment of inertia of the rotating pulley was identified |
| 1894 | Rayleigh, J.W.S. | He developed an approximate method to calculate the natural frequency of a continuous beam with distributed mass and flexibility using the energy method |
| 1895 | DeLaval, G. | He was responsible for the first experimental demonstration that a steam turbine is capable of sustained operation above the first critical speed |
| 1916 | Timoshenko, T. | He discovered the effects of transverse shear deflection on the natural frequency of a continuous beam and applied the principle to the case of the rotating shaft |
| 1919 | Jeffcott, H.H. | He examined the effect of unbalance on the whirl amplitudes and the forces transmitted to the bearings. The case of a light uniform shaft supported freely on bearings at its ends and carrying a thin pulley of mass $m$ at the center of the span was studied. He assumed the moment of inertia of the pulley to be negligible. Using this model, later known as the *Jeffcott model*, a comprehensive theory was developed to explain the behavior of the rotor as it passed through the critical speed |
| | | The effect of damping on the whirl amplitude, a phase change of angle $\pi$ as it passes through the critical speed, and the concept of synchronous rotor whirling (precession) were introduced and explained. He also recognized that with a separation margin of 10% on either side of a critical speed, the amplitude of vibration would not be excessive. He demonstrated that it is better from the vibration point of view to design the shaft with its critical speed below the working speed rather than to have a critical speed the same proportion above the working speed. Accordingly, he explained the behavior of the De Laval steam turbine and the economic advantages of operation above the critical speed |
| 1921 | Southwell, R.V. and Gough, B.S. | They found that a torque and an end thrust of constant magnitude lowers the critical speed of a rotating shaft, disproving Greenhill's earlier (1883) conclusions |

*(continued on next page)*

**TABLE 9.1** *(continued)*

| Year | Contributor | Description |
|------|-------------|-------------|
| 1921 | Holzer, H. | He developed a numerical method to calculate torsional critical speeds and mode shapes for a multidisk rotor system |
| 1924 | Newkirk, B.L. | He observed that a rotor operating at a speed above the first critical speed can enter into high, violent whirling and the center of the rotor will precess in the forward direction at a rate equal to that of the critical speed. Unlike in the case of synchronous whirling, if the speed is increased beyond the initial whirl speed, the whirl amplitude will continue to increase, eventually leading to failure. This was the first time that it was realized that nonsynchronous unstable motion can exist in a high-speed rotor |
| | | Based on experiments, he made the following key observations on nonsynchronous whirling. The amplitude and the onset speed of whirling are independent of the rotor balance. Whirling always occurs at speeds above the critical speed, and the whirl speed is always constant at the critical speed, regardless of the rotor speed. The whirl threshold speed can vary even for machines of similar construction. Whirling occurs only in built up rotors, and not in single piece constructions. Increasing the foundation flexibility, distortion or misalignment of the bearing housings, or introducing damping to the foundation or increasing the axial thrust bearing load, increased the threshold speed of whirling |
| 1924 | Kimball, A.T. | Suggested that internal friction or viscous action due to bending may cause a shaft to whirl when rotating at any speed above the first critical speed. He postulated that the nonsynchronous whirling observed by Newkirk was due to this phenomenon |
| 1924 | Newkirk, B.L. | Based on Mr Kimball's theory, he concluded that similar frictional forces are generated at the mating face between the shrunk on disk and the shaft of a built-up rotor, and the nonsynchronous whirling observed by him was due to this effect. However, he was unable to explain some of his experimental findings, in particular, the effects of bearing or foundation flexibility, damping, and misalignment |
| 1925 | Newkirk, B.L. | He experienced another form of nonsynchronous whirling, similar but different to that caused by the frictional effects of a shrink-fit disk. It occurred at rotor speeds just exceeding twice the first critical speed on shafts mounted on journal bearings. He recognized that the oil in the journal bearing was responsible for the violent motion and called it *oil whip*. The whirl speed and direction of whirling were the same as that for friction induced whirling, that is, the first critical speed in the forward direction. A theory to explain how the oil film can produce the whirling motion of a journal and to account for why it took the same direction as rotation of the shaft was proposed. However, the theory does not explain why whirling does not commence until the rotor speed reaches twice the critical speed value. The influence of foundation flexibility on the rotor stability was also found to be confusing to Newkirk. In the case of friction-induced whirl, he was able to totally eliminate the rotor instability by means of a flexibly mounted bearing. When this was tried with the journal bearings, the whirl amplitudes magnified. External damping at the bearing was found to have a favorable influence on whirl amplitudes |
| 1925 | Stodola, A. | He developed an iterative procedure to calculate the fundamental frequency of a vibrating system based on an assumed mode shape |
| 1927 | Stodola, A. | He provided an explanation and formulae for the gyroscopic moment effect on the critical speed of a rotor. He also introduced the notion of synchronous and nonsynchronous reverse precession of a rotor under specific conditions |

**TABLE 9.1**   (*continued*)

| Year | Contributor | Description |
|------|-------------|-------------|
| 1933 | Robertson, D. | In order to understand oil whip, he studied the stability of the ideal 360° infinitely long journal bearing, and erroneously concluded that the rotor will be unstable at all speeds and not only at speeds above twice the critical speed value |
| 1933 | Smith, D.M. | He studied the case of unsymmetrical rotors on unsymmetrical supports and obtained four different critical speed values in comparison to the single value for a symmetrical system. He also discussed the presence of additional critical speeds due to gyroscopic effects of large disks |
| 1944 | Myklestad, N. | A lumped parameter transfer matrix method to calculate natural frequencies for airplane wings was developed by him |
| 1945 | Prohl, M. | He developed a lumped parameter transfer matrix method for calculating critical speeds of flexible rotors |
| 1953 | Poritsky, H. | Using the small displacement theory, he derived a radial stiffness coefficient for the journal bearings and analyzed the rotor behavior under oil whirl conditions. He concluded that the rotor was stable below twice the critical speed and indicated that increasing the rotor or bearing flexibility will reduce the threshold speed of instability. He also proposed a stability criterion for a rotor based on the bearing and rotor stiffness |
| 1953 | Miller, D.F. | He introduced a solution to the steady-state forced vibration problem, for a beam or rotating shaft on damped, flexible end supports. The response of the rotor to an unbalance force and the damped resonance frequencies are calculated by this method |
| 1955 | Pinkus, O. | He investigated oil whirl in various journal bearing types and made the following major conclusions. The unbalance of the rotor has minimal effect on stability. The threshold of instability occurs at approximately twice the first critical speed of the rotor. In the unstable region, the whirl frequency remained constant at the first critical speed, irrespective of the shaft rotating speed. At speeds nearly equal to three times the first critical speed, whipping motion stops with a heavy shaft rotor, whereas with a light shaft rotor it does not cease. High loads, high viscosity, flexible mountings, and bearing asymmetry favor stability |
| 1958 | Lomakin, A. | The influence of the dynamic characteristics of seals on the critical speeds and stability of pump rotors were introduced by him |
| 1958 | Thomas, H. | He proposed that an eccentric turbine rotor would generate a destabilizing force due to the circumferential variation in clearance |
| 1966 | Gunter, E.J. Jr. | He combined the different theories on whirling developed by the rotor dynamist and the bearing specialist, and elegantly explained some of the conflicting experimental evidence gathered thus far. He emphasized the importance of considering the combined effects of rotor parameters and the bearing and foundation characteristics on rotor stability, and developed more comprehensive criteria for self-excited whirl instability |
| 1969 | Black, H.F. | He provided a comprehensive analysis of annular pressure seals on the vibrations of pump rotors |
| 1970 | Ruhl, R. | He introduced finite element models for flexible rotors for calculating rotor critical speeds and mode shapes. These models did not take into account gyroscopic effects and axial loading |
| 1974 | Lund, J. | A transfer matrix method to calculate damped critical speeds of a rotor taking into account the cross coupling terms as well were introduced by him |
| 1976 | Nelson, H. and McVaugh, J. | They extended the finite element model of a rotor to account for rotary inertia, gyroscopic effect, and axial loads |

(*continued on next page*)

**TABLE 9.1**    *(continued)*

| Year | Contributor | Description |
|------|-------------|-------------|
| 1978–1980 | Benckert, H. and Wachter, J. | A method to calculate flow induced spring constants for labyrinth gas seals and the use of *swirl breaks* to reduce the destabilizing force caused by tangential velocity in labyrinth seals was introduced by them |
| 1980 | Nelson, H. | He further developed a finite element model of a rotor to include shear deflection and axial torque effects |
| 1980 | Brennen, C. et al. | They recognized the presence of substantial shroud forces, which influences the rotor dynamics of a pump |
| 1986 | Muszynska, A. | She demonstrated that oil whirl occurs at about one-half the running speed in a vertical rotor. With further increase in speed, oil whip will commence when the whirl frequency approaches the critical speed of the rotor |

## 9.2    Vibration Basics

The vibration phenomena that manifest in rotating machinery can be divided into two major categories: *forced vibration* and *self-excited instability*. A stimulus or a source of excitation is required to initiate and sustain vibratory motion in a rotor. When the stimulus is a forcing phenomenon such as mass unbalance, it will produce forced flexural vibration in the rotor analogous to linear forced vibration response in a simple spring–mass system. On the other hand, self-excited vibration (instability) does not require a forcing phenomenon for its initiation or sustenance. A description of these phenomena is given next.

### 9.2.1    Forced Vibration

A rotating force vector (unbalance), a steady directional force (gravity), or a periodic force (pump impeller/diffuser interface action), will cause forced vibration in a rotating machine. The response of the rotor will depend on the nature of the forcing function and how it relates to rotor characteristics. The rotor responses to the most common excitation phenomena are examined below.

#### 9.2.1.1    Unbalance Response — Synchronous Whirling

As an introduction to the theory on rotating machinery vibration and understanding unbalance response, it is most appropriate to examine Jeffcott's (1919) rotor, which is a simple model that has many of the basic characteristics of more complex rotating machinery. The Jeffcott rotor represents a massless elastic shaft supported freely in bearings at its ends and carrying a disk of mass *m* at the center of its span. The mass center of the disk is eccentric to its geometric center by a distance *e*. Refer to Figure 9.1.

$C$ = geometric center of the disk
$\beta$ = phase angle

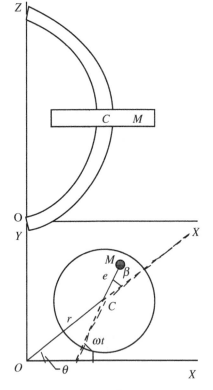

**FIGURE 9.1**    Jeffcott rotor.

$M$ = mass center of the disk
$c$ = viscous damping coefficient of on the rotor
$O$ = bearing center
$r$ = deflection of rotor from origin
$\theta$ = angle of precession
$k$ = shaft stiffness
$\omega$ = angular velocity of the rotor = $\dot{\theta} + \dot{\beta}$

*Whirling* is defined as the angular velocity of rotation of the rotor geometric center ($c$) or the time derivative ($\dot{\theta}$) of the angle of precession ($\theta$) (also see Chapter 7). *Synchronous whirling* is when the rate of whirling, $\dot{\theta}$, is equal to the total angular velocity, $\omega$, of the system.

Applying Newton's Laws of motion to the rotor, the differential equations of motion in polar coordinates ($r$, $\theta$) are obtained as

$$\ddot{r} + \frac{c}{m}\dot{r} + \left(\frac{k}{m} - \dot{\theta}^2\right)r = e\omega^2 \cos\beta \qquad (9.1)$$

$$r\ddot{\theta} + \left(\frac{c}{m}r + 2\dot{r}\right)\dot{\theta} = e\omega^2 \sin\beta \qquad (9.2)$$

For a steady-state condition, the values of $r$, $\beta$, $\dot{\theta}$, and $\omega$ are constant. For synchronous whirling, Equation 9.1 and Equation 9.2 reduce to

$$\left(\frac{k}{m} - \omega^2\right)r = e\omega^2 \cos\beta \qquad (9.3)$$

$$\frac{c}{m}\omega r = e\omega^2 \sin\beta \qquad (9.4)$$

From Equation 9.3 and Equation 9.4

$$r = \frac{e\omega^2}{\sqrt{\left(\frac{k}{m} - \omega^2\right)^2 + \left(\frac{c\omega}{m}\right)^2}} \qquad (9.5)$$

$$\beta = \tan^{-1}\frac{c\omega}{m\left(\frac{k}{m} - \omega^2\right)} \qquad (9.6)$$

$$F = \frac{kr}{2} = \frac{ke\omega^2}{2\sqrt{\left(\frac{k}{m} - \omega^2\right)^2 + \left(\frac{c\omega}{m}\right)^2}} \qquad (9.7)$$

Using the following relationships:

$$\omega_N = \sqrt{\frac{k}{m}} \text{ — Natural frequency of rotor without damping}$$

$$c_{cr} = 2\sqrt{km} \text{ — Critical damping coefficient}$$

$$\zeta = \frac{c}{c_{cr}} \text{ — Damping ratio}$$

Equation 9.6 and Equation 9.7 are reduced to the following nondimensional form:

$$\frac{r}{e} = \frac{2F}{ke} = \frac{(\omega/\omega_N)^2}{\sqrt{(1 - (\omega/\omega_N)^2)^2 + (2\zeta\omega/\omega_N)^2}} \qquad (9.8)$$

$$\beta = \tan^{-1}\frac{2\zeta(\omega/\omega_N)}{(1 - (\omega/\omega_N)^2)} \qquad (9.9)$$

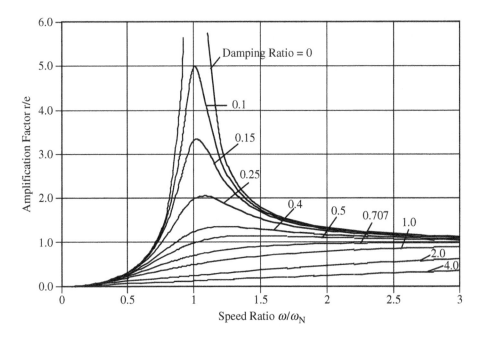

**FIGURE 9.2** Jeffcott rotor response with mass eccentricity — amplification vs. speed.

Figure 9.2 is a graphical representation of the unbalance response of the rotor as a function of rotating speed, $\omega$. Upon examination of the phase relationship, it is important to note that the phase angle, $\beta$, changes from approximately 0° at low speed to values approaching 180° at the higher speed. At $\omega_N$, $\beta = 90°$. A pictorial illustration of this phenomenon is given in Figure 9.3.

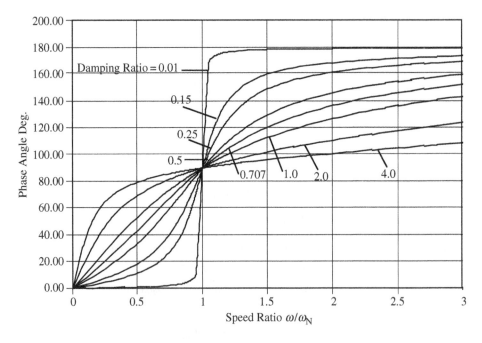

**FIGURE 9.3** Jeffcott rotor response to unbalance — phase angle vs. speed.

For the case of zero damping, when $\omega = \omega_N$, the rotor deflection and the bearing forces are unbounded. For all other cases, the rotor deflection and the bearing forces are bounded, and their amplitude depends on the damping ratio. If a shaft is quickly accelerated through its critical speed to a higher working speed, then there may not be enough time for large rotor deflection to take place. At high speeds, $\omega \gg \omega_N$, the amplitude of the rotor deflection decreases and approaches the value $e$, the eccentricity of the rotor.

The critical speed, $\omega_{cr}$, of a rotor in the general case, is the speed at which the rotor deflection amplitude or the force amplitude transmitted to the bearings is a maximum.

This implies that, at $\omega = \omega_{cr}$

$$\frac{dr}{d\omega} = \frac{dF}{d\omega} = 0$$

Using Equation 9.8, the following relationship between the natural frequency of the rotor and its critical speed is derived:

$$\omega_{cr} = \frac{\omega_N}{\sqrt{1 - 2\zeta^2}} \tag{9.10}$$

From Equation 9.10, it is evident that the critical speed of a rotor is not a fixed value and is dependent on the degree of rotor damping. When $\zeta = 1/\sqrt{2}$, the system is said to be critically damped.

It is important to note that rotor response to unbalance (or imbalance) is recognizable and controllable. The amplitude of the force transmitted to the bearing can be reduced by operation at speeds above the critical speed, reducing unbalance, increasing viscous damping, and avoiding operation close to critical speeds.

### 9.2.1.2 Shaft Bow

A rotor with a bent shaft will behave in a similar manner to a rotor with an eccentric mass (Ehrich, 1999). At high rotor speeds ($\omega \gg \omega_{cr}$), the shaft will tend to correct the bow as illustrated in Figure 9.4. When shaft bow is combined with mass eccentricity, unique behavior patterns are produced depending on the phase angle between the bow and the eccentric mass (Childs, 1993).

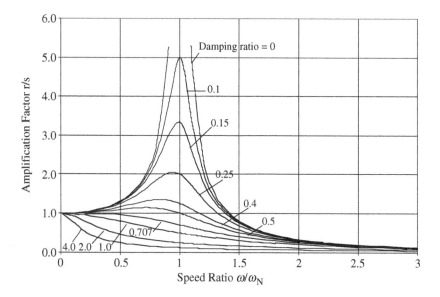

**FIGURE 9.4** Jeffcott rotor response with shaft bow — amplification vs. speed.

### 9.2.1.3  Gravity Critical

A special case of synchronous whirling may occur in certain types of horizontal rotors due to the gravitational force. It is a secondary critical speed commonly called the *gravity critical*, which can occur in a very heavy lightly damped rotor. The critical speed will occur at approximately half the natural frequency of the rotor and its amplitudes of deflection at the critical speed are bounded and approximately twice the static deflection of the rotor (Gunter, 1966).

### 9.2.1.4  The Influence of Rotor Inertia and Gyroscopic Action

The effect of rotor inertia is ignored in the Jeffcott model. However, in practice, it is recognized that rotor inertia and gyroscopic action has an influence on the natural frequencies, critical speeds, and unbalance response of the rotor, including reverse whirling. In the case of the natural frequency of the rotor (zero speed), the diametral or rotary inertia provides an additional natural frequency associated with the rotational degree of freedom (DoF). Also, the inertia effect lowers the first natural frequency (Childs, 1993). In the rotating case, the effect of inertia generates both forward and reverse whirling critical speeds (Childs, 1993). These forward whirling critical speeds tend to be higher (stiffening effect) and the reverse whirling critical speed lower than the natural frequency of the rotor. At the forward critical speeds, large amplitude whirling motion due to imbalance occurs, whereas the reverse critical speeds are insensitive to imbalance of the rotor.

### 9.2.1.5  Rotor Housing Response across an Annular Clearance

If the rotor deflection due to imbalance exceeds the uniform annular gap, continuous contact would occur between the rotor and stator resulting in coupled motion between the rotor and stator (Childs, 1993). For low contact frictional forces, synchronous forward whirling driven by the imbalance forces will occur. If the contact friction force is large enough to prevent slipping between the rotor and stator, reverse whirling will take place. For the case of synchronous forward whirling in a certain range of running speeds, instability will occur due to engagement between the rotor and stator (Black, 1968). The zones of instability depend on the coupled natural frequency of the rotor and stator and the degree of rotor deflection with respect to the annular gap.

### 9.2.1.6  Effect of Nonlinearity and Asymmetry on Forced Vibration Response

The foregoing analysis has assumed that stiffness and damping are linear and symmetric and the resulting forces are proportional to the deflection and velocity of the rotor. However, in reality, rotating machinery components have inherent nonlinearities and asymmetries that can have a profound influence on their rotordynamic behavior. At large amplitudes of motion, stiffness and damping coefficients become nonlinear and result in modifying the response amplitude and critical speeds of the rotor. Nonlinearity in the support stiffness will introduce considerable distortion to the otherwise simple harmonic vibration behavior of a purely linear system. The stiffness and damping coefficients of the bearings and their supports are asymmetric in most cases, in particular in horizontal machines. As a result, the forced vibratory responses in the two principal directions are different and can behave independent of each other. Each principal direction will display a critical speed unique to itself. Ehrich (1999) has presented a discussion on how nonlinearity and asymmetry of stator systems influence forced vibration response.

The influence of rotor stiffness asymmetry and inertia asymmetry on rotor stability is discussed in Section 9.2.3.

## 9.2.2  Self-Excited Vibration

*Instability* (nonsynchronous whirling) is a self-induced excitation phenomenon, sometimes described as *sustained transient motion*, that can occur in rotating machinery. At the inception of instability, the rotor deflection will continue to build up with increase in speed, whereas in the case of a critical speed resonance, the amplitude of the deflection reaches a maximum value and then decreases. If the rotor speed is increased above the instability threshold speed, the large amplitudes of motion will normally

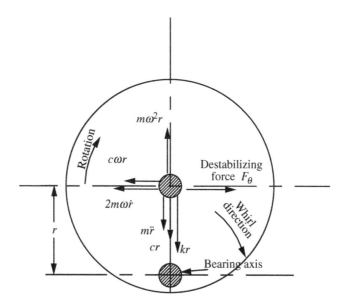

**FIGURE 9.5** Rotor instability — general case.

result in damage to the machine. Unlike forced flexural vibration, rotor instability is self-induced and does not require a sustained forcing phenomenon to initiate or maintain the motion. It is known to occur only in machines operating at speeds well above the critical speeds of the rotor. Furthermore, the rotor whirling speed is different to the rotor speed (nonsynchronous whirling), and it is identical to the critical speed irrespective of the rotor speed.

In general, the rotor instability is associated with the existence of a tangential force vector, $F_\theta$, acting at right angles to the deflection vector and directly opposing the damping force vector, as illustrated in Figure 9.5. The nature of $F_\theta$ is such that its magnitude increases proportionately with the rotor deflection. At the point where $F_\theta$ equals the external damping force, rotor instability will commence due to the nullification of the stabilizing force. This will produce a whirling motion of ever increasing amplitude. Several phenomena inherent in the rotor system that generates such tangential force vectors have been identified and are discussed below. Rotordynamists believe that there still remain more such phenomena to be discovered.

### 9.2.2.1 Internal Friction Damping

This type of instability was first experienced in the early 1920s in blast furnace compressors made by the General Electric Company. These machines were subject to occasional fits of violent vibration. Newkirk (1924) carried out a series of experiments to understand the unusual behavior of these machines. Based on the *internal friction theory* of Kimball (1924). Newkirk (1924) concluded that the interfacial friction damping forces at the disk shaft interface caused the subsynchronous whirling.

In order to understand the internal friction damping phenomena let us examine the shaft stresses in the whirling Jeffcott rotor (Figure 9.1). Figure 9.6 is a cross section of the shaft disk interface. Owing to its deflection, all of the fibers in the right half of the cross section are in tension, $T_e$, and those in the left half are in compression, $C_e$. These fiber stresses tend to straighten the shaft and produce a restoring force, $F_r$, which opposes the centrifugal force, $m\ddot{\theta}r$. Furthermore, a set of frictional forces are generated at the shaft disk interface due to stretching and compression of the fibers. The fibers in the bottom lower half will be stretched and are under frictional tension, $T_f$, and those in the upper half are being compressed under frictional compression $C_f$. Similarly to the reaction force, $F_r$, produced from right to left by $T_e$ and $C_e$, a reaction force, $F_\theta$, from bottom to top will be produced by the frictional stresses $T_f$ and $C_f$. The disturbing force, $F_\theta$, is in the same direction as the whirling motion and as a result will increase as

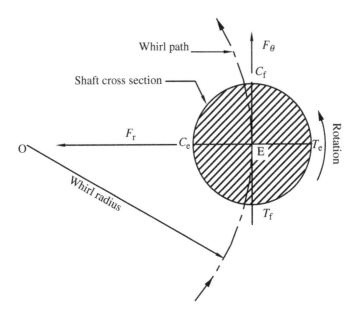

**FIGURE 9.6**   Internal friction damping forces acting on a rotor.

the whirling motion increases. The force, $F_\theta$, will oppose the external damping force, $c\dot\theta r$. At the threshold of instability, the two forces nullify each other. It is also know that the frequency of whirling, $\dot\theta$, at the threshold of instability equals the natural frequency, $\omega_n$, of the rotor. Mathematically it can be expressed as follows:

$$F_\theta = c_i r(\omega - \dot\theta) \tag{9.11}$$

where $c_i$ is the rotor internal damping coefficient:

$$c\dot\theta r = c_i r(\omega - \dot\theta) \tag{9.12}$$

$$\dot\theta = \omega_N \tag{9.13}$$

Equation 9.12 and Equation 9.13 yield the following relationship between the threshold speed of instability, the first critical speed, and the damping factors (both internal and external):

$$\frac{\omega}{\omega_N} = 1 + \frac{c}{c_i} \tag{9.14}$$

### 9.2.2.2   Tip Clearance Excitation (Alford's Force, Steam Whirl)

Thomas (1958) investigated the instability of steam turbines and suggested that nonsymmetric radial clearances caused by an eccentric rotor could result in destabilizing forces, and called them *clearance excitation forces*. Subsequently Alford (1965) discovered a similar phenomenon in aircraft gas turbines and, as a result, the destabilizing forces are sometimes referred to as *Alford forces* in North America.

The destabilizing force is created as a result of the variation in the gap between the blade tip and the stator. When the gap decreases, the leakage decreases and consequently the efficiency increases, resulting in a torque higher than the average torque produced by a uniform gap. When the gap increases, there is a corresponding decrease in the torque relative to the average. The variation in torque produced by the eccentricity results in a tangential force, which is normal to the radial deflection and is in the direction of the whirling motion as shown in Figure 9.7. Furthermore, it has been illustrated that the magnitude of the resulting force increases proportionally with the increase in rotor deflection, that is, the decrease

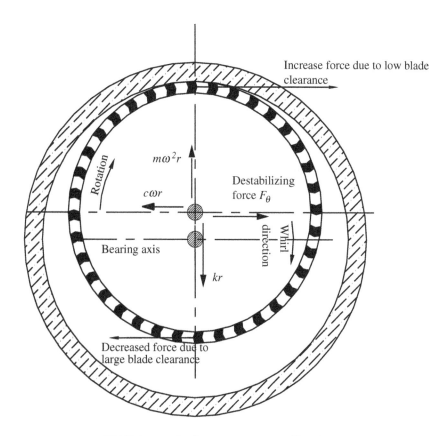

**FIGURE 9.7**   Tip clearance excitation (Alford forces).

in the gap. The resulting force is a destabilizing force that will oppose the external damping force, and at some point when they balance each other, rotor instability will occur. A detailed analysis of the tip clearance forces has been made by Urlichs (1977).

### 9.2.2.3   Impeller Diffuser Excitation Forces

Instability experienced in centrifugal compressors provides a strong suspicion that impeller–diffuser interaction phenomena are involved in the development of destabilizing forces. However, to date, no satisfactory destabilizing mechanism or source involving impellers has been identified. In the last two decades there have been several studies related to rotordynamic forces arising from shrouded centrifugal pump impellers, but very little work has been done on compressor impellers. The destabilizing force arising from the impeller–diffuser/volute interaction of a pump has been determined to be relatively small (Jery et al., 1984; Bolleter et al., 1985; Adkins and Brennen, 1986; Ohashi et al., 1986). The major portion of the destabilizing force is known to be generated in the narrow gap region between the casing and shroud of the impeller (Childs, 1986; Bolleter et al., 1989; Baskharone et al., 1994; Moore and Palazzolo, 2001). In the case of centrifugal compressors, an empirical method to determine stability of multistage machines has been proposed (Kirk and Donald, 1983). The stability maps proposed by them for flow through and back-to-back centrifugal compressors are shown in Figure 9.8 and Figure 9.9.

### 9.2.2.4   Propeller Whirl

Propeller whirl (Taylor and Browne, 1938; Houbolt and Reed, 1961) is another form of instability which occurs in aircraft rotors when there is a mismatch in the angular velocity vector of the propeller and the linear velocity vector of the aircraft. This angular mismatch results in the generation of a moment whose vector has a component of significant magnitude, which contributes to the instability of the

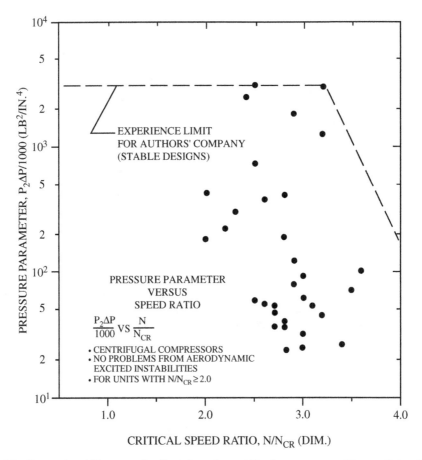

**FIGURE 9.8** Proposed stability map for flow through centrifugal compressors. (*Source:* Rotor Dynamical Instability, 1983. With permission.)

propeller (Vance, 1988). Its magnitude is proportional to both the angular mismatch and the linear speed of the aircraft. With increasing speed, the magnitude of the destabilizing moment will exceed the rotor viscous damping moment and result in propeller instability (refer to Figure 9.5). Since the propeller is supported only from one end, the whirling motion is conical and is found to be in the reverse direction to propeller rotation. The instability is sensitive to the velocity and density of the air and not a function of the torque of the machine.

### 9.2.2.5 Fluid Trapped in a Hollow Rotor

Wolf (1968) has demonstrated that trapped fluid inside a hollow rotor can produce a force component tangential to the whirl orbit due to viscous drag forces. Under subsynchronous whirling speeds, this force component acts in the same direction of whirling motion and its magnitude is proportional to the rotor deflection. With reference to Figure 9.5, this force has all the markings of a destabilizing force, which can produce instability in the rotor. The threshold speed of instability is reached when the whirling speed equals the first critical speed of the rotor. It has been shown (Ehrich, 1999) that, at the threshold of instability, the rotor speed is less than twice the first critical speed. This results in a ratio of whirl speed to rotor speed in the range of 0.5 to 1.0.

### 9.2.2.6 Dry Friction Rubs

In Section 9.2.1.5, a dry friction rub situation was identified, where slipping was prevented between the rotor and stator under contact conditions. The contact was made possible by the deflection of the rotor

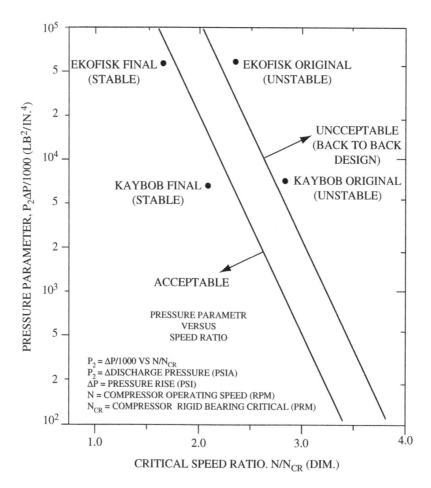

**FIGURE 9.9** Proposed stability map for back-to-back centrifugal compressors. (*Source:* Rotor Dynamical Instability, 1983. With permission.)

due to unbalance forces. When contact is made between the rotor and stator, Coulomb friction produces a tangential force in the direction opposite to shaft rotation. Since the frictional force prevents slipping, whirling in the reverse direction to rotation occurs. The whirling speed is equal to $r\omega/C$, where $r$ is the radius of the rotor, $C$ the radial gap, and $\omega$ the speed of the rotor. Since the frictional force is in the same direction of whirling, it will cause the magnitude of whirling to increase, resulting in further increase of the frictional force. When the magnitude of this force exceeds the viscous damping force, rotor instability will occur. Another possibility is for the dry friction whirling speed to approach the coupled natural frequency of the rotor and stator, in which case unstable motion termed *dry friction whipping* takes place (Ehrich, 1999).

In addition to the case described above, dry friction rubs can occur in journal bearings, seals, wear rings, or any situation where a small clearance between a rotor and stator exists. The inadvertent closure of the clearance due to unbalance or lack of proper lubrication can initiate dry friction rub induced instability in these cases as well.

### 9.2.2.7 Torque Whirl/Load Torque

When the rotor disk axis is not aligned with the bearing axis, as in the case with an overhung rotor, Vance (1988) has shown that nonsynchronous whirling (torque whirl) can occur as a result of the misalignment between the load torque and the driving torque. His findings are based on the analysis of a simple rotor model. It appears that torque whirl instability can occur only in the case of long slender shafts with high-load torque values. The practical implications of this theory are still to be fully explored.

### 9.2.2.8   Oil Whirl/Whip

Newkirk and Taylor (1925) first experienced shaft whipping due to oil action in journal bearings during their investigation into internal friction induced whirling of rotors. They found that under certain conditions, a rotor mounted on journal bearings whipped when the rotor was running at any speed above double the critical speed; the whirling motion was in the forward direction and its speed matched the critical speed of the rotor. He provided a qualitative explanation of the phenomenon based on the fact that the oil film rotates at half the velocity of the shaft due to friction drag. Hence, for rotational speeds near twice the critical speed, the oil film provides the stimulus as its speed matches the critical speed value resulting in large displacements and whipping. Others have also drawn similar conclusions based on the suggestion of an oil wedge rotating at half speed, or rotating fluid force fields at half shaft speed. However, the foregoing fails to explain why oil whip persists at speeds greater than twice the critical speed. Ehrich (1999) has also provided a qualitative explanation for oil whirl based on the general theory of rotor instability.

Although a comprehensive explanation of the physical phenomena of oil whirl is still outstanding, numerous analytical models to identify where it could be encountered have been suggested. Gunter (1966) has analytically demonstrated that the instability in a rotor supported on journal bearings can be attributed to the cross coupling bearing coefficients. As a result, most of the research on oil whirl instability has narrowed to accurate estimation of bearing cross coupling coefficients.

### 9.2.2.9   Influence of Bearings and Supports on Rotor Instability

The results of the Jeffcott model can be easily adopted to include bearing stiffness and bearing support stiffness effects, provided they are both linear and circumferentially symmetric (isotropic). For this particular case, the rotor stiffness, $k$, is the equivalent stiffness resulting from the series connection of the shaft, bearings, and support stiffness. The resulting values for $\omega_N$ and $\omega_{cr}$ will be less than those for the simply supported Jeffcott model. This will result in lowering the threshold speed of instability.

If the bearing stiffness or the bearing-support stiffness is not symmetric (orthotropic) then it can be shown (Childs, 1993) that the threshold speed of instability is increased and the maximum amplitude of deflection of the rotor is reduced in comparison to the case with symmetric bearings.

The effect of damping at the bearings or at the bearing-support is very similar to the influence of stiffness. It reduces the amplitude of the synchronous rotor response at the critical speed, and elevates the threshold speed of instability. However, there is a limit to the amount of damping that can be applied. Excessive damping causes a reduction in stability (Childs, 1993).

The mass of the bearings plays a significant role on rotor stability. If the bearing mass is significantly larger than the rotor mass, the threshold speed of instability is lowered.

## 9.2.3   Parametric Instability

The instability phenomena described in Section 9.2.2 can be represented by linear differential equations where the system parameters such as mass, inertia, stiffness, damping, and natural frequency are assumed to be constants. There is another subcategory of self-excited motion, referred to as *parametric instability*, since it is induced by the periodic variation of the system parameters such as inertia, mass, and stiffness. A discussion of the more common forms of this phenomenon follows.

### 9.2.3.1   Shaft Stiffness Asymmetry

If the shaft of a rotor contains a sufficient level of stiffness asymmetry in the two principal axis of flexure, rotor instability could occur. Smith (1933) investigated the rotor behavior under unsymmetrical flexibility of the bearing supports and unsymmetrical transverse flexibility of the shaft, taking into consideration the damping effects as well. The following conclusions were derived based on his investigation:

In the presence of stiffness asymmetry, the onset speed of internal-friction induced instability is lowered.

When there is no external damping, the rotor becomes unstable at all speeds between the two undamped natural frequencies in the two orthogonal directions. However, if external damping is significant, parametric instability may be eliminated.

Within the unstable range, the whirling motion is in the forward direction and is synchronous with the shaft speed. Further, unlike in the case of internal-friction induced instability; it is theoretically possible to run through parametric instability. This makes parametric instability quite similar to the case of unbalance response.

When the asymmetric rotor is acted upon by a transverse disturbing steady force such as gravitational force, the rotor whirls at twice the speed of the shaft. This motion exhibits a resonant increase in amplitude at a speed that is approximately half the mean of the two natural frequencies.

### 9.2.3.2   Rotor Inertia Asymmetry

Crandall and Brosens (1961) analyzed the parametric excitation of a rotor with nonsymmetrical principal moments of inertia. Their results indicate that the rotor behavior is very similar to the case of rotors with stiffness asymmetry described in Section 9.2.3.1, and parametric instability over similar speed ranges occurs.

### 9.2.3.3   Pulsating Torque

Constant torque acting on a rotor is known to lower its critical speeds because they effectively reduce the rotor's lateral stiffness. A pulsating torque introduces lateral vibrations and instabilities into a rotor. When a combination of a pulsating and constant torque is applied to a rotor, it will induce unstable lateral vibrations in a specific range of rotor speeds and certain combinations of torque amplitudes. In the region of unstable lateral motion, the whirling speed of the rotor will coincide with the first critical speed of the rotor regardless of the rotor speed or the frequency of the pulsating torque. At rotor speeds outside the unstable region, the whirling speed of the rotor will be coincident with the pulsating torque frequency.

### 9.2.3.4   Pulsating Longitudinal Loads

Pulsating axial forces on a shaft that are in the order of magnitude of the buckling load will effectively cause a periodic variation in its lateral stiffness. This will result in a proportionate reduction of the lateral natural frequency of the shaft. Therefore, pulsating axial loads are capable of inducing parametric instability in a shaft for both the rotating and the stationary cases.

### 9.2.3.5   Nonsymmetric Clearance Effects

Bentley (1974) recognized that large subsynchronous whirling can occur in rotating machinery due to certain types of nonsymmetric clearance conditions. One such condition is when a rotor's whirling motion causes rubbing with a stationary surface over a portion of the rotor orbit. This effectively results in an increase in the rotor stiffness during the contact portion of the orbit, producing a periodic variation in rotor stiffness during each cycle. Another situation that produces cyclic variation in rotor stiffness can occur in the case of a rotor supported on antifriction bearings mounted with a clearance fit to the housing. The cyclic variation of the effective rotor stiffness produces a subsynchronous whirling motion at exactly half the rotational speed. Occasionally, whirling at one third and one fourth the running speed has also been observed by Childs (1993). Instability will occur when the whirling speed corresponds to a critical speed of the rotor.

## 9.2.4   Torsional Vibration

Rotating machinery with rotors that have relatively large moments of inertia are susceptible to torsional vibration problems. Torsional vibration is an oscillatory angular motion that is superimposed on the steady rotational motion of the rotor. In practice it can easily go undetected, as standard vibration monitoring equipment is not geared to measure torsional vibration. Special equipment must be used to detect torsional vibration in rotating machinery. Since the introduction of electric motor variable frequency drives (VFD), the incidence of field torsional vibration problems has increased. This has been attributed to the inherent torsional excitation forces present in current designs of VFDs. An added complication when using VFDs, as compared with using fixed speed drives, is the requirement of eliminating torsional natural frequencies over a wider speed range. Large synchronous electric motors are known to produce a large pulsating torque at a frequency that changes from twice the line frequency at the start, down to zero at the synchronous operating speed. In this case, any torsional natural frequency between zero and twice the line frequency may be subject to excitation. Most industries have come to recognize that torsional vibration is a potential hazard, and therefore, needs to be investigated at the design stage of rotating machinery. Several standard design specifications now require that torsional analysis is part of the design procedure.

The standard design practice in modeling the system for torsional analysis has been to calculate the undamped eigenvalues of the rotor as a free body in space. This practice is acceptable for most types of rotating machinery since the torsional stiffness and damping of bearings is insignificant. Also, in most cases, the torsional damping of the rotors itself is extremely low. Although the absence of damping is favorable from an analysis point of view, it makes it extremely difficult to eliminate a torsional vibration problem when encountered. This deficiency has been partly addressed with the introduction of several new lines of couplings that have a significant degree of torsional damping. Although not commonly used in rotating machinery, several torsional dampers such as the Lanchester damper have been developed for use on reciprocating machines. Dampers of similar design could be developed for use in rotating machinery to solve torsional vibration problems.

The torsional critical speeds of a simple rotor with one or two DoFs can be calculated by analytical methods. Numerical methods are used to calculate critical speeds and mode shapes of more complex systems with higher DoF. The Holzer numerical method (Holzer, 1921) described in Section 9.3.1.11 is one of the common methods used for these analyses.

- The two major categories of vibration phenomena that occur in rotating machinery are forced vibration and self-excited instability.
- Parametric instability is a special case of self-excited instability where some of the normally constant parameters vary, influencing rotor motion.
- Torsional vibrations are similar to lateral vibrations but occur in the planes perpendicular to the shaft axis.

## 9.3 Rotordynamic Analysis

Rotordynamic analysis is a part of current design procedures for rotating machinery that is carried out to predict the vibration behavior during operation of the machine. Potential problems are identified and eliminated by analytical means well before the manufacturing of components is begun. Furthermore, when a machine in operation displays unusual vibration behavior, analytical means are employed to study, identify, and help resolve the problem. In order for the analysis to be useful, it must be accurate and cost-effective.

During the last 100 years, several analytical procedures have been developed to understand the vibration behavior of rotating machinery. Some of these techniques are of historical interest only, and their usefulness in practical systems is very limited. With the advent of computer technology and advanced modeling techniques, several computer-based procedures have been developed to predict the vibration behavior of rotating machinery quite precisely. Of the most commonly used procedures, two are based on the lumped-parameter model where the distributed elastic and inertial properties are represented as a collection of rigid bodies connected by massless elastic beam elements. These two procedures are the transfer matrix formulations introduced by Myklestad (1944) and Prohl (1945), and the direct stiffness matrix formulations proposed by Biezeno and Grammel (1954). Ruhl and Booker (1972) introduced the third commonly used method, based on the finite element analysis (FEA) model in which the rotor is represented as an assemblage of elements with distributed elastic and inertial properties. Several of the well-known procedures are discussed next, some of them for historical significance.

## 9.3.1 Analysis Methods

### 9.3.1.1 Rankine's Numerical Method

Rankine (1869) proposed that, for a shaft of a given length, diameter, and material, there is a limit of speed, and for a shaft of a given diameter and material, turning at a given speed, there is a limit of length, below which centrifugal whirling is impossible. The limits of length and speed depend on the way the shaft is supported. The critical speed of the shaft is given by the following equation:

$$\omega = \frac{k(Hg)^{1/2}}{b^2} \tag{9.15}$$

where

$\omega$ = critical speed in rad/sec
$k$ = radius of gyration of the cross section of the shaft
$g$ = acceleration due to gravity
$H$ = modulus of elasticity expressed in units of height of itself ($H = E/\rho$)
$E$ = Young's modulus
$\rho$ = density of the material
$l$ = shaft length
$b = l/\pi$ for a simply supported shaft
$b = l/0.595\pi$ for an overhanging shaft

### 9.3.1.2 Greenhill's Formulae

Greenhill (1883) introduced the following differential equation of motion for a uniform shaft slightly deformed from straightness by centrifugal whirling:

$$\frac{d^4y}{dx^4} - \frac{m\omega^2}{gEak^2}y = 0 \tag{9.16}$$

The general solution to Equation 9.16 is given by

$$y = B \cosh \mu x + A \cos \mu x \tag{9.17}$$

where

$\mu^4 = m\omega^2/gEak^2$
$m$ = weight of the shaft per unit length
$\omega$ = rotational speed
$a$ = cross-sectional area of the shaft

The constants $A$ and $B$ depend on the boundary conditions at the support locations.

### 9.3.1.3 Reynolds' Equations

Reynolds extended the differential equation of motion for a uniform rotating shaft (Equation 9.16) to include shafts loaded with pulleys (disks) and for multispan rotors (Dunkerly, 1894).

At a bearing support, the difference in shear force must equal the bearing load $P$:

$$\frac{dM_r}{dx} - \frac{dM_l}{dx} = P \tag{9.18}$$

where $M_r$ and $M_l$ are bending moments in the right (r) and left (l) sides of the load.

At a load consisting of a revolving weight, $W$, the above equation becomes

$$\frac{dM_r}{dx} - \frac{dM_l}{dx} = \frac{W}{g}\omega^2 y \tag{9.19}$$

A further equation may be obtained by considering the *centrifugal couple* (gyroscopic moment) as given by

$$M_r - M_l = \omega^2 I' \frac{dy}{dx} \tag{9.20}$$

where $I'$ = moment of inertia of the pulley.

The solution to Equation 9.20 is given by

$$y = A \cosh mx + B \sinh mx + C \cos mx + D \sin mx \tag{9.21}$$

The values of $A$, $B$, $C$, and $D$ will depend on the boundary conditions between any two singular points.

### 9.3.1.4  Dunkerley Method

When considering the effects of the pulleys and the shaft together, the formulae derived by Reynolds were found to be limited for practical purposes. Dunkerly (1894) proposed an empirical method to consider the effects of the shaft and each of the pulleys separately, and then combine them using the following formula to obtain the critical speed of the rotor:

$$\frac{1}{\omega_c^2} = \frac{1}{\omega_s^2} + \sum_{i=1}^{n} \frac{1}{\omega_i^2} \tag{9.22}$$

where

$\omega_c$ = critical speed of the rotor
$\omega_s$ = critical speed of the shaft alone
$\omega_i$ = critical speed of the $i$th disk on a weightless shaft

In the case of the unloaded shaft, the critical speed, $\omega_s$, is given by the following formula

$$\left(\frac{m\omega_s^2}{gEI}\right)^{1/4} l = a \tag{9.23}$$

where

$I$ = sectional inertia of shaft
$l$ = length of the span
$a$ = a coefficient dependent on the manner of support of the shaft

The critical speed of the rotor $\omega_i$ with a single disk of weight $W_i$ on a weightless shaft is given by

$$\omega_i = \theta\left(\frac{gEI}{W_i c^3}\right)^{1/2} \tag{9.24}$$

where

$c$ = distance of disk from nearest support
$\theta$ = a coefficient dependant on the manner in which the shaft is supported, the position of the disk within the span and the dimensions of the disk

### 9.3.1.5 Rayleigh Method

The Rayleigh method is based on the premise that, when a system vibrates at its natural frequency, the maximum potential energy stored in the elastic components is equal to the maximum kinetic energy stored in the masses (Rayleigh, 1945).

The first natural frequency of a vibrating uniform beam is given by the following equation:

$$\omega^2 = \frac{EI \int_0^l \left( \frac{d^2y}{dx^2} \right)^2 dx}{\mu \int_0^l y^2 \, dx} \tag{9.25}$$

The Rayleigh formula for a lumped mass system is

$$\omega^2 = \frac{\sum_{i=1}^n m_i y_i}{\sum_{i=1}^n m_i y_i^2} \tag{9.26}$$

where

$\omega$ = first natural frequency
$y$ = deflection of the beam
$x$ = distance along $x$-axis
$l$ = length of the beam
$m_i$ = $i$th lumped mass
$y_i$ = static deflection of $i$th mass

The accuracy of the Rayleigh method depends upon the selection of a suitable deflection curve that approximates the fundamental mode shape. If the assumed curve represents the true mode shape, then the correct fundamental natural frequency will result. All deviations from the true mode shape will yield frequencies that are higher than the correct value.

### 9.3.1.6 Ritz Method

The Ritz method is an improvement on the Rayleigh method (Timoshenko et al., 1974) where the mode shape is represented by several orthogonal functions with unknown coefficients that satisfy the boundary conditions. The orthogonal functions are represented by a series of functions, $\Phi_i(x)$, where $i$ varies from 1 to $n$. The mode shape is represented by the following expression:

$$y = \sum_{i=1}^n a_i \Phi_i(x) \tag{9.27}$$

In order for the coefficients $a_i$ in the above equation to yield minimum values when substituted in the energy balance equation proposed by Rayleigh, the following expression needs to be satisfied:

$$\frac{\partial}{\partial a_i} \frac{\int_0^l \left( \frac{d^2y}{dx^2} \right)^2 dx}{\int_0^l y^2 \, dx} = 0 \tag{9.28}$$

From Equation 9.25 and Equation 9.28, we find

$$\frac{\partial}{\partial a_i} \int_0^l \left[ \left( \frac{d^2y}{dx^2} \right)^2 - \frac{\omega^2 \mu}{EI} y^2 \right] dx = 0 \tag{9.29}$$

Substituting Equation 9.27 for $y$ in Equation 9.29 and performing the mathematical operations, a system of linear equations in $a_i$ is obtained. The number of such equations will be equal to $n$. These equations will yield solutions different from zero only if the determinant of the coefficients of $a_i$ is equal to zero. This condition yields the frequency equation, from which the frequency of each mode can be derived.

### 9.3.1.7 Stodola–Vianello Method

The Stodola–Vianello method is a numerical iterative process (Timoshenko et al., 1974) that can be used to calculate the natural frequencies and mode shapes of vibrating systems. An approximate mode shape is first assumed and by successive iterations it is refined until convergence is obtained to the desired level of accuracy. This method can be used to refine the assumed mode shape when using the Rayleigh formulae or in the more general case of the matrix iteration process illustrated below.

Using Newton's Second law, the equations of motion for a multi-DoF system in matrix notation are

$$\{Y\} = \omega_i^2 [A][m]\{Y\} \tag{9.30}$$

$$[A] = [K]^{-1} \tag{9.31}$$

where

[A] is the flexibility matrix
[m] is the mass matrix
[K] is the stiffness matrix

To start the iterative process a trial vector, $\{Y\}_1$, representing the mode shape is substituted to both sides of Equation 9.30 and solve for the natural frequency, $\omega_i$. For this reason, let the product of $[A]$, $[m]$, and $\{Y\}_1$ be $\{Y\}_2'$. The first approximation for $\omega_i$ may be obtained by dividing any one of the elements on $\{Y\}_1$ by $\{Y\}_2'$ (Note that, if $\{Y\}_1$ was the true mode shape, then the ratio for all such elements will be equal.) The vector $\{Y\}_2'$ is then normalized by dividing all the elements by the first element to produce $\{Y\}_2$. The vector $\{Y\}_2$ is premultiplied by $[m]$ and $[A]$ to produce $\{Y\}_3'$. Once again the ratio of corresponding elements of $\{Y\}_3'$ and $\{Y\}_2$ are compared for equality.

This procedure is repeated until the mode shape and the associated frequency is determined to the desired level of accuracy. In the above iteration procedure, the mode shape converges to the one corresponding to the lowest natural frequency. If the stiffness matrix had been used instead of the flexibility matrix, then convergence at the highest natural frequency is obtained. After the first mode of vibration is determined, it is removed from the system matrices by the use of a sweeping matrix so that higher modes can be obtained. This procedure is repeated until all the desired mode shapes and natural frequencies are determined.

### 9.3.1.8 Myklestad–Prohl Method (Transfer Matrix Method)

The Myklestad–Prohl transfer matrix formulation (Myklestad, 1944; Prohl, 1945) is commonly used to analyze lumped parameter models of rotating machinery. The distributed elastic and inertial properties of the rotor are represented as a collection of rigid bodies connected by massless elastic beam elements as illustrated in Figure 9.10. This method is best suited to calculate critical speeds and mode shapes of rotors neglecting the effects of viscous damping. The Myklestad–Prohl procedure can also be adopted to perform synchronous response and stability analysis, including for the effects of damping.

In order to demonstrate the transfer matrix procedure, an axisymmetric rotor is analyzed to determine its undamped critical speeds and mode shapes. Refering to Figure 9.10, the rotor is divided into $n$ nodes, and each node is connected to the adjacent node by a massless elastic beam with uniform cross-sectional properties. The mass of components such as disks, impellers, and so on, together with the mass of the adjacent portion of the shaft, is lumped at the nodes. The Myklestad–Prohl method is based on the solution of the Bernoulli–Euler equation and the variables of interest are displacement ($y$), slope ($\theta$), moment ($M$), and shear ($V$). The development of the following procedure follows Childs (1993).

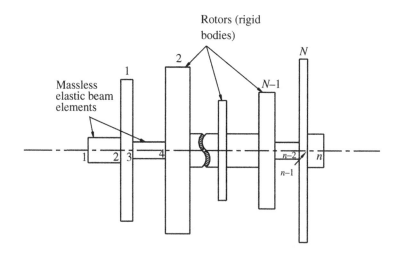

**FIGURE 9.10**  Lumped-parameter model of rotor.

At a typical nodal point ($n$), the variables on the left-hand side (l) are related to the variables on the right-hand side (r) by the following relationship:

$$\begin{Bmatrix} y_n^r \\ \theta_n^r \\ M_n^r \\ V_n^r \end{Bmatrix} = \begin{bmatrix} 1 & 0 & 0 & 0 \\ 0 & 1 & 0 & 0 \\ 0 & -J_n\omega^2 & 1 & 0 \\ -m_n\omega^2 & 0 & 0 & 1 \end{bmatrix} \begin{Bmatrix} y_n^l \\ \theta_n^l \\ M_n^l \\ V_n^l \end{Bmatrix} \qquad (9.32)$$

For the purpose of abbreviation:

$$(Q)_n^T = (y_n \quad \theta_n \quad M_n \quad V_n) \qquad (9.33)$$

and Equation 9.32 can be written in a more compact form as follows:

$$(Q)_n^r = [T_{mn}](Q)_n^l \qquad (9.34)$$

where $[T_{mn}]$ represents the transfer mass matrix at node $n$.

At a massless beam section, connecting node $n$ to node $n+1$ the transfer matrix is given by

$$\begin{Bmatrix} y_{n+1}^l \\ \theta_{n+1}^l \\ M_{n+1}^l \\ V_{n+1}^l \end{Bmatrix} = \begin{bmatrix} 1 & l_n & \dfrac{l_n^2}{2EI_n} & \dfrac{-l_n^3}{6EI_n} \\ 0 & 1 & \dfrac{l_n}{EI_n} & \dfrac{l_n^2}{2EI_n} \\ 0 & 0 & 1 & -l_n \\ 0 & 0 & 0 & 1 \end{bmatrix} \begin{Bmatrix} y_n^r \\ \theta_n^r \\ M_n^r \\ V_n^r \end{Bmatrix} \qquad (9.35)$$

Equation 9.35 may be written in a more abbreviated form as

$$(Q)_{n+1}^l = [T_{bn}](Q)_n^r \qquad (9.36)$$

where $[T_{bn}]$ represents the beam element transfer matrix connecting node $n$ to node $n+1$.

From Equation 9.34 and Equation 9.36, we obtain the combined transfer matrix for nodes $n$ and $n+1$:

$$(Q)_{n+1}^l = [T_{bn}][T_{mn}](Q)_n^l = [T_n](Q)_n^l \qquad (9.37)$$

Starting with node one, successive matrix multiplications are carried out until node $n + 1$ is reached. The last node $(n + 1)$ is a dummy node with the beam length, $l$, equal to zero, and the mass and inertias also equal to zero. This makes the nodal parameters on the left-hand side of node $n + 1$ equal to those on the right-hand side of node $n$. The result is as follows:

$$(Q)_n^r = [T_n][T_{n-1}]\cdots[T_1](Q)_1^l \quad \text{or} \quad (Q)_n^r = [T](Q)_1^l \tag{9.38}$$

The matrix $[T]$ is a function of the rotational speed, $\omega$. The Myklestad–Prohl method uses a trial and error solution to determine the values of $\omega$ which satisfy the boundary conditions and Equation 9.38 simultaneously. It is not necessary to store and multiply all the matrices together. The transfer matrix procedure is used to proceed from one end to the other without having to store all the nodal matrices. In all cases, two boundary conditions each are known at the two ends of the shaft, and the frequencies that satisfy these boundary conditions are the critical speeds of the rotor. Once the critical speeds are calculated, the corresponding mode shapes can also be determined using the transfer matrix procedure. It should be noted that other types of elements, such as elastic supports, flexible couplings, and so on, could also be introduced very conveniently.

### 9.3.1.9   Direct Stiffness Method

The direct stiffness method uses a lumped-parameter formulation to evaluate the dynamic characteristics of a flexible rotor. The general differential equation of motion that characterizes its behavior (less the damping and gyroscopic forces) is as follows:

$$\begin{bmatrix} [m] & 0 \\ 0 & [J] \end{bmatrix} \begin{Bmatrix} (\ddot{Y}) \\ (\ddot{\theta}) \end{Bmatrix} + [K] \begin{Bmatrix} (Y) \\ (\theta) \end{Bmatrix} = \begin{Bmatrix} (F) \\ (T) \end{Bmatrix} \tag{9.39}$$

where $[m]$ and $[J]$ are diagonal matrices which contains the nodal masses, $m_i$, and nodal moments of inertia, $J_i$, respectively. The stiffness matrix, $[K]$, contains the internal stiffness terms of the beam elements as well as any external spring stiffness at the supports. The vectors $(F)$ and $(T)$ represent external forces and moments acting on the system, respectively.

The stiffness matrix for a typical beam element based on the Bernoulli–Euler equations is as follows (Childs, 1993):

$$[K^i] = \frac{2EI_i}{l_i^3} \begin{bmatrix} 6 & 3l_i & -6 & 3l_i \\ 3l_i & 2l_i^2 & -3l_i & l_i^2 \\ -6 & -3l_i & 6 & -3l_i \\ 3l_i & l_i^2 & -3l_i & 2l_i^2 \end{bmatrix} \tag{9.40}$$

The overall stiffness matrix, $[K]$, has to be assembled by combining the individual component matrices in a systematic manner. The following procedure illustrates the process.

The stiffness matrix of the $i$th beam element in matrix notation is

$$[K^i] = [k_{j,k}^i] \tag{9.41}$$

where $j$ and $k$ vary from $(2i - 1)$ to $(2i + 2)$.

To form the overall stiffness matrix, the elements with the same subscripts of adjacent beam elements are added over $n$ beam elements as given by the following equation:

$$[K] = [K_{j,k}] = \sum_{i=1}^{n} \sum_{j=2i-1}^{2i+2} \sum_{k=2i-1}^{2i+2} k_{j,k}^i \tag{9.42}$$

Once the inertia matrix and the stiffness matrix for the entire system are assembled, the eigenvalues and eigenvectors can be evaluated by solving the following homogeneous equation derived from Equation 9.39:

$$[M](\ddot{Y}) + [K](Y) = 0 \tag{9.43}$$

There are numerous analysis procedures (Meirovitch, 1986) for the solution of Equation 9.43 that yield the eigenvalues and eigenvectors of the system. The method of choice will depend on the complexity and nature of the inertia and stiffness matrices. Perhaps the most widely known is the matrix iteration using the power method in conjunction with the sweeping technique. However, this method is not necessarily the most efficient, particularly for higher-order systems. The Jacobi's method, which uses matrix iteration to diagonalize a matrix by successive rotations, is more commonly used owing to its higher efficiency. Details of these techniques are given in the text by Meirovitch (1986).

When the damping matrix and the gyroscopic matrix is also included in Equation 9.39, the direct stiffness method can be used to calculate damped critical speeds, forced rotor response, and instability of the rotor in addition to the eigenvalues using similar methods of solution.

### 9.3.1.10  The Finite Element Analysis Method

The basis of the FEA method is to provide formulation for complex and irregular systems that can utilize the automation capabilities of computers (also see Chapter 9). The FEA method considers a rotordynamic system as an assemblage of discreet elements, where every such element has distributed and continuous properties, namely, the consistent representation of both mass and stiffness as distributed parameters. As illustrated in Section 9.3.1.9, the lumped-parameter method uses a consistent stiffness matrix equation (Equation 9.40) in its formulation, and therefore, the identical procedure can be adopted for the finite element method as well. For the distributed mass representation of an element, Archer (1963) procedure, which is based on the assumption that the mass distribution is proportional to the elastic distribution similar to the Rayleigh–Ritz formulation, is utilized. The resulting mass matrix is as follows:

$$[m^i] = \frac{m^i l_i}{420} \begin{bmatrix} 156 & 22l_i & 54 & -13l_i \\ 22l_i & 4l_i^2 & 13l_i & -3l_i^2 \\ 54 & 13l_i & 156 & -22l_i \\ -13l_i & -3l_i^2 & -22l_i & 4l_i^2 \end{bmatrix} \tag{9.44}$$

The overall stiffness matrix, $[K]$, for the entire system is assembled by combining the individual component matrices in a systematic manner according to Equation 9.41 and Equation 9.42. The overall mass matrix can also be assembled in precisely the same manner, as given by Equation 9.45 and Equation 9.46.

The mass matrix of the $i$th beam element in matrix notation can be represented as

$$[m^i] = [m^i_{j,k}] \tag{9.45}$$

where $j$ and $k$ varies from $(2i - 1)$ to $(2i + 2)$:

$$[M] = [M_{j,k}] = \sum_{i=1}^{n} \sum_{j=2i-1}^{2i+2} \sum_{k=2i-1}^{2i+2} m^i_{j,k} \tag{9.46}$$

Once the mass matrix and the stiffness matrix for the entire system are assembled, Equation 9.43 that describes the free vibration of the complete system can be solved. The solution methods of the eigenvalue problem, which can be utilized, are the same as those used for the direct stiffness method illustrated in Section 9.3.1.9 above. Details of the FEA methods are given in Ruhl and Booker (1972).

### 9.3.1.11 Torsional Analysis (Holzer Method)

The development of torsional analysis methods have gone through a similar evolutionary process to lateral vibration methods. Holzer (1921) first introduced the lumped-parameter numerical method to calculate torsional natural frequencies of a multi-DoF system. Even to-date, this is the most commonly used method because of its simplicity and reasonable degree of accuracy. The Holzer method is a transfer matrix formulation that uses a lumped parameter model similar to that used in the Myklestad–Phrol method described in Section 9.3.1.8. The only difference is that the transfer matrices represented by Equation 9.32 and Equation 9.35 are replaced by the equations

$$\left\{ \begin{array}{c} \theta \\ T \end{array} \right\}_n^r = \left[ \begin{array}{cc} 1 & 0 \\ -\omega^2 J & 1 \end{array} \right]_n \left\{ \begin{array}{c} \theta \\ T \end{array} \right\}_n^l \tag{9.47}$$

$$\left\{ \begin{array}{c} \theta \\ T \end{array} \right\}_{n+1}^l = \left[ \begin{array}{cc} 1 & \dfrac{1}{k} \\ 0 & 1 \end{array} \right]_{n+1} \left\{ \begin{array}{c} \theta \\ T \end{array} \right\}_n^r \tag{9.48}$$

Starting with node one, successive matrix multiplications are carried out until node $n + 1$ is reached. The result can be represented by Equation 9.38. The matrix $[T]$ is a function of the rotational speed, $\omega$. In all cases, one boundary condition at each end of the rotor is known. A trial-and-error solution to determine the values of $\omega$ which satisfy the boundary conditions and Equation 9.38 are simultaneously determined. These values are the torsional critical speeds of the rotor. Once the critical speeds are calculated, the corresponding torsional mode shapes can also be determined using the transfer matrix procedure.

In the case of branched systems and geared systems, particular attention has to be paid to the relative rotational speeds of the components. The rule is quite simple: multiply all stiffness and inertias of the geared shaft by $N^2$, where $N$ is the speed ratio of the geared shaft to the reference shaft.

Other methods such as the distributed mass matrix method, direct stiffness method, and finite element method can also be used to determine torsional critical speeds of rotors. These procedures are very similar to those for lateral critical speed analysis.

## 9.3.2 Modeling

The design and analysis of rotordynamic systems require the development of models that simulate the behavior of the physical system. In the past, the critical speed of the rotor was considered to be the main criterion for stable operation. Today, stable, well-damped rotordynamic response to the exciting forces within a machine is considered to be a necessary condition for high reliability. The accuracy and reliability of the results greatly depends on the credibility of the system model and its adaptability to the analytical procedure. Even the most accurate and efficient analytical method cannot produce good results from a bad model. The methods that are commonly used to model shaft sections and disks and other such elements attached to shafts have been discussed in the previous sections. Useful formulae for calculating critical speeds of simple systems are given in Table 9.2. Models to represent bearings, rotor dampers, seals, and rotor–stator interactions are discussed in the following sections.

### 9.3.2.1 Journal Bearings

Journal bearings were used in rotating machinery for a long time before their dynamic characteristics were fully understood. Considerable effort has been expended in the last few decades to understand and develop techniques for their accurate representation in rotordynamic analysis. A variety of bearing types with improved characteristics have been developed over the years. Figure 9.11 shows the most commonly used types in rotating machinery. Hagg and Sankey (1958) were amongst the first to provide dynamic stiffness and damping coefficients for a number of these bearing types. However, these coefficients are considered incomplete as cross-coupling terms were not considered. Soon after, there was a flurry of activity related to the analysis of journal bearings; Sternlicht (1959), Warner (1963),

**TABLE 9.2**   Useful Formulas in Vibration Analysis and Design

| | | |
|---|---|---|
| Rankine formula | $\omega = \dfrac{k(Hg)^{1/2}}{b^2}$ | *Note*: This formula is of historical interest only and has limited practical value |
| Greenhill formula | $\dfrac{d^4y}{dx^4} - \dfrac{m\omega^2}{gEak^2}y = 0$ | |
| Dunkerly equation | $\dfrac{1}{\omega_c^2} = \dfrac{1}{\omega_s^2} + \displaystyle\sum_{i=1}^{n}\dfrac{1}{\omega_i^2}$ | |

The above equation reduces to

$$\omega_c^2 = \frac{g}{\sum y_{stat}}$$

Formulas for natural
frequency calculation
(Blevins, 2001;
Gorman, 1975)

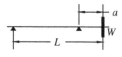

$$\omega_c = \frac{1}{ab}\sqrt{\frac{3EIL}{W}}$$

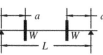

$$\omega_c = \frac{1}{a}\sqrt{\frac{6EI}{W(3L - 4a)}}$$

$$\omega_c = \frac{1}{a}\sqrt{\frac{3EI}{WL}}$$

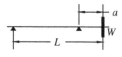

$$\omega_c = \left(\frac{12EIL^3}{Wa^3b^2(3L + b)}\right)^{1/2}$$

$$\omega_c = \left(\frac{3L^3EI}{Wa^3b^3}\right)^{1/2}$$

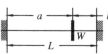

$$\omega_c = \left(\frac{3EI}{WL^3}\right)^{1/2}$$

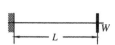

$$\omega_c = \left(\frac{98EI}{mL^4}\right)^{1/2}$$

$$\omega_c = \left(\left(\frac{64a}{9L}\right)^2 - \left(\frac{6a}{L}\right)\right.$$
$$\left. + \left(\frac{16}{3}\right)^2\left(\frac{EI}{mL^4}\right)\right)^{1/2},$$
$$\frac{a}{L} \geq 0.25$$

(*continued on next page*)

**TABLE 9.2**    (*continued*)

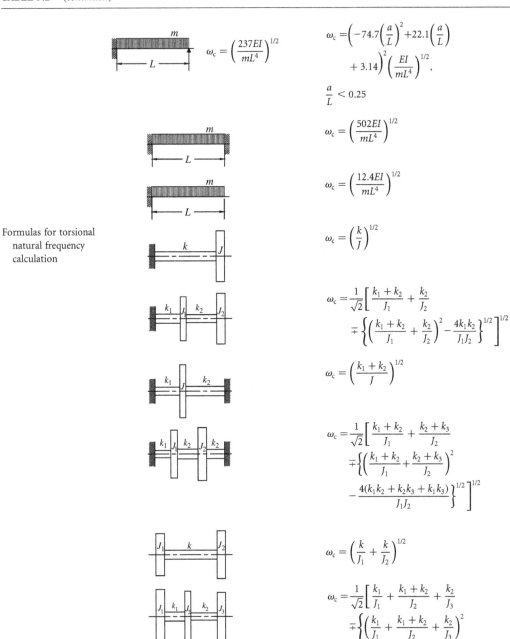

$$\omega_c = \left(\frac{237EI}{mL^4}\right)^{1/2}$$

$$\omega_c = \left(-74.7\left(\frac{a}{L}\right)^2 + 22.1\left(\frac{a}{L}\right) + 3.14\right)^2 \left(\frac{EI}{mL^4}\right)^{1/2},$$

$$\frac{a}{L} < 0.25$$

$$\omega_c = \left(\frac{502EI}{mL^4}\right)^{1/2}$$

$$\omega_c = \left(\frac{12.4EI}{mL^4}\right)^{1/2}$$

Formulas for torsional
natural frequency
calculation

$$\omega_c = \left(\frac{k}{J}\right)^{1/2}$$

$$\omega_c = \frac{1}{\sqrt{2}}\left[\frac{k_1 + k_2}{J_1} + \frac{k_2}{J_2} \right.$$
$$\left. \mp \left\{\left(\frac{k_1 + k_2}{J_1} + \frac{k_2}{J_2}\right)^2 - \frac{4k_1 k_2}{J_1 J_2}\right\}^{1/2}\right]^{1/2}$$

$$\omega_c = \left(\frac{k_1 + k_2}{J}\right)^{1/2}$$

$$\omega_c = \frac{1}{\sqrt{2}}\left[\frac{k_1 + k_2}{J_1} + \frac{k_2 + k_3}{J_2} \right.$$
$$\mp \left\{\left(\frac{k_1 + k_2}{J_1} + \frac{k_2 + k_3}{J_2}\right)^2 \right.$$
$$\left. \left. - \frac{4(k_1 k_2 + k_2 k_3 + k_1 k_3)}{J_1 J_2}\right\}^{1/2}\right]^{1/2}$$

$$\omega_c = \left(\frac{k}{J_1} + \frac{k}{J_2}\right)^{1/2}$$

$$\omega_c = \frac{1}{\sqrt{2}}\left[\frac{k_1}{J_1} + \frac{k_1 + k_2}{J_2} + \frac{k_2}{J_3} \right.$$
$$\mp \left\{\left(\frac{k_1}{J_1} + \frac{k_1 + k_2}{J_2} + \frac{k_2}{J_3}\right)^2 \right.$$
$$\left. \left. - \frac{4k_1 k_2(J_1 + J_2 + J_3)}{J_1 J_2 J_3}\right\}^{1/2}\right]^{1/2}$$

Rayleigh equations

$$\omega^2 = \frac{EI\int_0^l \left(\dfrac{\mathrm{d}^2 y}{\mathrm{d}x^2}\right)^2 \mathrm{d}x}{\mu\int_0^l y^2\, \mathrm{d}x}$$

**TABLE 9.2** *(continued)*

$$\omega^2 = \frac{\sum\limits_{i=1}^{n} m_i y_i}{\sum\limits_{i=1}^{n} m_i y_i^2}$$

Ritz method

$$y = \sum_{i=1}^{n} a_i \Phi_i(x)$$

$$\frac{\partial}{\partial a_i} \frac{\int_0^l \left(\dfrac{d^2 y}{dx^2}\right)^2 dx}{\int_0^l y^2 \, dx} = 0$$

$$\frac{\partial}{\partial a_i} \int_0^l \left[\left(\frac{d^2 y}{dx^2}\right)^2 - \frac{\omega^2 \mu}{EI} y^2\right] dx = 0$$

Stodola–Vianello method

$$\{Y\} = \omega_i^2 [A][m]\{Y\}, \ [A] = [K]^{-1}$$

Transfer matrix —
Myklestad–Phrol
method

$$\begin{Bmatrix} y_{n+1}^l \\ \theta_{n+1}^l \\ M_{n+1}^l \\ V_{n+1}^l \end{Bmatrix} = \begin{bmatrix} 1 & l_n & \dfrac{l_n^2}{2EI_n} & \dfrac{-l_n^3}{6EI_n} \\ 0 & 1 & \dfrac{l_n}{EI_n} & \dfrac{l_n^2}{2EI_n} \\ 0 & 0 & 1 & -l_n \\ 0 & 0 & 0 & 1 \end{bmatrix} \begin{Bmatrix} y_n^r \\ \theta_n^r \\ M_n^r \\ V_n^r \end{Bmatrix}$$

Stiffness matrix for a
beam element

$$[K^i] = \frac{2EI_i}{l_i^3} \begin{bmatrix} 6 & 3l_i & -6 & 3l_i \\ 3l_i & 2l_i^2 & -3l_i & l_i^2 \\ -6 & -3l_i & 6 & -3l_i \\ 3l_i & l_i^2 & -3l_i & 2l_i^2 \end{bmatrix}$$

Mass matrix for a beam
element

$$[m^i] = \frac{m^i l_i}{420} \begin{bmatrix} 156 & 22l_i & 54 & -13l_i \\ 22l_i & 4l_i^2 & 13l_i & -3l_i^2 \\ 54 & 13l_i & 156 & -22l_i \\ -13l_i & -3l_i^2 & -22l_i & 4l_i^2 \end{bmatrix}$$

Squeeze-film damper
coefficients

$$k = \frac{24R^3 L\mu\omega\varepsilon}{C_r^3(2+\varepsilon^2)(1-\varepsilon^2)}$$

$$c = \frac{12\pi R^3 L\mu}{C_r^3(2+\varepsilon^2)(1-\varepsilon^2)^{1/2}}$$

$$k = \frac{2RL^3 \mu\omega\varepsilon}{C_r^3(1-\varepsilon^2)^2}$$

$$c = \frac{\pi RL^3 \mu}{2C_r^3(1-\varepsilon^2)^{\frac{3}{2}}}$$

Unbalance sensitivity

$$\text{SF} = \frac{a}{U} M$$

Rolling element bearing
defect frequencies

$$f_{\text{bor}} = \frac{ND}{60d}\left[1 - \left(\frac{d}{D}\cos\theta\right)^2\right]$$

$N$ = rotational speed (rpm),
$D$ = rolling element pitch diameter

*(continued on next page)*

**TABLE 9.2**    (*continued*)

$$f_{\text{ir}} = \frac{Nn}{120}\left(1 + \frac{d}{D}\cos\theta\right)$$

$d$ = rolling element diameter,
$N$ = number of rolling elements

$$f_{\text{or}} = \frac{Nn}{120}\left(1 - \frac{d}{D}\cos\theta\right)$$

$\theta$ = contact angle with respect to axis, bor = ball or roller defect

$$f_c = \frac{N}{120}\left(1 - \frac{d}{D}\cos\theta\right)$$

ir = inner race defect,
or = outer race defect,
$c$ = cage defect

Lomakin formula for radial stiffness for a close clearance bushing

$$k = \frac{\pi}{8}(1 + \varsigma)\lambda\mu^4\left(\frac{l}{b_m}\right)^2 \Delta pD$$

$$\mu^2 = \frac{1}{1 + \varsigma + (\lambda l/2b_m)}$$

$b_m$ = radial clearance,
$l$ = length of bushing,

$\varsigma$ = inlet loss coefficient,

$D$ = diameter,
$\Delta p$ = differential pressure across bushing,
$\lambda$ = friction coefficient

Lund (1964), Lund (1965), Glienicke (1966), Orcutt (1967), Lund (1968), Someya et al. (1988), and several others provided complete bearing coefficients, including cross-coupling terms, for several bearing types. This information is considered to be a valuable resource for those engaged in rotordynamic analysis. The general form of the rotordynamic model for a journal bearing resulting from the above contributions is given by the following equation:

$$\begin{Bmatrix} F_X \\ F_Y \end{Bmatrix} = -\begin{bmatrix} k_{11} & k_{12} \\ k_{21} & k_{22} \end{bmatrix}\begin{Bmatrix} X \\ Y \end{Bmatrix} - \begin{bmatrix} c_{11} & c_{12} \\ c_{21} & c_{22} \end{bmatrix}\begin{Bmatrix} \dot{X} \\ \dot{Y} \end{Bmatrix} \tag{9.49}$$

Since the dawn of the digital computer era, several computer codes have been developed to analyze all aspects of journal bearings, including stiffness and damping coefficients. Many of these codes have been developed by equipment manufactures and research centers for their exclusive use. Several commercially available software codes popularized in North America are given in Table 9.3. Although bearing coefficients given in the form of charts and tables from the earlier studies are still in use, computer-based codes are growing in popularity.

### 9.3.2.2  Rolling Element Bearings

Rolling element bearings are used in numerous types of rotating machinery which are required to be compact, manage high loads, and have low heat rejection and simple lubrication systems. Unlike journal bearings, their load-carrying capacity is not speed-dependent and as a result is capable of full load capacity down to zero speed. Some of these salient features make rolling element bearings very attractive to many industries.

From a rotordynamic standpoint, rolling element bearings are modeled as linear spring elements with direct spring coefficients only. The damping terms are insignificant and as a result do not attenuate rotor deflections at critical speeds. A typical rolling element bearing is represented by the following equation:

$$\begin{Bmatrix} F_X \\ F_Y \end{Bmatrix} = -\begin{bmatrix} k & 0 \\ 0 & k \end{bmatrix}\begin{Bmatrix} X \\ Y \end{Bmatrix} \tag{9.50}$$

The absence of cross-coupling stiffness and damping terms signifies that bearing induced rotor instability will not occur. Although, for convenience of analysis, the spring stiffness is considered linear, its

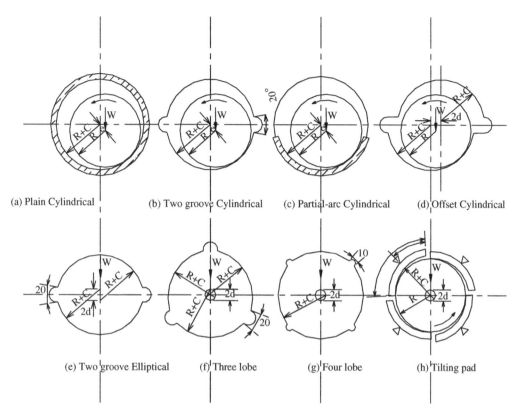

**FIGURE 9.11** Common types of journal bearings.

true behavior can be quite the opposite, leading to some calculation inaccuracies. The nonlinearities are most significant where the bearings have no preload and some internal clearance in the bearings exists. Preloaded bearings with little or no internal clearance behave quite linearly. Jones (1960), Harris (1991), and Kramer (1993) have analyzed the bearing stiffness coefficients for the common types of rolling element bearings, and this data can be utilized for rotordynamic study of rotating machinery.

### 9.3.2.3 Squeeze-Film Dampers

Squeeze-film dampers are used to introduce damping capacity to a rolling element bearing or, in the case of journal bearings, to provide additional damping and stiffness to eliminate rotor instability problems. Squeeze-film dampers have come into prominence through the modern aircraft gas turbine industry where the bearing of choice is the rolling element bearing. In the mid-1970s, several designs were introduced to add damping capacity and predictable stiffness to the rolling element bearings. In its basic form, a squeeze-film damper is very similar to a nonrotating cylindrical journal bearing where the outer race of the rolling element bearing forms the journal as illustrated in Figure 9.12. The addition of end seals (to control leakage) and centering springs are modifications that have been introduced to enhance its performance. The interactive force at a bearing using a squeeze-film damper can be represented by the following equation:

$$\begin{Bmatrix} F_X \\ F_Y \end{Bmatrix} = -\begin{bmatrix} k & 0 \\ 0 & k \end{bmatrix}\begin{Bmatrix} X \\ Y \end{Bmatrix} - \begin{bmatrix} c & 0 \\ 0 & c \end{bmatrix}\begin{Bmatrix} \dot{X} \\ \dot{Y} \end{Bmatrix} \tag{9.51}$$

The stiffness and damping coefficients for the squeeze-film dampers have been derived (Ehrich, 1999), from the solution of the Reynolds' equation for the case of a nonrotating journal bearing. For dampers with end seals, the long journal bearing theory is used to generate the following stiffness and damping

**TABLE 9.3** Rotordynamic Analysis Software

| Name of Software | Type of Analysis | Supplier |
|---|---|---|
| CAD20 | Lateral critical speeds of flexible rotors | CADENSE Programs, |
| CAD21 | Unbalance response of flexible rotors | Foster Miller Technologies |
| CAD21a | Response of flexible rotors to nonsynchronous sinusoidal excitation | Inc., Albany, NY, USA |
| CAD22 | Torsional critical speeds and response of geared systems | |
| CAD24 | Transient torsional critical speeds of geared system | |
| CAD25 | Dynamic stability of flexible rotors | |
| CAD25a | Transient response of flexible rotors | |
| CAD26 | Lateral critical speeds of multilevel rotors | |
| CAD27 | Unbalance response of multilevel rotors | |
| CAD30 | Dynamic coefficients of liquid lubricated journal bearings | |
| CAD30a | Dynamic coefficients of ball bearings | |
| CAD31 | Dynamic coefficients of liquid lubricated tilting pad journal bearings | |
| CAD32 | Dynamic coefficients of liquid lubricated axial-groove and single pad journal bearings | |
| CAD34a | Performance of tilting pad thrust bearings | |
| CAD34b | Performance of tapered-land thrust bearings | |
| CAD36 | Dynamic coefficients of liquid lubricated pressure dam journal bearings | |
| CAD38 | Dynamic coefficients of liquid lubricated deep-pocket hydrostatic journal bearings | |
| CAD40 | Dynamic coefficients of gas lubricated journal bearings | |
| CAD41 | Dynamic coefficients of gas lubricated tilting pad journal bearings | |
| CAD42 | Dynamic coefficients of gas lubricated spiral groove journal bearings | |
| CAD42i | Dynamic coefficients of liquid lubricated spiral groove journal bearings | |
| FEATURE | Rotor bearing system analysis | |
| COJOUR | Analysis of journal bearings | |
| DYNROT | A program designed to perform a complete study of the rotordynamic behavior of rotors. It is capable of linear, nonlinear and torsional analysis of rotors | Dipartimento di Meccanica, Politecnico di Torino, Torino, Italy |
| DyRoBeS | Comprehensive rotordynamic analysis software for lateral and torsional analysis, including bearing analysis of rotor-bearing systems | AGILE SOFTWARE CONCEPTS NREC White River Junction, |
| RotorLab | A software package for agile modeling of rotor systems, bearings, and seals. It combines the tasks of design, modeling, analysis, post processing, and data management into a consistent user interface | VT, USA |
| DAMBRG2 | Coefficients and rigid rotor stability information for two-lobe isoviscous bearings with a pressure dam in only one pad | ROMAC—Rotating Machinery and Controls Laboratory, University of Virginia, Charlottesville, VA, USA |
| HYDROB | Predicts the steady state and dynamic operating characteristics of hybrid journal bearings | |
| PDAM2D | This program can analyze stiffness and damping coefficients, and the rigid rotor stability threshold of multipad pressure dam bearings | |
| SQFDAMP | Determines stiffness and damping coefficients for short and long squeeze-film bearings with and without fluid film cavitation | |
| THBRG | Dynamic coefficients of multilobe journal bearings with incompressible fluid | |

**TABLE 9.3** *(continued)*

| Name of Software | Type of Analysis | Supplier |
|---|---|---|
| THPAD | Dynamic coefficients of tilting pad journal bearings with incompressible fluid | |
| THRUST | Predicts the steady-state operating characteristics of tilting-pad and fixed geometry fluid-film thrust bearings | |
| CRTSP2 | Undamped lateral critical speeds of dual-level rotor systems | |
| MODFR2 | Undamped lateral critical speeds of single or dual-level rotor systems | |
| TWIST2 | Undamped torsional critical speeds and mode shapes of rotor systems | |
| FRESP2 | Predicts the modal frequency forced response of dual rotor systems with a flexible substructure | |
| RESP2V3 | Nonplanar synchronous unbalance response of dual-level multimass flexible rotors | |
| HCOMB | Dynamic coefficients of straight-through honeycomb seals with a compressible gas | |
| LABY3 | Dynamic coefficients for straight-through and uniform interlocking type labyrinth seals with a compressible fluid | |
| SEAL2 | Stiffness and damping coefficients for plain and grooved seals with incompressible turbulent axial flow | |
| SEAL3 | Stiffness, damping and mass coefficients of both plain and circumferentially grooved seals | |
| TURSEAL | Stiffness and damping coefficients of turbulent flow annular seals or water lubricated bearings | |
| FSTB3 | Stability, damped critical speeds, and whirl mode shapes of multispool rotor systems | |
| ROTSTB | Stability, damped critical speeds, and whirl mode shapes of single spool rotor systems | |
| COTRAN | Nonlinear time transient analysis of multilevel rotors with substructure | |
| TORTRAN3 | Transient torsional rotor response | |
| hydrosealt | Stiffness and damping coefficients, and threshold speed of instability of cylindrical-pad journal bearings and pad-hydrostatic bearings of arbitrary arc lengths and preloads | Rotordynamics Laboratory, Texas A&M University, College Station, TX, USA |
| hydroflext | Stiffness and damping coefficients, and threshold speed of instability of a variety of bearing and seal types | |
| hydrotran | Predicts the transient force response of a rigid rotor supported on fluid film bearings | |
| hydrojet | Force coefficients for a variety of hybrid bearing and seal types handling process fluids | |
| hydroTRC | Stiffness and damping coefficients for a variety of bearing and seal types and for different types of fluids | |
| hseal2p | Stiffness and damping coefficients of seals that operate under two-phase flow conditions | |
| fembear | Stiffness and damping coefficients of cylindrical and fixed arc pad hydrostatic and hydrodynamic bearings for laminar and isothermal flow conditions | |
| sfdfem | Damping force coefficients of finite length squeeze-film dampers executing circular centered motion | |
| sfdflexs | Instantaneous fluid film forces for arbitrary journal motions and circular centered orbits in multiple pad integral squeeze-film dampers | |
| hsealm | Stiffness and damping coefficients of cylindrical annular pressure seals | |

*(continued on next page)*

**TABLE 9.3**   (*continued*)

| Name of Software | Type of Analysis | Supplier |
|---|---|---|
| lubsealn | Stiffness and damping coefficients of single-land and multiple-land high pressure oil seal rings and cylindrical journal bearings | |
| ROTECH | Lateral rotordynamic analysis for critical speeds; unbalance response, linear stability and nonlinear transient response of rotors. Also includes a torsional rotordynamic analysis program | ROTECH Engineering Services, Delmont, PA, USA |
| ROTOR-E | A comprehensive software package for lateral rotordynamic analysis of rotating equipment | Engineering Dynamics Inc., San Antonio, TX, USA |
| ROTORINSA | A software package devoted to the prediction of the steady-state lateral dynamic behavior of rotors | Laboratoire de Mecanique des Structures, LMST INSA Lyon, Lyon, France |
| TURBINE-PAK | A software package for rotordynamic analysis of nonlinear multibearing rotor-bearing-foundation systems | Scientific Engineering Research, Mt Best, Vic., Australia. |
| TURBINE-PAK NONLINEAR | Designed to study transient responses of rotor-bearing-foundation systems, including the loss of stability of the system | |
| XLrotor | A complete suite of analysis tools for rotating machinery dynamics. Handles both lateral and torsional analysis of rotors. Also includes codes for calculating coefficients for fluid film and antifriction bearings | Rotating Machinery Analysis Inc., Austin, TX, USA |
| XLTRC | A suite of codes for executing a complete lateral rotordynamic analysis of rotating machinery | The Turbomachinery Laboratory, Texas A&M University, College Station, TX, USA |
| XLAnSeal | Force and moment coefficients for annular turbulent seals in the laminar, turbulent, and transition flow regimes | |
| XLCGrv | Coefficients for centered grooved-stator, turbulent flow, annular pump seals | |
| XLLaby | Stiffness and damping coefficients for tooth-on-rotor or tooth-on-stator gas labyrinth seals | |
| XLIsotSL | Coefficients for smooth rotor/honeycomb stator annular seals | |
| XLLubGT | Coefficients for high-pressure oil bushing seals of compressors or smooth pump seals in the laminar flow regime | |
| XLJrnl | Stiffness and damping coefficients for fixed-arc and tilting-pad bearings | |
| HLHydPad | Stiffness and damping coefficients for hydrostatic and hybrid journal-pad bearings in the laminar flow regime | |
| XLTFPBrg | Stiffness and damping coefficients for fixed-arc, tilting-pad and flexure-pivot hydrostatic bearings | |
| XLPresDm | Stiffness and damping coefficients for multilobed, rigid-pad arc bearings with preload and pressure-dam bearings with relief tracks | |
| XLBalBrg | Stiffness coefficients for ball bearings | |
| XLLSFD | Damping and mass coefficients for locally sealed squeeze-film dampers | |
| XLOSFD | Damping and mass coefficients for open ended squeeze-film dampers | |
| XLSFDFEM | Damping coefficients for squeeze-film dampers with various types of end seals | |
| XLPIMPLR | Stiffness, damping, and mass matrices for centrifugal pump impellers | |
| XLWachel | Destabilizing cross-coupled force coefficients for impellers of centrifugal compressors | |
| XLClrEx | Destabilizing cross-coupled stiffness coefficients for unshrouded turbines | |

coefficients:

$$k = \frac{24R^3 L \mu \omega \varepsilon}{C_r^3 (2 + \varepsilon^2)(1 - \varepsilon^2)} \qquad (9.52)$$

$$c = \frac{12\pi R^3 L \mu}{C_r^3 (2 + \varepsilon^2)(1 - \varepsilon^2)^{1/2}} \qquad (9.53)$$

where

$R$ = the damper radius
$\omega$ = whirl speed
$L$ = length of damper
$\mu$ = viscosity of oil
$C_r$ = the radial clearance
$\varepsilon$ = eccentricity ratio (orbit radius/$C_r$)

Similarly, for a damper without end seals, the short journal bearing theory yields the following stiffness and damping coefficients:

$$k = \frac{2RL^3 \mu \omega \varepsilon}{C_r^3 (1 - \varepsilon^2)^2} \qquad (9.54)$$

$$c = \frac{\pi RL^3 \mu}{2C_r^3 (1 - \varepsilon^2)^{\frac{3}{2}}} \qquad (9.55)$$

Although the above equations, based on the Reynolds' equation, have been proposed to predict damper characteristics, the experimental evidence does not validate these equations. Therefore, these equations should be used with caution for practical purposes.

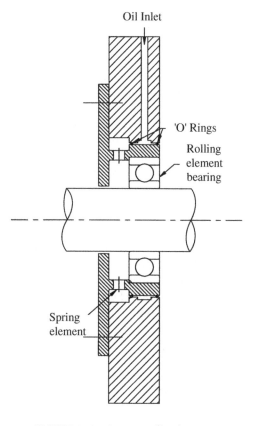

**FIGURE 9.12**  A squeeze-film damper.

### 9.3.2.4  Annular Seals

Annular seals are primarily used in pumps, compressors, gas turbines, and steam turbines to minimize leakage and thereby improve the volumetric efficiency of the machine. In addition to their basic function, they also play a vital role in the rotordynamics of the machine, especially in multistage machines, providing stiffness and damping and thereby enhancing high-speed operational capability. In fact, in the last few decades, most of the development work on seals has focused on understanding and improving their dynamic vibration characteristics rather than improving their efficiency in sealing.

Lomakin (1958) was the first to publish on the restoring forces in smooth annular clearances in pumps. However, it was more than a decade later that Black (1968) provided the major initial impetus for the understanding and development of seals. Childs (1993) provided an excellent compendium of the research work conducted in the area of seals. His book also provides the most comprehensive coverage of the subject of seal dynamics.

In the present context, seals are handled in the same manner as the stiffness and damping characteristics of journal bearings with some degree of modifications. In particular, fluid inertia effects are included, and it is assumed that the center of the shaft orbit is the same as the center of the stationary seal ring. Assuming rotational symmetry the reaction force-seal motion model can be represented by the following equation:

$$\begin{Bmatrix} F_X \\ F_Y \end{Bmatrix} = - \begin{bmatrix} k & k_c \\ -k_c & k \end{bmatrix} \begin{Bmatrix} X \\ Y \end{Bmatrix} - \begin{bmatrix} c & c_c \\ -c_c & c \end{bmatrix} \begin{Bmatrix} \dot{X} \\ \dot{Y} \end{Bmatrix} - \begin{bmatrix} m & 0 \\ 0 & m \end{bmatrix} \begin{Bmatrix} \ddot{X} \\ \ddot{Y} \end{Bmatrix} \qquad (9.56)$$

An added complexity is the predominance of turbulent flow in annular seals. This invalidates the use of Reynolds' equation for the derivation of seal coefficients. The highest degree of accuracy can be obtained by the direct solution of the Navier–Stokes and continuity equations. However, at the present moment, such methods are considered to be excessively costly and impractical. As a result, two practical semiempirical methods have been developed to derive seal coefficients. In the first approach, the semiempirical turbulent model is directly substituted in the Navier–Stokes equation and a numerical technique is used for its solution. The second, most commonly used technique uses a bulk flow model together with control volume formulations, namely, the continuity equation and momentum equation, to obtain the desired results. For a detailed discussion of these methods, solution techniques, the influence of various physical parameters on the coefficients, and an excellent compilation of computational and experimental results, the publication by Childs (1993) is recommended.

### 9.3.2.5 Impeller–Diffuser/Volute Interface

It is widely known that the flow fields within certain types of rotating machinery can significantly influence its vibration behavior. Thomas (1958) recognized and explained the presence of destabilizing clearance excitation forces in axial flow steam turbines. Black (1974) was the first to suggest that centrifugal pump impellers could also develop destabilizing forces. The nature of these forces and their influence on rotor instability has been explained in Section 9.2.2.2 and Section 9.2.2.3 of this chapter. The impeller–diffuser/volute forces assuming rotational symmetry can generally be modeled by an equation of the following form:

$$\begin{Bmatrix} F_X \\ F_Y \end{Bmatrix} = -\begin{bmatrix} k & k_c \\ -k_c & k \end{bmatrix}\begin{Bmatrix} X \\ Y \end{Bmatrix} - \begin{bmatrix} c & c_c \\ -c_c & c \end{bmatrix}\begin{Bmatrix} \dot{X} \\ \dot{Y} \end{Bmatrix} - \begin{bmatrix} m & m_c \\ -m_c & m \end{bmatrix}\begin{Bmatrix} \ddot{X} \\ \ddot{Y} \end{Bmatrix} \tag{9.57}$$

For analytical procedures for the derivation of impeller interaction coefficients and a comparison of experimental data, the work by Childs (1993) is recommended. It is well recognized that a considerable amount of work still needs to be done towards understanding the complex nature of impeller–diffuser/volute interactive forces, especially at off-design conditions.

## 9.3.3 Design

Since the real machine is not available for tests, at the preliminary design stage it is a common practice to develop an accurate mathematical model of the machine to predict its dynamic behavior in operation. It is also prudent to understand and estimate how the machine will interact with its operating environment and how the environment could influence the operation of the machine. A suitable model of the rotor can be developed using the techniques described in Section 9.3.2, and the rotordynamic characteristics of the machine can be analyzed using one of the methods described in Section 9.3.1, above. Based on these methods, numerous computer-based rotordynamic analysis programs have been developed. A listing of the most widely known computer programs in North America is given in Table 9.3. The objectives of the analysis are to predict the critical speeds, excitation frequencies, the amplitudes of deflection, and the magnitude of the forces of the rotor within its full operating range. In certain situations, evaluation of the energy content of the excitation may also be required.

Once the mathematical model is developed, the eigenvalues of the rotor and the mode shapes can be determined. The results can then be presented in the form of a Campbell diagram, where the eigenvalues along with the excitation frequencies are plotted as a function of rotor speed. Critical speeds occur at the speeds corresponding to the points of intersection of the excitation frequency lines and the eigenvalue lines. The Campbell diagram presentation (Figure 9.13) of the results is very useful since the influence of key parameters such as stiffness, damping, clearances (new and worn conditions), and so on can all be shown on the same diagram. A critical speed, although present, may be of little consequence if it is associated with sufficient damping. As illustrated in Figure 9.4, when the damping ratio $\zeta \geq 0.707$, the system is critically damped and above this level of damping there is no amplification of the rotor deflection. At or near a critical speed the amplification factor is $\approx 1/2\zeta$. Using this estimated value,

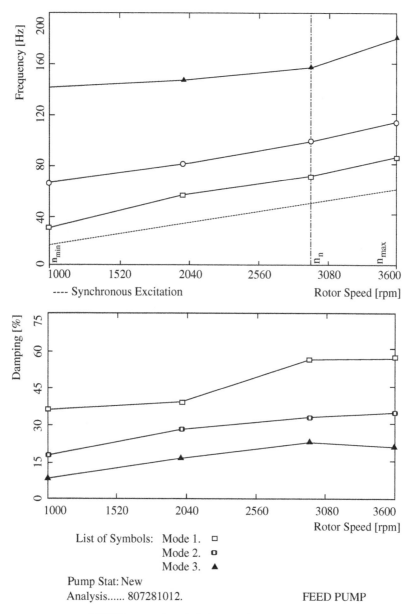

**FIGURE 9.13**  Campbell diagram for a multistage pump.

depending on the internal clearances of the machine, it is possible to assess if the rotor can pass through a critical speed without causing damage to the components. An amplification factor of 2.5 or below is a typical acceptance limit for centrifugal pumps, even for continuous operation at or near it. However, if the amplification factor exceeds the acceptable limits or the critical speeds are too close to the continuous operating speed, then design modifications have to be made to change the critical speed values. At the design stage, it is considered good practice to ensure that the critical speeds are not within ±10% of the continuous operating speed; these limits are sometimes referred to as *separation margins*. The mode shapes of the rotor are important from the standpoint of identifying where the maximum deflections occur. It also provides a good guide for assessing design modifications to improve damping or reduce sensitivity to unbalance forces.

The eigen analysis only provides relative deflections of the rotor. In order to estimate true deflections, a forced response analysis has to be made. Forcing functions of estimated magnitude are applied at selected locations to determine the resulting deflections at specific points on the rotor. This type of analysis is typically carried out for synchronous excitation forces only. The nature of the forcing function depends on the type of the machine; mechanical unbalance is common to all types of machines, whereas hydraulic unbalance is relevant to centrifugal pumps and electrical unbalance to electric motors. The challenge, of course, is to determine the magnitudes, directions, and locations of the forces to apply and how the resulting rotor response should be judged. Of course, these criteria are machine type-dependent and not necessarily applicable to all types of rotating machinery. An example of how forced response analysis on centrifugal pumps is evaluated is given below (Bolleter et al., 1992):

1. Maximum amplification factors and required separation margins are defined by specifications; example as shown in API 610, 8th edn., 1995.
2. Excitation forces are defined and the response is judged relative to admissible shaft vibration limits, and relative to clearances.
3. Apply unbalance forces of such a magnitude that maximum permissible vibration limits at the vibration probe locations are reached, and then evaluate if the deflections exceed the minimum clearances in the machine.
4. Apply an unbalance force of arbitrary magnitude and determine the resulting response at the same or another location, and calculate the sensitivity factor (SF) using the following formula:

$$SF = \frac{a}{U} M \tag{9.58}$$

where
$a$ = rotor deflection
$U$ = unbalance force
$M$ = rotor mass

The sensitivity factor should then be compared with experimental base values of similar machines for acceptance. The rotor responses to the applied forces can be further analyzed to extract other parameters of interest, such as phase angles and force magnitudes at the bearings, in order to evaluate the design.

In order to optimize the rotating machine design in terms of placement of critical speeds and control of deflections and forces, a parameter sensitivity coefficients analysis (Lund, 1979; Rajan et al., 1986; Rajan et al. 1987) may be carried out. For speed and convenience of analysis, the optimization routine can be automated.

---

- Rotordynamic analysis is a part of the current rotating machinery design practice used to predict their vibration behavior.
- The most current rotordynamic analytical procedures are computer-based and are derived from the lump-parameter model or the transfer matrix method.
- In the lumped-parameter model method, the distributed elastic and inertial properties of the rotor are represented as a collection of rigid bodies connected by massless elastic beams.
- In the transfer matrix method, commonly called the FEA method, the rotor is represented as an assembly of elements with distributed elastic and inertial properties.
- Accurate modeling and representation of rotor components is vital to the accuracy and reliability of analysis results. As a result, significant advancement in modeling shaft sections, disks, impellers, bearings, seals, rotor dampers, and rotor–stator interactions have been made.

# 9.4 Vibration Measurement and Techniques

## 9.4.1 Units of Measurement

The most commonly used units of measurement of vibration are as follows:

- Displacement (peak to peak): millimeter (mm) or micron ($10^{-6}$ m) in metric units, and *thou* or *mil* (0.001 in) in imperial units
- Velocity (peak, RMS or "true peak"): mm/sec or m/sec in metric units, and in./sec in imperial units
- Acceleration (RMS): $g$ or m/sec$^2$ in metric units, and $g$ or in./sec$^2$ in imperial units
- Frequency: hertz (Hz) or cycles/min (cpm) in both systems of units
- Phase angle: degrees in both systems of units

## 9.4.2 Measured Parameters and Methods

Under steady-state conditions, the vibration from a rotating machine is a periodic signal of a complex waveform. During unstable operation or upset conditions, the signal may become random in nature. In certain types of machines, transient signals that are nonperiodic could also be present due to internal impacts and damping. Based on the simple spring–mass system and the mathematical Fourier analysis procedure, all periodic complex waveforms can be reduced to the sum of a series of sinusoidal functions. In the case of random signals, averaging techniques are used to reduce them to periodic signals for convenience of analysis.

A quantitative assessment of the vibration can be made in terms of the amplitude, velocity, acceleration, or the magnitude of force of the motion. Other key parameters such as frequency, phase angle, and the time-varying nature of the signal are important in fully characterizing it. Because the true signal is not purely sinusoidal, it is important to identify its magnitude as, *peak, peak-to-peak* or *root mean square* (RMS). The preferred parameter of measurement varies throughout the industry and depends on the nature, complexity, and type of machine, and the purpose for which it is measured. A general classification of measured parameters and techniques based on industry and rotating machine type is given in Table 9.4.

---

- The displacement amplitude, velocity, acceleration, frequency, and phase angle of the vibration signal are some of the parameters that can be used to assess the condition of a rotating machine.
- The vibration signals generated by a typical rotating machine is complex in nature and, therefore, requires various mathematical analysis procedures and signal averaging techniques to reduce them to simple and interpretable forms.

---

# 9.5 Vibration Control and Diagnostics

## 9.5.1 Standards and Guidelines

Given the fact that some degree of vibration is always present in rotating machinery, some means of judging "how much is too much" has to be established so that vibrations can be controlled within reasonable limits. When such a judgment is not based on a scientific method, there is room for speculation and it will depend on the one making the decision, the manufacturer, end user, the governing authority, and so on. Since vibrations are a key indication parameter of the performance of rotating machinery, different interest groups monitor it for a variety of purposes. It can be used as a measure of

**TABLE 9.4**  Measurement Parameters and Techniques

| Measurement/Technique | Description | When/Where Used |
|---|---|---|
| Acceleration, RMS | When high frequency or force of the vibration is of interest | Gear boxes, rolling element bearings, gas/steam turbines. Mainly used by defense and aerospace industry |
| Bode/Nyquist plot | Plot of displacement amplitude and phase angle vs. speed | Observe critical speeds and instability in machines using journal bearings |
| Cepstrum analysis | Inverse Fourier transform of logarithmic power spectrum | To detect families of harmonics and sidebands in gearboxes, rolling element bearings, and electric motors |
| Condition monitoring | Analysis of signals generated by the machine to determine its condition on a continuous or periodic basis | On critical equipment where predictive maintenance programs are used. Can reduce equipment redundancy |
| *Displacement peak-to-peak* | | |
| (a) Absolute | Absolute displacement amplitude of rotor vibration | Where the rotor mass is very much larger than the stator mass. Large motors, generators, and fans |
| (b) Relative | Relative displacement amplitude of rotor vibration | Commonly used on machines with journal bearings or in close clearance seals |
| (c) External | Absolute displacement amplitude of stator component vibration | Low speed machines (less than 1000 rpm) |
| Modal analysis | To measure the vibration response of a structure to an applied force. The force can be periodic or an impact force | To determine the modal mass, stiffness and damping properties of a structure. Also used to measure structural natural frequencies |
| Orbit analysis | The path of the shaft centerline motion during rotation | Diagnostics of machines using journal bearings. Provides a picture of the motion of the journal in the bearing |
| Polar plots | A polar graph of amplitude versus phase at various machine speeds | Similar to Bode plots, can be used to detect critical speeds and instability. Modal properties can also be extracted from polar plots |
| Phase angle | Phase angle of vibration signal | Useful in balancing, diagnosing critical speeds, and misalignment problems |
| *Rolling element bearing analysis* | | |
| (a) Acceleration | Measure the amplitude of all pass and discreet frequency accelerations | When damage has progressed to generate audible noise amplitudes increase. Useful from 5–5 kHz |
| (b) Shock pulse method | A high frequency resonance technique tuned to the detector natural frequency | Early detection of failure, measures ultrasonic noise. Proprietary technique |
| (c) Envelope technique | Bearing defects cause periodic impacts, which make bearing components resonate. Demodulation and enveloping techniques are used to detect the impact (fault) frequencies | Used for early failure detection (ultrasonic noise) as well as for advanced stages of damage (audible noise) |
| (d) Spike energy method | Measure the broadband acceleration over the 5–45 kHz range | Used for early failure detection (ultrasonic noise) as well as for advanced stages of damage (audible noise) |
| (e) Kurtosis method | The normalized fourth moment of the probability distribution of acceleration over the 2–80 kHz range | Used for early failure detection (ultrasonic noise) as well as for advanced stages of damage (audible noise) |
| Run-up, run-down analysis (waterfall/ cascade plots) | Three-dimensional plot of frequency or time spectrum vs. time or speed | Used for diagnosing a variety of vibration problems. Helpful in analyzing transient signals |

**TABLE 9.4**    (*continued*)

| Measurement/Technique | Description | When/Where Used |
|---|---|---|
| Spectrum analysis | Plot of amplitude vs. frequency of vibration | Used for diagnostics, to determine frequency, harmonics, side bands, beats, transfer functions, etc., and to control of vibrations levels at discreet frequencies |
| Trend analysis | Vibration data collected periodically over an extended time domain | Useful in predictive maintenance programs in assessing machine conditions |
| Time averaging | Averaging of time records using triggering at the same point of the waveform of a repetitive signal | Used in the analysis of faulty gearboxes. Can reduce asynchronous components in the signal and improve signal-to-noise ratio |
| Time-domain analysis | Plot of amplitude versus time | To observe amplitude modulation, beats, impacts, transients, and phase angle. Very useful in diagnostics |
| Velocity-peak or RMS | Velocity amplitude of vibration signal | Parameter most commonly used in many industries to monitor vibrations. Peak readings relate to peak stress levels and rms to energy of vibrations |

quality and workmanship or the common basis for acceptance between the user and manufacturer. From a safety point of view, the operator can establish *normal, alarm,* and *shutdown levels* based on vibration limits. Vibration levels are also used in making maintenance decisions in rotating machinery.

Rathbone (1939) was the first to publish guidelines for vibration limits for machinery, based on his experience as an insurance agent. Since that time, numerous individuals, organizations, and governing bodies have developed a variety of guidelines and standards for vibration levels in rotating machinery. A listing of the more commonly used guidelines and standards is given in Table 9.5. It should be recognized that these are experience-based standards, and therefore, will grow and develop with technology.

ISO 10816 Part 1 to Part 5 is a comprehensive set of standards that has been developed for the evaluation of vibration of rotating machinery by measuring the vibration response on nonrotating, structural components such as bearing housings. Vibration measuring points as specified by these standards are shown in Figure 9.14. In a similar vein, ISO 7919 Part 1 to Part 5 has been developed for the evaluation of vibration by measuring the vibration on rotating shafts. These standards cover the most widely used types of rotating machinery and they relate to both operational monitoring and acceptance testing of equipment. Table 9.6 and Table 9.7 are derived from these standards and are presented as a general guideline for vibration limits of rotating machinery. For specific details, including the limitations of the standards, the reader is advised to refer to the relevant sections of ISO 10816 and ISO 7919 Standards.

## 9.5.2    Vibration Cause Identification

Vibrations are an inherent part of all rotating machinery. Vibration can be due to many causes: improper design, practical manufacturing limits, poor installation, the effect of system environment, component deterioration, operation outside of design limits, or a combination of the above. At times, finding the exact cause of vibration can be quite a challenge, as several of the causes have similar symptoms. Table 9.8 is a list of the more commonly known causes of vibration in rotating machinery and their symptoms.

## 9.5.3    Vibration Analysis — Case Study

In the past, it was common practice to operate centrifugal pumps at a fixed speed and attain required flow changes by means of throttling. This forces the pump to operate at low efficiency conditions

**TABLE 9.5** Vibration Guidelines and Standards

| Year | Author/ Organization | Reference Number | Title/Description |
|---|---|---|---|
| 2002 | AGMA | ANSI/AGMA 6000-B96 | Specification for Measurement of Linear Vibration on Gear Units |
| 2003 | API | ANSI/API std 541-2003 | Form-Wound Squirrel-Cage Induction Motors 500 hp and Larger |
| 1997 | API | API STD 546, second edition | Brushless Synchronous Machines, 500 kVA and Larger |
| 2004 | API | API STD 610/ISO 13709, ninth edition | Centrifugal Pumps for Petroleum, Petrochemical and Natural Gas Industries |
| 1997 | API | API STD 611, fourth edition | General Purpose Steam Turbines for Petroleum, Chemical and Gas Industry Services |
| 2005 | API | API STD 612/ISO 10437, sixth edition | Petroleum, Petrochemical and Natural Gas Industries – Steam Turbines – Special-Purpose Applications |
| 2003 | API | API STD 613, fifth edition | Special Purpose Gear Units for Petroleum, Chemical and Gas Industry Services |
| 1998 | API | API STD 616, fourth edition | Gas Turbines for the Petroleum, Chemical, and Gas Industry Services |
| 2002 | API | API STD 617, seventh edition | Axial and Centrifugal Compressors and Expander-compressors for Petroleum, Chemical and Gas Industry Services |
| 2000 | API | API STD 670, fourth edition | Mechanical Protection Systems |
| 2004 | API | API STD 672, fourth edition | Packaged Integrally Geared, Centrifugal Air Compressors for Petroleum, Chemical, and Gas Industry Services |
| 2001 | API | API STD 673, second edition | Special Purpose Fans |
| 1997 | API | API STD 677, second edition | General Purpose Gear Units for Petroleum, Chemical, and Gas Industry Services |
| 1996 | API | API STD 681, first edition | Liquid Ring Vacuum Pumps for Petroleum, Chemical, and Gas Industry Services |
| 2000 | API | API STD 685, first edition | Sealless Centrifugal Pumps for Petroleum, Heavy-Duty Chemical, and Gas Industry Services |
| 1965 | BDS | BDS 5626-65 | Measurement of Vibration on Electrical Rotating Machines |
| 1964 | Blake, M.P. | Hydrocarbon Processing, January 1964 | New Vibration Standards for Maintenance |
| 1963 | CAGI | | In-Service Standards for Centrifugal Compressors |
| 1975 | CAGI | | Standard for Centrifugal Air Compressors |
| 1971 | CSN | CSN 011410 | Permitted Limits for Unbalanced Solid Machine Elements |
| 1968 | Dresser Industrial | | General Guidelines for Vibration on Clark Centrifugal Compressors |
| 1966 | Gosstandart | GOST 12379-66 | Measurement of Vibration on Electrical Rotating Machines |
| 2002 | HI | ANSI/HI 9.6.4 | Centrifugal and Vertical Pumps — Vibration Measurement and Allowable Values |
| 1996 | IEC | IEC 60034-14 | Rotating Electrical Machines, Part 14: Mechanical Vibrations of Certain Machines with Shaft Heights 56 mm and Higher — Measurement, Evaluation and Limits of Vibration |
| 1964 | IRD | IRD #305D | General Machinery Vibration Severity Chart |
| 1995–2001 | ISO | | Mechanical Vibration — Evaluation of Machine Vibration by Measurements on Nonrotating Parts: |
| | | ISO 10816-1:1995 | Part 1: General Guidelines |
| | | ISO 10816-2:2001 | Part 2: Land-Based Steam Turbines and Generators in Excess of 50 MW with Normal Operating Speeds of 1500, 1800, 3000 and 3600 rpm |
| | | ISO 10816-3:1998 | Part 3: Industrial Machines with Nominal Power above 15 kW and Nominal Speeds between 120 and 15,000 rpm when Measured *In Situ* |
| | | ISO 10816-4:1998 | Part 4: Gas Turbine Driven Sets Excluding Aircraft Derivations |

**TABLE 9.5**     (*continued*)

| Year | Author/Organization | Reference Number | Title/Description |
|------|---------------------|------------------|------------------|
| | | ISO 10816-5: 2000 | Part 5: Machine Sets in Hydraulic Power Generating and Pumping Plants |
| 2002 | ISO | | Mechanical Vibration — Vibration of Active Magnetic Bearing Equipped Rotating Machinery |
| | | ISO 14839-1: 2002 | Part 1: Vocabulary |
| | | ISO/CD 14839-2:2004 | Part 2: Evaluation of Vibration |
| 1996–2001 | ISO | | Mechanical Vibrations of Nonreciprocating Machines — Measurement on Rotating Shafts and Evaluation Criteria |
| | | ISO 7919-1: 1996 | Part 1 (1996): General Guidelines |
| | | ISO 7919-2: 2001 | Part 2 (2001): Land-Based Steam Turbines and Generators in Excess of 50 MW with Normal Operating Speeds of 1500, 1800, 3000 and 3600 rpm |
| | | ISO 7919-3: 1996 | Part 3 (1996): Coupled Industrial Machines |
| | | ISO 7919-4: 1996 | Part 4 (1996): Gas Turbine Sets |
| | | ISO 7919-5: 1997 | Part 5 (1997): Machine Sets in Hydraulic Power Generating and Pumping Plants |
| 1993 | ISO | ISO 8579-2 | Acceptance Code for Gears, Part 2: Determination of Mechanical Vibration of Gear Units During Acceptance Testing |
| 2004 | ISO | | Rolling Bearings — Measuring Methods for Vibration |
| | | ISO 15242-1:2004 | Part 1: Fundamentals |
| | | ISO 15242-2:2004 | Part 2: Radial Ball Bearings with Cylindrical Bore and Outside surface |
| | | ISO/CD 15242-3 | Part 3: Spherical and Taper Radial Roller Bearings with Cylindrical Bore and Outside Diameter |
| 1959 | Kruglov, N.V. | Teplonerg, 8 (85), 1959 | Turbomachine Vibration Standards |
| 1967 | Maten, S | Hydrocarbon Processing, January 1967 | New Vibration Velocity Standards |
| 1983 | McHugh, J.D. | J. Lub. Tech., Trans. ASME, 1983, 105 | Estimating the Severity of Shaft Vibration within Fluid Film Journal Bearings |
| 1974 | MIL | MIL-STD-167-1 | Mechanical Vibration of Shipboard Equipment, Type I: Environmental, Type II: Internally Excited |
| 2003 | NEMA | NEMA MG 1-2003 | Motors and Generators, Part 7 — Mechanical Vibration — Measurement, Evaluation and Limits |
| 1991 | NEMA | NEMA SM 23-1991 | Steam Turbines for Mechanical Drive Service |
| 1991 | NEMA | NEMA SM 24-1991 | Land Based Steam Turbine Generator Sets 0 to 33,000 kW |
| 1965 | PKN | PN-65/E-04255 | Measurement of Vibration of Electrical Rotating Machines |
| 1939 | Rathbone, T.C. | Power Plant Engineering, November 1939 | Vibration Tolerances |
| 1964 | VDI | VDI 2056 | Evaluation Criteria for Mechanical Vibrations in Machines |
| 1982 | VDI | VDI 2059 P1 | Shaft Vibrations of Turbosets Principles for Measurement and Evaluation |
| 1990 | VDI | VDI 2059 P2 | Shaft Vibrations of Steam Turbosets for Power Station Measurement and Evaluation |
| 1985 | VDI | VDI 2059 P3 | Shaft Vibrations of Industrial Turbosets Measurement and Evaluation |
| 1981 | VDI | VDI 2059 P4 | Shaft Vibrations of Gas Turbosets Measurement and Evaluation |
| 1982 | VDI | VDI 2059 P5 | Shaft Vibrations of Hydraulic Machinesets Measurement and Evaluation |
| 1949 | Yates, H.G. | Trans. N.E. Coast Inst. Engrs Ship Builders, Vol. 65, 1949 | Vibration Diagnosis of Marine Geared Turbines |

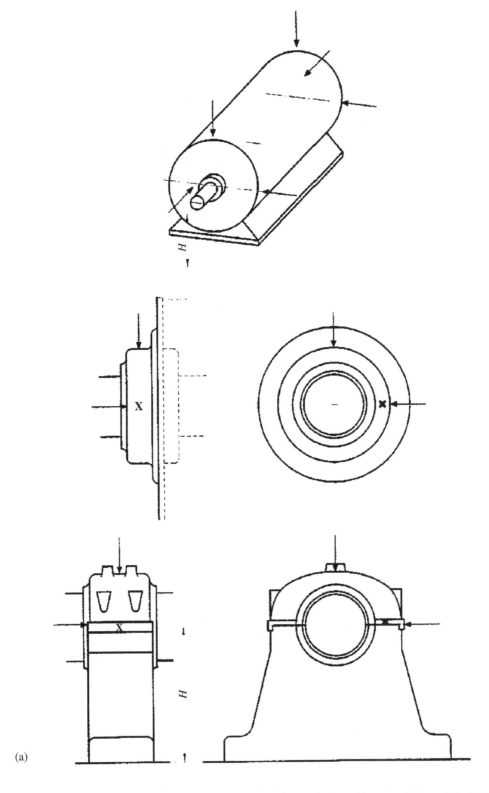

(a)

**FIGURE 9.14** (a) Measuring points; (b) measuring points for vertical machine sets. (*Source:* ISO 10816-3, 1998-05-15. With permission.)

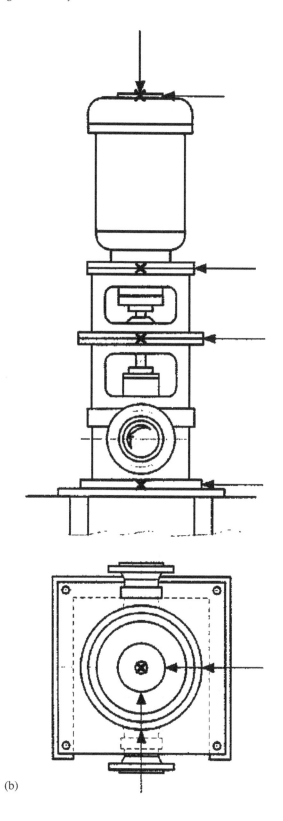

(b)

**FIGURE 9.14** (*continued*)

**TABLE 9.6A**  Acceptable Vibration Levels for Rotating Machinery Measured on Nonrotating Parts

| Machinery Type | Power Level | Speed Range (rpm) | Applicable Vibration Level | |
|---|---|---|---|---|
| | | | Rigid Support | Flexible Support |
| Steam turbines | $15 \leq P \leq 300$ kW | $120 \leq N \leq 15{,}000$ | V1 and D3 | V3 and D7 |
| | $300$ kW $\leq P \leq 50$ MW | $120 \leq N \leq 15{,}000$ | V3 and D5 | V6 and D8 |
| | $P > 50$ MW | $N < 1{,}500$ or $N > 3{,}600$ | V3 and D5 | V6 and D8 |
| | $P > 50$ MW | $N = 1{,}500$ or $1{,}800$ | V5 | V5 |
| | $P > 50$ MW | $N = 3{,}000$ or $3{,}600$ | V7 | V7 |
| Gas turbines | $15 \leq P \leq 300$ kW | $120 \leq N \leq 15{,}000$ | V1 and D3 | V3 and D7 |
| | $300$ kW $\leq P \leq 3$ MW | $120 \leq N \leq 15{,}000$ | V3 and D5 | V6 and D8 |
| | $P > 3$ MW | $3{,}000 \leq N \leq 20{,}000$ | V8 | V8 |
| Hydraulic turbines and | Horizontal machines | | | |
| pump turbine | $P > 1$ MW | $60 \leq N \leq 300$ | N/A | V4 |
| | $P > 1$ MW | $300 < N \leq 1{,}800$ | V2 and D6 | N/A |
| Vertical machines | $P > 1$ MW | $60 < N \leq 1{,}800$ | V2 and D6 | N/A |
| | $P > 1$ MW | $60 < N \leq 1{,}000$ | V2 and D6 | V4 and D9 |
| Centrifugal pumps | | | | |
| Separate driver | $P > 15$ kW | $120 \leq N \leq 15{,}000$ | V3 and D2 | V6 and D4 |
| Integral driver | $P > 15$ kW | $120 \leq N \leq 15{,}000$ | V1 and D1 | V3 and D2 |
| Electric motors | | | | |
| Shaft height $H \geq 315$ mm | $P > 15$ kW | $120 \leq N \leq 15{,}000$ | V3 and D5 | V6 and D8 |
| Shaft height $160 \leq H < 315$ mm | $P > 15$ kW | $120 \leq N \leq 15{,}000$ | V1 and D3 | V3 and D7 |
| Generators, excluding those used | $15 \leq P \leq 300$ kW | $120 \leq N \leq 15{,}000$ | V1 and D3 | V3 and D7 |
| in hydraulic power generation | $300$ kW $\leq P \leq 50$ MW | $120 \leq N \leq 15{,}000$ | V3 and D5 | V6 and D8 |
| | $P > 50$ MW | $N < 1{,}500$ or $N > 3{,}600$ | V3 and D5 | V6 and D8 |
| | $P > 50$ MW | $N = 1{,}500$ or $1{,}800$ | V5 | V5 |
| | $P > 50$ MW | $N = 3{,}000$ or $3{,}600$ | V7 | V7 |
| Generators and motors used in | Horizontal machines | | | |
| hydraulic power generation | $P > 1$ MW | $60 \leq N \leq 300$ | N/A | V4 |
| | $P > 1$ MW | $300 < N \leq 1{,}800$ | V2 and D6 | N/A |
| Vertical machines | $P > 1$ MW | $60 < N \leq 1{,}800$ | V2 and D6 | N/A |
| | $P > 1$ MW | $60 < N \leq 1{,}000$ | V2 and D6 | V4 and D9 |
| Compressors, rotary, blowers, | $15 \leq P \leq 300$ kW | $120 \leq N \leq 15{,}000$ | V1 and D3 | V3 and D7 |
| and fans | $300$ kW $\leq P \leq 50$ MW | $120 \leq N \leq 15{,}000$ | V3 and D5 | V6 and D8 |

resulting in wasted energy and premature failure of components due to high vibration. The current practice to obtain flow changes in the pump is by means of speed change. This eliminates flow throttling and allows the pump to operate close to its best efficiency point, where energy is not wasted and vibrations are a minimum. However, as illustrated below, variable speed operation of a pump-motor set over a wide speed range could pose several challenging problems.

**TABLE 9.6B**  Maximum Vibration Velocity Limits for Different Levels (mm/sec, RMS)

| Vibration Level | Zone A | Zone B | Zone C | Alarm | Trip |
|---|---|---|---|---|---|
| V1 | 1.4 | 2.8 | 4.5 | 3.5 | 5.6 |
| V2 | 1.6 | 2.5 | 4.0 | 3.1 | 5.0 |
| V3 | 2.3 | 4.5 | 7.1 | 5.6 | 8.9 |
| V4 | 2.5 | 4.0 | 6.4 | 5.0 | 8.0 |
| V5 | 2.8 | 5.3 | 8.5 | 6.6 | 10.6 |
| V6 | 3.5 | 7.1 | 11.0 | 8.9 | 13.8 |
| V7 | 3.8 | 7.5 | 11.8 | 9.4 | 14.8 |
| V8 | 4.5 | 9.3 | 14.7 | 11.6 | 18.4 |

**TABLE 9.6C**    Maximum Vibration Displacement Limits for Different Levels ($\mu$m, RMS)

| Vibration Level | Zone A | Zone B | Zone C | Alarm | Trip |
|:---:|:---:|:---:|:---:|:---:|:---:|
| D1 | 11 | 22 | 36 | 28 | 45 |
| D2 | 18 | 36 | 56 | 45 | 70 |
| D3 | 22 | 45 | 71 | 56 | 89 |
| D4 | 28 | 56 | 90 | 70 | 113 |
| D5 | 29 | 57 | 90 | 71 | 113 |
| D6 | 30 | 50 | 80 | 63 | 100 |
| D7 | 37 | 71 | 113 | 89 | 141 |
| D8 | 45 | 90 | 140 | 113 | 175 |
| D9 | 65 | 100 | 160 | 125 | 200 |

*Zone A*: Newly commissioned machines should fall within this zone. *Zone B*: Machines with vibrations within this zone are considered acceptable for long-term operation. *Zone C*: Machines with vibrations within this zone are normally considered unsatisfactory for long-term operation. Such a machine may be operated for a short period in this condition. *Alarm*: The values chosen will normally be set relative to a baseline value determined from experience. However, it is recommended that the alarm value shall not exceed those given herein. *Trip*: The values will generally relate to the mechanical integrity of the machine. They will generally be the same for all machines with similar design. It is recommended that the trip value shall not exceed those given herein. *Notes*: (1) The measured vibration is broadband, and the frequency range will depend on the type of machine being considered. A range from 2 to 1000 Hz is typical except for in high-speed machines, >10,000 rpm, where the upper limit should at least be six times the rotational frequency. (2) It is common practice to evaluate rotating machinery based on the broadband RMS vibration velocity, since it can be related to the vibration energy levels. However, other quantities such as vibration displacement or acceleration may be preferred. Especially low speed machines can have unacceptably large vibration displacements when the 1 × rpm component is dominant. Therefore, where specified, both the velocity and displacement criteria are met. (3) Since typical vibration waveforms measured on rotating machinery are complex in nature, there is no simple relationship between broadband velocity, displacement, and acceleration. (4) Vibration measurements shall be taken on bearing support housings, or other structural components, which adequately respond to the dynamic forces of the machine. Recommended locations for bearing housings are shown in Figure 9.14. (5) For certain types of machines, the axial vibration limits may differ from those for radial directions. Also, within the same machine set, in particular hydraulic power-generating sets, the applicable level may differ from bearing to bearing depending on its classification as a rigid or flexible support. (6) Above vibration limits apply to steady-state/normal operating conditions of the machine. If the vibration levels are sensitive to the operational conditions, then evaluation of the machine for operating conditions outside steady-state conditions will have to be based on different criteria. (7) The vibration limits specified herein should not be used to assess the condition of rolling element type bearings although it encompasses machines that may have these types of bearings. (8) It must be recognized that the vibration measurement on nonrotating parts alone does not form the only basis for judging the condition of a machine. In certain types of machines, it is common practice also to judge the vibration based on measurements taken on rotating shafts. (9) A support may be considered as rigid in a specific direction only if its natural frequency in that direction exceeds the main excitation frequency by at least 25%, otherwise it is considered to be flexible. In some cases, a support may be rigid in one direction and flexible in another. (10) In the case of hydraulic machine sets, major differences in radial bearing support arrangement can occur. For evaluation of the support type it is recommended that the reader refer to ISO 10816-5.

*Description.* The following case study is taken from a petroleum pipeline pump application where pump-motor sets with VFDs were installed in a new pipeline starting in Alberta, Canada and terminating in Minnesota, USA. VFDs are frequently used in the pipeline industry to power high horsepower pumps to eliminate power wasted by throttling, reduce inrush current at motor startup, and to provide greater operating flexibility. However, variable speed operation can cause vibration problems in the pump, motor, and the couplings that are not normally experienced with fixed speed pumps. Unexpected high torsional and lateral vibrations were experienced with these pumps and motors at certain operating speeds. A rotor torsional resonance, motor housing resonance, acoustic resonance in the internals of the pump, and discharge piping were identified to be the causes of the high vibration in the pump-motor set. Details on diagnosing the problems and the corrective measures taken to resolve them are given below:

*Pump type*: The pump was a centrifugal, two-stage, double volute horizontal pump, with six vane impellers, normally designed to operate at a fixed speed. Generation of pressure pulsations at the vane passing frequencies of 6 × and 12 × rotational speed is normally expected.

**TABLE 9.7A**  Acceptable Vibration Levels for Rotating Machinery, Measured on Rotating Shafts

| Machinery Type | Power Level | Speed Range (RPM) | Applicable Vibration Level | |
|---|---|---|---|---|
| | | | Relative Displacement | Absolute Displacement |
| Steam turbines | $P \leq 50$ MW | $1,000 \leq N \leq 30,000$ | D8 | — |
| | $P > 50$ MW | $N = 1,500$ | D5 | D7 |
| | $P > 50$ MW | $N = 1,800$ | D4 | D6 |
| | $P > 50$ MW | $N = 3,000$ | D2 | D5 |
| | $P > 50$ MW | $N = 3,600$ | D1 | D3 |
| Gas turbines | $P > 3$ MW | $3,000 \leq N \leq 30,000$ | D8 | — |
| | $P \leq 3$ MW | $1,000 \leq N \leq 30,000$ | D8 | — |
| Hydraulic turbines and pumps used in hydraulic power generation and pumping plants | $P > 1$ MW | $60 \leq N \leq 1,800$ | D9 | D9 |
| Centrifugal pumps | All | $1,000 \leq N \leq 30,000$ | D8 | — |
| Electric motors | All | $1,000 \leq N \leq 30,000$ | D8 | — |
| Generators, excluding those used in hydraulic power generation | $P \leq 50$ MW | $1,000 \leq N \leq 30,000$ | D8 | — |
| | $P > 50$ MW | $N = 1,500$ | D5 | D7 |
| | $P > 50$ MW | $N = 1,800$ | D4 | D6 |
| | $P > 50$ MW | $N = 3,000$ | D2 | D5 |
| | $P > 50$ MW | $N = 3,600$ | D1 | D3 |
| Generators and motors used in hydraulic power generation | $P > 1$ MW | $60 \leq N \leq 1,000$ | D9 | D9 |
| | $P > 1$ MW | $1,000 < N \leq 1,800$ | D8 | — |
| Compressors, rotary, blowers, and fans | All | $1,000 \leq N \leq 30,000$ | D8 | — |

*Motor*: The motor was a 3000 hp, two pole horizontal induction motor, designed to operate at 3600 rpm.

*Supply*: The supply was a VFD of the current source inverter type. These drives are known to generate an oscillatory torque at 6 × and 12 × the operating frequency.

*Coupling*: Flexible disc type coupling with a spacer was used. These couplings have very little torsional damping capacity.

*Speed range*: The speed range was from 1440 rpm (24 Hz) to 3900 rpm (65 Hz).

*Reference*: Refer to Figure 9.15 to Figure 9.18.

As for the resonance at second and third torsional critical speeds (Figure 9.15) the second and third torsional modes are excited when the 6 × component of rotational speed corresponds to the critical speeds of 92 and 268 Hz, respectively. The 6 × rpm torsional excitation is caused by the pressure pulsations in the pump. The waterfall plot (Figure 9.15a) was taken during a run down of the set with the power to the motor turned off.

A similar plot taken during run up of the motor (Figure 9.15b) shows excitations at the same frequencies but having different amplitude. Since both the pump and motor generate 6 × excitation, it suggests a phase difference between the excitation torques.

It is important to note that the conventional vibration monitoring devices cannot detect the torsional resonance problem. The only indication of a problem was the unusual chattering noise emitted by the coupling. Special techniques to measure dynamic torque using strain gauges had to be used to detect the torsional vibrations.

As for the motor housing resonance (Figure 9.16), the 2 × rotational speed vibration of the motor is dominant and peaks at 118 Hz corresponding to a natural frequency of the motor frame. In the waterfall plot, the natural frequencies corresponds to excitations that are parallel to the axis. Excitation at harmonics, including the 6 × component, is present but is not dominant.

As for the pump vibrations at the vane passing frequency (Figure 9.17), the 6 × vane passing frequency is dominant at all operating speeds. It peaks at 238 Hz, possibly due to an acoustic resonance in

**TABLE 9.7B**  Maximum Vibration Displacement $S_{p-p}$ Limits ($\mu$m) Peak-to-Peak Limits for Different Levels

|    | Zone A | Zone B | Zone C |
|----|--------|--------|--------|
| D1 | 75 | 150 | 240 |
| D2 | 80 | 165 | 260 |
| D3 | 90 | 180 | 290 |
| D4 | 90 | 185 | 290 |
| D5 | 100 | 200 | 320 |
| D6 | 110 | 220 | 350 |
| D7 | 120 | 240 | 385 |
| D8 | $4800/\sqrt{n}$ | $9000/\sqrt{n}$ | $13{,}200/\sqrt{n}$ |
| D9 | $10^{(2.3381-0.0704\log n)}$ | $10^{(2.5599-0.0704\log n)}$ | $10^{(2.8609-0.0704\log n)}$ |

*Zone A*: Newly commissioned machines should fall within this zone. *Zone B*: Machines with vibrations within this zone are considered acceptable for long-term operation. *Zone C*: Machines with vibrations within this zone are normally considered unsatisfactory for long-term operation. Such a machine may be operated for a short period in this condition. *Alarm*: The values chosen will normally be set relative to a baseline value determined from experience. However, it is recommended that the alarm value shall not exceed those given herein. *Trip*: The values will generally relate to the mechanical integrity of the machine. They will generally be the same for all machines with similar design. It is recommended that the trip value shall not exceed those given herein. *Notes*: (1) The measured vibration is broadband and is shaft vibration displacement peak to peak. Where applicable, vibration limits for both absolute and relative radial shaft vibrations are given in certain cases. (2) Relative displacement is the vibratory displacement between the shaft and an appropriate structural component such as the bearing housing. Absolute displacement is the vibratory displacement of the shaft with reference to an inertial frame of reference. (3) Relative measurements are carried out with a noncontacting transducer. Absolute readings are obtained by one of the following methods: by a shaft riding probe on which a seismic transducer is mounted so that it measures absolute shaft displacement directly, or with the combination of a noncontacting transducer which measures relative shaft displacement and a seismic transducer which measures support vibration. Their conditioned outputs are vectorially added to provide a measure of the absolute shaft motion. (4) The vibration evaluation criteria are dependent upon a variety of factors and the criteria adopted will vary for different types of machines. Some of these factors are the bearing type, clearance, and diameter. The adopted criteria have to be compared with the bearing diametral clearance ($C$) and adjusted to suit. Typical values are: Zone A $\leq 0.4C$; Zone B $\leq 0.6C$; and Zone C $\leq 0.7C$. (5) Above vibration limits apply to steady state/normal operating conditions of the machine. If the vibration levels are sensitive to the operational conditions then evaluation of the machine for operating conditions outside steady-state conditions will have to be based on different criteria. (6) It is recommended that vibration readings at each location be made with a pair of transducers and that the transducers are mounted perpendicular to the shaft axis and they are at an angle of 90° to one another. The vibration limits apply to each measured direction. (7) The mechanical and electrical run-out at each measurement location must be assessed and should be <25% of the allowable limit or 6 $\mu$m, whichever is greater. (8) It must be recognized that the vibration measurement on rotating shafts does not form the only basis for judging the condition of a machine. In certain types of machine, it is common practice also to judge the vibration based on measurements taken on nonrotating parts. (9) ALARM levels should be set relative to a baseline value determined from experience for the measurement position, direction and type of machine. It must provide a warning that a defined value, which is significantly above the baseline value, has been reached. The maximum ALARM setting should be $\leq 0.75C$. (10) The TRIP values should be based on protecting the mechanical integrity of the machine. Consideration of damage to bearings is typical; therefore, maximum TRIP setting should be $\leq 0.9C$.

the discharge pipe. It does not seem to correspond to a structural natural frequency due to the absence of excitations at 238 Hz at all speeds.

As for the acoustic resonance in the pump cross-over pipe (Figure 9.18), dynamic pressure pulsation measurements made on the pump cross-over pipe from the first stage discharge to the second stage suction show an acoustic resonance at 540 Hz. The consistent presence of some excitation at 540 Hz at all speeds confirms that it is an acoustic natural frequency of the cross-over pipe. When the 6 × rpm pressure pulsation frequency coincides with the acoustic natural frequency, a resonance condition occurs and the magnitude of the pressure pulsation increases by almost a factor of 30.

As for corrective action, for a pump that has to operate over a wide speed range, totally eliminating the coincidence of all the frequencies of exciting forces with the system natural frequencies is impractical. Therefore, the system has to be designed such that the resulting magnitudes of the forces are controlled to within tolerable levels so that safe and reliable operation can take place. This can be accomplished by a

**TABLE 9.8**  Vibration Cause Identification

| Cause | Dominant Frequency | Spectrum, Time Domain, Orbit Shape | Characteristics, Corrections, Comments |
|---|---|---|---|
| Mass unbalance | 1 × | High 1 × with much lower harmonics; circular or elliptic orbits | Corrected by shop or field balancing |
| Shaft bow | 1 × | Run down plot shows decrease of vibration at critical speed | The shaft has to be straightened using an acceptable method |
| Misalignment | 1 × and 2 × | Equally high 1 × and 2 × , figure 8 orbits | Realign at operating conditions; loads causing misalignment, such as nozzle loads, may have to be reduced |
| Worn journal bearings | 1 × , 1/2 × | Equally high 1 × and 1/2 × | Difficult to balance |
| Gravity critical | 2 × | Run down plot will show excitation at 1/2 critical speed | Can be corrected by balancing |
| Asymmetric shaft | 2 × | Run down plot will show excitation at 1/2 critical speed | Typically occurs on multistage machines when all the keyways lie in the same plane; correct by staggering them |
| Shaft crack | 1 × and 2 × | High 1 × and run down plots may show excitation at 1/2 critical speed | Confirmation and detection of location of the crack may require NDE techniques |
| Loose components | 1 × and higher orders plus fractional subharmonics | High 1 × with lower level orders and fractional subharmonics | Shimming and peening may be used as temporary methods to fix the problem |
| Coupling lockup | 1 × and 2 × | Equally high 1 × and 2 × , figure 8 orbits | Stop starts may change vibration pattern |
| Thermal instability | 1 × | High 1 × varies with temperature. Phase angle may change | Proper prewarming or compromise balancing can correct the problem |
| Oil whirl | < 1/2 × , typically 0.35 × to 0.47 × | Run-up plot will show 1/2 × increasing and locking into fixed value < 1/2 × | Temporary problem may be caused by excess clearances, oil viscosity, or unloading of the bearing; if it is a design problem, correct by changing to tilting pad bearings |
| Internal rubs | 1/4 × , 1/3 × , 1/2 × , 2 × , 3 × , 4 × , etc. | Run down plots may show decreasing amplitudes and disappearance; loops in orbits | May get progressively worse; galling between contact surfaces or heat build-up may cause seizure and shaft failure |
| Trapped fluids in rotor | 0.8 × to 0.9 × | Time domain signal will show beating | Balancing the rotor may reduce the vibration |
| Defective rolling element bearings | At bearing defect frequency | Peaks at defect frequencies in spectrum | Shock pulse measurements can also be used to detect problem |
| Damaged gears | Gear mesh frequency | High peaks at gear mesh frequency with side bands. Time domain may also show pulses | To determine exact nature of damage further analysis may be required |

**TABLE 9.8** (*continued*)

| Cause | Dominant Frequency | Spectrum, Time Domain, Orbit Shape | Characteristics, Corrections, Comments |
|---|---|---|---|
| Electric motor problems | 1 × (line frequency), 2 × (line frequency) | High peaks at 1 × and 2 × line frequency with side bands; disappears when power to motor is turned off | In the case of two pole motors, it can be confused with mechanical causes as the rotational speed is the same as line frequency |
| Casing distortion | 1 × | High 1 × , may change with time | Caused by high nozzle loads, casing not free to expand, soft foot or foundation distortion |
| Piping forces | 1 × , 2 × | Equally high 1 × and 2 × | Causes misalignment between bearings or between coupled equipment |
| Rotor and bearing critical | 1 × | High 1 × , on rundown plot 1 × decreases rapidly, may also show a large phase angle change | More common in machines originally designed for fixed speed operation, later converted to variable speed operation |
| Structural resonance | 1 × , 2 × | High 1 × and some 2 ×;can be easily identified on run down plot | Increase or decrease stiffness of structure or add or remove mass to change natural frequency |
| Rotor hysteresis | 0.65 × to 0.85 × | Spectrum will show high magnitudes at 0.65 × to 0.85 × | Occurs in built up rotors with transitional fits |
| Hydraulic causes | 1 × (vane pass frequency), 2 × (vane pass frequency) | High 1 × and 2 × vane pass frequency | Common in centrifugal pumps due to flow recirculation or inadequate gap between impeller and casing |

direct reduction of the exciting force or by means of increased damping. Based on these guidelines, the following modifications were proposed to correct the problem:

1. Torsional resonance
   - Use an electrometric type coupling that has a high degree of torsional damping to reduce the magnitude of the torsional excitation forces such that the torsional stresses within the rotors are within acceptable limits.
   - Since both the pump and VFD generate excitation at 6 × rpm, their effects could be compounding one another. Introducing either five vane or seven vane impellers into the pump will eliminate this possibility.
   - Additional filters could be introduced into the VFD to reduce the 6 × and 12 × component periodic torsional excitation.
   - Consider not operating (lock out) the pump within ±10% of the frequency at which torsional resonance occurs.
2. Motor housing resonance at 2 ×
   - Although the 2 × vibration is dominant, its magnitude is within tolerable levels. The fact that some 2 × vibration is also present in the pump indicates that the 2 × vibration is perhaps caused by misalignment between the pump and the motor. This can be corrected by proper

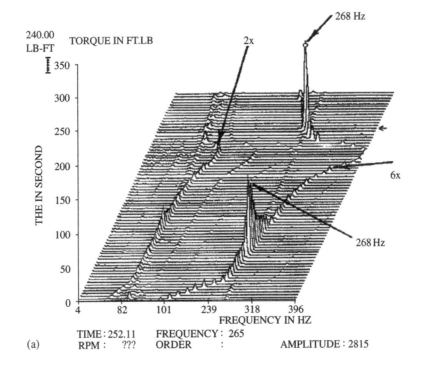

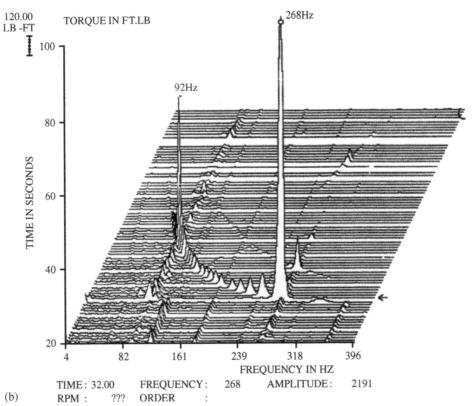

**FIGURE 9.15** (a) Torsional resonance run-up and run-down plots; (b) torsional resonance run-down plot. (*Source:* Private communique, Insight Engineering Services Ltd. Alta., Canada. With permission.)

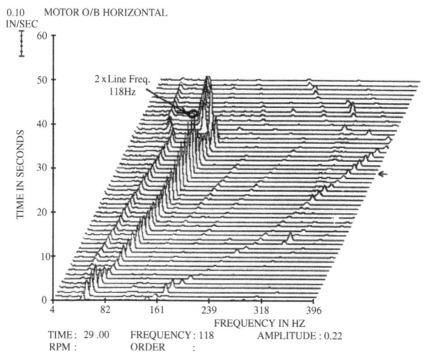

TIME : 29 .00       FREQUENCY : 118       AMPLITUDE : 0.22
RPM :       ORDER       :

**FIGURE 9.16**  Motor frame resonance. (*Source:* Private communique, Insight Engineering Services Ltd., Alta., Canada. With permission.)

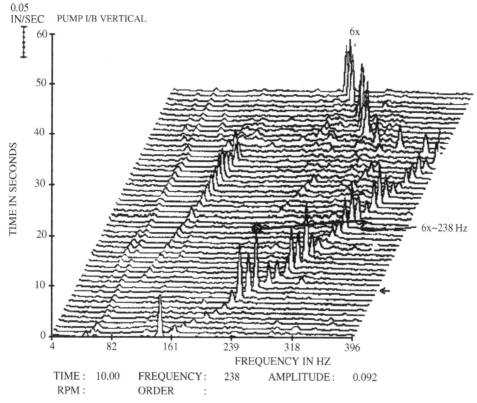

TIME :  10.00     FREQUENCY:  238     AMPLITUDE:  0.092
RPM :       ORDER       :

**FIGURE 9.17**  Pump bearing housing resonance. (*Source:* Private communique, Insight Engineering Services Ltd., Alta., Canada. With permission.)

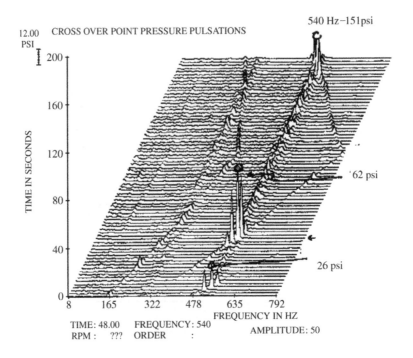

**FIGURE 9.18** Pump cross-over pipe acoustic resonance. (*Source:* Private communique, Insight Engineering Services Ltd., Alta., Canada. With permission.)

alignment and thus reducing the 2 × excitation forces. In some cases, due to an unequal air gap between the rotor and stator of the motor, the motor could generate the 2 × vibration. Under such conditions, accurate centering of the motor bearings will generally correct the problem.

3. Pump vibrations at vane passing frequency
   - Generally, high vibrations at vane passing frequency are caused by pressure pulsations generated at the discharge of the impeller. There are several hydraulic modifications that can be made to the pump to reduce the amplitude of these pulsations that occur at vane passing frequency. The most common method is to increase the gap between the impeller discharge vanes and diffuser/volute. Also, changing the ratio of the number of impeller vanes to diffuser/volute vanes can help in reducing vane passing frequency pressure pulsations and the resulting vibration.

4. Acoustic resonance in the pump cross-over pipe
   - Once the pump is constructed, it is not possible to change the acoustic natural frequency of the cross-over pipe. However, the excitation force, pressure pulsations generated at the impeller discharge, can be reduced by the methods outlined above.

---

- The root cause of a vibration problem in a rotating machine can be determined by careful study and analysis of the vibration signals.
- Industrial and international vibration standards and guidelines have been developed to ensure safe and reliable operation of rotating machinery.
- Equipment manufacturers, users, insurance companies, and public interest groups use vibration standards to control vibration to within acceptable levels.

# References

Adkins, D. and Brennen, C. 1986. Origins of hydrodynamic forces on centrifugal pump impellers, NASA CP No 2443, p. 467, In *Proceedings of a Workshop held at Texas A&M University*, Dallas, TX.

Alford, J., Protecting turbomachinery from self-excited rotor whirl, *Trans. ASME, J. Eng. Power*, 87, 333, 1965.

*API Standard 610*, 8th ed., Centrifugal Pumps for General Refinery Service, 1995.

Archer, J.S., Consistent mass matrix for distributed mass systems, *J. Struct. Div., Proc. ASCE*, 89, ST4, 161, 1963.

Baskharone, E.A., Daniel, A.S., and Hensel, S.J., Rotordynamic effects of the shroud-to-housing leakage flow in centrifugal pumps, *Trans. ASME, J. Fluid Eng.*, 116, 558, 1994.

Benckert, H., Wachter, J. 1980. Flow induced spring coefficients of labyrinth seals for application in turbomachinery, NASA CP No. 2133, p. 189, In *Proceedings of a Workshop held at Texas A&M University*, Dallas, TX.

Bentley, D. 1974. *Forced Subrotative Speed Dynamic Action of Rotating Machinery*, ASME, Dallas, TX, 74-PET-16.

Biezeno, C.B. and Grammel, R. 1954. Engineering Dynamics, Steam Turbines, Vol. III. D. Van Nostrand Co. Inc., New York (originally published in German in 1939 as Technische Dynamik by Julius Springer, Berlin, Germany).

Black, H.F., Interaction of a whirling rotor with a vibrating stator across a clearance annulus, *J. Mech. Eng. Sci.*, 10, 1, 1968.

Black, H.F. 1974. Lateral stability and vibrations of high speed centrifugal pump rotors, p. 56, In *Proceedings IUTAM Symposium on Dynamics of Rotors*, Lyngby, Denmark.

Blevins, Robert, D. 2001. Formulas for Natural Frequency and Mode Shapes, Krieger Publishing Co. Inc., Melbourne, FL.

Bolleter, U., Frei, A., Florjancic, S., Leibundgut, E., and Stürchler, R., *Rotordynamic Modeling and Testing of Boiler Feedpumps*, EPRI TR-100980, 1992.

Bolleter, U., Leibundgut, E., Stürchler, R., and McCloskey, T. 1989. Hydraulic interaction and excitation forces of high head pump impellers, p. 187, In *Proceedings of the Third ASCE/ASME Mechanical Conference*, La Jolla, CA.

Bolleter, U., Wyss, A., Welte, I., and Stürchler, R., Measurement of hydrodynamic interaction matrices of boiler feed pump impellers, *Trans. ASME, J. Vib. Stress Reliab. Des.*, 1985, SME, 85-DET-147, New York.

Brennen, C., Acosta, A., and Caughey, T. 1980. A test program to measure cross-coupling forces in centrifugal pumps and compressors, NASA CP No. 2133, p. 229, In *Proceedings of a Workshop held at Texas A&M University*, Dallas, TX.

Childs, D.W. 1986. Force and moment rotordynamic coefficients for pump-impeller shroud surfaces, NASA CP No. 2443, p. 467, In *Proceedings of a workshop held at Texas A&M University Dallas*, TX.

Childs, D. 1993. Turbomachinery Rotordynamics, Wiley, New York.

Chree, C., The whirling and transverse vibration of rotating shafts, *Phil. Mag.*, 7, 504, 1904.

COJOUR, User's Guide: Dynamic Coefficients for Fluid Film Journal Bearings, EPRI CS-4093.

Crandall, S.H. and Brosens, P.J., Whirling of unsymmetrical rotors, *J. Appl. Mech.*, 28, 567, 1961.

Dunkerly, S., On the whirling and vibration of shafts, *Phil. Trans. R. Soc., London A*, 185, 279, 1894.

Ehrich, F.F. 1999. Handbook of Rotordynamics, Revised ed., Krieger Publishing Co. Inc., Melbourne, FL.

Foppl, A., Das Problem der Laval'schen Turbinewelle, *Civilingenieur*, 41, 333, 1885.

Glienicke, J. 1966. Experimental investigation of the stiffness and damping coefficients of turbine bearings and their application to instability prediction, p. 122, In *Proceedings of the Journal Bearings for Reciprocating and Turbo Machinery Symposium*, Nottingham, UK.

Gorman and Daniel, J. 1975. *Free Vibration Analysis of Beams and Shafts*, Wiley, New York.

Greenhill, A.G., On the strength of shafting when exposed both to torsion and to end thrust, *Proc. I. Mech. Eng (London)*, 182, 1883.

Gunter, E.J. Jr., Dynamic stability of rotor-bearing systems, *NASA SP-113*, 1966.

Hagg, A.C. and Sankey, G.O., Elastic and damping properties of oil-film journal bearings for application to unbalance vibration calculations, *Trans. ASME, J. Appl. Mech.*, 25, 141, 1958.

Harris, T. 1991. Rolling Bearing Analysis, 3rd ed., Wiley, New York.

Holzer, H. 1921. Die Berechnung der Drehschwingungen, Springer, Berlin.

Houbolt, J.C. and Reed, W.H., Propeller nacelle whirl flutter, *Inst. Aerospace Sci.*, 1, 61, 1961.

ISO 10816-1. *Mechanical vibration—evaluation of machine vibration by measurements on non-rotating parts.* Part 1. General Guidelines, ISO, Geneva, Switzerland, 1995.

ISO 10816-2. *Mechanical vibration—evaluation of machine vibration by measurements on non-rotating parts.* Part 2. Land-based Steam Turbines and Generators in excess of 50 MW with normal operating speeds of 1500 r/min, 1800 r/min, 3000 r/min and 3600 r/min, ISO, Geneva, Switzerland, 2001.

ISO 10816-3. *Mechanical vibration—evaluation of machine vibration by measurements on non-rotating parts.* Part 3. Industrial machines with nominal power above 15 kW and nominal speeds between 120 r/min and 15 000 r/min when measured in situ, ISO, Geneva, Switzerland, 1998.

ISO 10816-4. *Mechanical vibration—evaluation of machine vibration by measurements on non-rotating parts.* Part 4. Gas Turbine Driven Sets Excluding Aircraft Derivations, ISO, Geneva, Switzerland, 1998.

ISO 10816-5. *Mechanical vibration—evaluation of machine vibration by measurements on non-rotating parts.* Part 5. Machine Sets in Hydraulic Power Generating and Pumping Plants, ISO, Geneva, Switzerland, 2000.

ISO 7919-1. *Mechanical vibrations of non-reciprocating machines—measurement on Rotating Shafts and Evaluation Criteria.* Part 1. General Guidelines, ISO, Geneva, Switzerland, 1996.

ISO 7919-2. *Mechanical vibrations of non-reciprocating machines—Measurement on Rotating Shafts and Evaluation Criteria.* Part 2. Land-Based Steam Turbines and Generators in Excess of 50 MW with Normal Operating Speeds of 1500 r/min, 1800 r/min, 3000 r/min and 3600 r/min, ISO, Geneva, Switzerland, 2001.

ISO 7919-3. *Mechanical vibrations of non-reciprocating machines—Measurement on Rotating Shafts and Evaluation Criteria.* Part 3. Coupled Industrial Machines, ISO, Geneva, Switzerland, 1996.

ISO 7919-4. *Mechanical vibrations of non-reciprocating machines—Measurement on Rotating Shafts and Evaluation Criteria.* Part 4. Gas Turbine Sets, ISO, Geneva, Switzerland, 1996.

ISO 7919-5. *Mechanical vibrations of non-reciprocating machines—Measurement on Rotating Shafts and Evaluation Criteria.* Part 5 Machine Sets in Hydraulic Power Generating and Pumping Plants, ISO, Geneva, Switzerland, 1997.

Jeffcott, H.H., The lateral vibration of loaded shafts in the neighbourhood of a whirling speed—the effect of want of balance, *Phil. Mag.*, 37, 304, 1919.

Jery, B., Acosta, A., Brennen, C., and Caughey, T. 1984. Hydrodynamic impeller stiffness, damping, and inertia in the rotordynamics of centrifugal flow pumps, NASA CP No. 2338, p. 137, In *Proceedings of a workshop held at Texas A&M University*, Dallas, TX.

Jones, A., A general theory for elastically constrained ball and radial roller bearings under arbitrary load and speed conditions, *Trans. ASME J. Basic Eng.*, 82, 309, 1960.

Kimball, A.L. Jr., Internal friction theory of shaft whirling, *Gen. Electr. Rev.*, 27, 244, 1924.

Kirk, R.G., Donald, G.N. 1983. Design Criteria of Improved Stability of Centrifugal Compressors, AMD-Vol. 55, Rotor Dynamical Instability. ASME, New York, p. 59.

Kramer, E. 1993. Dynamics of Rotors and Foundations, Springer, Berlin.

Lomakin, A.A., Calculating the critical speed and the conditions to ensure dynamic stability of the rotors in high pressure hydraulic machines, taking account of the forces in the seals, *Energomashinostroenie*, 4, 1, 1958.

Lund, J.W., Spring and damping coefficients for the tilting-pad journal bearing, *Trans. ASLE*, 7, 342, 1964.

Lund, J.W. 1965. Rotor-bearing Dynamics Design Technology. Part III. Design Handbook for Fluid-film Bearings, AFAPL-TR-64-45. Wright-Patterson Air Force Base, Dayton, OH.

Lund, J.W. 1968. Rotor-bearing Dynamics Design Technology. Part VII. The Three Lobe Bearing and Floating Ring Bearing, AFAPL-TR-65-45. Wright-Patterson Air Force Base, Dayton, OH.

Lund, J.W., Sensitivity of the critical speeds of a rotor to changes in the design, *Trans. ASME, J. Mech. Des.*, 102, 115, 1979.

Lund, J.W. and Orcutt, F.K., Calculation and experiments on the unbalance response of a flexible rotor, *Trans. ASME, J. Eng. Ind.*, 89, 785, 1967.

Meirovitch, L. 1986. Elements of Vibration Analysis, 2nd ed., McGraw-Hill, New York.

Miller, D.F., Forced lateral vibration of beams on damped flexible end supports, *Trans. ASME, J. Appl. Mech.*, 20, 167, 1953.

Moore, J.J. and Palazzolo, A.B., Rotordynamic force prediction of Whirling Centrifugal Impeller Shroud passages using Computational Fluid Dynamic techniques, *Trans. ASME, J. Eng. Gas Turbine Power*, 123, 910, 2001.

Muszynska, A., Whirl and whip—rotor/bearing stability problems, *J. Sound Vib.*, 110, 443, 1986.

Myklestad, N.O., A new method for calculating natural modes of uncoupled bending vibrations of airplane wings and other types of beams, *J. Aeronaut. Sci.*, 11, 153, 1944.

Nelson, H., A Finite rotating shaft element using Timoshenko beam theory, *Trans. ASME, J. Mech. Des.*, 102, 793, 1980.

Nelson, H. and McVaugh, J., The dynamics of rotor-bearing systems using finite elements, *Trans. ASME, J. Eng. Ind.*, 98, 593, 1976.

Newkirk, B.L., *Shaft Whipping Gen. Electr. Rev.*, 27, 169, 1924.

Newkirk, B.L. and Taylor, H.D., Shaft whipping due to oil action in journal bearings, *Gen. Electr. Rev.*, 28, 559, 1925.

Ohashi, H., Hatanaka, R., and Sakurai, A. 1986. Fluid force testing machine for whirling centrifugal impeller, In *Proceedings of the International Federation for Theory of Machines and Mechanisms, International Conference on Rotordynamics*, JSME, Tokyo, Japan.

Orcutt, F.K., The steady-state and dynamic characteristics of the tilting-pad journal bearing in laminar and turbulent flow regimes, *Trans. ASME, J. Lubricat. Technol.*, 89, 392, 1967.

Perera, L. 2002. Private communiqué, Insight Engineering Services Ltd, Alta., Canada.

Pinkus, O. and Sternlicht, B. 1961. Theory of Hydrodynamic Lubrication, McGraw-Hill, New York.

Poritsky, H., Contribution to the theory of oil whip, *Trans. ASME*, 75, 1153, 1953.

Prohl, M.A., A general method for calculating Critical Speeds of flexible rotors, *Trans. ASME, J. Appl. Mech.*, 12, A-142, 1945.

Rajan, M., Nelson, H.D., and Chen, W.J., Parameter sensitivity in the dynamics of rotor-bearing systems, *Trans. ASME, J. Vib. Acoust. Stress Reliab. Des.*, 108, 197, 1986.

Rajan, M., Rajan, S.D., and Nelson, H.D., and Chen, W.J., Optimal placement of critical speeds in rotor-bearing systems, *Trans. ASME, J. Vib. Acoust. Stress Reliab. Des.*, 109, 152, 1987.

Rankine, W.J.M., On the centrifugal force of rotating shafts, *Engineer*, 249, 9, 1869.

Rathbone, T.C., Vibration tolerances, *Power Plant Eng.*, November, 1939.

Rayleigh, J.W.S. 1945. Theory of Sound, Dover Publications, New York.

Robertson, D., Whirling of a journal in a sleeve bearing, *Phil. Mag.*, 15, 96, 113, 1933.

Robertson, D., Transient whirling of a rotor, *Phil. Mag.*, 20, 793, 1935.

Ruhl, R.L. and Booker, J.F., A finite element model for distributed parameter turborotor systems, *Trans. ASME, J. Eng. Ind.*, 94, 126, 1972.

Smith, D.M., The motion of a rotor carried by a flexible shaft in flexible bearings, *Proc. R. Soc. London A*, 142, 92, 1933.

Someya, T. 1989. *Journal Bearing Databook*, Springer, New York.

Southwell, R.V. and Gough B.S., 1921, *Complex Stress Distributions in Engineering Materials*, British Association For Advancement of Science Reports, 345.

Sternlicht, B., Elastic and damping properties of cylindrical journal bearings, *Trans. ASME, J. Basic Eng.*, 81, 101, 1959.

Stodola, A. 1927. *Steam and Gas Turbines*, Vol. I, McGraw-Hill, New York.

Taylor, E.S. and Browne, K.A., Vibration isolation of aircraft power plants, *J. Aeronaut. Sci.*, 6, 43, 1938.

Thomas, H.J., Unstable natural vibration of turbine rotors excited by the axial flow in stuffing boxes and blading, *Bull. AIM*, 71, 1039, 1958.

Timoshenko, S., Young, D.H., and Weaver, W. Jr. 1974. Vibration Problems in Engineering, 4th ed., Wiley, New York.

Urlichs, K., Leakage flow in thermal turbo-machines as the origin of vibration exciting lateral forces, NASA, TT-17409, 1977.

Vance, J.M. 1988. Rotordynamics of Turbomachinery, Wiley, New York.

Warner, P.C., Static and dynamic properties of partial journal bearings, *Trans. ASME, J. Basic Eng.*, 85, 247, 1963.

Wolf, J.A. 1968. Whirl Dynamics of a Rotor Partially Filled with Liquids, ASME, New York, 68-WA/APM-25.

# 10

# Regenerative Chatter in Machine Tools

Robert G. Landers
*University of Missouri at Rolla*

**Summary**

*Regenerative chatter, a result of unstable interactions between machining forces and structural deflections, is a great limitation in machining operations. This chapter describes the modeling, analysis, simulation, detection, and control of regenerative chatter in machining operations, and, in particular, turning and face milling. An analytical method is applied to calculate the limiting stable depth-of-cut and corresponding spindle speeds to generate stability lobe diagrams. The method is applied to both turning and face-milling operations. Time-domain simulation is described and applied to turning and face-milling operations. Methods for chatter detection are presented and experimental results from a face-milling operation are given. Chatter suppression techniques, namely spindle-speed selection, feed selection, depth-of-cut selection, and spindle-speed variation, are presented and two simulations of a turning operation are used to illustrate the spindle-speed selection and spindle-speed variation techniques. Finally, a case study of a face-milling operation is presented. The nomenclature used in the presentation is listed at the end of the chapter.*

## 10.1 Introduction

Regenerative chatter is a major limitation in machining operations. This phenomenon is a result of an unstable interaction between the machining forces and the structural deflections. The forces generated when the cutting tool and part come into contact produce significant structural deflections. These structural deflections modulate the chip thickness that, in turn, changes the machining forces. For certain cutting conditions, this closed-loop, self-excited system becomes unstable and regenerative chatter occurs. Regenerative chatter may result in excessive machining forces and tool wear, tool failure, and scrap parts due to unacceptable surface finish, thus severely decreasing operation productivity and part quality.

A typical chatter stability chart, the so-called stability lobe diagram, is shown in Figure 10.1. If the process parameters are above the stability borderline, chatter will occur, and if the process parameters are below the stability borderline, chatter will not occur. The asymptotic stability borderline is the depth-of-cut below which stable machining is guaranteed regardless of the spindle speed. The lobed nature of the stability borderline allows stable pockets to form; thus, at specific ranges of spindle speeds, the depth-of-cut may be substantially increased beyond the asymptotic stability limit. These pockets become smaller as the spindle speed decreases. The

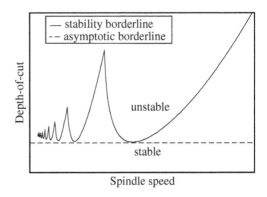

**FIGURE 10.1**  Stability lobe diagram.

stability borderline is "pulled up" for low spindle speeds due to process damping (i.e., the back side of the tool rubbing on the part surface). If accurate models of the structural components and the cutting process are available, the stability lobe diagram may be used to plan chatter-free machining operations.

The analysis of regenerative chatter as the interaction between the cutting forces and structural vibrations was established by Tobias (1965) and Koenigsberger and Tlusty (1971). Merritt (1965) used systems theory to determine stability and construct the stability lobe diagram by generating specialized plots from the harmonic solutions of the system's characteristic equation. Chatter analysis reveals a natural delay in the system leading many researchers to use Nyquist techniques to generate stability lobe diagrams (Minis et al., 1990a, 1990b; Lee and Liu, 1991a, 1991b; Minis and Yanushevsky, 1993). A set of process parameters is selected and the characteristic equation is formed. The Nyquist criterion is applied to determine if the system for this process parameter set is stable. The depth-of-cut is adjusted and the procedure is repeated until the critical depth-of-cut is determined. Another chatter analysis technique capable of generating stability lobe diagrams analytically for linear systems has recently been introduced (Altintas and Budak, 1995; Budak and Altintas, 1998a, 1998b). This technique is utilized in this chapter.

The theoretical analysis of regenerative chatter laid the foundation for developing techniques to automatically detect its occurrence and to automatically suppress it. Since there is a dominant chatter frequency, which is near a structural frequency, that occurs when chatter develops, most monitoring techniques analyze the frequency of a process variable, and chatter is detected when significant energy is present near a structural frequency. Most automatic chatter suppression routines either adjust the spindle speed to be in a pocket of the stability lobe diagram or vary the spindle speed to bring the current and previous tooth passes into phase. While automatic monitoring and control of regenerative chatter shows great promise, it has been mostly limited to laboratory applications. Therefore, commercial tools are not currently available.

While regenerative chatter in turning and face-milling operations is discussed in this chapter, this phenomenon is not limited to these specific manufacturing operations. Other machining operations for which chatter has been analyzed include end milling (Budak and Altintas, 1998a, 1998b), grinding (Inasaki et al., 2001), drilling (Tarng and Li, 1994), and so on. Also, the regenerative chatter phenomenon occurs in other manufacturing operations, most notably in rolling (Yun et al., 1998; Tlusty, 2000).

Section 10.2 and Section 10.3 present an analytical method to examine regenerative chatter in turning and face-milling operations, respectively. Section 10.4 discusses a numerical technique known as time domain simulation that may be used to analyze regenerative chatter for nonlinear systems. The subject of chatter detection is presented in Section 10.5, then methods to perform chatter suppression are discussed and illustrated in Section 10.6. Section 10.7 presents a case study of a face-milling operation.

## 10.2   Chatter in Turning Operations

A schematic of a turning operation is shown in Figure 10.2. The part structure is assumed to be perfectly rigid, while the cutting-tool structure is capable of vibrations in the longitudinal (i.e., the z) direction only. The machining force in the longitudinal direction is

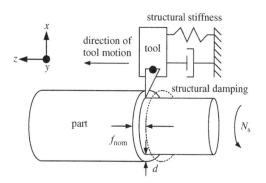

$$F(t) = Pdf(t) \qquad (10.1)$$

The depth-of-cut is assumed to be constant; however, the feed, and hence the machining force, is time-varying due to structural vibrations. It is assumed here that the machining force does not explicitly depend upon the cutting speed.

**FIGURE 10.2**   Turning operation schematic: current pass (solid line) and previous pass (dotted line).

The feed is the chip thickness in the longitudinal direction. The nominal feed is the distance the tool advances relative to the part each spindle revolution and is constant once the tool fully engages the part. However, the cutting tool vibrates, leaving an undulated surface on the part and, thus, modulates the feed. The instantaneous feed is

$$f(t) = f_{nom} + \Delta z(t) = f_{nom} + z(t) - z(t - T) \qquad (10.2)$$

The term $f_{nom}$ is the nominal feed, also known as the static feed. The term $\Delta z(t)$ is the feed due to the cutting-tool vibrations and is known as the dynamic feed. The parameter $T$ is the spindle-rotation period. The structural vibration, $z(t)$, known as the inner modulation, is the cutting-tool vibration at the current time. The delayed structural vibration $z(t - T)$, known as the outer modulation, is the cutting-tool vibration as of when the part was at the current angular during the previous spindle rotation. The modulation in feed due to structural vibrations is illustrated in Figure 10.3.

Inserting Equation 10.2 into Equation 10.1:

$$F(t) = Pdf_{nom} + Pd\Delta z(t) = F_{nom} + \Delta F(t) \qquad (10.3)$$

The force $F_{nom} = Pdf_{nom}$ is due to the nominal chip thickness and does not vary since the depth-of-cut and nominal feed are constant. The force $\Delta F(t) = Pd\Delta z(t)$ is due to changes in the nominal feed caused by structural vibrations.

The structural vibrations are related to the machining force by

$$z(s) = -g(s)F(s) \qquad (10.4)$$

where $g(s)$ is the transfer function relating the structural vibrations to the machining forces. Since $F(t) - F(t - T) = \Delta F(t) - \Delta F(t - T)$, the structural vibrations are related to the machining

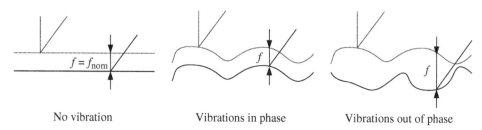

No vibration          Vibrations in phase          Vibrations out of phase

**FIGURE 10.3**   Modulation in feed due to structural vibrations in a turning operation: current pass (solid line) and previous pass (dotted line).

forces by

$$\Delta z(s) = -(1 - e^{-sT})g(s)\Delta F(s) \tag{10.5}$$

Substituting for $\Delta z$ in Equation 10.5 and rearranging:

$$\Delta F(s)\left\{1 + Pd[1 - e^{-sT}]g(s)\right\} = 0 \tag{10.6}$$

Equation 10.6 is now solved, based on the method presented by Budak and Altintas (1998a, 1998b), to determine the stability lobe diagram. Assuming the steady-state solution is a harmonic function at a single chatter frequency $\omega_c$, Equation 10.6 becomes

$$\Delta F(j\omega_c)e^{j\omega_c t}\left\{1 + Pd[1 - e^{-j\omega_c T}]g(j\omega_c)\right\} = 0 \tag{10.7}$$

where $j^2 = -1$. For nontrivial solutions of Equation 10.7, the following eigenvalue problem is derived:

$$\det\left\{1 + Pd[1 - e^{-j\omega_c T}]g(j\omega_c)\right\} = 0 \tag{10.8}$$

Since the structural dynamics are one-dimensional, Equation 10.8 reduces to

$$1 + Pd[1 - e^{-j\omega_c T}]g(j\omega_c) = 0 \tag{10.9}$$

The parameter $\Lambda$ is defined as

$$\Lambda = Pd[1 - e^{-j\omega_c T}] = \Lambda_R + j\Lambda_I \tag{10.10}$$

Using the Euler identity $1 - e^{-j\omega_c T} = 1 - \cos(\omega_c T) + j\sin(\omega_c T)$, the limiting stable depth-of-cut is

$$d_{\lim} = \left(\frac{1}{P}\right)\frac{\Lambda_R + j\Lambda_I}{1 - \cos(\omega_c T) + j\sin(\omega_c T)} \tag{10.11}$$

Equation 10.11 is rewritten as

$$d_{\lim} = \left(\frac{1}{2P}\right)\left\{\frac{\Lambda_R[1 - \cos(\omega_c T)] + \Lambda_I \sin(\omega_c T)}{1 - \cos(\omega_c T)} + j\frac{\Lambda_R \sin(\omega_c T) + \Lambda_I[1 - \cos(\omega_c T)]}{1 - \cos(\omega_c T)}\right\} \tag{10.12}$$

Since the limiting depth-of-cut must be a real number:

$$\Lambda_R \sin(\omega_c T) + \Lambda_I[1 - \cos(\omega_c T)] = 0 \tag{10.13}$$

The parameter $\kappa$ is defined as

$$\kappa = \frac{\Lambda_I}{\Lambda_R} = \frac{\sin(\omega_c T)}{1 - \cos(\omega_c T)} \tag{10.14}$$

The limiting stable depth-of-cut is solved explicitly as

$$d_{\lim} = \frac{\Lambda_R}{2P}(1 + \kappa^2) \tag{10.15}$$

Note that $\Lambda_R$ must be positive for $d_{\lim}$ to be positive. From Equation 10.9, the parameter $\Lambda$ is

$$\Lambda = -\frac{1}{g(j\omega_c)} \tag{10.16}$$

Equation 10.16 is used to determine $\Lambda_R$ and $\Lambda_I$, and these values are used to solve for $d_{\lim}$. Next, the spindle speed at which the limiting depth-of-cut occurs is determined. The trivial solution to Equation 10.14 is

$$\omega_c T = 0 + 2l\pi, \quad l = 0, 1, 2, \ldots \tag{10.17}$$

The quantity $\omega_c T$ may be interpreted as the number of vibration cycles during a spindle rotation. The trivial solution indicates that the successive vibrations are in phase (i.e., there is no regeneration). The nontrivial solution to Equation 10.14 is

$$\cos(\omega_c T) = \frac{\kappa^2 - 1}{\kappa^2 + 1} \tag{10.18}$$

and may be rewritten as

$$\omega_c T = \varepsilon + 2l\pi, \quad l = 0, 1, 2, \dots \tag{10.19}$$

where

$$\varepsilon = \cos^{-1}\left(\frac{\kappa^2 - 1}{\kappa^2 + 1}\right) \tag{10.20}$$

The parameter $\varepsilon$ is the fraction of the vibration cycles during a spindle rotation. The angle of $\Lambda$ in the complex plane is

$$\varphi = \tan^{-1}\left(\frac{\Lambda_I}{\Lambda_R}\right) = \tan^{-1}(\kappa) \tag{10.21}$$

Substituting $\kappa = \tan(\varphi)$ into Equation 10.18 yields

$$\cos(\omega_c T) = -\cos(2\varphi) \tag{10.22}$$

A solution to Equation 10.22 is

$$\omega_c T = \pi - 2\varphi + 2l\pi, \quad l = 0, 1, 2, \dots \tag{10.23}$$

Comparing Equation 10.19 and Equation 10.23, it is seen that the fraction of vibration cycles is $\varepsilon = \pi - 2\varphi$. Since $0 \le \varepsilon \le 2\pi$, one must ensure that $-\pi/2 \le \varphi \le \pi/2$ when computing $\varphi$. For example, if Equation 10.21 is solved using a four-quadrant inverse tangent function whose solution is bounded between $-\pi$ and $\pi$, then $-\pi/2 \le \varphi \le \pi/2$ since $\Lambda_R$ is positive. For milling applications, it will be seen that $\Lambda_R$ must be negative; therefore, the following conditions must be enforced to ensure $0 \le \varepsilon \le 2\pi$:

$$\begin{aligned} \text{if } \Lambda_I < 0 \quad &\text{then } \varphi \rightarrow \varphi + \pi \\ \text{if } \Lambda_I > 0 \quad &\text{then } \varphi \rightarrow \varphi - \pi \end{aligned} \tag{10.24}$$

The spindle speed is

$$N_s = \frac{60}{T} = \frac{60\omega_c}{\varepsilon + 2l\pi}, \quad l = 0, 1, 2, \dots \tag{10.25}$$

To construct a stability lobe diagram, the following steps are implemented:

1. Select a chatter frequency ($\omega_c$) near a dominant structural frequency.
2. Calculate $\Lambda_R$ and $\Lambda_I$ using Equation 10.16.
3. Calculate $d_{\lim}$ using Equation 10.15.
4. Select a stability lobe number ($l$) and calculate $N_s$ using Equation 10.25. The point $(N_s, d_{\lim})$ is the point on the stability lobe diagram corresponding to the chatter frequency, $\omega_c$, and the stability lobe number, $l$.
5. Repeat Step 4 for the desired number of stability lobes. The result is a vector of spindle speeds, $\vec{N}_s = \{N_{s_1} \quad N_{s_2} \quad \cdots \quad N_{s_n}\}$. Each point $\{(N_{s_1}, d_{\lim}) \quad (N_{s_2}, d_{\lim}) \quad \cdots \quad (N_{s_n}, d_{\lim})\}$. corresponds to a different stability lobe, and all of the points correspond to the chatter frequency $\omega_c$.
6. Select another chatter frequency and repeat Steps 2 to 5. In this manner, the stability lobe diagram is constructed. The smaller the difference between successive chatter frequencies, the greater the resolution of the stability lobe diagram. In general, the lobes will overlap. In this case, the

minimum limiting depth-of-cut is the smallest depth-of-cut. If the lobes do not overlap, then the range of chatter frequencies must be increased.

## 10.2.1 Example 1

The feed force for a turning operation is given by Equation 10.1 and the structural dynamics are given by Equation 10.26. An analytical expression for the limiting depth-of-cut and corresponding spindle speed for a given chatter frequency and stability lobe number is developed. The stability lobe diagram is plotted for $P = 0.6$ kN/mm$^2$, $\omega_n = 600$ Hz, $\zeta = 0.2$, and $k = 12$ kN/mm. The stability lobe diagram is compared to stability lobe diagrams for $\zeta = 0.1, 0.3$, and $0.4$. The stability lobe diagram is then compared with stability lobe diagrams for $\omega_n = 500, 700$, and $800$ Hz. The first ten lobes are included for all stability lobe diagrams:

$$\ddot{z}(t) + 2\zeta\omega_n\dot{z}(t) + \omega_n^2 z(t) = -\frac{\omega_n^2}{k}F(t) \tag{10.26}$$

The parameter $\Lambda$ is

$$\Lambda = \frac{-1}{g(j\omega_c)} = \Lambda_R + j\Lambda_I = \frac{k}{\omega_n^2}(\omega_c^2 - \omega_n^2) - j\frac{k}{\omega_n^2}(2\zeta\omega_c\omega_n) \tag{10.27}$$

The limiting depth-of-cut is

$$d_{\lim} = \frac{k(\omega_c^2 - \omega_n^2)}{2K\omega_n^2}\left[1 + \frac{4\zeta^2\omega_c^2\omega_n^2}{(\omega_c^2 - \omega_n^2)^2}\right] \tag{10.28}$$

The spindle speed is

$$N_s = \frac{60\omega_c}{\pi - 2\tan^{-1}\left(\dfrac{-2\zeta\omega_c\omega_n}{\omega_c^2 - \omega_n^2}\right) + 2l\pi}, \qquad l = 0, 1, 2, \ldots \tag{10.29}$$

where $l$ is the stability lobe number. Note that the chatter frequency must be greater than the structural natural frequency for the limiting depth-of-cut to be positive. The first ten lobes of the

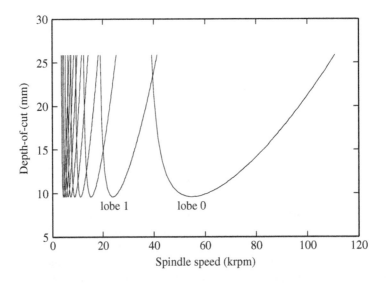

**FIGURE 10.4** Unprocessed stability lobe diagram for Example 1.

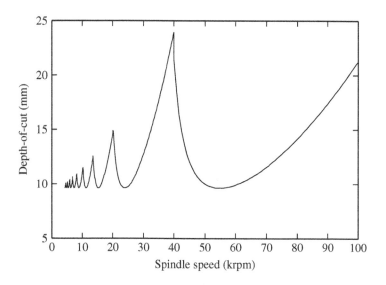

**FIGURE 10.5** Processed stability lobe diagram for Example 1.

stability lobe diagram are plotted in Figure 10.4 and Figure 10.5. In Figure 10.4, the entire solution for each of the ten stability lobes is shown. The largest stability lobe is the zeroth lobe on the right. The lobe number increases from right to left on the stability lobe diagram and successive lobes become closer together. The stability lobe diagram is processed in Figure 10.5 such that the minimum depth-of-cut is selected at each spindle speed showing the true stability borderline.

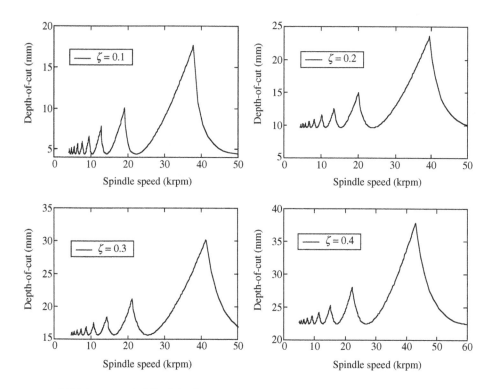

**FIGURE 10.6** Stability lobe diagrams for Example 1 with $\zeta = 0.1$, 0.2, 0.3, and 0.4.

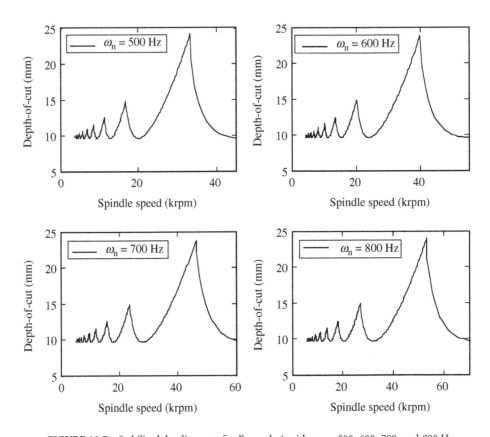

**FIGURE 10.7**  Stability lobe diagrams for Example 1 with $\omega_n = 500$, 600, 700, and 800 Hz.

In Figure 10.6, the effect of the structural damping ratio is illustrated: as the structural damping ratio increases, the lobes shift slightly to the left and the asymptotic stability boundary shifts up dramatically. The effect of the structural natural frequency is illustrated in Figure 10.7: as the structural natural frequency increases, the lobes shift to the right but the magnitude remains the same.

---

This section presented an analytical method to generate stability lobe diagrams for turning operations. The limiting depth-of-cut in a turning operation is given by

$$d_{\lim} = \frac{\Lambda_R}{2P}(1 + \kappa^2)$$

where $\Lambda_R$ is the real part of $-1/g(j\omega_c)$, $g(j\omega_c)$ is the structural transfer function evaluated at the chatter frequency, $\omega_c$, $\kappa = \Lambda_I/\Lambda_R$, and $\Lambda_I$ is the imaginary part of $-1/g(j\omega_c)$. The corresponding spindle speed is

$$N_s = \frac{60\omega_c}{\varepsilon + 2l\pi}$$

where $l = 0, 1, 2, \ldots$ is the stability lobe number and

$$\varepsilon = \cos^{-1}\left(\frac{\kappa^2 - 1}{\kappa^2 + 1}\right)$$

## 10.3    Chatter in Face-Milling Operations

A schematic of a face-milling operation is shown in Figure 10.8. In milling operations, multiple teeth may be in contact with the part simultaneously, the feed naturally varies as a function of the tooth angle even when structural vibrations are not present, and each tooth enters and leaves contact with the part every spindle revolution. The depth-of-cut is the chip thickness in the $z$ direction and is assumed to be constant, since the machine tool and part structures are typically much stiffer in the $z$ direction than in the $x$ and $y$ directions.

The instantaneous feed of the $i$th tooth, illustrated in Figure 10.9, is

$$f_i(t) = f_t \cos[\theta_i(t)] + \Delta x(t) \cos[\theta_i(t)] + \Delta y(t) \sin[\theta_i(t)] \tag{10.30}$$

where

$$\Delta x(t) = \{x_t(t) - x_t(t - T_t)\} - \{x_p(t) - x_p(t - T_t)\} \tag{10.31}$$

$$\Delta y(t) = \{y_t(t) - y_t(t - T_t)\} - \{y_p(t) - y_p(t - T_t)\} \tag{10.32}$$

The term $f_t \cos[\theta_i(t)]$ in Equation 10.30 represents the feed due to the distance the part advances relative to the cutting tool each tooth rotation and is known as the static feed. The terms $\Delta x(t) \cos[\theta_i(t)]$

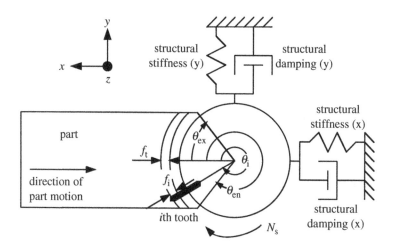

**FIGURE 10.8**    Face milling operation schematic: current pass (solid line), previous pass (dotted line), and depth-of-cut in $z$ direction.

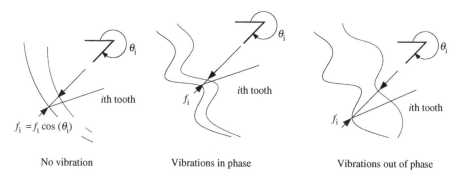

**FIGURE 10.9**    Modulation in feed due to structural vibrations in a face-milling operation: current pass (solid line) and previous pass (dotted line).

and $\Delta y(t) \sin[\theta_i(t)]$ in Equation 10.30 represent the feed due to tool and part vibrations in the $x$ and $y$ directions, respectively, at the tooth angle $\theta_i(t)$, and are known as the dynamic feed.

The machining forces in the $x$ and $y$ directions, respectively, are

$$F_x(t) = df_t \sum_{i=1}^{N_t} \left\{ -P_T \cos(\psi_r) \cos^2[\theta_i(t)] + P_C \cos[\theta_i(t)] \sin[\theta_i(t)] \right\} \sigma[\theta_i(t)]$$

$$+ d\Delta x(t) \sum_{i=1}^{N_t} \left\{ -P_T \cos(\psi_r) \cos^2[\theta_i(t)] + P_C \cos[\theta_i(t)] \sin[\theta_i(t)] \right\} \sigma[\theta_i(t)]$$

$$+ d\Delta y(t) \sum_{i=1}^{N_t} \left\{ -P_T \cos(\psi_r) \sin[\theta_i(t)] \cos[\theta_i(t)] + P_C \sin^2[\theta_i(t)] \right\} \sigma[\theta_i(t)] \tag{10.33}$$

$$F_y(t) = df_t \sum_{i=1}^{N_t} \left\{ -P_T \cos(\psi_r) \cos[\theta_i(t)] \sin[\theta_i(t)] - P_C \cos^2[\theta_i(t)] \right\} \sigma[\theta_i(t)]$$

$$+ d\Delta x(t) \sum_{i=1}^{N_t} \left\{ -P_T \cos(\psi_r) \cos[\theta_i(t)] \sin[\theta_i(t)] - P_C \cos^2[\theta_i(t)] \right\} \sigma[\theta_i(t)]$$

$$+ d\Delta y(t) \sum_{i=1}^{N_t} \left\{ -P_T \cos(\psi_r) \sin^2[\theta_i(t)] - P_C \cos[\theta_i(t)] \sin[\theta_i(t)] \right\} \sigma[\theta_i(t)] \tag{10.34}$$

where

$$\sigma[\theta_i(t)] = \begin{cases} 1 & \text{if } \theta_{en} \le \theta_i(t) \le \theta_{ex} \\ 0 & \text{if } \theta_{en} > \theta_i(t) > \theta_{ex} \end{cases} \tag{10.35}$$

The function $\sigma[\theta_i(t)]$ determines if the $i$th tooth is in contact with the part at the tooth angle, $\theta_i(t)$. The first terms in Equation 10.33 and Equation 10.34 are the machining forces acting on the tool in the $x$ and $y$ directions, respectively, due to the static feed. The second terms in Equation 10.33 and Equation 10.34 are the machining forces acting on the tool in the $x$ and $y$ directions, respectively, due to the dynamic feed resulting from structural vibrations in the $x$ direction. The third terms in Equation 10.33 and Equation 10.34 are the machining forces acting on the tool in the $x$ and $y$ directions, respectively, due to the dynamic feed resulting from structural vibrations in the $y$ direction.

The dynamic portion of the face milling force process model may be written compactly as

$$\begin{bmatrix} \Delta F_x(t) \\ \Delta F_y(t) \end{bmatrix} = dA(t) \begin{bmatrix} \Delta x(t) \\ \Delta y(t) \end{bmatrix} = d \begin{bmatrix} A_{11}(t) & A_{12}(t) \\ A_{21}(t) & A_{22}(t) \end{bmatrix} \begin{bmatrix} \Delta x(t) \\ \Delta y(t) \end{bmatrix} \tag{10.36}$$

where

$$A_{11}(t) = \sum_{i=1}^{N_t} \left\{ -P_T \cos(\psi_r) \cos^2[\theta_i(t)] + P_C \cos[\theta_i(t)] \sin[\theta_i(t)] \right\} \sigma[\theta_i(t)] \tag{10.37}$$

$$A_{12}(t) = \sum_{i=1}^{N_t} \left\{ -P_T \cos(\psi_r) \sin[\theta_i(t)] \cos[\theta_i(t)] + P_C \sin^2[\theta_i(t)] \right\} \sigma[\theta_i(t)] \tag{10.38}$$

$$A_{21}(t) = \sum_{i=1}^{N_t} \left\{ -P_T \cos(\psi_r) \cos[\theta_i(t)] \sin[\theta_i(t)] - P_C \cos^2[\theta_i(t)] \right\} \sigma[\theta_i(t)] \tag{10.39}$$

$$A_{22}(t) = \sum_{i=1}^{N_t} \left\{ -P_T \cos(\psi_r) \sin^2[\theta_i(t)] - P_C \cos[\theta_i(t)] \sin[\theta_i(t)] \right\} \sigma[\theta_i(t)] \tag{10.40}$$

These coefficients modulate the instantaneous feed as the tooth angular displacement changes. The summation from $i = 1$ to $N_t$ represents the contribution to this modulation for each of the $N_t$ teeth. Note the matrix $A(t)$ is time-varying and periodic with the tooth-passing period, $T_t$. For chatter analysis, the matrix $A(t)$ is typically expanded in a Fourier series using the zeroth term (Minis and Yanushevsky, 1993; Budak and Altintas, 1998a). The zeroth term of the Fourier expansion of the force process matrix $A(t)$ is

$$A^0 = \frac{N_t}{2\pi} \begin{bmatrix} A_{11}^0 & A_{12}^0 \\ A_{21}^0 & A_{22}^0 \end{bmatrix} \tag{10.41}$$

where

$$A_{11}^0 = \frac{1}{2}\left[ -P_T \cos(\psi_r)\left\{\theta + \frac{1}{2}\sin(2\theta)\right\} + P_C \sin^2(\theta) \right]_{\theta=\theta_{en}}^{\theta=\theta_{ex}} \tag{10.42}$$

$$A_{12}^0 = \frac{1}{2}\left[ -P_T \cos(\psi_r)\sin^2(\theta) + P_C\left\{\theta - \frac{1}{2}\sin(2\theta)\right\} \right]_{\theta=\theta_{en}}^{\theta=\theta_{ex}} \tag{10.43}$$

$$A_{21}^0 = \frac{1}{2}\left[ -P_T \cos(\psi_r)\sin^2(\theta) - P_C\left\{\theta + \frac{1}{2}\sin(2\theta)\right\} \right]_{\theta=\theta_{en}}^{\theta=\theta_{ex}} \tag{10.44}$$

$$A_{22}^0 = \frac{1}{2}\left[ -P_T \cos(\psi_r)\left\{\theta - \frac{1}{2}\sin(2\theta)\right\} - P_C \sin^2(\theta) \right]_{\theta=\theta_{en}}^{\theta=\theta_{ex}} \tag{10.45}$$

The dynamic force process is now approximated by the linear, time-invariant relationship:

$$\begin{bmatrix} \Delta F_x(t) \\ \Delta F_y(t) \end{bmatrix} = dA^0 \begin{bmatrix} \Delta x(t) \\ \Delta y(t) \end{bmatrix} \tag{10.46}$$

The tool and part vibrations, respectively, are related to the machining forces by

$$\begin{bmatrix} x_t(s) \\ y_t(s) \end{bmatrix} = G_t(s) \begin{bmatrix} F_x(s) \\ F_y(s) \end{bmatrix} = \begin{bmatrix} G_{t_{11}}(s) & G_{t_{12}}(s) \\ G_{t_{21}}(s) & G_{t_{22}}(s) \end{bmatrix} \begin{bmatrix} F_x(s) \\ F_y(s) \end{bmatrix} \tag{10.47}$$

$$\begin{bmatrix} x_p(s) \\ y_p(s) \end{bmatrix} = -G_p(s) \begin{bmatrix} F_x(s) \\ F_y(s) \end{bmatrix} = -\begin{bmatrix} G_{p_{11}}(s) & G_{p_{12}}(s) \\ G_{p_{21}}(s) & G_{p_{22}}(s) \end{bmatrix} \begin{bmatrix} F_x(s) \\ F_y(s) \end{bmatrix} \tag{10.48}$$

where $G_t(s)$ and $G_p(s)$ are the transfer functions relating the tool structural and part structural vibrations, respectively, to the machining forces. The negative sign in Equation 10.48 is due to the fact that the forces acting on the part are equal in magnitude and opposite in direction to the machining forces given in Equation 10.33 and Equation 10.34. Since

$$\begin{bmatrix} F_x(t) \\ F_y(t) \end{bmatrix} - \begin{bmatrix} F_x(t - T_t) \\ F_y(t - T_t) \end{bmatrix} = \begin{bmatrix} \Delta F_x(t) \\ \Delta F_y(t) \end{bmatrix} - \begin{bmatrix} \Delta F_x(t - T_t) \\ \Delta F_y(t - T_t) \end{bmatrix}$$

the structural vibrations can be related to the machining forces by

$$\begin{bmatrix} \Delta x \\ \Delta y \end{bmatrix} = (1 - e^{-sT_t})[G_t(s) + G_p(s)]\begin{bmatrix} \Delta F_x(s) \\ \Delta F_y(s) \end{bmatrix} \tag{10.49}$$

The machine tool and part vibrations are assumed to occur at a chatter frequency, $\omega_c$, when a marginally stable depth-of-cut is taken. Assuming the steady-state solution is a harmonic function at a chatter frequency, $\omega_c$, and substituting for the structural vibrations, Equation 10.49 becomes

$$\begin{bmatrix} \Delta F_x \\ \Delta F_y \end{bmatrix} e^{j\omega_c t} = \frac{dN_t}{2\pi}(1 - e^{j\omega_c T_t})G^0(j\omega_c)\begin{bmatrix} \Delta F_x \\ \Delta F_y \end{bmatrix} e^{j\omega_c t} \tag{10.50}$$

where the matrix $G^0$ is

$$G^0(j\omega_c) = \frac{2\pi}{N_t}A^0[G_t(j\omega_c) + G_p(j\omega_c)] \tag{10.51}$$

Equation 10.50 is now solved based on the method presented by Budak and Altintas (1998a, 1998b) to determine the stability lobe diagram. The characteristic equation of Equation 10.50 is

$$\det\left[I_2 - \frac{dN_t}{2\pi}(1 - e^{j\omega_c T_t})G^0(j\omega_c)\right] = 0 \tag{10.52}$$

where $I_2$ is the 2 × 2 identity matrix. The solution of Equation 10.52 yields the limiting stable depth-of-cut. The inverse of the eigenvalue of $G^0$ is defined as

$$\Lambda(j\omega_c) = \Lambda_R(j\omega_c) + j\Lambda_I(j\omega_c) = -\frac{dN_t}{2\pi}(1 - e^{j\omega_c T_t}) \tag{10.53}$$

Expanding the exponential term in Equation 10.53 and noting that the depth-of-cut must be a real number, the limiting stable depth-of-cut may be written as

$$d_{\lim} = -\frac{\pi\Lambda_R}{N_t}(1 + \kappa^2) \tag{10.54}$$

where the parameter $\kappa$ is defined by the transcendental equation

$$\kappa = \frac{\Lambda_I}{\Lambda_R} = \frac{\sin(\omega_c T_t)}{1 - \cos(\omega_c T_t)} \tag{10.55}$$

Equation 10.55 is solved for the tooth-passing period of the *l*th stability lobe and the tooth-passing period is related to the spindle speed to yield

$$N_s = \frac{60\omega_c}{N_t[\pi - 2\varphi + 2l\pi]}, \qquad l = 0, 1, 2, \dots \tag{10.56}$$

where, again, $\varphi = \tan^{-1}(\kappa)$. A chatter frequency is selected and the limiting stable depth-of-cut is calculated from Equation 10.54 corresponding to the spindle speed on the *l*th lobe as given by Equation 10.56.

### 10.3.1 Example 2

The cutting and thrust pressures in a face-milling operation are given by $P_C = 2.0$ kN/mm$^2$ and $P_T = 0.8$ kN/mm$^2$, respectively, and the lead angle is 45°. The part is assumed to be perfectly rigid and the tool structural dynamics for the x and y directions are given by Equation 10.57 and Equation 10.58, respectively. The nominal parameters are $\theta_{en} = -45°$, $\theta_{ex} = 45°$, $N_t = 4$, $k_x = 14$ kN/mm, $k_y = 17$ kN/mm, $\zeta_x = 0.15$, $\zeta_y = 0.1$, $\omega_x = 3000$ rad/s, and $\omega_y = 4000$ rad/s. Stability lobe diagrams are generated for the nominal parameters and $N_t = 1, 2$, and 8 teeth. Next, stability lobe diagrams are generated for the nominal parameters and $\theta_{ex} = 30°$, 60°, and 75°. The first 15 lobes are included for all stability lobe diagrams.

$$\ddot{x}_t(t) + 2\zeta_x\omega_x\dot{x}_t(t) + \omega_x^2 x_t(t) = \frac{\omega_x^2}{k_x}F_x(t) \tag{10.57}$$

$$\ddot{y}_t(t) + 2\zeta_y\omega_y\dot{y}_t(t) + \omega_y^2 y_t(t) = \frac{\omega_y^2}{k_y}F_y(t) \tag{10.58}$$

The effect of the number of teeth is illustrated in Figure 10.10: as the number of teeth increases, the lobes shift to the left and the asymptotic stability borderline decreases. In Figure 10.11, the effect of the exit angle is illustrated: as the exit angle increases, the asymptotic stability borderline decreases.

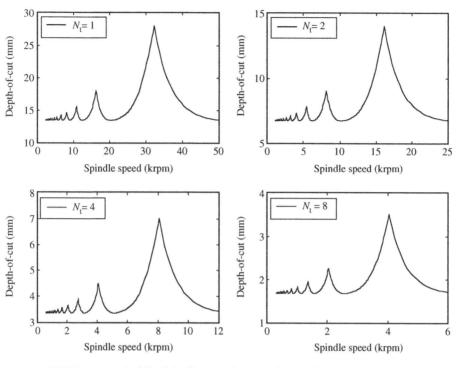

**FIGURE 10.10** Stability lobe diagrams for Example 2, with $N_t = 1, 2, 4,$ and 8.

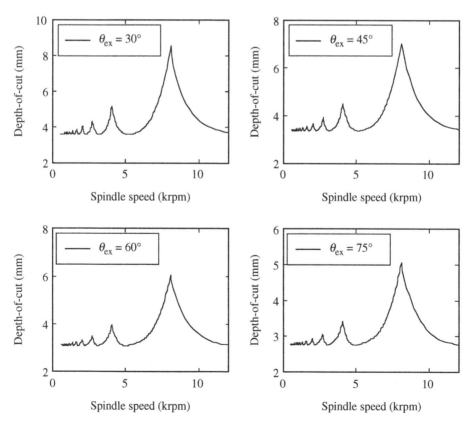

**FIGURE 10.11** Stability lobe diagrams for Example 2, with $\theta_{ex} = 30°, 45°, 60°,$ and $75°$.

This section presented an analytical method to generate stability lobe diagrams for face-milling operations. The limiting depth-of-cut in a face-milling operation is given by

$$d_{\lim} = -\frac{\pi \Lambda_R}{N_t}(1 + \kappa^2)$$

where $N_t$ is the number of teeth, $\Lambda_R$ is the inverse eigenvalue of $(2\pi/N_t)A^0[G_t(j\omega_c) + G_p(j\omega_c)]$, $A^0$ is the zeroth term of the Fourier expansion of the force process matrix, $G_t(j\omega_c)$ and $G_p(j\omega_c)$ are the transfer functions relating the tool structural and part structural vibrations, respectively, to the machining forces evaluated at the chatter frequency $\omega_c$, $\kappa = \Lambda_I/\Lambda_R$, and $\Lambda_I$ is the imaginary part of $(2\pi/N_t)A^0[G_t(j\omega_c) + G_p(j\omega_c)]$. The corresponding spindle speed is

$$N_s = \frac{60\omega_c}{N_t[\pi - 2\tan^{-1}(\kappa) + 2l\pi]}$$

where $l = 0, 1, 2, \ldots$ is the stability lobe number.

## 10.4 Time-Domain Simulation

Time-domain simulation (Tlusty and Ismail, 1981, 1983; Tlusty, 1986; Tsai et al., 1990; Lee and Liu, 1991a, 1991b; Smith and Tlusty, 1993; Elbestawi et al., 1994; Tarng and Li, 1994; Weck et al., 1994) is an alternative method for determining regenerative chatter. In a time-domain simulation, the machining forces and structural vibrations are simulated in the time domain for a specific set of process parameters and the resulting signals (i.e., forces and displacements) are examined to determine if chatter is present. The analyses presented above for the turning and face-milling operations assume that the tool always maintains contact with the part and that the cutting and thrust pressures are independent of the process parameters. Further, the face milling analysis approximated the time-varying force process matrix, $A(t)$, by the zeroth term of its Fourier expansion. With time domain simulations, nonlinear effects may be directly incorporated into the simulation; thus, more accurate stability prediction is possible. The disadvantage of time-domain simulations is the extreme computational cost that is required. For a specific spindle speed, several simulations must be conducted at different depths-of-cut; thus, the stability boundary for that spindle speed is determined iteratively. This procedure is repeated for a range of spindle speeds to construct a complete stability lobe diagram.

For turning operations, the machining force is calculated using Equation 10.1, the feed is calculated using Equation 10.2, and the tool displacement is calculated using Equation 10.4. For face-milling operations, the feed is calculated using Equation 10.30, the machining forces in the $x$ and $y$ directions are calculated using Equation 10.33 to Equation 10.35, and the tool and part displacements, respectively, are calculated using Equation 10.47 and Equation 10.48. To calculate the machining forces in the face-milling operation, the angular displacement of each tooth is required. The angular displacement of the $i$th tooth is

$$\theta_i(t) = \frac{2\pi}{60}N_s t + \frac{2\pi}{N_t}(i - 1) \tag{10.59}$$

The feed and force equations are static, while the structural displacement equations are dynamic and must be solved via a numerical integration technique. A sufficiently small time step must be utilized in the numerical integrations to account for the small system time constants associated with the large structural frequencies.

### 10.4.1 Example 3

The feed force for a turning operation is given by $F(t) = 0.6df^{0.7}(t)$. The structural dynamics are given by Equation 10.26 with the following parameters: $\omega_n = 600$ Hz, $\zeta = 0.2$, and $k = 12$ kN/mm.

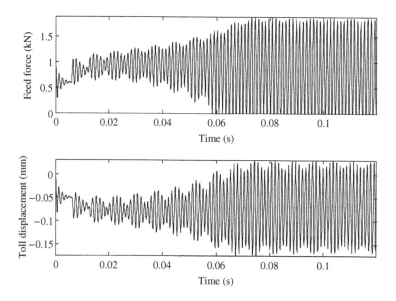

**FIGURE 10.12** Time-domain simulations for Example 3 with $f_{nom} = 0.1$ mm.

A time-domain simulation of the system including the effect of the tool disengaging from the part is constructed, and simulations for $N_s = 10,000$ rpm, $d = 8$ mm, and $f_{nom} = 0.1$ mm are conducted. The simulation is repeated for $f_{nom} = 0.2$ mm. For both simulations, the time history of the feed force and the tool displacement are plotted.

To account for the phenomenon of the tool disengaging from the part, the feed in Equation 10.2 must be modified as follows:

$$f(t) = \begin{cases} f_{nom} + z(t) - f_p(t - T) & \text{if } f_{nom} + z(t) - f_p(t - T) \geq 0 \\ 0 & \text{if } f_{nom} + z(t) - f_p(t - T) < 0 \end{cases} \qquad (10.60)$$

where

$$f_p(t) = \begin{cases} z(t) & \text{if } f_{nom} + z(t) - f_p(t - T) \geq 0 \\ -f_{nom} + f_p(t - T) & \text{if } f_{nom} + z(t) - f_p(t - T) < 0 \end{cases} \qquad (10.61)$$

If the feed at the current time is calculated to be negative, then the cutting tool has disengaged from the part and the feed is zero. The term $f_p(t)$ accounts for feed due to structural vibrations at the previous spindle rotation, even when the cutting tool disengages from the part. The results for $f_{nom} = 0.1$ mm and $f_{nom} = 0.2$ mm are shown in Figure 10.12 and Figure 10.13, respectively. As the nominal feed is increased, chatter is suppressed.

## 10.4.2 Example 4

The cutting and thrust forces in a face-milling operation are given by $F_C(t) = 1.4df^{0.6}(t)$ and $F_T(t) = 0.4df^{0.8}(t)$, respectively. The lead angle is 45°, the entry angle is $-60°$; the exit angle is 60°, the number of teeth is four, and the feed per tooth is $f_t = 0.15$ mm. The part is assumed to be perfectly rigid, and tool structural dynamics for the $x$ and $y$ directions are given by Equation 10.58 and Equation 10.59, respectively. A time-domain simulation is developed to determine the limiting stable depth-of-cut for spindle speeds of 1000 and 32,000 rpm. For both spindle speeds, the system is simulated for a depth-of-cut 10% below the limiting stable depth-of-cut and for a depth-of-cut 10% above the limiting stable depth-of-cut. The cutting force, thrust force, $x$ tool displacement, and $y$ tool displacement are plotted.

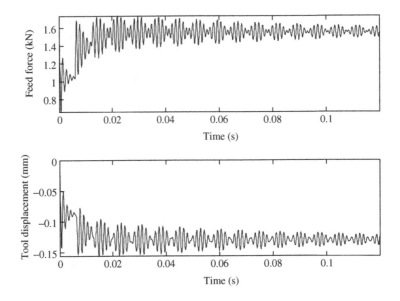

**FIGURE 10.13**  Time-domain simulations for Example 3 with $f_{nom} = 0.2$ mm.

The nonlinear effect of tooth disengagement is included:

$$\ddot{x}_t(t) + 2(0.15)(3000)\dot{x}_t(t) + 3000^2 x_t(t) = \frac{3000^2}{15} F_x(t) \tag{10.62}$$

$$\ddot{y}_t(t) + 2(0.1)(4000)\dot{y}_t(t) + 4000^2 y_t(t) = \frac{4000^2}{17} F_y(t) \tag{10.63}$$

To account for the phenomenon of the tool disengaging from the part, the feed in Equation 10.30 must be modified as follows:

$$f_i(t) = \begin{cases} f_{ci}(t) - f_{pi}(t - T_t) & \text{if } f_{ci}(t) \geq 0 \\ 0 & \text{if } f_{ci}(t) < 0 \end{cases} \tag{10.64}$$

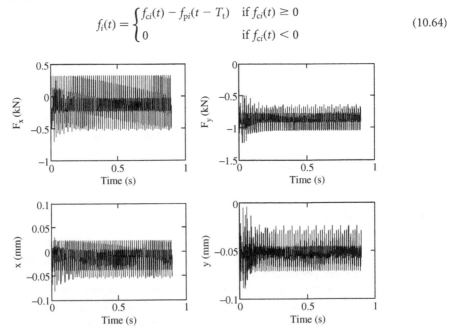

**FIGURE 10.14**  Time-domain simulation for Example 4 with $N_s = 1000$ rpm and $d = 2.025$ mm.

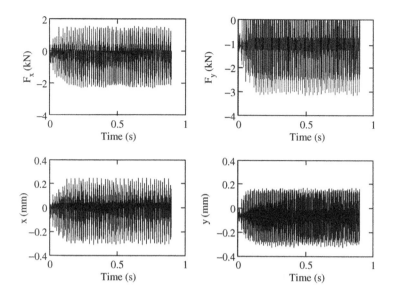

**FIGURE 10.15** Time-domain simulation for Example 4 with $N_s = 1000$ rpm and $d = 2.475$ mm.

where

$$f_{ci}(t) = f_t \cos[\theta_i(t)] + \{x_t(t) - x_p(t)\} \cos[\theta_i(t)] + \{y_t(t) - y_p(t)\} \sin[\theta_i(t)] \tag{10.65}$$

$$f_{pi}(t) = \begin{cases} \{x_t(t - T_t) - x_p(t - T_t)\} \cos[\theta_i(t)] + \{y_t(t - T_t) - y_p(t - T_t)\} \sin[\theta_i(t)] & \text{if } f_{ci}(t) \geq 0 \\ -f_t \cos[\theta_i(t)] + f_{pi}(t - T_t) & \text{if } f_{ci}(t) < 0 \end{cases} \tag{10.66}$$

Note that $\theta_i(t) = \theta_{i+1}(t - T_t)$ and $i - 1 \rightarrow N_t$ if $i = 1$. The term $f_{pi}(t)$ represents the contribution to the instantaneous feed when the previous tooth was at the same angular location as the $i$th tooth. If the tooth and part are in contact, this contribution is due to the tool and part vibrations. If the tooth and part are not in contact, this contribution is the previous contribution added to the static portion and the instantaneous feed is set to zero. Through time-domain simulations, the limiting stable depth-of-cut for

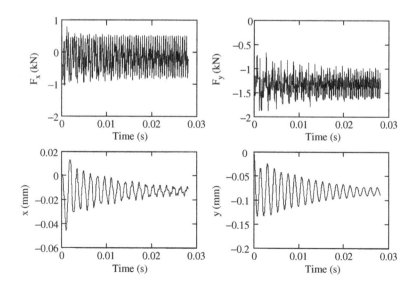

**FIGURE 10.16** Time-domain simulation for Example 4 with $N_s = 32,000$ rpm and $d = 3.105$ mm.

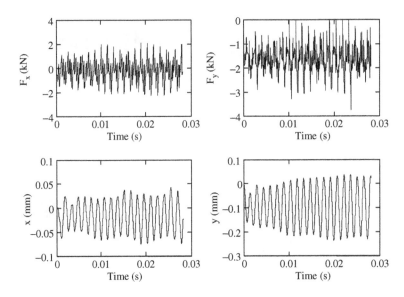

**FIGURE 10.17**   Time-domain simulation for Example 4 with $N_s = 32{,}000$ rpm and $d = 3.795$ mm.

$N_s = 1000$ rpm is found to be 2.25 mm and the limiting depth-of-cut for $N_s = 32{,}000$ rpm is found to be 3.45 mm. The results are shown in Figure 10.14 to Figure 10.17. The system is stable in Figure 10.14 and Figure 10.16, while instability is evidenced in Figure 10.15 and Figure 10.17 by the force in the $y$ direction saturating at 0 kN.

> This section presented the technique of time-domain simulation as an alternative means to analyze regenerative chatter. Time-domain simulations are the direct numerical simulations of the force process and structural vibrations. A process parameter is changed iteratively from simulation to simulation to determine the critical value at which chatter occurs. A sufficiently small time step must be utilized in the numerical integrations to account for the small system time constants associated with the large structural frequencies.

## 10.5   Chatter Detection

Regenerative chatter is easily detected by an operator due to the loud, high-pitched noise it produces and the distinctive "chatter marks" it leaves on the part surface. However, automatic detection is required for intelligent manufacturing (Cho and Ehmann, 1988; Delio et al., 1992). At the onset of chatter, process signals (e.g., force, vibration) contain significant energy at the chatter frequency. It is a well-known fact that the chatter frequency will be close to a dominant structural frequency. The most common method to detect the presence of chatter is to threshold the frequency signal of a process signal. To analyze the frequency content of a signal, a Fourier transform, or fast Fourier transform, is performed. If the frequency content of the resulting signal near a dominant chatter frequency is above a threshold value, then chatter is determined to be present. It should be noted that machining process signals also contain significant energy at the tooth-passing frequency. If the dominant structural frequencies and tooth-passing frequency are sufficiently separated, then the tooth-passing frequency may be ignored when determining the presence of chatter. If the dominant structural frequencies and tooth-passing frequency are close, then the signal must be filtered at the tooth-passing frequency using a notch filter. Also, forced vibrations, such as those resulting from the impact between the cutting tool and part, must not be allowed to falsely trigger the chatter detection algorithm. These thresholding algorithms all suffer from the lack of an analytical method of selecting a threshold value. This value is typically selected empirically and will not be valid over a wide range of cutting conditions and machining operations.

## 10.5.1 Example 5

An experimental face-milling operation, a complete description of which is given in Landers (1997), is conducted with a spindle speed of 1500 rpm and a tool with four teeth. The dominant structural frequencies are 334, 414, 653, and 716 Hz. The machining force $F_z$ is sampled at a frequency of 2000 Hz, and the time-domain signal is transformed into the frequency domain via a Fourier transform using 80 points (i.e., one spindle revolution). The power spectral density of the force signal is shown for depths-of-cut of 1.0 and 1.5 mm in Figure 10.18 and Figure 10.19, respectively. In Figure 10.18, there is significant energy at 100 Hz, which is the tooth-passing frequency. There is also significant energy at 750 Hz due to structural vibrations; however, the system did not chatter, as evidenced by the lack of chatter marks on the part and a high-pitched sound during machining. In Figure 10.19, there is significant energy at 665 Hz as well as 100 Hz. Chatter was evidenced by the chatter marks left on the part surface and the high-pitched sound during machining. The results demonstrate that the chatter frequency is 665 Hz, which is near the dominant structural frequency of 653 Hz. Note that the power spectral density at the frequency of 0 Hz is ignored in Figure 10.18 and Figure 10.19. This component is stronger than the components at all other frequencies since the machining force $F_z$ fluctuates about a static, nonzero value. In this application, a thresholding algorithm may ignore the low frequencies where the tooth-passing frequency is strong; however, if the operation

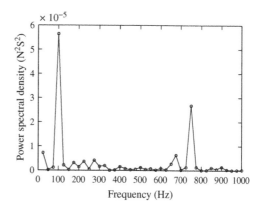

**FIGURE 10.18**   Power spectral density of $F_z$ in a face-milling operation with $d = 1.0$ mm.

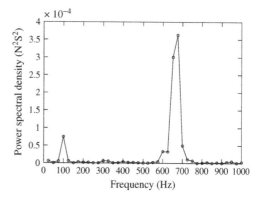

**FIGURE 10.19**   Power spectral density of $F_z$ in a face-milling operation with $d = 1.5$ mm.

were to be performed at a higher spindle speed, say 7500 rpm, or the number of teeth were increased from 4 to 20, the tooth-passing frequency would be 500 Hz, close to the structural frequencies. In this case, the force signal would have to be filtered at the tooth-passing frequency.

This section presented techniques to detect the occurrence of regenerative chatter. The phenomenon of regenerative chatter is easily detected by an operator due to the loud, high-pitched noise it produces and the distinctive "chatter marks" it leaves on the part surface. The most common method to detect the presence of chatter is to threshold the frequency signal of a process signal. In this case, one must be careful to separate out the spindle rotation and tooth-passing frequencies.

## 10.6  Chatter Suppression

Most machining process plans are derived from handbooks or from a database. Since these plans do not consider the physical machine that will be used, chatter-free operations cannot be guaranteed. Thus, multiple iterations, where the feed or spindle speed are adjusted using the operator's experience, are typically required. The tool position may also be adjusted (e.g., the depth-of-cut may be decreased) to suppress chatter and, while this is guaranteed to be effective due to the presence of the asymptotic stability borderline, this approach is typically not employed since part program must be rewritten to add multiple passes, thereby drastically decreasing productivity. The stability lobe diagram can be used as a tool to plan chatter-free machining operations and productivity can be greatly increased by selecting the process parameters to lie in a pocket between two lobes. A cutting tool design methodology (Altintas et al., 1999) has also been proposed for milling tools where the pitch is slightly adjusted such that the teeth are not evenly spaced. The variable pitch has the effect of changing the phase difference between successive teeth vibrations and, if designed properly, will suppress chatter. These design techniques are very sensitive to parameter variations and model uncertainty, and may not be used reliably for a large range of operating conditions. This section will describe methods for automatic chatter suppression.

### 10.6.1  Spindle-Speed Selection

For the stability lobe diagram generated from a system modeled as having a one-dimensional structure, it is seen that the maximum depths-of-cut are located at the tooth-passing frequencies (i.e., the number of teeth multiplied by the spindle speed) corresponding to the dominant structural frequency and integer fractions thereof. If the dominant structural frequency is known, it may be used as an aid in selecting spindle speeds; however, the structural dynamics are often unknown and may be determined only through costly testing. Further, structural dynamics change drastically over time.

It is known, however, that during chatter, the dominant frequency seen in the cutting-process output is close to a dominant structural frequency. This fact is used in Smith and Delio (1992) to suppress chatter automatically. The following steps are taken:

1. Implement a chatter detection routine to determine the presence of chatter.
2. If chatter is detected, determine the chatter frequency, $\omega_c$. This will be the frequency at which the process signal has the greatest energy.
3. Set the new spindle speed to be $N_s = \omega_c/[N_t(N + 1)]$, where $N$ is the smallest positive integer such that the new spindle speed does not violate the maximum spindle speed constraint.
4. Repeat Steps 1 to 3 until the chatter has been suppressed.

The equation $N_s = \omega_c/[N_t(N + 1)]$ may be interpreted as selecting the tooth-passing frequency, or an integer fraction thereof, corresponding to the approximate dominant structural frequency. Note that if the depth-of-cut is too large and the maximum spindle speed is too small, this technique will not be effective and the feed or depth-of-cut must be adjusted, or the spindle speed must be continuously varied.

### 10.6.2  Example 6

The feed force for a turning operation is given by Equation 10.1, and the structural dynamics are given by Equation 10.26. The system parameters are $P = 0.75 \text{ kN/mm}^2$, $f_{nom} = 0.1 \text{ mm}$, $\omega_n = 750 \text{ Hz}$, $\zeta = 0.1$, and $k = 15 \text{ kN/mm}$. The depth-of-cut is 5 mm. The spindle speed that should be selected to suppress chatter if the chatter frequency is 725 Hz, when the spindle speed is not constrained, is determined. The spindle speed that should be selected to suppress chatter if the maximum spindle speed is 15,000 rpm is also determined. The system is simulated for a spindle speed of 10,000 rpm for ten spindle revolutions and then for ten spindle revolutions for the spindle

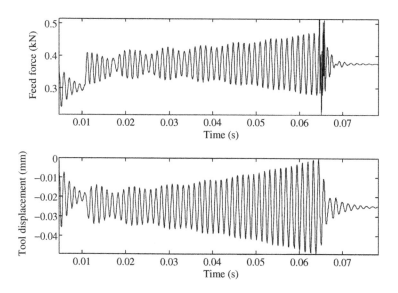

**FIGURE 10.20** Time-domain simulations using spindle speed selection with $N_s = 43,500$ rpm.

speed calculated when the spindle speed is not constrained. The simulation is then repeated for the spindle speed calculated when the spindle speed is constrained. Feed force and tool displacement are plotted for both cases.

For a chatter frequency of 725 Hz, the optimal spindle speed is $60(725) = 43,500$ rpm. Other possible spindle speeds are $43,500/2 = 21,750$ rpm, $43,500/3 = 14,500$ rpm, $43,500/4 = 10,875$ rpm, and so on. Therefore, when the maximum spindle speed is 15,000 rpm, a spindle speed of 14,500 rpm is used. The time domain simulations are in Figure 10.20 and Figure 10.21. The results illustrate that a depth-of-cut of 5.3 mm is stable at 43,500 rpm, but not at 14,500 rpm. Therefore, if the spindle speed is limited to 15,000 rpm, spindle-speed selection may not be used to suppress the chatter present in the machining operation.

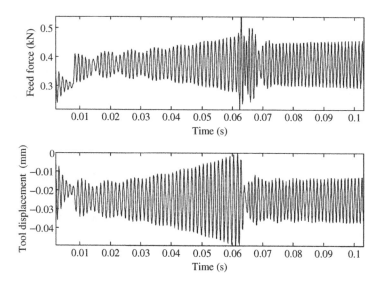

**FIGURE 10.21** Time-domain simulations using spindle speed selection with $N_s = 14,500$ rpm.

### 10.6.3   Feed and Depth-of-Cut Selection

When chatter occurs, operators will sometimes increase the feedrate via the feedrate override button on the machine tool control panel. This has the effect of increasing the feed, assuming the spindle speed remains constant. When linear chatter analysis techniques are employed, the force process gains are linearized about the nominal feed, and stability does not appear to be affected by the nominal feed. However, the stability results are only valid for a small region about the nominal feed. It is well known that there is a nonlinear relationship between the machining forces and the feed of the form $F = P(f)df$. The pressure can be expressed in the form $P(f) = Kf^\alpha$ where $\alpha < 0$; thus, the pressure decreases as the feed increases. Since the stable depth-of-cut is inversely proportional to the pressure, the stability limit will increase as the feed increases, assuming the spindle speed remains constant. An illustration of this phenomenon was shown in Example 3: when the feed was increased from 0.1 to 0.2 mm, chatter was suppressed. While increasing the feed can suppress chatter, the sensitivity of chatter to feed is limited and other adverse phenomenon, such as tooth chippage, may occur.

Another method to suppress chatter is to decrease the depth-of-cut (Weck et al., 1975). This method is guaranteed to work as evidenced by stability lobe diagrams. However, this method is typically not preferred as it dramatically decreases operation productivity by increasing the total number of tool passes that are required to complete the operation.

### 10.6.4   Spindle-Speed Variation

Spindle speed variation (SSV) is another technique that has shown the ability to suppress chatter (Inamura and Sata, 1974; Lin et al., 1990). The spindle speed is varied about some nominal value, typically in a sinusoidal manner. Although SSV is a promising technique, the theory required to guide the designer in the selection of suitable amplitudes and frequencies is in its infancy (Radulescu et al., 1997a, 1997b; Sastry et al., 2002). Also, in some cases, SSV may create chatter that would not occur when using a constant spindle speed.

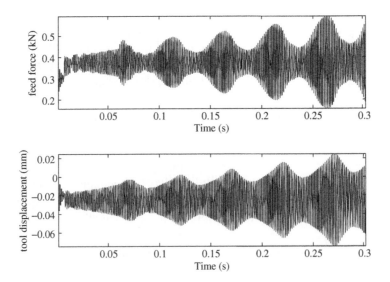

**FIGURE 10.22**   Time-domain simulations using spindle-speed variation with $A = 0.1$ and $\Omega = 20$ Hz.

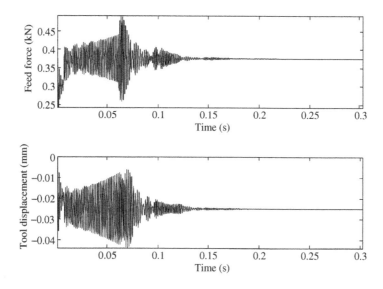

**FIGURE 10.23** Time-domain simulations using spindle-speed variation with $A = 0.25$ and $\Omega = 20$ Hz.

## 10.6.5 Example 7

The feed force for a turning operation is given by Equation 10.1 and the structural dynamics are given by Equation 10.26. The system parameters are $P = 0.75$ kN/mm$^2$, $f_{nom} = 0.1$ mm, $\omega_n = 750$ Hz, $\zeta = 0.1$, and $k = 15$ kN/mm. The depth-of-cut is 5 mm. The system is simulated for a nominal spindle speed of $N_{nom} = 10,000$ rpm for 10 spindle revolutions and then for 30 spindle revolutions for the spindle speed calculated from Equation 10.67 for the following three cases: $A = 0.1$ and $\Omega = 20$ Hz, $A = 0.25$ and $\Omega = 20$ Hz, and $A = 0.25$ and $\Omega = 160$ Hz. Feed force and tool displacement are plotted for all three cases.

$$N_s(t) = N_{nom}[1 + A \sin(\Omega t)] \tag{10.67}$$

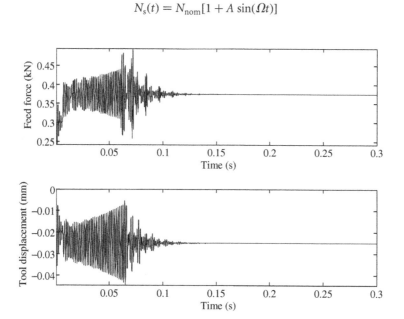

**FIGURE 10.24** Time-domain simulations using spindle-speed variation with $A = 0.25$ and $\Omega = 160$ Hz.

The time-domain simulations are in Figure 10.22 to Figure 10.24 for the respective cases. The results illustrate that SSV may be utilized to suppress chatter; however, the amplitude and frequency of the spindle speed vibration must be carefully chosen.

---

This section presented several techniques to suppress regenerative chatter. The three major techniques to suppress chatter are spindle-speed selection, feed selection, and SSV. In spindle-speed selection, the spindle speed is adjusted to be a multiple of the chatter frequency to place the spindle speed in a pocket of the stability lobe diagram. In feed selection, the feed is increased to suppress chatter. In SSV, the spindle speed is varied in a sinusoidal manner to decrease the phase difference between the current and previous tooth passes.

---

## 10.7   Case Study

A case study of regenerative chatter for a face-milling operation is now presented. Further details are presented in Landers (1997). The machine tool is a three-axis vertical milling machine (Figure 10.25). Each axis has a linear encoder with a resolution of 10 $\mu$m mounted on it. The axis motors (186 W) drive pulleys that rotate leadscrews and provide motion to the linear axes. The spindle (2240 W) drives the face mill (Carboloy R/L220.13-02.00-12, 50 mm diameter). The tool holds four carbide inserts (Carboloy SEAN 42AFTN-M14 HX, 45° lead angle). The part is 6061 aluminum. The spindle is run open-loop. A Kistler 9293 piezoelectric three-component dynamometer was utilized for force process modeling and chatter detection. The $x$ and $y$ channels have a natural frequency of 4.5 kHz, rigidity of 0.7 kN/$\mu$m, and range of $-20$ to 20 kN. The $z$ channel has a natural frequency of 5 kHz, rigidity of 7 kN/$\mu$m, and range of $-100$ to 200 kN. A Bently Nevada 3000 Series Type 190 proximity transducer was utilized to measure the static stiffnesses of the structural components. The sensor gain is 8 V/mm, the response is flat to 10 kHz, and the range is 1.02 mm. A Kistler Quartz Model #802A accelerometer (resonant frequency 36.7 kHz) was utilized to measure the dynamic characteristics of the structural components.

The cutting and thrust pressures, respectively, are

$$P_C = 0.29f^{-0.25}d^{-0.13}\left(\frac{V}{1000}\right)^{-0.72} \tag{10.68}$$

$$P_T = 0.16f^{-0.40}d^{-0.41}\left(\frac{V}{1000}\right)^{-0.58} \tag{10.69}$$

The transfer function matrices, respectively, between the tool structure and machining forces, and the part structure and machining forces are modeled as

$$\begin{bmatrix} x_t(s) \\ y_t(s) \end{bmatrix} = \begin{bmatrix} \dfrac{(4500^2/14)}{s^2 + 2(0.07)(4500)s + 4500^2} & 0 \\ 0 & \dfrac{(4100^2/14)}{s^2 + 2(0.11)(4100)s + 4100^2} \end{bmatrix} \begin{bmatrix} F_x(s) \\ F_y(s) \end{bmatrix} \tag{10.70}$$

$$\begin{bmatrix} x_p(s) \\ y_p(s) \end{bmatrix} = -\begin{bmatrix} \dfrac{(2600^2/9.5)}{s^2 + 2(0.09)(2600)s + 2600^2} & 0 \\ 0 & \dfrac{(2100^2/9.5)}{s^2 + 2(0.22)(2100)s + 2100^2} \end{bmatrix} \begin{bmatrix} F_x(s) \\ F_y(s) \end{bmatrix} \tag{10.71}$$

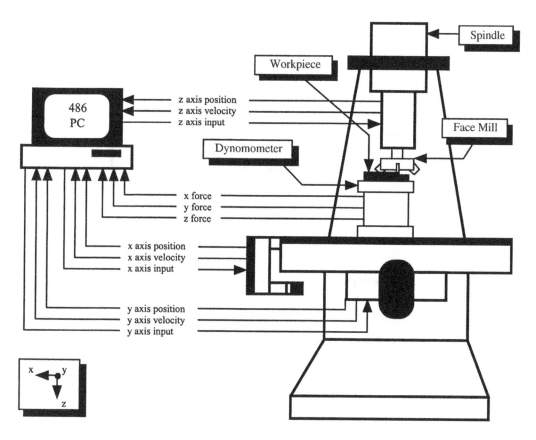

**FIGURE 10.25**  Three-axis vertical machine tool schematic.

Using the machining force and structural models, a stability lobe diagram was constructed using time domain simulations. Experimental data were collected by adjusting the depth-of-cut in increments of 0.1 mm until chatter occurred. The time-domain simulations and the experimental data are plotted in Figure 10.26. The cutting conditions were $f_t = 0.10$ mm/tooth, $N_t = 4$ teeth, $\theta_{en} = -90°$, and $\theta_{ex} = 90°$. The chatter detection methodology for this system was described in Example 5.

Spindle-speed adjustment is typically a more productive chatter suppression option. However, the machine tool in this case study is not equipped with automatic spindle speed control and, thus, the depth-of-cut is adjusted to suppress chatter.

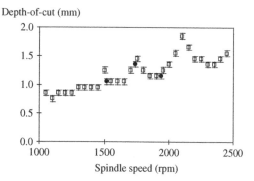

**FIGURE 10.26**  Stability lobe diagram for a face-milling operation — time-domain simulations (empty boxes) and experimental points (filled circles).

When chatter is detected, the chatter suppressor rewrites the part program to accommodate one additional tool pass (Figure 10.27). Therefore, the new operation depth-of-cut is

$$d_n = \frac{d_p}{1 + N_c} \qquad (10.72)$$

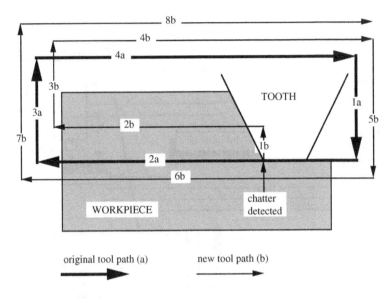

**FIGURE 10.27**  Original tool path (a) is rewritten when chatter occurs. New tool path (b) contains an additional tool pass.

where $d_p$ is the previous operation depth-of-cut and $N_c$ is the number of times the chatter suppression routine has been invoked. The new value may be well below the stability limit; however, making all passes an equal depth-of-cut provides a good balance between productivity and the search for a stable depth-of-cut. Results of this controller are presented in Landers and Ulsoy (1998, 2001).

# Nomenclature

| Symbol | Quantity | Symbol | Quantity |
|---|---|---|---|
| $d$ | depth-of-cut (mm) | $T$ | spindle rotational period (s) |
| $d_{lim}$ | limiting stable depth-of-cut (mm) | $T_t$ | tooth rotational period (s) |
| $f$ | feed (mm) | $x_p$ | part structural displacement in $x$ direction (mm) |
| $f_t$ | feed per tooth (mm) | | |
| $F$ | machining force (kN) | $x_t$ | cutting tool structural displacement in $x$ direction (mm) |
| $F_C$ | cutting force (kN) | | |
| $F_T$ | thrust force (kN) | $y_p$ | part structural displacement in $y$ direction (mm) |
| $F_x$ | force acting on cutting tool in $x$ direction (kN) | | |
| | | $y_t$ | cutting tool structural displacement in $y$ direction (mm) |
| $F_y$ | force acting on cutting tool in $y$ direction (kN) | | |
| | | $z$ | cutting tool structural displacement in $z$ direction (mm) |
| $F_z$ | force acting on cutting tool in $z$ direction (kN) | | |
| $k$ | structural stiffness (kN/mm) | $\theta$ | tooth angle (rad) |
| $N_s$ | spindle speed (rpm) | $\theta_{en}$ | tooth entry angle (rad) |
| $N_t$ | number of teeth | $\theta_{ex}$ | tooth exit angle (rad) |
| $P$ | machining pressure (kN/mm²) | $\omega_c$ | chatter frequency (rad/sec) |
| $P_C$ | cutting pressure (kN/mm²) | $\omega_n$ | structural natural frequency (rad/sec) |
| $P_T$ | thrust pressure (kN/mm²) | | |
| $t$ | time (sec) | $\psi_r$ | lead angle (rad) |
| | | $\zeta$ | structural damping ratio |

# References

Altintas, Y. and Budak, E., Analytical prediction of stability lobes in milling, *Ann. CIRP*, 44/1, 357–362, 1995.

Altintas, Y., Engin, S., and Budak, E., Analytical stability prediction and design of variable pitch cutters, *ASME J. Manuf. Sci. Eng.*, 121, 173–178, 1999.

Budak, E. and Altintas, Y., Analytical prediction of chatter stability in milling, part I: general formulation, *ASME J. Dyn. Syst. Meas. Control*, 120, 22–30, 1998a.

Budak, E. and Altintas, Y., Analytical prediction of chatter stability in milling, part II: application of the general formulation to common milling systems, *ASME J. Dyn. Syst. Meas. Control*, 120, 31–36, 1998b.

Cho, D.W. and Ehmann, K.F., Pattern recognition for on-line chatter detection, *Mech. Syst. Signal Process.*, 2, 279–290, 1988.

Delio, T., Tlusty, J., and Smith, S., Use of audio signals for chatter detection and control, *ASME J. Eng. Ind.*, 114, 146–157, 1992.

Elbestawi, M.A., Ismail, F., Du, R., and Ullagaddi, B.C., Modeling machining dynamics including damping in the tool–workpiece interface, *ASME J. Eng. Ind.*, 116, 435–439, 1994.

Inamura, T. and Sata, T., Stability analysis of cutting under varying spindle speed, *Ann. CIRP*, 23/1, 119–120, 1974.

Inasaki, I., Karpuschewski, B., and Lee, H.-S., Grinding chatter — origin and suppression, *Ann. CIRP*, 50/2, 515–534, 2001.

Koenigsberger, I., Tlusty, J. 1971. *Structures of Machine Tools*, Pergamon Press, New York.

Landers, R.G. 1997. Supervisory machining control: a design approach plus force control and chatter analysis components, Ph.D. Dissertation. Department of Mechanical Engineering and Applied Mechanics, University of Michigan, Ann Arbor.

Landers, R.G. and Ulsoy, A.G., Supervisory machining control: design and experiments, *Ann. CIRP*, 47/1, 301–306, 1998.

Landers, R.G. and Ulsoy, A.G., Supervisory control of a face-milling operation in different manufacturing environments, *Trans. Control Automat. Syst. Eng.*, 3, 1–9, 2001.

Lee, A.-C. and Liu, C.-S., Analysis of chatter vibration in a cutter–workpiece system, *Int. J. Mach. Tools Manuf.*, 31, 221–234, 1991a.

Lee, A.-C. and Liu, C.-S., Analysis of chatter vibration in the end milling process, *Int. J. Mach. Tools Manuf.*, 31, 471–479, 1991b.

Lin, S.C., DeVor, R.E., and Kapoor, S.G., The effects of variable speed cutting on vibration control in face milling, *ASME J. Eng. Ind.*, 112, 1–11, 1990.

Merritt, H.E., Theory of self-excited machine tool chatter: contribution to machine-tool chatter research — 1, *ASME J. Eng. Ind.*, 87, 447–454, 1965.

Minis, I., Magrab, E., and Pandelidis, I., Improved methods for the prediction of chatter in turning, Part III: a generalized linear theory, *ASME J. Eng. Ind.*, 112, 28–35, 1990a.

Minis, I. and Yanushevsky, R., A new theoretical approach for the prediction of machine tool chatter in milling, *ASME J. Eng. Ind.*, 115, 1–8, 1993.

Minis, I., Yanushevsky, R., and Tembo, A., Analysis of linear and nonlinear chatter in milling, *Ann. CIRP*, 39/1, 459–462, 1990b.

Radulescu, R., Kapor, S.G., and DeVor, R.E., An investigation of variable spindle speed face milling for tool work structures with complex dynamics, part 1: simulation results, *ASME J. Manuf. Sci. Eng.*, 119, 266–272, 1997a.

Radulescu, R., Kapor, S.G., and DeVor, R.E., An investigation of variable spindle speed face milling for tool work structures with complex dynamics, part 2: physical explanation, *ASME J. Manuf. Sci. Eng.*, 119, 273–280, 1997b.

Sastry, S., Kapor, S.G., and DeVor, R.E., Floquet theory based approach for stability analysis of the variable speed face-milling process, *ASME J. Manuf. Sci. Eng.*, 124, 10–17, 2002.

Smith, S. and Delio, T., Sensor-based chatter detection and avoidance by spindle speed selection, *ASME J. Dyn. Syst. Meas. Control*, 114, 486–492, 1992.

Smith, S. and Tlusty, J., Efficient simulation programs for chatter in milling, *Ann. CIRP*, 42/1, 463–466, 1993.

Tarng, Y.S. and Li, T.C., Detection and suppression of drilling chatter, *ASME J. Dyn. Syst. Meas. Control*, 116, 729–734, 1994.

Tlusty, J., Dynamics of high-speed milling, *ASME J. Eng. Ind.*, 108, 59–67, 1986.

Tlusty, J. 2000. *Manufacturing Processes and Equipment*, Prentice Hall, Upper Saddle River, NJ.

Tlusty, J. and Ismail, F., Basic nonlinearity in machining chatter, *Ann. CIRP*, 30/1, 299–304, 1981.

Tlusty, J. and Ismail, F., Special aspects of chatter in milling, *ASME J. Vib. Acoust. Stress Reliab. Des.*, 105, 24–32, 1983.

Tobias, S.A. 1965. *Machine Tool Vibration*, Wiley, New York.

Tsai, M.D., Takata, S., Inui, M., Kimura, F., and Sata, T., Prediction of chatter vibration by means of a model-based cutting simulation system, *Ann. CIRP*, 39/1, 447–450, 1990.

Weck, M., Altintas, Y., and Beer, C., CAD assisted chatter-free NC tool path generation in milling, *Int. J. Mach. Tools Manuf.*, 34, 879–891, 1994.

Weck, M., Verhagg, E., and Gather, M., Adaptive control of face-milling operations with strategies for avoiding chatter-vibrations and for automatic cut distribution, *Ann. CIRP*, 24/1, 405–409, 1975.

Yun, I.S., Wilson, W.R.D., and Ehmann, K.F., Review of chatter studies in cold rolling, *Int. J. Mach. Tools Manuf.*, 38, 1499–1530, 1998.

# 11

# Fluid-Induced Vibration

Seon M. Han
*Texas Tech University*

**Summary**

*This chapter gives an overview on the subject of fluid-induced vibration in an ocean environment. The main objective is to show how the fluid forces on an offshore structure due to current and random waves are modeled. The chapter is divided into three sections. The first section describes the ocean environment, especially the currents and random waves. The second section is dedicated to obtaining fluid forces utilizing the results from the first section and the third section gives some examples to show how the results from the first two sections can be used in practice. In the first section, the concept of spectral density is introduced. For a given spectrum, methods to obtain a sample time series are given. In the second section, the forces that the fluid can exert on a body are discussed. The regimes in which inertia, drag, or diffraction forces are dominant are shown in terms of the ratio of the wave height to the structural diameter and the ratio of the structural diameter to the wavelength. The Morison equation is extended to the case of a moving inclined cylinder. The Morison equation requires the use of experimentally determined fluid coefficients such as added mass, inertia, and drag coefficients. Plots of these fluid coefficients for various values of the fluid parameters are reproduced here. The vortex shedding force is discussed briefly. In the third section, four examples are given to show how fluid forces affect the static and dynamics of ocean structures, how the significant wave height can be chosen to represent the condition in a certain area for a long time, and how the time series can be constructed from a given spectrum. Finally, the available numerical codes for modeling slender flexible bodies in fluids are listed.*

## 11.1 Description of the Ocean Environment

In modeling offshore structures, one needs to account for the forces exerted by the surrounding fluid. In-depth studies are given in Kinsman (1965), Sarpkaya and Isaacson (1981), Wilson (1984), Chakrabarti (1987), and Faltinsen (1993). The vibration characteristics of a structure can be significantly altered when it is surrounded by water. For example, damping by the fluid (or the added mass) lowers the natural frequency of vibration. When considering the dynamics of an offshore structure, one must also consider the forces due to the surrounding fluid. The two important sources of fluid motion are ocean waves and ocean currents.

Most steady large currents are generated by the drag of the wind passing over the surface of the water, and they are confined to a region near the ocean surface. Tidal currents are generated by the gravitational attraction of the sun and the moon, and they are most significant near coasts. The ultimate source of the ocean circulation is the uneven radiation heating of the Earth by the Sun.

Isaacson (1988) suggested an empirical formula for the current velocity in the horizontal direction as a function of depth:

$$U_c(x) = (U_{\text{tide}}(d) + U_{\text{circulation}}(d))\left(\frac{x}{d}\right)^{1/7} + U_{\text{drift}}(d)\left(\frac{x - d + d_0}{d_0}\right) \tag{11.1}$$

where $U_{\text{drift}}$ is the wind-induced drift current, $U_{\text{tide}}$ is the tidal current, $U_{\text{circulation}}$ is the low-frequency long-term circulation, $x$ is the vertical distance measured from the ocean bottom, $d$ is the depth of the water, and $d_0$ is the smaller of the depth of the thermocline and 50 m. The value of $U_{\text{tide}}$ is obtained from tide tables, and $U_{\text{drift}}$ is about 3% of the 10 min mean wind velocity at 10 m above the sea level.

It should be noted that these currents evolve slowly compared with the time scales of engineering interests. Therefore, they can be treated as a quasisteady phenomenon. Waves, on the other hand, cannot be treated as a steady phenomenon. The underlying physics that govern wave dynamics are too complex and, therefore, waves must be modeled stochastically. The subsequent section discusses the concept of the spectral density, available ocean wave spectral densities, a method to obtain the spectral density from wave time histories, methods to obtain a sample time history from a spectral density, the short-term and long-term statistics, and a method to obtain fluid velocities and accelerations from wave elevation using linear wave theory.

## 11.1.1 Spectral Density

Here, we will consider only surface gravity waves. Let us first consider a regular wave in order to familiarize ourselves with the terms that are used to describe a wave. The wave surface elevation is denoted as $\eta(x, t)$ and can be written as $\eta(x, t) = A \cos(kx - \omega t)$, where $k$ is the wave number, and $\omega$ is the angular frequency. Figure 11.1 shows the surface elevation at two time instances ($t = 0$ and $t = \tau$) and the surface elevation at a fixed location ($x = 0$). $A$ is the amplitude, $H$ is the wave height or the distance between the maximum and minimum wave elevation or twice the amplitude, and $T$ is the period given by $T = 2\pi/\omega$.

In practice, waves are not regular. Figure 11.2 shows a schematic time history of an irregular wave surface elevation. The wave height and frequency are not easy to find. Therefore, we rely on a statistical description for the wave elevation such as the wave spectral density. The spectral

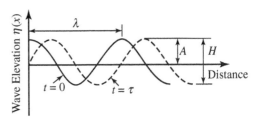

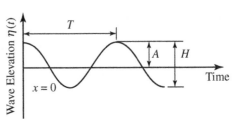

**FIGURE 11.1**   Regular wave.

density tells us how the energy of the system is distributed among frequencies. The random surface elevation $\eta(t)$ can be thought of as a summation of regular waves with different frequencies. The surface elevation $\eta(t)$ is related to its Fourier transform $X(\omega)$ by

$$\eta(t) = \frac{1}{2\pi} \int_{-\infty}^{\infty} X(\omega) \exp(-i\omega t) d\omega$$

Suppose that the energy of the system is proportional to $\eta^2(t)$ so that we can write the energy as

$$E = \frac{1}{2}C\eta^2(t)$$

where $C$ is the proportionality constant.

Let us assume that the expected value of the energy is given by

$$E\{E\} = \frac{1}{2}CE\{\eta^2(t)\}$$

**FIGURE 11.2** Time history of random wave.

where $E\{\eta^2(t)\}$ is the mean square of $\eta(t)$. If $\eta(t)$ is an *ergodic process*, then the mean square of $\eta(t)$ can be approximated by the time average over a long period of time:

$$E\{\eta^2(t)\} = \lim_{T_s\to+\infty}\frac{1}{T_s}\int_{-T_s/2}^{T_s/2}\eta^2(t)dt = \lim_{T_s\to+\infty}\frac{1}{T_s}\frac{1}{2\pi}\int_{-\infty}^{\infty}|X(\omega)|^2\,d\omega \tag{11.2}$$

where we have used Parseval's theorem

$$\int_{-\infty}^{\infty}\eta^2(t)dt = \frac{1}{2\pi}\int_{-\infty}^{\infty}|X(\omega)|^2\,d\omega \tag{11.3}$$

where

$$|X(\omega)|^2 = X(\omega)X^*(\omega),\quad X(\omega) = \int_{-\infty}^{\infty}\eta(t)\exp(-i\omega t)dt,\quad X^*(\omega) = \int_{-\infty}^{\infty}\eta(t)\exp(i\omega t)dt$$

We define the power spectral density (or simply the spectrum) as

$$S_{\eta\eta}(\omega) \equiv \frac{1}{2\pi T_s}|X(\omega)|^2 \tag{11.4}$$

so that $E\{\eta^2(t)\}$ is given by

$$E\{\eta^2(t)\} = \int_{-\infty}^{\infty}S_{\eta\eta}(\omega)d\omega \tag{11.5}$$

For a zero-mean process, $E\{\eta^2(t)\}$ is also the variance $\sigma_\eta^2$. The spectral density has units of $\eta^2 t$. Where $\eta$ is the wave elevation, the spectral density has a unit of $m^2$ sec.

It can also be shown that $S_{\eta\eta}(\omega)$ is related to the autocorrelation function, $R(\tau)$, by the Wiener–Khinchine relations (Wiener, 1930; Khinchine, 1934):

$$S_{\eta\eta}(\omega) = \frac{1}{2\pi}\int_{-\infty}^{\infty}R_{\eta\eta}(\tau)\exp(-i\omega\tau)d\tau,\quad R_{\eta\eta}(\tau) = \int_{-\infty}^{\infty}S_{\eta\eta}(\omega)\exp(i\omega\tau)d\omega \tag{11.6}$$

It should be noted that, in some textbooks, the factor $1/2\pi$ appears in the second equation instead of the first. Figure 11.3 shows some important pairs of $S_{\eta\eta}(\omega)$ and $R_{\eta\eta}(\tau)$.

There are a few properties of the spectral density that readers should become familiar with. The first property is that the spectral density function of a real-valued stationary process is both real and symmetric. That is, $S_{\eta\eta}(\omega) = S_{\eta\eta}(-\omega)$ (Equation 11.4). Secondly, the area under the spectral density is equal to $E\{\eta^2(t)\}$ (Equation 11.5) and is also equal to $R_{\eta\eta}(0) = \sigma_\eta^2 - \mu_\eta^2$, where $\sigma_\eta^2$ is the variance and $\mu_\eta^2$ is the mean of $\eta(t)$. In most cases, we only consider a zero-mean process so that the area under the spectral density is just $\sigma_\eta^2$. If the process does not have a zero mean, the mean can be subtracted from it so that the process has a zero mean.

For ocean applications, a one-sided spectrum in terms of cycles per second (cps) or hertz is often used. We will denote the one-sided spectrum with a superscript "o". The one-sided spectrum can be obtained from the two-sided spectrum by

$$S_{\eta\eta}^o(\omega) = 2S_{\eta\eta}(\omega),\quad \omega \ge 0$$

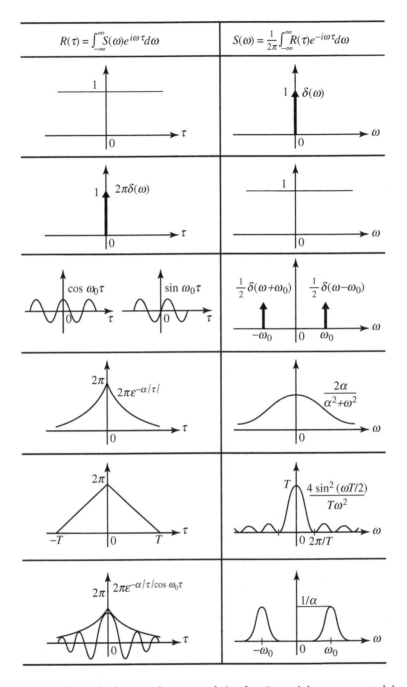

**FIGURE 11.3** Relationship between the autocorrelation function and the power spectral density.

The two-sided spectrum in terms of $\omega$ can be transformed to the spectrum in terms of $f$ (where $\omega = 2\pi f$) by

$$S_{\eta\eta}(f) = 2\pi S_{\eta\eta}(\omega), \quad f, \omega \geq 0$$

Then, the two-sided spectrum in terms of $\omega$ can be transformed to the one-sided spectrum in terms of cps (or hertz) by

$$S^{\circ}_{\eta\eta}(f) = 4\pi S_{\eta\eta}(\omega), \quad f, \omega \geq 0$$

It should be noted that the spectral density that we have defined here is the *amplitude half-spectrum*. The amplitude, height, and height double spectra are related to the amplitude half-spectrum by

$$S^A(\omega) = 2S(\omega), \quad S^H(\omega) = 8S(\omega), \quad S^{2H}(\omega) = 16S(\omega)$$

## 11.1.2 Ocean Wave Spectral Densities

In this section, we will discuss spectral density models to describe a random sea. An excellent review of existing spectral density models is given in Chapter 4 of Chakrabarti (1987).

The ocean wave spectrum models are semiempirical formulas. That is, they are derived mathematically but the formulation requires one or more experimentally determined parameters. The accuracy of the spectrum depends significantly on the choice of these parameters.

In formulating spectral densities, the parameters that influence the spectrum are *fetch limitations, decaying* vs. *developing seas, water depth, current,* and *swell.* The fetch is the distance over which a wind blows in a wave-generating phase. Fetch limitation refers to the limitation on the distance due to some physical boundaries so that full wave development is prohibited. In a developing sea, the sea has not yet reached its stationary state under a stationary wind. In contrast, a wind has blown for a sufficient time in a fully developed sea, and the sea has reached its stationary state. In a decaying sea, the wind has dropped off from its stationary value. Swell is the wave motion caused by a distant storm and persists even after the storm has died down or moved away.

The Pierson–Moskowitz (P–M) spectrum (Pierson and Moskowitz, 1964) is the most extensively used spectrum for representing a fully developed sea. It is a one-parameter model in which the sea severity can be specified in terms of the wind velocity. The P–M spectrum is given by

$$S^o_{\eta\eta}(f) = \frac{8.1 \times 10^{-3} g^2}{\omega^5} \exp\left(-0.74\left(\frac{g}{U_{w,19.5\,m}}\right)^4 \omega^{-4}\right)$$

where $g$ is the gravitational constant and $U_{w,19.5\,m}$ is the wind speed at a height of 19.5 m above the still water. The P–M spectrum is also called the wind-speed spectrum because it requires wind data. It can also be written in terms of the modal frequency $\omega_m$ as

$$S^o_{\eta\eta}(f) = \frac{8.1 \times 10^{-3} g^2}{\omega^5} \exp\left(-1.25\left(\frac{\omega_m}{\omega}\right)^4\right) \tag{11.7}$$

Note that the modal frequency is the frequency at which the spectrum is the maximum.

In some cases, it may be more convenient to express the spectrum in terms of significant wave height rather than the wind speed or modal frequency. For a narrowband Gaussian process[1], the significant wave height is related to the standard deviation by $H_s = 4\sigma_\eta$. The standard deviation is the square root of the area under the spectral density, $\int_{-\infty}^{\infty} S_{\eta\eta}(\omega)d\omega = \sigma_\eta^2$. Then, the spectrum can be written as

$$S^o_{\eta\eta}(f) = \frac{8.1 \times 10^{-3} g^2}{\omega^5} \exp\left(-\frac{0.0324 g^2}{H_s^2} \omega^{-4}\right) \tag{11.8}$$

and the peak frequency and the significant wave height are related by

$$\omega_m = 0.4\sqrt{g/H_s} \tag{11.9}$$

The P–M spectrum is applicable for deep water, unidirectional seas, fully developed and local-wind-generated seas with unlimited fetch, and was developed for the North Atlantic. The effect of swell is not accounted for. Although it was developed for the North Atlantic, the spectrum is valid for other locations. However, the limitation that the sea is fully developed may be too restrictive because it cannot model the

---

[1]See Section 11.1.5 for details.

effect of waves generated at a distance. Therefore, we consider a two-parameter spectrum, such as the Bretschneider spectrum, in order to model a sea that is not fully developed as well as a fully developed sea.

The Bretschneider spectrum (Bretschneider, 1959, 1969) is a two-parameter spectrum in which both the sea severity and the state of development can be specified. The Bretschneider spectrum is given by

$$S^o_{\eta\eta}(f) = 0.169 \frac{\omega_s^4}{\omega^5} H_s^2 \exp\left(-0.675\left(\frac{\omega_s}{\omega}\right)^4\right)$$

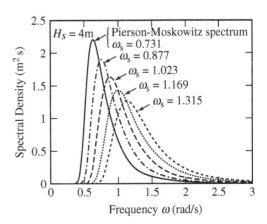

FIGURE 11.4 Bretschneider spectrum with various values of $\omega_s$.

where $\omega_s = 2\pi/T_s$ and $T_s$ is the significant period. The sea severity can be specified by $H_s$ and the state of development can be specified by $\omega_s$. It can be shown that the relationship $\omega_s = 1.167\omega_m$ (equivalent to $\omega_s = 1.46/\sqrt{H_s}$) renders the Bretschneider spectrum and the P–M spectrum equivalent. Figure 11.4 shows the Bretschneider spectra for $H_s = 4$ m. When $\omega_s = 0.731$ rad/sec, the P–M and the Bretschneider spectra are identical. It should be noted that the developing sea will have a slightly higher modal frequency than the fully developed sea, and can be described by $\omega_s$ greater than $1.46/\sqrt{H_s}$.

Other two-parameter spectral densities that are often used are the International Ship Structures Congress (ISSC) and the International Towing Tank Conference (ITTC) spectra. The ISSC spectrum is written in terms of the significant wave height and the mean frequency, where the mean frequency is given by

$$\bar{\omega} = \sqrt{\frac{\int_0^\infty \omega S(\omega)d\omega}{\int_0^\infty S(\omega)d\omega}} = 1.30\omega_m$$

Thus, the ISSC spectrum is given by

$$S^o_{\eta\eta}(f) = 0.111 \frac{\bar{\omega}^4}{\omega^5} H_s^2 \exp\left(-0.444\left(\frac{\bar{\omega}}{\omega}\right)^4\right)$$

The ITTC spectrum is based on the significant wave height and the zero crossing frequency and is given by

$$S^o_{\eta\eta}(f) = 0.0795 \frac{\omega_z^4}{\omega^5} H_s^2 \exp\left(-0.318\left(\frac{\omega_z}{\omega}\right)^4\right)$$

where the zero crossing frequency, $\omega_z$, is given by

$$\omega_z = \sqrt{\frac{\int_0^\infty \omega^2 S(\omega)d\omega}{\int_0^\infty S(\omega)d\omega}} = 1.41\omega_m$$

The Bretschneider, ITTC, and ISSC spectra are called two-parameter spectra, and they can be written as

$$S^o_{\eta\eta}(f) = \frac{A}{4} \frac{\bar{\omega}^4}{\omega^5} H_s^2 \exp\left(-A\left(\frac{\bar{\omega}}{\omega}\right)^4\right)$$

with $A$ and $\omega$ given in Table 11.1.

**TABLE 11.1** Two-Parameter Spectrum Models
$S^0_{\eta\eta}(\omega) = (A/4)H_s^2\,\tilde\omega^4/\omega^5\,\exp(-A(\omega/\tilde\omega)^{-4})$

| Model | $A$ | $\tilde\omega$ |
|---|---|---|
| Bretschneider | 0.675 | $\omega_s$ |
| ITTC | 0.318 | $\omega_z$ |
| ISSC | 0.4427 | $\tilde\omega$ |

The spectra that we have discussed so far do not allow us to generate spectra with two peaks to represent local or distant storms or to specify the sharpness of the peaks. The Ochi–Hubble (O–H) spectrum (Ochi and Hubble, 1976) is a six-parameter spectrum with the form:

$$S^0_{\eta\eta}(\omega) = \frac{1}{4}\sum_{i=1}^{2}\frac{((4\lambda_i+1)\omega_{mi}^4/4)^{\lambda_i}}{\Gamma(\lambda_i)}\frac{H_{si}^2}{\omega^{4\lambda_i+1}}\exp\left(-\left(\frac{4\lambda_i+1}{4}\right)\left(\frac{\omega_{mi}}{\omega}\right)^4\right)$$

where $\Gamma(\lambda_i)$ is the Gamma function, $H_{s1}$, $\omega_{m1}$, and $\lambda_1$ are the significant wave height, modal frequency, and shape factor for the lower frequency components, respectively, and $H_{s2}$, $\omega_{m2}$, and $\lambda_2$ are those for the higher frequency component. Assuming that the entire spectrum is that of a narrow band, the equivalent significant wave height is given by

$$H_s = \sqrt{H_{s1}^2 + H_{s2}^2}$$

For $\lambda_1 = 1$ and $\lambda_2 = 0$, the spectrum reduces to the P–M spectrum. With the assumption that the entire spectrum is narrowband, the value of $\lambda_1$ is much higher than $\lambda_2$. The O–H spectrum represents unidirectional seas with unlimited fetch. The sea severity and the state of development can be specified by $H_{si}$ and $\omega_{mi}$, respectively. In addition, $\lambda_i$ can be selected to control the frequency width of the spectrum. For example, a small $\lambda_i$ (wider frequency range) describes a developing sea, and a large $\lambda_i$ (narrower frequency range) describes a swell condition. Figure 11.5 shows the O–H spectrum with $\lambda_1 = 2.72$, $\omega_{m1} = 0.626$ rad/sec, $H_{s1} = 3.35$ m, $\lambda_2 = 2.72$, $\omega_{m2} = 1.25$ rad/sec, and $H_{s2} = 2.19$ m.

Finally, another spectrum that is commonly used is the Joint North Sea Wave Project (JONSWAP) spectrum developed by Hasselmann et al. (1973). It is a fetch-limited spectrum because the growth over a limited fetch is taken into account. The attenuation in shallow water is also taken into account. The JONSWAP spectrum is written as

$$S^0_{\eta\eta}(\omega) = \frac{\alpha g^2}{\omega^5}\exp\left(-1.25\left(\frac{\omega_m}{\omega}\right)^4\right)\gamma^{\exp(-(\omega-\omega_m)/2\tau^2\omega_m^2)}$$

where $\gamma$ is the peakedness parameter and $\tau$ is the shape parameter. The peakedness parameter $\gamma$ is the ratio of the maximum spectral energy to the maximum spectral energy of the corresponding P–M spectrum. That is, when $\gamma = 7$, the peak spectral energy is seven times that of the P–M spectrum.

$\gamma = \begin{cases} 7.0 & \text{for very peaked data} \\ 3.3 & \text{for mean of selected JONSWAP data} \\ 1.0 & \text{for P–M spectrum} \end{cases}$

$\tau = \begin{cases} 0.07 & \text{for } \omega \leq \omega_m \\ 0.09 & \text{for } \omega > \omega_m \end{cases}$

$\alpha = 0.076(\bar{X})^{-0.22}$ or 0.0081 if fetch independent

$\bar{X} = gX/U_w^2$

$X =$ fetch length (nautical miles)

$U_w =$ wind speed (knots)

$\omega_m = 2\pi \times 3.5(g/U_w)\bar{X}^{-0.33}$

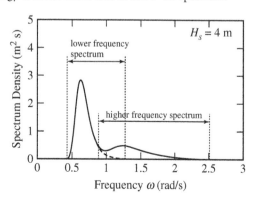

**FIGURE 11.5** Ochi–Hubble spectrum.

Figure 11.6 shows the JONSWAP spectrum when $\alpha = 0.0081$ and $\omega_m = 0.626$ rad/sec for three peakedness parameters.

### 11.1.3 Approximation of Spectral Density from Time Series

From the time history of the wave elevation, the spectral density function can be obtained by two methods.

The first method is to use the autocorrelation function $R_{\eta\eta}(\tau)$, which is related to the spectral density function $S_{\eta\eta}(\omega)$ by the Wiener–Khinchine relations (Equation 11.6).

The autocorrelation $R_{\eta\eta}(\tau)$ is the expected value of $\eta(t)\eta(t+\tau)$ or $R_{\eta\eta}(\tau) = E\{\eta(t)\eta(t+\tau)\}$, where $t$ is an arbitrary time and $\tau$ is the time lag. For a weakly stationary process, the autocorrelation is a function of the time lag only.

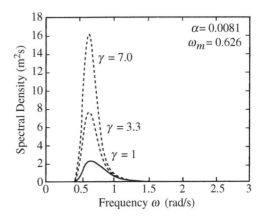

**FIGURE 11.6**   JONSWAP spectrum for $\gamma = 1.0$, 3.3, and 7.0.

Assuming that the process is ergodic, the autocorrelation function for a given time history of length $T_s$ can be approximated as

$$\hat{R}_{\eta\eta}(\tau) = \lim_{T_s \to \infty} \frac{1}{T_s - \tau} \int_0^{T_s - \tau} \eta(t)\eta(t+\tau)dt \quad \text{for } 0 < \tau < T_s$$

Note that the superscript $\wedge$ is used to emphasize that the variable is an approximation based on a sample time history of length $T_s$. The spectral density is then obtained by taking the Fourier cosine transform of $\hat{R}_{\eta\eta}(\tau)$,

$$\hat{S}_{\eta\eta}(\omega) = \frac{1}{\pi} \int_0^{T_s} \hat{R}_{\eta\eta}(\tau) \cos \omega\tau \, d\tau \tag{11.10}$$

The second method for obtaining the spectral density function is to use the relationship between spectral density and the Fourier transform of the time series. They are related by

$$\hat{S}_{\eta\eta}(\omega) = \lim_{T_s \to \infty} \frac{1}{2\pi T_s} |\hat{X}(\omega)\hat{X}^*(\omega)| \tag{11.11}$$

where $\hat{X}(\omega)$ is given by

$$\hat{X}(\omega) = \int_0^{T_s} \eta(t) \exp(-i\omega t)dt$$

and $\hat{X}^*(\omega)$ is the complex conjugate given by

$$\hat{X}^*(\omega) = \int_0^{T_s} \eta(t) \exp(i\omega t)dt$$

In order to obtain the Fourier transforms of the time series, the discrete Fourier transform (DFT) or the fast Fourier transform (FFT) procedure can be used. For detailed descriptions of how this is done, see Appendix 1 in Tucker (1991). Nowadays, spectral analysis is almost always carried out via FFTs because it is easier to use and faster than the formal method via correlation function.

It should be noted that the length of the sample time history only needs to be long enough so that the limits converge. Taking a longer sample will not improve the accuracy of the estimate. Instead, one should take many samples or break one long sample into many parts. For $n$ samples, the spectral densities are obtained for each sample time history using either Equation 11.10 or Equation 11.11, and they are averaged to give the estimate.

The determination of the spectral density from wave records depends on the details of the procedure such as the length of the record, sampling interval, degree and type of filtering and smoothing, and time discretization.

## 11.1.4  Generation of Time Series from a Spectral Density

In a nonlinear analysis, the structural response is found by a numerical integration in time. Therefore, one needs to convert the wave elevation spectrum into an equivalent time history. The wave elevation can be represented as a sum of many sinusoidal functions with different angular frequencies and random phase angles. That is, we write $\eta(t)$ as

$$\eta(t) = \sum_{i=1}^{N} \cos(\omega_i t - \varphi_i)\sqrt{2S_{\eta\eta}(\omega_i)\Delta\omega_i} \tag{11.12}$$

where $\varphi_i$ is a uniform random number between 0 and $2\pi$, $\omega_i$ are discrete sampling frequencies, $\Delta\omega_i = \omega_i - \omega_{i-1}$, and $N$ is the number of partitions. Recall that the area under the spectrum is equal to the variance, $\sigma_\eta^2$. The incremental area under the spectrum, $S_{\eta\eta}(\omega_i)\Delta\omega_i$, can be denoted as $\sigma_i^2$ such that the sum of all the incremental area equals the variance of the wave elevation or $\sigma_\eta^2 = \sum_{i=1}^{N} \sigma_i^2$. The time history can be written as

$$\eta(t) = \sum_{i=1}^{N} \cos(\omega_i t - \varphi_i)\sqrt{2}\sigma_i$$

The sampling frequencies, $\omega_i$, can be chosen at equal intervals such that $\omega_i = i\omega_1$. However, the time history will then have the lowest frequency of $\omega_1$ and will have a period of $T = 2\pi/\omega_1$. In order to avoid this unwanted periodicity, Borgman (1969) suggested that the frequencies are chosen so that the area under the spectrum curve for each interval is equal or $\sigma_i^2 = \sigma^2 = \sigma_\eta^2/N$. The time history is written as

$$\eta(t) = \sqrt{\frac{2}{N}}\sigma_\eta \sum_{i=1}^{N} \cos(\bar{\omega}_i t - \varphi_i) \tag{11.13}$$

where $\bar{\omega}_i = (\omega_i + \omega_{i-1})/2$. The discrete frequencies, $\omega_i$, are chosen such that the area between the interval $0 < \omega < \omega_i$ is equal to $i/N$ of the total area under the curve between the interval $0 < \omega < \omega_N$ or

$$\int_0^{\omega_i} S_{\eta\eta}(\omega)d\omega = \frac{i}{N} \int_0^{\omega_N} S_{\eta\eta}(\omega)d\omega \quad \text{for } i = 1, \ldots, N$$

where it is assumed that the area under the spectrum beyond $\omega_N$ is negligible. If $\eta(t)$ is a narrowband Gaussian process, the standard deviation can be replaced by $\sigma_\eta = H_s/4$, and the time history can be written as

$$\eta(t) = \frac{H_s}{4}\sqrt{\frac{2}{N}} \sum_{i=1}^{N} \cos(\bar{\omega}_i t - \varphi_i)$$

Shinozuka (1972) proposed that the sampling frequencies, $\bar{\omega}_i$, in Equation 11.13 should be randomly chosen according to the density function, $f(\omega) \equiv S_{\eta\eta}^o(\omega)/\sigma_\eta^2$. This is equivalent to performing an integration using the Monte Carlo method. The random frequencies $\omega$ distributed according to $f(\omega)$ can be obtained from uniformly distributed random numbers, $x$, by $\omega = F^{-1}(x)$, where $F(\omega)$ is the cumulative distribution of $f(\omega)$.

The random frequencies obtained this way are used in Equation 11.13 to generate a sample time series. It should be noted that many sample time histories should be obtained and averaged to synthesize a time history for use in numerical simulations.

## 11.1.5  Short-Term Statistics

In discussing wave statistics, we often use the term *significant wave* to describe an irregular sea surface. The significant wave is not a physical wave that can be seen but rather a statistical description of random waves. The concept of significant wave height was first introduced by Sverdrup and Munk (1947) as the average height of the highest one third of all waves. Usually, ships co-operate in programs to find sea statistics by reporting a rough estimate of the storm severity in terms of an observed wave height.

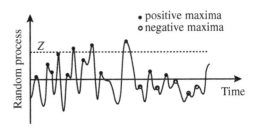

**FIGURE 11.7**  A sample time history.

This observed wave height is consistently very close to the significant wave height.

Stationarity and ergodicity are two assumptions that are made in describing short-term waves statistics. These assumptions are valid only for "short" time intervals — approximately two hours or the duration of a storm — but not for weeks or years. The wave elevation is assumed to be weakly stationary so that its autocorrelation is a function of time lag only. As a result, the mean and the variance are constant, and the spectral density is invariant with time. Therefore, the significant wave height and the significant wave period are constant when we consider short-term statistics. In this case, the individual wave height and wave period are the stochastic variables. We then need to determine certain statistics for the analysis and design of offshore structures when we consider short time intervals.

Consider a sample time history of a zero-mean random process, as shown in Figure 11.7. The questions that we ask are how often is a certain level (e.g., $z$ in the figure) exceeded, and how are the maxima distributed? Likewise, we can ask when we can expect to see that a certain level is exceeded for the first time, and what are the values of the peaks of a random process? The first question is important when a structure may fail due to a one-time excessive load, and the second question is important when a structure may fail due to cyclic loads.

It is found that the rate at which a random process $X(t)$ crosses $Z$ with a positive slope (zero up-crossing) may be calculated from

$$\nu_{z^+} = \int_0^\infty \nu f_{X\dot{X}}(z, \nu) d\nu$$

where $f_{X\dot{X}}(x, \dot{x})$ is the joint probability density function of $X$ and $\dot{X}(t)$. The expected time of the first up-crossing is then the inverse of the crossing rate or

$$E\{T\} = 1/\nu_{z^+}$$

The probability density function of the maxima, $A$, can be calculated from

$$f_A(a) = \frac{\displaystyle\int_{-\infty}^0 -\omega f_{X\dot{X}\ddot{X}}(a, 0, \omega) d\omega}{\displaystyle\int_{-\infty}^0 -\omega f_{X\ddot{X}}(0, \omega) d\omega}$$

where $f_{X\dot{X}\ddot{X}}(x, \dot{x}, \ddot{x})$ is the joint probability density function of $X$, $\dot{X}$, and $\ddot{X}$.

If $X(t)$ is a Gaussian process, then we can write the joint probability density functions as

$$f_{X\dot{X}}(x, \dot{x}) = \frac{1}{2\pi\sigma_X\sigma_{\dot{X}}} \exp\left[ -\frac{1}{2}\left(\frac{x}{\sigma_X}\right)^2 - \frac{1}{2}\left(\frac{\dot{x}}{\sigma_{\dot{X}}}\right)^2 \right], \quad -\infty < x < \infty, \quad -\infty < \dot{x} < \infty$$

and

$$f_{X\dot{X}\ddot{X}}(x, \dot{x}, \ddot{x}) = \frac{1}{(2\pi)^{3/2}|M|^{1/2}} \exp\left[ -\frac{1}{2}(\{x\} - \{\mu_X\})^{\mathrm{T}}[M]^{-1}(\{x\} - \{\mu_X\}) \right]$$

where

$$[M] = \begin{bmatrix} \sigma_X^2 & 0 & \sigma_{\ddot{X}}^2 \\ 0 & \sigma_{\dot{X}}^2 & 0 \\ \sigma_{\ddot{X}}^2 & 0 & \sigma_{\ddot{X}}^2 \end{bmatrix} \quad \text{and} \quad \{x\} - \{\mu_X\} = \begin{bmatrix} x - \mu_X \\ \dot{x} - \mu_{\dot{X}} \\ \ddot{x} - \mu_{\ddot{X}} \end{bmatrix}$$

Then, for a stationary Gaussian process, the up-crossing rate is given by

$$\nu_z^+ = \int_0^\infty f_{X\dot{X}}(Z, \dot{x}) \dot{x} \, d\dot{x} = \frac{1}{2\pi\sigma_X\sigma_{\dot{X}}} \exp\left[ -\frac{1}{2}\left(\frac{Z}{\sigma_X}\right)^2 \right] \int_0^\infty \exp\left[ -\frac{1}{2}\left(\frac{\dot{x}}{\sigma_{\dot{X}}}\right)^2 \right] \dot{x} \, d\dot{x}$$

$$= \frac{\sigma_{\dot{X}}}{2\pi\sigma_X} \exp\left[ -\frac{1}{2}\left(\frac{Z}{\sigma_X}\right)^2 \right] \tag{11.14}$$

and the probability density function of maxima is given by the Rice density function (Rice, 1954)

$$f_A(a) = \frac{\sqrt{1-\alpha^2}}{\sqrt{2\pi}\sigma_\eta} \exp\left[ -\frac{1}{2}\frac{\alpha^2}{\sigma_\eta^2(1-\alpha^2)} \right] + a\frac{\alpha}{\sigma_\eta^2} \Phi\left( \frac{a\alpha}{\sigma_\eta\sqrt{\alpha^2-1}} \right) \exp\left( -\frac{1}{2}\frac{a^2}{\sigma_\eta^2} \right) \quad \text{for} \ -\infty < a < \infty$$

where $\Phi(x)$ is the cumulative distribution function of standard normal random variable

$$\Phi(x) = \frac{1}{\sqrt{2\pi}} \int_{-\infty}^x \exp(-z^2/2) dz$$

and $\alpha$ is the irregularity factor equivalent to the ratio of the number of zero up-crossings (number of times that $\eta[t]$ crosses zero with a positive slope) to the number of peaks. $\alpha$ ranges from 0 to 1, and it is also equal to

$$\alpha = \frac{\sigma_{\dot{\eta}}^2}{\sigma_\eta^2\sigma_{\ddot{\eta}}^2}$$

If $X(t)$ is a broadband process, $\alpha = 0$ and the Rice distribution is reduced to the Gaussian probability density function given by

$$f_A(a) = \frac{1}{\sqrt{2\pi}\sigma_\eta} \exp\left[ -\frac{1}{2}\frac{\alpha^2}{\sigma_\eta^2} \right] \quad \text{for} \ -\infty < a < \infty$$

If $X(t)$ is a narrowband process, it is guaranteed that it will have a peak whenever $\eta(t)$ crosses its mean. In this case, the irregularity factor is close to unity, and the Rice distribution is reduced to the Rayleigh probability density function given by

$$f_A(a) = \frac{a}{\sigma_\eta^2} \exp\left[ -\frac{1}{2}\frac{\alpha^2}{\sigma_\eta^2} \right] \quad \text{for } 0 < a < \infty$$

In other words, the amplitudes of a narrowband stationary Gaussian process are distributed according to the Rayleigh distribution.

Figure 11.8 shows the Rice distribution for various values of $\alpha$. Note that the Rice distribution includes both positive and negative maxima except when $\alpha = 1$, in which case all the maxima are positive. The positive maxima are the local maxima that occur above the mean of $X(t)$, and the negative maxima are the local maxima that occur below the mean, as shown in Figure 11.7. In some cases, the negative maxima may not mean

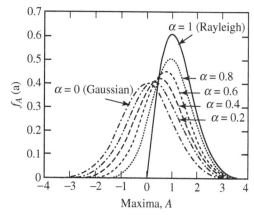

**FIGURE 11.8** Rice distribution for maxima.

much physically. In those cases, we can use the truncated Rice distribution, where only the positive portion of $f_A(a)$ is used. $f_A(a)$ is normalized by the area under the probability density for positive maxima (Longuet-Higgins, 1952; Ochi, 1973):

$$f_A^{\text{trunc}}(a) = \frac{f_A(a)}{\displaystyle\int_0^\infty f_A(a)\,da}, \quad a \geq 0$$

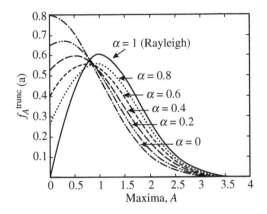

**FIGURE 11.9**   Truncated Rice distribution.

The truncated Rice distribution is shown in Figure 11.9.

If $X(t)$ is the wave elevation, its maxima, $A$, are the amplitudes of the wave elevation. The wave height, $H = 2A$, is then distributed according to

$$f_H(h) = f_A(H/2)\frac{dA}{dH} = \frac{h}{4\sigma_\eta^2}\exp\left[-\frac{1}{2}\frac{h^2}{4\sigma_\eta^2}\right] \quad \text{for } 0 < h < \infty$$

For any given wave, the probability that the height is less than $h$ (the cumulative distribution) is

$$F_H(h) = 1 - \exp\left[-\frac{1}{2}\frac{h^2}{4\sigma_\eta^2}\right] \quad \text{for } 0 < h < \infty$$

If $\eta(t)$ is a stationary narrowband process so that the peaks are distributed according to the Rayleigh distribution, we find that the root-mean-square wave height, $\sqrt{E\{H^2\}}$, is given by

$$\sqrt{E\{H^2\}} = \int_0^\infty h^2 f_H(h)\,dh = 2\sqrt{2}\sigma_\eta$$

In addition, it can be shown that the average and the significant wave heights are given by

$$H_0 \equiv E\{H\} = \sqrt{2\pi}\sigma_\eta, \quad H_s \equiv E\{H_{1/3}\} = 4\sigma_\eta \qquad (11.15)$$

where $E\{H_{1/3}\}$ means that it is the expectation of the highest one third of the waves.

## 11.1.6   Long-Term Statistics

Because offshore structures are designed for long life spans, we must also consider long-term wave statistics. Previously, when we considered the short-term statistics, the significant wave height and spectrum were assumed to be invariant with time. This assumption is valid only over time periods of days at most. For longer time periods, the significant wave height has its own statistics and is a random variable.

When one uses short-term statistics to describe long-term events, improbable events seem unjustifiably probable. For example, let us consider the probability that the wave height, distributed according to the Rayleigh distribution, exceeds a certain extreme value. Let us assume that the mean period of this wave is 10 sec and the probability that the height of any given wave is greater than 300 ft is $10^{-10}$. The value is small and the occurrence of a 300 ft wave seems improbable. However, the probability that the height will exceed 300 ft at least once in 10 years ($3 \times 10^8$ sec) is given by

$$1 - (1 - 10^{-10})^{3\times10^8/10} = 0.997$$

Thus, the statistical description states that it is almost certain that the wave height will exceed 300 ft at least once in 10 years. This prediction is a shortcoming of the short-term statistics since waves of this magnitude do not arrive at this probability.

In order to compute the probability that a wave height will exceed a certain extreme value, we require statistics for these extreme events. The actual maximum amplitude in a sequence of random amplitudes is a random variable itself. It has a probability distribution with mean value, standard deviation and other statistical properties. In fact, the distributions of these maximum values are called the extreme value distributions (EVDs). Gumbel (1958) obtained three methods of extrapolation known as three asymptotes. They are the Gumbel, Fretchet, and Weibull distributions. We will discuss the Gumbel and Weibull distributions in the next section. For the moment, we will discuss the concept of the *N*-year storm.

In long-term statistics, we often speak of an *N*-year storm. It means that, for any given year, the probability that we will have an *N*-year storm is

$$p = \frac{1}{N}$$

It follows that the probability that we will have *m* storms in *n* years is given by

$$\Pr\{mN\text{-year storms in } n \text{ years}\} =_n P_m \left(\frac{1}{N}\right)^m \left(1 - \frac{1}{N}\right)^{n-m}$$

where $_n P_m$ is the permutation given by

$$_n P_m = \frac{n!}{(n-m)!}$$

The probability that we will have at least one *N*-year storm in *n* years is

$$\Pr\{\text{at least one } N\text{-year storms in } n \text{ years}\} = 1 - \left(1 - \frac{1}{N}\right)^n$$

For a large *N*, the probability can be approximated as $1 - \exp(n/N)$. It should be noted that the probability that we will have exactly one *N*-year storm in *N* years is not one, but

$$\Pr\{\text{one } N\text{-year storm in } N \text{ years}\} = \left(1 - \frac{1}{N}\right)^{M-1}$$

As $N \to \infty$, we find that

$$P = 1/e \approx 0.3679$$

The probability that we will have at least one *N*-year storm in *N* years is

$$\Pr\{\text{at least one } N\text{-year storms in } n \text{ years}\} = 1 - \left(1 - \frac{1}{N}\right)^N$$

As $N \to \infty$, we find that

$$P = 1 - 1/e \approx 0.6321 \tag{11.16}$$

### 11.1.6.1 Weibull Distribution

The Weibull distribution fits probabilities of extremes quite satisfactorily. In long-term statistics, the significant wave height follows the Weibull distribution closely. The probability density and the cumulative distribution are given by

$$f(h) = \frac{m}{\beta} \left(\frac{h-\gamma}{\beta}\right)^{m-1} \exp\left(-\left(\frac{h-\gamma}{\beta}\right)^m\right), \quad F(h) = 1 - \exp\left(-\left(\frac{h-\gamma}{\beta}\right)^m\right) \quad \text{for } \gamma < h \tag{11.17}$$

where $m$ is called the shape parameter. Manipulating the cumulative distribution, we can write

$$\ln(-\ln\{1 - F(h)\}) = m\{\ln(h - \gamma) - \ln(\beta)\}$$

where the left-hand side is known from data. If we let $y = \ln(-\ln\{1 - F(h)\})$ and $x = \ln(h - \gamma)$, $y$ is a straight line with slope of $m$ and a $y$-intercept of $-m \ln \beta$:

$$y = mx - m \ln \beta$$

Suppose we have significant wave height data over a long period of time, and our goal is to find the Weibull parameters, $\gamma$, $\beta$, and $m$ that best fit the distribution of the significant wave heights. These parameters can be determined by the least-squares method or using the Weibull paper. Using the latter method, we first guess $\gamma$ so that the discrete points $(x, y)$ or $(\ln(h - \gamma), \ln(-\ln\{1 - F(h)\}))$ form a straight line. The slope of this line is $m$, and the value of $y$ when the line intersects the $y$ axis is $-m \ln \beta$. This method will be illustrated in Section 11.3.3.

The Gumbel distribution is given by

$$f(h) = \alpha \exp(-\alpha(h - \beta))\exp\{\exp[-\alpha(h - \beta)]\}$$
$$F(h) = \exp\{-\exp[-\alpha(h - \beta)]\} \quad \text{for } -\infty < h < \infty \tag{11.18}$$

When $\ln[-\ln F(h)]$ is plotted against $h$, the result is a line with a slope of $-\alpha$ and $y$ intercept of $\alpha\beta$.

Another distribution that may be used is the lognormal distribution given by (Jasper, 1954)

$$f(h) = \frac{1}{\sqrt{2\pi}\sigma h} \exp\left[ -\frac{1}{2} \frac{(\ln h - \mu)^2}{\sigma^2} \right], \quad F(h) = \Phi\left( \frac{\ln h - \mu}{\sigma} \right) \quad \text{for } 0 \le h \tag{11.19}$$

### 11.1.6.2  Wave Velocities *via* Linear Wave Theory

The wave velocities that correspond to the wave elevation given in Equation 11.12 can be obtained by *linear wave theory*. Linear wave theory, also called airy wave theory, sinusoidal wave theory, and small-amplitude theory, is the simplest wave theory. It is also the most important wave theory because it forms the basis for the probabilistic spectral description of waves.

Linear wave theory assumes that the wave height is small compared with the wavelength and wave depth. In addition, fluid particles are assumed to follow a circular orbit. The readers should refer to Kinsman (1965) and LeMehaute (1976) for detailed descriptions.

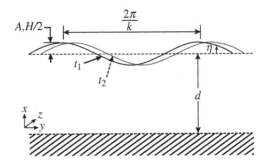

FIGURE 11.10   A schematic of a simple sinusoidal wave shown at two different times.

In linear wave theory, the surface elevation is given by

$$\eta(y, t) = A \cos(\omega t - ky) \tag{11.20}$$

which is a plane wave traveling to the right in Figure 11.10. Linear wave theory relates this sinusoidal surface elevation to the wave velocities given by

$$w_y(x, y, t) = A\omega \frac{\cosh kx}{\sinh kd} \cos(\omega t - ky), \quad w_y(x, y, t) = A\omega \frac{\sinh kx}{\sinh kd} \sin(\omega t - ky) \tag{11.21}$$

where $k$, $\omega$, and $A$ are wave number, angular frequency, and amplitude of a surface wave, respectively. The velocities vary with time, horizontal coordinate $y$, and depth $x$ measured from the ocean floor. The wave velocities are sinusoidal in $y$ and $t$, but exponentially decrease with the distance from the surface.

The frequency $\omega$ is related to the wave number $k$ by the dispersion relation given by

$$\omega^2 = gk \tanh kd$$

where $d$ is the water depth. For deep water, $\tanh kd$ approaches unity and the frequency is given by

$$\lim_{d \to \infty} \omega^2 = gk$$

For the surface elevation given in Equation 11.12, the surface elevation and the wave velocities are given by

$$\eta(y, t) = \sum_{i=1}^{N} \cos(\omega_i t - k_i y - \varphi_i)\sqrt{2S_{\eta\eta}(\omega_i)\Delta\omega_i}$$

$$w_y(x, y, t) = \sum_{i=1}^{N} \omega_i \frac{\cosh k_i x}{\sinh k_i d} \cos(\omega_i t - k_i y - \varphi_i)\sqrt{2S_{\eta\eta}(\omega_i)\Delta\omega_i} \qquad (11.22)$$

$$w_y(x, y, t) = \sum_{i=1}^{N} \omega_i \frac{\sinh k_i x}{\sinh k_i d} \sin(\omega_i t - k_i y - \varphi_i)\sqrt{2S_{\eta\eta}(\omega_i)\Delta\omega_i}$$

The wave accelerations can be obtained by differentiating the wave velocities with respect to time. Sample time histories of the wave velocity and acceleration can be obtained using either Borgman's or Shinozuka's method.

## 11.1.7 Summary

In this section, the concept of spectral density is introduced. It is then shown how the concept is used to describe the ocean wave heights. The spectral density or spectrum is related to the autocorrelation function by the Wiener–Khinchine relations:

$$S_{\eta\eta}(\omega) = \frac{1}{2\pi} \int_{-\infty}^{\infty} R_{\eta\eta}(\tau)\exp(-i\omega\tau)d\tau$$

$$R_{\eta\eta}(\tau) = \int_{-\infty}^{\infty} S_{\eta\eta}(\omega)\exp(-i\omega\tau)d\omega$$

In addition, the spectral density function of a real-valued stationary process is also real and symmetric, or

$$S_{\eta\eta}(\omega) = S_{\eta\eta}(-\omega) \qquad (11.23)$$

and the area under the spectral density is given by

$$\int_{-\infty}^{\infty} S_{\eta\eta}(\omega)d\omega = R_{\eta\eta}(0) = \sigma_\eta^2 - \mu_\eta^2 \qquad (11.24)$$

If the spectral density is given only for $\omega \geq 0$, then this one-sided spectrum is related to the two-sided spectrum by

$$S_{\eta\eta}^o(\omega) = 2S_{\eta\eta}(\omega), \quad \omega \geq 0$$

When the frequency is given in Hertz instead of in rad/sec, the spectra are related by

$$S_{\eta\eta}(f) = 2\pi S_{\eta\eta}(\omega), \quad \omega \geq 0$$

The spectra that are often used to describe wave heights are the P–M, Bretschneider, ITTC, ISSC, O–H, and JONSWAP spectra. The most widely used spectrum is the P–M spectrum, which is a single parameter spectrum. The P–M spectrum is applicable for deep water, unidirectional seas, fully developed and local-wind-generated sea with unlimited fetch, and was originally developed for the North Atlantic. The single parameter for this spectrum can be expressed as the wind velocity at 19.5 m above sea level or the significant wave height that specifies the sea severity. When it is written in terms of the significant wave height, it is given by

$$S^o_{\eta\eta}(f) = \frac{8.1 \times 10^{-3}g^2}{\omega^5} \exp\left(-\frac{0.0324g^2}{H_s^2}\omega^{-4}\right)$$

Bretschneider, ITTC, and ISSC spectra are two-parameter spectra in which the state of development as well as the sea severity can be specified. The O–H spectrum is a six-parameter spectrum that allows us to represent local and distant storm effects and to specify sharpness of the peaks as well as to specify the sea severity and the state of development. The JONSWAP spectrum allows us to account for growth over a limited fetch.

For a given ocean spectrum, a sample time history can be obtained by

$$\eta(t) = \sqrt{\frac{2}{N}}\sigma_\eta \sum_{i=1}^{N} \cos(\bar{\omega}_i t - \varphi_i)$$

In Borgman's method, the sampling frequencies $\bar{\omega}_i = (\omega_i + \omega_{i-1})/2$ are chosen so that the area between $\omega_{i-1}$ and $\omega_i$ are equal. In Shinozuka's method, the sampling frequencies are chosen randomly. The traditional method of choosing the sampling frequency at even intervals is not recommended.

When a relatively short interval of time is considered, for example, about two hours or the duration of a storm, it can be assumed that the spectrum and its statistics are invariant with time. In this case, the distribution of local maxima or peaks of a stationary Gaussian process is described by the Rice distribution. The Rice distribution can be reduced to the Rayleigh distribution when the process is narrowband and to the Gaussian distribution when the process is broadband. In the long-term statistics, the spectrum and its statistics may vary with time. In this case, the term "N-year storm" is often used to indicate the sea severity, and the significant wave heights closely follow the Weibull distribution.

Finally, the wave velocities and accelerations are related to the wave velocities using linear wave theory.

## 11.2 Fluid Forces

The following is a list of several types of forces that the fluid can exert on a body:

1. *Drag force.* This is due to the pressure difference between the downstream and upstream flow region. It can be thought of as the force required to hold a body stationary in a fluid of constant velocity. The drag force is proportional to the square of the velocity of the fluid relative to the structure.
2. *Inertia force.* This is the force exerted by the fluid while it accelerates and decelerates as it passes the structure. It is also the force required to hold a rigid structure in a uniformly accelerating flow, and it is proportional to the fluid acceleration. The concept of the inertia force in an inviscid flow was first formulated by Lamb (1945).
3. *Added mass.* As the body accelerates or decelerates in a stationary fluid, the body carries a certain amount of the surrounding fluid along with it. This entrained fluid is called the *added, apparent,* or *virtual mass.* In order to accelerate the body, additional force is required to accelerate or decelerate the added mass.
4. *Diffraction force.* This is due to the scattering of an incident wave on the surface of the structure. It is important when the body is large compared with the wavelength of the incident wave.
5. *Froude–Kryloff force.* This is the pressure force on the structure due to the incident wave, assuming that the structure does not exist and does not interfere with the incident wave.
6. *Lift force.* This is due to nonsymmetrical separation of the fluid or due to vortices that are shed in a nonsymmetrical way. The component of the force perpendicular to the flow direction is the lift force.

7. *Wave slamming force.* This is due to a single occasional wave with a particularly high amplitude and energy, and it may be important at the free surface. Sarpkaya and Isaacson (1981) reviewed the research on slamming of water against circular cylinders. Miller (1977, 1980) found that the peak wave slamming force on a rigidly held horizontal circular cylinder is proportional to the square of the horizontal water particle velocity.

## 11.2.1   Wave Force Regime

Previously, we discussed various types of forces caused by waves and currents. In some cases, one type of force may be dominant. Hogben (1976) gave a literature review of the fluid force in various regimes. The load regime of importance can be demonstrated for the case of a vertical cylinder in Figure 11.11 in terms of $H/D$ and $\pi D/\lambda$, where $H$ is the wave height, $D$ is the cylinder diameter, and $\lambda$ is the wavelength. When linear wave theory is used, $H/D$ is related to the Keulegan–Carpenter number by

$$K = \pi H/D$$

The Keulegan–Carpenter number gives a measure of the importance of drag force relative to the inertia force. The term $\pi D/\lambda$ is called the diffraction parameter, and it determines the importance of the diffraction effect. As $H/D$ increases, the drag force becomes more important and the inertia force becomes less important. As $\pi D/\lambda$ increases, the diffraction force becomes important.

Using linear wave theory, the maximum drag force to the maximum inertia force can be written as

$$\frac{f_{\text{drag}}}{f_{\text{inertia}}} = \frac{1}{2\pi}\frac{H}{D} = \frac{K}{2\pi^2}$$

From the last relation, we find that the drag force is 5% of the inertial force when $H/D = 0.314$. The Morison equation may be used for $D/\lambda < 0.2$ and $f_{\text{drag}}/f_{\text{inertia}} > 0.1$ or thereabouts. It should be noted that Figure 11.11 is valid only near the surface. The drag force is predominant for a cylinder that extends from the bottom to the near surface, so that the Morison equation may be used.

For example, consider a fixed jacket platform with legs with a diameter of 10 m and bracings with a diameter of 0.8 m. For a 10-year storm with $\lambda = 100$ m and $H = 8$ m, the ratios $H/D$ and $D/\lambda$ for the leg are 0.8 and 0.1, respectively. Similarly, the ratios $H/D$ and $D/\lambda$ for the bracings are 10 and 0.08, respectively. Figure 11.11 shows that the inertia force is dominant for the legs, and both inertia and the drag forces are important for the bracings.

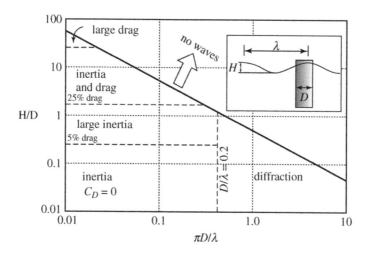

**FIGURE 11.11**   Load regimes near surface.

## 11.2.2   Wave Forces on Small Structures — Morison Equation

The added mass, $M_A$, can be written as

$$M_A = C_A M_{\text{disp}}$$

where $C_A$ is called the *added mass coefficient* and $M_{\text{disp}}$ is the mass of the fluid displaced by the structure. For a cylinder with a diameter, $D$, and height, $h$, the displaced fluid mass is $\pi D^2 h/4$. It should be noted that the added mass is a tensor quantity. That is, we can speak of the added mass force in the $x_i$ direction due to the acceleration of the body in the $x_j$ direction, denoted as $M_{ij}^A$. $M_{ij}^A$ is symmetric so that the added mass force in the $x_i$ direction due to the acceleration in the $x_j$ direction is equal to the added mass force in the $x_j$ direction due to the acceleration in the $x_i$ direction. The off-diagonal terms are not zero if the cross-section is not symmetric.

Similarly, the inertia force can be written as

$$F_M = C_M M_{\text{disp}} \dot{w} \tag{11.25}$$

where the proportionality constant, $C_M$, is called the *inertia coefficient*.

It should be noted that the added mass and the inertia effects are often neglected for a body vibrating in air since the displaced air mass is negligible.

The drag force is proportional to the square of the fluid velocity, $w$, the density of the fluid, $\rho$, and the area of the body projected onto the plane perpendicular to the flow direction, $A_f$,

$$F_D = \frac{1}{2} C_D \rho A_f w |w|$$

where $C_D$ is the *drag coefficient*. The absolute value sign is used to ensure that the drag force always acts in the direction of the flow. For a cylinder with a diameter $D$ and height $h$, the projected area $A_f$ is $Dh$.

For a body with nonzero velocity, the drag force is given by

$$F_D = \frac{1}{2} C_D \rho A_f (w - v) |w - v| \tag{11.26}$$

where $w - v$ is the velocity of the fluid relative to the body.

Morison et al. (1950) combined the inertia and drag terms (Equation 11.25 and Equation 11.26) so that the fluid force on a body is given by

$$f = \frac{1}{2} C_D \rho A_f w |w| + C_M M_{\text{disp}} \dot{w}$$

For a cylinder, the fluid force per unit length can be written as

$$f = \frac{1}{2} C_D \rho D w |w| + C_M \rho \pi \frac{D^2}{4} \dot{w}$$

For a moving cylinder with velocity $v$, the Morison force is given by

$$f = \frac{1}{2} C_D \rho D (w - v) |w - v| + C_M \rho \pi \frac{D^2}{4} \dot{w}$$

### 11.2.2.1   Inclined Cylinder

Let us now consider the inclined cylinder shown in Figure 11.13. The direction of the flow makes an angle of $\theta$ with the cylinder. Often, only the fluid force in the normal direction is considered. The normal component is given by

$$f^n = \frac{1}{2} C_D \rho D (w^n - v^n) |w^n - v^n| + C_M \rho \pi \frac{D^2}{4} \dot{w}^n \tag{11.27}$$

where the superscript is used for the normal component. The term, $w^n - v^n$, is the normal component of the relative velocity of the fluid with respect to the structure. Suppose that fluid is flowing to the right, and the cylinder is also moving to the right, as shown in Figure 11.12. The normal components of the fluid and cylinder velocities are

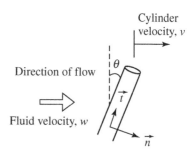

$$w^n = |w| \cos \theta, \quad v^n = |v| \cos \theta$$

In three dimensions, it may be difficult to picture what the normal component should be. Here, we can find the normal component using the formula

$$(w^n - v^n)\vec{n} = \vec{t} \times (\vec{w} - \vec{v}) \times \vec{t} \qquad (11.28)$$

**FIGURE 11.12** Inclined cylinder.

where $\vec{t}$ is the unit vector tangent to the cylinder and $\vec{n}$ is the unit vector normal to the cylinder. Note that the normal direction depends on the direction of the flow as well as the inclination of the cylinder.

In some cases, the tangential drag force may be included, and it can be written as

$$f^t = \frac{1}{2} C_T \rho D (w^t - v^t) |w^t - v^t| \qquad (11.29)$$

where $C_T$ is the tangential drag coefficient. Note that $C_T$ is usually a very small number.

The normal component of the fluid force is more dominant than the tangential component. It may seem strange that the fluid force does not act in the direction of the fluid motion. Instead, the force is predominantly in the normal direction defined by Equation 11.28. In Section 11.3.1, we will demonstrate what this means by considering a towing cable.

### 11.2.2.1.1 Determination of Fluid Coefficients

The drag, inertia, and added mass coefficients must be obtained by experiment. However, for a long cylinder, $C_M$ approaches its theoretical limiting value (uniformly accelerated inviscid flow) of 2, and $C_A$ approaches unity (Lamb, 1945; Wilson, 1984). In reality, the inertia and drag coefficients are functions of at least three parameters (Wilson, 1984):

$$C_M = C_M(\text{Re}, K, \text{cylinder roughness})$$
$$C_D = C_D(\text{Re}, K, \text{cylinder roughness})$$

where Re is the Reynolds number and $K$ is the Keulegan–Carpenter number given by

$$Re \equiv \frac{\rho_f U D}{\mu}, \quad K \equiv \frac{UT}{D} \qquad (11.30)$$

where $\rho_f$ is the density of the fluid, $U$ is the free stream velocity, $D$ is the diameter of the structure, $\mu$ is the dynamic or absolute viscosity, and $T$ is the wave period.

Sarpkaya looked at the variation of these hydrodynamic coefficients extensively and obtained the plots shown in Figure 11.13 to Figure 11.15 (Sarpkaya, 1976; Sarpkaya et al., 1977). Figure 11.13 shows the inertia and drag coefficients for a smooth cylinder as a function of $K$ for various values of Re and the reduced frequency $\beta$, defined by $\beta = \text{Re}/K$. From this figure, we find that for low Re and $\beta$, the inertial coefficient decreases and the drag coefficient increases at about $10 < K < 15$. It is found that the drop and the increase in these coefficients are due to shedding vortices, which also exert forces perpendicular to the structure and the flow.

Figure 11.14 and Figure 11.15 show the inertia and drag coefficients for a rough cylinder, whose roughness is measured by $k/D$. Figure 11.14a shows a drop in the drag coefficient for Re between $10^4$ and $10^5$, and this is called the "drag crisis." For a larger Re, the drag coefficient stays constant. As the surface becomes rougher, the drop occurs at lower Re and the drag coefficients for the larger Re increases.

Figure 11.14 to Figure 11.16 can be used to obtain proper values of the drag and inertia coefficients for fluid with known Re, Keulegan–Carpenter number, and cylinder roughness.

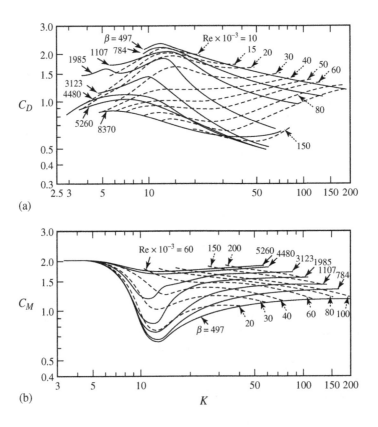

**FIGURE 11.13** Drag and inertia coefficients as functions of $K$ for various values of Re and $\beta$. (*Source*: Sarpkaya, 1976, *Proceedings of the Eighth Offshore Technology Conference.* With permission.)

### 11.2.3 Vortex-Induced Vibration

When the flow passes around a fixed cylinder, for a very low Re ($0 < \text{Re} < 4$), the flow separates and reunites smoothly. When the Re is between 4 and 40, eddies are formed and are attached to the downstream side of cylinder. They are stable and there is no oscillation in the flow. For a flow with a Reynolds number greater than about 40, the fluid near the cylinder starts to oscillate due to shedding vortices. These shedding vortices exert an oscillatory force on the cylinder in the direction perpendicular to both the flow and the structure. The frequency of oscillation is related to the nondimensionalized parameter, the Strouhal number, defined by

$$St = \frac{f_v D}{U} \tag{11.31}$$

where $f_v$ is the frequency of oscillation, $U$ is the steady velocity of the flow, and $D$ is the diameter of the cylinder. For circular cylinders, the Strouhal number stays roughly at 0.22 for laminar flow ($10^3 < \text{Re} < 2 \times 10^5$) and 0.3 for turbulent flow (Patel, 1989).

The lift force due to these shedding vortices can be written as

$$f_L = \frac{1}{2} C_L \rho A_f U^2 \cos 2\pi f_v t \tag{11.32}$$

where $C_L$ is the lift coefficient, which is also a function of Re, $K$, and the surface roughness. The experimental data of the lift coefficients show considerable scatter with typical values ranging from 0.25 to 1. For smooth cylinders, the lift coefficient approaches about 0.25 as Re and $K$ increase.

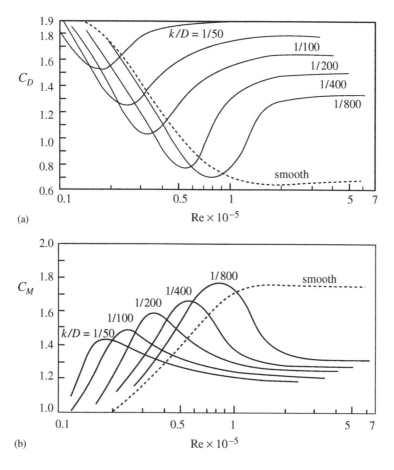

**FIGURE 11.14** Drag and inertia coefficients for a rough cylinder as functions of Re for various values of cylinder roughness (as measured by *k/D*) for *K* = 20. (*Source*: Sarpkaya et al., 1977, *Proceedings of the Ninth Offshore Technology Conference*. With permission.)

It should be noted that the vortex forces are not generally correlated on the entire cylinder length. That is, the phase of the vortex shedding forces varies over the length. The correlation length — the length over which vortex shedding is synchronized — for a stationary cylinder is about three to seven diameters for laminar flow. If sectional forces are randomly phased, the net effect will be small. The total force on a cylinder of length $L$ will be only a fraction of $Lf_{\mathrm{L}}$. This fraction is called the joint acceptance and depends on the ratio of the correlation length to the total length.

When the flow passes by a cylinder that is free to vibrate, the shedding frequency is also controlled by the movement of the cylinder. When the shedding frequency is close to the first natural frequency of the cylinder ($\pm 25$ to 30% of the natural frequency [Sarpkaya and Isaacson, 1981]), the cylinder takes control of the vortex shedding. The vortices will shed at the natural frequency instead of at the frequency determined by the Strouhal number. This is called lock-in or synchronization, which is a result of nonlinear interaction between the oscillation of the body and the action of the fluid. Figure 11.16 shows the shedding frequency, as a function of flow velocity in the presence of a structure. $f_1$ and $f_2$ are the natural frequencies of the structure.

The amplitude of the structural response and the range of the fluid velocity over which the lock-in phenomenon persists are functions of a reduced damping parameter — the ratio of the damping force to the exciting force (Vandiver, 1985, 1993). If the reduced damping parameter is small, the lock-in can persist over a greater range of flow velocity.

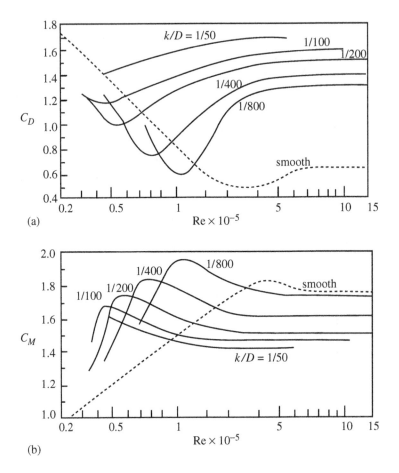

(a)

(b)

**FIGURE 11.15**  Drag and inertia coefficients for a rough cylinder as functions of *Re* for various values of cylinder roughness (as measured by *k/D*) for *K* = 60. (*Source*: Sarpkaya et al., 1977, *Proceedings of the Ninth Offshore Technology Conference.* With permission.)

The existing models for vortex-induced oscillation for a rigid cylinder include single-degree-of-freedom models and coupled models. The single-DoF models assume that the effect of vortex shedding is an external forcing function, which is not affected by the motion of the body. The coupled models assume that the equations that govern the motion of the structure and the lift coefficients are coupled so that the fluid and the structure affect each other (Billah, 1989).

### 11.2.4  Summary

Some of the fluid forces are discussed briefly, and the regimes where inertia, drag, and diffraction forces are important are shown as functions of the ratio of the structural diameter to the wave

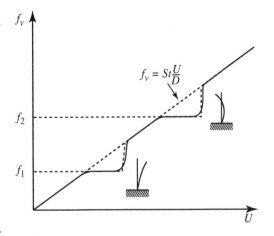

**FIGURE 11.16**  An example of fluid elastic resonance.

length, $D/\lambda$, and the ratio of the wave height to the structural diameter, $H/D$. The wave forces on small structures are modeled by the Morison equation, and it is valid for $D/\lambda < 0.2$ and $H/D > 0.63$ or thereabouts. The Morison equation includes the effects of added mass, inertia, and drag. The added mass is simply

$$M_A = C_A M_{disp}$$

For a cylinder with transverse velocity, $v$, the normal and the tangential components of the drag and the inertia forces are given by

$$f^n = \frac{1}{2} C_D \rho D (w^n - v^n)|w^n - v^n| + C_M \rho \pi \frac{D^2}{4} \dot{w}^n$$

$$f^t = \frac{1}{2} C_T \rho D (w^t - v^t)|w^t - v^t|$$

The fluid coefficients are at least functions of three parameters: the Reynolds number, the Keulegan–Carpenter number, and the cylinder roughness. The plots of these coefficients are reproduced in Figure 11.13 to Figure 11.15.

The frequency of the lift force that is exerted by shedding vortices is closely related to the Strouhal number given by

$$St = \frac{f_v D}{U}$$

The lift force due to these shedding vortices can be written as

$$f_L = \frac{1}{2} C_L \rho A_f U^2 \cos 2\pi f_v t$$

If the structure is free to vibrate, then lock-in or synchronization may occur when the shedding frequency is close to the structure's natural frequency. The structure takes control of the vortex shedding. Many nonlinear models are available to capture this phenomenon.

## 11.3 Examples

Four examples are given in this section. The first example illustrates the roles of the normal and the tangential components of the drag force in the static configuration of a towing cable. The second example shows how the equation of motion of an articulated tower can be formulated in the presence of surrounding fluid. The third example shows how to choose a single significant wave height to represent a certain condition from significant wave height data over a long period of time. The final example shows how to reconstruct time series data from a given spectrum.

### 11.3.1 Static Configuration of a Towing Cable

For the purpose of ocean surveillance, oceanographic or geographic measurements, or ocean exploration, marine cables with instrument packages or Remotely Operated Vehicles are often towed behind ships or submarines. For example, the goal of the VENTS program by the National Oceanic and Atmospheric Administration (NOAA) is to conduct research on the impacts and consequences of submarine volcanoes and hydrothermal venting on the global ocean. In attempts to locate and map the distributions of hydrothermal plumes in the Mid-Ocean Ridge system, an instrument package called a CTD (Conductivity, Temperature and Depth Sensors) is towed behind a ship.

Let us consider a cable and a body towed behind a ship at a constant velocity with no current as shown in Figure 11.17. What kind of shape will the cable take? What will be the distance between the ship and the towed body?

We immediately recognize that this is equivalent to having a stationary ship with a steady current in the opposite direction. The equation of motion is given by

$$\sum \vec{F} = m\vec{a}(s,t) = \vec{0} = \frac{\partial}{\partial s}(T\vec{t}) + f^{n}\vec{n} + f^{t}\vec{t} + mg\vec{k}$$

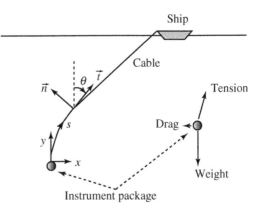

**FIGURE 11.17**  Towed system in equilibrium and the forces acting on the towed body.

where $m$ is the mass of the cable per unit length, $\vec{a}(s,t)$ is the acceleration of the cable, $s$ is the coordinate along the cable, $T$ is the tension which is a function of $s$, $(\vec{t}, \vec{n}, \vec{b})$ is the set of unit vectors of the curvilinear coordinate system, $\vec{k}$ is the unit vector downward in the direction of gravity, $g$ is the gravitational acceleration, $f^{n}$ is the normal drag force, and $f^{t}$ is the tangential drag force. The added mass and the inertial terms are zero because the fluid acceleration and the cable acceleration are zero. The normal and tangential drag forces are given in Equation 11.27 and Equation 11.29. In our case, they are given by

$$f^{n} = C_{D}\rho \frac{D}{2} U^{2} \cos^{2}\theta, \quad f^{n} = -C_{T}\rho \frac{D}{2} U^{2} \sin^{2}\theta$$

The corresponding scalar equations are given by

$$\frac{dT}{ds} - C_{T}\rho \frac{D}{2} U^{2} \sin^{2}\theta - mg \cos\theta = 0$$

$$-T\frac{d\theta}{ds} + C_{D}\rho \frac{D}{2} U^{2} \cos^{2}\theta - mg \sin\theta = 0$$

(11.33)

where $\theta$ is the angle that the tangential vector makes with the vertical and measured positive clockwise. Note that we have used $\partial \vec{t}/\partial s = (-\partial \theta/\partial s)\vec{n}$ and $\vec{k} = -\cos\theta\vec{t} - \sin\theta\vec{n}$. Equation 11.33 shows that the tangential components of the external forces act to increase the tension, while the normal components cause the towline to bend. Because the normal component of the drag force is much larger than the tangential component, most of the fluid force is used to turn the cable.

From the force diagram (in Figure 11.17), the angle that the cable makes with the vertical where it is connected to the towed body is given by

$$T(0)\cos\theta(0) = W, \quad T(0)\sin\theta(0) = \text{Drag}$$

Once we know the weight and the drag force on the towed body, the tension and the angle at $s = 0$ can be found. If the drag is negligible compared with the weight, then the cable must be near vertical and the tension must be equal to the weight of the towed body at $s = 0$:

$$T(0) \approx W \quad \text{and} \quad \theta(0) \approx 0$$

For now, let us assume that this is the case. Then, with these initial conditions, the system of ordinary differential equations (Equation 11.33) can be solved numerically for $T(s)$ and $\theta(s)$. For example, even very simple finite difference equations will work. A set of equations

$$T_{i+1} = T_i + \left( mg \cos \theta_i - C_T \rho \frac{D}{2} U^2 \sin^2 \theta_i \right) \Delta s \tag{11.34}$$

$$\theta_{i+1} = \theta_i - \left( mg \sin \theta_i + C_D \rho \frac{D}{2} U^2 \cos^2 \theta_i \right) \Delta s / T_i \tag{11.35}$$

where $T_i = T(i \Delta s)$, are used here, and it works very well for $\Delta s = 0.05$.

The Cartesian coordinates, $x$ and $y$, are related to $\theta$ by

$$\frac{dx}{ds} = \sin \theta \quad \text{and} \quad \frac{dy}{ds} = \cos \theta$$

and can also be obtained by integrating them numerically.

Figure 11.18 shows the results when $mg = 1.5$ N/m, $C_D \rho D U^2/2 = 10$ N/m, $C_T \rho D U^2/2 = 0.1$ N/m, $W = 100$ N, and the cable is 100 m long. Care is taken so that the ship is located at $x = 0$ and $y = 0$.

It is interesting to note that $\theta$ approaches a critical value, and the shape gradually becomes linear toward the ship. Mathematically, $d\theta/ds$ becomes zero. This is when the drag force is completely balanced by the normal component of the cable weight. The angle at which this occurs, $\theta_{cr}$, can be obtained from the second governing equation and

$$mg \sin \theta_{cr} = -f^n, \quad \frac{\sin \theta_{cr}}{\cos \theta_{cr}} = C_D \rho \frac{D}{2} U^2 \frac{1}{mg}$$

In our case, $\theta_{cr} = 1.184$ rad, and this value agrees with Figure 11.18.

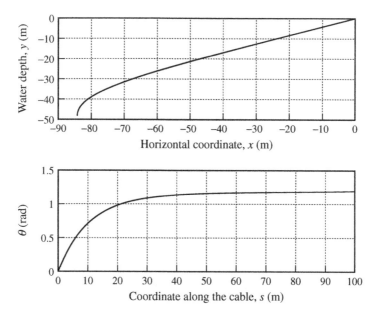

**FIGURE 11.18** The equilibrium configuration of a towed cable and the angle that the cable makes with the vertical when $mg = 1.5$ N/m, $C_D \rho D U^2/2 = 10$ N/m, $C_T \rho D U^2/2 = 0.1$ N/m, and $W = 100$ N.

## 11.3.2   Fluid Forces on an Articulated Tower

Offshore structures are used in the oil industry as exploratory, production, oil storage, and oil landing facilities. They are designed to be self-supporting and sufficiently stable for offshore activities such as drilling and production of oil. An articulated tower as seen in Figure 11.19 is an example of an offshore platform that consists of a base, shaft, universal joint that connects the base and the shaft, ballast chamber, buoyancy chamber, and deck. The ballast chambers provide the extra weight so that the tower's bottom stays on the ocean floor, and the buoyancy chamber adds the necessary buoyancy so that the tower does not fall.

An articulated tower can be effectively modeled as a rigid inverted pendulum, where the deck is modeled as a point mass, the shaft as a uniform rigid bar, and the buoyancy chamber by a point buoyancy. In two dimensions, motion of the tower can be described with a single DoF (Chakrabarti and Cottor, 1979; Bar-Avi, 1996). The equation of motion in terms of the tower's deflection angle is obtained by summing the moment about the point O in Figure 11.20 and is given by

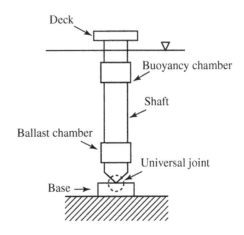

**FIGURE 11.19**   Schematic of an articulated tower.

$$I\frac{d^2\theta}{dt^2} = \sum M_O = mg\frac{L}{2}\sin\theta$$
$$+ MgL\sin\theta - Bl\sin\theta + \int_0^L f^n x\,dx$$

where $I$ is the mass moment of inertia about the point O given by $I = mL^2/3 + ML^2$, $m$ is the mass of the shaft, $g$ is the gravitational acceleration, $L$ is the length of the shaft, $M$ is the point mass at the top, $B$ is the buoyancy provided by the buoyancy chamber, $l$ is its moment arm, $f^n$ is the normal fluid force per unit length, and $x$ is the coordinate along the shaft from O.

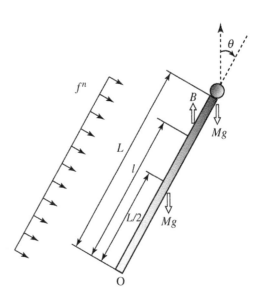

**FIGURE 11.20**   Free-body diagram.

The fluid force per unit length in the normal direction is given by

$$f^n = C_D \rho \frac{D}{2}(w^n - v^n)|w^n - v^n| + C_M \rho \pi \frac{D^2}{4}\dot{w}^n - C_A \rho \pi \frac{D^2}{4}a^n$$

where the last term is the force in the normal direction due to the added mass. $v^n$ and $a^n$ are the velocity and the acceleration of the body in the normal direction and are given by

$$v^n = x\frac{d\theta}{dt} \quad \text{and} \quad a^n = x\frac{d^2\theta}{dt^2}$$

If we assume that the surrounding fluid is stationary, then the normal velocity and the acceleration of the fluid ($w$ and $\dot{w}$) are zeros. Thus, the moment due to the fluid force is given by

$$\int_0^L f^n x \, dx = \int_0^L \left( -C_D\rho\frac{D}{2}x^2\left(\frac{d\theta}{dt}\right)^2 \text{sign}\left(\frac{d\theta}{dt}\right) + C_A\rho\pi\frac{D^2}{4}x\frac{d^2\theta}{dt^2} \right)x \, dx$$

$$= -C_D\rho\frac{D}{2}\frac{L^4}{4}\left(\frac{d\theta}{dt}\right)^2 \text{sign}\left(\frac{d\theta}{dt}\right) + C_A\rho\pi\frac{D^2}{4}\frac{L^3}{3}\frac{d^2\theta}{dt^2}$$

and the equation of motion is given by

$$\left(m\frac{L^2}{3} + ML^2 + C_A\rho\pi\frac{D^2}{4}\frac{L^3}{3}\right)\frac{d^2\theta}{dt^2} = \left(mg\frac{L}{2} + MgL - Bl_b\right)\sin\theta - C_D\rho\frac{D}{2}\frac{L^4}{4}\text{sign}\left(\frac{d\theta}{dt}\right)\left(\frac{d\theta}{dt}\right)^2$$

Note that the normal fluid drag force adds directly to the restoring moment in the case of a rigid bar. The equation of motion can be solved numerically once the initial conditions ($\theta[0]$ and $d\theta/dt[0]$) are given.

The equation of motion can be simplified if we assume that the angle of rotation $\theta$ is small. More specifically, if we assume that $\theta^2$ is negligible when compared with 1, then we find that[2]

$$\sin\theta \approx \theta$$

The equation of motion can be simplified to

$$\left(m\frac{L^2}{3} + ML^2 + C_A\rho\pi\frac{D^2}{4}\frac{L^3}{3}\right)\frac{d^2\theta}{dt^2} - \left(mg\frac{L}{2} + MgL - Bl_b\right)\theta + C_D\rho\frac{D}{2}\frac{L^4}{4}\text{sign}\left(\frac{d\theta}{dt}\right)\left(\frac{d\theta}{dt}\right)^2 = 0$$

which resembles the equation for a linear oscillator with a nonlinear damping term. Note that the system becomes unstable when the stiffness term (the coefficient of $\theta$) becomes negative. This occurs when the buoyancy is not sufficient or

$$B < \frac{1}{l_b}\left(mg\frac{L}{2} + MgL\right)$$

## 11.3.3 Distribution of Significant Wave Heights — Weibull and Gumbel Distributions

The National Buoy Data Center (NBDC) run by NOAA collects ocean data such as wind, current, wave, pressure, and temperature data in various locations and the records are made public. Let us say that we are to design an articulated tower (in Section 11.3.2) in one of these locations where the data are available. The first task is to characterize the environment. Using all of the information that is collected is inefficient and impractical. Instead, we are interested in choosing a single number that can represent typical and extreme situations such as 10- and 50-year storms. For now, let us only consider random waves. We are then interested in finding the significant wave heights representing 10- and 50-year storms.

From NBDC data for a buoy outside Monterey Bay, the number of occurrences for ranges of significant wave heights is constructed in Table 11.2. The measurements were taken every hour for about 12 years. We first construct the corresponding Weibull distribution using the method described in Section 11.1.6. We first guess $\gamma$ so that a pair of $\ln(-\ln\{1 - F(h)\})$ and $\ln(h - \gamma)$ form a

---

[2]This is called the small angle assumption.

**TABLE 11.2**    Number of Occurrences of Various Sea States

| Significant Wave Height, $h$ (m) | Number of Occurrences | Sum |
|---|---|---|
| <1 | 2,367 | 2,367 |
| 1–2 | 46,353 | 48,720 |
| 2–3 | 3,4285 | 83,005 |
| 3–4 | 1,3181 | 96,186 |
| 4–5 | 3,813 | 99,999 |
| 5–6 | 716 | 100,715 |
| 6–7 | 145 | 100,860 |
| 7–8 | 32 | 100,892 |
| 8–9 | 8 | 100,900 |
| 9–10 | 2 | 100,902 |
| Total | 100,902 | |

straight line. Figure 11.21 shows that the pair yields nearly a straight line when $\gamma \approx 0.84$. The slope and the $y$ intercept of this line are 1.6 and $-0.78$, respectively. The Weibull parameters are then $m = 1.6$ and $\beta = 1.6$.

Similarly, we can find the corresponding Gumbel probability density function by plotting pairs of $(h, \ln(-\ln\{F(h)\}))$ to form a line. For the data shown in Table 11.2, the line has a slope of $-1.52$ and $y$ intercept of 2.84 so that $\alpha = -1.52$ and $\beta = 1.87$.

Figure 11.22 shows the Weibull probability density and the cumulative distribution (Equation 11.17) in solid lines, the Gumbel probability density and the cumulative distribution in dotted lines (Equation 11.18), and the discrete probability density and the cumulative distribution derived from Table 11.2 in symbols.

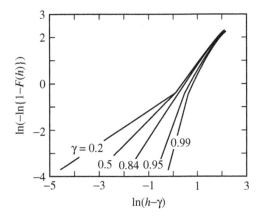

**FIGURE 11.21**    Plots of $(\ln(h - \gamma), \ln[-\ln\{1 - F(h)\}])$ for various values of $\gamma$.

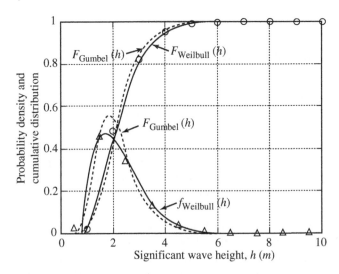

**FIGURE 11.22**    Weibull approximations of the probability density and cumulative distribution of significant wave heights measured in the outer Monterey Bay area. The symbols are the values given in Table 11.2.

**TABLE 11.3** Comparison of Representative Significant Wave Heights for Long-Term Predictions from Gumbel and Weibull Distributions

|  | 5-Year (m) | 10-Year (m) | 50-Year (m) |
|---|---|---|---|
| Weibull | 7.84 | 8.15 | 8.79 |
| Gumbel | 8.83 | 9.33 | 10.4 |

The next step is to find a significant wave height that can represent an $N$-year storm, $h_N$. The probability that we will not have an $N$-year storm in any given year is $1 - 1/N$ and is equivalent to the probability that the significant wave height will not exceed $h_N$ in the same year. The probability that $h < h_N$ in a single measurement is $F(h_N)$, and the probability that $h < h_N$ in every measurement taken in a year is $F(h_N)^{24 \times 365}$. Then, we have

$$1 - \frac{1}{N} = F(h_N)^{24 \times 365}$$

Table 11.3 shows significant wave heights that represent 5-, 10-, and 50-year storms obtained using the Weibull and Gumbel distributions.

The Gumbel probability distribution gives higher significant wave heights. For this particular set of data, the Weibull distribution seems to fit the data better (Figure 11.22), and the Weibull distribution is the most often used distribution in the offshore industry.

## 11.3.4 Reconstructing Time Series for a Given Significant Wave Height

Previously, we found significant wave heights that could represent 5-, 10-, and 25-year storms for a given site. Recall that the significant wave height can entirely characterize the Pierson–Moskowitz spectra. Once the spectral density is determined, a sample time history of the wave profile, $\eta(t)$, can be determined using either Borgman's or Shinozuka's method (Section 11.1.4). Here, Shinozuka's method is used to generate the random wave elevations.

Let us first find the random frequencies distributed according to $S^o_{\eta\eta}(\omega)/\sigma^2_\eta$. The P–M spectrum in terms of the significant wave height is given by Equation 11.8.

$$S^o_{\eta\eta}(\omega) = 0.7795\omega^{-5}\exp\left(-\frac{3.118}{H^2_s}\omega^{-4}\right)$$

The variance is given by

$$\sigma^2_\eta = \int_0^\infty S^o_{\eta\eta}(\omega)d\omega = \frac{H^2_s}{16}$$

The probability density and the cumulative distribution functions are given by

$$f(\omega) = S^o_{\eta\eta}(\omega)/\sigma^2_\eta = \frac{12.472}{H^2_s}\omega^{-5}\exp\left(-\frac{3.118}{H^2_s}\omega^{-4}\right), \quad F(\omega) = 1 - \exp\left(-\frac{3.118}{H^2_s}\omega^{-4}\right)$$

The inverse of the cumulative distribution function is given by

$$F^{-1}(x) = \left(-\frac{H^2_s}{3.118}\ln(1-x)\right)^{-1/4}$$

The random frequencies distributed according to $f(\omega)$ can be obtained from uniformly distributed random numbers $x$ from 0 and 1. Table 11.4 shows uniform random numbers between

**TABLE 11.4**   Generation of Random Frequencies Distributed According to $f(\omega)$ from Uniform Random Numbers

| Uniform Random Numbers $0 < x < 1$ | Random Frequencies $\omega$ Distributed According to $f(\omega)$ |
|---|---|
| 0.950 | $(-19.713\ln[1-0.950])^{-1/4} = 0.360$ |
| 0.231 | $(-19.713\ln[1-0.231])^{-1/4} = 0.662$ |
| 0.606 | $(-19.713\ln[1-0.606])^{-1/4} = 0.483$ |
| $\vdots$ | $\vdots$ |

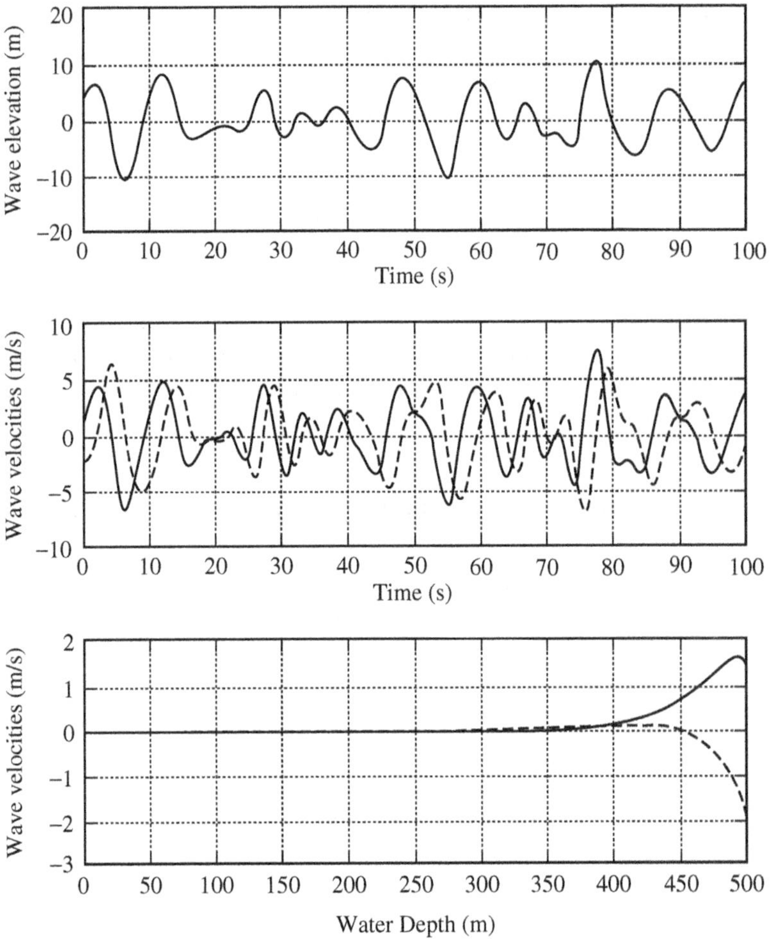

**FIGURE 11.23**   Wave elevation and velocities.

0 and 1 and the random frequencies distributed according to $f(\omega)^3$. The significant wave height of 7.84 m is used.

We can obtain 100 in this way, and the wave elevation is also obtained using Equation 11.13. The random phase $\varphi_i$ is obtained by multiplying uniform random numbers (different from the ones used to generate the random frequencies) by $2\pi$.

Figure 11.23 shows the surface elevation as a function of time, the corresponding wave velocities at the water surface (Section 11.1.7) as functions of time, and the wave velocities at $t = 0$ as functions of the water depth. Note that the wave velocities decay with depth.

---

[3]The uniform random numbers can be generated by the MATLAB® rand function.

## 11.3.5 Available Numerical Codes

Many numerical codes are available for modeling the dynamics of slender structures such as risers, tether, umbilicals, and mooring lines. The first example in this section was solved by a numerical code, WHOI Cable, developed at Woods Hole Oceanographic Institution. WHOI Cable is a time-domain program that can be used for analyzing the dynamics of towed and moored cable systems in both two and three dimensions. It takes into account bending and torsion as well as extension.

Comparative studies investigating flexible risers were carried out by ISSC Committee V7 from computer programs developed by 11 different institutions in the period between 1988 and 1991, and the results were reported by Larsen (1992). More recently, Brown and Mavrakos (1999) conducted a comparative study on the dynamic analysis of suspended wire and chain mooring lines and reported results from 15 different numerical codes. The participants included engineering consultancies, and academic and research institutions involved in marine technology. Some of the time-domain programs that were included in the comparative study are MODEX by Chalmers University of Technology, FLEXAN-C by Institute Francais du Petrole, DYWFLX95 by MARIN, R.FLEX by MARINTEK, CABLEDYN by National Technical University of Athens, DMOOR by Noble Denton Consultancy Services Ltd, V.ORCAFLEX by Orcina Ltd Consulting Engineers, ANFLEX by Petrobras SA, TDMOOR-DYN by University College London, FLEXRISER by Zentech International. Some of these programs are available to academic institutions and government laboratories at no cost.

# Acknowledgments

The author wishes to express gratitude for the funding from the Woods Hole Oceanographic Institution and the Department of Mechanical Engineering at Texas Tech University.

# References

Bar-Avi, P. 1996. Dynamic response of an offshore articulated tower, Ph.D. thesis, The State University of New Jersey, Rutgers, May 1996.

Billah, K. 1989. A study of vortex induced vibration, Ph.D. thesis, Princeton University, May 1989.

Borgman, L., Ocean wave simulation for engineering design, *J. Waterway Harbors Div.*, 95, 557–583, 1969.

Bretschneider, C. 1959. *Wave variability and wave spectra for wind-generated gravity waves.* Technical Memorandum No.118, Beach Erosion Board, U.S. Army Corps of Engineers, Washington D.C.

Bretschneider, C. 1969. Wave forecasting. In *Handbook of Ocean and Underwater Engineering*, Ed. J.J. Myers, McGraw-Hill, New York.

Brown, D. and Mavrakos, S. Comparative study on mooring line dynamic loading, *Marine Structures*, 12, 131–151, 1999.

Chakrabarti, S.K. 1987. *Hydrodynamics of Offshore Structures*, Computational Mechanics Publications, Southampton, U.K.

Chakrabarti, S.K. and Cottor, D., Motion analysis of articulated tower, *J. Waterway Port Coast. Ocean Div.*, 105, 281–292, 1979.

Faltinsen, O.M. 1993. *Sea Loads on Ships and Offshore Structures*, Cambridge University Press, Cambridge.

Gumbel, E. 1958. *Statistics of Extremes*, Columbia University Press, New York.

Hasselmann, K., Barnett, T., Bouws, E., Carlson, H., Cartwright, D., Enke, K., Ewing, J., Gienapp, H., Hasselmann, D., Kruseman, P., Meerburg, A., Muller, P., Olbers, D., Richter, K., Sell, W., and Walden, H. 1973. Measurement of wind-wave growth and swell decay during the joint North Sea wave project (JONSWAP), *Technical Report 13 A*. Deutschen Hydrographischen Zeitschrift.

Hogben, N. 1976. Wave loads on structures, *Behavior of Offshore Structures (BOSS)*, Oslo, Norway.

Isaacson, M., Wave and current forces on fixed offshore structures, *Can. J. Civil Eng.*, 15, 937–947, 1988.

Jasper, N., Statistical distribution patterns of ocean waves and of wave induced ship stresses and motions with engineering applications, *Trans. Soc. Nav. Arch. Mar. Engrs*, 64, 375–432, 1954.

Khinchine, A., Korrelations theorie der stationaren stochastischen prozesse, *Math. Ann.*, 109, 604–615, 1934.

Kinsman, B. 1965. *Wind Waves*, Prentice Hall, Englewood Cliffs, NJ.

Lamb, H. 1945. *Hydrodynamics*, 6th Ed., Cambridge University Press, New York.

Larsen, C. Flexible riser analysis — comparison of results from computer programs, *Marine Structure*, 5, 103–119, 1992.

LeMehaute, B. 1976. *Introduction to Hydrodynamics and Water Waves*, Springer-Verlag, New York.

Longuet-Higgins, M., On the statistical distribution of the height of sea waves, *J. Mar. Res.*, 11, 3, 245–266, 1952.

Miller, B., Wave slamming loads on horizontal circular elements of offshore structures, *Nav. Arch.*, 3, 81–98, 1977.

Miller, B. 1980. Wave slamming on offshore structures, *Technical Report No. NMI-R81*. National Maritime Institute.

Morison, J., O'Brien, M., Johnson, J., and Schaaf, S., The force exerted by surface waves on piles, *Pet. Trans., AIME*, 189, 149–157, 1950.

Ochi, M., On prediction of extreme values, *J. Ship Res.*, 17, 29–37, 1973.

Ochi, M. and Hubble, E. 1976. Six parameter wave spectra. ASCE pp. 301–328. *Proceedings of the Fifteenth Coastal Engineering Conference*, Honolulu, HI.

Patel, M. 1989. *Dynamics of Offshore Structures*, Butterworths, London.

Pierson, W. and Moskowitz, L., A proposed spectral form for fully developed wind seas based on the similarity theory of S.A. Kitaigorodskii, *J. Geophys. Res.*, 69, 24, 5181–5203, 1964.

Rice, S.O. 1954. Mathematical analysis of random noise. In *Selected Papers on Noise and Stochastic Processes*, N. Wax, Ed., Dover Publications, New York.

Sarpkaya, T. 1976. In-line and transverse forces on cylinders in oscillating flow at high Reynolds numbers, OTC 2533, pp. 95–108. *Proceedings of the Eighth Offshore Technology Conference*, Houston, TX.

Sarpkaya, T., Collins, and N., Evans, S. 1977. Wave forces on rough-walled cylinders at high Reynolds numbers, OTC 2901, pp. 175–184. In *Proceedings of the Ninth Offshore Technology Conference*, Houston, TX.

Sarpkaya, T., and Isaacson, M. 1981. *Mechanics of Wave Forces on Offshore Structures*, Van Nostrand Reihold, New York.

Shinozuka, M., Monte Carlo Solution of structural dynamics, *Comput. Struct.*, 2, 855–874, 1972.

Sverdrup, H., and Munk, W. 1947. Wind, sea, and swell: theory of relations for forecasting, *Technical Report 601*. U.S. Navy Hydrographic Office.

Tucker, M. 1991. *Waves in Ocean Engineering: Measurements, Analysis, and Interpretation*, Ellis Horwood, Chichester, U.K.

Vandiver, J. 1985. Prediction of lockin vibration on flexible cylinders in sheared flow, May 1985. In *Proceedings of the 1985 Offshore Technology Conference*, Paper No. 5006, Houston, TX.

Vandiver, J., Dimensionless parameters important to the prediction of vortex-induced vibration of long, flexible cylinders in ocean currents, *J. Fluids Struct.*, 7, 5, 423–455, 1993.

Wiener, N., Generalized harmonic analysis, *Acta Math.*, 55, 117–258, 1930.

Wilson, J. 1984. *Dynamics of Offshore Structures*, Wiley, New York.

# 12

# Sound Levels and Decibels

S. Akishita
*Ritsumeikan University*

**Summary**

*In this chapter, the basic characteristics of sound and sound propagation are described. Levels and decibels, which represent the magnitude of sound waves, are defined and explained.*

## 12.1 Introduction

Sound is related to vibration, and is described as a propagating perturbation through a fluid, which is air or water in most cases. A very wide variety of noise sources exists. Each source is peculiar to its generation mechanism, which may cover a wide range of phenomena including fluid mechanics and the vibration of structures. Sound is perceived by the ear of the listener as a pressure wave superimposed upon the ambient air pressure. The *sound pressure* is the incremental variation about the ambient atmospheric pressure. Generally, it is detected by a microphone and expressed as oscillatory electric signal output from an audio measurement instrument. We shall present a mathematical description of these pressure waves that are known as sound. The field of acoustics concerns sound and vibration, and is treated in Chapter 12 to Chapter 20 of this book.

## 12.2 Sound Wave Characteristics

The characteristics of a sound wave are described by a pressure oscillation of a pure tone. A "pure tone" is a sinusoidal pressure wave of a specific frequency and amplitude, propagating at a velocity determined by the temperature and pressure of the medium (air).

Let us consider a hypothetical sound field in a duct with constant cross-sectional area, as shown in Figure 12.1a. A reciprocating piston at the left end emits the sound wave and it propagates toward the right-side end along the indicated axis. It is detected by a microphone at the right end. Figure 12.1b shows the instantaneous pressure distribution in a duct at time $t = t_0$. Figure 12.1c shows the pressure variation of the time history detected by the microphone at $x = x_0$.

The wavelength, $\lambda$, is the distance between successive two peaks in the waveform in Figure 12.1b. Wavelength is related to the frequency, $f$, and the velocity of wave propagation, $c$, by

$$\lambda = \frac{c}{f} \quad \text{(ft or m)} \quad (12.1)$$

The period, $T$, of the sinusoidal wave is the time interval required for one complete cycle, as depicted in Figure 12.1b. The period, $T$, is related to the frequency, $f$, by

$$T = \frac{1}{f} \quad \text{(s)} \quad (12.2)$$

## 12.2.1 Velocity of Sound

The velocity of sound is identical to the velocity of wave propagation, $c$, and in air it is given by

$$c = \sqrt{\frac{\gamma p_0}{\rho}} \quad \text{(ft/s or m/s)} \quad (12.3)$$

where $\gamma$ denotes the ratio of specific heat, $p_0$ denotes the ambient or equilibrium pressure, and $\rho$ denotes the ambient or equilibrium density. For air, $\gamma$ is taken as 1.4. Equation 12.3 then becomes

$$c = \sqrt{\frac{1.4 p_0}{\rho}} \quad \text{(ft/s or m/s)} \quad (12.4)$$

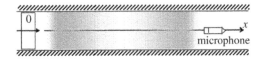

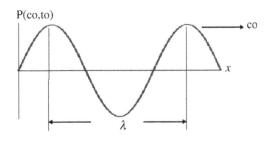

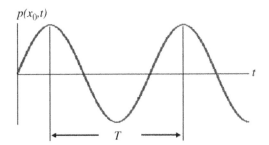

**FIGURE 12.1** (a) Propagating sound wave in a duct; (b) instantaneous pressure distribution; (c) pressure variation in time history detected by a microphone at $x = x_0$.

which can be further simplified by the fact that the ratio $p_0/\rho$ is related to the temperature of the gas. On assuming that the air behaves virtually as an ideal gas, the velocity, $c$, is related to the absolute temperature in degrees Kelvin (K) by

$$c = 20.05\sqrt{T} \quad \text{(m/s)} \quad (12.5)$$

where $T$, the temperature in degrees Kelvin, is

$$T = 273.2° + (°C) \text{ K} \quad (12.6)$$

## Example 12.1

Calculate the velocity of sound, $c$, giving the temperature of 15°C.

## Solution

$T = 273.2° + 15° = 288.2$ K, then
$c = 20.05\sqrt{288.2} = 340.4$ m/s

is obtained. This value means a typical velocity of sound in the air.

## 12.3   Levels and Decibels

Sound pressure and power are commonly expressed in terms of *decibel levels*. This allows us to use a logarithmic rather than a linear scale. It provides the distinct advantage of allowing accurate computations using small numerical values, while accommodating a wide range of numerical values.

### 12.3.1   Sound Power Level

*Sound power level* describes the acoustical power radiated by a given source with respect to the international reference of $10^{-12}$ W. The sound power level, $L_W$, is defined as

$$L_W = 10 \log\left(\frac{W}{W_{re}}\right) \ \text{(dB)} \tag{12.7}$$

where $W$ denotes sound power in question and $W_{re} = 10^{-12}$ W (reference).

### Example 12.2

Determine the sound power level of a small ventilation fan that generates 10 W of sound power.

### Solution

$$L_W = 10 \log\left(\frac{W}{W_{re}}\right) = 10 \log\left(\frac{10}{10^{-12}}\right) = 130 \ \text{dB}$$

### 12.3.2   Sound Pressure Level

*Sound pressure levels* are expressed in decibels, as are sound power levels. The sound pressure level, $L_p$, is defined as

$$L_p = 10 \log\left(\frac{\bar{p}^2}{p_{re}^2}\right) = 20 \log\left(\frac{\bar{p}}{p_{re}}\right) \ \text{(dB)} \tag{12.8}$$

where $\bar{p}$ denotes root-mean-square (RMS) sound pressure in question Pa or N/m$^2$ and $p_{re} = 20 \times 10^{-6}$ Pa $= 0.0002 \ \mu$bar. The pressure of $20 \times 10^{-6}$ Pa has been chosen as a reference because it has been found that the average young adult can perceive a $10^3$ Hz tone at this pressure. This reference is often referred to as the threshold of hearing at $10^3$ Hz.

### Example 12.3

Giving $L_p = 50$ dB for the Aeolian tone of 200 Hz, determine the RMS pressure of the tone.

### Solution

Given $L_p$ as 50 dB, then $\bar{p}$ is determined by using Equation 12.8.

$$50 = 20 \log\left(\frac{\bar{p}}{p_{re}}\right)$$

then

$$\bar{p} = 10^{50/20} \ p_{re} = 316.2 p_{re}$$

$$\bar{p} = 6.32 \times 10^{-3} \ \text{Pa} = 0.0632 \ \mu\text{bar}$$

Note that this value is very small, contradicting the magnitude of the sensory impression of the human ear.

### 12.3.3 Overall Sound Pressure Level

The sound pressure level is defined assuming "pure tone" sound. However, practically any real sound contains various components of pure tone sound. Let us consider a set of $n$ components of pure tone, denoted by

$$\begin{cases} p_1(t) = a_1 \sin(2\pi f_1 t + \phi_1) \\ \qquad\qquad \vdots \\ p_n(t) = a_n \sin(2\pi f_n t + \phi_n) \end{cases} \qquad (12.9)$$

$$p(t) = p_1(t) + \cdots + p_n(t) \qquad (12.10)$$

If $L_p$ of $p(t)$ is evaluated in RMS pressure, $\bar{p}$, we have

$$\bar{p} = \left[ \lim_{T\to\infty} \frac{1}{T} \int_0^T p^2(t) dt \right]^{1/2} = \left[ \lim_{T\to\infty} \frac{1}{T} \int_0^T (p_1(t) + \cdots + p_n(t))^2 dt \right]^{1/2} \qquad (12.11)$$

Since

$$\lim_{T\to\infty} \frac{1}{T} \int_0^T p_i(t) p_j(t) dt = 0, \quad i \neq j$$

is valid, $\bar{p}$ is obtained as

$$\bar{p} = \left[ \overline{p_1(t)}^2 + \cdots + \overline{p_n(t)}^2 \right]^{1/2} = \left[ \bar{p}_1^2 + \cdots + \bar{p}_n^2 \right]^{1/2} \qquad (12.12)$$

where

$$\bar{p}_i^2 \equiv \overline{p_i(t)}^2 \equiv \lim_{T\to\infty} \frac{1}{T} \int_0^T p_i^2(t) dt = \frac{1}{2} a_i^2 \qquad (12.13)$$

Let us define $L_{pi} \equiv 10 \log(\bar{p}_i^2/p_{re}^2)$ $(i = 1, 2, \ldots, n)$. Then the overall sound pressure level, $L_p$, of $p(t)$ is expressed by

$$L_p \equiv 20 \log \frac{\bar{p}}{p_{re}} = 10 \log \frac{1}{p_{re}^2} (\bar{p}_1^2 + \cdots + \bar{p}_n^2)$$

or $L_p$ is expressed by $L_{pi}$ $(i = 1, 2, \ldots, n)$ as follows:

$$L_p = 10 \log(10^{L_{p1}/10} + 10^{L_{p2}/10} + \cdots + 10^{L_{pn}/10}) \qquad (12.14)$$

### Example 12.4

Determine the overall sound pressure level of the combination of three pure tones, the sound pressure levels of which are expressed by

$$L_{p1} = 60 \text{ dB } (f_1 = 250 \text{ Hz}), \qquad L_{p2} = 65 \text{ dB } (f_2 = 500 \text{ Hz}), \qquad L_{p3} = 55 \text{ dB } (f_3 = 1000 \text{ Hz})$$

### Solution

We have $10^{L_{p1}/10} = 10^6$, $10^{L_{p2}/10} = 10^{6.5}$, and $10^{L_{p3}/10} = 10^{5.5}$. Then the overall level, $L_p$, is determined by using Equation 12.14 as follows:

$$L_p = 10 \log(10^6 + 10^{0.5} \times 10^6 + 10^{-0.5} \times 10^6)$$

$$= 10 \log 10^6 (1 + 10^{0.5} + 10^{-0.5}) = 60 + 10 \log 4.479 = 66.5 \text{ dB}$$

Note that the sum of 65, 60 and 55 dB is just 66.5 dB.

# 13

# Hearing and Psychological Effects

S. Akishita

*Ritsumeikan University*

## Summary

*In this chapter, first the characteristics of human hearing are discussed, including a brief description of the anatomy and function of the hearing mechanisms. Next, the frequency and loudness responses of the human hearing are explained, and then hearing loss causing permanent damage is described. Finally, the psychological response to noise is discussed by defining the indices, loudness (sones), noise-criteria curves, and sound level.*

## 13.1 Introduction

This chapter considers the characteristics of human hearing. After a brief description of the anatomy and function of the hearing mechanism, those aspects of hearing that are important in noise control are discussed. The perception of sound by the human ear is a complicated process, dependent both on the frequency and pressure amplitude of the sound. We shall consider the structure of the ear and hearing mechanism. We will also briefly discuss various means of measuring the psychological effects of noise.

## 13.2 Structure and Function of the Ear [1]

The main components of the human ear are depicted in Figure 13.1(a). The ear is commonly divided into three main components: (1) the outer ear, (2) the middle ear, and (3) the inner ear.

The visible portion of the ear is called the *pinna*. Because of its small size compared with the primary wavelengths that we hear, the pinna serves only to produce a small enhancement of the sounds that arrive from the front of the listener as compared to those which arrive from behind; that is, the human sound reception system has a small frontal directivity. The remainder of the outer ear, which consists of the ear canal terminated in the ear drum, forms a resonant cavity at about 3 kHz. This resonant or near-resonant condition allows for a nearly reflection-free termination of the *ear canal* and thus a good impedance match of the *ear drum* to the air in which the sound wave was propagated.

The middle ear consists of three small *ear bones*, the hammer, anvil, and stirrup. The middle ear serves as an impedance transformer, which matches the low impedance of the air in which sound travels and in

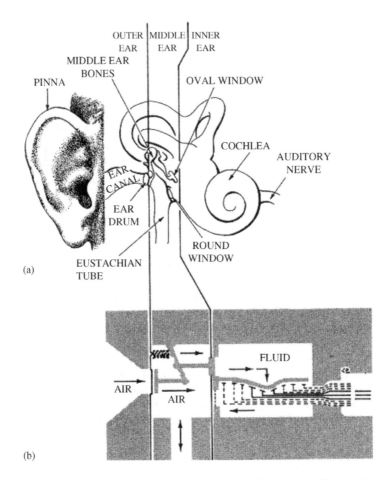

**FIGURE 13.1**   (a) Main components of the human ear; (b) functional diagram of the ear.

which the ear drum is located to the high impedance of the lymphatic fluid of the *cochlea* beyond the *oval window*. Without this impedance-matching transformation, a mismatch would occur, resulting in a loss of approximately 30 dB.

In the inner ear, the cochlea is the main component where the actual reception of sound takes place. The schematic extended structure of the cochlea is depicted in Figure 13.1(b). The cochlea, which is located in extremely hard temporal bone, is divided almost its entire length by the *basilar membrane*. At the end of the cochlea, the two canals are connected by the *helicotrema*, which allows for the flow of the *lymphatic fluid* between the two sections. The basilar membrane, which is about 3 cm long and 0.02 cm wide, has about 24,000 nerve ends terminated in *hair cells* located on the membrane. The motion of the oval window is transmitted to the basilar membrane and its associated sensing cells. This motion is sensed as sound.

## 13.3   Frequency and Loudness Response

The threshold of hearing, defined for binaural listening, is that sound pressure in the free field which one can just still hear as the signal is decreased. The threshold of hearing, for what is considered normal hearing, is shown in Figure 13.2. As seen from the curve, human hearing is most sensitive in the range of 2000 to 5000 Hz; furthermore, we note that the response in this range is very close to 0 dB, or $20 \times 10^{-6}$ Pa. At the other end of the scale, there is the threshold of pain, which is usually taken as about 135 to 140 dB. Thus, there is a dynamic range of normal hearing of approximately 140 dB.

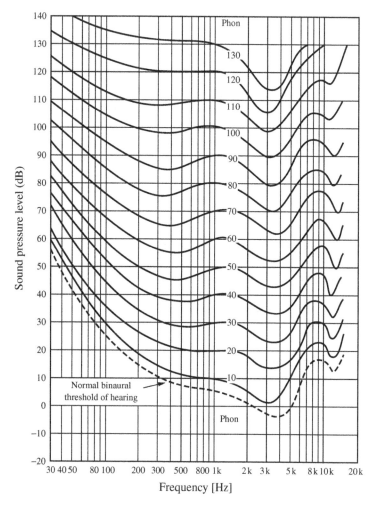

**FIGURE 13.2**  Equal-loudness contour for free-field binaural listening.

One also readily notes from the curve of Figure 13.2 that the threshold of hearing is a function of frequency. For example, with normal hearing, one would just be able to hear a 2000 Hz tone at a 0 dB level. However, one would require a pressure level of about 15 dB to be able to barely hear a 200 Hz tone. Thus, in describing the subjective loudness of sound, it is necessary to consider the characteristics of the human ear. This concept of loudness is quantized by the loudness level.

The *loudness* level of a particular sound is determined by the subjective comparison of the loudness of the sound to that of a 1000 Hz pure tone. The level, measured in *phons*, is equal numerically to the sound pressure level, in dB, of the 1000 Hz tone, which was regarded to be of equal loudness. A set of internationally standardized equal loudness contours is plotted in Figure 13.2. In keeping with the definition of loudness level, note that at 1000 Hz, all the equal loudness contours are equal in phons to the sound pressure level in dB.

## Example 13.1

Determine the sound pressure level of a 100 Hz tone with a loudness level of 30 phon.

## Solution

From Figure 13.2, we find the sound pressure level to be 44 dB.

## 13.4   Hearing Loss

Excessive and prolonged noise exposure causes permanent hearing loss. Various theories have been put forth in an effort to characterize and predict the possible damage that might be caused by a given exposure. Absolute proof of any theory concerning such a complex biological phenomenon is virtually impossible to achieve. However, reliable data have been collected, which deal with situations where workers have been continuously exposed to more or less the same noise environment for many years.

It is well established that excessive noise exposure causes permanent hearing damage by destroying the auditory sensor cells. These cells are hair cells located on the basilar membrane. Furthermore, other types of inner ear damage include harm to the auditory neurons, as well as damage to the structure of the organ of Corti. In all, the various theories and data have been taken advantage of establishing the noise-exposure criteria set forth in the noise exposure regulations.

## 13.5   Psychological Effects of Noise

In this section, certain generally accepted aspects of the psychological effects of noise will be discussed and quantified. Various indexes have been proposed that quantify the psychological effects of noise. However, only a few of indices, *loudness (sones)*, *noise-criteria* (NC) *curve*, and *sound level*, are introduced in the following presentation.

### 13.5.1   Loudness Interpretation

As was discussed relating to Figure 13.2, loudness level is measured in phons, and the related quantity, loudness, is measured in *sones*. A sone is defined as the loudness of a 1000 Hz pure tone with a sound pressure level of 40 dB. On recalling the definition of loudness level, or by referring to Figure 13.2, one notes that 40 phon have a loudness equal to 1 sone. This relationship may be simply expressed as

$$S = 2^{(L_L - 40)/10} \text{ sone} \tag{13.1}$$

where $S$ = loudness (sones), $L_L$ = loudness level (phons), or conversely

$$L_L = 33.2 \log S + 40 \text{ phon} \tag{13.2}$$

### Example 13.2

Make the following two conversions using the appropriate equation (Equation 13.1 or Equation 13.2): (1) convert 80 phon to sone, (2) convert 100 sone to phon.

### Solution

1. To convert phons to sones, use Equation 13.2:

$$S = 2^{(L_L - 40)/10} = 2^{(80 - 40)/10} = 2^4 = 16 \text{ sone}$$

2. To convert sones to phons, use Equation 13.2:

$$L_L = 33.2 \log S + 40 = 33.2 \log 100 + 40 = 66.4 + 40 = 106.4 \text{ phon}$$

How should we determine the "total loudness" (sones), when the sound is composed of multiple frequency components? Probably the most widely used method for establishing the loudness of a complex noise is that developed by Stevens [2]. The method is based on the measurement of the 1-octave, 1/3-octave, or 1/2-octave band pressure levels. The measured band pressure levels are used

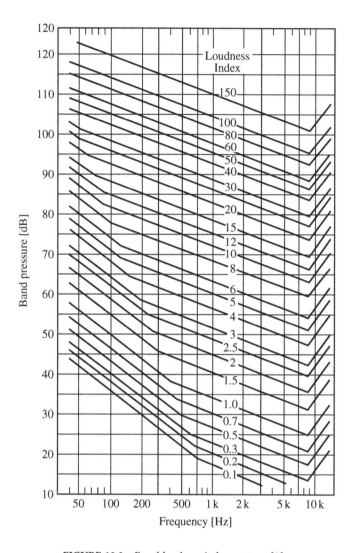

**FIGURE 13.3** Equal loudness index contour [2].

in conjunction with the equal loudness index contours shown in Figure 13.3 to determine the loudness or loudness level by means of a simple calculation.

A step-by-step outline of the procedure is as follows:

1. Measure the band pressure levels (1-octave, 1/2-octave, or 1/3-octave) over the frequency range of interest. Usually, the range chosen is from about 50 to 10,000 Hz.
2. Enter the center frequency and band pressure level for each band in the contour of Figure 13.3, and determine the loudness index for each band.
3. Calculate the total loudness, $S_t$, in sones, by using

$$S_t = I_m(1 - K) + K \sum_{i=1}^{n} I_i \text{ (sone)} \tag{13.3}$$

where $S_t$ = the total loudness (sones), $I_m$ = the largest of the loudness indices, $I_i$ = the loudness indices, including $I_m$, $K$ = weighting factor for the bands chosen. $K = 0.3$ for 1-octave bands, $K = 0.2$ for 1/2-octave bands, $K = 0.15$ for 1/3-octave bands.

4. If so desired, one may calculate the loudness level in phons using Equation 13.2, or one may convert to loudness level by means of the conversion curve of Figure 13.3.

## Example 13.3

A particular complex noise was measured to yield the one-octave band pressure given in the following table

| Center Frequency (Hz) | Band Pressure Level (dB) | Loudness Index (sone) |
|---|---|---|
| 63 | 66 | 2.5 |
| 125 | 63 | 3.2 |
| 250 | 65 | 4.8 |
| 500 | 70 | 7.5 |
| 1000 | 73 | 10.6 |
| 2000 | 76 | 15.2 |
| 4000 | 81 | 25.1 |
| 8000 | 79 | 29.0 |

Compute the loudness level using the procedure described before.

## Solution

As a first step, the loudness indices are determined from Figure 13.3 and recorded in tabular form with the band pressure levels. Next, we note that one-octave bands have been used. Therefore $K = 0.3$ in Equation 13.3, and

$$S_t = I_m(1 - 0.3) + 0.3 \sum_{i=1}^{8} I_i$$

From the table above, we find that $I_m = 29.0$ sone and, summing up, find $\sum I_i = 97.9$ sone. Therefore,

$$S_t = 29(1 - 0.3) + 0.3(97.9) = 49.67 \text{ sone}$$

We find that the loudness, $S_t \cong 50$ sone. The loudness level may now be calculated by means of Equation 13.2 as

$$L_L = 33.2 \log S_t + 40 = 33.2 \log 50 + 40 = 96.4 \text{ phon}$$

Therefore, $L_L$, the loudness level, is 96 phon.

### 13.5.2  Noise-Criteria Curves

*Noise-criteria* curves, which are neglected here, were established in 1957 for rating indoor noise. The curves have been utilized as one method of rating background noise level in a room. Each curve specifies the maximum octave-band sound pressure level for a given NC rating. If the octave band levels for a given noise spectrum are known, the rating of that noise in terms of the NC curves is given by plotting the noise spectrum on the set of NC curves to determine the point of highest penetration.

In 1971, some objections to the NC curves led to their modification. The new curves, which are shown in Figure 13.4, are called the *preferred noise-criteria* (PNC) *curves*. Although these curves differ from the NC curves, they are used in exactly the same manner.

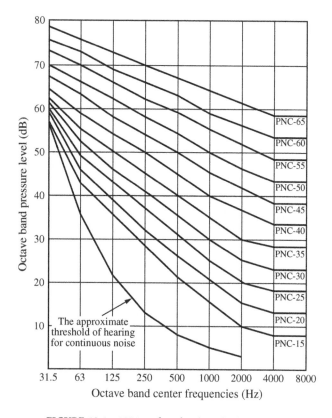

**FIGURE 13.4** 1971 preferred noise-criteria curves.

## Example 13.4

Determine the PNC rating for the octave-band noise spectrum tabulated below.

| Center frequency (Hz) | 63 | 125 | 250 | 500 | 1000 | 2000 | 4000 | 8000 |
|---|---|---|---|---|---|---|---|---|
| Band pressure level (dB) | 65 | 60 | 60 | 63 | 55 | 50 | 45 | 40 |

## Solution

The highest penetration is found at 500 Hz on PNC-60. Hence, the answer is PNC-60.

## 13.5.3 Sound Level

*Sound levels* are sound pressure levels that have been weighted according to a particular weighting curve. Three weightings, A, B, and C, and associated sound levels, have been developed as a method to subjectively evaluate the impact of noise upon the human ear, in a proper manner. The frequency response and decibel conversions from a flat response for each of these weightings are given in Figure 13.5 and Table 13.1, respectively.

The A-weighting network is now used almost exclusively in measurements that relate directly to the human response to noise, both from the viewpoint of hearing damage and of annoyance. Such measurements are referred to as *sound level measurements*. Sound level is designated by **L** and the designated unit is the dBA. Similarly, dBB and dBC are used to designate sound level weighted by B weighting and C weighting networks, respectively.

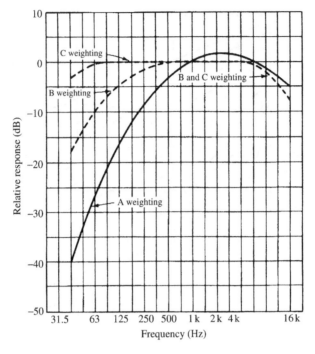

**FIGURE 13.5** Frequency response for the A, B, and C weighting networks.

**TABLE 13.1**   Sound Level Conversion Chart from Flat
Response to A Weighting

| Frequency (Hz) | A Weighting (dB) |
|---|---|
| 50 | − 30.2 |
| 63 | − 26.2 |
| 80 | − 22.5 |
| 100 | − 19.1 |
| 125 | − 16.1 |
| 160 | − 13.4 |
| 200 | − 10.9 |
| 250 | − 8.6 |
| 315 | − 6.6 |
| 400 | − 4.8 |
| 500 | − 3.2 |
| 630 | − 1.9 |
| 800 | − 0.8 |
| 1,000 | 0 |
| 1,250 | +0.6 |
| 1,600 | +1.0 |
| 2,000 | +1.2 |
| 2,500 | +1.3 |
| 3,150 | +1.2 |
| 4,000 | +1.0 |
| 5,000 | +0.5 |
| 6,300 | − 0.1 |
| 8,000 | − 1.1 |
| 10,000 | − 2.5 |
| 12,500 | − 4.3 |
| 16,000 | − 6.6 |

**TABLE 13.2**  Octave-Band Sound Pressure Levels

| $f_c$ (Hz) | $L_{flat}$ (dB) | $\Delta L_A$ (dB) | $L_A = L_{flat} + \Delta L$ (dB) | $I_{iA}$ |
|---|---|---|---|---|
| 63 | 74 | −26.2 | 47.8 | $0.60 \times 10^5$ |
| 125 | 71 | −16.1 | 54.9 | 3.09 |
| 250 | 61 | −8.6 | 52.4 | 1.74 |
| 500 | 60 | −3.2 | 56.8 | 6.31 |
| 1000 | 62 | 0 | 62.0 | $1.585 \times 10^6$ |
| 2000 | 60 | 1.2 | 61.2 | 1.318 |
| 4000 | 62 | 1.0 | 63.0 | 1.995 |
| 8000 | 69 | −1.1 | 67.9 | 6.166 |
| Sum | | | | $12.238 \times 10^6$ |

*Note*:

- $f_c$: band center frequency
- $L_{flat}$: sound pressure level with flat weighting
- $\Delta L_A$: A-weighting level
- $I_{iA}$: sound pressure intensity with A weighting.

In noise-abatement problems, it is often necessary to convert calculated a 1-octave-band or 1/3-octave-band sound pressure level to a total sound level in dBA. Table 13.1 gives sound level conversion by A weighting from flat response pressure.

## Example 13.5

Determine the total A weight sound level, $L$, of the set of octave-band sound pressure levels given in Table 13.2.

## Solution

Refer Table 13.1 for the dB conversion from a flat response level, $L_{flat}$, for each of the octave bands to a sound pressure intensity with A weighting $I_{iA}$, and then the sum of $I_{iA}$. Finally, the total sound level with A weighting $L_{total,A}$ is given by

$$L_{total,A} = 10 \log \sum_{i=1}^{n} I_{iA} = 10 \log 12.238 \times 10^6 = 70.9 \text{ (dB)}$$

## References

1. Irwin, J.D. and Graf, E.R. 1979. *Industrial Noise and Vibration Control*, Prentice Hall, Englewood Cliffs, NJ.
2. American National Standard USAS S3.4-1968. 1968. *Procedure for the Computation of Loudness of Noise*, America National Standards Institute, New York, NY.

# 14

# Noise Control Criteria and Regulations

S. Akishita

*Ritsumeikan University*

**Summary**

*In this chapter, the basic ideas behind the development of noise control criteria and regulations are discussed, taking into consideration that the standards and criteria vary from country to country and depend on governments in power. Legislations in the European Union and regulations in Japan are introduced as typical examples. Some indexes as measures of noise evaluation are described.*

## 14.1  Introduction

In order to protect people from being exposed to excessive noise, different communities have implemented different types of legislative control. While the controls vary in scope, control mechanisms, and technical requirements, and are based on different control philosophies, they are intended to achieve a balance between the demand for a tranquil environment and the need for maintaining economic and social activities. In general, the noise standards vary according to the time of day and the use of the land concerned, with the more stringent standards applied to rest periods and areas where the noise sensitivity is high, such as those with schools and hospitals, and exclusive residential areas. Different countries have adopted different noise standards and regulations to meet their local situations and requirements. This chapter cannot describe all major control criteria and regulations in the world, or even in the major industrialized countries. Only the main issues of legislation on noise emission and reception are briefly introduced in the chapter. More details in the on-going noise control issues are found in Refs. [2,3].

## 14.2  Basic Ideas behind Noise Policy

Every noise policy originates from the idea of protecting the quality of life from noise pollution of all kinds. When establishing a noise policy, it is useful to consider the distinction between noise *emission* and *immission* (or *reception*) [6]. The former means literally emitting or radiating sound energy or power from a noise source, whereas the latter means receiving, perceiving, or observing radiated noise, which leads to the extent of the noise exposure at a position near the noise source. Therefore, noise emission is controlled with noise regulation law by the government, whereas noise immission is legislated with environmental quality standards. The measure of the extent describing the former is the "sound power level," and that describing the latter is the sound pressure level.

The global professional organization on noise control, the International Institute of Noise Control Engineering (I-INCE) recently started its activities to develop a global noise policy [5]. In response to the question "is noise policy a global issue, or is it a local issue?", I-INCE had a common theme presented in special session. It was felt that noise is primarily a global policy issue, but many noise problems can only be solved with the active participation of local authorities. The task of the technical study group is to take a global approach to noise in order to define the requirements for an international noise control policy to be effective, stated as follows:

> All *vehicles, devices, machinery,* and *equipment* that emit audible sound are manufactured products; most are entered into world trade and many are produced in two or more different countries by companies with worldwide operations. The *noise emission* of these products is an appropriate subject of international agreements and regulations. The *noise immisions* resulting from the operation of these products are growing in severity as traffic flow and the pace of industrialization continues to increase in many parts of the globe.

The technical study group reports the classification of noise areas as follows:

1. OCCUPATION NOISE—noise received at the workplace, indoors and outdoors, caused by all noise sources in the vicinity of the workplace.
2. ENVIRONMENTAL NOISE—noise perceived by individuals in the domestic environment, indoors and outdoors, caused by sources controlled by others.
3. CONSUMER PRODUCT NOISE—noise perceived by users and bystanders of noise generating products over which the individual has some control, including noise in the passenger compartment of vehicles, excluding occupational and environmental noise.

## 14.3 Legislation

The World Health Organization (WHO) published the historic "Guidelines for Community Noise" in 2000, which has been accepted as the most significant recommendations for noise exposure criteria. The bodies that are responsible for enacting the regulations as law include the Federal Government in the USA, the European Union (EU) in Europe, and the Japanese Government in Japan. In the following, the EU's legislation on noise immission is shown, as an example of the flow of legislation process [8].

On July 18th, 2002, a European Directive on the assessment and management of environmental noise was published in the *Official Journal of European Communities*. It was required to be implemented in the national legislation of the EU Member States no later than July 18th, 2004. From then on, a program was to start, containing periodic noise mapping, the making of action plans, and information of the public. The directive also has strengthened the position of the European Commission regarding the reduction of noise emission.

In 2002, the development of the European Directive on environmental noise resulted in an approved directive relating to the assessment and management of environmental noise, for which the acronym "DAMEN" is used. According to Article 1 of the DAMEN its objective is to "define a common approach to avoid, prevent or reduce harmful effects, including annoyance, due to exposure to environmental noise." A rough description of actions in the DAMEN is shown in Figure 14.1. Brief notes are given next to supplement Figure 14.1.

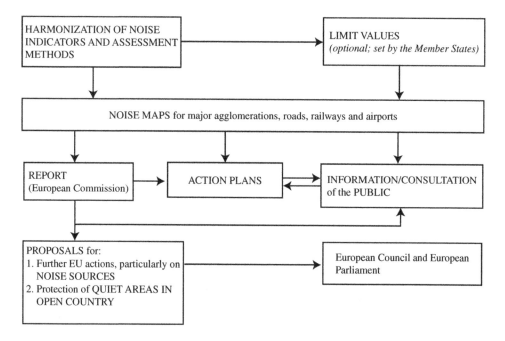

**FIGURE 14.1**　Overview of the DAMEN. (*Source*: Wolde, T.T. 2003. The European Union's legislation on noise immission, pp. 4367–4371. In *Proceedings of Inter-noise 2003* (N832). With permission.)

## 14.3.1　Action Plans

In a case where the mapping results are such that they violate the local or national limit value, or are found unsatisfactory for other reasons, action plans shall be developed for the improvement of the situation. These action plans shall be discussed with the citizens involved. A summary of the action plans shall be sent to the European Commission.

## 14.3.2　Publication of Data by the European Commission

Every five years, starting in 2009, the Commission shall publish a summary report from the noise maps and the action plans.

## 14.3.3　Proposal for Further European Union Action

In 2004, the European Commission was to submit a report to the European Parliament and the Council containing a review of existing EU measures relating to sources of environmental noise and present proposals for improvement, if appropriate. In 2009, the European Commission will submit to the European Parliament and the Council a report on the implementation of the directive. That report will in particular assess the need for further EU action and, if appropriate, propose implementing strategies on aspects such as:

- Long-term and medium-term goals for the reduction of the number of persons harmfully affected by environmental noise
- Additional measures on noise emission by specific sources
- The protection of quiet areas on the open country

## 14.4   Regulation

Noise regulation is executed by local governments once the central government enacts a noise regulation law. The law is considered the "national minimum." For example, factory noise, construction work noise, and road traffic noise are under the purview of the Noise Control Act, which means the central government is responsible of regulating these kinds of noise. On the other hand, community noise and factory noise are under the purview of the original regulation of local governments. It can be said that local governments are responsible for a great part of the noise policy, although they may not always fully understand the situations concerned. In what follows, an outline of the legal system for environmental noise problem in Japan is given as an example of a typical legal system for noise regulation [6].

In Japan, the "Environmental Quality Standards for Noise" was revised in 1999 after 27 years with the old law. Figure 14.2 outlines of the legal system in Japan.

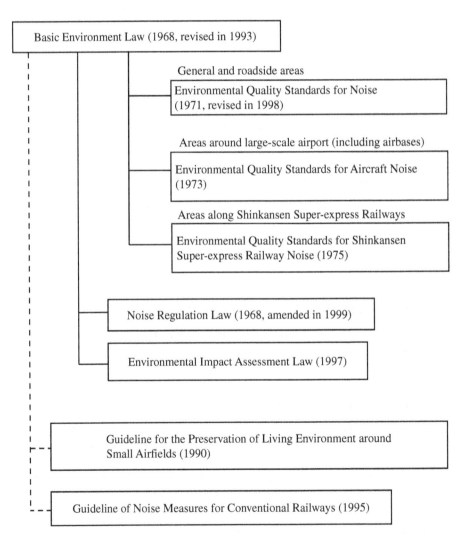

**FIGURE 14.2**   Legal system for environmental noise in Japan. (*Source:* Tachibana, H. and Kaku, J. 2003. Acoustic measures for the environmental noise assessment in Japan, pp. 3317–3322. In *Proceedings of Inter-noise 2003* (N1007). With permission.)

Each of these laws and standards is legislated for a specific noise problem (a noise source), and therefore different noise indices are specified according to the respective noise problems. To review this situation from a historical viewpoint, it can be said that each law or standard responds to a specific noise problem promptly, through the use of available measurement technology at that time. However, some laws and standards have become outdated since their establishment, when considering the current situation, international dynamics, and the current acoustic measurement technology.

## 14.5 Measures of Noise Evaluation

Basically, the A weighting networks are applied to obtain a measure of noise evaluation. As the measures are legislated by governments, they are dependent on the legislative regulations and standards. In what follows, the measure of noise evaluation legislated in Japan is given to show the concepts behind the legislation [6].

In the regulations and standards for environmental noise problems, a variety of noise measures are used. In order to improve these legislative regulations and standards in the future, the present measures shall be reviewed considering the difference between noise emission and immission and the difference between noise measurement and monitoring, and impact assessment and prediction. These measures legislated in Japan are listed and classified in Table 14.1 by considering the difference between noise characteristics.

**TABLE 14.1** Assessment Methods Specified in Laws and Standards for Environmental Noises in Japan

| Noise Sources | Law and Standards | Noise Indices | Assessment Time |
|---|---|---|---|
| Roads | Environmental Quality Standards for Noise | $L_{Aeq,T}$ [a] | Daytime (6:00–22:00); nighttime (22:00–6:00) |
| Shinkansen superexpress railways | Environmental Quality Standards for Shinkansen Superexpress Railways | $L_{A,Smax}$ [b] | Every event |
| Conventional railways | Guideline of Noise Measures for Conventional Railways | $L_{Aeq,T}$ | Daytime (7:00–22:00); nighttime (22:00–7:00) |
| Aircrafts | Environmental Quality Standards for Aircrafts Noise | WECPNL [c] | Time weighting |
| | Guideline for the Preservation of Living Environment around Small Airfields | $L_{den}$ [d] | |
| Construction works | Noise Regulation Law | According to | Not specified; |
| Factories | (specific noise sources) | time variation: | every event |
| Large-scale retail stores | Law concerning the measures by large scale retail stores for preservation of living environment | $L_A$ [e]; $L_{A,Fmax}$ [f]; $L_{A5}$ [g]; $L_{A,Fmax,5}$ [h] | |

[a] $L_{Aeq,T}$, equivalent continuous A-weighted sound pressure level.
[b] $L_{A,Smax}$, SLOW maximum value of A-weighted sound pressure level.
[c] WECPNL, weighted equivalent continuous perceived noise level(calculated from $L_{A,Smax}$).
[d] $L_{den}$, day/evening/night equivalent continuous A-weighted sound pressure level.
[e] $L_A$, FAST maximum value of A-weighted sound pressure level.
[f] $L_{A,Fmax}$, A-weighted sound pressure level.
[g] $L_{A5}$, upper value of the 90% range of A-weighted sound pressure level.
[h] $L_{A,Fmax,5}$, upper value of the 90% range of the FAST maximum Aweighted sound pressure level.
*Source*: Tachibana, H. and Kaku, J. 2003. Acoustic measures for the environmental noise assessment in Japan, pp. 3317–3322. In *Proceedings of Inter-noise 2003* (N1007). With permission.

When considering the consistency between noise measurement and monitoring, and noise prediction for impact assessment, it is most reasonable to use energy based indices such as $L_{Aeq}$. Of course, $L_{Aeq}$ is not a panacea and some secondary adjustment may be needed for the exact assessment of environmental noise with different characteristics. Nevertheless, the possibility of unification by $L_{Aeq}$ should be considered in the near future in Japan. Although $L_{Aeq}$ is now being widely used for the assessment of aircraft noise in almost all countries, WECPNL is still being used in Japan. WECPNL is very close to $L_{Aeq}$ in concept and it is not difficult to change the assessment index from WECPNL to $L_{Aeq}$.

The aim of the laws and standards shown in Figure 14.2 is to measure and assess the environmental noise for prevention or maintenance of the present situation. Therefore, any noise index should be appropriately used for each of noise problems, as shown in Table 14.1, which presents assessment methods specified in laws and standards for environmental noise in Japan. In particular, when predicting the future noise situation in environmental impact assessments, the indices should be suitable for theoretical calculation. The statistical noise indices such as the percentile level ($L_{A5}$) and maximum level ($L_{A,Fmax}$ or $L_{A,Smax}$) specified in the laws and standards have to be predicted statistically. It is difficult to predict these quantities by a simple physical calculation model, in principle. In this respect, the energy-based noise indices such as $L_{Aeq}$ can be easily treated in energy based calculation, and the prediction model becomes simple and clear in physical meaning. In an environmental impact assessment, the predicted results are to be compared with the related laws or standards. In the case of road traffic noise, $L_{Aeq}$ has been adopted in the new environmental quality standards, and therefore prediction has become very simple in theory, founded on energy-based indices.

In the prediction of road traffic noise, a motor vehicle as the noise source can be treated as a stationary sound sources of a constant sound power for a limited path. On the other hand, in the case of predicting construction noise, there are many complicated problems because various kinds of machines and equipment with various temporal variations of characteristics must be treated. Therefore, in the construction noise prediction method given in the "Acoustic Society of Japan CN-model 2002" [7], various noise indices for describing the acoustic output of various types of noise sources are specified as given in Table 14.2, which presents classification of noise sources and indices for expressing their acoustic output. Finally, Table 14.3 presents definitions and indices of measurement for acoustical output of noise sources.

**TABLE 14.2** Classification of Noise Sources and Indices for Expressing Their Acoustic Output

| Temporal Variation | Indices for Expressing Acoustic Output Sign | Terms |
|---|---|---|
| Stationary | $L_{WA}$ | A-weighted sound power level |
| | $L_A(r_0)$ | A-weighted sound power level at the reference distance ($r_0 = 1$ m) |
| Fluctuating randomly and widely | $L_{WAeff}$ | Effective A-weighted sound power level |
| | $L_{Aeff}(r_0)$ | Effective A-weighted sound pressure level at the reference distance ($r_0 = 1$ m) |
| | $L_{A,Fmax,5}(r_0)$ | 5% value of A-weighted sound pressure level at the reference distance ($r_0 = 1$ m) |
| Intermittent impulsive | $L_{JA}$ | A-weighted sound energy level |
| | $L_{WAeff}$ | Effective A-weighted sound power level |
| | $L_{AE}(r_0)$ | Single event sound exposure level at the reference distance ($r_0 = 1$ m) |
| | $L_{A,Fmax}(r_0)$ | FAST max. of A-weighted sound pressure level at the reference distance ($r_0 = 1$ m) |

*Source*: Tachibana, H. and Kaku, J. 2003. Acoustic measures for the environmental noise assessment in Japan, pp. 3317–3322. In *Proceedings of Inter-noise 2003* (N1007). With permission.

**TABLE 14.3** Definitions and Measurements of Indices for Acoustical Output of Noise Sources

| Indices | Definition | Measurement Method |
|---|---|---|
| $L_{WA}$ | $L_{WA} = 10 \log \dfrac{P_A}{P_0}$ (1) <br><br> Here, $P_0 = 1\,pW$ <br><br> | $L_{WA} = L_A(r) + 20 \log \dfrac{r}{r_0} + 8$ (2) <br><br> Here, $L_A(r)$ is the A-weighted sound pressure level measured at a distance of $r$, $r_0 = 1$ m |
| $L_{WAeff}$ | Effective A-weighted sound power level applied to fluctuating, intermittent and impulsive sounds <br><br> | $L_{WAeff} = L_{Aeff}(r) + 20 \log \dfrac{r}{r_0} + 8$ (3) <br><br> Here, $L_{Aeff}$ is the A-weighted sound pressure level measured at a distance of $r$ <br><br> $L_{Aeff} = 10 \log \left[ \dfrac{1}{T} \displaystyle\int_1^2 \dfrac{p_A^2(t)}{p_0^2} dt \right]$ (4) <br><br> Here, $T(t_1 - t_2)$ is averaging time (s), $p_0 = 20\,\mu Pa$ |
| $L_{JA}$ | $L_{JA} = 10 \log \dfrac{E_A}{E_0}$ (5) <br><br> Here, $E_0 = 1\,pJ$ <br><br> | $L_{JA} = L_{AE}(r) + 20 \log \dfrac{r}{r_0} + 8$ (6) <br><br> Here, $L_{AE}$ is the single event sound exposure level measured at a distance of $r$ <br><br> $L_{AE} = 10 \log \left[ \dfrac{1}{T} \displaystyle\int_1^2 \dfrac{p_A^2(t)}{p_0^2} dt \right]$ (7) <br><br> Here, $T_0 = 1$ s, $t_1 - t_2$ is the time including the event (s) |
| $L_A(r_0)$ <br> $L_{A,Fmax}(r_0)$ | A-weighted sound pressure level converted to the value at the reference distance $(r_0) = 1$ m | $L_A(r_0) = L_A(r) + 20 \log \dfrac{r}{r_0} + 8$ (8) <br><br> Here, $L_A(r)$ is the A-weighted sound pressure level measured at a distance of $r$ |

*Source:* Tachibana, H. and Kaku, J. 2003. Acoustic measures for the environmental noise assessment in Japan, pp. 3317–3322. In *Proceedings of Inter-noise 2003* (N1007). With permission.

## References

1. Fahy, F. 1985. *Sound and Structural Vibration, Radiation, Transmission and Response*, Academic Press, New York, chap. 2.
2. Fields, J.M. and de Jong, R.G., Standardized general-purpose noise reaction questions for community noise survey: research and a recommendation, *J. Sound Vib.*, 242, 641–679, 2001.
3. Harris, C.M., Ed. 1979. *Handbook of Noise Control*, 2nd ed., McGraw-Hill, New York, chap. 37.
4. Irwin, J.D. and Graf, E.R. 1979. *Industrial Noise and Vibration Control*, Prentice Hall, New York, chap. 5.
5. Lang, W.W. and Wolde, T.T. 2003. Progress report for TSG#5 'Global Noise Policy', pp. 98–101. In *Proceedings of Inter-noise 2003* (N872).

6.  Tachibana, H. and Kaku, J. 2003. Acoustic measures for the environmental noise assessment in Japan, pp. 3317–3322. In *Proceedings of Inter-noise 2003* (N1007).

7.  Tachibana, H. and Yamamoto, K. 2003. *Construction Noise Prediction Model*, ASJ CN-Model 2002, proposed by the Acoustical Society of Japan, EURONOISE, in Naples (2003.5).

8.  Wolde, T.T. 2003. The European Union's legislation on noise immission, pp. 4367–4371. In *Proceedings of Inter-noise 2003* (N832).

# 15

# Instrumentation

Kiyoshi Nagakura

*Railway Technical Research Institute*

**Summary**

*This chapter describes some measuring methods for the identification and ranking of noise source that are of benefit in noise control projects. Sound intensity measurement and directional measuring devices such as the mirror–microphone system and microphone array are introduced and their principles and applications are described.*

## 15.1 Sound Intensity Measurement

Every noise control project starts with the identification and ranking of the noise sources. Several methods have been proposed for the purpose and have proved to be useful and widely utilized. In this chapter, sound intensity measurement and directional measuring devices such as the mirror–microphone system and microphone array are introduced and their principles and applications are described. Other useful measurements, such as acoustic holography method [1,2] and spatial transformation of sound fields [3], are described in the literature.

### 15.1.1 Theoretical Background

Sound intensity is a measure of the magnitude and direction of the flow of sound energy. The instantaneous intensity vector, $\mathbf{I}(t)$, is given by the product of the instantaneous sound pressure, $p(t)$, and the corresponding particle velocity, $\mathbf{u}(t)$, that is, $\mathbf{I}(t) = p(t)\mathbf{u}(t)$.

In practice, the time-averaged intensity, $\bar{\mathbf{I}}$, is more important, and is given by the equation:

$$\bar{\mathbf{I}} = \lim_{T \to \infty} \frac{1}{T} \int_{-T/2}^{T/2} p(t)\mathbf{u}(t)\mathrm{d}t \qquad (15.1)$$

The intensity vector denotes the net rate of flow of energy per unit area (watts/m$^2$). Thus, the acoustic power, $W$, of the source located in a closed surface, $S$, is given by the integral of the intensity passing through the surface, $S$, as

$$W = \iint_s \bar{\mathbf{I}} \cdot \mathrm{d}S \qquad (15.2)$$

Equation 15.2 indicates that the measurement of sound intensity over a surface enclosing a source enables the estimation of its sound power, which shows the usefulness of the sound intensity concept.

## 15.1.2  Measurement Method

The principle of intensity measurement systems in commercial production employs two closely spaced pressure microphones [4,5], as shown in Figure 15.1.

The particle velocity, $u_r(t)$, in a particular direction, $r$, can be approximated by integrating over time the difference of sound pressures at two points separated by a distance $\Delta r$ in that direction:

$$u_r(t) = -\frac{1}{\rho_0} \int_{-\infty}^{t} \frac{p_2(\tau) - p_1(\tau)}{\Delta r}\,d\tau \quad (15.3)$$

where $p_1$ and $p_2$ are the sound pressure signals from the two microphones. The sound pressure at the center of two microphones is approximated by

$$p(t) = \frac{p_1(t) + p_2(t)}{2} \quad (15.4)$$

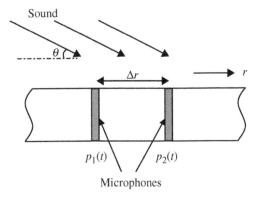

**FIGURE 15.1**  Microphone arrangement used to measure sound intensity.

Thus, the intensity in the direction $r$ can be calculated as

$$I_r(t) = -\frac{1}{2\rho_0 \Delta r}[p_1(t) + p_2(t)] \int_{-\infty}^{t} [p_2(\tau) - p_1(\tau)]d\tau \quad (15.5)$$

Some commercial intensity analyzers use Equation 15.5 to measure the intensity. Another type of analyzer uses the equation in the frequency domain:

$$I_r(\omega) = -\frac{\text{Im}[G_{12}]}{\omega \rho_0 \Delta r} \quad (15.6)$$

where $G_{12}$ is the cross spectrum between the two microphone signals. Equation 15.6 makes it possible to calculate sound intensity with a dual-channel fast fourier transform (FFT) analyzer.

## 15.1.3  Errors in Measurement of Sound Intensity

The principal systematic error of the two-microphone method is due to the approximation of the pressure gradient by a finite pressure difference. When the incident sound is a plane wave, the ratio of the measured intensity, $\hat{I}_r$, and the true intensity, $I_r$, is given by

$$\hat{I}_r/I_r = \frac{\sin(k\Delta r \cos\theta)}{k\Delta r \cos\theta} \quad (15.7)$$

where the angle $\theta$ is as defined in Figure 15.1 and $k$ is the wave number. Equation 15.7 indicates that the upper frequency limit is inversely proportional to the distance between the microphones.

Another serious error is caused by the phase mismatch between the two measurement channels. In the calculation of intensity from Equation 15.5, the phase difference, $\varphi$, between the two microphone signals, $p_1$ and $p_2$, is very important. Hence, the phase mismatch between the two measurement channels, $\Delta\varphi$, must be much smaller than $\varphi$. Since $\varphi$ increases with frequency, this error is serious in lower frequencies. Other possible errors, such as in the sensitivity of microphones and random errors associated with a given finite averaging time, are usually less serious.

## 15.1.4   Applications

One important application of sound intensity measurement is the determination of the sound power level using Equation 15.2. Furthermore, measurement of the intensity in the very near field of a source surface makes it possible to identify and rank the noise-sources. Plots of the sound intensity measured on a surface near a sound source are useful for investigating noise source distributions. Figure 15.2 shows sound intensity of noise from a wheel of a railway car. An intensity probe is located in the vicinity of the wheel and the normal component of sound intensity is measured by traversing the probe on a plane 100 mm away from the side surface of the wheel. These figures show a free vibration behavior of the wheel at each frequency; the wheel vibrates with one nodal diameter at 700 Hz and with three nodal diameters at 1150 Hz. Visualization by intensity vectors also gives valuable information about a noise source. Figure 15.3 shows the sound intensity vectors at each octave band measured in the vicinity of a railway car running at 120 km/h. These results suggest that the main radiator of rolling noise is the rail at the 500 Hz to 1 kHz band and the wheels at the 2 to 4 kHz band.

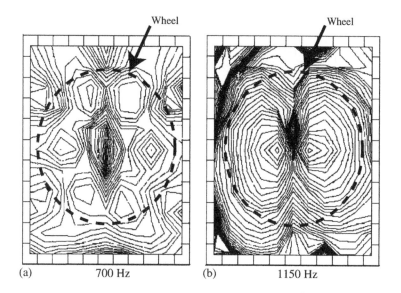

**FIGURE 15.2**   Measurements of the sound intensity radiated by a wheel of a railway car (1 dB contour).

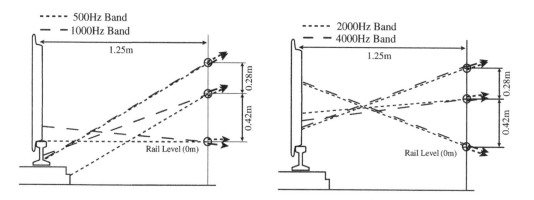

**FIGURE 15.3**   Sound intensity vectors measured in the vicinity of a railway car running at 120 km/h.

## 15.2 Mirror–Microphone System

### 15.2.1 Principle of Measurement

A mirror–microphone system consists of a reflector of elliptic or parabolic shape and an omnidirective microphone located at its focus [6,7]. Figure 15.4 shows the layout of a reflector of elliptic shape, an omnidirective microphone, and a noise source. Here, $S$ and $S'$ denote the front and back surfaces of the mirror, respectively; $P(\mathbf{r})$ denotes the pressure field on this configuration; $P_i(\mathbf{r})$ denotes the pressure field of free space; $\mathbf{r}_m$ is the position of the microphone; $\mathbf{r}$ is a point on the mirror surface. The normal, $\mathbf{n}_0$, directs toward the medium.

Using Green's theorem, the pressure at the microphone position $P(\mathbf{r}_m)$ is obtained by

$$P(\mathbf{r}_m) = P_i(\mathbf{r}_m) + \iint_{(s+s')} P(\mathbf{r}) \frac{\partial}{\partial \mathbf{n}_0}$$

$$\times \left[ \frac{e^{ikR_m}}{4\pi R_m} \right] d^2\mathbf{r} \tag{15.8}$$

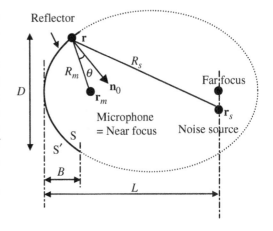

**FIGURE 15.4** Layout of a reflector, microphone, and noise source.

where $k = 2\pi f/c_0$ is the wave number, $f$ is the frequency of sound, $c_0$ is the speed of sound, and $R_m = |\mathbf{r} - \mathbf{r}_m|$ is the distance between the microphone and the mirror surface. If the wavelength is sufficiently smaller than the diameter of the reflector, the pressure field $P(\mathbf{r})$ is approximated by $2P_i(\mathbf{r})$ on the front surface, $S$, and by zero on the back surface, $S'$. In such a frequency range, the incident field term $P_i(\mathbf{r}_m)$ can be ignored. With these approximations, assuming that the noise source is a monopole type point source located at a position, $\mathbf{r}_s$, Equation 15.8 reduces to

$$P(\mathbf{r}_m) = -\frac{m(f)}{8\pi^2} \iint_s \frac{e^{ik(R_m+R_s)}}{R_m R_s}\left(ik - \frac{1}{R_m}\right)$$

$$\times \cos\theta(\mathbf{r})d^2\mathbf{r} \tag{15.9}$$

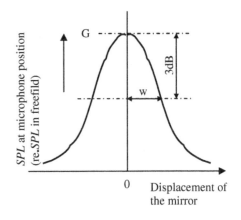

**FIGURE 15.5** Directivity pattern of a mirror–microphone system.

Here, $m(f)$ is the amplitude of the mass-flux rate of the source, $R_s = |\mathbf{r} - \mathbf{r}_s|$ is the distance between the sound source and the mirror surface and the angle $\theta(\mathbf{r})$ is defined in Figure 15.4. When the noise source is located at the far focus of the mirror, the sound pass length $R_m + R_s$ is constant with respect to $\mathbf{r}$, and a strong signal is obtained. As the noise source is moved away in the direction perpendicular to the mirror axis, the variance of the sound pass length, $R_m + R_s$, due to the position $\mathbf{r}$ increases, and thus the microphone signal drops off due to interference (see Figure 15.5, which we call the "directivity pattern"). The ratio of the peak level to the free field level at the microphone, $G$, is referred to as the "gain factor." The spatial resolution of the mirror is characterized by the displacement of the mirror position, $w$, at which the microphone signal drops off by a given relative amount, such as 3 dB. The quantities $G$ and $w$

can be related to the mirror geometries in Figure 15.4 by

$$G \approx 10 \log(CD^4/\lambda^2 B^2) \quad (C = \text{const.}) \tag{15.10}$$

$$w \propto \lambda L/D \tag{15.11}$$

The gain factor, $G$, increases with frequency at the rate of 6 dB per octave, and the spatial resolution, $w$, is inversely proportional to the frequency. The lower frequency limit is decided by the size of the mirror. On the other hand, there is no higher frequency limit, except for the capacity of an omnidirectional microphone itself. Thus, measurements with the mirror–microphone system are more suited to a scaled model test.

## 15.2.2 Applications

The mirror–microphone system has proved useful for identification of a noise source because of its directional property [8–10]. A scan of the source region produces a noise source map. It has an advantage in that the measurement is possible at a far field and it needs only one sensor, but has a disadvantage in that the measuring process is a time-consuming task.

Figure 15.6 shows an example of source maps of aerodynamic noise generated by a one-fifth scale high-speed train model, obtained from measurements by a mirror–microphone system, in a wind tunnel test. The surface of the car model is divided into several noise-source areas and the noise-source distribution in each area is measured by traversing the mirror–microphone system over the surface. The diameter and focal distance of the reflector are 1.7 and 3 m, respectively. Detailed maps of noise-source strength are obtained, which show that aerodynamic noise from high-speed trains is generated in relatively localized areas, namely, the local surface structures. The mirror–microphone system can be used for the measurement of the source distribution of a moving noise source. Figure 15.7 gives a time

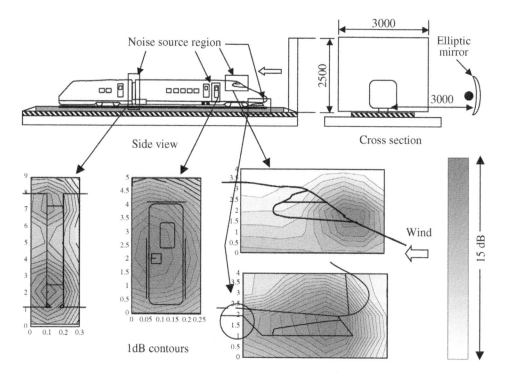

**FIGURE 15.6** Noise-source distribution of a one-fifth scale Shinkansen car model in a wind tunnel test measured with an elliptic mirror–microphone system.

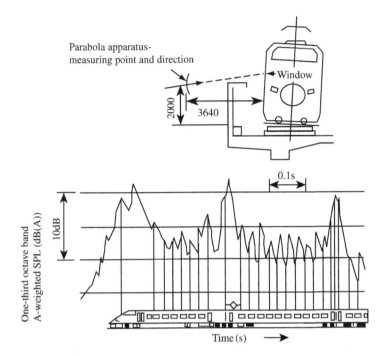

**FIGURE 15.7** Time history of the A-weighted one-third octave band ($f_0 = 8$ kHz) sound pressure level measured with a parabolic mirror–microphone system ($D = 1$ m, train speed = 274 km/h).

history of noise from a high-speed train measured with a parabolic mirror–microphone system, the diameter of which is 1 m. Peaks of the time history correspond to pantographs, doors, gaps between cars and the step-up of windows, which shows that they are main noise sources.

## 15.3 Microphone Array

### 15.3.1 Principle of Microphone Array

A microphone array [11] consists of several microphones distributed spatially to measure an acoustic field. The time signals from each microphone are added, accounting for the time delay between sound sources and microphones, and a directional output signal can be obtained as a result. The algorithm is called "beamforming." Now, consider $M$ omnidirectional microphones distributed in a far field of noise sources. The output signal of the array focused to a particular location in the source region, $\mathbf{r}$, and $z(\mathbf{r}, t)$, is calculated as a sum of delayed and weighted signals of each microphone:

$$z(\mathbf{r}, t) = \sum_{m=1}^{M} w_m p_m(t - \Delta_m) \tag{15.12}$$

Here, $p_m(t)$ is the signal from the $m$th microphone, $w_m$ is a weighting factor, and $\Delta_m$ is a time delay applied to signal of the $m$th microphone, as given by

$$\Delta_m = \frac{r_o - r_m}{c_0} \tag{15.13}$$

where $r_o$ and $r_m$ are the distances from the focus point to the reference point o and the $m$th microphone, respectively. When the focus location coincides with the source location, a strong signal is obtained (see Figure 15.8). If this process is repeated for various focus locations, $\mathbf{r}$, on the source surface, then a noise-source map can be obtained.

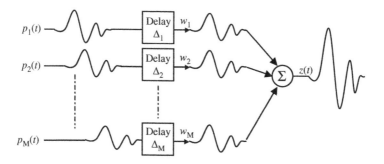

**FIGURE 15.8** Principle of a microphone array. Individual time delays are chosen such that signals arriving from a given point will be added up coherently.

## 15.3.2 Array's Directivity Pattern

The performance of a microphone array is characterized by the spatial resolution and signal-to-noise ratio. For simplicity, consider a linear array of $M = 2N + 1$ microphones spaced equally by $d$. When a harmonic plane wave is propagating with an incident angle $\theta$, and weighting factors all equal $1/M$, the ratio of the output signal of the array to that of the center microphone is computed using

$$W(\theta) = \frac{1}{M} \frac{\sin((M/2)kd \sin \theta)}{\sin((1/2)kd \sin \theta)} \tag{15.14}$$

where $k$ is the wave number. Figure 15.9 shows the directivity patterns for different values of the product $kd$ based on Equation 15.14. The highest peak appears at $\theta = 0$, which we call a "main lobe," and lower peaks also appear at some locations that are separate from a true source direction, which we call "side lobes." The width of the main lobe decides the performance of the array to separate two closely lying sources (which we call spatial resolution), and the ratio of main lobe to side lobe decides the signal-to-noise ratio of the array. The spatial resolution improves as $kd$ increases, that is, in proportion to the ratio of the array length to the wavelength. However, when $kd = 2\pi$, a peak of the same strength as the true source appears due to a spatial aliasing at $\theta = 90°$, which occurs when $d > \lambda/2$, where $\lambda$ is the wavelength. Thus, the acoustic frequency, $f$, is restricted by $f < c_0/2d$, to avoid aliasing.

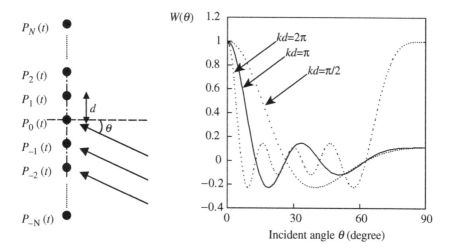

**FIGURE 15.9** Directivity patterns of a linear array for different values of the product $kd$ ($M = 2N + 1 = 9$).

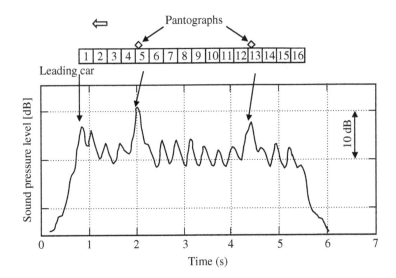

**FIGURE 15.10**   Time history of sound pressure level for a passing train measured with a linear microphone array located at a point 25 m away from a track (train velocity = 285 km/h).

In the above case of the linear array, the directivity exists only in the direction of the array (one-dimensional). If microphones are arranged in a two-dimensional plane, a two-dimensional directivity can be obtained. Recently, many microphone arrangements have been proposed that obtain better spatial resolution and to reduce side lobes [12–15].

### 15.3.3  Applications

Microphone arrays have been used for identification of the noise source in various situations, for example, in wind tunnel tests. Many actual examples can be found in published literature [10,16–18]. The measurement with a microphone array has the advantage of much shorter measuring time than that of a mirror–microphone system. Furthermore, the lower frequency limit is not so serious because the size of the apparatus can be easily extended.

Another fundamental example is given now. Nine microphones are arranged equally spaced by $d = \lambda/2$ for each one-third octave band, and their signals are summed without time delay. In this case, the array is focused to a fixed direction, perpendicular to the array axis. Figure 15.10 shows a time history of noise generated by a high-speed train, measured with a linear microphone array located at a point 25 m away from the track. It is found that pantographs, the leading car, and gaps between cars are the main noise sources in this example.

### References

1.   Ferris, H.G., Computation of far field radiation patterns by use of a general integral solution to the time independent scalar wave equation, *J. Acoust. Soc. Am.*, 41, 1967.
2.   Maynard, J.D., Nearfield acoustic holography: theory of generalized holography and the development of NAH, *J. Acoust. Soc. Am.*, 78, 1985.
3.   Ginn, K. and Hald, J., The effect of bandwidth on spatial transformation of sound field measurements, *Inter-Noise*, 87, 1987.
4.   Fahy, F.J., Measurement of acoustic intensity using the cross-spectral density of two microphone signals, *J. Acoust. Soc. Am.*, 62, 1977.

5. Chung, J.Y., Cross-spectral method of measuring acoustic intensity without error caused by instrument phase mismatch, *J. Acoust. Soc. Am.*, 64, 1978.

6. Grosche, F.R., Stiewitt, H., and Binder, B., On aero-acoustic measurements in wind tunnels by means of a highly directional microphone system, *Paper AIAA-76-535*, 1976.

7. Sen, R., Interpretation of acoustic source maps made with an elliptic-mirror directional microphone system, *Paper AIAA-96-1712*, 1996.

8. Blackner, A.M. and Davis, C.M., Airframe noise source identification using elliptical mirror measurement techniques, *Inter-Noise 95*, 1995.

9. Dobrzynski, W., Airframe noise studies on wings with deployed high-lift devices, *Paper AIAA-98-2337*, 1998.

10. Dobrzynski, W., Research into landing gear airframe noise reduction, *Paper AIAA-2002-2409*, 2002.

11. Johnson, D.H. and Dudgeon, D.E. 1993. *Array Signal Processing*, Prentice Hall, Englewood Cliffs, NJ.

12. Elias, G., Source localization with a two-dimensional focused array: optimal signal processing for a cross-shaped array, *Inter-Noise 95*, 1995.

13. Dougherty, R.P. and Stoker, R.W., Sidelobe suppression for phased array aeroacoustic measurements, *Paper AIAA-98-2242*, 1998.

14. Nordborg, A., Optimum array microphone configuration, *Inter-Noise 2000*, 2000.

15. Hald, J. and Christensen, J.J., A class of optimal broad band phased array geometries designed for easy construction, *Inter-Noise 2002*, 2002.

16. Piet, J.F. and Elias, G., Airframe noise source localization using a microphone array, *Paper AIAA-97-1643*, 1997.

17. Hayes, J.A., Airframe noise characteristics of a 4.7% scale DC-10 model, *Paper AIAA-97-1594*, 1997.

18. Stoker, R.W., Underbrink, J.R., and Neubert, G.R., Investigation of airframe noise in pressurized wind tunnels, *Paper AIAA-2001-2107*, 2001.

# 16

# Source of Noise

S. Akishita
*Ritsumeikan University*

**Summary**

*In this chapter, a mathematical description of sound radiation is briefly presented accompanied by an introduction to sound sources, monopole, and dipole. The modeling of a simple source of noise is discussed, introducing Green's function. As an example of simple sound radiation, the sound field generated by a source embedded in a plane surface is described. Finally, an estimation of noise-source sound power is presented by introducing the power conversion factor of actual machinery.*

## 16.1 Introduction

A careful examination of noise measurement data reveals that there exists a very wide variety of noise sources. Each source is peculiar to its generation mechanism, which can be any of a wide range of phenomena including fluid mechanics and the vibration of structures. However, in analysis, sources are normally simplified to rather simple and typical models in their generation mechanism.

The vibration of a solid body, which may be in contact with the fluid medium, generates sound waves or vibratory forces acting directly on a fluid, will result in the emission of acoustic energy in the medium. In the next section, an expression for an idealized sound source is introduced. We will assume that the fluid medium outside the source region is initially uniform and at rest. Also, we will concentrate on wave propagation in an infinite medium.

Generally, acoustic waves sensed as a sound represent a very small energy density in the medium. Only a very small fraction of the mechanical energy of a source body is converted into acoustic energy. The conversion factor, defined as the ratio of sound power to the mechanical power of the source, is in the order of $10^{-7}$ to $10^{-5}$. Some examples of estimated sound power conversion factor are given for typical common noise sources.

## 16.2 Radiation of Sound

### 16.2.1 Point Source

#### 16.2.1.1 Simple Source: Spherical Wave by a Monopole

Propagation of sound pressure wave $p(x, y, z, t)$ is described by the following partial differential equation for a medium where a field point is expressed by orthogonal coordinate system $O\text{-}xyz$, as shown in Figure 16.1(a):

$$\nabla^2 p - \frac{1}{c^2}\frac{\partial^2 p}{\partial t^2} = 0,$$

$$\nabla^2 \equiv \left(\frac{\partial^2}{\partial x^2} + \frac{\partial^2}{\partial y^2} + \frac{\partial^2}{\partial z^2}\right) \qquad (16.1)$$

where $c$ denotes the velocity of sound propagation. If the source region is compact and the generating motion has no preferred direction, it will produce a wave, which spreads spherically outwards. As the medium is assumed infinite in extent, the waveform will depend on the distance, $r$, from the center of the source. The wave equation in this case is

$$\frac{1}{r^2}\frac{\partial}{\partial r}\left(r^2\frac{\partial p}{\partial r}\right) - \frac{1}{c^2}\frac{\partial^2 p}{\partial t^2} = 0 \qquad (16.2)$$

When a monopole source of angular frequency $\omega$ is assumed, the simplest solution for the outward propagating waveform is expressed as

$$p(r,t) = p_\omega(\mathbf{r})e^{-i\omega t},$$

$$p_\omega(\mathbf{r}) = \frac{-\omega\rho}{4\pi r}S_\omega\, e^{ikr}, \quad k = \frac{\omega}{c} = \frac{2\pi}{\lambda} \qquad (16.3)$$

where $\rho$ denotes the density of the medium and $k$ denotes the wave number. Here, $p_\omega$ is used to denote the sinusoidal component of the sound pressure with angular frequency, $\omega$. The subscript $\omega$ on a variable typically indicates the sinusoidal component of a variable, but the variables related to sound energy and sound power, such as $w$, $I$, $W$,

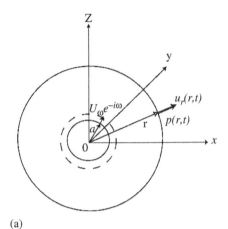

(a)

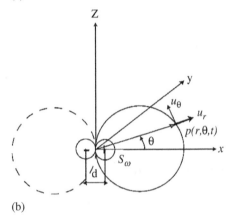

(b)

**FIGURE 16.1** Directivity of monopole and dipole, (a) Spherical sound field by a monopole; (b) Axisymmetric sound field by a dipole.

do not have the subscript $\omega$ even if they mean the sinusoidal component. In this case, the source of sound is taken as a "pulsating globe" of radius $a$ and radial velocity $U_\omega$ on the surface. Therefore, the flow outward from the origin, $S_\omega$, is related to $U_\omega$ as follows, and as shown in Figure 16.1(a):

$$S_\omega = 4\pi a^2 U_\omega \qquad (16.4)$$

We should note that, while a pulsating globe with a finite radius $a$ is assumed as the physical sound source, the sound field by a monopole with infinitesimal small size and finite magnitude, as expressed mathematically in Equation 16.3, is used.

The other quantities related to the spherical wave are described next [1]:

$$\begin{cases} u_{r\omega} = \dfrac{-1}{4\pi r^2}(ikr - 1)S_\omega\, e^{ik(r-ct)}; & \text{radial velocity} \\[3mm] w = \rho\left(\dfrac{1}{4\pi r^2}\right)^2 |S_\omega|^2\left[(kr)^2 + \dfrac{1}{2}\right]; & \text{energy density} \\[3mm] I = \rho c\left(\dfrac{k}{4\pi r}\right)^2 |S_\omega|^2 = \dfrac{|p|^2}{\rho c}; & \text{energy flux intensity} \\[3mm] W = (4\pi r^2)I = \rho c\dfrac{\pi}{\lambda^2}|S_\omega|^2 = \dfrac{\rho\omega^2}{4\pi c}|S_\omega|^2; & \text{total power} \end{cases} \qquad (16.5)$$

We should note that the first term in the parentheses, $kr$, is negligible in the region where $|kr| \ll 1$ is valid. Hence,

$$w = \frac{\rho}{2}(|S_\omega|/4\pi r^2)^2 = \frac{\rho}{2}u_{r\omega}^2 = \left(\frac{\lambda}{\omega}\right)\frac{I}{4\pi r^2}$$

is reduced. Conversely, when $|kr| \gg 1$ is valid, then $w = \rho(k/4\pi r)^2|S_\omega|^2 = I/c$ is valid.

#### 16.2.1.2 Simple Source: Plane Wave by an Alternating Piston

Another example of simple sound wave is generated in the one-dimensional field of fluid medium, as shown in Figure 16.1. Let us set the coordinate $x$ along the axis of wave propagation, for example, the axis of duct with a constant cross-sectional area. Then the wave equation is

$$\frac{\partial^2 p}{\partial x^2} - \frac{1}{c^2}\frac{\partial^2 p}{\partial t^2} = 0 \qquad (16.6)$$

The solution for a periodic source is given by

$$p(x,t) = p_\omega(x)e^{-i\omega t}, \quad p_\omega(x) = \rho c u_\omega\, e^{ikx} \qquad (16.7)$$

This is known as a *plane wave*, which is generated by the piston motion at the origin, the velocity of which is expressed by

$$u(t) = u_\omega(x)e^{-i\omega t} \qquad (16.8)$$

The other quantities related with the plane wave are given below:

$$\begin{cases} u_\omega(x) = u_\omega\, e^{ikx}; & \text{particle velocity} \\ w = I = \rho c |u_\omega|^2 = |p_\omega|^2/\rho c^2; & \text{energy density, sound intensity} \\ W = SI\ (S;\ \text{cross-sectional area}); & \text{total power} \end{cases} \qquad (16.9)$$

A plane wave is generated in very limited situations, but its utility is rather wide since the sound wave propagating through a duct or duct-like space with a gradually varying cross section is approximated as the plane wave. Network theory is applied to the sound wave propagating through a branch and junction by using the description of a plane wave.

#### 16.2.1.3 Dipole and Multipoles and Their Sound Field

Let us return to the three-dimensional sound field. The second simple solution to Equation 16.1 is the "dipole" sound field. Suppose that a pair of monopoles, close together, opposite in sign, and equal in magnitude, $S_\omega$, are located along the $x$-axis as shown in Figure 16.1(b). Since only a preferred direction is assigned along the $x$-axis, the sound field is axisymmetric as represented by

$$p(r,\theta,t) = -k^2 D_\omega\frac{\rho c\cos\theta}{4\pi r}\left(1 + \frac{i}{kr}\right)e^{-i(\omega t - kr)} \qquad (16.10)$$

$D_\omega$ is defined by

$$D_\omega = S_\omega d \qquad (16.11)$$

where $d$ denotes the separation of the monopoles as shown in Figure 16.1(b). Mathematically, $d$ tends to zero, keeping $D_\omega$ finite, and the preferred axis is the $x$-axis. Physically, a sound field is commonly realized by a pair of monopoles with a finite separation that is short compared with the wavelength, $\lambda$, as illustrated next with realistic examples.

The characteristic quantities relating with dipole sound field are described below [1]:

$$
\begin{cases}
u_{r\omega} = -\dfrac{k^2 D_\omega \cos\theta}{4\pi r}\left(1 + \dfrac{2i}{kr} - \dfrac{2}{k^2 r^2}\right)e^{-i\omega t + kr}; & \text{radial velocity} \\[2ex]
u_{\theta\omega} = i\dfrac{k^2 D_\omega \sin\theta}{4\pi r^2}\left(1 + \dfrac{i}{kr}\right)e^{-i\omega t + kr}; & \text{peripheral velocity} \\[2ex]
w = \rho\left(\dfrac{k^2 |D_\omega|}{4\pi r}\right)^2\left[\cos^2\theta + \dfrac{1}{2}\left(\dfrac{1}{kr}\right)^2 + \dfrac{1}{2}\left(\dfrac{1}{kr}\right)^4(1 + 3\cos^2\theta)\right]; & \text{energy density} \\[2ex]
I_r = \rho c\left(\dfrac{k^2 |D_\omega|}{4\pi r}\right)^2\cos^2\theta, \quad I_\theta = 0; & \text{sound intensity} \\[2ex]
W = \dfrac{\rho\omega^4}{12\pi c^3}|D_\omega|^2; & \text{total power}
\end{cases}
\tag{16.12}
$$

We should add the following notes on the dipole sound field.

When $|kr| \gg 1$ is assumed, the second term in parentheses in Equation 16.10 is negligible. Then the directivity for $p(r, \theta, t)$ is expressed by $\cos\theta$. A similar directivity is found on $I_r$ and on $u_{r\omega}$, $u_{\theta\omega}$, and $w$ with the assumption $|kr| \gg 1$.

A pair of dipoles produces a *quadrupole*, a pair of quadrupoles produce an *octopole*, and so on. These are called *multipole* in general. Out of multipoles, the quadrupole is common in representing a sound field generated by mixing fluid flow, especially jet flow. More details on multipoles are found in Ref. [1].

## 16.2.2 Sources of Finite Volume

### 16.2.2.1 Description of Sound Field by Green's Function

In order to describe the sound field from distributed sources, source terms are introduced to the right side of Equation 16.1. The partial differential equation with source term is derived from the equation system representing the dynamics of fluid flow with periodic motion at angular frequency $\omega$:

$$
\nabla^2 p_\omega + k^2 p_\omega = -m_\omega + \text{div } \mathbf{F}_\omega
\tag{16.13}
$$

where $p_\omega$ denotes the acoustic pressure amplitude according to $p(x, y, z, t) = p_\omega(x, y, z)e^{-i\omega t}$, $m_\omega$ denotes the effective monopole source density expressed by $-i\omega\rho s_\omega$, $s_\omega$ denotes the generalization of the point source strength, $S_\omega$, of Equation 16.4 for a distributed source, and $\mathbf{F}_\omega$ denotes the vector representation of point force-density in the fluid. Introduction of Green's function, $g_\omega$, of angular frequency, $\omega$, satisfying the following equation is useful in general:

$$
\nabla^2 g_\omega + k^2 g_\omega = -\delta(x - x_0)\delta(y - y_0)\delta(z - z_0)e^{i\omega t_0}
\tag{16.14}
$$

Here, $\delta(z)$ denotes the Dirac impulse (delta) function of the variable $z$; the coordinate $(x_0, y_0, z_0)$ denotes the position of unit source, $r_0$, with periodic angular velocity, $\omega$, and $t_0$ denotes the time pertaining to the source. The solution of Equation 16.14 is

$$
g_\omega(\mathbf{r}, \mathbf{r}_0) = \frac{1}{4\pi R}e^{ikr}, \quad |\mathbf{r} - \mathbf{r}_0| = \sqrt{(x - x_0)^2 + (y - y_0)^2 + (z - z_0)^2}
\tag{16.15}
$$

where $\mathbf{r}$ denotes the position vector of sound field with coordinates $(x, y, z)$. The solution of Equation 16.13 is described by using $g_\omega(\mathbf{r}, \mathbf{r}_0)$ as follows:

$$
p_\omega(\mathbf{r}) = \int\int\int [m_\omega(\mathbf{r}_0) - \text{div } \mathbf{F}_\omega(\mathbf{r}_0)]g_\omega(\mathbf{r}, \mathbf{r}_0)dx_0 dy_0 dz_0
\tag{16.16}
$$

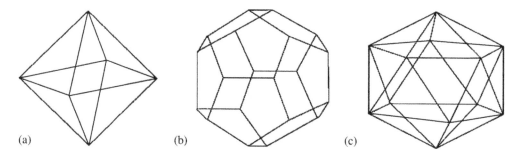

**FIGURE 16.2** Practical monopole sources.

### 16.2.2.2 Radiation from Vibrating Small Body

A pulsating globe, as the simplest sound source, is rarely realized in the real world. The approximated monopole source required in most measurements is an eight-sided polyhedron, as depicted in Figure 16.2, where a loud speaker is installed on each of the surfaces. The sound field radiated by a thus approximated source is almost the same as that by a pulsating globe at the far field, where $kr \gg 1$ is valid.

In this case, the sound field is represented by Equation 16.16, assuming that $F_\omega(r_0) = 0$, $m_\omega(r_0) = -i\omega\rho U_\omega \delta(n)$, and $dx_0 dy_0 dz_0 = dn dS$, where the source $m_\omega(r_0)$ is distributed on a thin layer of thickness, $dn$, on the spherical surface element, $dS$. Substitution of these relationships into Equation 16.16 yields the formula

$$p_\omega(\mathbf{r}) = -\iint_S \frac{i\omega\rho U_\omega}{4\pi R} e^{ikR} dS$$

$$= -\frac{i\omega\rho}{4\pi R_0}(4\pi a^2) U_\omega e^{ikR_0} \quad (16.17)$$

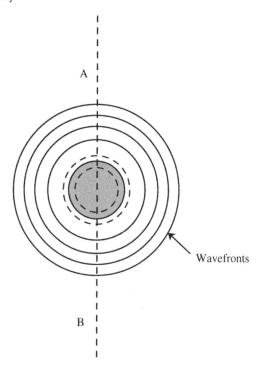

**FIGURE 16.3** Field of a pulsating spherical source.

where $\iint_S dS$ means integration on the approximately spherical surface, and then $R \equiv |\mathbf{r} - \mathbf{r}_0| \cong R_0 = \sqrt{x^2 + y^2 + z^2}$ is applied. The final reduction of Equation 16.17 gives the same expression as is deduced from Equation 16.3 and Equation 16.4.

The sound field generated by a monopole source is illustrated in Figure 16.1. This field is unchanged with the presence of a rigid plane AB in Figure 16.3. It is clear, on account of symmetry, that the presence of the rigid plane does not alter the sound field in any way, because only the tangential component of particle velocity is induced on the plane. This utilization of the symmetry and construction of the semi-infinite field is applied to the expression of the sound field generated by the baffled structure, as shown in Figure 16.4.

Let us imagine a sound field generated by an oscillating small body in an infinite medium, as shown in Figure 16.4(a). Since the oscillation is caused by an external force, the sound source is modeled by the force $-F_\omega(r_0)$. Therefore, the sound field is described by Equation 16.16, assuming $m_\omega(r_0) = 0$. After applying a law of vector analysis, such as $g_\omega(\mathbf{r}, \mathbf{r}_0) \text{div } F_\omega(r_0) \cdot \text{grad}_0 g_\omega + \text{div}_0(g_\omega F_\omega)$, the sound field is

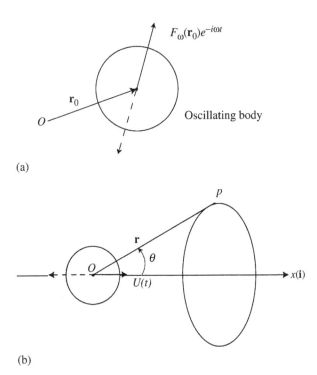

**FIGURE 16.4** Dipole field generated by an oscillating small body: (a) oscillating small body; (b) oscillating sphere along x-axis.

represented by

$$p_\omega(\mathbf{r}) = \iiint \mathbf{F}_\omega(\mathbf{r}_0) \cdot \nabla_0 g_\omega(\mathbf{r}, \mathbf{r}_0) dx_0 dy_0 dz_0 \tag{16.18}$$

where $\iiint \mathrm{div}_0(g_\omega \mathbf{F}_\omega) dx_0 dy_0 dz_0 = \iint_s g_\omega \mathbf{F}_\omega dS_n(x_0, y_0, z_0) = 0$ is applied. Assume that a small sphere of radius $a_0$ is oscillating along the x-axis with angular frequency $\omega$. Instead of an oscillating sphere with velocity $U_t = U_\omega e^{-i\omega t}$, the sound field is generated by the concentrated body force, $F(t) = -m(d/dt)U(t) = i\omega m U_\omega e^{-i\omega t}$, ($m$ is the mass of the sphere), at the origin, $\mathbf{r}_0 = 0$. Then $\mathbf{F}_\omega(\mathbf{r}_0)$ is expressed by $F_\omega(\mathbf{r}_0) = (F_{x\omega}\mathbf{i})\delta(x_0)\delta(y_0)\delta(z_0)$; $F_{x\omega} = im\omega U_\omega$. This approximate reduction is appropriate when $ka_0 \ll 1$ is valid. Substituting the approximation, Equation 16.18 is rewritten as

$$p_\omega(\mathbf{r}) \cong F_{x\omega} \frac{\partial}{\partial x_0} g_\omega(\mathbf{r}, \mathbf{r}_0, \theta) = -k^2 D_\omega \frac{\rho c}{4\pi r} \cos\theta \left(1 + \frac{i}{kr}\right) e^{ikr} \tag{16.19a}$$

$$D_\omega = \frac{i}{k\rho c} F_{\omega x} \tag{16.19b}$$

As the sound field is axisymmetric about x-axis, the sound pressure, $p_\omega(\mathbf{r}, \theta)$, depends only on $\mathbf{r}$ and $\theta$, where $\mathbf{r}$ and $\theta$ are defined in Figure 16.4(b). The expression is applicable to the sound field generated by an oscillating small body in a free space.

## 16.2.3   Radiation from a Plane Surface

### 16.2.3.1   Radiation from a Small Body in Infinite Plane Surface

The introduction of an infinite rigid plane surface to the sound field, as shown in Figure 16.3, simplifies the formulation of the sound field generated by an oscillating body adjacent to a large plane.

The configuration discussed in this section relates to a source in the presence of an infinite plane barrier, so that the medium is confined to one side of the plane.

By taking the effect of the image caused by the rigid plane surface, the Green's function, $g_\omega(\mathbf{r}, \mathbf{r}_0)$, in this case simplifies as given below, to what is called Rayleigh's formula [2]:

$$g_\omega(\mathbf{r}, \mathbf{r}_0) = \frac{1}{2\pi R} e^{ikr}, \quad R = |\mathbf{r} - \mathbf{r}_0| \tag{16.20}$$

where $\mathbf{r}_0$ denotes the projection position of the source on the surface. Therefore, Equation 16.16 is rewritten as follows:

$$p_\omega(\mathbf{r}) = \int\int\int \frac{e^{ikR}}{2\pi R}[m_\omega(\mathbf{r}_0) - \text{div } \mathbf{F}_\omega(\mathbf{r}_0)]dx_0 dy_0 dz_0 \tag{16.21}$$

### 16.2.3.2 Radiation from a Circular Piston

Let us consider the sound field generated by a rigid circular piston of radius $a$ mounted flush with the surface of an infinite baffle and vibrating with simple harmonic motion of angular frequency, $\omega$. The solution of this example is applicable to a number of related problems, including the radiation from the open end of a flanged organ pipe.

The coordinate system is shown in Figure 16.5, where the infinite baffle and the piston are placed in the $Oxy$-plane. Note that the observation point, $P$, is denoted by $\mathbf{r} = \overrightarrow{OP}$, while the source point, $Q$, is denoted by $\mathbf{r}_0 = \overrightarrow{OQ}$.

Since sound is observed only in the semi-infinite plane where $z > 0$ is valid, only $m_\omega(\mathbf{r}_0)$ remains nonzero in Equation 16.21. As the velocity of the piston is denoted by $U_\omega e^{-i\omega t}$ along the $z$-axis, $m_\omega(\mathbf{r}_0) = -i\omega\rho U_\omega \delta(z_0)$, is distributed only on the circular piston. Finally, Equation 16.21 is rewritten as

$$p_\omega(\mathbf{r}) = \frac{-i\omega\rho U_\omega}{2\pi}\int\int_S \frac{e^{ikR}}{R}dx_0 dy_0 \tag{16.22}$$

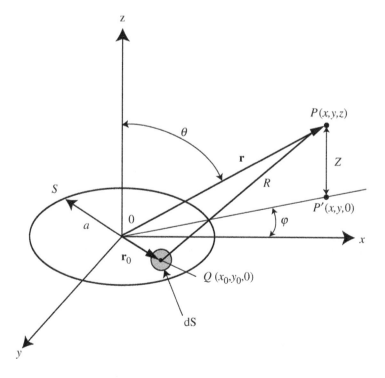

**FIGURE 16.5** The rigid circular piston and the coordinate system.

Assuming $a \ll r$, the following approximation can be made:

$$R \cong r - (x_0 \cos \varphi + y_0 \sin \varphi) \sin \theta,$$

$$\frac{1}{R} \cong \frac{1}{r} \tag{16.23}$$

where $\varphi$ and $\theta$ denote the angles defining the observation point, $P$, as shown in Figure 16.5. By changing from Cartesian coordinates, $x_0$, $y_0$, to polar coordinates, $\rho$, $\varphi_0$, such that $x_0 = \rho \cos \varphi_0$, $y_0 = \rho \sin \varphi_0$ in the integral above, we rewrite Equation 16.22 as

$$p_\omega(\mathbf{r}) = -\frac{i\omega\rho U_\omega}{2\pi R} e^{ikr} \int_0^{2\pi} d\varphi_0$$

$$\times \int_0^a \exp[-k\rho \cos(\varphi_0 - \varphi) \sin \theta]\rho d\rho \tag{16.24}$$

The integration is performed by introducing the Bessel function of the first order, $J_1(z)$ as follows:

$$p_\omega(\mathbf{r}) = -\frac{i\omega\rho \, e^{ikr}}{2\pi r} f_\omega(\theta),$$

$$f_\omega(\theta) = \pi a^2 U_\omega \left[ \frac{2J_1(ka \sin \theta)}{ka \sin \theta} \right] \tag{16.25}$$

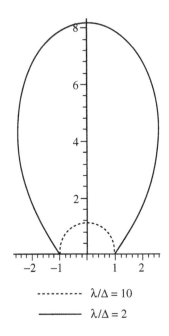

- - - - - - - - $\lambda/\Delta = 10$

———— $\lambda/\Delta = 2$

**FIGURE 16.6**  Directivity of sound intensity generated by an oscillating piston.

Note that dependency on $\varphi$ has disappeared in the integration process, which follows from the axisymmetry of the sound field. The corresponding intensity at $\mathbf{r}$ is given by

$$I_r(\theta) \cong \frac{|p_\omega|}{\rho c} = \frac{\rho c U_\omega^2 a^2}{4r^2} (ka)^2 \left[ \frac{2J_1(ka \sin \theta)}{ka \sin \theta} \right]^2 \tag{16.26}$$

Figure 16.6 illustrates the dependency of $I_r$ on $\theta$ for two cases of $ka$. Note that, for the smaller $ka = 2\pi a/\lambda$, $I_r$ is almost independent on $\theta$, which is similar to the dependence of the monopole.

### 16.2.3.3  Radiation from a Rectangular Plate

An normal velocity distribution, $u_\omega(x_0, y_0)$, is prescribed over a baffled planar radiator located in the plane $z_0 = 0$ in the region $-L_x \le x_0 \le L_x$, $-L_y \le y_0 \le L_y$, as shown in Figure 16.7.

In this case, the sound pressure field is represented by the following equation, similar to the previous section:

$$p_\omega(\mathbf{r}) = \frac{-i\omega\rho}{2\pi} \int_{-L_y}^{L_y} dy_0 \int_{-L_x}^{L_x} \frac{U_\omega(x_0, y_0)e^{ikR}}{R} dx_0 \tag{16.27}$$

Assuming $2L_x$, $2L_y \ll r$, the same approximation in as Equation 16.23 is acceptable. Therefore, $p_\omega(\mathbf{r}) = p_\omega(r, \theta, \phi)$ takes the form [2]:

$$p_\omega(r, \theta, \phi) = \frac{-i\omega\rho \, e^{ikr}}{2\pi r} \int_{-L_y}^{L_y} dy_0 \int_{-L_x}^{L_x} U_\omega(x_0, y_0) \exp[-ik \sin \theta(x_0 \cos \phi + y_0 \sin \phi)]dx_0 \tag{16.28}$$

The result of the above integration, assuming $U_\omega(x_0, y_0) = U_\omega = const$, which means the rigid rectangular piston oscillates with amplitude $U_\omega$ along the $z$-axis, is

$$p_\omega(r, \theta, \phi) = \frac{i\omega\rho U_\omega}{2\pi r} (4L_x L_y)e^{ikr} S(kL_x \sin \theta \cos \phi)S(kL_y \sin \theta \sin \phi) \tag{16.29a}$$

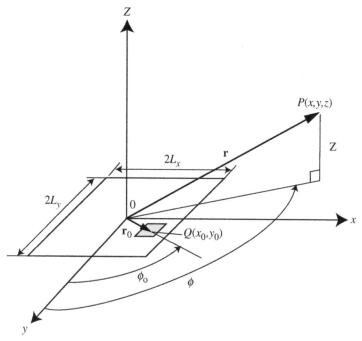

**FIGURE 16.7** A flexural rectangular plate and the coordinate system.

where the function $S(z)$ is defined by

$$S(z) = \frac{\sin z}{z} \tag{16.29b}$$

Now, let us discuss the case of flexural plate, which has the same lengths, $2L_x$ and $2L_y$, but $U_\omega(x_0, y_0)$ represents the flexural vibration mode. Flexural mode patterns of rectangular panels take the general form of contiguous regions of roughly equal area and shape, which vary alternately in vibration phase, and are separated by nodal lines of zero vibration.

For simply supported edges, the normal vibration velocity distribution is

$$U_\omega(x_0, y_0) = U_{mn} \sin\left(\frac{m\pi x}{2L_x}\right) \sin\left(\frac{n\pi y}{2L_y}\right), \quad (0 \le x \le 2L_x,\ 0 \le y \le 2L_y) \tag{16.30}$$

Note that the coordinate system $O\text{-}x_0 y_0 z_0$ shifts its origin from the center point of the panel to the edge at the leftmost and frontmost point in Figure 16.7.

The approximation represented by Equation 16.23 is valid as well in this case. The result of integration is rather simple as given below [3]:

$$p_\omega(r, \theta, \phi) = \frac{-i\omega\rho U_{mn}\, e^{kr}}{2\pi r}\, \frac{4L_x L_y}{mn\pi^2} \left[\frac{(-1)^m\, e^{i\alpha} - 1}{(\alpha/m\pi)^2 - 1}\right]\left[\frac{(-1)^n\, e^{i\beta} - 1}{(\beta/n\pi)^2 - 1}\right] \tag{16.31}$$

where $\alpha$ and $\beta$ are defined by

$$\alpha = 2kL_x \sin\theta\cos\phi, \quad \beta = 2kL_y \sin\theta\sin\phi \tag{16.32}$$

The corresponding intensity, $I_r(\theta, \varphi)$, at $\mathbf{r}$ is expressed by

$$I_r(r, \theta, \phi) = \frac{|p_\omega|^2}{\rho c} = 4\rho c|U_{mn}|^2\left(\frac{4kL_x L_y}{\pi^3 rmn}\right)^2 \left\{\frac{\cos\left(\frac{\alpha}{2}\right)\cos\left(\frac{\beta}{2}\right)}{\sin\left(\frac{\alpha}{2}\right)\sin\left(\frac{\beta}{2}\right)}{[(\alpha/m\pi)^2 - 1][(\beta/n\pi)^2 - 1]}\right\} \tag{16.33}$$

where $\cos(\alpha/2)$ is used when $m$ is an odd integer, and $\sin(\beta/2)$ is used when $m$ is an even integer; $\cos(\beta/2)$ is used for even $n$ and $\sin(\beta/2)$ for odd $n$.

## 16.2.4 Estimation of Noise-Source Sound Power

### 16.2.4.1 Power Conversion Factor of Machinery

It is often necessary to estimate the expected sound power that a particular machine might introduce into an environment [4]. One way where such an estimate may be approached for a particular class of machine is by means of the sound power conversion factor, $\eta_n$. This factor is defined as

$$\eta_n = \frac{P}{P_m} \tag{16.34}$$

where $P$ = sound power of the machine (W), and $P_m$ = power of the machine (W). This relationship is valid for both mechanical and electrical machinery. The conversion factors for some common noise sources are given in Table 16.1.

## Example 16.1

Estimate the sound power level of a typical 1-kW electric motor that operates at 1200 rpm.

## Solution

From Table 16.1, we find that for typical electric motors, $\eta_n = 1 \times 10^{-7}$. Thus, using Equation 16.34, we obtain

$$P = \eta_n P_m = (1 \times 10^{-7}) \times 1000 = 10^{-4}(W)$$

as the total sound power of motor. Then using Equation 37.7, $L_W$ is given by

$$L_W = 10 \log\left(\frac{10^{-4}}{10^{-12}}\right) = 80(dB)$$

### 16.2.4.2 Fan Noise

We are familiar with noise nuisance caused by a domestic ventilating fan. The mechanical power of the fan is expressed by

$$P_m = p_T Q \tag{16.35}$$

where $p_T$ denotes the total pressure rise through the fan and $Q$ denotes the volumetric flow rate. According to the law of sixth power of flow velocity deduced by the aeroacoustics theory, sound power of the fan is proportional to $p_T^{2.5} Q$. Therefore, the specific ratio $k_T$ defined below is more useful than $\eta_n$ for the fan:

$$k_T = \frac{P_m}{p_T^{2.5} Q} \tag{16.36}$$

**TABLE 16.1**　Estimated Sound Power Conversion Factors for Common Noise Sources[a]

| Noise Source | Conversion Factor | | |
|---|---|---|---|
| | Low | Midrange | High |
| Compressor, air (1–100 hp) | $3 \times 10^{-7}$ | $5.3 \times 10^{-7}$ | $1 \times 10^{-6}$ |
| Gear trains | $1.5 \times 10^{-8}$ | $5 \times 10^{-7}$ | $1.5 \times 10^{-6}$ |
| Loud speakers | $3 \times 10^{-2}$ | $5 \times 10^{-2}$ | $1 \times 10^{-1}$ |
| Motors, diesel | $2 \times 10^{-7}$ | $5 \times 10^{-7}$ | $2.5 \times 10^{-6}$ |
| Motors, electric (1200 rpm) | $1 \times 10^{-8}$ | $1 \times 10^{-7}$ | $3 \times 10^{-7}$ |
| Pumps, over 1600 rpm | $3.5 \times 10^{-6}$ | $1.4 \times 10^{-5}$ | $5 \times 10^{-5}$ |
| Pumps, under 1600 rpm | $1.1 \times 10^{-6}$ | $4.4 \times 10^{-6}$ | $1.6 \times 10^{-5}$ |
| Turbines, gas | $2 \times 10^{-6}$ | $5 \times 10^{-6}$ | $5 \times 10^{-5}$ |

[a] Total sound power for the four octave bands from 500 to 4000 Hz.

*Source:* Irwin, J. D. and Graf, E. R. 1979. *Industrial Noise and Vibration Control*, Prentice Hall, Englewood Cliffs, NJ.

**TABLE 16.2**  Specific Fan Noise Level for Low-Pressure Fans

| Type | $K_T$ (dB) |
|------|------------|
| Axial | $-87.5$ to $-70.7$ |
| Centrifugal/cirroco | $-87.5$ to $-85.7$ |
| Centrifugal/radial | $-98.7$ to $-84.7$ |
| Centrifugal/turbo | $-104.7$ to $-89.7$ |
| Cross-flow | $-76.7$ to $-68.7$ |

Furthermore, usually the specific fan noise level, $K_T$, as given below is more useful than $k_T$.

$$K_T = L_W - 10 \log(p_{T^{2.5}} Q)\mathrm{dB} \tag{16.37}$$

where $p_T$ denotes the total pressure rise in mmAg, $Q$ denotes the flow rate in m$^3$/min, $L_W$ denotes the total sound power level in dB, and $K_T$ represents the radiation efficiency of the fan noise. This efficiency varies with the type of the fan and with the flow rate when the model is assigned. In particular, $K_T$ will be the lowest at the flow rate at which the aerodynamic power efficiency is the highest. Therefore, we will have the best advantage on the sound environment when the fan is operated at the highest efficiency. For convenience of design, the $L_W$ is often evaluated by A-weighted total sound level (dB-A) not by linear total level (dB). Table 16.2 gives the specific fan noise level evaluated with dB-A for five types of the low-pressure fans.

## Example 16.2

Consider a ventilating axial-flow fan, the specifications of which are: $p_T = 10$ mmAq, $Q = 30$ m$^3$/min, $D = 30$ cm (diameter of the duct containing fan rotor). Estimate the directional distribution of intensity $I_r(\theta)$, assuming $r = 3$ m and $f = 150$ Hz for the main component of the fan noise, and $K_T = -79.0$ dB-A from Table 16.2. The fan noise is assumed to be radiated as a plane sound wave at the mouth of the duct.

## Solution

First, we modify the $K_T$ in dB-A to that of linear scale. From the frequency response for A-weighting network shown in Figure 2.5, for $f = 150$ Hz, the modification is found as $\Delta K_T = 15$ dB. Then, the modified specific fan noise level is $K_T = -79.0 + 15.0 = -64.0$ dB. By using Equation 16.37, $L_W = K_T + 25 \log P_T + 10 \log Q = -64 + 25 + 15 = -24$ dB is obtained. This means the emitted total sound power is $W = 10^{-24/10} = 3.98 \times 10^{-3}$ W. Since we assume a plane sound wave at the mouth of the duct for the fan noise, we can use the sound radiation model of a circular piston with $a = 15$ cm for the radiated sound wave. In our case $k_a = 2\pi a/(c/f) = 0.415$, or $\lambda/a = 15.1$. A monopole model will be valid for the directional distribution of intensity from Figure 16.6. Therefore, $I_r = W/2\pi r^2 = 0.704 \times 10^{-4}$ W/m$^2$ or $|p|^2 = I_r \rho c = 2.92 \times 10^{-2}$(Pa$^2$). This is the same as the sound pressure level $L_p = 10 \log|p|^2/p_{\mathrm{ref}} = 78.6$ dB or $L_p(A) = L_p - 15 = 63.6$ dB-A.

## References

1. Morse, P.M. and Ingard, K.U. 1986. *Theoretical Acoustics*, Princeton University Press, Princeton, NJ, chap. 7.
2. Junger, M.C. and Feit, D. 1986. *Sound, Structures, and Their Interaction*, 2nd ed., MIT Press, Cambridge, chap. 4.
3. Fahy, F. 1985. *Sound and Structural Vibration. Radiation, Transmission and Response*, Academic Press, New York, chap. 2.
4. Irwin, J.D. and Graf, E.R. 1979. *Industrial Noise and Vibration Control*, Prentice Hall, Englewood Cliffs, NJ, chap. 5.

# 17

# Design of Absorption

Teruo Obata
*Teikyo University*

**Summary**

*This chapter presents the basics of designing devices for sound absorption. The absorption coefficient and acoustic impedance are introduced. Characteristic properties and parameters of sound absorption material and basic elements are presented. Acoustic modeling, analysis, and design considerations of components, such as ducts, and noise attenuation devices, such as mufflers, are presented. A practical design example is presented for illustration of the concepts presented in the chapter.*

## 17.1 Introduction

Sound-absorption equipment is used for multiple purposes in architectural acoustics, mechanical noise countermeasures, and so on. In this context, the necessity for designing sound-absorption equipment from the viewpoint of noise control is explained. Architectural acoustics is an important area of study, which involves architecture, sound-absorption design, and sound-measurement facility.

In the area of noise reduction, the characteristics of sound are important, and proper sound-absorbing material should be selected based on how much attenuation is necessary for each frequency of sound. In particular, acoustic characteristics of sound-absorbing material such as the type of material and the sound-absorption mechanism are important. In this chapter, the basics of sound absorption are given, and the prediction and calculation methods for attenuation of lined or dissipative mufflers are outlined.

## 17.2 Fundamentals of Sound Absorption

### 17.2.1 Attenuation of Sound

When an acoustic wave propagates in a medium, the sound energy attenuates due to such reasons as viscosity, heat conduction, and the effects of molecular absorption. In a medium of small volume surrounded by a boundary surface, the attenuation is particularly considerable, for example, when the medium is a thin tube. This is because there is the dissipation of the energy controlled by the viscosity of the medium and heat conduction between the material and the medium of tube wall. A sound-absorbing material may be utilized to adjust such dissipation of acoustic energy.

#### 17.2.1.1 Absorption Coefficient and Normal Acoustic Impedance

Some amount of energy is lost when an acoustic wave hits the surface of a sound-absorbing material. Figure 17.1 illustrates an infinite medium of absorbing material separated by air and the reflected wave (sound pressure $p_r$) from the boundary surface with the air where a plane wave of sound pressure $p_i$ is emitted in the direction indicated by an arrow, at an angle $\theta$. When $\theta = 0$, sound pressure $p$ in air is given by

$$p = p_i + p_r = (Ae^{-jkx} + Be^{jkx})e^{j\omega t} \qquad (17.1)$$

where

$A, B =$ the amplitude of sound pressure of incident and reflected waves (in Pa),
$j = \sqrt{-1}$,
$k = 2\pi f/c$; wave number (1/m),
$\omega =$ angular frequency (rad/sec).

The sound pressure, $p_m$, in the absorbing material may be expressed using a complex propagation constant, by the equation:

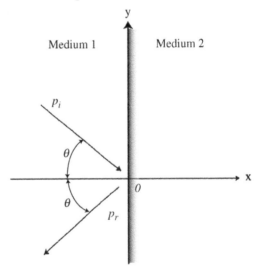

**FIGURE 17.1** Plane wave incidence on an infinite absorbing material.

$$p_m = p_{x=0}e^{-\gamma x}e^{j\omega t} \qquad (17.2)$$

where

$\gamma =$ the propagation constant in the absorbing material ($m^{-1}$). *Note:* $\gamma = \delta + j\beta$. $\gamma$ is a property of the material itself and is not dependent on the mounting conditions when large areas of material are considered.
$\delta =$ attenuation constant. *Note:* $\delta$ tells us how much of the sound wave will be reduced as it travels through the material.
$\beta =$ phase constant. *Note:* $\beta$ is a measure of the velocity of propagation of the sound wave through the material.

The relation for determining the velocity of sound in the material is given by

$$c_m = \omega/\beta \qquad (17.3)$$

Boundary conditions must be satisfied on the boundary surface. The acoustic impedance of a unit area of air and of absorbing material are, respectively, denoted by $z$ and $z_a$. The pressure and the particle velocity on both sides of the boundary are equal. We have

$$\left.\begin{array}{c} p_i + p_r = p_{x=0} \\ \dfrac{p_i - p_r}{z_a} = \dfrac{p_{x-0}}{z} \end{array}\right\} \tag{17.4}$$

The amplitude of reflectance of sound pressure, $r$, is obtained from Equation 17.4, and is given by

$$r = \frac{p_r}{p_i} = \frac{z_a - z}{z_a + z} \tag{17.5}$$

The reflectivity is the energy reflection rate. The absorption coefficient, $\alpha$, of an absorbing material is defined as

$$\alpha = 1 - |r^2| \tag{17.6}$$

The impedance, $z_n$, through a surface is the quantity that represents the dissipation of energy of sound as well as the absorption coefficient. It is given as a ratio between sound pressure and particle velocity on boundary surface in the reflecting acoustic wave:

$$z_n = \left(\frac{p}{u}\right)_{x=0} = \left(\frac{\rho c}{\cos\theta}\right)\frac{p_i + p_r}{p_i - p_r} \tag{17.7}$$

Note that $z_n$ is a complex quantity and involves both amplitude and phase, both of which depend on the sound pressure at the boundary surface in the reflecting acoustic wave.

In the case of oblique incidence, the surface impedance can be expressed by following equation:

$$z_n = Z\gamma z/q \tag{17.8}$$

where $z =$ the acoustic impedance (Pa sec/m$^3$). Here,

$$Z = \frac{z_1 \cosh(ql) + (\gamma z/q)\sinh(ql)}{z_1 \sinh(ql) + (\gamma z/q)\cosh(ql)}$$

$$q = \sqrt{\gamma^2 + k^2 \sin^2\theta}$$

The absorption coefficient, $\alpha(\theta)$, for an oblique incidence with angle $\theta$ may be expressed by

$$\alpha(\theta) = 1 - \left|\frac{z_n \cos\theta - \rho c}{z_n \cos\theta + \rho c}\right|^2 \tag{17.9}$$

## 17.3 Sound-Absorbing Materials

### 17.3.1 Porous Material

Porous acoustical materials are a special category of a more general class of gas–solid mixtures. They range from porous solids, for example, porous rocks, fibrous granular solids, expanded plastics, and form materials, to porous or turbid gases, for example, suspensions and emulsions. Sound is attenuated in a gas-saturated porous solid due to the restriction on the gas movement within it. A convenient microstructure model for such materials is one of a rigid solid matrix through which run cylindrical, capillary pores (tubing) with constant radius, normal to its surface. This model enables the use of Kirchoff's theory of sound propagation in narrow tubes with rigid walls. Accordingly, this mechanism of dissipation may be identified as (1) a viscous loss in the boundary layer at the wall of each capillary tube associated with the relative motion between the viscous gas and the solid wall, or (2) heat conduction between compressions and rarefactions of the gas and the conducting solid walls.

### 17.3.2  Tubular Material

Consider the absorption of low-frequency sound using the tubular absorbing material. By itself, sound absorption is not satisfactory with the tubular absorbing material. The material produces bending vibration due to an acoustic wave through it, and sound absorption occurs by the internal friction of the material. For hard plywood and gypsum boards, there is a natural frequency in the range 100 to 200 Hz, and the absorption coefficient ranges from 0.3 to 0.5. It is possible to increase the absorption coefficient by coating the board surface with fibrous absorbing material.

### 17.3.3  Membrane Material

For membrane material, the sound-absorption mechanism makes use of resonant vibration. Hence, resonant frequency is a governing parameter. The imaginary part (the reactance term) of the acoustic impedance of a membrane gives rise to a resonance. The associated natural frequency is given by

$$f_r = \frac{1}{2}\left\{ \frac{1}{m}\left( \frac{1.4 \times 10^5}{L} \right) + K_m \right\} \tag{17.10}$$

where

$f_r$ = natural frequency (Hz)
$m$ = surface density (kg/m$^2$)
$L$ = thickness of air space (m)
$K_m$ = board rigidity (kg/m$^2$ sec$^2$)

The $K_m$ values of some boards are shown in Figure 17.2. The absorption coefficient is approximately 0.3 to 0.4 in the frequency range of 300 to 1000 Hz, when the thickness of the air space between the membrane and the rigid wall behind it is 50 to 100 mm.

### 17.3.4  Perforated Plate

A perforated board of sound absorbing material (i.e., a board with holes) is placed over a rigid wall at a fixed clearance, as shown in Figure 17.3. The sound-absorption characteristics depend on the board thickness, $t$, the hole diameter of the perforations, $d$, the clearance, $L$, between the perforated board and the rigid wall, and so on. The absorption coefficient becomes a maximum at resonant frequency. In the present case, the resonant frequency is given by

$$f_r = \frac{c}{2\pi}\sqrt{\frac{\varepsilon}{(t + 0.8d)L}} \tag{17.11}$$

where sound speed is $c$, the airspace thickness is $L$ (typically, 300 mm or less), and the ratio of the total area of holes to the total area of the board is $\varepsilon$. The absorption coefficient is approximately 0.3 to 0.4.

### 17.3.5  Acoustic Resonator

Yet another method of achieving sound absorption is using an acoustic resonator of Helmholtz type, which consists of a vessel of any shape containing a volume air, as shown in Figure 17.4. The air volume is in direct communication with the ambient air in the room through an interconnecting tube, which may be long or short and of any cross-sectional shape. An example of a resonator of Helmholtz type may be a 1 gal jar. When a sound wave impinges on the aperture of neck of the jar, the air in the neck will be set in oscillation, periodically expanding and compressing the air in the vessel.

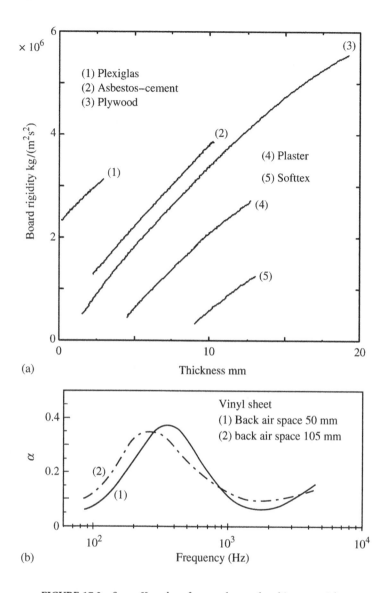

**FIGURE 17.2** Some $K_m$ values for membrane absorbing materials.

The resulting amplified motion of the air particles in the neck of the jar, due to phase cancellation between the air plug in the neck and the air volume in the vessel, causes energy dissipation due to friction in and around the neck. This type of absorber can be designed to produce maximum absorption over a very narrow frequency band or even a wide frequency band. The resonant frequency of a Helmholtz resonator may be expressed as

$$f_r = \frac{c}{2m}\sqrt{\frac{\varepsilon}{(t + 0.8d)L}} \tag{17.12}$$

where

$c$ = speed of sound (m/s)
$S_n$ = cross-sectional area of neck of jar (m$^2$)
$d_n$ = diameter of neck of jar (m)
$V$ = volume of vessel (m$^3$)

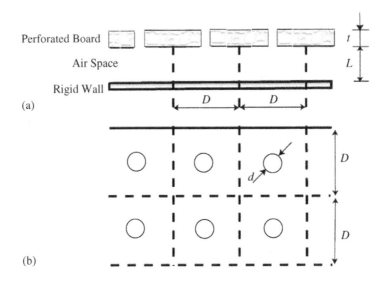

FIGURE 17.3 Sound-absorption characteristics of a perforated plate structure: (a) cross sectional view; (b) plan view.

## 17.4 Acoustic Characteristic Computation of Compound Wall

### 17.4.1 Absorption Coefficient of Combined Plate with Porous Blanket

A common form of problem in noise control is the need to reduce the sound radiated from a duct or some other object. A way to achieve this is by lining the duct with several centimeters of porous acoustic material, and covering it with a solid plate of some type, as indicated in Figure 17.5.

Consider the case of normal incidence with the sound-absorbing structure of Figure 17.5. Assume that the boundary conditions for the sound pressure and the volume flow-rate are identical. For plane wave incidence on the hard wall, the magnitude of reflection coefficient is $-1$ [1]. The following equation is obtained:

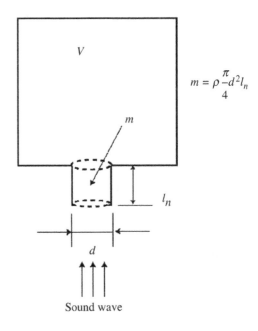

$$m = \rho \frac{\pi}{4} d^2 l_n$$

$$
\begin{bmatrix}
1 & -1 & -1 & 0 \\
-1 & -m_1 & m_1 & 0 \\
0 & e^{-\gamma l_1} & e^{\gamma l_1} & -(1+e^{-2jkl_2}) \\
0 & m_2 e^{-\gamma l_1} & m_2 e^{\gamma l_1} & -(1-e^{-2jkl_2})
\end{bmatrix}
\begin{bmatrix}
B_1 \\
A_1 \\
B_1 \\
B_2
\end{bmatrix}
$$

$$
=
\begin{bmatrix}
-1 \\
-1 \\
0 \\
0
\end{bmatrix}
\qquad (17.13)
$$

FIGURE 17.4 Geometry of a Helmholtz resonator. Volume, $V$, is connected to an infinitely open area by a neck tube of diameter $d$ and length $l_n$.

where

$j = \sqrt{-1}$

$m_1 = z_0/z_1,\ m_2 = z_1/z_2$

$z_0, z_1, z_2$: acoustic impedance of each medium
    (Pa s/m³)

$\gamma$ = complex propagation constant (1/m)

The absorption coefficient for normal incidence is
given by the following equation:

$$\alpha_0 = 1 - \left|\frac{B_0}{A_0}\right|^2 = 1 - |B_0|^2 \qquad (17.14)$$

The absorption coefficient for random incidence
may be approximated by

$$\alpha = \frac{1}{n}\sum_{i=1}^{n}\alpha(\theta)_i \qquad (17.15)$$

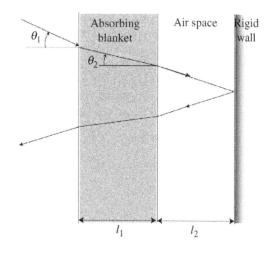

FIGURE 17.5   Structure for sound absorption using a
blanket and an air space showing angles $\theta_1$ in the air and
$\theta_2$ in the blanket.

where $\theta$ = the incident angle of sound, $0 < \theta < \pi/2$.

It is known that the propagation speed of the sound in fibrous materials changes with air, and the
following equation holds on the boundary surface:

$$\sin\theta/\sin\theta' = c/c_m \qquad (17.16)$$

Here, $c_m$ is the sound speed in fibrous materials, which is calculated from the imaginary part of Equation
17.18, given later. The angle of reflection, $\theta'$, in the boundary surface of the back air space is obtained in a
similar way. Hence, the following equation is substituted in Equation 17.13 instead of the thickness of the
absorber, $l_1$, and the thickness of the air space, $l_2$, to obtain the absorption coefficient in oblique
incidence:

$$l_1' = l_1/\cos\theta',\quad l_2' = l_2/\cos\theta'' \qquad (17.17)$$

The complex propagation constant, $\gamma$, is an important physical quantity in absorbing material of
propagated sound, which is given per unit length of acoustic attenuations, and phase changes. Between
the aeroelasticity rate, $K_a$, of absorbing material and the bulk modulus, $Q$, of absorbing material, $\gamma$ is
given by the following equation, for $K_a > 20\,Q$ [2,3]:

$$\gamma = j\omega\sqrt{Y/K}\sqrt{\langle\rho_1\rangle - j\langle R_1\rangle/\omega} \qquad (17.18)$$

$$\langle R_1\rangle = \frac{R_1[1 - \rho_0(1 - Y)/\rho_m]}{\left[1 + \dfrac{\rho_0(\kappa - 1)}{\rho_m}\right]^2\left[1 + \dfrac{R_1^2}{\rho_m^2\omega^2[1 + \rho_0(\kappa - 1)/\rho_m]^2}\right]}$$

$$\langle\rho_1\rangle = \rho_0\kappa - \frac{\dfrac{R_1^2(Y/\kappa + \rho_m/\rho_0\kappa)}{\rho_m^2\omega^2[1 + \rho_0(\kappa - 1)/\rho_m]^2} + \dfrac{1 + \rho_0 Y(\kappa - 1)/\rho_m\kappa}{1 + \rho_0(\kappa - 1)/\rho_m}}{1 + \dfrac{R_1^2}{\rho_m^2\omega^2[1 + \rho_0(\kappa - 1)/\rho_m]^2}}$$

where

$\rho_m$ = density of acoustical material (kg/m³)

$\rho_0$ = density of air (kg/m³)

$c_0$ = speed of sound in air (m/s)

$K$ = volume coefficient of elasticity of air (N/m$^2$)

$R_1$ = alternating flow resistance for unit thickness of material due to the difference between the velocity of the skeleton and the velocity of air in the interstices (Pa s/m$^2$). $R_1$ values are given in Table 17.1

$Y$ = porosity = the ratio of the volume of the voids in the material to the total volume; porosity equals the total volume minus the fiber volume, all divided by total volume

$\kappa = 5.5 - 4.5Y$, the structure factor of the interstices in the skeleton

$\omega = 2\pi f$, the angular frequency (radians/s)

The acoustic impedance, $z_1$, of absorbing material is given by

$$z_1 = R + jX = -\frac{jK\gamma}{\omega Y} \tag{17.19}$$

in which

$$R = \rho_0 c_0 \left\{ 1 + 0.0571(\rho_0 f/R_f)^{-0.754} \right\}$$

$$X = -\rho_0 c_0 \left\{ 0.0870(\rho_0 f/R_f)^{-0.732} \right\}$$

## 17.4.2 Transmission Loss through a Single Porous Board

Assume that a sound wave impinges on the left side of a porous board at normal incidence and emerges with a reduced amplitude from the right side. The associated transmission loss of the porous board is obtained from

$$TL_0 = 10 \log_{10}(X + Y)$$

$$X = \left\{ 1 + \frac{\omega^2 m^2 PR_f}{2\rho_0 c_0 (\omega^2 m^2 P^2 + R_f^2)} \right\}^2$$

$$Y = \left\{ \frac{\omega m R_f^2}{2\rho_0 c_0 (\omega^2 m^2 P^2 + R_f^2)} \right\}^2 \tag{17.20}$$

where

$m$ = surface density of the blanket (kg/m$^2$)

$P$ = porosity of the blanket (porosity = the total volume minus the fiber volume, all divided by the total volume)

$R_f$ = specific flow resistance of material (Pa s/m)

**TABLE 17.1**   Flow Resistance Values of Glass-Wool Board (Quality Regulation Range by JIS)

| Board Type | K value | Gross Specific Gravity (kg/m$^3$) | Specific Flow Resistance ($\times 10^{-3}$ N s/m$^4$) | Standard of JIS for Glass Wool |
|---|---|---|---|---|
| #1 Glass-wool board | 8 | 8 ± 2 | 1.5 ~ 7.0 | JIS A 9505-A |
| | 12 | 12 ± 2 | 2.5 ~ 12.0 | |
| | 16 | 16 ± 2 | 4.7 ~ 17.0 | |
| | 20 | 20 ± 3 | 5.0 ~ 22.0 | |
| | 24 | 24 ± 3 | 6.5 ~ 27.0 | |
| #2 Glass-wool board | 12 | 12 ± 2 | 1.5 ~ 7.0 | JIS A 9505-B |
| | 16 | 16 ± 2 | 2.5 ~ 10.0 | |
| | 20 | 20 ± 3 | 3.0 ~ 13.0 | |
| | 24 | 24 ± 3 | 4.0 ~ 16.0 | |
| | 32 | 32 ± 4 | 6.0 ~ 22.0 | |
| | 48 | 48 ± 5 | 11.0 ~ 38.0 | |
| | 64 | 64 ± 6 | 18.0 ~ 60.0 | |
| | 96 | 96 ± 10 | 27.0 ~ 95.0 | |
| #3 Glass-wool board | 96 | 96 ± 10 | 15.0 ~ 40.0 | JIS A 9505-C |

$\rho_0$ = density of air (kg/m³)
$c_0$ = sound speed in air (m/s)

### 17.4.3 Transmission Loss through a Sandwich Board

Consider a wide wall formed by two panels (sheets) of infinite area separated with a homogeneous filling of fibrous acoustical material, as shown in Figure 17.6. Suppose that a plane wave impinges at an angle $\theta$. As the pressure of both sides of the wall is equal with regard to the amplitude of the progressing wave and the reflected wave in each boundary surface, the following result may be established [4]:

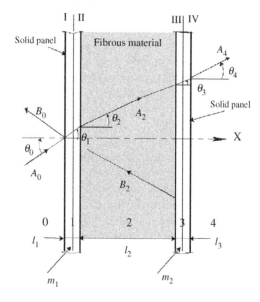

**FIGURE 17.6** Cross-sectional view of a sandwich panel.

$$\left.\begin{aligned}
A_0 + B_0 &= A_1 + B_1 \\
(A_0 - B_0)/z_0 &= (A_1 - B_1)/z_1 \\
A_1 e^{-jkl_1'} + B_1 e^{jkl_1'} &= A_2 + B_2 \\
(A_1 e^{-jkl_1'} - B_1 e^{jkl_1'})/z_1 &= (A_2 - B_2)/z_2 \\
A_2 e^{-\gamma l_2'} + B_2 e^{\gamma l_2'} &= A_3 + B_3 \\
(A_2 e^{-\gamma l_2'} - B_2 e^{\gamma l_2'})/z_2 &= (A_3 - B_3)/z_3 \\
A_3 e^{-\gamma l_3'} + B_3 e^{\gamma l_3'} &= A_4 + B_4 \\
(A_3 e^{-\gamma l_2'} - B_3 e^{\gamma l_2'})/z_3 &= (A_4 - B_4)/z_0
\end{aligned}\right\} \quad (17.21)$$

where $A$ and $B$ are the amplitude of sound pressures.

From Equation 17.17, $l_1' = l_1/\cos\theta_1$, $l_2' = l_2/\cos\theta_2'$, and $l_1'$ and $l_2'$ may be calculated.

The speed of sound in the walls is given by the following equation in terms of the modulus of longitudinal elasticity, $E_i$:

$$c_i = \sqrt{E_i/\rho_i} \tag{17.22}$$

The real part of acoustic impedance, $z_i$ ($i = 1, 3$), is given by $R_i = r_i/\cos\theta$, and of the imaginary part is given at $X_i = m_i\omega$. The internal resistances, $r_i$, are functions of such factors as the material, frequency, temperature, and density. Some typical values are given in Table 17.2.

If the space of the transmission side is infinite, $B_4$ in Equation 17.21 becomes equal to zero. Then, the transmission loss is given is given by

$$TL(\theta) = 10 \log_{10}\left|\frac{A_4}{A_0}\right|^2 \tag{17.23}$$

**TABLE 17.2** Internal Resistance Values of Several Useful Materials

| Material | Thickness (mm) | Internal Resistance (Pa sec/m³) |
|---|---|---|
| Aluminum | 0.4 | 3.0 |
| Plywood | 3.0 | 7.5 |
| Plaster board | 7.0 | 15.0 |

## 17.5    Attenuation of Lined Ducts

### 17.5.1    Computation of Attenuation in a Lined Duct

A lined duct is an air passage with one or more of the interior surfaces covered with an acoustical material such as a glass or mineral fiber blanket. The parallel baffles are merely a series of side-by-side ducts that generally have a rectangular or round cross section. If the walls are covered with absorptive material, attenuation will occur because of the viscous motion of the air in and out of the porous of blanket.

Figure 17.7 shows an isometric illustration of a lined duct. The attenuation of sound for a lined duct is dependent primarily on the duct length, $l_e$, the thickness of the lining, $b$, the density of the lining, $\rho$, the width of the air passage, $l$, and the wavelength of sound, $\lambda$. At low frequencies ($l/\lambda < 0.1$), the attenuation of sound in a lined duct may be calculated from the following empirical formula:

$$\text{ATT} = K_1 P/S \qquad (17.24)$$

where

$K_1 = $ the coefficient, which is determined from the random incidence absorption coefficient of lined material, given in the chart of Figure 17.8
$P = $ acoustically lined perimeter of duct (m)
$S = $ cross-sectional open area of duct (m$^2$)

If the absorbing material is lined in the rectangular cross section as shown in Figure 17.9 to Figure 17.11, the attenuation can be estimated using the formulas given in Table 17.3 [5].

### 17.5.2    Attenuation in a Lined Bend

A lined bend duct is shown in Figure 17.12. The insertion loss, IL, of a lined bend results from two mechanisms: the reflection of sound back toward the source side, and the scattering of sound energy into the high-frequency region is rapidly attenuated by the lining beyond the bend. Higher-frequency modes will be attenuated by even an unlined duct for frequencies below the ratio of the air passage between the linings to the wavelength of sound equal to 0.5. At frequencies well above this ratio, the insertion loss of a lined bend is expected to be comparable to the reverberant-field

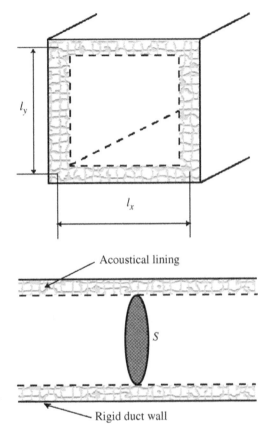

**FIGURE 17.7**   Illustration of a lined duct.

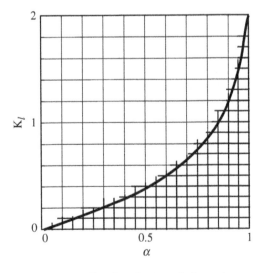

**FIGURE 17.8**   $K_1$ value for sound-absorption coefficient by reverberation room method.

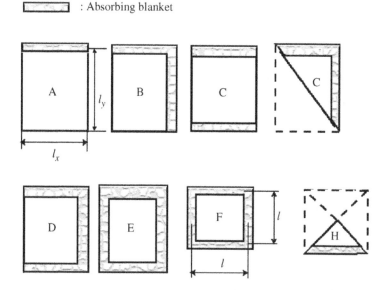

: Absorbing blanket

**FIGURE 17.9**   Duct-liner configurations corresponding to Table 42.3.

end correction derived for the duct. The insertion loss of a lined bend may be obtained as following equation [6]:

$$\mathrm{IL} = \frac{K_1 P}{S} + (l_1 + l_2) + \varPhi \tag{17.25}$$

where $\varPhi$ is obtained from Figure 17.13.

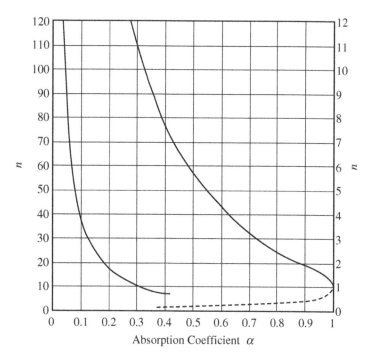

**FIGURE 17.10**   Relationship between absorption coefficient and stationary wave factor, $n$.

The total insertion loss for a lined bend is given in Figure 17.13 along with the attenuation of the lining beyond the bend.

### 17.5.3 Attenuation in Splitter Lined Duct

The use of parallel or zigzag baffle-type separators (splitters) to increase the perimeter–area ratio results in more compact attenuators. In rock-wool blankets, the attenuation of a parallel type splitter duct may be obtained directly from Figure 17.14. The peak value of the attenuation is related to wavelength of sound and the splitter interval. With the zigzag arrangement of acoustic blankets, the attenuation of high frequencies is improved over that of the parallel splitter [7].

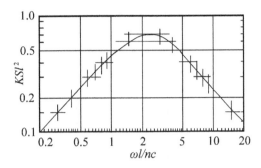

**FIGURE 17.11** Damping function $KSl^2$ as a function of dimensionless frequency, $\omega l/nc$.

## 17.6 Attenuation of Dissipative Mufflers

### 17.6.1 Transmission Loss of Lined Expansion Chamber

The geometry and nomenclature for a dissipative muffler are given in Figure 17.15. For $f < 1.2c/D$, the assumption of plane wave is acceptable where $D =$ the diameter of the muffler.

The transmission loss for the light lining in the chamber may be obtained using [8,9]:

$$TL = 10\log_{10}\left[\left\{\cosh(\delta_e l_e/2) + \frac{m+1}{2m}\sinh(\delta_e l_e/2)\right\}^2\cos^2 kl_e + \left\{\sinh(\delta_e l_e/2) + \frac{m+1}{2m}\cosh(\delta_e l_e/2)\right\}^2\sin^2 kl_e\right]$$

$$(17.26)$$

**TABLE 17.3** Formulas for Attenuation of Several Lined Ducts

| See Figure 17.7 | Low-Frequencies Range: $\dfrac{\omega l}{nc} < 1$ | Middle-Frequencies Range $(K_y S_y l_y$; see Figure 17.7) | High-Frequencies Range: $\dfrac{\omega l}{nc} > 5$ |
|---|---|---|---|
| (A) | $\beta = \dfrac{4.34}{nl_y}$ | $\beta = \dfrac{8.7c}{l_y^2\omega}(K_y S_y l_y^2)$ | $\beta = 21.4\dfrac{c^2 n}{\omega^2 l_y^3}$ |
| (B) | $\beta = 4.34\left(\dfrac{1}{n_y l_y} + \dfrac{1}{n_x l_x}\right)$ | $\beta = \dfrac{8.7c}{\omega}\left(\dfrac{K_y S_y l_y^2}{l_y^2} + \dfrac{K_x S_x l_x^2}{l_x^2}\right)$ | $\beta = 21.4\dfrac{c^2}{\omega^2}\left(\dfrac{n_y}{l_y^3} + \dfrac{n_x}{l_x^3}\right)$ |
| (C) | $\beta = \dfrac{8.7}{nl_y}$ | $\beta = \dfrac{34.7c}{l_y^2\omega}\left(\dfrac{K_y S_y l_y^2}{4}\right)$ | $\beta = 171\dfrac{c^2}{\omega^2}\dfrac{n_y}{l_y^3}$ |
| (D) | $\beta = 4.34\left(\dfrac{2}{n_y l_y} + \dfrac{1}{n_x l_x}\right)$ | $\beta = \dfrac{8.7c}{\omega}\left(\dfrac{K_y S_y l_y^2}{l_y^2} + \dfrac{K_x S_x l_x^2}{l_x^2}\right)$ | $\beta = 21.4\dfrac{c^2}{\omega^2}\left(\dfrac{8n_y}{l_y^3} + \dfrac{n_x}{l_x^3}\right)$ |
| (E) | $\beta = 8.7\left(\dfrac{1}{n_y l_y} + \dfrac{1}{n_x l_x}\right)$ | $\beta = \dfrac{34.7c}{\omega}\left(\dfrac{K_y S_y l_y^2}{4l_y^2} + \dfrac{K_x S_x l_x^2}{4l_x^2}\right)$ | $\beta = \dfrac{171c^2}{\omega^2}\left(\dfrac{n_y}{l_y^3} + \dfrac{n_x}{l_x^3}\right)$ |
| (F) | $\beta = \dfrac{17.4}{nl}$ | $\beta = \dfrac{69.5c}{4l^2\omega}KSl^2$ | $\beta = \dfrac{341c^2 n}{\omega^2 l^3}$ |

$\beta$ is attenuation (dB/m), $n$ is absorbing factor plotted in Figure 17.7, $K_y S_y l_y$ is damping function, plotted in Figure 17.8, $c$ is sound speed, $l$ is the width of the duct, $\omega = 2\pi f$: angular frequency, $x, y$: coordinates, see Figure 17.7.

*Source*: Brüel, P.V. 1951. *Sound Insulation and Room Acoustics*, Chapman & Hall, London, p.159. With permission.

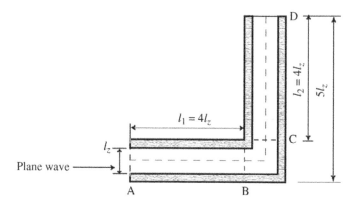

**FIGURE 17.12**  Sketch of a typical lined bend with plane wave incidence. (*Source*: Beranek, L.L. *Noise Reduction*, McGraw-Hill, 1960. With permission.)

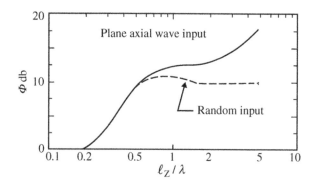

**FIGURE 17.13**  Insertion loss for lined bend. (The lining must extend two to four duct widths beyond the bend for this data to be valid.) (*Source*: Beranek, L.L. *Noise Reduction*, McGraw-Hill, 1960. With permission.)

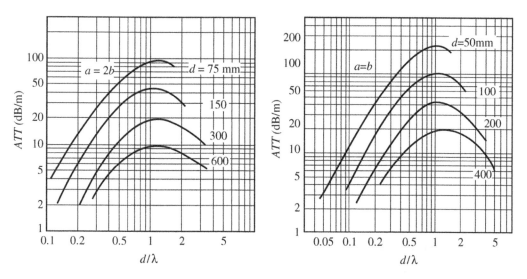

**FIGURE 17.14**  Sound attenuation for a splitter duct. Each baffle is constructed with two sheets of perforated metal filled with mineral wool, with about 100 to 140 kg/m$^3$ gross density; $a$ = the width of the open space, $b$ = the width of the baffle, $d$ = the center-to-center distance of baffles, $\lambda$ = the wavelength of the sound.

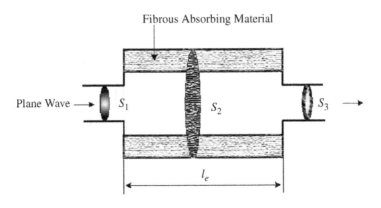

**FIGURE 17.15**   A dissipative muffler.

**TABLE 17.4**   Filled up Factor of Glass Wool, $\alpha_g$

| $V_g/V$ | 0.05 | 0.10 | 0.15 | 0.20 | 0.30 | 0.40 | 0.50 | 0.60 | 0.70 | 0.80 | 0.90 | 1.00 |
|---|---|---|---|---|---|---|---|---|---|---|---|---|
| $\alpha_g$ | 0.106 | 0.124 | 0.288 | 0.365 | 0.529 | 0.677 | 0.794 | 0.885 | 0.935 | 0.960 | 0.987 | 1.0 |

$V_g$ = Filled up volume (factors of 100 kg/m³), $V$ = Volume of chamber

in which $\delta_e$ = the attenuation per unit length for the lined duct, which is given by the following equation:

$$20 \log_{10}(\delta_e l_e) = \frac{K_1 P l_e}{S} \tag{17.27}$$

The $K_1$ values are obtained from the absorption coefficient, as shown earlier (see Figure 17.8). In particular, $\delta_e$ is given by

$$\delta_e = \frac{1}{l_e} 10^{0.05 K_1 P l_e / S} \tag{17.28}$$

where $m$ = the ratio of the area of expanded or lined sections to the area of inlet or outlet sections of muffler; $k = 2\pi f/c$, and $l_e$ = the length of the muffler.

The transmission loss for the case of a thick lining of glass wool in the chamber is obtained using the empirical formula [10]

$$\text{TL} = 10 \log_{10}\left[ 1 + \left\{ \frac{1}{2} \alpha_g mkl_e \right\}^2 \right] \tag{17.29}$$

where

$\alpha_g$ = the coefficient, which is obtained from Table 17.4, using the filling volume and the density of glass wool

$m$ = the ratio of the area of expanded or lined sections to the area of inlet or outlet sections of muffler

$k = 2\pi f/c$

$l_e$ = the length of muffler

## 17.6.2   Transmission Loss of a Plenum Chamber

The geometry and nomenclature for a plenum chamber are given in Figure 17.16. A plenum chamber is similar in many ways to a lined expansion chamber. The main difference is that the inlet and outlet of a plenum chamber are not located in line. Generally, there is an offset to direct transmission of sound. Sound is reflected at the square-cornered bend as the cross section dimension of the duct is sufficiently large. Particularly at high frequencies, almost all of the sound energy may reflect many times off the lined sides when propagating from the inlet to the outlet. The transmission loss of a single plenum chamber can be obtained approximately from [11]:

$$TL = 10 \log_{10}\left\{S_w\left(\frac{\cos\theta}{2\pi d^2} + \frac{1}{R}\right)\right\} \quad (17.30)$$

where

$S_w = lW$ = area of the inlet and outlet
$d = \{(L - l)^2 + H^2\}^{1/2}$ = the slant distance
     from inlet to outlet
$\cos\theta = H/d$
$R = a/(1 - \alpha_m)$
$a$ = the total lined area in chamber times absorption coefficient
$\alpha_m$ = the statistical absorption coefficient of the lining

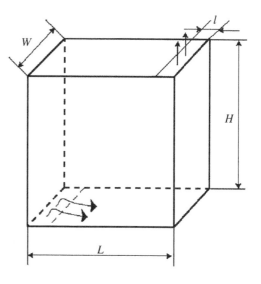

**FIGURE 17.16**  A single-plenum chamber showing the nomenclature used in Equation 17.26.

# 17.7  General Considerations

In order to carry out the design of noise-control measures for a particular problem, we must consider not only the fundamental acoustical properties of the material as discussed before, but also such practical aspects of the problem as (1) gas flow velocities, (2) temperature of gas, (3) moisture exposure, and (4) head losses for gas-flow. The client depends heavily on the expertise of the designer to realize adequate protection of the noise-control equipment under operating conditions.

## 17.7.1  Surface Treatment with Lining of Acoustic Material

Fibrous material in the market has some form of resin binder. Comparatively long fiber flocculent and comparatively short fiber are available. The packaging density of flocculent is about 60 to 100 kg/m³. It is necessary to cover with perforated thin metal or wire netting so that an arbitrary shape may be maintained in the absorbing material. The perforated metal does not take into account the numerical aperture, hole shape, hole diameter, and metal thickness. From the acoustic viewpoint, a suitable numerical aperture is given in Table 17.5.

## 17.7.2  Gas Flow Velocity

Noise control problems often involve the use of an acoustical material in high-velocity gas-flow such as those found in the exhaust of engines or ventilating systems. Deterioration of the acoustical

**TABLE 17.5**  Perforated Metal for Treatment of Absorbing Material (Gas Flow Velocity is 25 m/sec or Less)

Perforation rate: 30 to 50%
Hole diameter: 5 to 10 mm
Hole shape: round, plus, slit and interminglement
Metal: iron, stainless steel (used in case of the corrosive gas)

material due to high-velocity gas flowing past it can be a serious problem. In addition, turbulence in the gas flow subjects the materials to vibration and can cause further deterioration. One solution to this problem is to install the acoustical material behind some form of protective facing, which will vary in complexity depending on the gas velocity.

A limited amount of information on this subject is available through field experience, as shown in Figure 17.17 [12]. However, the parameters of the treatment structure are not well established, for example, those concerning perforated metal, wire net, absorbing material, and gas flow. Multiple layers are used under conditions of flow velocity exceeding 25 m/sec, and the associated performance analysis can become rather complex.

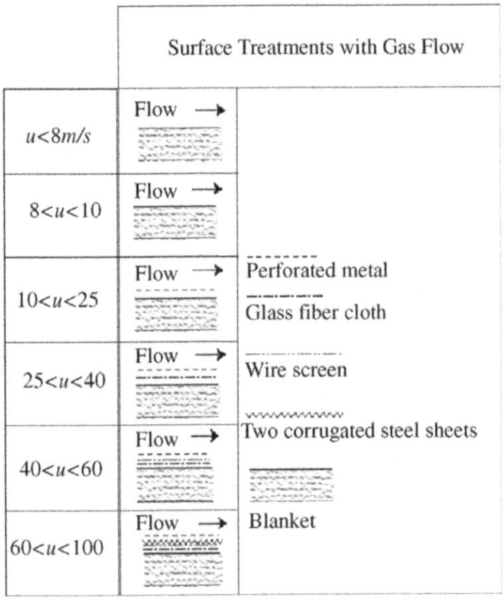

FIGURE 17.17 Protective surface for absorbing material subjected to high-velocity gas flow.

### 17.7.3 Gas Temperature

In many noise-control problems, temperature is a very important factor. Sometimes high-temperature ducts that are radiating noise, for example, in diesel engines, and induced draft fans, must be wrapped. With a proper choice, it is possible to combine thermal and acoustical insulations using one single material. Under extremely high temperatures, the tensile strength of materials tends to decrease, and the material may be subjected to thermal shock.

Examples of absorbing materials that are currently available for use where temperature is an important consideration are given in Table 17.6.

### 17.7.4 Dust and Water Exposure

The holes of perforated metal can be blocked if a dust treatment is not carried out, and the sound absorption performance will deteriorate with adhesion to the surface of the absorbing material. Methods of dust accumulation and removal may be designed into cavity type mufflers used on the sound absorption equipment.

A fan of a cooling tower, for example, experiences a considerable amount of moisture. Precautions must be taken so that water droplets are not deposited on the sound absorbing material. The underside of the equipment should be treated with rust prevention material. Figure 17.18 shows the degradation

TABLE 17.6   Fibrous Materials of Use in Hot Gas Flows

| Materials | Maximum Allowable Temperature (°C) |
| --- | --- |
| Glass fibers with binder | 320 ~ 360 |
| Glass wool | 960 ~ 1060 |
| Mineral wool felts | 1160 |
| Mineral wool | 1660 |
| Asbestos fibers | 760 |
| Alumina-silica | 1900 |

of the acoustic characteristic of absorbing material due to moisture, using the normal incidence absorption coefficient [13].

## 17.8 Practical Example of Dissipative Muffler

An example is given on the design of a dissipative muffler for noise reduction in an axial-flow fan for a ventilation system.

1. Specification of the axial-flow fan

   - Volume flow rate: $Q = 125$ m$^3$/min
   - Wind pressure: $p = 80$ mm Aq
   - Rotor blade number: $Z = 10$
   - Stationary blade number: $Z_s = 5$
   - Rotational speed: $N = 2580$ rpm
   - Shaft horsepower: $P = 3.75$ kW

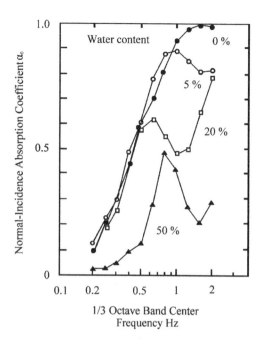

**FIGURE 17.18** Degradation of the absorption coefficient by water content.

2. The desired values of attenuation and head loss with the muffler installation follow. The noise of the fan propagates both intake and discharge sides. A performance level (noise reduction) of about 37.5 dB is required, when specific sound level $K_s$ is obtained on the basis of the axial-flow fan specification given by

$$K_s = L_A - 10 \log_{10}(p^2 Q) \tag{17.31}$$

3. Figure 17.19 gives the noise spectrum for the axial-flow fan. The blade passing frequency, BPF, is a fundamental component of the velocity fluctuation as the flow passes the blades. It is seen in the noise spectrum in Figure 17.19 at 430 Hz ($Q \times Z = (2580/60) \times 10$). By adding the background noise spectrum to this spectrum, it is seen that a muffler that provides an attenuation over 20 dB near 430 Hz, and about 15 dB in the frequency range of 800 to 1000 Hz is necessary. The head loss value of the muffler is to be maintained within 4 mm Aq.

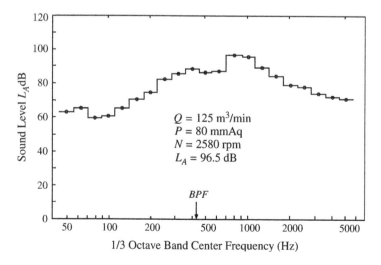

$Q = 125$ m$^3$/min
$P = 80$ mmAq
$N = 2580$ rpm
$L_A = 96.5$ dB

**FIGURE 17.19** Noise spectrum of the axial flow fan of a factory ventilation system.

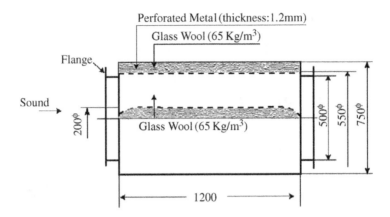

**FIGURE 17.20**   Half cross-sectional view of dissipative muffler for the axial flow fan.

4. The structure of the dissipative muffler is shown in Figure 17.20. The maximum value of outer diameter of the muffler is 750 mm, and the length chosen to optimize the performance.
5. The packing density of glass wool is chosen as 65 kg/m³. The surface treatment of glass wool uses perforated metal with 1 mm thickness, 36% open area with 6 mm hole diameter. A sound absorption body of 200 mm diameter is supported in the center part, and it is welded to the outside cylinder by three props in the flow direction, and two in the circumferential direction.
6. The attenuation characteristics of the dissipative muffler may be calculated using Equation 17.27

$$\text{TL} = 10 \log_{10}\left[ 1 + \left\{\frac{1}{2}\alpha_g mkl_e\right\}^2 \right]$$

The proportion of the volume of glass wool filled into the muffler is approximately $(0.75^2 - 0.55^2 + 0.2^2)/0.75^2 = 0.53$. For a packing density of 100 kg/m³, we have $0.53 \times 65/100 = 0.347$. The value of $\alpha_g \approx 0.6$ is obtained from Table 17.2.

The required expansion ratio, $m$, length, $l_e$, and wave number, $k$, are given by

$$m = (750/500)^2 = 2.25, \quad l_e = 1.2 \text{ m}, \quad k = 2\pi f/c$$

The speed of sound $c$ depends on the environmental temperature. For a temperature of 25°C, we get 346.5 m/sec ($= 331.5 + 0.6 \times 25$). The TL values at 100 to 1000 Hz are calculated. We have

$$f = 100 \text{ Hz}; \quad \text{TL} = 5.0 \text{ dB}$$

$$f = 430 \text{ Hz}; \quad \text{TL} = 16.1 \text{ dB}$$

$$f = 1000 \text{ Hz}; \quad \text{TL} = 23.4 \text{ dB}$$

The flow velocity satisfies desired value of head loss $p_{\text{loss}}$ (in mm Aq), and is calculated by following empirical equation [12]:

$$p_{\text{loss}} = \left\{0.142 m_f^{-0.1}\left(\frac{l_e}{d_1}\right)^{3/4}\left(\frac{d_m}{d_1}\right)^{-1/3}\right\}\frac{u^2}{g} \tag{17.32}$$

Use the numerical values as follows:

- $m_f = (550/500)^2 = 1.21$, ratio of cross-sectional area between air passage and muffler.
- $d_1 = 500$ mm, diameter of inlet.
- $l_e = 1.2$ m, length of muffler.
- $d_m = 200$ mm, diameter of absorption body.
- $u = 10$ m/sec or less, flow velocity at inlet.
- $g = 9.8$ m/sec², acceleration of gravity force.

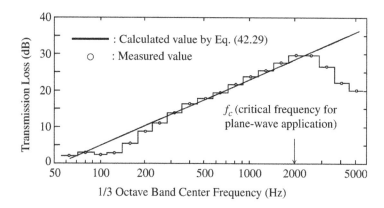

**FIGURE 17.21**  Noise-reduction characteristics of the designed dissipative muffler.

The corresponding head loss is 3.25 mm Aq, which corresponds to JIS B 833, and nearly agrees with the predicted value. Specifically, the condition of 4 mm Aq or less of the designed value is satisfied. The connection of axial-flow fan and the muffler uses vibration isolation, using the thick synthetic rubber.

7. The result of the attenuation realized from the spectrum after the muffler installation is shown in Figure 17.21. It is proven that the attenuation characteristics almost parallel the designed value. The frequency range where the approximation is valid is given by

$$f < c/d = 346.5/0.175 = 1980 \text{ Hz}$$

For frequencies below 250 Hz, the estimated result of the attenuation becomes slightly overestimated.

## References

1.  Obata, T., Hirata, M., Nishiwaki, N., Ohnaka, I., and Kato, K., Noise reduction characteristics of dissipative mufflers, 1st report, acoustical characteristics of fibrous materials, *Trans. Japan Soc. Mech. Eng.*, 42, 363, 3500, 1976.
2.  Zwikker, C. and Kosten, C.W. 1949. *Sound Absorbing Materials*, Elsevier, New York.
3.  Beranek, L.L., Acoustical properties of homogeneous isotropic rigid tiles and flexible blankets, *J. Acoust. Soc. Am.*, 19, 4, 556, 1947.
4.  Obata, T. and Hirata, M., Estimation of acoustical transmission loss for combined walls, *Proc. Japan Soc. Mech. Eng. Annu. Meet.*, 780, 1, 42, 1978.
5.  Brüel, P.V. 1951. *Sound Insulation and Room Acoustics*, Chapman & Hall, London, p. 159.
6.  Beranek, L.L. 1971. *Noise and Vibration Control*, McGraw-Hill, New York, chap. 17, p. 390.
7.  King, A.J., Attenuation of lined ducts, *J. Acoust. Soc. Am.*, 30, 6, 505, 1958.
8.  Davis, D.D. Jr. and Stokes, G.M., 1954. *Natl. Advisory Comm. Aeronaut. Ann. Rept.*, 1192.
9.  Davis, D.D. Jr. 1957. *Acoustical Filters and Mufflers. Handbook of Noise Control*, C.M. Harris, Ed., McGraw-Hill, New York, chap. 21.
10. Hagi, S., Studies on Silencer for Ventilating System, A Doctoral Thesis of University of Tokyo, 1961.
11. Wells, R.J., Acoustical plenum chambers, *Noise Control*, 4, 4, 9, 1958.
12. Obata, T. and Hirata, M., Estimation of acoustic power of flow-generated noise within silencer and head losses, *J. Acoust. Soc. Japan*, 34, 9, 532, 1978.
13. Koyasu, M., Acoustical properties of fibrous materials, personal letter, RC-SC35, *Japan Soc. Mech. Eng. Div. Meet.*, 15, 1975.

# 18

# Design of Reactive Mufflers

Teruo Obata
*Teikyo University*

**Summary**

*This chapter concerns the design of noise suppression devices such as mufflers. In particular, reactive mufflers that are inserted into long ducts are considered in detail. Analytical and empirical equations and information that are useful in the modeling and analysis of mufflers are presented, with an indication of their application ranges and limitations. A design procedure, complete with the necessary computations, is given. Methodologies and parameters of the performance analysis of acoustic systems with mufflers are indicated. An illustrative example of a muffler for a double-acting reciprocating compressor is presented.*

## 18.1 Introduction

In noise-reduction applications, the need for a reactive muffler usually arises when transporting gas through a duct. For sound transmission through the duct to be minimized, an acoustic suppression device must be incorporated into the duct system. For example, in internal-combustion engines, it is required to reduce the intake and exhaust noise to acceptable levels. This may be accomplished by inserting a muffler in the intake and exhaust ducting to attenuate the pressure pulsations before they reach the environment.

A successful muffler design must satisfy at least the following three criteria: (1) muffler performance as a function of frequency (the maximum permissible noise generated by the gas flow through the muffler may have to be specified as well); (2) the maximum permissible average pressure drop through the muffler at a given temperature and mass flow; (3) the maximum allowable volume and restrictions on space utilization.

The customer may ask for a muffler with unrealistically high noise attenuation, virtually no backpressure, and very small size. In addition, it is important to the customer that the muffler is inexpensive and durable, and presents no maintenance problems. Needless to say, in practice, these criteria for muffler design are unrealistic, and have to be modified to practical levels. In this chapter, we will present some of the analytical and empirical tools that are helpful in muffler design.

## 18.2 Fundamental Equations

### 18.2.1 Analytical Model

The physical behavior of a reactive muffler may be adequately modeled by linear differential equations. The law of conservation of mass must hold, while three simultaneous equations, Newton's, Boyle–Charles, and that of conservation, must be satisfied. When these equations are combined, we obtain the wave equation for the plane, one-dimensional sound-pressure wave:

$$\frac{\partial^2 \xi}{\partial t^2} = c^2 \frac{\partial^2 \xi}{\partial x^2} \tag{18.1}$$

$$p = \rho c^2 \frac{\partial \xi}{\partial x} \tag{18.2}$$

where

$\xi =$ displacement of particle motion (m)
$c =$ velocity of sound (m/s)
$p =$ sound pressure (Pa)
$\rho =$ density of air (kg/m$^3$)
$t =$ time (s)
$x =$ coordinate system along which wave travels (m)

The stationary solutions for angular frequency $\omega$ of Equation 18.1 and Equation 18.2 are given by

$$\xi = (A\, e^{-jkx} - B\, e^{jkx}) e^{j\omega t} \tag{18.3}$$

$$p = -\rho c^2 k (A\, e^{-jkx} + B\, e^{jkx}) e^{j\omega t} \tag{18.4}$$

where $A$, $B =$ amplitudes of sound pressure or particle motion for traveling and reflecting waves, $k = 2\pi f/c$, wave number, and $j = \sqrt{-1}$.

### 18.2.2 Boundary Conditions

The boundary conditions are given below.

(1) *Sound source.* The sound source is assumed to be independent of the existence of the muffler, and the volume rate of the particles is assumed constant, as given by

$$S\dot{\xi} = \text{const} \tag{18.5}$$

in which · denotes the time derivative.

(2) *Open end of duct.* The reflection coefficient, $R$, at the open end of an unflanged circular pipe is available, and is given by

$$R = \frac{B\, e^{jkx}}{A\, e^{-jkx}} \tag{18.6}$$

The magnitude of the reflection coefficient, $|R|$, is shown in Figure 18.1a as a function of $ka$, where $a$ is the pipe radius. The phase shift can be determined from Figure 18.1b, which is a plot of $\alpha/a$ as a function of $ka$. Also, the reflection coefficient is [1]:

$$R = -|R| e^{-2jk\alpha} \tag{18.7}$$

For the small values of $ka$ that are most often encountered in reactive muffler design, $|R| \approx 1$ and $\alpha/a = 0.613$.

(3) *Closed end.* The displacement of particle motion is zero at a rigid wall. Hence, we have

$$\xi = 0 \tag{18.8}$$

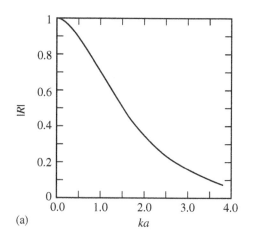

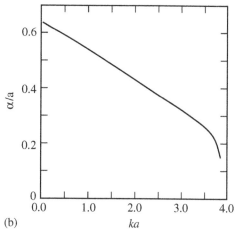

**FIGURE 18.1** (a) Magnitude of the reflection coefficient at open end of an unflanged circular pipe; (b) End correction for an unflanged circular pipe. (*Source:* Levine, H. and Schwinger, J., On the radiation of sound from an unflanged circular pipe, *J. Phys. Rev.*, 73, 383, 1948. With permission.)

(4) *Junction conditions.* The following equations correspond to the continuity of volume flow rate of the particles and the continuity of pressure, even if the cross section changes suddenly:

$$S_i \dot{\xi}_i = S_{i+1} \dot{\xi}_{i+1} \tag{18.9}$$

$$p_i = p_{i+1} \tag{18.10}$$

## 18.3 Effects of Reactive Mufflers

The acoustic behavior of a reactive muffler may be expressed in term of the insertion loss, the difference in the noise levels measured at some external point with and without the muffler in the system. The transmission loss is defined as the insertion loss for a nonreflecting source and the end of exhaust duct.

### 18.3.1 Insertion Loss

A single expansion-type muffler installation is shown schematically in Figure 18.2. At the open end of a pipe, as in Figure 18.2, the traveled and reflected waves of the source become $A_0 \, e^{-jkl}$, $B_0 \, e^{jkl}$, over a length $l$, where the amplitudes are denoted by $A_0$, $B_0$. The reflective coefficient for length $l_0$ is given by

$$R_0 = \frac{B_0 \, e^{jkl_0}}{A_0 \, e^{-jkl_0}} \tag{18.11}$$

This is obtained from Equation 18.6 with $x = l_0$.

The energy, $W_0$, of the acoustic wave escaping from the open end of the pipe is given by

$$W_0 \propto \frac{S_0 A_0^2 (1 - R_0^2)}{\rho_0 c_0} \tag{18.12}$$

in which $\rho_0$ = density of air, and $c_0$ = speed of sound in air. The equation of the sound-pressure level measured at an open point at some distance is given by

$$p_0 = 10 \log_{10} \left( \frac{Q_d}{4\pi r_0^2} + \frac{4}{R_r} \right) + PWL_{r_0} \tag{18.13}$$

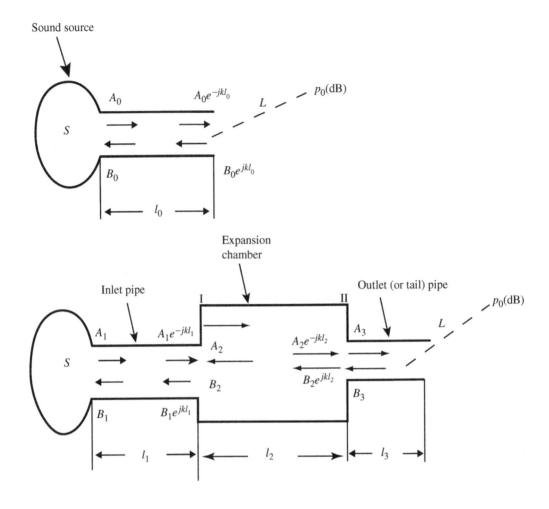

**FIGURE 18.2** Measurement of insertion loss.

where

$r_{0,}$ = distance
$R_r = A/(1 - \alpha)$, room constant
$A$ = the indoor sound absorbing power (indoor surface area times indoor average absorption coefficient)
$\alpha$ = the indoor average absorption coefficient
$Q_d$ = the directivity factor from the open end

Therefore, the measured value of insertion loss can be obtained from Equation 18.14, when $Q_d$ values are equal. Power level is defined as

$$PWL = 10 \log_{10}\left(\frac{W}{10^{-12}}\right)$$

Now,

$$IL = PWL_{r_0} - PWL_r \tag{18.14}$$

Using Equation 18.14, it can be shown that IL can be expressed by

$$IL = 10 \log_{10}\left|\frac{S_0}{S_3}\left|\frac{A_0}{A_3}\right|^2\right|\left|\frac{1 - R_0^2}{1 - R_3^2}\right| \tag{18.15}$$

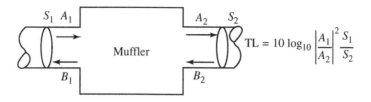

**FIGURE 18.3**  Definition of transmission loss for a muffler.

## 3.2  Transmission Loss

It is desirable to eliminate the source and radiation characteristics from the system in Figure 18.3, and to look only at some property of the muffler itself. This may be accomplished by defining a quantity called "transmission loss" (TL) as follows:

$$\text{TL} = 10 \log_{10} \left| \frac{A_1}{A_2} \right|^2 \frac{S_1}{S_2} \tag{18.16}$$

In the measurement of TL, it is difficult to separate the reflected wave. The theoretical calculation is easy and useful.

# 18.4  Calculation Procedure

For the reactive muffler shown in Figure 18.2, the following equations are obtained from Equation 18.5 to Equation 18.7 and Equation 18.9. The inlet pipe, cavity, and tail pipe are denoted by suffix in the figure [2].

(1)  The sound source:

$$S_0(A_0 - B_0) = S_1(A_1 - B_1) \tag{18.17}$$

(2)  The open end of pipes:

$$\left. \begin{aligned} B_0\, e^{jkl_0} &= A_0\, e^{-jkl_0} R_0 \\ B_3\, e^{jkl_3} &= A_3\, e^{-jkl_3} R_3 \end{aligned} \right\} \tag{18.18}$$

(3)  The sudden expansion (junction I):

$$\left. \begin{aligned} A_1\, e^{-jkl_1} + B_1\, e^{jkl_1} &= A_2 + B_2 \\ S_1(A_1\, e^{-jkl_1} - B_1\, e^{jkl_1}) &= S_2(A_2 - B_2) \end{aligned} \right\} \tag{18.19}$$

(4)  The sudden contraction (junction II):

$$\left. \begin{aligned} A_2\, e^{-jkl_2} + B_2\, e^{jkl_2} &= A_3 + B_3 \\ S_2(A_2\, e^{-jkl_2} - B_2\, e^{jkl_2}) &= S_3(A_3 - B_3) \end{aligned} \right\} \tag{18.20}$$

From these linear equations, the following equation can be obtained:

$$\begin{aligned} \frac{A_0}{B_0} = \frac{m_{10}}{1 - R_0\, e^{-2jkl_0}} &\Big\{ j\Big(1 + R_3\, e^{-2jkl_3}\Big)(\sin kl_1 \cos kl_2 + m_{21}kl_1 \sin kl_2) \\ &+ m_{32}\Big(1 - R_3\, e^{-2jkl_3}\Big)(m_{21} \cos kl_1 \cos kl_2 - \sin kl_1 \sin kl_2) \Big\} \end{aligned} \tag{18.21}$$

When the reflection coefficients are absent, $R_0 = R_3 = 0$, we have

$$IL = 10 \log_{10} m_{03} \{ m_{10}^2 (\sin kl_1 \cos kl_2 + m_{21} \cos kl_1 \sin kl_2)^2 + m_{10}^2 m_{32}^2$$
$$\times (m_{21} \cos kl_1 \cos kl_2 - \sin kl_1 \sin kl_2)^2 \} \tag{18.22}$$

where $m_{10} = S_0/S_1$, $m_{21} = S_2/S_1$, $m_{32} = S_3/S_2$, $m_{03} = S_0/S_3$, which are ratios of cross-sectional areas of the pipes.

For the reflection factors, $R_0 = R_3 = -1$, $S_0 = S_1 = S_3$, and $m_{21} = m$, the equation of insertion loss becomes

$$IL = 20 \log_{10} \left\| \left[ \frac{\cos kl_1}{\cos kl_0} \left\{ \cos kl_{21} \cos kl_3 - m \sin kl_2 \sin kl_3 - \tan kl_1 \left( \cos kl_2 \sin kl_3 + \frac{1}{m} \sin kl_2 \cos kl_3 \right) \right\} \right] \right\| \tag{18.23}$$

The ratio $|A_1/A_3|$ may also be obtained from Equation 18.17 to Equation 18.20. When the magnitude of reflection coefficients $R_0 = R_3 = 0$, $S_0 = S_1 = S_3$, and expansion ratio of the cross-sectional open area of the pipes $m_{21} = m$, the transmission loss is given by

$$TL = 10 \log_{10} \left\{ 1 + \frac{1}{4} \left( m - \frac{1}{m} \right)^2 \sin^2 kl_2 \right\} \tag{18.24}$$

When $R_0 = R_3 = -1$, $S_0 = S_1 = S_3$, and $m_{21} = m$, TL is obtained from the following equation:

$$TL = 10 \log_{10} \left| 1 + \frac{1}{m^2} (m^2 - 1) \{ (m^2 + 1) \sin^2 kl_3 - 1 \} \sin^2 kl_2 + \frac{1}{2} \left( -m + \frac{1}{m} \right) \sin 2kl_2 \sin 2kl_3 \right| \tag{18.25}$$

Computation formulas of insertion loss and transmission loss for the case of an expansion chamber with insertion pipe and resonator are shown in Table 18.1.

The principal structures of several reactive mufflers are shown in Figure 18.4.

## 18.5 Application Range of Model

### 18.5.1 Condition for Approximation of Plane Wave

The frequency range where the approximation of a plane wave is valid is given by

$$f_c < 1.22 \frac{c}{D} \tag{18.26}$$

where

$f_c$ = critical frequency of plane wave (Hz)
$c$ = speed of sound (m/s)
$D$ = diameter of muffler (m)

It is seen that the expansion ratio of an open area of a pipe increases with IL or TL. However, the application range of the analytical model decreases with increasing diameter of chamber.

### 18.5.2 Effect of Temperature

Under conditions of high-temperature and high-speed gas flow, as in an engine exhaust system, the primary effect of a change in pipe temperature is the corresponding change in the speed of sound, which is proportional to the square root of the absolute temperature. In the design of a reactive muffler, it is necessary to use the actual speed of sound in the gas inside the pipe. The most accurate values available for density ($\rho_0$) and the speed of sound ($c_0$) at each element should be used in calculating the impedance

**TABLE 18.1** Transmission Loss of Reactive Mufflers and Insertion Loss of Reactive Mufflers

| Muffler (see Figure 18.4) | TL (dB) | Application Limits and Comments |
|---|---|---|
| *Transmission loss of reactive mufflers* | | |
| (a) | $\text{TL} = 10 \log_{10}\left\{1 + \left(\dfrac{kS}{4a}\right)^2\right\}$ <br><br> $a$: radius of orifice | $a/\lambda < 0.1,\ R = 0$ |
| (b) | $\text{TL} = 20 \log_{10}\left|\dfrac{(1 + R_3\, e^{-2jkl_3})(\cos kl_2 + jm_{21}\sin kl_2)}{+\, m_{32}(1 - R_3\, e^{-2jkl_3})(j\sin kl_2 + m_{21}\cos kl_2)}\right|$ <br><br> $m_{10} = S_1/S_0,\ m_{21} = S_2/S_1,\ m_{32} = S_3/S_2,\ R_0,\ R_3$: the reflection coefficient of the open end, <br> $l_3$ = length of the tail pipe <br> (1) when $R_0 = R_3 = 0,\ S_0 = S_1 = S_3$ <br><br> $\text{TL} = 10 \log_{10}\left\{1 + \dfrac{1}{4}\left(m_{21} - \dfrac{1}{m_{21}}\right)^2 \sin kl_2\right\}$ <br><br> (2) when $R_0 = R_3 = -1,\ S_0 = S_1 = S_3$ <br><br> $\text{TL} = 10 \log_{10}\left[1 + \left(\dfrac{m_{21}^2 - 1}{m_{21}^2}\right)\left\{(m_{21}^2 + 1)\sin^2 kl_3 - 1\right\}\sin^2 kl_2 + \dfrac{1}{2}\left(\dfrac{1}{m_{21}} - m_{21}\right)\sin 2kl_2 \sin 2kl_3\right]$ | $f < 1.22c/D$ |
| (c) | $\text{TL} = 10 \log_{10}\left[\left\{2\cos k(l_1 - l_{11} - l_{22}) - \dfrac{m-1}{m}\sin k(l_1 - l_{11} - l_{22})\right\}^2 \right.$ <br><br> $+ \left\{\left(m + \dfrac{1}{m}\right)\sin k(l_1 - l_{11} - l_{22}) + (m - 1)\cos(l_1 - l_{11} - l_{22})(\tan kl_{11} + \tan kl_{22})\right\}^2$ <br><br> $\left. - \dfrac{(m-1)^2}{m}\tan kl_{11}\tan kl_{22}\sin k(l_1 - l_{11} - l_{22})\right]$ <br><br> $M = S_1/S_0$ | $R \approx 0$ |
| (d) | $\text{TL} = 10 \log_{10}\left\{1 + \dfrac{1}{4}\left(\dfrac{m}{\dfrac{kS_2}{C_0} - \cot kl}\right)^2\right\}$ <br><br> $C_0 = NC_i;\ N$: number of holes, $C_i = 2\pi a_i^2/(l_b + \pi a_0),\ l_b,\ l_b$: thickness of the pipe, $a_i$: radius of a hole, <br> $m = S_{12}/S$ | $R \approx 0$ |

*(continued on next page)*

**TABLE 18.1**  *(continued)*

| Muffler (see Figure 18.4) | TL (dB) | Application Limits and Comments |
|---|---|---|
| (e) | $$TL = 10\log_{10}\left[1 + \frac{1}{4}\left|\frac{\dfrac{\sqrt{C_0 V}}{S}}{\dfrac{f}{f_r} - \dfrac{f_r}{f}}\right|^2\right]$$ $$f_r = \frac{c}{2\pi}\sqrt{\frac{C_0}{V}},$$ $$C_0 = \frac{2\pi a^2}{2l_b + \pi a}$$ $l_b$: length of the neck or thickness of the pipe, $a$: radius of the neck or hole | $R \approx 0, l_b \ll \lambda$ Resonator size $\ll \lambda$ |
| (f) | $$TL = 10\log_{10}\left\{1 + \frac{m^2}{4}\left(\frac{\tan kl_b - \dfrac{S_b}{kV}}{\dfrac{S_b}{kV}\tan kl_b + 1}\right)\right\}$$ $m = S_b/S$ | $R \approx 0, l_b \ll \lambda$ Resonator size $\ll \lambda$ |
| (g) | $$TL = 20\log_{10}\frac{1}{16m^2}\left|[4m(m+1)^2\cos 2k(l+l_c) - 4m(m-1)^2\cos 2k(l-l_c)]\right.$$ $$+ j\{2(m^2+1)(m+1)^2\sin 2k(l+l_c) - 2(m^2+1)(m-1)^2\sin2k(l-l_c)$$ $$\left.- 4(m^2-1)^2\sin 2kl_c\}\right|$$ $m = S_2/S_1$ | $R \approx 0$ |
| (h) | $$TL = 10\log_{10}\left|\{\cos 2kl - (m-1)\sin 2kl_c \tan kl_c\}^2 + \left\{\frac{j}{2}\left(m + \frac{1}{m}\right)\sin 2kl\right.\right.$$ $$\left.\left.+ (m-1)\tan kl_c\left(\left(m+\frac{1}{m}\right)\cos 2kl - \left(m - \frac{1}{m}\right)\right)\right\}^2\right|$$ $m = S_2/S_1$ | $R \approx 0$ |

(i)

$$TL = 10\log_{10}\left\{\frac{1}{4}\left|\frac{A_1 + jB_1}{A_2 + jB_2}\right|^2\right\}$$

$$A_1 = Y_3 X_1^2 + Z_0 Y_3^2 + Z_0(X_1 + X_3)^2$$

$$B_1 = X_1 Y_3^2 + X_1 X_3(X_1 + X_3)$$

$$A_2 = Y_3 X_1^2 \cos kl + Z_0 X_1 Y_3 \sin kl$$

$$B_2 = X_1 Y_3^2 + X_1 X_3(X_1 + X_3)\cos kl - Z_0 X_1(X_1 + X_3)\sin kl$$

$$X_1 = \frac{\omega\rho}{C_0} - \frac{\rho c^2}{\omega V_1},$$

$$X_2 = \frac{\omega\rho}{C_0} - \frac{\rho c^2}{\omega V_2}$$

$$X_3 = \frac{Z_0^2\left(X_2\cos 2kl + \frac{1}{2}Z_0\sin^2 kl\right)}{(X_2\sin kl - Z_0\cos kl)^2 + X_2^2\cos^2 kl}$$

$$Y_3 = \frac{Z_0 X_2^2}{(X_2\sin kl - Z_0\cos kl)^2 + X_2^2\cos^2 kl}$$

$$Z_0 = \frac{\rho c}{S_0},$$

$$C_0 = \frac{2\pi a^2}{2l_b + \pi a}$$

$R \approx 0$ Resonator size $\ll \lambda$

$l_b$: thickness of the pipe, $a$: radius of hole

(j)

$$TL = 10\log_{10}\left\{\left(\cos kl + \frac{\rho c}{4S_0 X}\left(m + \frac{1}{m}\right)\sin kl - \frac{\rho c}{4S_0 X}\left(m - \frac{1}{m}\right)\cos 2kl_b\,\sin kl\right)^2\right.$$
$$\left. + \left(\frac{1}{2}\left(m + \frac{1}{m}\right)\sin kl + \frac{\rho c}{4S_0 X}\left(m - \frac{1}{m}\right)\sin 2kl_b\,\sin kl - \frac{\rho c}{2S_0 X}\cos kl\right)^2\right\}$$

$$X = \frac{\omega\rho}{C_0} - \frac{\rho c^2}{\omega V},$$

$$S = \frac{S}{S_0}$$

$$m = \frac{S}{S_0}$$

$R \approx 0$

*(continued on next page)*

**TABLE 18.1**  (*continued*)

| Muffler (see Figure 18.4) | IL (dB) | Application Limits and Comments |
|---|---|---|
| *Insertion loss of reactive mufflers* | | |

(b)

$$IL = 10 \log_{10} \frac{1}{m_{30}}$$

$$\times \left| \frac{m_{10}}{1 - R_0\, e^{-2jkl_0}} \{j(1 + R_3\, e^{-2jkl_4})(\sin kl_1 \cos kl_2 + m_{21} \cos kl_1 \sin kl_2)) \right.$$

$$\left. + m_{32}(1 - R_3\, e^{-2jkl_5})(m_{21} \cos kl_1 \cos kl_2 - \sin kl_1 \sin kl_2)\} \right|^2$$

Application: $f < 1.22c/D$, $R$ is plotted in Fig.18.1

(1)

$$IL = 10 \log_{10} \left| \{1 + (m_{21}^2 - 1)\} \left(1 - \frac{m_{21}^2 + 1}{m_{21}^2} \sin^2 kl_1 \right) \sin^2 kl_2 \right.$$

$$\left. + \frac{1}{2}\left(m_{21} - \frac{1}{m_{21}}\right) \sin 2kl_1 \sin 2kl_2 \right|$$

Application: $R_0 = R_3 = 0$, $S_0 = S_1 = S_3$

(2)

$$IL = 10 \log_{10} \left| \left( \frac{\cos kl_1}{\cos kl_0} \{\cos kl_2 \cos kl_3 - m_{21} \sin kl_2 \sin kl_3 - \tan kl_1 (\cos kl_2 \sin kl_3 \right.\right.$$

$$\left.\left. + \frac{1}{m_{21}} \sin kl_2 \cos kl_3)\} \right)^2 \right|$$

Application: $R_0 = R_3 = -1$, $S_0 = S_1 = S_3$

(c)

$$IL = 20 \log_{10} \left| \frac{\cos kl_1 \cos kl_2 - m \sin kl_1 \sin kl_2}{\cos kl_{11} \cos kl_{22}} \right|$$

$$m = S_1/S_0$$

Application: $R = -1$

(d)

$$IL = 20 \log_{10} \left[ \cos^2 kl_2 + \frac{m}{\dfrac{kS_2}{C_0} - \cot kl} \sin 2kl_2 + \left( \frac{m}{\dfrac{kS_2}{C_0} - \cot kl} \right)^2 \sin^2 kl_2 \right]$$

Application: $R = -1$, $kl_0 \ll 1$

$C_0 = NC_i$; $C_i = 2\pi a_i^2/(l_b + \pi a_i)$, $N$: number of holes, $l_b$: thickness of the pipe, $a_i$: radius of a hole, $m = S_{12}/S$

(e)

$$IL = 10 \log_{10} \left| \frac{\frac{\sqrt{C_0}\,V}{S}\sin 2kl_2 + \frac{C_0 V}{S^2}\sin^2 kl_2 + \cos^2 kl_2}{\frac{f}{f_r}-\frac{f_r}{f}} \left(\frac{f}{f_r}-\frac{f_r}{f}\right)^2 \right|$$

$$f_r = \frac{c}{2\pi}\sqrt{\frac{C_0}{V}},$$

$$C_0 = \frac{2\pi a^2}{2l_b + \pi a}$$

$l_b$: length of the neck or thickness of pipe, $a$: radius of the neck or hole

$$R = -1, \; kl_0 \ll 1$$

(f)

$$IL = 10 \log_{10} \left[ \cos^2 kl_2 + m \sin 2kl_2 \frac{\frac{S_b}{kV}\tan kl_b + 1}{\tan kl_b - \frac{S_b}{kV}} + m^2 \left(\frac{\frac{S_b}{kV}\tan kl_b + 1}{\tan kl_b - \frac{S_b}{kV}}\right)^2 \sin^2 kl_2 \right]$$

$$f_r = \frac{c}{2\pi}\sqrt{\frac{C_0}{V}},$$

$$C_0 = \frac{2\pi a^2}{2l_b + \pi a}$$

$l_b$: length of the neck or thickness of the pipe, $a$: radius of the neck or hole

$$R = -1, \; kl_0 \ll 1$$

(k)

$$IL = 20 \log_{10} |(\cos kl_{11} - m_1 \sin kl_1 \sin kl_{11}) + (\cos kl_2 \cos kl_{22} - m_2 \sin kl_2 \sin kl_{22}) + \cdots + (\cos kl_i \cos kl_{ii} - m_i \sin kl_i \sin kl_{ii})|$$

$$R = -1, \; kl_0 \ll 1$$

(l)

$$IL = 20 \log_{10} \left| \left\{ \frac{\cos kl_1 \cos kl_1 - m \sin kl_1 \sin kl_1}{\cos kl_{11}} + \frac{\cos kl_2 \cos kl_2 - m \sin kl_2 \sin kl_2}{\cos kl_{12} \cos kl_{21}} + \cdots + \frac{\cos kl_n \cos kl_n - m \sin kl_n \sin kl_n}{\cos kl_{(n-1)2}} \right\} \right|$$

$$R = -1, \; kl_0 \ll 1$$

$A$ is the radius of tube in orifice hole or diameter of side branch, $c$ is the sound speed, $C_0$ is the conductivity, $D$ is the diameter of chamber, $f$ is the frequency, $f_r$ is the resonant frequency of the resonator, $k = 2\pi f/c$ is the wave number, $L$ is the length, $m = S_i/S_{i+1}$ is the ratio of the cross section, $N$ is the number of holes, $R$ is the reflection coefficient, $S$ is the cross section, IL is the insertion loss, TL is the transmission loss, $V$ is the volume of chamber, $Z$ is the acoustic impedance, $\rho$ is the density, $\lambda$ is the wavelength, $\omega = 2\pi f$, angular frequency.

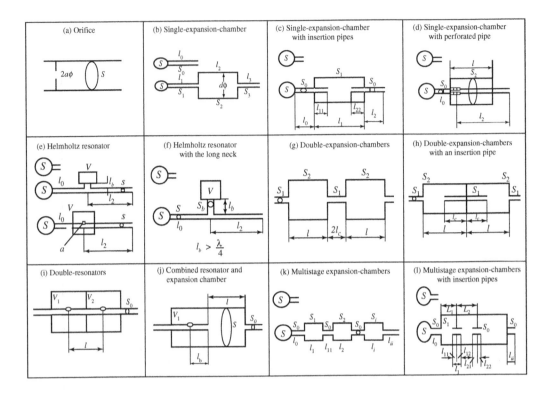

**FIGURE 18.4**   Sketches of the 12 principal structures of reactive mufflers.

of the elements. The impedance, $z$, is given by

$$z = -j\frac{\rho_0 c_0}{S}\frac{1}{kl} \tag{18.27}$$

where

$S$ = the cross-sectional open area of pipe
$k = 2\pi f/c_0$, wave number
$l$ = the length of the pipe element

Note that the impedance of the resonator chamber is proportional to $\rho_0 c_0^2$. However, $c_0^2$ is proportional to the absolute temperature of gas ($T$) and $\rho_0$ is proportional to $1/T$. Hence, the chamber impedance is independent of temperature. The connector impedance is a function of $T$, but in most cases the connector will be at the pipe temperature. For a resonator-type muffler, a temperature difference between the pipe and chamber is expected to have little effect on the performance of the muffler.

### 18.5.3   Effect of Gas Flow in Pipe

Under conditions of high-temperature and high-speed gas flow in a pipe, the pressure amplitude in the pipe is large, and is larger than what is predicted by theory. Analysis by the characteristic curve method is desirable under such conditions.

In a reactive muffler where the pipe flow passes through a sudden pipe expansion or an orifice, the computed transmission loss or insertion loss tends to be an overestimate because of new noise that is generated due to the resulting irregular air-flow within the muffler.

### 18.5.4 Effect of Friction Loss in Pipe

When an acoustic wave propagates in a pipe, it will attenuate due to viscous friction. The effect is large for long pipes of small diameter. Friction damping in a pipe may be incorporated into the propagation constant, $\gamma$, such that

$$\gamma = \delta + jk \tag{18.28}$$

where $\delta$ is the attenuation constant per unit length of pipe. By substituting Equation 18.28 into Equation 18.3 and Equation 18.4, we obtain

$$\xi = (A\,e^{-\gamma l} - B\,e^{\gamma l})e^{j\omega t} \tag{18.29}$$

$$p = -\rho c^2 k(A\,e^{-\gamma l} + B\,e^{\gamma l})e^{j\omega t} \tag{18.30}$$

Empirical formulas are given below for two cases of the attenuation coefficient $\delta$ [3].

(1) The formula for seamless steel or chloride-ethylene pipes (regression formula when the inside roughness is 4 to 8 $\mu$m and length under 3 m) is

$$\delta = 26,100\lambda^{-0.5}\frac{\mu}{\rho c d} \tag{18.31}$$

where

$\lambda$ = wavelength of sound (m)
$\mu$ = viscosity of gas in the pipe (Pa s)
$\rho$ = density of gas (kg/m$^3$)
$d$ = diameter of the pipe (m)

(2) The equations for lining with glass wool are

$$\left.\begin{aligned}
\delta_2 &= 2491\lambda^{-0.476}\left(\frac{\rho c d}{\mu}\right)^{-1.068}\\[2mm]
\delta_6 &= 5175\lambda^{-0.476}\left(\frac{\rho c d}{\mu}\right)^{-1.303}\\[2mm]
\delta_6 &= 11596\lambda^{-0.476}\left(\frac{\rho c d}{\mu}\right)^{-1.270}
\end{aligned}\right\} \tag{18.32}$$

The suffix of $\delta$ gives the thickness of absorbing material in mm.

## 18.6 Practical Example

### 18.6.1 Expansion-Type Muffler for Reciprocating Compressor

Consider a double-acting (i.e., fluid on both sides of the piston in the cylinder) reciprocating compressor for supplying high-pressure air to a machine shop of a factory, for example.

The specifications of the reciprocating compressor follow:

- Delivery pressure: $6.9 \times 10^5$ Pa
- Rotational speed of driving shaft: 600 rpm
- Power of driving shaft: 450 kW
- Diameter of inlet pipe: 380 mm

Pressure pulsations of 10 and 20 Hz are produced by the compressor due to the rotational speed, as seen in Figure 18.5. The pressure wave from the inlet propagates the free space and can the damage nearby private houses. Wooden doors with glass paneling, wooden sliding-doors, and leaves of plants and foliage, and have been found to vibrate due to low-frequency audible sound. An attenuation of 15 to 20 dB was

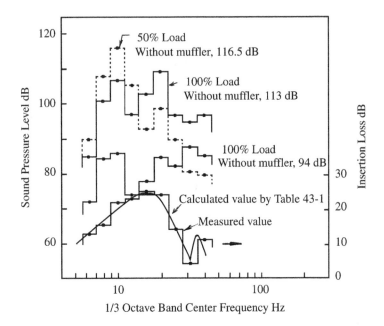

**FIGURE 18.5** Noise spectrum at inlet of reciprocating compressor and insertion loss of designed muffler.

necessary at the frequencies 10 and 20 Hz. A muffler using an expansion and tail pipe type was suggested to handle the problem. The reflection coefficient of the tail pipe is approximately $R = -1$ and $ka = 2\pi fa/c = 2\pi \times 20 \times 0.19/345 = 0.0692$.

$$IL = 20 \log_{10}|\cos kl_2 \cos kl_3 - m \sin kl_2 \sin kl_3|$$

where

$k = 2\pi f/c$; wave number (1/m)
$l_2 =$ the length of the chamber (m)
$l_3 =$ the length of the tail pipe (m)
$m =$ the expansion ratio of the cross section between the chamber and inlet

With $kl_2 = kl_3 = \pi/2$, we have

$$IL = 20 \log_{10}|m|$$

We need $m > 10$ in order to satisfy the desired value of IL. For 20 Hz, we use $kl_2 = kl_3 = \pi$. When $kl_2 = kl_3 = \pi/2$ at $f = 10$ Hz, we have IL $= 0$. Then, using $kl_2 = kl_3 = \pi/2$ at frequency 15 Hz, we can satisfy the IL condition of 20 dB at both frequencies. Hence, $l_2 = l_3 = 345/(4 \times 15) = 5.75$ (m) is chosen at a speed of sound $c = 345$ m/s.

The noise spectrum at the inlet of the reciprocating compressor under study and insertion loss of the muffler design in this example are shown in Figure 18.5.

The diameters or the lengths of the chamber and the tail pipe are properly selected, as shown and in Figure 18.6. At 10 Hz, IL is determined as

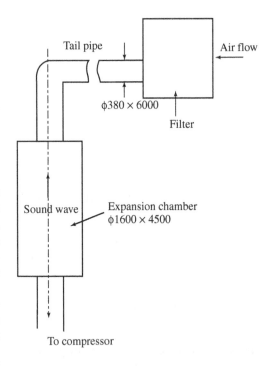

**FIGURE 18.6** The muffler designed for the noise control of a reciprocating compressor.

indicated below:

$$IL = 20 \log_{10} |\cos(2\pi \times 10 \times 4.5/345)\cos(2\pi \times 10 \times 6/345) - (1.6/0.38)^2 \sin(2\pi \times 10 \times 4.5/345)$$
$$\times \sin(2\pi \times 10 \times 6/345)|$$
$$= 20 \log_{10} |0.6825 \times 0.4600 - 17.728 \times 0.7308 \times 0.8879| = 21.0 \text{ (dB)}$$

Similarly, at 20 Hz, we have

$$IL = 20 \log_{10} |\cos(2\pi \times 20 \times 4.5/345)\cos(2\pi \times 20 \times 6/345) - (1.6/0.38)^2 \sin(2\pi \times 20 \times 4.5/345)$$
$$\times \sin(2\pi \times 20 \times 6/345)|$$
$$= 20 \log_{10} |(-0.0682) \times (-0.5767) - 17.728 \times 0.9977 \times 0.8170| = 23.2 \text{ (dB)}$$

Clearly, the attenuation at both frequencies satisfies the desired lower limit of 20 dB. Calculated values of IL at low frequencies are shown by a curved continuous line in Figure 18.5.

## References

1. Levine, H. and Schwinger, J., On the radiation of sound from an unflanged circular pipe, *J. Phys. Rev.*, 73, 383, 1948.
2. Ohnaka, I., Lecture for noise reduction of machines, no. 2, *J. Marine Eng. Soc. Jpn.*, 4, 179, 1969.
3. Suyama, E. and Hirata, M., Attenuation constant of plane wave in a tube, *J. Acoust. Soc. Jpn.*, 35, 152, 1979.

# 19

# Design of Sound Insulation

Kiyoshi Okura
*Mitsuboshi Belting Ltd.*

## Summary

*This chapter presents useful theory and design procedures for sound insulation. Related concepts and representations of transmission loss, the transmission coefficient, and impedance are given. Analysis and design procedures for sound insulation structures such as single and multiple panels and walls with sound absorption material are presented. Practical applications for the design of sound insulation components and systems are described.*

## 19.1 Theory of Sound Insulation

### 19.1.1 Expressions of Sound Insulation [1]

#### 19.1.1.1 Transmission Coefficient

Let us denote by $I_i$ the acoustic energy incident on a wall per unit area and unit time. Some energy is dissipated in the wall, and, apart from the energy that is reflected by the wall, the rest is transmitted through the wall. Using $I_t$ to denote the transmitted acoustic energy, the transmission coefficient of the wall is defined as

$$\tau = \frac{I_t}{I_i} \tag{19.1}$$

#### 19.1.1.2 Transmission Loss

As an expression for sound insulation performance, we may use transmission loss (TL), which is defined as (also see Chapter 17 and Chapter 18)

$$\text{TL} = 10 \log\left(\frac{1}{\tau}\right) = 10 \log\left(\frac{I_i}{I_t}\right) \tag{19.2}$$

### 19.1.2 Transmission Loss of a Single Wall

Consider a plane sound wave incident on a impermeable infinite plate at angle $\theta$, which is placed in a uniform air space as shown in Figure 19.1. The sound pressure of the incident, reflected, and transmitted waves, denoted by $p_i$, $p_r$, and $p_t$, respectively, are given by

$$p_i = P_i e^{j\omega t - jk(x \cos \theta + y \sin \theta)}$$

$$p_r = P_r e^{j\omega t - jk(-x \cos \theta + y \sin \theta)} \qquad (19.3)$$

$$p_t = P_t e^{j\omega t - jk(x \cos \theta + y \sin \theta)}$$

where $P_i$, $P_r$, and $P_t$ are the sound pressure amplitudes of incident, reflected, and transmitted waves, respectively; $\omega$ is angular frequency; $k$ is the wave number of the sound wave; $c$ is the speed of sound, respectively in the air. Assuming that the plate is sufficiently thin compared with the wavelength of the incident sound wave, the vibration velocities on the incident and transmitted surfaces of the plate are equal. Then vibration velocity, $u$, of the plate in the $x$ direction is equal to the particle velocity of the incident and transmitted sound waves, and we obtain relations

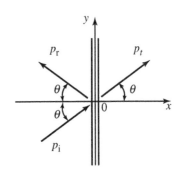

**FIGURE 19.1**   Plane sound wave incidence on an infinite plate.

$$u = -\frac{1}{j\omega\rho}\frac{\partial(p_i + p_r)}{\partial x} = -\frac{1}{j\omega\rho}\frac{\partial p_t}{\partial x} \qquad (19.4)$$

$$\frac{p_i + p_r - p_t}{u} = Z_m \qquad (19.5)$$

where $\rho$ is the air density and $Z_m$ is the *mechanical impedance* of the plate per unit area. From these equations, the transmission coefficient, $\tau_\theta$, and then the transmission loss, $\mathrm{TL}_\theta$, at the incident angle, $\theta$, are obtained according to

$$\mathrm{TL}_\theta = 10 \log\frac{1}{\tau_\theta} = 10 \log\left|\frac{p_i}{p_t}\right|^2 = 10 \log\left|1 + \frac{Z_m \cos \theta}{2\rho c}\right|^2 \qquad (19.6)$$

### 19.1.2.1   Coincidence Effect

Consider the vibration of the plate in the $x$–$y$ plane shown in Figure 19.1. Denoting by $m$ the surface density, and by $B$ the bending stiffness per unit length of the plate, the equation of motion of the plate is given by

$$m\frac{\partial^2 \xi}{\partial t^2} + B(1 + j\eta)\frac{\partial^4 \xi}{\partial y^4} = p_i + p_r - p_t, \quad B = \frac{Eh^3}{12(1 - \nu^2)} \cong \frac{Eh^3}{12} \qquad (19.7)$$

where

$\xi =$ displacement in the $x$ direction
$E =$ Young's modulus of the plate
$h =$ thickness of the plate
$\eta =$ loss factor of the plate
$\nu =$ Poisson's ratio of the plate

The plane sound wave of angular frequency, $\omega$, and of incidence angle, $\theta$, causes a bending wave in the plate where displacement is assumed to be $\xi = \xi_0 e^{j(\omega t - k_1 y)}$, as a solution of Equation 19.7. Hence, the mechanical impedance per unit area is obtained:

$$Z_m = \frac{p_i + p_r - p_t}{\partial \xi/\partial t} = \eta\frac{Bk_1^4}{\omega} + j\left(\omega m - \frac{Bk_1^4}{\omega}\right) \qquad (19.8)$$

where $k_1 = k \sin \theta (k = \omega/c)$ is the wave number of the bending wave in $y$ direction caused by the incident sound wave. Propagation speed of the forced bending wave, $c_1$, and a free bending wave of the plate, $c_B$, are given by

$$c_1 = \omega/k_1, \quad c_B = \left(\frac{\omega^2 B}{m}\right)^{1/4} \tag{19.9}$$

Equation 19.9 reduces Equation 19.8 to

$$Z_m = \eta\omega m\left(\frac{c_B}{c_1}\right)^4 + j\omega m\left[1 - \left(\frac{c_B}{c_1}\right)^4\right] \tag{19.10}$$

When the speed of forced bending wave, $c_1$, and the speed of free bending wave, $c_B$, are equal in Equation 19.10, the imaginary part of $Z_m$ becomes 0, and a form of "resonance" occurs. Then the transmission loss decreases rapidly. This phenomenon is called the *coincidence effect*, and the resonant frequency dependent on the incident angle is given by

$$f = \frac{c^2}{2\pi \sin^2\theta}\sqrt{\frac{m}{B}} \tag{19.11}$$

The minimum of the resonant frequency is called coincidence critical frequency, or critical frequency for short, and it reduces to

$$f_c = \frac{c^2}{2\pi}\sqrt{\frac{m}{B}} \cong \frac{c^2}{1.8 c_L h} \tag{19.12}$$

where $c_L = \sqrt{E/\rho_P}$ is the speed of longitudinal wave in the plate, and $\rho_P$ denotes the density of the plate.

Let us show the relations of the critical frequency and the plate thickness of typical material of sound insulation shown in Figure 19.2. Using the relation $c_B/c_1 = \sqrt{f/f_c} \sin\theta$, Equation 19.10

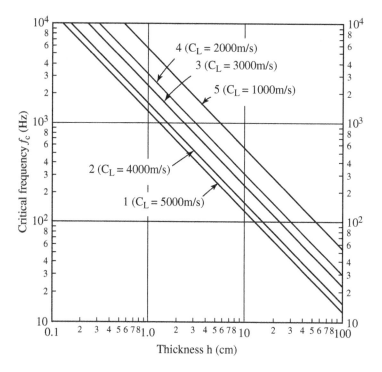

**FIGURE 19.2** Critical frequency vs. plate thickness of typical sound insulation materials: (1) aluminum, steel, or glass; (2) hardboard or copper; (3) dense concrete, plywood, or brick; (4) gypsum board; (5) lead or light weight concrete. (*Source*: Beranek, L.L. 1988. *Noise and Vibration Control*, INCE/USA. With permission.)

becomes

$$Z_m = \eta \omega m \left(\frac{f}{f_c}\right)^2 \sin^4\theta + j\omega m \left[1 - \left(\frac{f}{f_c}\right)^2 \sin^4\theta\right] \tag{19.13}$$

### 19.1.2.2   Mass Law of Transmission Loss

When $f \ll f_c$, Equation 19.13 becomes $Z_m \cong j\omega m$. Then, the transmission loss depends on the incident angle, the frequency and the surface density of the plate. This is called the mass law of transmission loss.

*Mass law of normal incidence* represents the transmission loss at the incident angle $\theta = 0$, as given by

$$TL_0 = 10 \log\left[1 + \left(\frac{\omega m}{2\rho c}\right)^2\right] \tag{19.14}$$

For $\omega m \gg 2\rho c$, it becomes

$$TL_0 \cong 10 \log\left(\frac{\omega m}{2\rho c}\right)^2 = 20 \log mf - 42.5; \quad \text{for air} \tag{19.15}$$

*Mass law of random incidence* represents the transmission loss at the angle averaged over a range of $\theta$ from 0 to 90°, which is realized for perfectly diffused sound field. We have

$$TL_r = 10 \log(1/\tau_r) \cong TL_0 - 10 \log(0.23TL_0) \tag{19.16}$$

where the random incident transmission coefficient, $\tau_r$, is defined as

$$\tau_r = \int_0^{\pi/2} \tau_\theta \cos\theta \sin\theta \, d\theta \bigg/ \int_0^{\pi/2} \cos\theta \sin\theta \, d\theta \tag{19.17}$$

An approximation for Equation 19.16, as given below, is generally used for a practical use and this is often useful.

$$TL_r = 18 \log mf - 44 \tag{19.18}$$

*Mass law of field incidence* represents the transmission loss at the angle averaged over a range of $\theta$ from 0 to about 78°, which is said to agree with actual sound field. We have

$$TL_f = TL_0 - 5 \tag{19.19}$$

The three types of transmission loss presented above are compared in Figure 19.3.

### 19.1.2.3   Stiffness Law of Transmission Loss [2]

The plate described above is assumed to be infinite. However, an actual plate is always supported by some structures at its boundaries and the plate size is finite. Transmission loss of a finite plate is considered to be related to the nature of excitation of vibration in the plate, for example, sound wave incidence, modes of vibration and characteristics of sound radiation. Therefore, the governing relationships become very complex. However, in the following frequency range, it is known that the transmission loss conforms to the mass law

$$\frac{c}{2a} < f \ll f_c \tag{19.20}$$

where $a$ is length of shorter edge for rectangular plate.

When $f < c/2a$, the whole plate is excited in phase, and stiffness effects from the supports of its edges will appear. If we denote the equivalent stiffness of the plate as $K$ and assume a loss factor of 0,

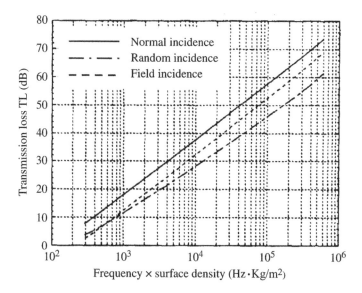

**FIGURE 19.3**  Theoretical transmission loss based on mass law.

the mechanical impedance of the plate is obtained by using Equation 19.8; thus

$$Z_\mathrm{m} = j\left(\omega m - \frac{K}{\omega}\right)$$ (19.21)

The frequency at $Z_\mathrm{m} = 0$ corresponds to the first mode natural frequency, $f_{11}$, of the plate and consequently, the equivalent stiffness of rectangular plate with simple edge-support is given by

$$f_{11} = \frac{1}{2\pi}\sqrt{\frac{K}{m}} \equiv \frac{\pi}{2}\sqrt{\frac{B}{m}}\left[\left(\frac{1}{a}\right)^2 + \left(\frac{1}{b}\right)^2\right]$$ (19.22)

Then,

$$K = B\pi^4\left(\frac{1}{a^2} + \frac{1}{b^2}\right)^2$$

where $a$ and $b$ are the length of the short and long edges for the rectangular plate, respectively. When $f \ll f_{11}$ is assumed in Equation 19.21, the mass term can be neglected, and from Equation 19.6 the normal incidence transmission loss, $\mathrm{TL}_{S0}$, is given by

$$\mathrm{TL}_{S0} = 10\log\left|1 - j\frac{K}{2\omega\rho c}\right|^2$$

$$= 10\log\left[1 + \left(\frac{K}{2\omega\rho c}\right)^2\right]$$

$$\cong 20\log(K/f) - 74.5$$ (19.23)

This is called the stiffness law of the transmission loss, and it shows a 6 dB decay per octave.

The characteristics mentioned above for single wall transmission loss are shown in Figure 19.4 and summarized below.

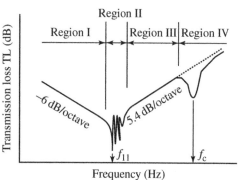

**FIGURE 19.4**  Transmission loss characteristics of a single wall.

1. Region I ($f \ll f_{11}$). Transmission loss is controlled by the stiffness of the panel:

$$\mathrm{TL} = \mathrm{TL}_0 - 40 \log\left(\frac{f}{f_{11}}\right) \tag{19.24}$$

2. Region II ($f \approx f_{11}$). Transmission loss is controlled by the lower-mode natural frequencies of the panel, and the estimation becomes very complex.
3. Region III ($f_{11} \ll f \leq f_c/2$). Transmission loss is controlled by the mass (surface density) of the panel:

$$\mathrm{TL} = 18 \log mf - 44 \tag{19.25}$$

4. Region IV ($f > f_c/2$). Transmission loss is controlled by the mass and the damping of the panel, and it is reduced by coincidence effects.

For $f_c/2 < f \leq f_c$:
    TL is represented by a straight line connecting the value at $f = f_c/2$ of Equation 19.25 and the value at $f = f_c$ of Equation 19.26.
For $f > f_c$:

$$\mathrm{TL} = \mathrm{TL}_0 + 10 \log\left(\frac{2\eta}{\pi}\frac{f}{f_c}\right) \tag{19.26}$$

### 19.1.3 Transmission Loss of Multiple Panels

To realize sound insulation of high performance, we often use a double wall or a multiple panel composed of insulation materials like steel plates and absorbing materials like fiber-glass. In this subsection, transmission loss of a multiple panel is described [3].

#### 19.1.3.1 Calculation Method

Consider a multiple panel of infinite lateral extent as shown in Figure 19.5, which is composed of $n$ acoustic elements, each element consisting of three basic materials, an impermeable plate, air space, and an absorption layer. Furthermore, consider a plane wave incident on the left-hand side surface of the $n$th element at angle $\theta$. Let the sound pressure of the incident wave be $p_i$, and of the reflected wave be $p_r$, and the wave be propagating through the structure, and then radiating from the right-hand side of the first element as a plane wave of pressure $p_t$ into a free field at transmission angle $\theta$.

In the analysis, we append the subscript $k(= 1, 2, \ldots, n)$ to the physical parameters of the $k$th element, and "2" and "1" to the left- and right-hand side values of these parameters, respectively, as shown in Figure 19.5. The ratio of the sound pressure at the incident surface, $p_{n2}$, to the incident wave, $p_i$, is given by

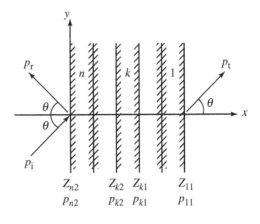

**FIGURE 19.5** Calculation model of $n$-element multiple panel.

$$\frac{p_{n2}}{p_i} = \frac{p_i + p_r}{p_i} = \frac{2Z_{n2}}{Z_{n2} + \rho c/\cos\theta} \tag{19.27}$$

where $Z_{n2}$ is the acoustic impedance of the left-hand side normal to the surface of the $n$th element and $\rho c/\cos\theta$ is the acoustic impedance normal to the surface, which is equal to the radiation impedance of the first element, $Z_{11}$, shown in Figure 19.5. Using the usual condition of pressure matching at each interface, we can write the expression for the oblique incidence transmission coefficient as

$$\tau(\theta) = \left|\frac{p_t}{p_i}\right|^2 \equiv \left|\frac{p_{11}}{p_i}\right|^2 = \left|\frac{p_{n2}}{p_i}\right|^2 \cdot \left|\frac{p_{n1}}{p_{n2}} \cdots \frac{p_{k1}}{p_{k2}} \cdots \frac{p_{11}}{p_{12}}\right|^2 \tag{19.28}$$

Hence, we obtain the following expression for the random incidence transmission loss:

$$\mathrm{TL} = 10 \log \left( \frac{\displaystyle\int_0^{\theta_1} \cos\theta \sin\theta \, d\theta}{\displaystyle\int_0^{\theta_1} \tau(\theta) \cos\theta \sin\theta \, d\theta} \right) \tag{19.29}$$

where $\theta_1$ is the limiting angle above which no sound is assumed to be received, and it varies between 78° and 85°.

If we know $Z_{n2}$ in Equation 19.27 and the pressure ratio across each of the single elements in Equation 19.28, we can calculate the TL using Equation 19.29. We can obtain $Z_{n2}$ by using the conditions of impedance matching at each interface from the rightmost to the leftmost element in order, if we know the impedance relations across each of the single elements.

Now, we present the pressure ratios and the acoustic impedance relations across three basic elements.

### 19.1.3.2 Impermeable Plate

Consider the vibration of an infinite impermeable plate of thickness, $h$, induced by the sound pressure difference on each side of the plate, as illustrated in Figure 19.6. In this case, the particle velocity on both sides of the plate must be the same as the plate vibration velocity. Then, from Equation 19.8, the following expressions are obtained:

$$Z_2 = Z_1 + Z_m \tag{19.30}$$

$$\frac{p_2}{p_1} = \frac{Z_2}{Z_1} \tag{19.31}$$

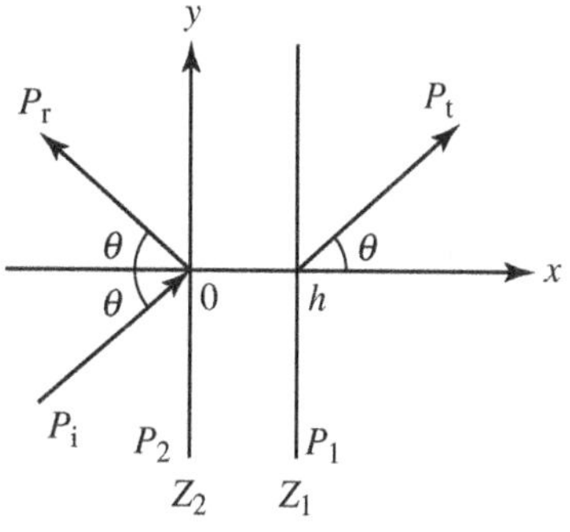

FIGURE 19.6 Excitation of infinite plate by a plane sound wave.

where $p_2, p_1$ are the sound pressure at the incident surface $x = 0$ and at the transmitted surface $x = h$, respectively, $Z_m$ is the mechanical impedance of the plate, and $Z_2, Z_1$ are the acoustic impedance normal to the incident surface at $x = 0$ and the transmitted one at $x = h$, respectively.

### 19.1.3.3 Sound Absorbing Material

For a sound absorbing material layer of thickness $d$ and infinite lateral extent, consider a plane wave incident at an angle $\theta$ to the normal, as shown in Figure 19.7. Deriving the wave equation in the sound absorbing material and applying the continuity conditions of the sound pressure across the surface at $x = 0$ and $x = d$, with some mathematical manipulation we get the following results:

$$Z_2 = \frac{\gamma Z_0}{q} \coth(qd + \varphi) \tag{19.32}$$

$$\frac{p_2}{p_1} = \frac{\cosh(qd + \varphi)}{\cosh\varphi} \tag{19.33}$$

$$q = \gamma\sqrt{1 + \left(\frac{k}{\gamma}\right)^2 \sin^2\theta}, \quad \varphi = \coth^{-1}\left(\frac{qZ_1}{\gamma Z_0}\right) \tag{19.34}$$

where $\gamma$ is a propagation constant and $Z_0$ is a characteristic impedance of a homogeneous, isotropic absorbing material.

If porous material is used as the absorbing material, the following relations are applicable for $\gamma$ and $Z_0$ [4]:

$$Z_0 = R + jX$$

$$R/\rho c = 1 + 0.0571(\rho f/R_1)^{-0.754} \quad (19.35)$$

$$X/\rho c = -0.0870(\rho f/R_1)^{-0.732}$$

$$\gamma = \alpha + j\beta$$

$$\alpha/k = 0.189(\rho f/R_1)^{-0.595} \quad (19.36)$$

$$\beta/k = 1 + 0.0978(\rho f/R_1)^{-0.700}$$

$$(0.01 \le \rho f/R_1 \le 1)$$

where $\rho$ is the air density, $f$ is the frequency, and $R_1$ is the flow resistivity, respectively. Specifically, note that $R_1$ is defined as the flow resistance of the porous absorbing material per unit thickness. With data measured with a measuring tube of flow resistance, we can write

$$R_1 = \frac{\Delta p}{l \cdot u} \quad (19.37)$$

where $\Delta p$ is pressure difference between the inlet and the outlet of the absorbing material in the tube, $u$ is the mean flow velocity in the tube, and $l$ is the thickness of the absorbing material. It is known that the flow resistivity of porous absorbing material such as fiber-glass or rock wool is related to the bulk density, as shown in Figure 19.8.

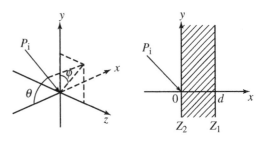

**FIGURE 19.7**   Schematic relation of sound wave directions.

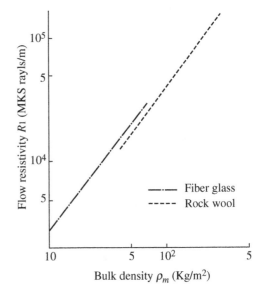

**FIGURE 19.8**   Flow resistivity vs. bulk density for porous, sound absorbing materials.

#### 19.1.3.4   Air Space

For an air space, $Z_0 = \rho c$ and $\gamma = jk$. Hence, Equation 19.32 to Equation 19.34 reduce to

$$Z_2 = \frac{\rho c}{\cos \theta} \coth (jkd \cos \theta + \delta) \quad (19.38)$$

$$\frac{p_2}{p_1} = \frac{\cosh (jkd \cos \theta + \delta)}{\cosh \delta} \quad (19.39)$$

$$\delta = \coth^{-1} \left( \frac{Z_1 \cos \theta}{\rho c} \right) \quad (19.40)$$

#### 19.1.3.5   Double Wall [2]

Applying the theory formulated above, we can easily obtain the transmission loss of a double wall composed of the three elements: impermeable plate, air space, and impermeable plate, as shown in Figure 19.9. Assume that the two impermeable plates have the same surface density, $m$, and the mechanical impedance of the plates is $j\omega m$. Then, we can obtain following equations for element one and element three:

$$Z_{12} = Z_{11} + Z_m = \frac{\rho c}{\cos \theta} + j\omega m,$$

$$\frac{p_{12}}{p_{11}} = \frac{Z_{12}}{Z_{11}} = 1 + j\frac{\omega m \cos \theta}{\rho c} \tag{19.41}$$

$$Z_{32} = Z_{31} + Z_m = Z_{22} + j\omega m,$$

$$\frac{p_{32}}{p_{31}} = \frac{Z_{32}}{Z_{22}} = 1 + j\frac{\omega m}{Z_{22}} \tag{19.42}$$

For element two:

$$Z_{22} = \frac{\rho c}{\cos \theta} \coth\left( jkd \cos \theta + \delta' \right),$$

$$\frac{p_{22}}{p_{21}} = \frac{\cosh\left( jkd \cos \theta + \delta' \right)}{\cosh \delta'} \tag{19.43}$$

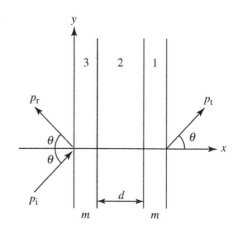

**FIGURE 19.9** Calculation model of a double wall with air space.

where, by applying impedance matching conditions at the interface of element one and element two, the following definition is introduced:

$$\delta' = \coth^{-1}\left( \frac{Z_{21} \cos \theta}{\rho c} \right) = \coth^{-1}\left( \frac{Z_{12} \cos \theta}{\rho c} \right) = \coth^{-1}\left( 1 + j\frac{\omega m \cos \theta}{\rho c} \right) \tag{19.44}$$

In this case, Equation 19.27 reduces to

$$\frac{p_i}{p_{32}} = \frac{Z_{32} + \rho c/\cos \theta}{2Z_{32}} = \frac{Z_{22} + j\omega m + \rho c/\cos \theta}{2(Z_{22} + j\omega m)} \tag{19.45}$$

Substituting Equation 19.41 to Equation 19.45 into Equation 19.28, we obtain the transmission loss of the double wall:

$$\text{TL}_\theta = 10 \log[1/\tau(\theta)] = 10 \log[1 + 4a^2 \cos^2 \theta (\cos \beta - a \cos \theta \sin \beta)^2] \tag{19.46}$$

$$a = \omega m/2\rho c, \quad \beta = kd \cos \theta$$

In Equation 19.46, the transmission loss is zero, and full passage (i.e., "all-pass" in the filter terminology) of sound occurs when the following equation holds:

$$\cos \beta - a \cos \theta \sin \beta = 0 \tag{19.47}$$

When $\beta \ll 1(kd \ll 1)$, the frequency of full passage for normal incidence is given by

$$f_r = \frac{1}{2\pi}\sqrt{\frac{2\rho c^2}{md}} \tag{19.48}$$

This is the natural frequency of a vibrating system consisting of two masses, $m$, connected by a spring of spring constant, $\rho c^2/d$.

When $\beta \gg 1(kd \gg 1)$, the solution of Equation 19.47 for $\beta$ is $\beta \cong n\pi$, and the frequency of all passage for normal incidence is given by

$$f_n = \frac{nc}{2d} \ (n = 1, 2, 3, \ldots) \tag{19.49}$$

These are the acoustic resonant frequencies of the air space $d$.

Characteristics of the transmission loss given by Equation 19.46, in case of normal incidence ($\theta = 0$), are as follows:

1. $f < f_r\left(\beta < \sqrt{2\rho d/m}\right)$

$$\text{TL} \cong 10 \log(4a^2) = \text{TL}_0 + 6 \tag{19.50}$$

This is equal to the transmission loss of a single wall of surface density $2m$.

2. $f_r \leq f < f_1/\pi \left( \sqrt{2\rho d/m} \leq \beta < 1 \right)$

$$TL \cong 10 \log(4a^4\beta^2) = 2TL_0 + 20 \log(2kd) \qquad (19.51)$$

This transmission loss indicates an 18 dB increase per octave.

3. $f = (2n - 1)c/4d(\beta = n\pi - \pi/2)$

$$TL \cong 10 \log(4a^4) = 2TL_0 + 6 \qquad (19.52)$$

A straight line connecting the transmission losses at these frequencies in Figure 19.10 indicates a 12 dB increase per octave. When the two impermeable plates have different surface densities, $m_1$ and $m_2$, Equation 19.41 to Equation 19.52 reduce to

1. $f < f_r \left( \beta < \sqrt{2\rho d/m} \right)$

$$TL = 20 \log[\omega(m_1 + m_2)/2\rho c] \qquad (19.53)$$

2. $f_r \leq f < f_1/\pi \left( \sqrt{2\rho d/m} \leq \beta < 1 \right)$

$$TL = TL_1 + TL_2 + 20 \log(2kd) \qquad (19.54)$$

3. $f = (2n - 1)c/4d(\beta = n\pi - \pi/2)$

$$TL = TL_1 + TL_2 + 6 \qquad (19.55)$$

In these equations, $TL_1$ and $TL_2$ are the transmission losses of each plate, which are given by Equation 19.15.

The transmission loss of a double wall, as mentioned above, is shown schematically in Figure 19.10. An actual double wall, however, is finite in size and the air space forms a closed acoustic field, which

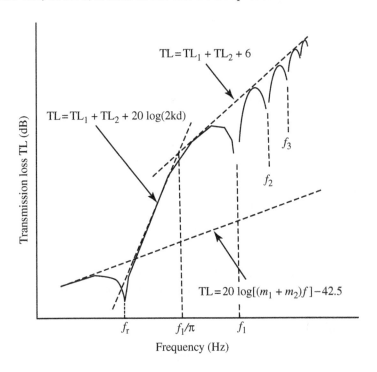

**FIGURE 19.10**   Transmission loss of a double wall with air space.

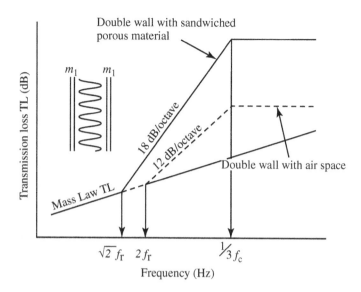

**FIGURE 19.11**  Design chart for estimating the transmission loss of a double wall with sandwiched porous material or air space.

makes the transmission loss deviate from the theoretical value. Figure 19.11 gives a design chart of an actual double wall, which is based on theory and experiments.

## 19.1.4  Transmission Loss of Double Wall with Sound Bridge [5]

In the previously presented theory, each plate of the multiple panel is considered to be structurally independent. In actual multiple panels, such as partitions of a building or sound insulation laggings of a duct, however, each plate is connected with steel sections, stud bolts, and the like, which are called sound bridges. This is illustrated in Figure 19.12.

The sound pressure of the transmitted wave through a double wall with sound bridges is given by the summation of radiated sound pressure from the vibration of the transmitted side plate excited by the sound in the air space and that mechanically excited by the sound bridges.

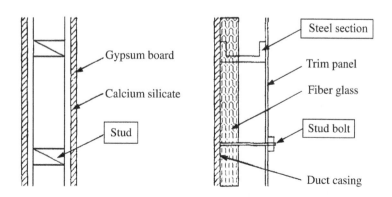

**FIGURE 19.12**  Examples of actual double wall with sound bridges.

The acoustic power radiated from the area, $S$, of an infinite plate excited by sound pressure is given by

$$W_P = \rho c S v_2^2 \tag{19.56}$$

where $v_2^2$ is the space averaged mean square vibration velocity over the plate. The acoustic power radiated from the plate mechanically excited by a point force or a line force is

$$W_B = \rho c \chi v^2 (f << f_c) \tag{19.57}$$

$$\chi = \begin{cases} \dfrac{8}{\pi^3} \lambda_c^2 & \text{(point force excitation)} \\[2mm] \dfrac{2}{\pi} l \lambda_c & \text{(line force excitation)} \end{cases} \tag{19.58}$$

where $v^2$ is the mean square vibration velocity of the plate at the excitation point, $\lambda_c = c/f_c$ is the wavelength of the bending wave at the critical frequency, and $l$ is the length of the line force. By comparing Equation 19.56 and Equation 19.57, it is noted that $\chi$ is the effective area of the acoustic power radiated from the infinite plate excited by the point or line force. Acoustic power, $W_B$, is the power radiated from a small area near the excitation point, because a free bending wave propagating in an infinite plate can radiate little sound when $f < f_c$.

From the equations given above, the total acoustic power radiated from the transmitted side plate is obtained as

$$W_T = W_P + W_B = \rho c S v_2^2 \left[ 1 + \frac{n\chi}{S} \left( \frac{v}{v_2} \right)^2 \right] \tag{19.59}$$

where $n$ is the number of excitation forces applied to the area, $S$. Then, transmission loss $\text{TL}_T$ of the double wall with sound bridges is given by

$$\text{TL}_T = 10 \log\left( \frac{W_I}{W_T} \right) = 10 \log\left( \frac{W_I}{W_P} \cdot \frac{W_P}{W_T} \right) = \text{TL} - \text{TL}_B \tag{19.60}$$

where $W_I$ is the acoustic power incident on the double wall, TL is the transmission loss of the double wall without a sound bridge, and $\text{TL}_B$ denotes the transmission loss reduction by the sound bridges, and is given by

$$\text{TL}_B = 10 \log\left( \frac{W_T}{W_P} \right) = 10 \log\left[ 1 + \frac{n\chi}{S} \left( \frac{v}{v_2} \right)^2 \right] \tag{19.61}$$

We assume the following:

1. The vibration velocity of the incident side plate is not affected by the sound bridges.
2. The vibration velocity of the transmitted side plate at the excitation points (connecting points) is equal to the velocity $v_1$ of the incident side plate, and consequently, the next equation holds

$$\frac{v}{v_2} \cong \frac{v_1}{v_2}$$

With these assumptions, we apply the method presented in section 19.1.3, to determine $v/v_2$ as

$$\frac{v}{v_2} \cong \frac{v_1}{v_2} = \frac{\omega^2 m_2 d}{\rho c^2} \quad (f_r < f < f_1/\pi) = \frac{\omega m_2}{\rho c} \quad (f > f_1/\pi) \tag{19.62}$$

Using Equation 19.60 and Equation 19.53, we can obtain the increase in transmission loss, $\Delta$TL, from the transmission loss of mass law based on the total mass of the double wall, as presented below.

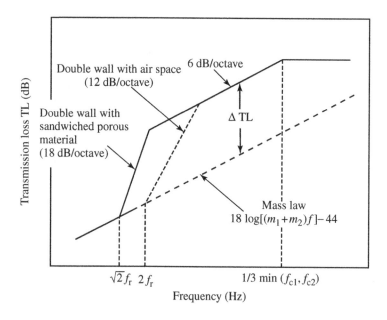

**FIGURE 19.13**  Design chart for estimating the transmission loss of a double wall with sound bridges.

1. Point connection

$$\Delta TL = TL - TL_B - 20 \log\left[ \frac{\omega(m_1 + m_2)}{2\rho c} \right]$$

$$= 20 \log(ef_c) + 20 \log\left( \frac{m_1}{m_1 + m_2} \right) + 10 \log\left( \frac{\pi^3}{8c^2} \right) \qquad (19.63)$$

2. Line connection

$$\Delta TL = TL - TL_B - 20 \log\left[ \frac{\omega(m_1 + m_2)}{2\rho c} \right]$$

$$= 10 \log(bf_c) + 20 \log\left( \frac{m_1}{m_1 + m_2} \right) + 10 \log\left( \frac{\pi}{2c} \right) \qquad (19.64)$$

where $e = \sqrt{S/n}$ is the distance between point forces and $b = S/nl$ is the distance between line forces.

Figure 19.13 presents a practical and useful design chart of the transmission loss for a double wall with sound bridges, which is based on Figure 19.11 and Equation 19.63 and 19.64.

## 19.2  Application of Sound Insulation

### 19.2.1  Acoustic Enclosure

Performance of an enclosure may be represented by the insertion loss (IL), which is the difference of acoustic power level before and after installation of the enclosure. When we assume a noise source and also an enclosure with one-dimensional model as shown in Figure 19.14a, the insertion loss through frequency is shown in Figure 19.14b. It is divided into the following four regions:

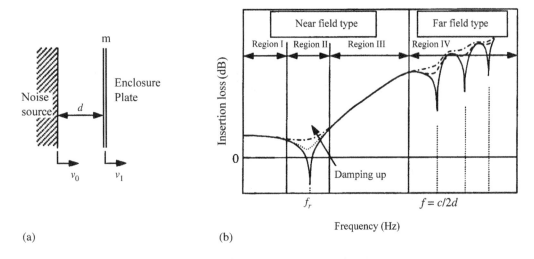

<table>
<tr><td></td></tr>
</table>

(a)                                               (b)

**FIGURE 19.14**  One-dimensional model for calculating the insertion loss characteristics of an acoustic enclosure: (a) One-dimensional model; (b) insertion loss of an enclosure.

1. Region I ($f < f_r$) is controlled by the stiffness of the enclosure plate and air space.
2. Region II ($f \cong f_r$) is the resonance region for a vibrating system consisting of the mass, the stiffness of the enclosure plate, and the capacitance of air space.
3. Region III ($f_r < f < c/2d$) is controlled by the mass of the enclosure plate.
4. Region IV ($f > c/2d$) is controlled by the diffused sound field. This region cannot be represented by a one-dimensional model.

### 19.2.1.1  Near Field Type [4]

When the distance between the noise source and the enclosure is less than half of a wavelength of the emitted sound from the source, insertion loss corresponds to the characteristics of the regions I to III, and we can represent them with a one-dimensional acoustic model, as shown in Figure 19.14a.

Consider an infinite flat enclosure plate with distance $d$ from a noise source of plane sound wave, as shown in Figure 19.14a. The insertion loss of the plate is given by

$$IL = 10 \log\left(\frac{v_0}{v_1}\right)^2 = 10 \log\left[ 1 - \frac{2 \sin \theta(X \cos \theta - R \sin \theta)}{\rho c} + \frac{\sin^2 \theta(X^2 + R^2)}{\rho^2 c^2} \right] \tag{19.65}$$

$$\theta = kd = \omega d/c, \quad R = \eta \omega m, \quad X = (\omega m - K/\omega) = \omega m[1 - (\omega_{11}/\omega)^2], \quad \omega_{11} = \sqrt{K/m}$$

where $m$ and $K$ are the density and equivalent stiffness of the plate per unit area, respectively, and $\eta$ is the loss factor of the plate. If the enclosure is a rectangular plate of size $a \times b$ and is simply supported at its edges, the equivalent stiffness is given by Equation 19.22, and $\omega_{11}$ is the natural (angular) frequency of the first mode.

In Equation 19.65, the conditions in which the brackets of the right-hand side are equal to zero or $IL = -\infty$ are satisfied by following frequencies:

(1) $\theta \ll \pi$

$$f_r = \frac{1}{2\pi} \sqrt{\frac{K + \rho c^2/d}{m}} = \frac{1}{2\pi} \sqrt{\omega_{11}^2 + \frac{\rho c^2}{md}} \tag{19.66}$$

This is the natural frequency of vibration of the one-degree-of-freedom (one-DoF) system determined by the stiffness of the plate, the spring constant of the air space, and the surface density of the plate, as shown in Figure 19.15.

(2) $\theta = n\pi$ $(n = 1, 2, ...)$

$$f_n = \frac{nc}{2d} \quad (n = 1, 2, ...) \quad (19.67)$$

These are the resonant frequencies of the air space.

The frequency characteristics of the IL given by Equation 19.65 are shown in Figure 19.16, where the normal incidence transmission losses are shown by broken lines, as a reference. Equation 19.65 is approximated by

1. $f < f_r$

$$IL \cong 10 \log\left(1 + \frac{2Kd}{\rho c^2}\right) \quad (19.68)$$

2. $f_r \leq f < f_1$

$$IL \cong 20 \log(mdf^2) + 20 \log\left(\frac{4\pi^2}{\rho c^2}\right)$$
$$= 20 \log(mdf^2) - 71 \quad (19.69)$$

### 19.2.1.2  Far Field-Type (Absorption Type) Enclosure [1]

When the distance between the noise source and the enclosure is larger than half of a wavelength of the emitted sound, insertion loss may be represented by the characteristics of Region IV, and it can be analyzed using the theory of room or hall acoustics.

Consider the enclosure shown in Figure 19.17, with a noise source of power level $L_{W0}$.

From the theory of room acoustics, the average sound pressure level, $L_{P0}$, on the inner surface of the enclosure plate is obtained as the sum of the direct and reverberant sound pressures:

$$L_{P0} = L_{W0} + 10 \log\left(\frac{1}{S} + \frac{4}{R}\right)$$
$$= L_{W0} - 10 \log S + 10 \log\left(1 + \frac{4S}{R}\right) \quad (19.70)$$

$$R = \frac{\bar{\alpha} S}{(1 - \bar{\alpha})} \quad (19.71)$$

where $S$ is the inner surface area of the enclosure and $\bar{\alpha}$ is the average absorption coefficient on the inner surface of the enclosure. In Equation 19.70, the first and the second terms of the right-hand side represent the influence of the direct sound field, and the third term represents the buildup caused by the covering.

When we use $S_P$ for the area of the enclosure plate and $S_O (= S - S_P)$ for the area of the enclosure opening, and assume diffusing condition for the sound field in the enclosure, acoustic power levels

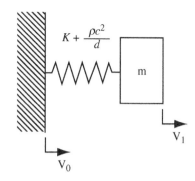

**FIGURE 19.15**  One-degree-of-freedom vibrating system.

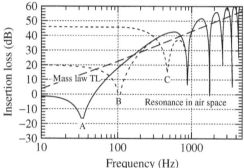

$m = 16$ Kg/m², $d = 0.19$ m
Curve A: $f_{11} = 0$ Hz, $\eta \cong 0.033$ (at 33 Hz)
Curve B: $f_{11} = 100$ Hz, $\eta \cong 0.033$ (at 105 Hz)
Curve C: $f_{11} = 475$ Hz, $\eta \cong 0.033$ (at 475 Hz)

**FIGURE 19.16**  Example of theoretical insertion loss of a near field-type enclosure.

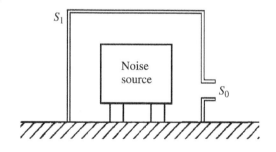

**FIGURE 19.17**  Calculation model of a far field type enclosure.

radiated from the plate and the opening are given, respectively, by

$$L_{WP} = L_{P0} - (TL + 6) + 10 \log S_P, \tag{19.72}$$

$$L_{W0} = L_{P0} - 6 + 10 \log S_O$$

where TL is the random incident transmission loss of the enclosure plate. Then, the insertion loss of the enclosure is

$$IL = L_{W0} - 10 \log\left(10^{L_{WP}/10} + 10^{L_{W0}/10}\right) \tag{19.73}$$

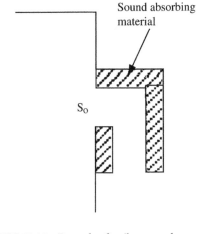

**FIGURE 19.18**   Example of a silencer at the opening.

In the design of an acoustic enclosure, special attention should be paid to the following points:

1. *Buildup* increases the sound pressure in the enclosure and also the power level radiated from the enclosure. We must treat the inner surface of the enclosure with sound absorbing materials to reduce the absorption coefficient and to decrease the buildup.

2. *Opening* radiates more acoustic power than the enclosure plate by TL, as is clear from Equation 19.72. It is desirable to make the opening as small as possible within the range given by

$$\frac{S_O}{S_P} \leq 10^{-TL/10} = \tau \tag{19.74}$$

   This equation means that the acoustic power from the opening is less than that from the enclosure plate. If the relation in Equation 19.74 cannot be satisfied because of ventilation requirements, and so on, some type of silencers should be provided at the opening, as shown in Figure 19.18.

3. *Structure-borne noise*, which is caused by the vibration propagating from the base of the machine (noise source) to the enclosure plate, significantly decreases the insertion loss of the enclosure. In this case, some means of noise/vibration suppression should be provided, for example, the following:
   - Place supporting structures of the enclosure at the points of lowest vibration level, and the vibrations of the machine should be prevented from propagating to the enclosure plate, using vibration isolation materials.
   - Add damping materials to the enclosure plate so as to reduce the vibration level of the plate.

## 19.2.2   Sound Insulation Lagging

In electric power plants and chemical plants, for example, piping for high-pressure water or steam, and ducts for air or gas flow form major noise sources. For controlling these noise sources, we usually use sound insulation laggings, which cover the noise sources with heavy and impermeable plates or sheets with sound absorbing materials, as shown in Figure 19.19.

### 19.2.2.1   Pipe Lagging [2]

Approximate the cylindrical piping and pipe lagging with a one-dimensional model as shown in Figure 19.20. The insertion loss of one-layered lagging approximated by a one-DoF system is given by Equation 19.69 in the frequency region $f_r \leq f < f_1$, as mentioned before. It is not practical, however, to directly apply Equation 19.69 to actual laggings, and we approximate the insertion loss of actual laggings by

$$IL = a \log(mdf^2) + b$$

where $a$ and $b$ are constants.

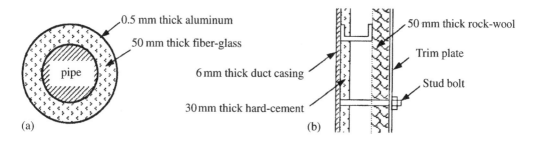

**FIGURE 19.19** Examples of typical sound insulation laggings: (a) pipe lagging, (b) duct lagging.

By taking $mdf^2$ as the horizontal axis and plotting the insertion loss data from laboratory tests and field tests (the vertical coordinates), we obtain Figure 19.21. Apply regression analysis to the data in Figure 19.21 to obtain the insertion loss of one layered lagging as

$$\text{IL} = 11.7 \log(mdf^2) - 43.3 \ (5 \times 10^3 \le mdf^2 \le 10^8) \tag{19.75}$$

Applying the same method to double layered lagging, approximated by a two-DoF vibrating system, we get Figure 19.22, and the insertion loss

$$\text{IL} = 6.9 \log(m_1 m_2 d_1 d_2 f^4) - 40.3 \ (10^6 \le m_1 m_2 d_1 d_2 f^4 \le 10^{15}) \tag{19.76}$$

where the subscripts "1" and "2" denote the first layer and the second layer, respectively.

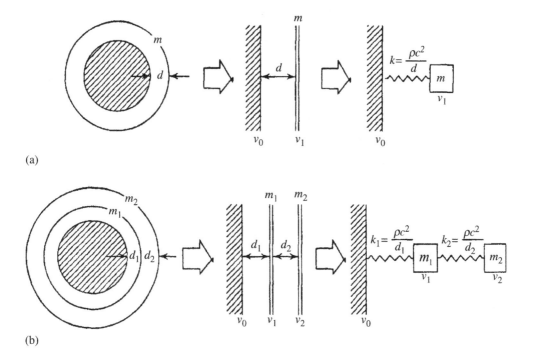

**FIGURE 19.20** Examples of pipe laggings and calculation model: (a) one-layered lagging; (b) double-layered lagging.

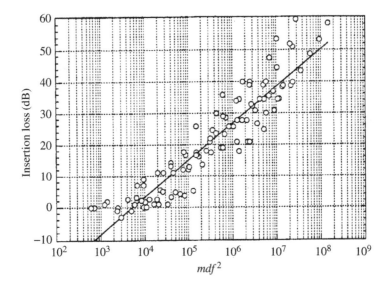

**FIGURE 19.21** Measured insertion loss data obtained from laboratory and field tests for one-layered laggings and regression analysis.

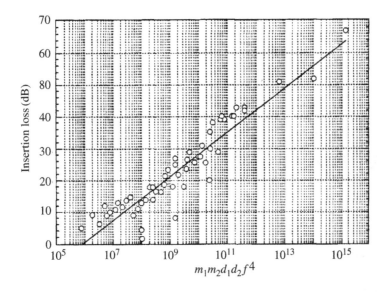

**FIGURE 19.22** Measured data of insertion loss obtained from laboratory and field tests for double-layered laggings and regression analysis.

### 19.2.2.2 Duct Lagging [4]

Various types of the duct laggings are used according to the need, for example, as shown in Figure 19.19. A simpler and more practical approach is to place a thin plate on the duct casing through absorbing materials, as shown in Figure 19.23. In this case, assuming that vibration of the duct casing is not affected by the placed plate, the insertion loss of the duct lagging is obtained by following equations, which can be deduced from the method used in the transmission loss of double wall with sound bridges.

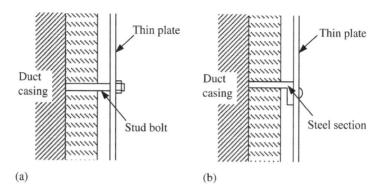

**FIGURE 19.23** Examples of duct laggings and connection types of thin plate to the duct casing: (a) point connection; (b) line connection.

1. Point connection

$$\text{IL} = -10 \log\left[ \beta^2 \frac{8}{\pi^3} n_P \frac{c^2}{f_c^2} + \left(\frac{f_r}{f}\right)^4 \right] + 10 \log \sigma \tag{19.77}$$

2. Line connection

$$\text{IL} = -10 \log\left[ 0.64 n_L \frac{c}{f_c} + \left(\frac{f_r}{f}\right)^4 \right] + 10 \log \sigma \tag{19.78}$$

where
$\beta$ = vibration isolation factor of the flexible support ($\beta = 1$ for rigid support)
$n_P$ = number of attachment points per unit area
$n_L$ = number of studs per unit length

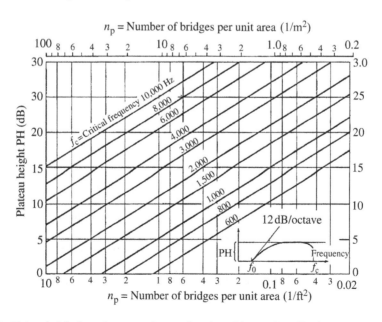

**FIGURE 19.24** Plateau height for point connection as a function of the number of bridges per unit area [4]. (*Source:* Beranek, L. L. 1988. *Noise and Vibration Control*, INCE/USA. With permission.)

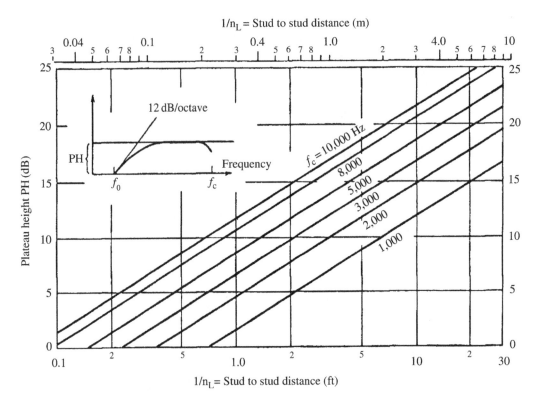

**FIGURE 19.25** Plateau height for line connection as a function of the number of studs per unit length [4]. (*Source*: Beranek, L. L. 1988. *Noise and Vibration Control*, INCE/USA. With permission.)

$f_c$ = critical frequency of the thin plate given by Equation 19.12
$f_r$ = resonant frequency given by Equation 19.66
$\sigma$ = sound radiation efficiency of the thin plate

Knowing the critical frequency, $f_c$, and the resonant frequency, $f_r$, we can obtain the insertion loss from the charts given in Figure 19.24 and Figure 19.25 instead of using Equation 19.77 and Equation 19.78. In Figure 19.24 and Figure 19.25, it is assumed that $\sigma = 1$. Note from Equation 19.77 and Equation 19.78 that we must consider the following measures to obtain a higher insertion loss.

1. Make the distance between attachment points or studs as large as possible (decrease $n_P$ and $n_L$).
2. Make the air space as large as possible (decrease $f_r$).

## References

1.    Shiraki, K., ed. 1987. *From Designing of Noise Reduction to Simulation* (in Japanese), Ouyou-gijutsu Shuppan, Chiyoda-ku, Tokyo.
2.    Tokita, Y., ed. 2000. *Sound Environment and Control Technology, Vol. I, Basic Engineering* (in Japanese), Fuji-techno-system, Bunkyo-ku, Tokyo.
3.    Okura, K. and Saito, Y., Transmission loss of multiple panels containing sound absorbing materials in a random incidence field, *Inter-noise*, 78, 637, 1978.
4.    Beranek, L.L., ed. 1988. *Noise and Vibration Control*, INCE/USA, Ames, IA.
5.    Sharp, B.H. 1973. A study of techniques to increase the sound insulation of building elements, Wyle Laboratory Report WR 73-5, El Segundo, CA.

# 20

# Statistical Energy Analysis

Takayuki Koizumi
*Doshisha University*

## Summary

*This chapter presents the basics of statistical energy analysis (SEA) as applied to acoustic problems in structural systems. Power flow equations for structures consisting of two or more subsystems are described. The modeling and analysis procedures for the structural subsystems and acoustic subsystems are given. An estimation procedure of the necessary SEA parameters is given. The practical application of the SEA procedure in structures is illustrated using a two-story building as an example.*

## 20.1 Introduction

This chapter describes the basic concepts of the method of statistical energy analysis (SEA) and presents its application to structures. The analysis and computation techniques for vibration response and radiating sound in instruments and structures vary according to the characteristics of the physical object and the frequency range of interest. Here, we analyze vibration and noise in relation to a rather large-scale structure over a wide frequency band. Extensive computations are usually required, when, for example, the finite element method is used for the computations, with respect to a given oscillation mode. In particular, when the computations must be performed in the high-frequency range and when many modes are included in the frequency band, the level of computation becomes considerable, generally resulting in reduced computational accuracy. To supplement the weak point of the traditional approach, it is necessary to redistribute statistically the energy equally from all modes in the analytical frequency band. This allows computed results to be compared with experimental results for a structure across a wide frequency band. This is the SEA method [1]. Early in its development, the objective of this analytical method was to predict the vibration response of artificial satellites and rockets that receive sound excitation when the jet discharges, and to predict the response of vibration stress in the boundary layer noise of an aircraft's airframe. It also became a model that allows an exciting force to be statistically (randomly) diffused (distributed) over a wide frequency band. This technique considers energy of excitation of a diffused (distributed) sound field and its variables that represent the sound pressure, acceleration, and force. Thus, it can be applied to problems of solid-borne sound in which vibration propagates through each element [2] and problems of air-borne sound in which multiple barriers exist [3], even when more excitation points than one are present.

## 20.2 Power Flow Equations

With the SEA method, we do not deal with specific characteristic modes of the analyzed structure. Instead, we consider the structural components as a set of equivalent vibrating elements, and evaluate the vibration condition of the components as a macroscopic quantity averaged statistically over the frequency band and space (by describing the energy). We assume that the vibration modes within a given frequency band are distributed uniformly and are excited to the same degree.

Using the SEA method, we can formalize the relationships of power flows between subsystems, and by solving these relationships, we can compute the energy stored in each subsystem. Next, the equations of such basic power flow [4] are explained.

### 20.2.1 Power Flow Equations of a Two-Subsystem Structure

The power flow relationships of a structure consisting of a two-subsystem structure are shown in Figure 20.1. The equations for the power flows between subsystem 1 and subsystem 2 under typical conditions are expressed as

$$\text{Subsystem 1: } P_{i1} = P_{11} + P_{12} \qquad (20.1)$$

$$\text{Subsystem 2: } P_{i2} = P_{12} + P_{21} \qquad (20.2)$$

where $P_{i1}$ is the input power to subsystem 1 from outside, $P_{11}$ is the internal power loss of subsystem 1, and $P_{12}$ is the transmitted power from subsystem 1 to subsystem 2.

The internal power loss, $P_{11}$, is written as

$$P_{11} = \omega \eta_1 E_1 \qquad (20.3)$$

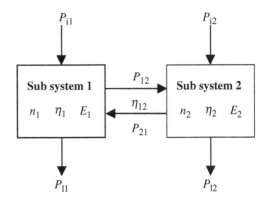

**FIGURE 20.1**  Power flow relationships between two subsystems.

where $\omega$ is the central angular frequency in the band, $E_1$ is the energy in the bandwidth $\Delta\omega$ of subsystem 1, and $\eta_1$ is the internal loss factor (ILF).

The average modal energy $E_{m1}$ in subsystem 1, and $E_{m2}$ in subsystem 2, are given by

$$E_{m1} = \frac{E_1}{N_1}, \quad E_{m2} = \frac{E_2}{N_2} \qquad (20.4)$$

where $N_1$ is number of modes in the bandwidth $\Delta\omega$ of subsystem 1, and $N_2$ is number of modes in the bandwidth $\Delta\omega$ of subsystem 2.

The transferred power, $P_{12}$, between subsystems 1 and 2 is expressed as

$$P_{12} = -P_{21} = P'_{12} - P'_{21} \qquad (20.5)$$

$$P'_{12} = \omega \eta_{12} E_1 = \omega \eta_{12} N_1 E_{m1} \qquad (20.6)$$

$$P'_{21} = \omega \eta_{21} E_2 = \omega \eta_{21} N_2 E_{m2} \qquad (20.7)$$

where $\eta_{12}$ and $\eta_{21}$ are the coupling loss factors (CLFs) from subsystem 1 to subsystem 2, and from subsystem 2 to subsystem 1. They satisfy the reciprocity relationship $\eta_{12}n_1 = \eta_{21}n_2$. Therefore, transferred power, $P_{12}$, becomes

$$P_{12} = \omega\eta_{12}N_1(E_{m1} - E_{m2}) = \omega\eta_{12}N_1\left(\frac{E_1}{N_1} - \frac{E_2}{N_2}\right) \tag{20.8}$$

Consequently, the power flow equations (Equation 20.1 and Equation 20.2) can be expressed as follows:

$$P_{i1} = \omega\eta_1 E_1 + \omega\eta_{12}N_1\left(\frac{E_1}{N_1} - \frac{E_2}{N_2}\right) \tag{20.9}$$

$$P_{i2} = \omega\eta_2 E_2 + \omega\eta_{21}N_2\left(\frac{E_2}{N_2} - \frac{E_1}{N_1}\right) \tag{20.10}$$

If the SEA parameters (i.e., the modal density, intrinsic loss factor, CLF, and input power) are given, then each subsystem's energy condition can be easily computed.

## 20.2.2 Power Flow Equations of a Multiple Subsystem Structure

By expanding the formulation in the previous section, it is possible to formalize the power flow relationships of a structure composed of multiple subsystems in the same way. The power flow equation for a structure composed of $N$ subsystems is expressed by the following equation in the matrix form:

$$\omega \begin{bmatrix} \left(\eta_1 + \sum_{i\neq1}^{N}\eta_{1i}\right)n_1 & -\eta_{12}n_1 & \cdots & -\eta_{1N}n_1 \\ -\eta_{21}n_2 & \left(\eta_2 + \sum_{i\neq2}^{N}\eta_{2i}\right)n_2 & \cdots & -\eta_{2N}n_2 \\ \vdots & \vdots & \ddots & \vdots \\ -\eta_{N1}n_n & \cdots & \cdots & \left(\eta_N + \sum_{i\neq N}^{N-1}\eta_{Ni}\right)n_N \end{bmatrix} \times \begin{bmatrix} E_1/n_1 \\ E_2/n_2 \\ \vdots \\ E_N/n_N \end{bmatrix} = \begin{bmatrix} P_{i1} \\ P_{i2} \\ \vdots \\ P_{iN} \end{bmatrix} \tag{20.11}$$

From Equation 20.11, if the SEA parameters are given in the same way as for the structure of two subsystems, then the energy equation of each subsystem can be obtained.

The average energy of a subsystem is expressed by the following equations by using the vibration velocity and sound pressure:

$$E = M\langle v^2\rangle \tag{20.12}$$

$$E = \frac{M\langle p^2\rangle}{Z_0^2} \tag{20.13}$$

where $M$ is the mass of the subsystem, $\langle v^2\rangle$ is the average spatial square of the vibration velocity, $\langle p^2\rangle$ is the average spatial square of the sound pressure, and $Z_0$ is the specific acoustic impedance of air.

Accordingly, if each condition of component's energy is determined from Equation 20.11, it is possible to compute the vibration variable and the sound pressure with Equation 20.12 and Equation 20.13.

## 20.3  Estimation of SEA Parameters

To solve the power flow equations, it is necessary to determine the SEA parameters (i.e., the modal density, ILF, CLF, and input power). In this subsection, a method is given for computing the SEA parameters.

### 20.3.1  Modal Density

#### 20.3.1.1  Structural Subsystem

Modal density is a key parameter for determining the dynamic characteristic of a structure. The number of modes, $N$, included in the frequency band (for estimation), is a factor denoting how easily energy, in transferring between subsystems, can be obtained. To determine $N$ in the prescribed frequency band, it is first necessary to determine the modal density $n(f)$, that is, the gradient of $N$ in the frequency band.

The modal density of a structural subsystem is computed by using the following equation [4,5]:

$$n(f) = \frac{dN}{df} = \frac{1}{f_0} = \frac{A}{2t}\sqrt{\frac{12\rho(1-\nu^2)}{E}} \tag{20.14}$$

where $A$ is the area of cross section, $t$ is the thickness of the structural subsystem, $\rho$ is the mass density, $\nu$ is the Poisson's ratio, $E$ is the Young's modulus, and $f_0$ is the fundamental natural frequency of the structural subsystem.

#### 20.3.1.2  Acoustic Subsystem

The modal density of an acoustic subsystem is determined by the following analytical equation [6]:

$$n(f) = \frac{dN}{df} = \frac{4\pi V}{c^3}f^2 \tag{20.15}$$

where $c$ is the speed of sound propagation within the acoustic subsystem, and $V$ denotes the volume of the acoustic subsystem.

Modal density of the cavity in the low frequency band is deduced in a similar manner to that in the two-dimensional space. Define the depth of the cavity by $d$, and the frequency of the standing wave in the cavity by $f_d = c/2d$. If $f < f_d$, then the modal density is assumed to be uniformly distributed, and is estimated by

$$n(f) = \frac{2\pi S}{c^2}f \tag{20.16}$$

where $S$ is the area of the cavity.

If $f > f_d$, modal density can be estimated using Equation 20.15, because the cavity is designated as acoustically three dimensional.

### 20.3.2  Internal Loss Factor

#### 20.3.2.1  Structural Subsystem

The ILF, $\eta_1$, of a subsystem gives the loss ratio when the input power to the subsystem from the outside is converted to kinetic energy of the subsystem. An excitation test to measure the damping ratio is employed to estimate the ILF of the structural subsystem. There are several methods for estimating the internal loss factor. The ILF applied in the SEA method is estimated by measuring input energy and output energy simultaneously, or by measuring the attenuation ratio within a given period of time. Both methods require the same setup to conduct an excitation test, and one is able to improve the measurement precision by conducting both methods.

With the energy measuring methods mentioned above, the ILF can be estimated by

$$\eta = \frac{\int_{f_1}^{f_2} \text{Re}(Y)F^2 df}{\omega_0 M \left\langle \int_{f_1}^{f_2} v^2 df \right\rangle} \tag{20.17}$$

where $Y$ is the complex mobility at the driving point in the range of $f_1$ to $f_2$, $F^2$ is the power spectrum of the input vibration force, and $v^2$ is the power spectrum of the response speed. In addition, $\langle \, \rangle$ denotes the space average operator.

### 20.3.2.2 Acoustic Subsystem

The ILF of an acoustic subsystem is determined by [7]

$$\eta = \frac{cS\bar{\alpha}}{4V\omega} \tag{20.18}$$

where $\bar{\alpha}$ is the average acoustic absorption coefficient, $V$ is the volume of the acoustic subsystem, and $S$ denotes the surface area. The acoustic absorption coefficient can be estimated by measuring the reverberation time.

Both the ILF of the cavity in the low-frequency band and the modal density are deduced similar to that in two-dimensional spaces. For $f < f_d$, the ILF is estimated using

$$\eta = \frac{cS_p \alpha_p}{\pi \omega V_c} \tag{20.19}$$

where $\alpha_p$ is the acoustic absorption coefficient in the cavity, $S_p$ is the peripheral area of the cavity, and $V_c$ is the volume of the cavity.

For $f > f_d$, the modal density can be estimated using Equation 20.18 because the cavity is taken as an acoustic subsystem.

## 20.3.3 Coupling Loss Factor

### 20.3.3.1 Between Structural Subsystems

The CLF $\eta_{ij}$ gives the loss ratio when power transmits between two subsystems [4]. For example, the CLF between two flat plates can be estimated using

$$\eta_{ij} = \frac{c_{gi} L_c \tau}{\pi \omega S_i} \tag{20.20}$$

where $c_{gi}$ is the group velocity of the bending waves, $L_c$ is the coupled length, $S_i$ is the surface area, and $\tau_{ij}$ is the energy transmission factor from subsystem $i$ to subsystem $j$. The transmission factor varies with the type of coupling, for example, I-type, L-type, or T-type shown in Figure 20.2. In this section, we use energy transmission efficiency of vertical incidence, reported by Cremer [7].

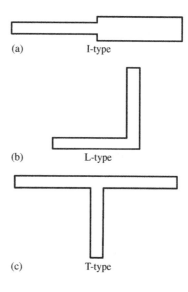

(a)       I-type

(b)       L-type

(c)       T-type

**FIGURE 20.2** Coupled type.

### 20.3.3.2 Between a Structural Subsystem and an Acoustic Subsystem

Coupling power between a structural subsystem and an acoustic subsystem is the power flow based on resonance at transmission. The CLF between a structural subsystem and an acoustic subsystem is given by

$$\eta_{ij} = \frac{Z_0 S_c \sigma}{\omega M_i} \qquad (20.21)$$

where $Z_0$ is the specific acoustic impedance of air, $S_c$ is the surface area of coupling, $\sigma$ is the acoustical radiation efficiency, and $M_i$ is the mass of the structural subsystem.

### 20.3.3.3 Between an Acoustic Subsystem and a Cavity

Coupling power between an acoustic subsystem and a cavity is power flow based on resonance in transmission. The CLF between an acoustic subsystem and a cavity is given by [8]

$$\eta_{sc} = \frac{c_s S_{cs} \tau_m}{4\omega V_s} \qquad (20.22)$$

where $c_s$ is the sound velocity in the acoustic space, $V_s$ is the volume in the acoustic space, $\tau_m$ is the transmission factor at random incidence depending on mass flow through the partition, and $S_{cs}$ is the coupling area.

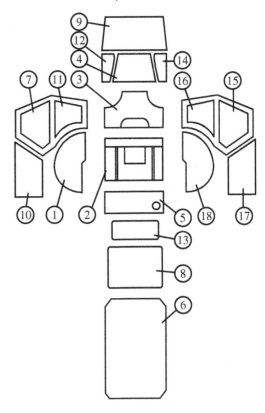

**FIGURE 20.3** Modeling of the tractor cabin.

## 20.3.4 Input Power

### 20.3.4.1 Vibration Input Power

The vibration input power $P_{iN}$ is given by

$$P_{iN} = \omega M_i \langle v_i^2 \rangle \qquad (20.23)$$

where $M_i$ is the equivalent mass, $\langle v_i^2 \rangle$ is the spatial average of square of the vibration velocity, and $\omega$ is the central angular frequency.

### 20.3.4.2 Acoustical Input Power

The acoustical input power $P_s$ is given by

$$P_s = \frac{\langle p^2 \rangle S^2 n(f)}{4M} \sigma_{rad} \frac{c^2}{2\pi S f^2} \qquad (20.24)$$

where $\langle p^2 \rangle$ is the square average of the input sound pressure, $S$ is the surface area of the component, $M$ is the mass of the component, $n(f)$ is the modal density, and $\sigma_{rad}$ is the sound radiation factor [2].

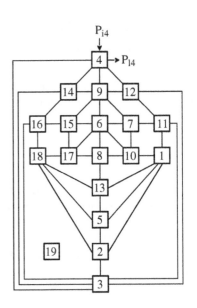

**FIGURE 20.4** Power flows in the tractor cabin.

## 20.4 Application in Structures

In this section, we present an example of modeling with significant analytical accuracy, and discuss the application of the SEA method for structures.

### 20.4.1 Application for Prediction of Noise in a Tractor Cabin

Figure 20.3 shows a model of the tractor cabin. This figure shows that the cabin consists of a floor, a door, a ceiling, and other components. Figure 20.4 presents the power flow relationships within the cabin [9,10].

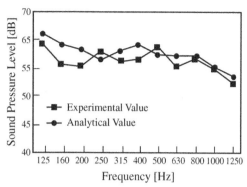

**FIGURE 20.5**   Results of estimating the sound pressure level in a cabin.

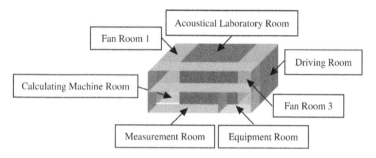

**FIGURE 20.6**   The configuration of the building.

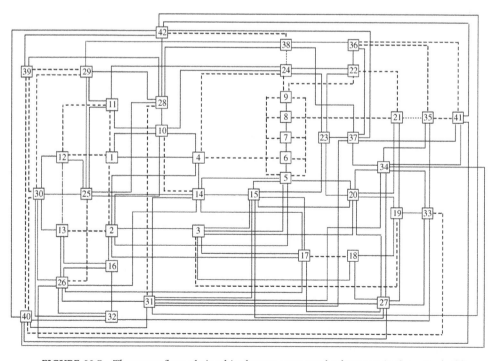

**FIGURE 20.7**   The power flow relationships between structural subsystems in the entire building.

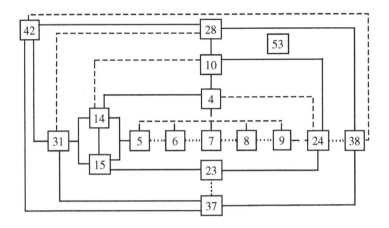

**FIGURE 20.8**   The power flow relationships in the acoustical laboratory room.

The results obtained for the cabin are shown in Figure 20.5. According to this figure, the disagreement between the computation and the measurement was found about 2 dB in the medium- to high-frequency band.

## 20.4.2   Application for Prediction of Noise and Vibration in a Building

Consider a two-story laboratory reinforced with concrete [11]. The building configuration is shown in Figure 20.6. This building comprises a driving room, an acoustical laboratory room, a computer room, a measurement room, an equipment room, and others.

We modeled the structural subsystem using I-, L-, or T-type connected points, and the acoustic subsystem as an element shown in Figure 20.6.

The SEA model constructed in this manner is composed of 61 elements, and has 244 connecting points. Subsystems 1 to 17 and subsystems 19 to 42 are concrete components. Subsystem 18 and subsystems 43 to 48 are plasterboard components;

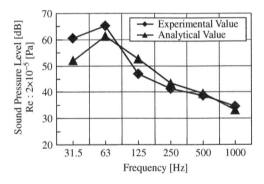

**FIGURE 20.9**   Estimated sound pressure level results for other rooms.

subsystems 49 to 55 are room components; and subsystems 56 to 61 are cavity components. For example, Figure 20.7 shows the power flow relationships between structural subsystems in the entire building, while Figure 20.8 shows them in the acoustical laboratory room. Here, the thin-dotted, dotted, and solid lines indicate the I-, L-, and T-type combinations, respectively. Subsystem 53 is the room component, and it is connected with all structural components shown in Figure 20.8. The plasterboards located between the computer room and the measurement room are considered as a partition; therefore, connections between subsystem 49 and subsystems 56 to 59 (cavity components), and subsystem 50 and subsystems 60 and 61 (cavity components) are derived from nonresonant modes.

The results obtained for some other rooms are shown in Figure 20.9. Computing accuracy in this building is worse than in the cabin because the structure of this building is complicated, although the differences between the computed values and the measured values were approximately 4 dB in the medium- to high-frequency band.

The computations take approximately 10 sec, so the workload on the personal computer is quite light.

# References

1.  Lyon, R.H. 1975. *Statistical Energy Analysis of Dynamical Systems: Theory and Applications*, MIT Press, Cambridge, MA.
2.  Irie, Y., Solid propagation sound analysis by SEA, *Mitsubishi Heavy Ind. Tech. Rev.*, 21, 4, 571–578, 1984.
3.  Crocker, M.J., Sound transmission using statistical energy method, *AIAA J.*, 15, 2, 75–83, 1977.
4.  Irie, Y., Solid propagation sound analysis by SEA, *Nihon Onkyo Gakkai-shi*, 48, 6, 433–444, 1992.
5.  Lyon, R.H. and Dejong, R.G. 1995. *Theory and Applications of Statistical Energy Analysis*, 2nd ed., Butterworth-Heinemann, Oxford.
6.  Craik, R.J.M. 1996. *Sound Transmission Through the Buildings Using Statistical Energy Analysis*, Gower Press, England.
7.  Cremer, L., Heckle, M., and Unger, E.E. 1973. *Structure Borne Sound*, Springer, New York, pp. 347–370.
8.  *Mechanical Noise Handbook*, JSME, Japan Society of Mechanical Engineers, pp. 179–181, 1991 (in Japanese).
9.  Koizumi, T., Tsujiuchi, N., Kubomoto, I., and Ishida, E., Estimation of the noise and vibration response in a tractor cabin using statistical energy analysis, *JSAE Paper*, 28, 4, 49–54, 1997.
10. Koizumi, T., Tujiuchi, N., Kubomoto, I., and Ishida, E., Estimation of the noise and vibration response in a tractor cabin using statistical energy analysis, *SAE Paper*, 1999-01-2821, 1999.
11. Koizumi, T., Tujiuchi, N., Tanaka, H., Okubo, M., and Shinomiya, M., Prediction of the vibration in building using statistical energy analysis, *IMAC Paper*, 7–13, 2002.

# Index

## A

**Absorber**
broadband excitation, **5**-5–6
control systems, **7**-2
  versus control, **5**-2, 3
  damped absorber, **7**-51–57
  versus isolation, **5**-2
  undamped absorbers, **7**-46–51
design and implementation of systems, **5**-12–24
  active resonator system, **5**-18–24
  delayed resonator system, **5**-13–18
primary system vibration suppression, **5**-2
**Absorption**
coefficient, sound, **17**-3, 4
  combined plate with porous blanket, **17**-6–8
  stationary wave factor and, **17**-11
**Acceleration**
limits specification, **7**-4
seismic vibration-induced; *see also* Seismic vibration
  sliding isolation system simulations, equipment, **4**-35, 36
  sliding isolation system simulations, structure, **4**-33
**Accelerometer**, **3**-7, 8
control systems, **7**-62
helicopter rotor tuning, *see* Helicopter rotor tuning
**Acoustic enclosure**, **19**-13–16
**Acoustic impedance**, **17**-2, 3
**Acoustic resonance**, rotating machinery, **9**-54, 55
**Acoustics**, **13**-1–9
hearing, **13**-1–9
  frequency and loudness response, **13**-2–4
  hearing loss, **13**-4
  psychological effects of noise, **13**-4–9
  structure and function of ear, **13**-1–2
noise control criteria and regulations
  evaluation, **14**-5–7
  legislation, **14**-2–4
  policy concepts, **14**-1–2
  regulation, **14**-4–5
noise sources, **16**-1–11
  estimation of noise-source sound power, **16**-10–11
  point source, **16**-1–4
  radiation from plane surface, **16**-6–10
  sources of finite volume, **16**-4–6
sound control design, **17**-1–19
  compound wall, computation of acoustic characteristics, **17**-6–9
  dissipative mufflers, attenuation of, **17**-12, 14–15
  dissipative mufflers, practical example, **17**-17–19
  dust and water exposure, **17**-16–17
  fundamentals of sound absorption, **17**-2–3
  gas flow velocity, **17**-15–16
  gas temperature, **17**-16

  insulation, *see* Insulation, sound
  lined ducts, **17**-10–12, 13
  materials, **17**-3–6
  reactive mufflers, *see* Mufflers, reactive
  surface treatment with lining of acoustic material, **17**-15
sound levels and decibels, **12**-1–4
  levels and decibels, **12**-3–4
  sound wave characteristics, **12**-1–3
sound measurement instrumentation
  microphone array, **15**-6–8
  mirror-microphone system, **15**-4–6
  sound intensity measurement, **15**-1–4
statistical energy analysis, **20**-1–8
  applications in structures, **20**-7–8
  estimation of parameters, **20**-4–6
  power flow equations, **20**-2–3
**Acoustic sensors**, **3**-7, 8
**Active control systems**, **5**-2, 7–8; **7**-2
design, **7**-58, 61–67
  systems, **7**-61–62
  techniques, **7**-62–67
**Active damping/dampers**, **1**-2
damping theory, **2**-13–14
resonator absorbers, **5**-8, 18–24
**Adaptive-passive control systems**, **5**-4
**Adaptive tuning**, helicopter rotor, **6**-8–12
  estimation of feasible region selection of blade adjustments, **6**-9–10
  interval model, **6**-8–9
  learning, **6**-10–12
**Added mass**, fluid forces, **11**-16, 18
**Adjustable elements**, control systems, **5**-8–12
**Aircraft**
propeller whirl, **9**-13–14
**Air damping**, **2**-30–31
  brass and solid rod pendula, **3**-31–34
  damping theory, **2**-9–10
**Air flow**, self-excited vibrations, **7**-39
**Alford's force**, **9**-12–13
**Algebra**, *see* Linear algebra
**Amplifers**
damping sensor, **2**-24–25
**Analog-digital conversion**
analog-to-digital conversion, **2**-3
  damping studies, **2**-3
control systems, **7**-62
**Analysis/analytical methods and models**
rotordynamic, **9**-18–38
  case study, **9**-46–49, 51–54
  design, **9**-36–38
  methods of determining, **9**-18–26
  modeling, **9**-26–36
  software packages, **9**-32–34

## Related Titles

*Vibration Simulation Using MATLAB and ANSYS*
Michael R. Hatch
ISBN: 1584882050

*Vibration: Fundamentals and Practice, Second Edition*
Clarence W. de Silva
ISBN: 0849319870

Printed and bound by CPI Group (UK) Ltd, Croydon, CR0 4YY

23/10/2024

01778259-0007